4th International Conference on Operational Research (InteriOR 2017)

IOP Conference Series: Materials Science and Engineering
Volume 300

Medan, Indonesia
21 - 23 August 2017

Editor:

Herman Mawengkang

ISBN: 978-1-5108-5885-5
ISSN: 1757-8981

Printed from e-media with permission by:

Curran Associates, Inc.
57 Morehouse Lane
Red Hook, NY 12571

Some format issues inherent in the e-media version may also appear in this print version.

Printed by Curran Associates, Inc. (2018)

For permission requests, please contact the Institute of Physics
at the address below.

Institute of Physics
Dirac House, Temple Back
Bristol BS1 6BE UK

Phone: 44 1 17 929 7481
Fax: 44 1 17 920 0979

techtracking@iop.org

Additional copies of this publication are available from:

Curran Associates, Inc.
57 Morehouse Lane
Red Hook, NY 12571 USA
Phone: 845-758-0400
Fax: 845-758-2633
Email: curran@proceedings.com
Web: www.proceedings.com

Table of contents

Volume 300

4th International Conference on Operational Research (InteriOR)21–23 August 2017, Medan, Indonesia

Accepted papers received: 10 January 2018Published online: 15 February 2018

Preface

Papers

OR

ST

IOP Conf. Series: Materials Science and Engineering **300** (2018) 011001 doi:10.1088/1757-899X/300/1/011001

The 4th International Conference on Operational Research

August 21st-23rd, 2017 Grand Aston City Hall Hotel, Medan-Indonesia

The 4th International Conference on Operational Research 2017 (InteriOR 2017) is organized by the Department of Mathematics Universitas Sumatera Utara, Medan Indonesia. Our co-organizer is The Indonesian Mathematical Society Section Aceh and Sumatera Utara. This conference is a biannual conference, started at 2011. The topics of the conference are divided into four categories: they are Operational Research, Statistics, Applied Mathematics, and Pure Mathematics including their Applications.

The keynote talks are focused especially on the advanced achievement on Operational Research and their Applications. This year theme is "Better Living with Operational Research". We have four keynote speakers for the conference:

- Prof. Dr. Gerhard W Weber, Middle East Technical University, Turkey
- Prof. Dr. Abdel Salhi, Essex University, UK
- Prof. Dr. Yong Hong Wu, Curtin University, Australia
- Prof. Dr. Saib Suwilo, Universitas Sumatera Utara, Indonesia.

InteriOR 2017 was a success event, attracting researchers from many countries and provided a great academic experience and social activities for the participants. I would like to extend my deepest appreciation to all keynote speakers, participants and organizers. Without their support this conference was not a success. Finally, I wish you all a great success in the years to come. Hope to see you again at The InteriOR 2019.

Editor in Chief

Prof. Dr. Herman Mawengkang

IOP Conf. Series: Materials Science and Engineering **300** (2018) 011001 doi:10.1088/1757-899X/300/1/011001

IOP Conf. Series: Materials Science and Engineering **300** (2018) 011001 doi:10.1088/1757-899X/300/1/011001

IOP Conf. Series: Materials Science and Engineering **300** (2018) 011001 doi:10.1088/1757-899X/300/1/011001

IOP Conf. Series: Materials Science and Engineering **300** (2018) 011002 doi:10.1088/1757-899X/300/1/011002

The 4th International Conference on Operational Research (InteriOR)

"Better Living with Operations Research"

August 21 - 23, 2017
Grand Aston City Hall Hotel
Medan, Sumatera Utara - Indonesia

INTRODUCTION

The Department of Mathematics University of Sumatera Utara is pleased to announce the 4th International Conference on Operational Research (InteriOR 2017). The InteriOR is held every two years. The first InteriOR was held in 2011. The InteriOR 2017 will be held at Grand Aston City Hall Hotel, Medan, Indonesia on August 21-23, 2017 with the theme "Better Living With Operations Research". The objective of the conference is to provide a forum for researchers, lecturers, educators, students and practitioners in all related disciplines to discuss, share, exchange and to extend their knowledge of new advances on theories and practices in mathematics, particularly in operations research.

IMPORTANT DATES

Registration	:	April 01, 2017
Abstracts Submission Deadline	:	April 30, 2017
Notification of Acceptance (Abstracts)	:	May 31, 2017
Full Paper Submission Deadline	:	June 17, 2017
Notification of Acceptance (Full Paper)	:	July 17, 2017
Conference	:	August 21 - 23, 2017

TOPICS

We invite the submission of all papers related to operations research (O.R.). The topics are (but not limited to):

1. Algebra
2. Applied Mathematics
3. Business Analytics
4. Combinatorics and Graph Theory
5. Data Mining
6. Decision Analysis
7. Financial Engineering
8. Forecasting
9. Fuzzy theory
10. Game Theory
11. Heuristics
12. Industrial Engineering
13. Logistics
14. Mathematical Computer Sciences
15. Mathematical Finance & Actuary
16. Mathematical Optimization
17. Numerical Analysis
18. Probability and Statistics
19. Simulation
20. Social Network
21. Forecasting Model
22. Supply Chain Management

KEYNOTES

Prof. Dr. Gerhard Wilhelm Weber
Middle East Technological University, Turkey

Prof. Dr. Katsunori Ano
Shibaura Institute of Technology, Japan

Prof. Dr. Abdel Salhi
University of Essex, United Kingdom

Prof. Dr. Erwin Pesch
Universität Siegen, Germany

Prof. Dr. Yong Hong Wu
Curtin University, Australia

Dr. Elise del Rosario
IFORS, Philippines

REGISTRATION

To participate in InteriOR 2017, you are asked to register for the conference via our official website : *http://ocs.usu.ac.id/interior.*

REGISTRATION FEE

International Participants

Early Bird	:	USD 350
Reguler	:	USD 450

Domestic (Local) Participants

Early Bird	:	IDR 2.500.000,-
Reguler	:	IDR 3.000.000,-
Student	:	IDR 1.000.000,-

FOR MORE INFORMATION

interior@usu.ac.id *http://ocs.usu.ac.id/interior*

Supported by
IndoMS IORA

Indexed by
Scopus THOMSON REUTERS INIS IET Inspec Compendex Polymer Library NASA Astrophysics Data System Chemical Abstract Service

Proceeding Publisher
IOP Conference Series conferenceseries.iop.org
Material Science & Engineering (MSE)

IOP Conf. Series: Materials Science and Engineering **300** (2018) 011003 doi:10.1088/1757-899X/300/1/011003

Conference Information

Date	: August 21 – 23, 2017
Organizer	: Department of Mathematics Faculty of Mathematics and Natural Sciences University of Sumatera Utara
Venue	: Grand Aston City Hall Hotel Jalan Balai Kota No. 1, Medan, Sumatera Utara, Indonesia, 20122
Phone	: +62 61 4557 000
Email	: info@grandastonmedan.com
Hotel Website	: https://www.astonhotelsinternational.com/
Official Language	: English
Secretariat	: Graduate School of Mathematics Faculty of Mathematics and Natural Sciences, University of Sumatera Utara Jalan Bioteknologi No. 1, Kampus USU, Medan 20155, Sumatera Utara
Email	: interior@usu.ac.id
Website	: http://ocs.usu.ac.id/interior

IOP Conf. Series: Materials Science and Engineering **300** (2018) 011003 doi:10.1088/1757-899X/300/1/011003

Committee

IOP Conf. Series: Materials Science and Engineering **300** (2018) 011003 doi:10.1088/1757-899X/300/1/011003

Dr. Fahmi F. (University of Sumatera Utara, Indonesia)

Dr. Tulus Ikhsan Nasution (University of Sumatera Utara, Indonesia)

Dr. Intan Muchtadi-Alamsyah (Institut Teknologi Bandung, Indonesia)

Dr. MDH. Gamal (Universitas Riau, Indonesia)

Dr. Elvina Herawati (University of Sumatera Utara, Indonesia)

ORGANIZING COMMITTEE

Chairman : Prof. Dr. Saib Suwilo

Co-chair : Prof. Dr. Herman Mawengkang

LOCAL ORGANIZING COMMITTEE

Director : Dr. Maya Silvi Lydia

Secretary : Dr. Syahril Efendi

Treasurer : Dr. Elly Rosmaini

Executive Members :
1. Dr. Ariswoyo
2. Dr. Mardiningsih
3. Dr. Esther Sorta Mauli Nababan
4. Dr. F. Buulolo
5. Dr. Suyanto
6. Dr. Sawaluddin
7. Dr. Sutarman

IOP Conf. Series: Materials Science and Engineering **300** (2018) 011003 doi:10.1088/1757-899X/300/1/011003

Venue Map

Grand Aston City Hall Hotel

Address : Jalan Balai Kota No. 1, Medan, Sumatera Utara, Indonesia, 20122
Phone : +62 61 4557 000

Website : https://www.astonhotelsinternational.com/

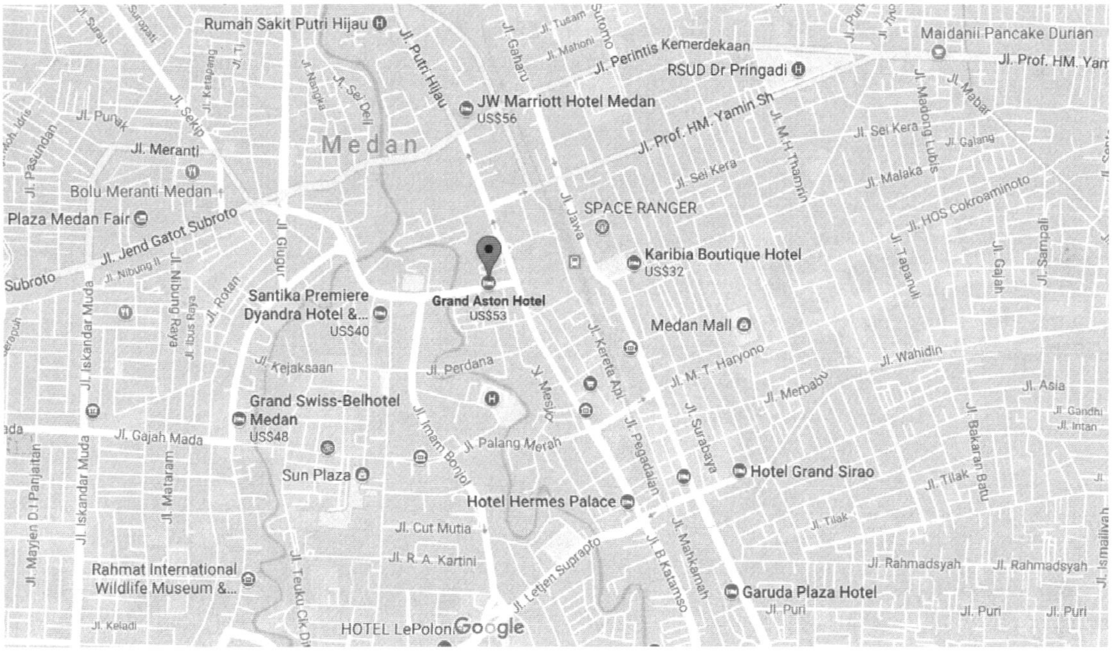

IOP Conf. Series: Materials Science and Engineering **300** (2018) 011004 doi:10.1088/1757-899X/300/1/011004

Peer review statement

All papers published in this volume of *IOP Conference Series: Materials Science and Engineering* have been peer reviewed through processes administered by the proceedings Editors. Reviews were conducted by expert referees to the professional and scientific standards expected of a proceedings journal published by IOP Publishing.

Portfolio optimization by using linear programing models based on genetic algorithm

Sukono[1], Y. Hidayat[2*], E. Lesmana[3*], A. S. Putra[4], H. Napitupulu[5], S. Supian[6]

[1,3,4,5,6]Department of Mathematics, Faculty of Mathematics and Natural Sciences, Universitas Padjadjaran, INDONESIA

[2]Department of Statisticss, Faculty of Mathematics and Natural Sciences, Universitas Padjadjaran, INDONESIA

Email: sukono@unpad.ac.id; yuyunhidayat@unpad.ac.id; man.msie@gmail.com; napitupuluherlina@gmail.com; sudradjat@unpad.ac.id

*Corresponding author's e-mail: yuyunhidayat@unpad.ac.id; man.msie@gmail.com

Abstract. In this paper, we discussed the investment portfolio optimization using linear programming model based on genetic algorithms. It is assumed that the portfolio risk is measured by absolute standard deviation, and each investor has a risk tolerance on the investment portfolio. To complete the investment portfolio optimization problem, the issue is arranged into a linear programming model. Furthermore, determination of the optimum solution for linear programming is done by using a genetic algorithm. As a numerical illustration, we analyze some of the stocks traded on the capital market in Indonesia. Based on the analysis, it is shown that the portfolio optimization performed by genetic algorithm approach produces more optimal efficient portfolio, compared to the portfolio optimization performed by a linear programming algorithm approach. Therefore, genetic algorithms can be considered as an alternative on determining the investment portfolio optimization, particularly using linear programming models.

1. Introduction

Investing in stocks is very attractive to many investors, even if investors have to face risks in their investments. The strategy in dealing with these risky investments is to form a portfolio of several stocks [14]. In the formation of a portfolio, investors always want to maximize the expected return with a certain level of risk they are willing to bear, or look for a portfolio that offers the lowest risk with a certain rate of return [15,16]. This portfolio characteristic, referred to as an efficient portfolio. To establish an efficient portfolio, it must adhere to assumptions about how investors behave in making investment decisions to be taken [14]. One of the most important assumptions is that all investors do not like risk (risk averse). Investors like this if faced with two investment options that offer the same returns with different risks, will tend to choose investments with lower risk [3]. While the optimal portfolio, is a portfolio selected an investor of the many options that exist in the collection of efficient portfolio [2,15]. Surely the portfolio selected by the investor is a portfolio that corresponds to the investor's preference, to the return or risk that it is willing to bear [3].

In the formation of a portfolio, an important thing to be noticed by investors is how to allocate capital resources, and the applicable provisions in investing, all of which are limited. This is with the aim that the portfolio is formed, resulting in an optimal portfolio. To find the solution is can be done by using linear programming (LP) optimization model. Such ways have been widely practiced by previous researchers. For example Mansini et al. [10], reviewing various portfolio optimization models that can be solved by LP. Presented in the literature of real features that have been modeled, and the solution approach to the resulting model. In most cases the model is integer linear programming (MILP). Mbah and Onwukwe [9], conducting an analysis of linear programming methods for the issue of portfolio selection and optimal financial investment in economic development. Liu [7], conducted a study that presents three possible models for the problem of portfolio selection with minimum transaction lots. As well as designing the appropriate genetic algorithm to get the solution. El hachlou et al. [1] optimizing stock portfolios using classification and genetic algorithm. Kapiamba et al. [4], performing a comparison between simulated annealing and genetic algorithm in optimal portfolio selection. Sadaf and Ghodrati [11], undertook a development of genetic algorithm methods for the selection and optimization of stock portfolios.

Based on the above description, there is a very interesting thing to investigate, that is how to optimize investment portfolio involving the use of linear programming model and genetic algorithm. Therefore, in this paper, research on investment portfolio optimization using linear programming model based on genetic algorithms is conducted. As the object of his research are some stocks that are traded on the capital market in Indonesia. The goal is to obtain a proportion of investment capital allocation, in order to obtain an optimal portfolio. The models used to achieve these objectives are discussed in the following sections.

2. Model of linear programming on investment portfolios and genetic algorithms

In this section we aim to discuss the models used in portfolio optimization, using linear programming based on genetic algorithms. The discussions include: investment portfolio, linear programming model, and genetic algorithm. It starts with a discussion of the investment portfolio as follows.

2.1. Portfolio investment in stocks

In this section it is intended to describe the investment portfolio in stocks. Investments in shares can be understood as a means of investment, either directly or indirectly, aimed at obtaining certain benefits, as a result of investments made [14]. In every investment decision on stock, as a rational investor, his attention is always aimed at the return of his investment. Investors will choose investments in stocks that give the highest expected return (return). However, investments in stocks that are carried contain an element of uncertainty, so investors should consider the risk factor [3].

Expected return is basically the weighted average of various historical returns, with the probability of each stock return as its weighting factor. Expected return is reflected by the mean of the profitability probability distribution. While the risk of an investment is measured by the amount of variance or standard deviation from expected return [15,16]. The greater the degree of spread (variance), illustrates that investments are becoming increasingly risky. A common strategy used in risky investment conditions, is to form a portfolio of some stocks [2]. The essence of forming a portfolio is to allocate funds to various investment alternatives on stocks, so the risk of investing in stocks as a whole can be minimized, and the profit rate of return can be maximized 13].

To minimize the risk and maximization of the expected return can be done using linear programming, as discussed in the following sections.

2.2. Linear programming model on investment portfolio

This section aims to discuss the formulation of linear programming model for optimization of investment portfolio in stocks.

Suppose that R_i is the i-th random variable of stock return, $i = 1,...,N$ with N is the number of stocks, and μ_i is the return expectation of R_i. Suppose also that W_0 is the initial capital owned by the investors, and π is the return desired by investors. Meanwhile, notation δ_i is the maximum value that investors want to invest in stocks i, $i = 1,...,N$ with N is the numbers of stocks. It is assumed that short selling is not allowed, thus the proportion $w_i \geq 0$, $i = 1,...,N$ with N is the number of stocks. Thus, suppose that the short selling is denoted by following equation (1) [17].

$$S = \{\mathbf{w'} = (w_1,...,w_N): \sum_{j=1}^{N} w_j \mu_j \geq \pi W_0, \sum_{j=1}^{N} w_j = W_0, 0 \leq w_j \leq \delta_j, j = 1,...,N\}. \quad (1)$$

Konno's formulated the investment portfolio optimization model as given in equation (2) [17].

$$\text{Minimum} \quad L_1 = E\left|\sum_{j=1}^{N} w_j R_j - E\left(\sum_{j=1}^{N} w_j R_j\right)\right|, \quad (2)$$

Subject to $\mathbf{w} \in S$.

Since the objective function is nonlinear, then according to Konno's and Yamazaki's methods, the model is constructed in the following way [17]:

$$\text{Minimum} \quad \frac{1}{T}\sum_{t=1}^{T} z_t, \quad (3)$$

$$\text{Subject to} \quad z_t \geq -\sum_{j=1}^{N} w_j (R_{jt} - \mu_j), \ t = 1,...,T,$$

$$z_t \geq \sum_{j=1}^{N} w_j (R_{jt} - \mu_j), \ t = 1,...,T,$$

$$\mathbf{w} \in S.$$

Here μ_j is the return expectation of j-th stocks, R_{jt} is the return of j-th stocks during periode t. Note that in this model there is no need to estimate the variance-covariance matrix.

Meanwhile, according to Cai's, the above investment portfolio optimization can be formulated as follows [17]:

$$\text{Minimum} \quad L_\infty(\mathbf{w}) = \underset{j}{\text{Maksimum}} E\,|w_j R_j - w_j \mu_j|, \quad (4)$$

Subject to $\mathbf{w} \in S$.

This model can also be transformed into the following linear form [17]:

$$\text{Minimum} \quad z, \quad (5)$$

$$\text{Subject to} \quad w_j q_j \leq z, \ j = 1,...,N,$$

$$\mathbf{w} \in S.$$

where $q_j = E|R_j - \mu_j|$, $j = 1,...,N$, is the absolute deviation of expectations of R_j from the mean return. Obviously, that if the distribution of each variable R_j given is random, this function is explicitly defined. Historical data can also be used to estimate μ_j and q_j. Model of L_∞ and related techniques are easy to be manipulated and applied practically. In addition, optimal portfolio selection does not contain correlation between stocks, which is the same as Konno's model and the constant number for this model is determined by the number of stocks.

Furthermore, Teo's model formulates optimization of investment portfolio as follows [6]:

$$\text{Minimum} \quad H_\infty^T = \frac{1}{T}\sum_{t=1}^{T} \underset{j}{\text{Maksimum}} E\,|w_j R_{jt} - E(R_{jt}) w_j|,$$

Subject to $\mathbf{w} \in S$.

For this model, each capital asset pricing model (CAPM) between the market portfolio and each stock return is determined by the use of an optimization method. This model can be transformed into the following linear form [17]:

$$\text{Minimum} \quad \frac{1}{T}\sum_{t=1}^{T} z_t ,$$

$$\text{Subject to} \quad w_j a_{jt} \le z_t ,\ t = 1,...,T ,\ j = 1,...,N ,$$

$$\mathbf{w} \in S .$$

where $a_{jt} = E\,|\,R_{jt} - E(R_{jt})\,|$, $t = 1,...,T$, $j = 1,...,N$. It is clear that the size of a constant is determined by the number of stocks and the number of periods. Therefore, if N and T become large, the computational speed of the completion of this model can be relatively slowly.

Recall the classic model of Markowitz's, as follows [13]:

$$\text{Minimum} \quad Var(\mathbf{w}) = \sum_{i=1}^{N}\sum_{j=1}^{N} w_i w_j \sigma_{ij} ,$$

$$\text{Subject to} \quad \mathbf{w} \in S .$$

The objective function in the Markowitz model is the quadratic form in the variance-covariance matrix. This means optimal allocation of portfolios can be solved using standard quadratic programming algorithms. The use of this quadratic programming algorithm needs to be assured that the elements of the variance-covariance matrix have an inverse. The allocation of portfolio proportions may be different from the previous three models, since the previous three models do not require the variance-covariance matrix estimation [17].

2.3. Genetic Algorithm

This section aims to discuss genetic algorithms in linear programming models for optimizing investment portfolios of stocks.

Genetic algorithms can be understood as a search method based on the mechanisms of scientific genetics and natural selection [5]. The mechanisms of scientific genetics reflect the ability of individuals to marry, and produce offspring that have characteristics similar to their parents. While the mechanism of natural selection states that living things can survive, if able to adapt to the environment [8]. Therefore, it is expected that the resulting offspring have the best combination of characteristics from their parents, and can sustain future generations.

The general structure of a genetic algorithm can be expressed as the following steps [12]:

a) Generating the initial population, this initial population is generated randomly so as to obtain an initial solution;
b) The population itself consists of a number of chromosomes that present the desired solution;
c) Forming a new generation, in forming a new generation used three operators, namely reproduction / selection, cross over and mutation;
d) Evaluation of the solution, this process will evaluate each population by calculating the fitness value of each chromosome, and evaluating it until the stop criteria are met. If the stopping criteria have not been met, it will be formed again new generation by repeating steps b).

Based on the general structure of the genetic algorithm, the genetic algorithm to solve the linear programming model of investment portfolio optimization issues such as equations (3), (4), and (5), can generally be composed as follows:

1) Determine the initial population. The N number of initial population, are generated randomly. This initial random population number is then converted into the form of decimal values $\mathbf{w}$, where $\mathbf{w} \in S$;

2) Evaluate chromosomes. The fitness values of chromosomes are the objective function values (3), (4), and (5). Based on fitness values, the smallest value is selected for the minimization program.
3) Calculation of population convergent percentage. Percent of population convergence p_c, is a percentage of the number of individuals who have the same fitness value and also the best. This p_c number is calculated using the following formula:

   $$p_c = \frac{n}{pop} \times 100\%,$$

   where n is the number of individuals who have the same fitness and also the most, and pop is the number of population.
4) Checking stopping condition. The process of genetic algorithm will stop when the generation counter has reached the number of defined generations c_g that is c_g =1000, or the convergent percentage of the population p_c reach the defined threshold limit that is $\theta = 90\%$.
   1) Chromosome selection. Selection process is done using roulette wheel selection. Due to the minimization program, it evaluates the fitness $eval(v_i)$, $i = 1,...,N$, done based on equations (3), (4), and (5), by the formula:

      $$eval(v_i) = \frac{1}{f(\mathbf{w})},$$

      where $f(\mathbf{w})$ is the fitness values referring to equations (3), (4), and (5).
   2) Crossbreeding. The new population of the selection results is cross-breeded, using the Single-Point Crossover (SPX) method.
   3) Mutation. Mutations of each generation are obtained by using $m \times pop\_size \times p_m$, with m as a number of mutation, pop_size is a population size, and p_m probability of mutation (its value is determined randomly).
4) Decoding. It is the process of encoding the genes in a chromosome to get its value back as it was, changing the coding to decimal values.

Furthermore, this genetic algorithm is used to analyze the case of investment portfolio optimization as follows.

3. Result and discussion

In this section it is aimed at modeling linear programming, and finding the optimal solution of the investment portfolio optimization problem. The discussions include: data being analyzed, linear programming modeling, and finding the optimal solution. It starts with the discussion of analyzed data as follows.

3.1. Data analyzed

This section is intended to explain the data being analyzed. The data analyzed in this paper is the closing price of stock from five companies listed on a stock market, with a period of one year (12 months), including stocks of: A, B, C, D, and E. Each stocks has a monthly return as given in Table 1. While the difference of monthly data to its average, are given in Table 2.

Toward these five stocks, investors provide budget of W_0 at IDR 100,000,000 to be invested. However, this investor expects a minimum monthly return π at 3% (or equivalent to IDR 3,000,000). Investors also want no stocks to be proportioned δ_i $(i = 1,...,N)$ greater 75% than its budget (or equivalent to IDR 75,000,000).

Table 1. Montly stocks return

Month	A	B	C	D	E
t	R_{1t}	R_{2t}	R_{3t}	R_{4t}	R_{5t}
1	0.0062	-0.0153	0.0175	0.0068	-0.0220
2	0.0000	0.0308	0.0044	0.0023	0.0073
3	0.0272	-0.0408	-0.0433	0.0000	0.0073
4	0.0026	0.0253	0.0477	0.0114	-0.0037
5	-0.0635	0.0051	0.0179	-0.0023	0.0000
6	0.0188	0.0000	0.0000	-0.0023	-0.0253
7	0.0439	0.0261	0.0091	-0.0023	0.0180
8	0.0183	-0.0209	-0.0181	0.0023	0.0110
9	0.0079	0.0356	0.0135	-0.0181	0.0261
10	0.0044	-0.0149	-0.0269	-0.0045	0.0190
11	0.0023	0.0245	0.0139	0.0152	0.0254
12	-0.0062	-0.0049	0.0000	0.0158	0.0800
μ_j	0.0052	0.0042	0.0030	0.0020	0.0119

Table 2. The difference of return and expected return

Month	A	B	C	D	E
t	$R_{1t}-\mu_1$	$R_{2t}-\mu_2$	$R_{3t}-\mu_3$	$R_{4t}-\mu_4$	$R_{5t}-\mu_5$
1	0.00105	-0.01949	0.01449	0.01564	-0.03391
2	-0.00515	0.02655	0.00143	0.00259	-0.00466
3	0.02203	-0.04504	-0.04628	-0.04512	-0.00461
4	-0.00256	0.02110	0.04475	0.04591	-0.01560
5	-0.06867	0.00092	0.01496	0.01611	-0.01193
6	0.01364	-0.00422	-0.00298	-0.00182	-0.03725
7	0.03870	0.02189	0.00611	0.00727	0.00609
8	0.01312	-0.02516	-0.02108	-0.01992	-0.00096
9	0.00278	0.03141	0.01056	0.01172	0.01414
10	-0.00072	-0.01911	-0.02989	-0.02873	0.00712
11	-0.00285	0.02028	0.01092	-0.02773	0.01347
12	-0.01136	-0.00913	-0.00298	-0.00182	0.06811

Based on the return data from the five stocks and the information described above, optimization is done by using linear programming based on genetic algorithm, as follows.

3.2. *Modeling of Konno's and Yamazaki's optimization methods*

To optimize the investment portfolio in the above five stocks, this paper is done by using Konno's and Yamazaki's linear programming model. Referring to equation (3), using data in Table 1 and Table 2, the Konno's and Yamazaki's linear programming model is structured as given in (6).

$$\text{Minimum} \quad \frac{1}{12}\sum_{t=1}^{12} z_t = \frac{1}{12}(z_1+z_2+z_3+z_4+z_5+z_6+z_7+z_8+z_9+z_{10}+z_{11}+z_{12}), \tag{6}$$

Subject to:

$$z_1 - 0.00105w_1 + 0.01949w_2 - 0.01449w_3 - 0.01564w_4 + 0.03391w_5 \geq 0$$
$$z_1 + 0.00105w_1 - 0.01949w_2 + 0.01449w_3 + 0.01564w_4 - 0.03391w_5 \geq 0$$
$$z_2 + 0.00515w_1 - 0.02655w_2 - 0.00143w_3 - 0.00259w_4 + 0.00466w_5 \geq 0$$
$$z_2 - 0.00515w_1 + 0.02655w_2 + 0.00143w_3 + 0.00259w_4 - 0.00466w_5 \geq 0$$
$$z_3 - 0.02203w_1 + 0.04504w_2 + 0.04628w_3 + 0.04512w_4 + 0.00461w_5 \geq 0$$
$$z_3 + 0.02203w_1 - 0.04504w_2 - 0.04628w_3 - 0.04512w_4 - 0.00461w_5 \geq 0$$
$$z_4 + 0.00256w_1 - 0.02110w_2 - 0.04475w_3 - 0.04591w_4 + 0.01560w_5 \geq 0$$
$$z_4 - 0.00256w_1 + 0.02110w_2 + 0.04475w_3 + 0.04591w_4 - 0.01560w_5 \geq 0$$
$$z_5 + 0.06867w_1 - 0.00092w_2 - 0.01496w_3 - 0.01611w_4 + 0.01193w_5 \geq 0$$
$$z_5 - 0.06867w_1 + 0.00092w_2 + 0.01496w_3 + 0.01611w_4 - 0.01193w_5 \geq 0$$
$$z_6 - 0.01364w_1 + 0.00422w_2 + 0.00298w_3 + 0.00182w_4 + 0.03725w_5 \geq 0$$

$$z_6 + 0.01364w_1 - 0.00422w_2 - 0.00298w_3 - 0.00182w_4 - 0.03725w_5 \geq 0$$
$$z_7 - 0.03870w_1 - 0.02189w_2 - 0.00611w_3 - 0.00727w_4 - 0.00609w_5 \geq 0$$
$$z_7 + 0.03870w_1 + 0.02189w_2 + 0.00611w_3 + 0.00727w_4 + 0.00609w_5 \geq 0$$
$$z_8 - 0.01312w_1 + 0.02516w_2 + 0.02108w_3 + 0.01992w_4 + 0.00096w_5 \geq 0$$
$$z_8 + 0.01312w_1 - 0.02516w_2 - 0.02108w_3 - 0.01992w_4 - 0.00096w_5 \geq 0$$
$$z_9 - 0.00278w_1 - 0.03141w_2 - 0.01056w_3 - 0.01172w_4 - 0.014146w_5 \geq 0$$
$$z_9 + 0.00278w_1 + 0.03141w_2 + 0.01056w_3 + 0.01172w_4 + 0.014146w_5 \geq 0$$
$$z_{10} + 0.00072w_1 + 0.01911w_2 + 0.02989w_3 + 0.02873w_4 - 0.00712w_5 \geq 0$$
$$z_{10} - 0.00072w_1 - 0.01911w_2 - 0.02989w_3 - 0.02873w_4 + 0.00712w_5 \geq 0$$
$$z_{11} + 0.00285w_1 - 0.02028w_2 - 0.01092w_3 + 0.02773w_4 - 0.01347w_5 \geq 0$$
$$z_{11} - 0.00285w_1 + 0.02028w_2 + 0.01092w_3 - 0.02773w_4 + 0.01347w_5 \geq 0$$
$$z_{12} + 0.01136w_1 + 0.00913w_2 + 0.00298w_3 + 0.00182w_4 - 0.06811w_5 \geq 0$$
$$z_{12} - 0.01136w_1 - 0.00913w_2 - 0.00298w_3 - 0.00182w_4 + 0.06811w_5 \geq 0$$
$$0.0052w_1 + 0.0042w_2 + 0.0030w_3 + 0.0020w_4 + 0.0119w_5 \geq 3{,}000{,}000$$
$$w_1 + w_2 + w_3 + w_4 + w_5 \leq 100{,}000{,}000$$
$$0 \leq w_1, w_2, w_3, w_4, w_5 \leq 75{,}000{,}000$$

3.3. *Optimal solution using genetic algorithm*

In this section it seeks to solve the linear programming model of investment portfolio optimization, set up in section 3.2. Completion is done using genetic algorithm optimization. The algorithm used is referring to what has been described in section 2.3. The genetic algorithm process is dismissed when the counter generation has reached the number c_g =1000, or percent of population conventions p_c has reached the number $\theta = 90\%$. Based on the optimization process using genetic algorithm, optimal solution obtained when the allocation of capital is as follows: w_1 = 14,340,000.00, w_2 = 8,970,000.00, w_3 = 6,760,000.00 , w_4 = 66,560,000.00 , and w_5 = 3,370,000.00. The allocation of such investment capital gets the values from z_1 = -80,080.58; z_2 = 527,066.25; z_3 = -584,574.40; z_4 = 982,734.59; z_5 = -503,510.46; z_6 = -103,807.88; z_7 = 1,116,909.92; z_8 = -88,601.07; z_9 = 766,340.08; z_{10} = -268,632.90; z_{11} = 957,989.80; dan z_{12} = 998,166.65. So it gets the minimum magnitude at $z = (1/12)\sum_{i=1}^{5} z_i = 310{,}000.00$.

By considering the values of z_1 to z_{12}, it is shown that the income earned monthly, from January to December is not always the same, but varies even some of which are negative or causing loss. The largest positive income occurred in July, which was IDR1,116,909.92, and the smallest positive income occurred in May at IDR503,510.46. While the largest negative income occurred in March at IDR584,574.40, and the smallest negative income occurred in January at IDR80,080.58. The average income in one year amounted IDR310,000.00.

Although this formulation seems very simple because it does not require an estimate of variance-covariance matrix, it does however lead to a rather pleasant portfolio. This formulation will produce an unlikely solution, if all shares during a given month generate a negative return. If for example during August, even stock A has a negative income, just like the others, there will be no portfolio with positive income. One possibility to reach a viable solution is to let z have a negative value as well. For example, if stock a returns in May is -0.0635, it will usually change in the next month.

4. Conclusions

In this paper we have discussed about the investment portfolio optimization using linear programming model based on genetic algorithms. Based on the results of the discussion can be concluded as follows. The investment portfolio optimization model that has an objective function in the quadratic form can be transformed into a linear form based on Konno's and Yamazaki's methods. The process of optimizing linear programming model can be solved by using genetic algorithm. Such investment portfolio optimization techniques are applied for stock analysis: A, B, C, D, and E. The optimal portfolio is obtained when the capital allocation is as follows: w_1 = 14,340,000.00, w_2 = 8,970,000.00, w_3 = 6,760,000.00, w_4 = 66,560,000.00, and w_5 = 3,370,000.00. Optimizations result obtained that the income monthly, from January to December is not always the same, but varied even some of which are negative or bring loss. However, in this case, with a capital investment of IDR100,000,000.00, a positive income of IDR310,000.00 is earned. Such optimization techniques are one way that can be used in investment portfolio analysis.

Acknowledgment

Further thanks to the Rector, Director of DRPMI, and Dean of FMIPA, Universitas Padjadjaran, which has a grant program of the Academic Leadership Grant (ALG) under the coordination of Prof. Dr. Sudradjat, and a grant program of Competence Research of Lecturer of Unpad (Riset Kompetensi Dosen Unpad/RKDU) under the coordination of Dr. Sukono, which is a means to increase research activities and publications to researchers at Universitas Padjadjaran.

References

[1] El hachlou, M, Guennoun, Z, and Hamza, F 2012 Stocks Portfolio Optimization Using Classification and Genetic Algorithms. *Appl. Math. Sci.* **6** 94 4673 – 84

[2] Gambrah P S N and Pirvu T A 2014 Risk Measures and Portfolio Optimization *J. Risk Financial Manag* 2014, 7 pp 113-129; doi:10.3390/jrfm7030113. ISSN 1911-8074. http: www.mdpi.com/journal/jrfm.

[3] Hult H, Lindskoq F, Hammarlid O, and Rehn, C J 2012 *Risk and Portfolio Analysis: Principles and Method.* (NEW YORK: Springer Series in Operations Research and Financial Engeneering)

[4] Kapiamba J N, Ulungu B E L, and Mubenga P K 2015 Simulated Annealing vs Genetic Algorithm to Portfolio Selection. *Int. J. of Sci. and Innovative Mathematical Research (IJSIMR)* **3** 5, May 2015, pp 18-30. ISSN 2347- 307X (Print) & ISSN 2347-3142 (Online). http://www.arcjournals.org

[5] Kim J S, Kim Y C, and Shin K Y 2005 An Algorithm for Portfolio Optimization Problem. *INFORMATICA* **16** 1 pp 93–106 2005 Institute of Mathematics and Informatics, Vilnius

[6] Konak F and Bagci B 2016 Fuzzy Linear Programming on Portfolio Optimization: Empirical Evidence from FTSE 100 Index. Global Journal of Management and Business Research: *C. Finance* , **16** 2 Version 1.0, Year 2016, Type: Double Blind Peer Reviewed, International Research Journal. (USA: Global Journals Inc.)

[7] Liu H W 2011 Linear Programming for Portfolio Selection Based on Fuzzy Decision-Making Theory *The Tenth Int. Symp. on Operations Research and Its Applications* (ISORA 2011) Dunhuang, China, August 28–31, 2011 Copyright © 2011 ORSC & APORC pp 195–202

[8] Lin C C and Liu Y T 2008 Genetic Algorithms for Portfolio Selection Problems with Minimum Transaction Lots. *European J. of Operational Research* 185 (2008), pp 393–404

[9] Mbah G C E and Onwukwe L C 2016 Linear programming Method for Portfolio Selection and Optimal Financial Investment in a Developing Economy. *Int. J. of Applied Research* 2016 **2** 12 pp 338- 342

[10] Mansini R, Ogryczak W, and Speranza M G 2014 Twenty Years of Linear Programming Based Portfolio Optimization *European J. of Operational Research* 234 (2014) pp 518–535 journal homepage: www.elsevier.com/locate/ejor

[11] Sadaf A and Ghodrati H 2015 An Improved Genetic Algorithm Method for Selection and Optimizing the Share Portfolio *Int. J. of Computer Science and Mobile Computing*, **4** 1, January- 2015, pp 342-353 Available Online at www.ijcsmc.com.

[12] Sefiane S and Benbouziane M 2012 Portfolio Selection Using Genetic Algorithm. *MPRA* Paper No. 41783, posted 12, October 2012 08:08 UTC, Online at https://mpra.ub.uni-muenchen.de/41783/

[13] Sarker M R 2013 Markowitz Portfolio Model: Evidence from Dhaka Stock Exchange in Bangladesh *IOSR J. of Business and Management (IOSR-JBM.* e-ISSN: 2278- 487X **8** 6 (Mar. - Apr. 2013), pp 68-73. http://www.iosrjournals.org.

[14] Sukono, Hidayat Y, Bon A T B, and Supian S. 2016. Modelling of Capital Asset Pricing Model with Considering the lagged Effects *IOP Conf. Series: Materials Science and Engineering* 166 (2017) 012001 doi:10.1088/1757-899X/166/1/012001. pp 1-9.

[15] Sukono, Sidi P, Bon A T, and Supian S Modeling of Mean-VaR Portfolio Optimization by Risk Tolerance When the Utility Function is Quadratic. *Statistics and its Applications.* AIP Conf. Proc. 1827, 020035-1–020035-9; doi: 10.1063/1.4979451 Published by AIP Publishing. 978-0-7354-1495-2/. pp 1-9

[16] Wang, X 2013 Theoretical Analysis of Financial Portfolio Model *iBusiness* 2013 5 pp 69-73. http://dx.doi.org/10.4236/ib.2013.53B015 Published Online September 2013 (http://www.scirp.org/journal/ib).

[17] Yu M, Inove H, and Shi J None Portfolio Optimization Problems with Linear Programming Models. *Paper.* E-mail: yumei@amss.ac.cn. Downloaded on May 5, 2017, via google.co.id.

Solving Quadratic Assignment Problem with Fixed Assignment (QAPFA) using Branch and Bound Approach

Sharifah Shuthairah Syed-Abdullah[1], Syariza Abdul-Rahman[2], Aida Mauziah Benjamin[3], Antoni Wibowo[4], Ku-Ruhana Ku-Mahamud[5]
[1,2,3]School of Quantitative Sciences (SQS)
[5]School of Computing (SOC)
College of Arts and Sciences Universiti Utara Malaysia
06010 Sintok, Kedah, Malaysia
[4]Computer Science Department-Binus Graduate Program
Bina Nusantara University
Anggrek Campus Jl. Kebon Jeruk Raya No. 27, Kebon Jeruk,
West Jakarta 11530, Indonesia

Corresponding author's e-mail: s_shuthairah89@yahoo.com

Abstract. Quadratic Assignment Problem (QAP) has been a very popular problem to be solved among researchers due to its practical applications. Several variants of QAP have been proposed by researchers in the past in order to reflect the real situations of QAP. One of the real problems of QAP is related with facilities which are required to be assigned to certain locations due to its function. In solving this problem, a fixed assignment has to be made thus allowing for the complexity of the problem to be reduced. Hence, this study introduces Quadratic Assignment Problem with Fixed Assignment (QAPFA) with the objective to minimize the assignment cost between the facility and location. This assignment takes into account the flow and distance between facility and location. QAPFA represents the real-world situation of the problem especially in dealing with specific requirement of some facilities to specific locations. Dataset of QAPFA is introduced and is solved using branch and bound approach. As for validation, the results of QAPFA are compared with QAP in terms of objective function and running time. The computational results show that the solution quality of QAPFA is lower when compared with the QAP, while the running time for QAPFA is lower than the QAP. Since the complexity of the problem is reduced by fixing the assignment, thus there is possibility that QAPFA has lower quality than QAP due to the fixed assignment. Nevertheless, in terms of running time QAPFA is better than QAP. It can be concluded that this problem reflect the real problem and practical to be used.

1. Introduction

Quadratic Assignment problem (QAP) is an NP-hard problem which has been used in modeling various types of problem such as facility layout [1-3], wiring problem in electronics [4], scheduling [5], transportation [6], computer manufacturing [7, 8] and sports [9]. QAP is defined as a problem of assigning a set of facilities to a set of locations with a flow and distance matrices while minimizing the cost of assignment of each facility to each location. QAP extension has been evolved throughout the years alongside with the research development. Variants of QAP which represent real world applications has been introduced by several researchers such as Quadratic Bottleneck Assignment Problem (QBAP) [11, 12], Quadratic 3-dimensional Assignment Problem (Q3AP) [13, 14], Quadratic Semi-Assignment Problem (QSAP) [15, 16], Biquadratic Assignment Problem (BiQAP) [17, 18], Multi-objective Quadratic Assignment Problem (mQAP) [19, 20] and Subset Quadratic Assignment Problem (SQAP) [21].

In dealing with real situation, there are some situations where some facilities need to be allocated at some specific locations due to their needs. For instance, some facilities in a hospital such as operating theater and emergency room should not be too far from each other and the flow of product (patients) between the two of the facilities should be minimized. Hence, fixed assignment of QAP is introduced in this study which is called as Quadratic Assignment Problem with Fixed Assignment (QAPFA)

QAP has been solved by researchers using various approaches to produce an optimal or near optimal solution. Approaches in solving QAP can be classified into three types which are exact, metaheuristic and hybrid methods. Exact method is used in searching an optimal solution but the method is limited to solve only small data sets [22]. An example of the exact methods for solving QAP are Branch and Bound approach (BB) [23, 24], Dynamic Programming [25] and Cutting Plane Method [26]. Nevertheless, metaheuristic and hybrid methods are used to solve large and complex problem of QAP in order to produce a best solution or near optimal solution. Example of such methods are Tabu search algorithm [27], Simulated Annealing algorithms [28], Genetic Algorithms [29] and Ant Colony Optimization [30].

This study proposed a variant of QAP which is called as Quadratic Assignment Problem with Fixed Assignment (QAPFA). In QAPFA, some assignments of facility and location are fixed due to its special requirement. This would represent the real-world situation. The benchmark dataset of QAPFA is proposed and solved using branch and bound approach. The computational results of the QAPFA are compared with the QAP.

The rest of the paper is organized as follows. Section 2 discusses on the literature review of the approaches in solving QAP and its variant, studies related to fixed assignment in QAP and in other types of fixed assignment. Section III presents the mathematical model of QAPFA, data representation and Branch and Bound approach in solving QAPFA. Several experimental results of QAPFA are presented and compared with QAP in Section IV. Finally, we concluded and give some future directions of this study in Section V.

2. Literature review

2.1. Quadratic Assignment Problem (QAP)

QAP is considered as an NP-hard problem [31, 33] which represents the facility layout problem mathematically. This problem considers the assignment of facilities to sites or location whenever there is an exchange between facilities with the objective of minimizing the cost. According to Kratica, Tosic, Filipovic, and Dugosija [32], QAP can be explained as the problem of assigning facilities n to locations n with given flows between the facilities and given distances between the locations, placing the facilities on locations in such a way that the sum of the product between flows and distances is minimized. QAP is a very interesting and challenging problem that can model many real-life problems. Solving this problem is very important for effective and efficient production. Thus, an efficient layout can reduce the cost and thus increase productivity. Another definition of QAP is given by Lim, Wibowo, Desa and Haron [33] where QAP is an assignment model involving a set of facilities that need to be allocated to a set of locations, with each location only will be having one facility. While, the objective is to minimize the total cost of the assignment involve multiply with the flow and the distances between facilities where, n can be the number of facilities and also the number of locations. Then, given two $n \times n$ matrices as which is F refer to flow and f_{ik} is the flow from facility i to facility k. Another matrix is matrix for distance or D, where d_{jl} is the distance from location j to location l.

In 1963, Lawler [34] recreated and proposed a general mathematical model originally introduced by Koopman and Beckman [10]. The mathematical model of the problem is as follow:

$$Min\sum_{j=1}^{n}\sum_{i=1}^{n} C_{ikjl} X_{ij} X_{kl} \tag{1}$$

Subject to;

$$\sum_{j=1}^{n} x_{ij} = 1, i \in N \tag{2}$$

$$\sum_{j=1}^{n} x_{ij} = 1, j \in N \tag{3}$$

$$x_{ij} \in \{0,1\}, i, j \in N \tag{4}$$

Equation (1) is an objective function where C_{ikjl} is equal to f_{ik} multiply by d_{jl}. That is, $f(i,j)$ represents the amount of flow between facilities i and j, $d(j,l)$ represents the distance between locations i and j. Equation (2) shows that each facility need to be assigned to each location. Equation (3) shows that only one facility can be assigned to one location and all constrains need to be fulfilled. While, Equation (4) shows that x_{ij} = 1 if facility i is assigned to location j and if otherwise, x_{ij} = 0.

2.2. Approaches in solving QAP

Numerous algorithms have been proposed for both exact and approximate solutions such as heuristics and metaheuristics for solving QAP. Exact algorithms are limited to solving small data sets of the QAP with huge parallel computers, whereas metaheuristics is used for a larger size of dataset and can offer near-optimal solutions within reasonable optimization times [22]. Branch and Bound (BB) algorithms are the most ideal approach in exact methods in solving the QAP [23, 24][33].

The first study in solving QAP using BB was proposed by Gilmore in 1962 [35]. BB was used to find an optimal solution for a problem in assigning a single facility to a single location. Duffuaa and Fedjki [36] modified the BB proposed by Gilmore [35] to solve generalized quadratic semi-assignment problem (GQSAP) which allow several facilities to be allocated in a single location. Another exact method used in solving QAP is Dynamic Programming. It has been used by [25] in solving one of the QAP applications which related with facility layout problem. The incomplete dynamic programming concept was proposed in solving the problem. Cutting Plane Method is one of the earliest method used in solving QAP.

Bazaraa and Sherali [26] used cutting plane procedure in solving quadratic assignment problem formulation. This study used QAP formulation for minimizing a concave quadratic function over the assignment polytope. A cutting plane procedure requires a huge computational effort for solving quadratic assignment problem because of the calculation of the lower bound that derived on the number of cuts needed for termination. However, the result gained from the cutting planes produced optimal or good quality solutions early on in the search process [26].

The capability to use QAP in producing many solution of many different problems has made it the subject of wide research area for comprehensive and metaheuristic strategies. Small size QAP datasets are suitable for exact solutions but the larger dataset cannot be solved by using exact solution in reasonable times due to the computational limits [37]. Therefore, metaheuristic approaches have grew a reputation for their ability to produce high-quality solutions within the computational limitations. A survey done by [38] shows that hybrid procedures from different metaheuristic arrangements are the most utilized solution procedure in solving QAP such as Morita and Shio [39] used a hybrid brand and bound method with Genetic algorithm in solving the multi-stage flexible flowshop-scheduling problem.

2.3. Studies related with fixed assignment

Jiang and Hu [21] introduced subset quadratic assignment problem by giving an idea of application situations. The idea is to place facilities to a subset of locations. For example, the hospital proposed plans, which are to assigning departments at suitable units. The strategy is to minimize the overall cost of time or distance concerning the movement of nurses, doctors and patients between the hospital departments. Besides, the locations should also suitable with the assigned department, for example, the wards need to be allocated at the location that have plenty of sunshine and the emergency rooms should be closed to the front gate.

Other similar applications also can be found in the wireless sensor deployment process. For example, body sensor network has usually used to detect the body functionalities of patients. Some fixed position should be determined in order to place different sensor to detect certain body functionality status and place some information flow around deployed sensor nodes. The location for the sensor nodes must be suitable depending on the function, for example in order to detect body temperature the sensors must be placed inside the mouth or under the armpit, meanwhile to detect the hard functionalities the sensors must be place near the pulse or heart [40, 41].

In order to obtain an optimal solution, the most suitable method to use is exact method which is branch and bound for solving a small data. This is because branch and bound make an overall searching without leaving one value in order to search an optimal result value. Since, this study considers solving small dataset of QAPFA, thus exact method is the most suitable for solving the problem. BB is chosen as an approach for solving the proposed problem.

3. Methodology

3.1. Quadratic Assignment Problem with Fixed Assignment

This study proposed a model of QAP with fixed assignment which is called QAPFA. The model considers multiple sets of facilities and locations where some facilities must be fixed allocated at some locations. The mathematical model of the QAPFA is developed by considering the requirement of the facilities to the locations. The mathematical formulation of quadratic assignment problem with fixed assignment is presented as shown below:

$$Min \sum_{i=1}^{n} \sum_{j=1}^{n} \sum_{k=1}^{n} \sum_{q=1}^{n} f_{ik} d_{jq} x_{ij} x_{kq} \quad (5)$$

Subject to;

$$\sum_{j=1}^{n} x_{ij} = 1, i \in N \quad (6)$$

$$\sum_{j=1}^{n} x_{ij} = 1, j \in N \quad (7)$$

$$x_{ij} = 1, \ i, j \in DMpref \quad (8)$$

$$DMpref \subseteq Allpref \ ,$$

$$Allpref = \{(f_i, d_j) \mid f_i \in F, d_j \in L\}$$

$$x_{ij} \in \{0,1\}, i, j \in N \quad (9)$$

Equation (5) is an objective function to minimize the cost of assignment of facility to location by taking into consideration the cost of flow and distance between them. Where n represent the number of total number of facility and location, while, $f(i,k)$ represents the amount of flow between facilities i and k, $d(j,q)$ represents the distance between locations j and q. Equations (6), (7), (8) and (9) are constraints for the problem and each one of the constrains need to be fulfil. Equation (6) shows that each facility need to be assign to each location. Equation (7) shows that only one facility can be assign to one location. Equation (8) indicates fixed assignment where the value of *i* and *j* is determine by decision maker preference or *DMpref. DMpref* refers to the preference of the decision maker on the fixed assignment of facility to location where the *DMpref* is the subset of the *Allpref. Allpref* represent all preferences of the allocation of the facility to the location and *F* represent the facility and *L* represent the location. Equation (9) shows that $x_{ij} = 1$ if facility i is assigned to location j and if otherwise, $x_{ij} = 0$. The illustration of the assignment is show below:

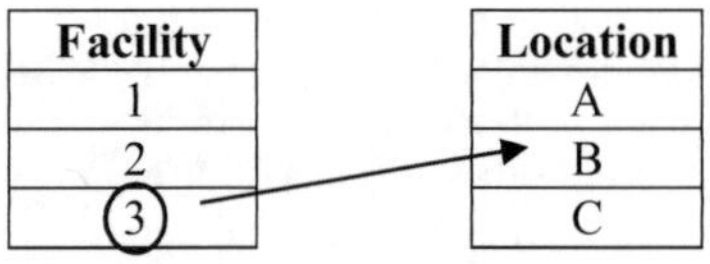

Figure 1. Illustration of QAPFA

$$Min \sum_{i=1}^{n} \sum_{j=1}^{n} \sum_{k=1}^{n} \sum_{q=1}^{n} f_{ik} d_{jq} x_{ij} x_{kq} \qquad (10)$$

Subject to;

$$\sum_{j=1}^{n} x_{ij} = 1, i \in N \qquad (11)$$

$$\sum_{j=1}^{n} x_{ij} = 1, j \in N \qquad (12)$$

$$x_{3B} = 1 \qquad (13)$$

$$x_{ij} \in \{0,1\}, i, j \in N \qquad (14)$$

Figure. 1 is an example for 3 facilities to be assigned to 3 locations with the objective of minimizing cost of allocation by taking into consideration the flow-distance cost. Equation (13) represents the Equation (8) which is *DMpref.* Equation (13) shows the fixed assignment of facility 3 where it must be assigned to location *B*. This is may be due its special requirement that need to be fulfilled where only location *B* can fulfil the requirement of facility 3. Therefore, fixed assignment is written as a constraint as shown in Equation (13).

3.2. Branch and Bound approach

Branch and Bound (BB) is used in solving QAPFA. The first step is to determine the upper bound and the lower bound of the problem solving. The lower and upper bound are used to close up the region of searching the solution value. When the bound value is found, the value is used to help to determine which of the subproblem should be eliminated. Given node S of the tree and the subproblem of the original problem is a child of the S. Usually, each subproblem is derived from the S through intruding a single new constraint and the child of the S are those subproblem. The feasible solution represents the leaves of the tree. An exponential number of the leaves can determine the instance exist in the search tree. The search is continued until the optimal solution is obtained. This branch and bound approach is tested on dataset ranged from size 3 until 15. The dataset size 12 to 15 is from QAPLIB (website) with some modifications on fixed assignment, while other dataset were generated using dummy data.

$$\text{Flow } (f_{ik}) = \begin{matrix} 0 & 5 & 7 & 9 \\ 5 & 0 & 4 & 6 \\ 7 & 4 & 0 & 3 \\ 9 & 6 & 3 & 0 \end{matrix} \qquad \text{Distance } (d_{jq}) = \begin{matrix} 0 & 6 & 8 & 9 \\ 6 & 0 & 5 & 1 \\ 8 & 5 & 0 & 2 \\ 9 & 1 & 2 & 0 \end{matrix}$$

Figure 2. Example of flow and distance matrices of $n = 4$.

Fig. 2 shows an example of the flow and distance matrices. The matrix used is in symmetric where the f_{ik} mean f represent flow, i is row facility and k is column facility. While, d_{jk} means that d represent distance, j is a row distance and q is a column distance. The number in the matrix represent the cost between the facility i to k.

4. Computational results

The QAPFA using BB methods was programmed in Lingo 16.0 on Acer Intel®Core i7 2.00 GHZ computer with 8 GB RAM. In order to validate the QAPFA, the instances with different sizes of QAP are also tested using BB. The computational results and running time for both problems are compared as presented in Table II. The bold font shows the best objective function value, while the italic font represents the minimum running time.

Table 1. Computational results of QAP and QAPFA

Size (*n*)	QAP	Running time (sec)	QAPFA	Running time (sec)	No. Of fixed assignment
3	**116**	0.26	118	*0.05*	1
4	**294**	*0.07*	348	0.11	2
5	**31 968**	*0.18*	42 002	0.20	2
6	**57 720**	0.60	71 902	*0.56*	3
8	**312**	4.98	344	*3.60*	3
10	**174 220**	217.41	196 056	*17.57*	3
12	**9 552**	85.76	23 452	*62.76*	5
14	**1 014**	75957.40	1 064	*197.77*	5
15	**9 896**	17098.87	23 508	*309.22*	5

Table 1 shows the comparisons of the objective function value and the running time between QAP and QAPFA on sizes from $n = 3$ until 15. The objective function value shows that the optimal solutions of QAP are better than the QAPFA. The fixed assignment of QAPFA has made the objective function value to increase. Since some facilities required to have fixed assignment to certain locations, thus the assignment which can give minimum cost never been considered. Hence, it limits the solution search and increased the objective function value of QAPFA compared to QAP. Nevertheless, the running time of QAPFA is shorter than the QAP. Due to fixed assignment of certain facilities to certain locations, thus allowing for the complexity to be reduced significantly compared with QAP. The last column of Table 1. shows the number of assignment that have been fixed for the QAPFA.

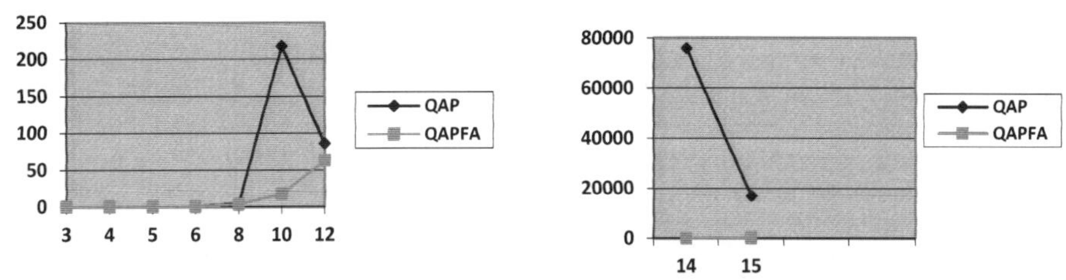

Running Time Comparison between QAP and QAPFA

Figure 3. Line plot for running time for both QAP and QAPFA.

Figure 3 show the comparison of running between QAP and QAPFA using line plot. The Figure 3 shows that QAP has longer running time due to its complexity. However, the dataset 12 and 15 shows a decreasing in the running time. Fig. 4 shows the comparison of the objective function values for dataset 3 to 15 between QAP and QAPFA. The bar chart is divided into two based on the size of the dataset. The bar chat show that QAPFA have a higher value than the QAP. This mean that QAP result produce lower cost that meet the objective of the research which is a minimum cost. The fixes assignment that has been done makes the minimum cost never been considered. The second bar chart represent QAP and QAPFA with high objective function results. The result are higher is because of the value in the dataset itself.

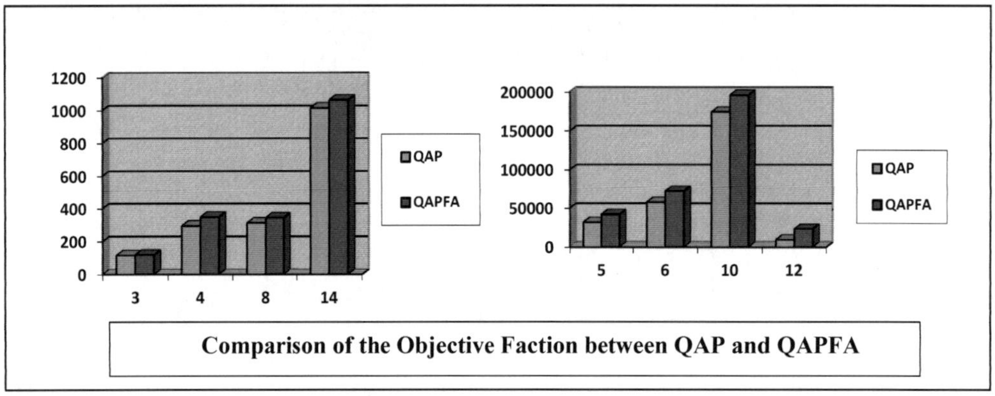

Figure 4. Comparison of result for the of QAP and QAPFA.

5. Conclusion and Future Work

In this study, a QAPFA is introduced which represent a real situation of QAP. Fixed assignment of certain facility to certain location according to their requirement is considered. The mathematical formulation of QAPFA is presented and various sizes of the QAPFA instances is introduced. The problem was solved using Branch and Bound approach. Finally, the results of QAPFA and QAP were compared in terms of its objective function value and running time. Findings showed that the QAPFA produced higher objective function values compared to the QAP because when the fixed assignment is made and make the it less difficult. However, the running time for QAPFA is less than QAP because of the fixed assignment of certain facilities to certain locations that also reduce the complexity compared to QAP. For the future research, we intend to use a larger size of benchmark data and solve it by using a metaheuristic method so see the result on the objective function.

6. Acknowledgement

The authors wish to thank the Ministry of Higher Education Malaysia for funding this study under the Fundamental Research Grant Scheme, S/O code 13586 and RIMC, Universiti Utara Malaysia, Malaysia, for the administration of this study..

7. References

[1] Benjaafar S 2002 *Management Science* **5** p 679.
[2] Duman, E and Or I. 2007 *Computers & Operations Research* **1,** p 163.
[3] Rabak C S and Sichman, J S 2003 *Advanced* Engineering Informatics **2** p 95
[4] Steinberg L 1961 *Siam Review* vol 1 p 37-50
[5] Geoffrion A and Graves G W 1976 *Operations Research* **4** p 595
[6] Malucelli F 1993 *Bulletin-European Association For Theoretical Computer Science* p 533
[7] Grötschel M 1992 *ICIAM* p 119
[8] Jünger M, Martin A, Reinelt G and Weismantel R 1994 *Mathematical Programming* **3** 257
[9] Heffley D R 1977 *Optimal strategies in sports* p 169
[10] Koopmans T and Beckmann M 1957 *Econometrica* **25** p 53
[11] Steinberg L 1961 *Siam Review* **1** p 37
[12] Burkard R E and Rissner R 2011 *Journal of combinatorial optimization* **4** p 845
[13] Pierskalla W F 1967 *Technical Memorandum No. 93, Operations Research Department, CASE Institute of Technology, September*. Available from: http://www.anderson.ucla.edu/faculty/william.pierskalla/Chronological_Bank /Research_and_Publication_Chro.html#Mathematical
[14] Mittelmann H D and Salvagnin D 2015 *Mathematical Programming Computation* **2** 219-234,
[15] Domschke W, Forst P and Voß S 1992 *In New directions for operations research in manufacturing*, p 389
[16] Drwal M 2014 *Optimization* Letters **3** p 1183
[17] Burkard R E, Çela E and Klinz B 1994 On the biquadratic assignment problem In *Quadratic Assignment and Related Problems* [edited by P.M. Pardalos and H. Wolkowicz], DIMACS Series in Discrete Mathematics and Theoretical Computer Science **16** p 117
[18] Mavridou T Pardalos P M Pitsoulis L S and Resende M G 1998 *European Journal of Operational Research* **3** p 613
[19] Knowles J and Corne D 2003 *International Conference on Evolutionary Multi-Criterion Optimization* 295
[20] Eleyan H M 2015 Artificial Bee Colony Optimization for Multiobjective Quadratic Assignment Problem (Doctoral dissertation, Eastern Mediterranean University (EMU)-Doğu Akdeniz Üniversitesi (DAÜ)),.
[21] Jiang H and Hu Y 2012 *In International Conference on Industrial, Engineering and Other Applications of Applied Intelligent Systems* 226 June.
[22] Dokeroglu T 2015 *Computers & Industrial Engineering.* doi: http://dx.doi.org/10.1016/j.cie.2015.03.001.
[23] Hahn P M and Saltzman M J 2010 *Stud. Inform. Univ.* **2** p 97
[24] Duffuaa S O and Fedjki C A 2012 *International Journal of Operational Research* **2** p 185
[25] Urban T L 1998 *Annals of operations research* **76** p 323
[26] Bazaraa M S and Sherali H D 1982 *Journal of the Operational Research Society* **11** p 991
[27] Acan A and Ünveren A 2015 *Applied Soft Computing* 185
[28] Ku M Y, Hu M H and Wang M J 2011 *International Journal of Production Research* **6** p 1801
[29] Ahmed Z H 2015 *IEEE Telecommunications and Signal Processing (TSP)* 5
[30] Sahu S K and Pandey M 2014 *International Journal of Computer Science and Information Technologies* **4** p 5956
[31] Garey M R and Johnson D S 1979 Computers and intractability: A guide to the theory of NP-completeness *New York: WH Freeman*
[32] Kratica J, Tosic D, Filipovic V and Dugosija D 2011 *Yugoslav Journal of Operations Research* **2** p 225

[33] Lim W L, Wibowo A, Desa M I and Haron H 2016 *Computational intelligence and neuroscience* **27** p 1
[34] Lawler E L 1963 *Management Science* **9** p 586
[35] Gilmore P C 1962 *Journal of the society for industrial and applied mathematics* **2** p 305
[36] Duffuaa S O and Fedjki C A 2012 *International Journal of Operational Research* **2** p 185
[37] Loiola E M, Abreu N M M D, Boaventura-Netto P O Hahn P and Querido T 2007 *European Journal of Operational Research* **176** p 657
[38] Zaied A N H and Shawky L A 2014 *International Journal of Computer Applications* **101** p 28
[39] Morita H and Shio N 2005 *International Journal Series C Mechanical Systems, Machine Elements and Manufacturing JSME* **1** p 46
[40] Pottie G J and Kaiser W J 2000 *Communications of the ACM* **5** p 51
[41] Akyildiz I F and Kasimoglu I H 2004 *Ad hoc networks* vol **4** p 351

IOP Conf. Series: Materials Science and Engineering **300** (2018) 012003 doi:10.1088/1757-899X/300/1/012003

Integer programming of cement distribution by train

Indarsih

Department of Mathematics, Universitas Gadjah Mada, Yogyakarta

E-mail: indarsih@ugm.ac.id

Abstract. Cement industry in Central Java distributes cement by train to meet daily demand in Yogyakarta and Central Java area. There are five destination stations. For each destination station, there is a warehouse to load cements. Decision maker of cement industry have a plan to redesign the infrastructure and transportation system. The aim is to determine how many locomotives, train wagons, and containers and how to arrange train schedules with subject to the delivery time. For this purposes, we consider an integer programming to minimize the total of operational cost. Further, we will discuss a case study and the solution the problem can be calculated by LINGO software.

1. Introduction

Railroad transportation has a good role in transportation. Many researchers work in this area. Newman and Yano [2] consider rail transportation of intermodal container. Intermodal transportation consists of combining modes, usually ship, truck, or rail to transport freight. They formulate the problem as an integer programming and develop a novel decomposition procedure to find near optimal solution. Ahuja et.al. [1] study the locomotive scheduling problem with multiple types of locomotives. The locomotives have different pulling and cost characteristic. Locomotives pull a set of trains from their origin to their destinations. The mixed integer is formulated to minimize fixed cost and operational cost.

In Central Java, cement distribution from factory to big city is handled by train. For long distance, train transportation is more efficient than truck transportation. Cement Industry will redesign transportation system with container system. Transport surveys have been conducted on alternative locations loading and unloading to provide an overview of location development unloading and redesigning the loading process. For business feasibility of transportation development then calculation of transport operation pattern is considered that oriented to utilities the means. Design of operation pattern and investment is ideal transport. These design structure are the estimation of volume transportation, design of operation pattern, the number of transportation facility, design of facilities and supporting facilities and finance. The manufactured cement should distribute within a day. Many things need to be reviewed to support the distribution system. Surveys conducted against feasibility of facilities and infrastructure, including station, unloading location, unloading time, warehouse, carriage capacity and long hopper.

2. Model

Based on assesment by Pustral [3], we construct the integer programming model with refer to [4] to help Cement Industry. To meet daily demand in Central Java areas, Cement Industry

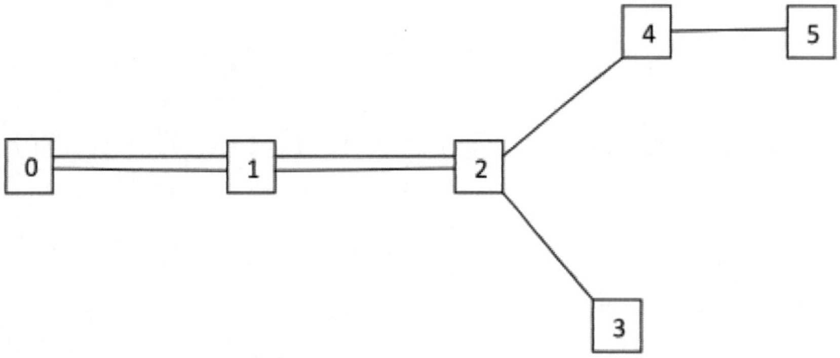

Figure 1. Stations Network

plans the container system to distribute cement. Here, we want to determine the number of locomotive, train wagon, container to support distribution system that minimize fixed cost. And then we want to determine distribution's pattern to minimize the operational cost that is equivalent to minimize the total of delivery times. Integer programming is based on the trans-shipment principle. Output of this model can be used to determine the number of locomotive, train wagon, container and train routes. A distribution networks in Central Java is given in the Figure 1. Node 0 represents a source station and nodes 1,2,3,4,5 represent five destination cities.

In our model, we have some assumptions in the system as follows:

(a) There are two tracks in the rail way i.e. departure and arrival, respectively.

(b) The loading time is not depend on the time distribution. It means that the number of container is enough to support distribution system. Whereas, the time distribution depends on time of departure preparation in source station. They are the same for each train.

(c) The maximum of train's velocity is suitable with the rail way condition. This is influenced by the travel time between two stations.

(d) Locomotive pulls train wagons that contain containers to distribute cement to the destination station. The container is unloaded in the destination station. Then the train continue to delivery to the other destinations. In the way back to the source station, train takes empty container that was unloaded in the destination stations. Except, since train delivers to the station 3, then train should take container after containers are unloaded. After that train continues to deliver to other destination or going back to source station. This policy is applied to efficiency delivery times. So, there is no train that go to station 3 only to take empty containers. Assume that unloading time for containers from train wagon are the same, such that loading time for empty container to train wagon are assumed the same.

(e) Pattern of distribution system is per day. So, in one cycle (one day), each train operate less than 24 hours. Although, to keep a trouble in the trip, we give slack time.

(f) Consider the delivery times from source station to station 1(the longest distance). We assume that there is no train takes a trip twice in one day.

(g) Containers capacity are equal.

(h) Train capacity is equal.

(i) Power of pull locomotive is equal.

According to our assumptions, we let the parameters in the following:

(a) Power of pull locomotive : G train wagons.
(b) Train wagon capacity : K containers.
(c) Container capacity : C tons.
(d) Demand for station j : D_j container. Measure of demand is container, if demand D in ton, we let $D_j = \lceil \frac{D}{C} \rceil$ container.
(e) Locomotive availability: L . Let $L = \frac{\sum_{j=1}^{5} D_j}{G.K}$.
(f) Travel time from station i to station j : $t_{ij}(t_{ij} = t_{ji})$. We have $t_{ij} = \frac{S_{ij}}{V_{ij}^{max}}$, where S_{ij} is distance from station i to station j and V_{ij}^{max} is maximum velocity for each train.
(g) Unloading time of container from train wagon per container :TB.
(h) Loading time of blank container to train wagon per container :TM.
(i) Slack time of each train : TS.
(j) Preparation time in source station :TP.

Furthermore, we let the decision variables in the following :

(a) Decision variable $0 - 1$,

$$L_{lij} = \begin{cases} 1, & \text{if train } l \text{ travels from station } i \text{ to station } j \\ 0, & \text{if train } l \text{ does not travel from station } i \text{ to station } j \end{cases}$$

(b) Non negative integer variable d_{lj}. It means that the number container that unloaded in station j by train l.

To summarize, our problem is to determine the pattern of daily delivery for each train subject to delivery times to minimize the total of operational cost. Because the number of containers is fixed corresponding to demand and each train unload container at delivery and load again empty container to back to source station then TB and TM are not influence in objective function. But this condition influence in the delivery times constrain. Here, we have total time of delivery and going back is the same. So, in the objective function we want to minimize total delivery time of trains. The objective function is formulated in equation 1,

$$\min T = t_{01}\Sigma_{l=1}^{L} L_{l01} + t_{12}\Sigma_{l=1}^{L} L_{l12} + t_{23}\Sigma_{l=1}^{L} L_{l23} + t_{24}\Sigma_{l=1}^{L} L_{l24} + t_{45}\Sigma_{l=1}^{L} L_{l45} \quad (1)$$

The constrains of the model are formulated in the following:

(i) Locomotive capacity constrain(L constrains),
$\Sigma_{j=1}^{5} d_{lj} \leq G.K, \quad l = 1, \ldots, L.$
(ii) Demand constrain for station j (5 constrains),
$\Sigma_{l=1}^{K} d_{lj} \geq D_j, \quad j = 1, \ldots, 5.$
(iii) If train l delivers cement to station j, it should travel the railway ij, where station i exactly after station j. There are $L \times 5$ constrains,
$d_{lj} \leq D_j L_{lij}, \quad \forall l, j.$
(iv) Trip constrain ($4 \times L$ constrains)
Given arrange of station i, j, k. If train travels from station j to station k, It is clear that train travels from station i to station j. Due to Figure 1, we formulate these constrains,
$L_{l12} \leq L_{l01}, \quad \forall l$
$L_{l23} \leq L_{l12}, \quad \forall l$
$L_{l24} \leq L_{l12}, \quad \forall l$
$L_{l45} \leq L_{l24}, \quad \forall l.$

IOP Conf. Series: Materials Science and Engineering **300** (2018) 012003 doi:10.1088/1757-899X/300/1/012003

(v) Travel time constrain for each train (L constrains)
Total of preparation time, slack time, travel time, unloading time, loading time of empty container is less than 24 hours. This constrain can be formulated,

$$TP+TS+2t_{01}L_{l01}+2t_{12}L_{l12}+2t_{23}L_{l23}+2t_{24}L_{l24}+2t_{45}L_{l45}+(TB+TM)\Sigma_{j=1}^{5}d_{lj} \leq 24, \ \forall l.$$

3. Case Study

In this section we give a case study that employe daily demand of cement industry in Central Java. Given parameters in the model. Travel times between two stations is given in Figure 2. Total of demand in ton is given in Table 1.

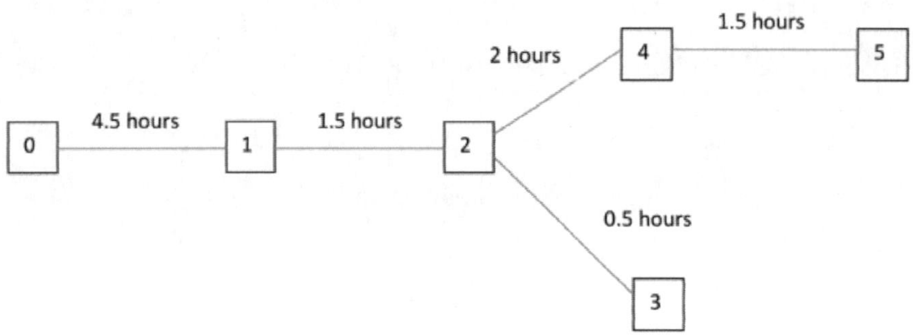

Figure 2. Station Network with Travel Time

Table 1. Demand

Station	Demand(in ton)
Station1	600
Station2	800
Station3	400
Station4	300
Station5	400

Another parameters are $TP = 30$ minutes (0.5 hours), $TS = 60$ minutes(1 hour) and $TB = TM = 3$ minutes (0.05 hours). Based on emplacement data in station, locomotive is permitted to pull 16 train wagons. Container capacity is 20 ton. Train wagon capacity is 2 containers or 40 tons. It means that locomotive capacity is 32 containers. So cement industry should equip $\lceil \frac{120}{32} \rceil = 4$ locomotives. The integer programming model has 40 variables and 49 constrains. We solve the model by LINGO software. The optimal solution has interpretation in the following:

(i) Locomotive 1 pulls 16 train wagons with 32 containers. The number of containers that deliver to station 3 and station 4 are 20 and 12 containers, respectively. The route for this train is 0-1-2-3-2-4-2-1-0 with travel times 20.2 hours.

(ii) Locomotive 2 pulls 16 train wagons with 32 containers. The number of containers that deliver to station 2, station 4 and station 5 are 9, 3 and 20 containers, respectively. The route for this train is 0-1-2-4-5-4-2-1-0 with travel times 22.2 hours.

(iii) Locomotive 3 pulls 16 train wagons. This train deliver to station 1 and station 2 a number of 1 and 31 containers, respectively. The route for this train is 0-1-2-1-0 with time travel 15.2 hours.

(iv) Locomotive 4 pulls 15 train wagons with 29 containers for station 1. Travel times of this train is 21.2 hours.

4. Conclusion

In this research, we construct a model based on station network in Central Java. Furthermore, the model can be developed for another area.

Acknowledgements. The author would like to express deepest gratitude to the Department of Mathematics, Faculty Mathematics and Natural Sciences, Universitas Gadjah Mada, for the financial support in my research.

5. References

[1] Ahuja R K, Liu J, Orlin J B, Sharma D and Shughart L A 2005 Solving Real Life Locomotive Scheduling Problems, *Transportation Science* **39** 503-517.

[2] Newman A M and Yano A Y 2000 Scheduling Direct and indirect Trains and Containers in an Intermodal Setting *Transportation Science* **34** 256-270.

[3] Tim PUSTRAL 2008 *Laporan Hasil Survai Lokasi Pengembangan Angkutan Semen PT. HOLCIM*, PUSTRAL UGM.

[4] Winston W L 2004 *Operation Research Applications and Algorithms* (California:Duxbury Press).

Analysis of portfolio optimization with lot of stocks amount constraint: case study index LQ45

Liem Chin[1], Erwinna Chendra[2], Agus Sukmana[3]
[1, 2, 3] Mathematics Department, Parahyangan Catholic University, Bandung

[1] chin@unpar.ac.id, [2] erwinna@unpar.ac.id, [3] asukmana@unpar.ac.id

Abstract. To form an optimum portfolio (in the sense of minimizing risk and / or maximizing return), the commonly used model is the mean-variance model of Markowitz. However, there is no amount of lots of stocks constraint. And, retail investors in Indonesia cannot do short selling. So, in this study we will develop an existing model by adding an amount of lot of stocks and short-selling constraints to get the minimum risk of portfolio with and without any target return. We will analyse the stocks listed in the LQ45 index based on the stock market capitalization. To perform this analysis, we will use Solver that available in Microsoft Excel.

1. Introduction

Portfolio is defined as a collection of investments that can be composed of various types of assets, such as bonds and stocks. Portfolio optimization is a process of selecting the proportion of assets in a portfolio that make the portfolio better than others based on several criteria, such as: minimum risk and / or maximum return.

Currently, the LQ45 stock index that was launched in February 1997 [6] includes indicators of stocks in the capital market in Indonesia. LQ45 uses 45 preferred stocks with criteria determined by Indonesia Stock Exchange, including liquidity and market capitalization. According to " Buku Panduan Indeks Harga Saham Bursa Efek Indonesia (2010) [3], the value of transactions in the regular market is the main measure of liquidity and since January 2005, the number of trading days and the frequency of transactions as a measure of liquidity has been added. Stocks that enter into the LQ45 index will be evaluated once every 3 months and the replacement of stocks into the LQ45 index is done once every six months, i.e in early February and August.

The problem that investors often encounter in investing, especially with stock instruments, is how to optimize their portfolio. Of course, many criteria can be used to optimize the portfolio. In [3], several criteria are discussed to optimize a portfolio, including minimizing risk, maximizing return, minimizing risk with a certain return target, and maximizing return with a certain target return. In a previous study [4], we have modelled the problem of portfolio optimization consisting of LQ45 group shares with the criteria of minimizing risk and minimizing risk with a certain return target. Both criteria that we use are also added to the terms short-selling is not allowed.

In our previous portfolio model, if the portfolio consists of all shares, then the number of shares purchased based on the proportion of funds obtained can be a positive real number. In fact, retail investors in Indonesia can buy shares in units of lot, where 1 lot equal to 100 shares. Thus, we need to develop an existing model by including the amount of lot of stocks constraint that need to be purchased in order to minimize the portfolio risk, with or without the target return.

2. The Models

Suppose that in a portfolio there are n assets and r_{ij}denotes the percentage return of the asset-i in the period j with $i = 1,2, \dots, n$ and $j = 1,2, \dots, m$ and assumed $m > n$. Let y_i $(i = 1,2, \dots, n)$ denotes the proportion of investment for the asset-i with $\sum_{i=1}^{n} y_i = 1$. The risk of portfolio risk (V) on the Markowitz model [5] can be defined as the variance of the portfolio [1, 2], i.e.

$$V = \boldsymbol{y}^T Q \boldsymbol{y}$$

with

$$\boldsymbol{y} = \begin{pmatrix} y_1 \\ y_2 \\ \vdots \\ y_n \end{pmatrix} \text{ and } Q = \begin{pmatrix} \sigma_{11} & \sigma_{12} & \cdots & \sigma_{1n} \\ \sigma_{21} & \sigma_{22} & \cdots & \sigma_{2n} \\ \cdots & \cdots & \cdots & \cdots \\ \sigma_{n1} & \sigma_{n2} & \cdots & \sigma_{nn} \end{pmatrix}$$

The matrix Q is the variance-covariance matrix of historical stock returns. Biggs [1] has been discussed how to optimize the portfolio with several criteria, including minimizing risk, maximizing return, minimizing risk with specific target returns, and maximizing returns with certain target returns. Some of the criteria were also added to the terms without short-selling.

In the previous study, we have discussed how to minimize the risk with and without a certain return target with the absence of short selling terms. Then, the model we have obtained, is applied to a portfolio of stocks included in the LQ45 index under various scenarios [4].

In this study, we will include the amount of lot of stocks constraint that need to be purchased so that the portfolio risk is minimum. The number of stock lots is a nonnegative integer. Then we will conduct a risk analysis and portfolio return consisting of several stocks with the best average return based on market capitalization whose shares are in the LQ45 group. The market capitalization value of a firm represents the price a person pays to buy the entire company. The value of capitalization can be calculated by multiplying the share price by the number of shares outstanding. We divide the stocks into LQ45 index groups into three groups based on their capitalization value. Group I are companies with a capitalization value greater than 100 trillion, Group II are companies with a capitalization value of 21 trillion to 100 trillion, and Group III are companies with a capitalization value smaller than or equal to 20 trillion. We also analyse the risk and return of a portfolio consisting of several stocks with the best average returns for each group based on this market capitalization value. The purpose of our study is to provide learning for retail investors in preparing an optimal portfolio in which this portfolio consists of only stocks.

In [4], we have discussed two models, they are

1. Minimize the risk of portfolio without a target return and short selling is not allowed.

   Model01

$$\min V = y^T Q y = \sum_{i=1}^{n} \sigma_i^2 \, y_i^2 + 2 \sum_{i=1}^{n-1} \sum_{j=i+1}^{n} \sigma_{ij} \, y_i y_j$$

   with constraints

$$\sum_{i=1}^{n} y_i = 1$$

   and $y_i \geq 0$ for every i.

2. Minimize the risk of portfolio with a particular target return and short selling is not allowed.

   Model02

$$\min V = y^T Q y = \sum_{i=1}^{n} \sigma_i^2 y_i^2 + 2\sum_{i=1}^{n-1}\sum_{j=i+1}^{n} \sigma_{ij} y_i y_j$$

with constraints

$$\sum_{i=1}^{n} y_i = 1$$

$$\sum_{i=1}^{n} \bar{r}_i y_i = R_p$$

and $y_i \geq 0$ for every i. The $\bar{r}_i$ is the mean return for each asset.

Both models are transformed by utilizing penalty functions. Then, the model we have obtained, is applied to a portfolio consisting of stocks included in the LQ45 index with various scenarios [4].

The next problem is the purchase of shares must be in lot (1 lot = 100 shares). The number of shares i need to buy in forming a portfolio is

$$\frac{y_i M}{P_i}$$

with P_i: the current price of stock i and M denoting the amount of funds to be invested. Thus, the number of stock to buy (in lots) is

$$z_i = \frac{y_i M}{100 P_i} \Leftrightarrow y_i = 100\frac{z_i P_i}{M}$$

with z_i: the number of lots of stock i. Since this lot number must be a nonnegative integer ($\mathbb{Z}^+$)), then **Model01** is modified to

Model01M

$$\min V = \boldsymbol{y}^T Q \boldsymbol{y}$$

with constraints (short selling is prohibited)

$$100\sum_{i=1}^{n} z_i P_i = M$$

$$z_i \in \mathbb{Z}^+$$

In a similar way, **Model02** is modified to

Model02M

$$\min V = \boldsymbol{y}^T Q \boldsymbol{y}$$

with constraints (short selling is prohibited)

$$100\sum_{i=1}^{n} z_i P_i = M$$

$$\frac{100}{M}\sum_{i=1}^{n} \bar{r}_i z_i P_i = R_p$$

$$z_i \in \mathbb{Z}^+$$

In both **Model01M** and **Model02M**, all entries of y_i's are replaced with $100\frac{z_i P_i}{M}$. Since the result is a positive integer, the two models are not transformed as we did in [4]. However, we use the Solver which is a Microsoft Excel tool that is quite easy to use.

3. The Results

In this study, we will determine the number of shares in the LQ45 index to be selected along with the number of lots of z_i that need to be purchased from each of these stocks so that portfolio has the least risk. We use historical stock price data from July 1, 2015 to June 30, 2016. We download this data on July 10, 2016 at yahoo finance [7]. From this data we calculate the average return of each stock. To form a variance-covariance matrix Q, the formula [1, 2] is used:

$$\sigma_{ii} = \sigma_i = \frac{1}{m}\sum_{j=1}^{m}\left(r_{ij} - \bar{r}_i\right)^2$$

and

$$\sigma_{ij} = \frac{1}{m}\sum_{k=1}^{m}\left(r_{ik} - \bar{r}_i\right)\left(r_{jk} - \bar{r}_j\right)$$

with $\bar{r}_i$ representing the average return of the stock i. The parameters we use for the following results are $M =$ Rp 100 million and the accuracy of each constraint is 10^{-5}.

From the historical data of stocks price, we get that there are 14 stocks whose average return is negative. Therefore, the 14 stocks we do not use in the formation of portfolio. Thus, we only use 31 shares of 45 stocks in the LQ45 group.

The following is given two tables of selected portfolio consisting of stocks in the LQ45 index. Table 1 and Table 2 are each portfolios formed using **Model01M** and **Model02M**. From both tables it is seen that the target return constraints play a role in the formation of portfolio. In Table 2, the portfolio consists of more stocks (13 stocks) than Table 1 (only 7 stocks). For a portfolio with a 0.02% return target provides greater risk than a portfolio with no target return.

Table 1. Selected portofolio using Model01M

Stock	Lots
ADHI	83
WSKT	25
AKRA	19
MYRX	353
PTBA	3
LPKR	68
BBCA	16
$V = 0.017\%$	
$R = 0.092\%$	

Table 2. Selected portofolio using Model02M

Stock	Lots	Stock	Lots
WKST	159	TLKM	3
PWON	89	SCMA	1
CPIN	57	SMRA	1
HMSP	1	PTPP	1
BBTN	56	MYRX	3
ANTM	10	SRIL	1
ICBP	7		
$V = 0.026\%$			
$R_p = 0.02\%$			

Next, we divide the 31 stocks by market capitalization. Market capitalization is total market value of all of company's outstanding stocks. This value is calculated by multiplying a company's stocks outstanding by the current market price of one stock. We group the stocks in the LQ45 index into 3 groups based on market capitalization, that is, Group 1 is the company with market capitalization larger than 100 trillion Rupiahs (9 stocks), Group 2 is the company with market capitalization between 20 and 100 trillion Rupiahs (15 stocks), and Group 3 is the company with market capitalization under 20 trillion Rupiahs (7 stocks).

Similar as we do before, the following five tables will be displayed which state the number of lots needed to form the portfolio by using the **Model01M** and **Model02M** consisting of small, medium and big market capitalization stocks, respectively. Nevertheles, there is no feasible solution for selected portfolio using the Model02M for big market capitalization.

Table 3. The selected portfolio consists of small cap stocks using Model01M

Stock	Lots
ANTM	2
BBTN	59
MYRX	355
PTBA	2
PTPP	58
SRIL	40
SSMS	73
WIKA	82
$V = 0.016\%$	
$R = 0.074\%$	

Table 4. The selected portfolio consists of small cap stocks using Model02M

Stock	Lots
ADHI	247
ANTM	14
BBTN	125
MYRX	33
SRIL	305
$V = 0.06\%$	
$R_p = 0.2\%$	

Table 5. The selected portfolio consists of medium cap stocks using Model01M

Stock	Lots
ADRO	105
AKRA	9
ICBP	15
JSMR	25
LPKR	117
PWON	40
WSKT	123
$V = 0.019\%$	
$R = 0.136\%$	

Table 6. The selected portfolio consists of medium cap stocks using Model02M

Stock	Lots
ADRO	6
BSDE	27
ICBP	18
PWON	148
SMRA	2
WSKT	216
$V = 0.06\%$	
$R_p = 0.2\%$	

Table 7. The selected portfolio consists of big cap stocks using Model01M

Stock	Lots
BBCA	14
BBRI	54
UNVR	1
TLKM	48
$V = 0.033\%$	
$R = 0.055\%$	

A moderate investor (who can accept the risk with the middle level) can form a portfolio without a target return with reference to Table 5. In this table, the investor can expect a return of 0.136% with the risk of 0.019%. The expected return is greater than the portfolio of Table 7 and slightly smaller than the portfolio Table 3. Whereas if an investor wants to expect a return of 0.2%, then the investor can form a portfolio of Table 4 or Table 6 which both provide the same risk, i.e. 0.06%.

4. Conclusions and Further Research

From the above discussion, we can conclude that the portfolio consisting of LQ45 index with market capitalization of between Rp 20 and Rp 100 trillion is relatively better (in terms of less risk and bigger returns) than other portfolios. For further research, we will include the cost of stock purchase transactions in the formation of the portfolio. In addition, portfolio rebalancing also seems to be considered to keep the portfolio set up by investors to provide a return that matches their expectations.

References

[1] Biggs M C B 2005 *Nonlinear optimization with financial applications* (London: Kluwer Academic Publisher).

[2] Biggs M C B and Kane S J 2009 A global optimization problem in portfolio selection *Computational Management Science* **6** (3) 329-345.

[3] Bursa Efek Indonesia 2010 *Buku panduan indeks harga saham Bursa Efek Indonesia.*

[4] Chin L, Chendra E, and Sukmana A 2015 Analysis of portfolio optimization consisting of stocks in the LQ45 index *ICMAME Proc. Sanata Dharma University.*

[5] Markowitz H 1952 Portfolio selection *The Journal of Finance* **7** (1) 77-91.

[6] www.idx.co.id.

[7] http://finance.yahoo.com

IOP Conf. Series: Materials Science and Engineering **300** (2018) 012005 doi:10.1088/1757-899X/300/1/012005

Optimization of Personnel Assignment Problem Based on Traveling Time by Using Hungarian Methods: Case Study on the Central Post Office Bandung

Sudradjat Supian[1], Sri Wahyuni[2], Julita Nahar[3], Subiyanto[4*]

[1,2,3]Department of Mathematics, Faculty of Mathematics and Natural Science, Universitas Padjadjaran, Indonesia

[4]Department of Marine Science, Faculty of Fishery and Marine Science, Universitas Padjadjaran, Indonesia

*Corresponding author: subiyanto17@gmail.com

Abstract. In this paper, traveling time workers from the central post office Bandung in delivering the package to the destination location was optimized by using Hungarian method. Sensitivity analysis against data changes that may occur was also conducted. The sampled data in this study are 10 workers who will be assigned to deliver mail package to 10 post office delivery centers in Bandung that is Cikutra, Padalarang, Ujung Berung, Dayeuh Kolot, Asia-Africa, Soreang, Situ Saeur, Cimahi, Cipedes and Cikeruh. The result of this research is optimal traveling time from 10 workers to 10 destination locations. The optimal traveling time required by the workers is 387 minutes to reach the destination. Based on this result, manager of the central post office Bandung can make optimal decisions to assign tasks to their workers.

1. Introduction

The personnel assignment problem is a special case of the transportation problem. It arises in a variety of decision-making situations. A distinguishing feature of the personnel assignment problem is that one worker is assigned to one and only one task [1].

In general, the assignment problem includes n tasks that must be assigned to n workers where each worker has different competencies in completing each task [8]. The purpose of the assignment problem is to assign each task appropriate to the worker so that the total expenditure of resources to complete all tasks can be optimized. The optimized resource can be an assignment fee, time spent on completing tasks, mileage and so on.

Many researchers have been applied assignment problem to solve decision-making situations. A new approach to one sided assignment problems was developed by Sasaki [6]. Maxon and Bhadury [3] proposed assignment problem with repetitive tasks and tried to introduce a human element into the analysis. A simple random assignment problem with a unique solution was proposed by Bogomolnaia and Moulin [2]. Nuass [4] described a special purpose branch-and-bound algorithm for solving assignment problems. Sourd [7] studied the continuous assignment problem with the aim of solving scheduling problems with irregular cost functions. Odior et. al. [5] addressed a problem of effectiveness of feasible solutions of assignment problems.

This research discusses how to optimize personnel assignment problem then solved by using Hungarian method. This optimization process is applied to case study of the central post office Bandung in assigning employees to deliver the packet to the destination location based on several criteria owned by each employee and also conducted a sensitivity analysis of data changes that may occur so as not to change the optimal assignment from the initial problem.

2. Materials and Methods

The concept of assignment problem based on traveling time by using Hungarian methods was applied to solve a problem for case study on the central Post Office Bandung. Data collected (estimated time requirements in minutes) from the central Post Office Bandung is shown in Table 1 below.

Table 1. Traveling time (in minutes)

Worker	Area									
	I	II	III	IV	V	VI	VII	VIII	IX	X
1	26	40	27	35	39	70	30	75	63	30
2	30	45	30	32	35	75	30	79	62	32
3	29	30	20	32	30	83	34	80	62	30
4	26	32	25	30	36	80	29	75	60	29
5	29	37	27	31	35	85	27	76	60	30
6	27	37	30	30	36	90	30	69	57	35
7	25	42	30	35	30	69	34	66	58	36
8	30	41	26	28	30	87	28	67	58	37
9	25	41	29	36	40	90	30	78	60	34
10	25	40	23	30	38	88	32	68	62	28

Table 1 shows that the traveling time of 10 employees to deliver mailing packets to 10 post office delivery centers in Bandung that is Cikutra (I), Padalarang (II), Ujung Berung (III), Dayeuh Kolot (IV), Asia-Africa (V), Soreang (VI), Situ Saeur (VII), Cimahi (VIII), Cipedes (IX) and Cikeruh (X).

3. Result and Discussion

The goal of the assignment problem is to minimize the traveling time of of 10 employees to deliver mailing packets to 10 post office delivery centers in Bandung. An important characteristic of this assignment problem is the number of workers is equal to the number of destinations. It is explained in the following way.

- Only one job is assigned to worker.
- Each worker is assigned with exactly one destination.

Based on the data collected, manager of the Central Post Office Bandung has ten workers for ten separate destinations and the traveling time of assigning each destination to each worker is given. His goal is to assign one and only destination to each worker in such a way that the total traveling of assignment is minimized.

Balanced assignment problem: The number of workers is equal to the number of destinations.
The mathematical model of the assignment problem in this case can be written as follows:

$$\text{Min T} = \sum_{i=1}^{10}\sum_{j=1}^{10} c_{ij}x_{ij}$$

$$s.t\sum_{j=1}^{10} x_{ij} = 1 \quad ; i = 1,2,..,10$$

$$\sum_{i=1}^{10} x_{ij} = 1 \quad ; j = 1,2,..,10$$

$$x_{ij} = 0 \text{ or } 1, \text{ for } i \text{ and } j$$

where:

$$x_{ij} = \begin{cases} 1, \text{if worker } i \text{ go to destination } j \\ 0, \text{if worker } i \text{ does not go to destination } j \end{cases}, i = 1,2,\ldots,10 \text{ and } j = 1,2,\ldots,10.$$

Based on data in the Table 1, the optimal assignments matrix was produced by applied the Hungarian method in this case as shown in Table 2 below.

Table 2. Optimal Assignment Problem Matrix

Worker	Area									
	I	II	III	IV	V	VI	VII	VIII	IX	X
1	3	7	4	7	9	0	4	8	7	2
2	3	8	3	0	1	1	0	8	2	0
3	9	0	0	7	3	16	11	16	9	5
4	4	0	3	3	7	11	4	9	5	2
5	5	3	3	2	4	14	0	8	3	1
6	3	3	6	1	5	19	3	1	0	6
7	3	10	8	8	1	0	9	0	3	9
8	7	8	3	0	0	17	2	0	2	9
9	0	6	4	6	8	18	2	9	2	4
10	2	7	0	2	8	18	6	1	6	0

Based on the optimal solution in Table 2, the assignment of workers from Central Post Office Bandung to each post office delivery centers in Bandung can be seen in Table 3 below.

Table 3. Workers assignment determination

Worker	Post office delivery centers in Bandung
1	Soreang
2	Dayeuh Kolot
3	Ujung Berung
4	Padalarang
5	Situ Saeur
6	Cipedes
7	Cimahi
8	Asia-Afrika
9	Cikutra
10	Cikeruh

By adjusting the decision variables with the initial table (Table 1) then the total traveling time (T) is 387 minutes.

In order to determine the data changes range of traveling time that do not change the assigned optimal assignment of workers, the sensitivity analysis was done. Sensitivity analysis to change of objective function coefficient from traveling time data was obtained as shown in Table 4.

Table 4. Range of Travel Time Coefficient

Worker	Area									
	I	II	III	IV	V	VI	VII	VIII	IX	X
1	$[23,\infty)$	$[33,\infty)$	$[23,\infty)$	$[28,\infty)$	$[30,\infty)$	$[-\infty,71]$	$[26,\infty)$	$[67,\infty)$	$[56,\infty)$	$[28,\infty]$
2	$[27,\infty)$	$[37,\infty)$	$[27,\infty)$	$[-\infty,33]$	$[34,\infty)$	$[74,\infty)$	$[29,31]$	$[71,\infty)$	$[60,\infty)$	$[31,33]$
3	$[20,\infty)$	$[29,31]$	$[-\infty,21]$	$[25,\infty)$	$[27,\infty)$	$[67,\infty)$	$[23,\infty)$	$[64,\infty)$	$[53,\infty)$	$[25,\infty)$
4	$[22,\infty)$	$[-\infty,33]$	$[22,\infty)$	$[27,\infty)$	$[29,\infty)$	$[69,\infty)$	$[25,\infty)$	$[66,\infty)$	$[55,\infty)$	$[27,\infty)$
5	$[24,\infty)$	$[34,\infty)$	$[24,\infty)$	$[29,\infty)$	$[31,\infty)$	$[71,\infty)$	$[-\infty,28]$	$[68,\infty)$	$[57,\infty)$	$[29,\infty]$
6	$[24,\infty)$	$[34,\infty)$	$[24,\infty)$	$[29,\infty)$	$[31,\infty)$	$[71,\infty)$	$[27,\infty)$	$[68,\infty)$	$[-\infty,58]$	$[29,\infty)$
7	$[22,\infty)$	$[32,\infty)$	$[22,\infty)$	$[27,\infty)$	$[29,\infty)$	$[68,70]$	$[25,\infty)$	$[-\infty,67]$	$[55,\infty)$	$[27,\infty)$
8	$[23,\infty)$	$[33,\infty)$	$[23,\infty)$	$[27,29]$	$[-\infty,31]$	$[70,\infty)$	$[26,\infty)$	$[66,68]$	$[56,\infty)$	$[28,\infty)$
9	$[-\infty,26]$	$[35,\infty)$	$[25,\infty)$	$[30,\infty)$	$[32,\infty)$	$[72,\infty)$	$[28,\infty)$	$[69,\infty)$	$[58,\infty)$	$[30,\infty)$
10	$[23,\infty)$	$[33,\infty)$	$[22,24]$	$[28,\infty)$	$[30,\infty)$	$[70,\infty)$	$[26,\infty)$	$[67,\infty)$	$[56,\infty)$	$[-\infty,29]$

Based on Table 4, it is obtained the range of coefficients changes from the objective function allowed for each worker to reach the destination location does not change the initial assignment.

4. Conclusion

In this paper, the concept of assignment problem has been applied to solve a problem for of the central post office Bandung which had a difficulty in assigning 10 workers to 10 destination locations of post office delivery centers. Based on the data collected, Hungarian Method was used to solve the problem. Optimal assignments of the cases were obtained for the central post office Bandung. It was obtained that, the optimal assignment of workers from Central Post Office Bandung to each post office delivery centers as follows: worker 1 to Soreang, worker 2 to Dayeuh Kolot, worker 3 to Ujung Berung, worker 4 to Padalarang, worker 5 to Situ Saeur, worker 6 to Cipedes, worker 7 to Cimahi, worker 8 to Asia-Afrika, worker 9 to Cikutra, worker 10 to Cikeruh. The total optimal number of travelling time of the cases is 387 minutes. The Sensitivity Analysis for assignment issues can illustrate the range of coefficient value changes for the allowed objective function so that this does not change the optimal assignment of the initial problem.

Acknowledgment

We would like to thank the financial support from Universitas Padjadjaran through the grant program of the Academic Leadership Grant (ALG) and Central Post Office Bandung for providing travelling time data.

References

[1] Boah, D.K., Adu, I. K., and Gyebil, F. J. 2015. Assignment problem of a legal firm in Kumasi, Ghana. *International Journal of Computing and Optimization*, 2(1),1-5.

[2] Bogomolnaia, A. and Moulin, H. 2002. A Simple Random Assignment Problem with a Unique Solution, *Economic Theory*, 19, 623 - 636

[3] Maxon, S. L. and Bhadury J. 2001. An Ms-excel implementation of a multi-period assignment problem with repetitive tasks, *Proceedings of the 13th Annual CSUPOM Conference,*

IOP Conf. Series: Materials Science and Engineering **300** (2018) 012005 doi:10.1088/1757-899X/300/1/012005

California State University San Bernardino, 39–48.

[4] Nauss, R. M. 2003. Solving the generalized assignment problem: an optimizing and heuristic approach, *INFORMS Journal on Computing*. 15(3), 249 – 266.

[5] Odior, A.O., Owaba, C. and Oyawale, O.E. 2010. Determining Feasible Solutions of a Multicriteria Assignment Problem, *Journal of Applied Sciences and Environmental Management*, 14(1), 35-38.

[6] Sasaki, H. 1995. Consistency and Monotonicity in Assignment Problems, *International Journal of Game Theory*, 24, 373-397.

[7] Sourd F. 2004. The continuous assignment problem and its application to pre-emptive and non-pre-emptive scheduling with irregular cost functions, *Informs Journal on Computing*, 16(2), 198-208.

[8] Taha, H. A.1996. *Riset Operasi, Jilid 1, Edisi ke-5*. Jakarta: Binarupa Aksara.

IOP Conf. Series: Materials Science and Engineering **300** (2018) 012006 doi:10.1088/1757-899X/300/1/012006

Stochastic Optimization in The Power Management of Bottled Water Production Planning

Budi Antoro[1], Esther Nababan[2] and Herman Mawengkang[3]
Department of Mathematics, Universitas Sumatera Utara, Medan, Indonesia
E-mail: [1]budiantoro01@gmail.com; [2]esther@usu.ac.id; [3]hmawengkang@yahoo.com

Abstract. This paper review a model developed to minimize production costs on bottled water production planning through stochastic optimization. As we know, that planning a management means to achieve the goal that have been applied, since each management level in the organization need a planning activities. The built models is a two-stage stochastic models that aims to minimize the cost on production of bottled water by observing that during the production process, neither interfernce nor vice versa occurs. The models were develop to minimaze production cost, assuming the availability of packing raw materials used considered to meet for each kind of bottles. The minimum cost for each kind production of bottled water are expressed in the expectation of each production with a scenario probability. The probability of uncertainly is a representation of the number of productions and the timing of power supply interruption. This is to ensure that the number of interruption that occur does not exceed the limit of the contract agreement that has been made by the company with power suppliers.

Keyword: Optimization, Stochastic, Production Planning, Power Supply, Water Supply

1. Introduction

Population growth is rapidly increasing life these days that led to once of dependence on the higher basic needs. Humans who are part of the high growth of popolasi on earth to provide many impacts for the existing ecosystem sustainability. Its need for basic needs such as water encourage its procurement for the fulfillment of daily consumption (Strang, 2004). The high level of human activity makes the need for water both as drinking water and the fulfillment of other need is higher. Drinking water is the main need among the various needs that must always be sought to support the existence of existing existence. Various efforts were developed by humans to be able continue to meet the need of drinking water, such as water management through the bottled water industry (Ahmad and Bajahlan, 2009).

The rapid development of the drinking water industry lately, encouraging each company to develop its competitive ability. This is because the high level of demand by consumers of different types of bottled drinking water. Based on that, it is necessary to have a separate planning system of production. Planning is a very important activity in the production process. Meanwhile, production planning can be defined as a planning of what products and how many products will be produced within a certain period that will be related to the availability of raw materials, labor, machinery and other equipment. Planning is also the main managerial function for a company, which is the direction or direction to coordinate and cooperate in the company's overall operation. Various things can happen in a productivity. Certainty in production planning is defined as a deterministic, meaning that production planning of a goods is a thing that is certain and inevitable. Community needs of certain goods become the basis of a company to produce the goods. A high level of demand is a benchmark for companies to be confident that the provision of such goods will provide substantial benefit. This is what is mean by the provider of the goods.

Risk and uncertainty are factors that arise from situations that need to be considered and foreseen to overcome the unexpected later. Uncertainty defined by Galbraith (2005) as the difference between the amount of information needed to perform a particular task in production with the amount of information that has been previously owned. Realy, there are many forms of uncertainty affecting the production process: environmental uncertainty in the form of demand uncertainty and supply uncertainty, and system uncertainties such as uncertain operational results, production system failures, product quality uncertainty and product structure changes (Ho, C, 1989). While the risk is one of the impact of uncertainty that will affect the demand for production planning for the next period. Disruption (*interruption*) in the production process is often a constraint of a company in the production process. Many factors blamed as a disruption in the production process can be discussed in production planning such as natural conditions that allow disasters to occur during disasters (earthquakes, hurricanes and storms), unforeseen possibilities in operation (accidents in logistics and human error), instability political, security and public health (Yang, et al, 2005), even the most common is a power outage.

2. Literature Review

2.1. Prodction planning

Planning is the most basic and very extensive management functions including estimates and calculations of the activities to be carrie out in the future following a certain rule. Planning is a means of management to achieve the goals that have been applied therefore every level of management in the organization is in dire need of planning activities (Ivancevich and Thomas, 2007). In many organizations, production planning is part of hierarchical planning, capacity / having allocation, scheduled and controlled resources that take into account the value of capacity, time period, supply and demand and so on. Production planning also talks about the product and the amount of production that the company will produce. Therefore, the interaction between production planning and scheduling is inevitable, not only scheduling / controlling of decisions that constitute a separate constraint in production decisions, but also because a disruption or constraint its implementation which may result in the optimization and / or likely of both plan and scheduling (Cai, Yiwei, et, al, 2011). In production planning, inventory of goods and raw materials into part that are not separated in attention. In conducting inventory management, the type of goods or products must also be a concern in determining the optimal policy. This is because not all types of raw materials or products can be last a long or have expiration limi that can cause harm to the company or buyer later. It is called deterioration. Forms of deterioration can be various kinds, such as the loss (damage), product failure / defect (spoilage) or drynase. Inventory control is the activities and techniques of stockpiling goods at a certain level, both in the form of raw materials and materials in the process and finished products.

2.2. Bottled Water

Water is the most important substance in life after air. Three-quarters of the human body consists of water. Water is also the most severe substance due to contamination. Diseases that attack humans can be transmitted and distributed through water. Water has been regarded as one of the most important natural resources in the life on earth and its availability has been condition for everything first in economic activity (Gibbons, 1986). At the same time, it should be remember that water is a limited resource that has generated many conflicts over the allocation of its use and its hygienic water availability (Lee, 1999). Judging from this complex role, the availability of water as the main consumption of production factors in the industrial or economic field then obliges the government to invest in infrastructure that ensures its availability which is often risky to the many disruption of water supply. Drinking water is water used for human consumption. The provision of clean water for drinking water needs to meet the requirements set by the government. Drinking water is safe for health when it meets the requirements Physics, microbiology, chemistry, and radioactivity. The mandatory parameter determining the microbiological quality of drinking water is total coliform bacteria and Escherichia coli. According to the health department, the terms of drinking water is a tasteless, odorless, colorless, does not contain microorganisms that are harmful, and do not contain heavy metals. Drinking water is water that through the process of processing or without processing that meet health requirements and can be directly in drinking. Currently there are the drinking water crisis in many developing countries in the world due to the number of people who are too many and water pollution. Demographic pressures and economic development led to a continued increase in demand for water consumption, while at the same time climatic conditions are also factors that limit water availability in many geographic zones that are increasingly seen worsening as a result of the drastic climate change process lately (Doria, 2010). Many factors that cause water supply disruption include the lack of care of the community to protect the environment so as to pollute the ecosystem that preserves the cleanliness suppy, natural conditions dipasing-masig areas tend to be different, the use of uncontrolled water and so forth. It will unwittingly trigger the emergence of policies governing the use of which limits water consumption. As it is said that water is important in human life, its disruption will also have an impact on human activities. In this case will be examined further impact on the industrial field. It is well known that the quality of clean water supply is important to maintain the stability of the production system in the field of economy and industry. Due to the high disruption of water supply in the industrial field, forcing the company to make the best decision in looking at and solve the problem. It is intended to suppress the monopoly in industrial competition (Tschirhart and Jen, 2006). In the meantime, the disruption of water supply that occurs resulting in a decrease in the production power of each company that impacts on the unfulfilled number of requests. Another case with the damage that may occur due to power failure with a sudden disconnection.

2.3. Power Supply

Electricity is a part that can not be separated from human life especially in the globalization and the rapid development of the curren. According to Culter-Hemmer (1999) electricity is a charge consisting of positive charge and negative charge by jumping or moving electrons from its orbit to produce electron flow. The development of science requires the fulfillment of electricity consumption in every joint of life, whether on a small scale such as households as well as a wider scope such as factories and other industries. Therefore, the empowerment of natural resources continue to be improved along with the increasing power consumption. Starting from the development power of solar energy, power of air and water energy, power of diesel energy, to the most recent is the use of nuclear. As it is said that electricity is an important thing in the joints of human life, its interference will also have an impact on human activities. In this case will be examined further impact on the industrial field. It is well known that the quality of electric power is essential for maintaining the stability of the power system. A production process in the industry rely heavily on electrical equipment to maintain the continuity of the production process (Borgeson et al, 2015). The problem of power failure is a very complicated problem because it involves several factors, such as the lack of experts in handling the disrupton, the maintenance of the generator machine, the poor quality of equipment, the theft of electric power and so forth. Efforts to overcome these disorders continue to be pursued. This is according to Borgeston (2015) due to the magnitude of the role of electricity in human life which is also the

IOP Conf. Series: Materials Science and Engineering **300** (2018) 012006 doi:10.1088/1757-899X/300/1/012006

most important contributor in the field of industry for the economic progress of a region and the state demanded fulfillment maximally.

2.4. Production Planning Model With Stochastic Two Stage

According Barik (et al 2013), the problem of stochastic two-stage program was formulated to several optimization decisions which obtained in the two stages are not the same. First-stage decisions are obtained before the realization of random events and second-stage decisions are obtained after realization. "The two-stage stochastic program model is the conversion of stochastic problems into an equivalent deterministic problem. In the two-stage stochastic program model, the decision variable is divided into two parts. The first stage is to determine the decision variable of the uncertainty parameter. Next make an objective function to minimize cost and expectation value in the first stage. The market situation is relatively volatile followed by the economic development of each different country is very influential on the demand for certain goods products. Uncertainty is what makes a service provider company should see all possible risks that must be faced. The production planning model with stochastic optimization is designed to contribute to the uncertainty. The two-stage stochastic model can be written as follows.

Set:

C : Set of Contract Type
T : Set of Time
M : Set of Many Production Machines
L : The Hour of May Workers
U : Set of Uncertainly
I : Set of Power Supply Disruption

Index:

c : Contracts Type
t : Time
l : Workers
i : Power Supply Disruption
Pro : Production

Deterministic Parameter:

$\prod^s$: Function Opportunities of Scenario is s
A_t^{pro} : Total Production time t (unit)
C_t^{pro} : Production Cost while t (Rp)
t : Time to Producing per Unit (Hour/Unit)
$A_{m,t}^{pro}$: Number of Machines at t (Unit)
C_l : Worker‘s Salary without overtime (Rp)
$A_{l,t}^{pro}$: Number of Employed for Producing the Time t (People)
C_{cl} : Cost of Power Contract (Rp/KWh)
$K_{m,t}^{pro}$: Machine Capacity to Producing Per Production Time (Unit/Hour)
C_m^{rev} : Machine Maintenance Cost (Rp)
d_{Pro} : Demand of the Product at t (Unit)
$C_{Gen,i}$: Operational Cost of Generator when Power Supply Disruption (Rp/Hour)
$C_{l,c2,i}$: Overtime Worker‘s Salary when Power Supply Disruption (Rp/Hour)
inv_0^{Pro} : Early Production Inventory
$K_{Inv,t}^{Pro}$: Inventory Production Capacity at t
K_i : Maximum Power Supply Disruption Happens

Variable:

$Z(\cdot)$: Objective Function
X_t^{Pro} : Total Production at t
$inv_{pro,t}$: Inventory of Currennt Production t

Production planning models with Two Stage Stochastic

Minimize :

$$Z(.) = \sum_{s=1}^{s} \pi^s \left[A_t^{pro}.t. \sum_{m\in M, l\in L, t\in T} A_{l,t}^{pro} \cdot A_{m,t}^{pro} \cdot K_{m,t}^{pro} + \sum_{l\in L, m\in M, t\in T} C_t^{pro} \cdot C_{l,t} \cdot C_m^{rev}.C_{c1} \right]$$
$$+ \left[A_{t(i)}^{pro} \cdot t_i \cdot \sum_{m\in M, l\in L, t\in T} A_{l,t(i)}^{pro} \cdot A_{m,t(i)}^{pro} \cdot K_{m,t(i)}^{pro} + \sum_{l\in L, t\in T} C_{Gen,(i)} \cdot C_{l,c2,(i)} \cdot K_{(i)} \right] \quad (1)$$

subject to:

$$\sum_{t=1}^{T} (c_{t,(i)}^{Pro} \cdot A_{m,t,(i)}^{Pro}) \cdot A_{l,t,(i)}^{Pro} \cdot C_{l,c2,(i)} + \sum_{m=1}^{M} (A_{m,t,(i)}^{Pro} \cdot K_{m,(i)}^{Pro}) \cdot C_{gen,(i)} \cdot K_{(i)}$$
$$\leq \sum_{t=1}^{T} (c_{t,}^{Pro} \cdot A_{m,t,}^{Pro}) \cdot A_{l,t}^{Pro}) \cdot C_l + \sum_{m=1}^{M} (C_m^{rev} \cdot K_m^{Pro}) \cdot C_{c1} \quad (2)$$

$$A_{l,t}^{Pro} A_{m,t}^{Pro} > 0 \ \forall l \in L, m \in M$$

$$C_t^{Pro} C_l C_{c1} C_{gen} > \ \forall l \in L, t \in T$$

$$X_{t(i)}^{Pro} \geq \sum_{l\in L, t\in T} C_t^{Pro} \cdot C_l \cdot A_l^{Pro} \cdot C_{c1} + \sum_{m\in M} A_m^{Pro} \cdot K_{m,t}^{Pro} \quad (3)$$

$$X_{t(i)}^{Pro} \geq \sum_{l\in L, t\in T} C_t^{Pro} \cdot C_{lc2(i)} \cdot A_l^{Pro} \cdot C_{Gen(i)} + \sum_{m\in M} A_m^{Pro} \cdot K_{m,t}^{Pro} \quad (4)$$

$$A_l^{Pro}, X_0^{Pro}, A_m^{Pro} > 0 \ \forall l \in L, t \in T, m \in M$$

$$inv_{pro,t} = inv_0^{Pro} + \sum_{t\in T} X_t^{Pro} - \sum d_{Pro} \quad (5)$$

$$X_t^{Pro}, d_{Pro} > 0 \ \forall t \in T, Pro$$

Equation (4.1) is a two-stage stochastic model for the bottled water production planning by looking at the lowest cost of the two production planning processes ie production planning normally or without any disruption of power supply (term 1) and production planning in the event of power supply disruption during processing Production (term 2). The above built model is intended to minimize the cost of production when there is a power supply disruption occurring at any time through a power supply termination scenario. This variable is assumed by the availability of packing raw materials that are considered to meet to produce each type of beverage packaging. While Equations (4.2) to (4.4) are constraints for the equation model (4.1). Where equation (4.2) shows the financing for production planning in the presence of power supply disruptions must be smaller / lower than the financing of production planning without any disruption of the power supply that occurs. This is based on a contract made by a bottled water company with a power supply company with the initial assumption of no interruption of power supply occurring during the agreed time. This is in contrast to the uncertainty that occurs with the disruption of power supplies that can occur at any time due to certain factors. Equations (4.3) to (4.4) indicate the large amount of bottled water production during uninterrupted power supply and with the disruption of power supply must be greater than the amount of financing and machinery. Equation (4.5) is an equation that shows the current production inventory t with the sum of the initial inventory of production and amount of production time t minus the number of market demand for products.

3. Conclusion

The built model is a two-stage stochastic model for the bottled water production planning by looking at the lowest cost of two production planning processes ie normal production planning or without any disruption of power supply and production planning in the event of power supply disruption during production process to minimize production cost. When there is a power supply interruption that occurs at any time through a power supply termination scenario. Expectations of each type of production that is expressed through the scenario π^s opportunities to contribute financing minimum values for each respective type of bottled water production. The difficulties encountered in this calculation are related to the unnecessary demand that is necessary to be estimated.

The probability $\xi \in U$ represents the amount of production and the timing of a power supply disruption. The set of uncertainties to ensure that the amount of disruption that occurs does not exceed the limit of the contract agreement that has been made by the company with the power supply supplier. Other variables that may be added may be concerning the financing of the electrical contracts that have been made in advance in order to reduce the risk of large losses for the bottled water company.

References:

[1] Ahmad M and Bajahlan A S (2009) Quality comparison of tapwater vs bottled water in the industrial city of Yanbu (Saudi Arabia). *Journal Environmental Monit Assess.* **Vol:159**, 1-14.

[2] Barik S K, Biswal M P and Chakravarty D (2013) Two-Stage Stochastic Programming Problem Involving Interval Discrite Random Variables, *OPRESEARCH*, **Vol.49**,PP-280-298.

[3] Borgeson S, Haley B, Hart E, Mahone A, Price S, Ryan N and Williams J (2015) *Results GHG Scenario Results. Energy + Environmental Economics.* (California: PATHWAYS PUBLISHING).

[4] Cai, Yiwei, Kutanoglu E and Hasenbein J (2011) Production Planning and Scheduling Onterraction and Coordination. *International Series in Operation Research & Management Science* **152**, Springer Science+ Business Media.

[5] Cutler-Hemmer (1999) *Learning Module 2: Fundamentals of Electricity. Milwaukee* (Wisconsin U.S.A: Eaton Corp).

[6] Doria M F (2010) Factors influencing public perception of drinking water quality. Division of Water Sciences, United Nations Educational, Scientific and Cultural Organization (UNESCO), *Journal Water Policy* **10** page 1-19.

[7] Galbraith J (2005) *Designing Complex Organizations.* (Buston, MA, USA: Addison-Wesley Logman Publishing Co., Inc)

[8] Gibbons D C (1986) *The Economic Value of Water.* (Washington D.C: Resources for the Future).

[9] Ho C(1989) Evaluating the impact of operating environments on MRP system nervousness. *International Journal of Production Research* **2**, 11151135.

[10] Ivancevich J M and Thomas N (2007) *Business: Principles, Guidelines, and Practices.* (Atomic Dog Publishing) page:172.

[11] Lee T R (1999) *Water Management in the 21st Century : the Allocation Imperative.* (Cheltenham: Edward Elgar Publishing).

[12] Strang V (2004) *The Meaning of Water.* (Oxford: Berg Publisher).

[13] Tschirhart and Jen F (2006) Behavior of a monopoly offering interruptible service. *The Bell Journal of Economics*, **10(1)**:244258, 1979. ISSN 0361915X.Van Valkenburg. Electricity theorema and Applied, Muskogee. Oklahoma, United States.

[14] Yang J, Qi X, and Yu G (2005). Disruption management in production planning. *Naval Research Logistics* NRL**52(5)**, 420-442.

An Optimization Model For Strategy Decision Support to Select Kind of CPO's Ship

Siti Suaibah Nst[1], **Esther Nababan**[2] **and Herman Mawengkang**[3]

Department of Mathematics, Universitas Sumatera Utara, Medan, Indonesia

E-mail: [1]ssuaibahnst@gmail.com; [2]esther@usu.ac.id; [3]hmawengkang@yahoo.com

Abstract. The selection of marine transport for the distribution of crude palm oil (CPO) is one of strategy that can be considered in reducing cost of transport. The cost of CPO's transport from one area to CPO's factory located at the port of destination may affect the level of CPO's prices and the number of demands. In order to maintain the availability of CPO a strategy is required to minimize the cost of transporting. In this study, the strategy used to select kind of charter ships as barge or chemical tanker. This study aims to determine an optimization model for strategy decision support in selecting kind of CPO's ship by minimizing costs of transport. The select of ship was done randomly, so that two-stage stochastic programming model was used to select the kind of ship. Model can help decision makers to select either barge or chemical tanker to distribute CPO.

Keywords: *Optimization, Cost, Two – Stage Stochastic Programming Model*

1. Introduction

Some of alternatives distribute CPO needs strategy for determine kind of ship and giving effect for determining cost to every companies. Ships as a maritime transportstion is more efficiently and environmentally friendly [14]. The modelling is dificult factors to affect of characteristics every kind of transport but The plan transportation is an activity of the selecting alternatives procurement of transport facilities to optimal [9].

One of problems in the distribute CPO has the cost of transporting which relatively large. Cost will affect to determine of cost production and the competitiveness of production. Distribute problems will be complex when it was uncertain problems. Solution to reduce of risk is minimizying cost of distribute CPO. The select either barge charters or chemical tanker charters are a startegy of decision which taken at random. Therefore, this study was used two-stage stochastic programming model.

2. Planning of CPO's Transport

Transportation services related to safety, where the goods are a transported must remain infact until at destination. Parties involved in this study as CPO's factory, shipowner, oil's factory. Transport process of CPO's ship was begins from CPO's factory as shipper, and then CPO will transported to port by truck for loading process at terminal handling. Loading process was used pipeline system, where Pipeline system is a activity flow of CPO from storage tanks to ship. After ship was load at port, tenant is take care of all the technical requirements, administrative, cost of marine transportation service, and have obtained a permit to sailling. If all matters have been fulfilled, then the company ship will dispatch ships to destination. Arriving at destination, cargo will be unloaded with equipment available. Unload Activity was pipeline system or truck lossing. In event of damage or loss to cargo, then consequences occur will transferred in form of unexpected costs.

Voyage charter can be used as agreement, where in a voyage charter the shipowner gets fee based on the amount of cargo transported from L_1 to L_2 [6], therefore The cost of charter was calculated to number ships of cargo ships that transported and frequency of shipping. Shipowner will determine cost of charter based on time for one cruise. This study was used chemical tanker (9000-15.000 DWT) and barge (2000-8000 DWT). As Consideration, Chemical tanker has greater speed than by barge, but it needs fuel more. Otherwise barge required less fuel bacause it has lack of speed. The speed and amount of fuel required ship into consideration in minimize cost. Possible risks responsed in the form of unexpected costs. Risks that may occur such as delay in arrival of goods, reduced volume of payload and damage of quality to charge.

3. Two-Stage Stochastic Programming Model

The purpose of model is determining a feasible solution model to the problem of uncertain. In this study, uncertainty is a plan of require time and transportation costs. The decision Stochastic programming model is an optimization method for modeling related to uncertain problems [2]. This study, two-stage stochastic programming model is formulated for two different stages. The first stage is analysis before the realization of uncertain events, while the second stage after uncertain events is realized. Mathematically, two-stage stochastic programming problem can be stated as :

$$min : z = \sum_{j=1}^{n} c_j x_j + E(\sum_{i=1}^{m} P_i|y_i|) \tag{1}$$

subyek to :

$$y_i = b_i - \sum_{j=1}^{n} a_{ij}x_j \qquad i = 1, 2, ..., m$$

$$\sum_{j=1}^{n} r_{sj}x_j \geq h_s \qquad s = 1, 2, ..., l$$

$$y_i \geq 0, \;\; x_j \geq 0, \qquad i = 1, 2, ..., m, \;\; j = 1, 2, ..., n$$

where the model was assumed that the first stage of decision variables $x_j, j = 1, 2, ..., n$ and the second stage decision variables $y_i, i = 1, 2, ..., m$ are stochastic in the problem, otherwise $p_i, i = 1, 2, ..., m$ are probabilistic to select of ships.

4. The Model

The Model is a strategy to plan of distribute CPO from CPO's factory to oil's factory. The magnitude of CPO's ships and the complexity of CPO production could affect to number of demands and price of CPO in Indonesian, so that decision makers has difficult to respon problem.

PARAMETERS

α	:	The total cost to every ships (Rp)
P	:	Probability of unexpected costs to every ships
V	:	The set kind of CPO's ships
Q_v	:	The amount of CPO's cargoes to every ships (MT)
C^{VC}	:	The cost of charter ship per tonnage (Rp)
C^{Tb}	:	The cost of charter tugboat (Rp)
T^{VC}	:	Time of charter ships for a single voyage (days)
C^{TVC}	:	The total cost of charter ship at beginning of caharter party (Rp)
b_1	:	The total cost charged to tenant after ship arrives at port of destination (Rp)
C^s	:	Cost of fuel per tonnage to every ships (Rp)
H^s	:	The amount of fuel needed to every ships (Rp)
C^{FT}	:	The total cost of fuel to every ships (Rp)
b_2	:	Budget of fuel costs to every ships (Rp)
T^{Port}	:	Time for loading and unloading at port (days)
C^{Port}	:	The cost of port to every ship per day (Rp)
C^{PT}	:	The total cost of port to every ships (Rp)
b_3	:	Budget of port to every ships (Rp)
C^{TT}	:	The total cost of unexpected costs to every ships (Rp)

INDEX

I	:	Set of barge ($i = 1, 2, ..., n$)
J	:	Set of Chemical Tanker ($j = 1, 2, ..., m$)

DECISION VARIABLES

X_{v_1}	:	Barge
X_{v_2}	:	Chemical Tanker
Y_i	:	Binarry variable 1 if barge has unexpected costs
Y_j	:	Binarry variable 1 if chemical tanker has unexpected costs
δ_{v_1}	:	Binarry variable 1 if barge was selected
δ_{v_2}	:	Binarry variable 1 if chemical tanker was selected

In the above section, it has been described that selection of ship is done randomly, so two stage stochastic model as follows:

Objective Function :

$$min : z = [\sum_{i=1}^{n} \alpha_i X_{v_1} + E(\sum_{i=1}^{n} P|y_i|)]\delta_{v_1} + [\sum_{j=1}^{m} \alpha_j X_{v_2} + E(\sum_{j=1}^{m} P|y_j|)]\delta_{v_2} \tag{2}$$

Subyek to :

$$C_i^{TT} = b_1 - \sum_{i=1}^{n} [(Q_i C_i^{VC} T_i^{VC}) + C_i^{Tb}] \quad (3)$$

$$C_j^{TT} = b_1 - \sum_{j=1}^{m} (Q_j C_j^{VC} T_j^{VC})$$

$$i = 1, 2, ..., n \ and \ j = 1, 2, ..., m \ \forall \ i \in I, j \in J$$

The above cost is calculation of determine following Y value:

$$C_i^{TT} \in Y_i$$

$$C_j^{TT} \in Y_j$$

Variabe Y is binarry variable which determining as decision support on describe that presence or absence of unexpected costs to every ships.

$$Y = \begin{cases} 1 & \text{if ship has unexpected costs} \\ 0 & \text{if ship hasn't unexpected costs} \end{cases}$$

Table 1. Strategy of determining presence or absence of unexpected costs

Calculations result	Y
$C^{TT} < 0$	0
$C^{TT} > 0$	1

$$\sum_{i=1}^{n} C^s H_i^s \leq b_2 \quad (4)$$

$$\sum_{j=1}^{m} C^s H_j^s \leq b_2$$

$$H_i^s > 0, \ H_j^s > 0 \qquad \forall \ i \in I, j \in J$$

$$\sum_{i=1}^{n} T_i^{Port} C_i^{Port} \leq b_3 \quad (5)$$

$$\sum_{j=1}^{m} T_j^{Port} C_j^{Port} \leq b_3$$

$$T_i^{Port} < T_i^{VC}, \ T_j^{Port} < T_j^{VC} \qquad \forall \ i \in I, j \in J$$

$$Q_{v_1} = Q_{v_2} \qquad \forall \ v_1, v_2 \in V \quad (6)$$

$$X_{v_1} > 0,\ \ X_{v_2} > 0 \qquad \forall\ v_1, v_2 \in V \tag{7}$$

$$\delta_{v_1} + \delta_{v_2} \leq 1 \qquad \forall\ i \in I, j \in J. \tag{8}$$

δ is a decision variable to indicate kind of ship to be selected or not selected by comparing sum of every cost and expectation value.

$$\delta = \begin{cases} 1 & \text{if ship was selecting} \\ \\ 0 & \text{if ship wasn't selecting} \end{cases}$$

Determination is one kind of the ships, it can be selected is assumed as follows:

Assumed:

$$u = [\sum_{i=1}^{n} \alpha_i X_{v_1} + E(\sum_{i=1}^{n} P|Y_i|)]$$

$$w = [\sum_{j=1}^{m} \alpha_j X_{v_2} + E(\sum_{j=1}^{m} P|Y_j|)]$$

Assignment of δ 1 or 0 was determinning by analyze difference between two costs charged to ship with above assumptions.

$$z = u - w$$

Table 2. Strategy select kind of ship

	δ_{v_1}	δ_{v_2}
$z < 0$	1	0
$z > 0$	0	1

Values of δ_{v_1} and δ_{v_2} will be projected on objective function. Subsequently obtained decision kind of ship to be used at lowest cost, where components of above model are limited to a single voyage.

Based on equation (2), Objective function is minimize cost for transporting CPO with strategy to select kind of ship. Decision to be taken is whether to barge charter or chemical tanker charter. Every ships has same of probability, so $P_1 + P_2 + ... + P_n = 1$. α is total cost of charter ship, cost of port, and cost of fuel ship.

$$\alpha = [C^{TVC} + C^{PT} + C^{FT}]$$

Based on equation (3), unexpected costs occur if total cost is greater than planned lease budget and agreed upon by tenants and shipowners ($b_1 > C^{TVC}$). Unexpected cost of model was shown in form of a binarry variable $\{0, 1\}$, provided that it is shown in table 1. Based on equations (4) and (5) shows respective total cost of fuel required and total cost of port. For the total cost of port is some of the total costs incurred while at port such as navigation costs, loading and unloading costs, water costs, crew and skipper wages costs, etc. Based on equation (6) show total CPO will be distributed by barge must be equal to total CPO charge that will be distributed by chemical tanker. Based on equation (7) show that every kind of ship will represent calculation to minimize cost of transporting CPO and determine only one kind of ship charter for distribute CPO. Based on equation (8) aims at deciding what kind of ship to be used for transporting of CPO. Decision will be taken through provisions shown in table 2. The cost of truck is not included in model because cargo is transported in same volume, so the cost was same for distrubute CPO. For route problem is not included in model, because model was restricted only to routes which both ship can pass.

5. Conclusion

Based on after chapters, the first stsge of this study is analyzing all cost components before the realization of uncertain events, while in the second stsge is analyzing unexpected cost as respon of the risk on this study. The strategy of select one kind of ship was using binarry variable $\{0, 1\}$, so as to produce decision to select is one kind of CPO's ships. If $z < 0$ then $\delta_i = 1$ and $\delta_j = 0$, otherwise if $z > 0$ then $\delta_i = 0$ and $\delta_j = 1$. Result of projected value δ will be multiplied by every components of cost, then it will be seen what kind of ship will be selected with lowest cost. The lowest cost can be seen when The total cost of one ship to one kind of ship will be multiplied by number of CPO's ship, and it was sum with the average value of unexpected cost. The average of value was determined after multiplication between probability of slected ship to same kind of ship with the unexpected cost to every ship.

References

[1] Alatas, A. (2015). Trend Produksi dan Ekspor Minyak Sawit (*CPO*) Indonesia. *Jurnal Gadjah Mada.*, Vol 1, No 2.
[2] Barik, S.K., Biswal M.P., and Chakravarty D. (2013). Two-Stage Stochastic Programming Problems Involving Some Continuous Random Variables, *Journal of Uncertain System.*, Vol 7, No 4, 277-288.
[3] Basirzadeh, H. (2011). An Approach for Solving Fuzzy Transportation Problem. *Applied Mathematical Sciences.*, Vol 5, No 32, 1549-1566.
[4] Christiansen, M., Fagerholt, K., Nygreen, B., dan Ronen, D. (2007). Maritime Transportation. *Transportation.*, 14, 189-284.
[5] Gannett, E. (1969). *Tanker Performance and Cost.* Cornell Maritime Press, Inc. Cambridge.
[6] Laake, J. (2010). An Optimization Model For Strategic Decision Support In Maritime Transportation. [Thesis]. Norwegian: University of Science and Technology.
[7] Lasse. (2012). *Manajemen Muatan.* PT Rajagrafindo Persada. Jakarta.
[8] Mihajlovic, M., Jovanovic, M., Pesic, R., Jovanovic, J., and Milanovic, Z. (2016). Volatile Organic Compounds (VOC) Policy Innovation in Petrochemicals River Barge Transportation. *Journal of Cleaner Production.*, 112, 1559-1567.
[9] Miro, F. (2005). *Perencanaan Transportasi.* Erlangga. Padang.
[10] Mulder, J., Dekker, R. (2016). Optimization in Container Liner Shipping. *Econometric Institute* 05.
[11] Nur. (2004). *Manajemen Transportasi.* Ghalia Indonesia. Jakarta.
[12] Powell, W. B., dan Topaloglu, H. (2002). Stochastic Programming In Transportation and Logistic, *Handbook of Operation Research and Management Science.* Vol 10, 555-626.
[13] Radiks. (1997). *Angkutan Muatan Laut 2.* PT Rineka Cipta, Jakarta.
[14] Rodrigue, J. P., Comtois, C., and Slack, B. (2013). *The Geography of Transport Systems.* 3rd. Routledge.
[15] Santoso, T., Ahmed, S., Goetschalckx, A., and Shapiro, A. (2005), A Stochastic Programming Approach for Supply Chain Network Design Under Uncertainty, *European J. of Oper. Res 167.*, 96-115.
[16] Siddiqui, A. W., and Verma, M. (2015). A Bi-Objective Approach to Routing and Scheduling Maritime Transportation of Crude Oil. *Transportation Research Part D.*, 37, 65-78.
[17] UNCTAD. (2016). *Review of Maritime Transport.* United Nations, New York.

Data Envelopment Analysis (DEA) Model in Operation Management

Meilisa Malik[1], **Syahril Efendi**[2], **Muhammad Zarlis**[3]
Department of Mathematics, Universitas Sumatera Utara, Medan, Indonesia
E-mail: [1]malikmeilisa@gmail.com, [2]syahnyata1@gmail.com, [3]m.zarlis@usu.ac.id

Abstract. Quality management is an effective system in operation management to develops, maintains, and improves quality from groups of companies that allow marketing, production, and service at the most economycal level as well as ensuring customer satisfication. Many companies are practicing quality management to improve their bussiness performance. One of performance measurement is through measurement of efficiency. One of the tools can be used to assess efficiency of companies performance is Data Envelopment Analysis (DEA). The aim of this paper is using Data Envelopment Analysis (DEA) model to assess efficiency of quality management. In this paper will be explained CCR, BCC, and SBM models to assess efficiency of quality management.
Keywords : Data Envelopment Analysis, Quality Management, performance measurement

1. Introduction

Operation management is a management system maximally of production factors in transformation process to be various products and services. Operation managemet is an effort to convert inputs into outputs by using physical resources so give benefits customers to meet effectiveness, the aim, compatibility and efficiency of companies [11].

The aim of operation management is to producing products or services with right quality and quantity at the time and right production cost so can be result maximum profit. Quality control is one of the important factors in operation management. Quality control is the way that used to improve and result level of products and services quality which are desired. This is systematic control from many factors that influence products or services quality.

Various the quality measurement incluce performance, features, reliability, durability, commitment to due date, etc. Now, customers demand products and services with greater durability and reability at the most economical price. this forces manufacturer to follow procedure of right quality start from design until delivery and installation of product. So, the aim of competitive industries are producing products or services at the economical price and can used customers satisfication. This can be achieved through total quality management [11].

Total quality management is effective system in operation management to develops, maintains, and improves quality of groups in company so allows marketing, production, and services at the most economical level and ensure customers satisfication. Many companies that practicing quality management to improve bussiness performance and still competitive in industries.

Uyar [14] show that financial and nonfinancial measures must be used in a balanced way to evaluate quality performance of companies. Therefore need a the tool can be used to asses quality performance of companies based on financial and nonfinancial measurements. One of the companies perfoemance measurement level is efficiency assessment. There are several methods can be used to assess efficiency, such as : Stochastic Frontier Analysis (SFA) and Data Envelopment Analysis (DEA)

SFA method were parametric approach that allow to give greater specification, especially in the case of panel data. SFA can be used to statistical test and construction confidence intervals. However, while a statistical stochastic approach allow result frontier for all data and this is not consistent to tecnological development. Therefore, the choice between different approaches must be based on trade-off concerning the purpose of the study, type of data, technology characteristics [12].

DEA method were nonparametric approach to asses performance set of entities which are called Decision Making Units (DMUs) that change inputs into outputs. Since DEA the first time introduced until current formulation, the researchers various fields had admitted that DEA method excellent and easy used to modeling operational process in assess performance. This had been accompanied by other developments. The examples, Zhu [15] presented a spreedsheet DEA model can be used to assess and became a benchmark of performance. Pournader [13] developed DEA model to outsourcing performance quality assessment in services supply chain. Use of this model in manufacturing requires a review to inputs, outputs, and intermediate products such as production and purchase of facilities, market speed, security, etc.

DEA can be used to assess efficiency level with existence of several inputs and outputs entities. Efficiency measurement quality performance of companies very needed to compare performance of companies with their competitors, and developed their business based on performance improvement plan. Therefore, this study is focused on capability Data Envelopment Analysis to assess efficiency of quality management.

2. Measurement Efficiency of Quality

Critical Factors in total quality management that found in literature vary from several writer, however there were common factors [5], such as :

- customer based approach
- management leadership and commitment
- quality planning
- management based on facts
- continious improvement
- Human resource management
- learning
- process management
- cooperation with suppliers

Performance measurement system, measuring inputs and outputs on operation include explain good or bad operation by use that system. Performance measurement important for all functional division in companies, especially operation management which directly impact on bussiness in the process efficiency include purcase acquisition and raw material inventory. Simple efficiency ratio [1], that is : $\frac{Output}{Input}$ be measured on individual variable that is employee, machine, and room.

Two important things in quality management were improvement process management and production quality. Improvement reliability and conformity in production process can improve internal process quality. Improvement of product quality can increase market share, higher prices, and better reputation. Therefore both of these things can increase profit and operational performance. Relationship between quality management and performance of companies, that is: quality performance, financial performance, and operational performance.

3. Data Envelopment Analysis (DEA)

Data Envelopment Analysis (DEA) was introduced by Charnes, Cooper, dan Rhodes [2]. DEA method was made as tool to evaluate performance an activity in entity units or organization. DEA was nonparametric approach which is basically a technique linear programming. The researchers on several fields quickly admitted that DEA was excellent methodology to modelling operational process.

On DEA, Companies were called as Decission Making Units (DMUs). Term Decission Making Units (DMUs) was introduced in the same way as entities, each of entities be evaluated as part of group that utilize inputs to producing outputs. The result of evaluation made in score efficiency about between 0 until 1 and represent degree of efficiency which is obtained from entities assessed. Whit the score, DEA also identified source and number of not efficient DMUs on every input and output. It also identified DMU (that placed on efficient frontier) that included actively in the appearance of these results. Assessment entities were all efficient DMUs and hence can fuction as benchmarking to effective improvement at future performance from DMUs be evaluated [4].

Some basic DEA models and expansion DEA models used in this paper that is: CCR model, BCC model, and Slack-Based (SBM) model.

3.1. CCR Model

CCR DEA model was the first DEA model which was introduced by Charnes, Cooper dan Rhodes (1978). Consider there are n DMUs, that are: DMU_1, DMU_2, $\mathrm{DMU}_3, \ldots, \mathrm{DMU}_n$. Each DMU_j, $(j = 1, ..., n)$ use m inputs x_{ij} $(i = 1, \ldots, m)$ dan generates s outputs y_{rj} $(r = 1, \ldots, s)$. Let the input weights v_i $(i = 1, 2, \ldots, m)$ and the output weight u_r $(r = 1, 2, \ldots, s)$ as variables. Let the DMU_j to be evaluated on any trial be designated as DMU_o $(o = 1, 2, \ldots, n)$. The efficiency of each DMU_o, e_o, is thus found by solving the linear programming below, which is known as multiplier form in DEA [10]:

$$e_o = \max \sum_r u_r y_{ro} \tag{1}$$

$$s.t \quad \sum_r u_r y_{ro} - \sum_i v_i x_{io} \leq 0$$

$$\sum_i v_i x_{io} - 1$$

$$u_r, v_i \geq 0$$

The model is run n times in identifying the relative efficiency scores of all the DMUs. Each DMU selects a set of input weights v_i and output weights u_r that maximize its efficiency score. Generally, a DMU is efficient if it obtains the maximum score of 1, else a DMU is inefficient.

Model dikerjakan n kali untuk mengidentifikasi skor efisiensi relatif dari seluruh DMU. Setiap DMU memilih sekumpulan bobot input v_i dan bobot output u_r yang memaksimumkan skor efisiensinya. Secara umum DMU efisien jika memperoleh skor 1 jika tidak maka DMU tidak efisien. CCR model is that it assumes constant return to scale (CRS), which may not be true for some applications. To address this issue, researchers have implemented variable return to scale (VRS) into the original DEA model.

3.2. BCC Model

DEA model that implemented Variable Return to Scale (VRS) was known as model BCC, Banker, Charnes dan Cooper (1984). In BCC model, VRS is assumed and the efficient frontier is formed by the convex hull of the existing DMUs. The envelopment form of BCC [10] is

$$\min \theta_o \tag{2}$$

$$s.t \quad \sum_j \lambda_j x_{ij} - \theta_o x_{io} \leq 0$$

$$\sum_j \lambda_j y_{rj} - y_{ro} \geq 0$$

$$\sum_j \lambda_j = 1$$

$$\lambda_j \geq 0$$

Note that BCC differs from CCR in that it has the additional convexity constraint, $\sum_j \lambda_j = 1$ A DMU is BCC efficient if it has an optimal solution of $\theta = 1$ $\lambda = 1$, and $\lambda \neq 0$.

3.3. Slack Based Model (SBM)

A variation model was proposed by name slack based model (SBM) that representing mesurement efficiency [10]:

$$\min \rho = \frac{1 - \frac{1}{m}\sum_i s_i^- / x_{io}}{1 + \frac{1}{s}\sum_r s_r^+ / y_{ro}} \tag{3}$$

$$s.t \quad \sum_j \lambda_j x_{ij} - x_{io} + s_i^- = 0$$

$$\sum_i \lambda_i y_{rj} - y_{io} - s_r^+ = 0$$

$$\sum_j \lambda_j = 1$$

$$\lambda_j, s_i^-, s_r^+ \geq 0$$

which s_i^- and s_r^+ are slacks, s_i^- is input excesses dan s_r^+ is output shortfalls. The optimum solution of SBM is 1, which can only be achieved when all the slacks equal to zero. This is consistent with CCR and BCC models.

4. Efficiency of Quality Management Using Data Envelopment Analysis (DEA) Model

Performance measurement in this paper include measuring quality, operational and financial. Then divided into specific variables like the following table:

In this study given an illustration example about quality management, data is taken from research had developed earlier by Kuah, Wong, and Behrouzi [9]. Ilustration problem consist of 10 small companies as DMUs with 2 inputs and 4 outputs. The inputs and outputs data: inputs:

- Quality costs (x_1);
- Employees involved in quality(x_2).

Table 1. Performance Measures dan output variables

Performance Measures	output variables
Quality Performance Measures	Quality products (%)
	Customer satisfaction rate (%)
Operational Performance Measures	On-time delivery rate (%)
Financial Performance	Measures Revenue (million USD)

Output:

- Quality products (y_1);
- Customer satisfaction rate (y_2);
- On-time delivery rate (y_3);
- Revenue (y_4).

Table 2 show inputs and outputs data from 10 DMUs. These data should be transformed into CCR, BCC, and SBM models, then determined which companies are already efficient. Linear Programming models are run by software LINDO.

Table 2. Data for illustration example

DMU	X_1	X_2	Y_1	Y_2	Y_3	Y_4
1	18	39	98	92	91	17
2	27	40	92	93	98	22
3	30	50	95	92	81	33
4	19	44	93	92	82	40
5	31	35	95	96	97	37
6	17	33	93	92	95	31
7	23	47	91	93	81	32
8	30	46	90	94	92	29
9	17	38	95	99	96	16
10	22	42	96	96	84	14

5. Conclusion

One of methods can be used to solve efficiency in operation management by Data Envelopment Analysis (DEA). By focusing on measure quality performance, operational performance, and financial performance can be evaluated what extent system of quality management can be repaired to development bussiness performance of companies and still competitive in industries. This paper extends the allocation resources of literature about application DEA in a reciprocal context between efficiency and effectiveness in quality management of companies. From comparison result on illustration example were done by DEA CCR, BCC, and SBM obtained the same result to DMU 4, 6, and 9. Whereas for other DMUs there were difference of results so the next research should be done improvement or review of data inputs to produce appropriate outputs, so that DMUs were efficient.

Table 3. Efficiency of 10 DMU

DMU	θ^*_{CCR}	θ^*_{BCC}	ρ^*_{SBM}
1	0.98	1	1
2	0.85	1	1
3	0.68	1	1
4	1	1	1
5	1	0.94	0.76
6	1	1	1
7	0.75	1	1
8	0.73	1	1
9	1	1	1
10	0.82	1	1

References

[1] Brown S, Blackmon K, Cousins P, and Maylor H 2001 *Operations Management.* Butterworth Heinemann, Italy.

[2] Charnes A, Cooper W W, and Rhodes E 1978 Measuring the efficiency of decision making units. *European Journal of Operational Research*, Volume **2**, 429–444.

[3] Cooper W W, Seiford L M, and Zhu J 2011 *Handbook on Data Envelopment Analysis*, Vol. **164** second edition ed F.S. Hillier. Springer.

[4] Cooper W W, Seiford L M, and Tone K 2006 *Introduction to Data Envelopment Analysis and Its Uses : with DEA Solver Software and Reference.* Springer, New York.

[5] Conca F J, Llopis J, and Tari J J 2004 Development of a measure to asses quality management in certified firm. *European Journal of Operational Research*, Volume **156**, 683–697.

[6] Dean J W, and Bowen D E 1994 Management theory and total Quality: improving research and practice through theory development. *The Academy of Management Review*, Volume **19**.

[7] Halkos G, Tzeremes N, and Kourtzidis S 2011 The use of supplay chain DEA models in operations management : A survey. *Munich Personal RePEc Archive*, No. 31846.

[8] Knowles G 2011 Quality Management. *Graeme Knowles and Ventus Publishing APS.*

[9] Kuah C T, Wong K Y, and Behrouzi F 2010 Application of Data Envelopment Analysis to asses quality management efficiency. *International Journal of Mechanical, Aerospace, Industrial, Mechatronic, and Manufacturing Engineering*, Volume **4**, No. 10

[10] Kuah C T, Wong K Y, and Behrouzi F 2010 A review on Data Envelopment Analysis (DEA). *Fourth Asia International Conference on Mathematical / Analytical Modelling and Computer.*

[11] Kumar S A, and Suresh N 2008 *Production and Operations Management.* New Age International Publisher, New Delhi.

[12] Kumbhakar S C, and Heshmati A 1996 DEA, DFA, and SFA : a comparison. *Journal of Productivity Analysis.*

[13] Pournader M , Kach A, Fahimnia B, and Sarkis J 2015 Outsorcing Performance quality assesment using DEA analytics. *Int. J. Production Economics.*

[14] Uyar A 2009 Quality performance measurement practice in manufacturing companies. *The TQM Journal*, Volume **21**, Issue 1, 72–86.

[15] Zhu J 2003 Quantitative models for performance evaluation and benchmarking : DEA with spreadsheet and DEA excel solver. *Springer Since + Bussiness Media.*

IOP Conf. Series: Materials Science and Engineering **300** (2018) 012009 doi:10.1088/1757-899X/300/1/012009

Expected value based fuzzy programming approach to solve integrated supplier selection and inventory control problem with fuzzy demand

Sutrisno[1], Widowati[2], Sunarsih[3], Kartono[4]

[1,2,3,4]Dept. of Mathematics, Diponegoro University, Jalan Prof. Soedarto, SH., Semarang, Indonesia, 50275

[1]tresno.math@undip.ac.id

Abstract. In this paper, a mathematical model in quadratic programming with fuzzy parameter is proposed to determine the optimal strategy for integrated inventory control and supplier selection problem with fuzzy demand. To solve the corresponding optimization problem, we use the expected value based fuzzy programming. Numerical examples are performed to evaluate the model. From the results, the optimal amount of each product that have to be purchased from each supplier for each time period and the optimal amount of each product that have to be stored in the inventory for each time period were determined with minimum total cost and the inventory level was sufficiently closed to the reference level.

1. Introduction

In logistic and supply chain management (LSCM), commonly a manufacturer faces supplier selection problem which can be presented as an optimal supplier finding to supply some raw material or product to the manufacturer with minimal total cost [1]. The most developed method to find the optimal supplier is mathematical model method in a mathematical optimization form such as mixed-integer linear programming that was developed in [2, 3] . To meet the demand for the future period, the manufacturer can store some product in the storage where the holding cost is charged. Commonly, the stored product is to be minimized in order to minimize the holding cost, but for some cases like retail manufacture, the decision maker is deciding to store the product as many he wants which can be called set point inventory level. To minimize the total cost so that the inventory level as close as possible to a set point level, a reference tracking control method is needed.

The most published articles was developed a method to solve the supplier selection with known demand value i.e. the demand value of all product at each time period is certainly measured. Unfortunately, there are many case that the demand value is unknown which can be called uncertain demand. To solve this problem, the uncertainty theory is needed. Since it was introduced in [4], possibility theory was being powerful tool to solve many problems in uncertain environment mainly in fuzzy optimization. Inspired by expected value of a random variable in probabilistic theory, the expected value of a fuzzy variable was introduced in [5] to solve a fuzzy programming. Many researchers were successfully utilized fuzzy expected value approach to solve many problems like data envelopment analysis [6 - 8], industrial grinding process [9, 10], portfolio optimization [11-13] and many more.

In this paper, we propose a mathematical model in quadratic programming with fuzzy parameter to determine the optimal strategy for integrated inventory control and supplier selection problem with fuzzy demand where the corresponding optimization will be solved by using expected value based fuzzy programming approach. Numerical experiment will be performed to evaluate the proposed model and to analyze the optimal strategy for this problem.

2. Problem Definition and Notations

Suppose that a manufacturer will solve supplier selection integrated with inventory control problem. The condition of the problem is assumed to be multi-product, multi-supplier and multi-period problem. To formulate the mathematical model, symbol of variables and parameters are introduced as follows:

T : Set of time period;
S : Set of supplier;
P : Set of product;
X_{tsp} : Volume of product $p \in P$ from supplier $s \in S$ at time period $t \in T$;
UP_{tsp} : Unit price of product $p \in P$ from supplier $s \in S$ at time period $t \in T$;
TC_{ts} : Transportation cost of all product from supplier $s \in S$ at time period $t \in T$;
Y_{ts} : Supplier assignment $s \in S$ at time period $t \in T$ (1 if $X_{tsp} > 0$ exist, 0 if none);
SC_{tsp} : Supplying capacity of supplier $s \in S$ for product $p \in P$ at time period $t \in T$;
UPC_{tsp} : Unit penalty cost for defect product unit $p \in P$ from supplier $s \in S$ at time period $t \in T$;
UDC_{tsp} : Unit delay cost for delayed product unit $p \in P$ from supplier $s \in S$ at time period $t \in T$;
DLT_{tsp} : Delay lead time of product $p \in P$ from supplier $s \in S$ at time period $t \in T$;
Q_{tsp} : Quality level of product $p \in P$ at period $t \in T$ from supplier $s \in S$;
I_{tp} : Inventory level of product $p \in P$ at time period $t \in T$;
M_p : Storage capacity of product $p \in P$;
B_t : Cost budget at time period $t \in T$;
HC_{tp} : Holding cost of product $p \in P$ at time period $t \in T$.

3. Expected Value Based Fuzzy Programming Model

The general form of an expected value based fuzzy programming with crisp objective function and fuzzy constraint can be expressed as follows

$$\begin{cases} \min f(x) \\ \text{s.t. } g_i(x,\xi) \geq 0, i = 1,2,\ldots,p. \end{cases} \tag{1}$$

where x is the decision vector, ξ is a fuzzy vector, $f(x)$ is the objective/return function and $g_i(x,\xi) \geq 0$ are constraint functions. Since the constraint do not produce a crisp feasible set, model (1) is not well-defined. Then we use expected value to approach the crisp value of the constraint. There are several ways to define the expected value of a fuzzy variable. In this paper, we use the expected value that was defined by Liu and Liu [5]. For any t and p, the expected value of fuzzy demand $\tilde{D}_{tp}$ is defined by

$$E\left[\tilde{D}_{tp}\right] = \int_0^{\infty} Cr\left\{\tilde{D}_{tp} \geq r\right\} dr - \int_{-\infty}^{0} Cr\left\{\tilde{D}_{tp} \leq r\right\} dr \tag{2}$$

provided that at least one of the two integrals is finite where $Cr[\cdot]$ denotes the credibility value. By using (2), it can be proven that the expected value of a triangle fuzzy variable (a,b,c) is

$E[(a,b,c)] = \frac{a+2b+c}{4}$ and the expected value of a trapezoidal fuzzy variable (a,b,c,d) is $E[(a,b,c,d)] = \frac{a+b+c+d}{4}$. For a discrete fuzzy variable ξ whose membership function is given by

$$\mu_\xi(x) = \begin{cases} \mu_1, \text{ if } x = x_1 \\ \mu_2, \text{ if } x = x_2 \\ \vdots \\ \mu_m, \text{ if } x = x_m \end{cases} \tag{3}$$

where $x_1, x_2, \ldots, x_m$ are distinct numbers and $x_1 < x_2 < \cdots < x_m$, then the expected value of ξ is

$$E[\xi] = \sum_{i=1}^{m} w_i x_i \tag{4}$$

where

$$w_i = \tfrac{1}{2}\left(\max_{1\le j\le i} \mu_j - \max_{1\le j< i} \mu_j + \max_{i\le j\le m} \mu_j - \max_{1< j\le m} \mu_j \right) \tag{5}$$

for $i = 1, 2, \ldots, m$.

The mathematical model for integrated supplier selection and inventory control of a multiproduct inventory system with fuzzy demand is formulated as follows. The objective function is the total procurement and inventory cost i.e.

$$\begin{aligned} Z = &\sum_{t=1}^{T}\sum_{s=1}^{S}\sum_{p=1}^{P} X_{tsp}\cdot UP_{tsp} + \sum_{t=1}^{T}\sum_{s=1}^{S} TC_{ts}\cdot Y_{ts} + \sum_{t=1}^{T}\sum_{s=1}^{S}\sum_{p=1}^{P}(1-Q_{tsp})\cdot UPC_{tsp}\cdot X_{tsp}\cdot Y_{ts} \\ &+ \sum_{t=1}^{T}\sum_{s=1}^{S}\sum_{p=1}^{P} UDC_{tsp}\cdot DLT_{tsp}\cdot X_{tsp}\cdot Y_{ts} + \sum_{t=1}^{T}\sum_{p=1}^{P} HC_{tp}\cdot I_{tp} + \sum_{t=1}^{T}\sum_{p=1}^{P}\left(I_{tp} - r_{tp}\right)^2 \end{aligned} \tag{6}$$

where the terms in Z representing the product buying cost for all product from all supplier for all time period, the transportation cost, suppliers service level cost, product defect cost, holding cost, and inventory level tracking objective respectively. The constraints of the model are formulated as follows starting from the constraints who's containing fuzzy variable. The first constraint is the inventory management which is

$$\sum_{s=1}^{S} X_{tsp} - I_{tp} \ge \tilde{D}_{tp}, \text{ for } t = 1, \forall p \in P, \text{ and } V_{t-1,p} + \sum_{s=1}^{S} X_{tsp} - V_{tp} \ge \tilde{D}_{tp}, \text{ for } t > 1, t \in T, \forall p \in P. \tag{7}$$

The second constraint is presenting to determine the supplier's assignment to supply a product i.e.

$$\left(\sum_{p=1}^{P} \tilde{D}_{tp}\right) Y_{ts} \ge \sum_{p=1}^{P} X_{tsp}, \forall t \in T, \forall s \in S. \tag{8}$$

The remaining constraints are formulated as follows

$$X_{tsp} \le SC_{tsp}, \forall t \in T, \forall s \in S, \forall p \in P, \tag{9}$$

$$Y_{ts} = \begin{cases} 1, & \text{if } \sum_{p=1}^{P} X_{tsp} > 0 \\ 0, & \text{others} \end{cases}, \forall t \in T, \forall s \in S \quad (10)$$

$$I_{tp} \leq M_{tp}, \forall t \in T, \forall p \in P, \quad (11)$$

$$\begin{aligned} &\sum_{s=1}^{S}\sum_{p=1}^{P} X_{tsp} \cdot UP_{tsp} + \sum_{s=1}^{S} TC_{ts} \cdot Y_{ts} + \sum_{s=1}^{S}\sum_{p=1}^{P} (1-Q_{tsp}) \cdot UPC_{tsp} \cdot X_{tsp} \cdot Y_{ts} \\ &+ \sum_{s=1}^{S}\sum_{p=1}^{P} UDC_{tsp} \cdot DLT_{tsp} \cdot X_{tsp} \cdot Y_{ts} + \sum_{p=1}^{P} HC_{tp} \cdot I_{tp} + \sum_{t=1}^{T}\sum_{p=1}^{P} \left(I_{tp} - r_{tp}\right)^2 \leq B_t, \forall t \in T, \end{aligned} \quad (12)$$

$$X_{tsp} \geq 0, \forall t \in T, \forall s \in S, \forall p \in P, \quad (13)$$

$$X_{tsp}, I_{tp}, \forall t \in T, \forall s \in S, \forall p \in P \text{ integer}, \quad (14)$$

where constraints (9)-(14) are presenting the supplier capacity satisfying, product quality level, supplier's assignment, non-negativity, storage capacity satisfying, budget satisfying, and integer constraint respectively. By applying expected value (2) to the fuzzy demand parameter $\tilde{D}_{tp}$, then we formulate the model as minimizing total cost (6) subject to (7)-(14) i.e.

$$\min Z \quad (15)$$

subject to:

$$\sum_{s=1}^{S} X_{tsp} - I_{tp} \geq E\left[\tilde{D}_{tp}\right], \text{ for } t=1, \forall p \in P, \text{ and } V_{t-1,p} + \sum_{s=1}^{S} X_{tsp} - V_{tp} \geq E\left[\tilde{D}_{tp}\right], \text{ for } t>1, t \in T, \forall p \in P;$$

$$\left(\sum_{p=1}^{P} E\left[\tilde{D}_{tp}\right]\right) Y_{ts} \geq \sum_{p=1}^{P} X_{tsp}, \forall t \in T, \forall s \in S;$$

and (9)-(14). The expected value $E\left[\tilde{D}_{tp}\right]$ can be calculated according to the definition of a membership function. The authors deal with the definition (2) to define the expected value of a fuzzy variable.

4. Numerical Examples

In this section, we evaluate the proposed model with randomly generated data which available in appendix. Suppose that we have four suppliers S1, S2, S3, S4 to supply three kind of products P1, P2, P3. The decision maker is deciding that the inventory level of each product must follow a set point as close as possible with minimum total cost.

Let the initial inventory level is 0 and the decision maker wants to determine the optimal strategy by using the model (15) for 5 time periods while the parameter values of the problem is available in appendix. We simulate the model with two different membership functions of the fuzzy demand parameter which are discrete membership function (discussed in sub-subsection 4.1) and triangular membership function (discussed in sub-subsection 4.2).

4.1. Discrete fuzzy demand

For the first example, suppose that the demand for any time period and any product is assumed to be a fuzzy variable with discrete membership function in the form of (3) that is defined by the following membership function

$$\mu_{\tilde{D}_{tp}} = \begin{cases} 0.45 \text{ if } \tilde{D}_{tp} = 300;\ 0.50 \text{ if } \tilde{D}_{tp} = 310;\ 0.65 \text{ if } \tilde{D}_{tp} = 320;\ 0.80 \text{ if } \tilde{D}_{tp} = 330;\ 0.95 \text{ if } \tilde{D}_{tp} = 340; \\ 1.00 \text{ if } \tilde{D}_{tp} = 350;\ 0.85 \text{ if } \tilde{D}_{tp} = 360;\ 0.75 \text{ if } \tilde{D}_{tp} = 370;\ 0.60 \text{ if } \tilde{D}_{tp} = 380;\ 0.55 \text{ if } \tilde{D}_{tp} = 390 \end{cases}$$

for all t and p. By using (4), the expected value of $\tilde{D}_{tp}$ is $E\left[\tilde{D}_{tp}\right] = \sum_{i=1}^{10} w_i \left(\tilde{D}_{tp}\right)_i$ where $w_1 = 0.225$, $w_2 = 0.025$, $w_3 = 0.075$, $w_4 = 0.0.075, w_5 = 0.1$, $w_6 = 0.05, w_7 = 0.075$, $w_8 = 0.025, w_9 = 0.275$, and $w_{10} = 0.275$. We solve optimization (15) by using LINGO® 16.0 in Windows Operating System with 2 GB of memory and Dual Core 1.5 GHz of processor. The solution is given in Fig. 1 and Fig. 2.

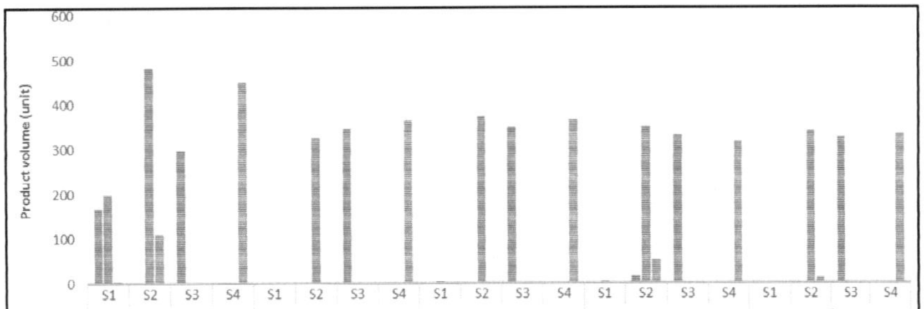

	S1	S2	S3	S4	S1	S2	S3	S4	S1	S2	S3	S4	S1	S2	S3	S4	S1	S2	S3	S4
	1				2				3				4				5			
Xtsp P1	167	2	298	0	0	2	347	0	0	1	349	0	0	17	330	0	0	0	325	0
Xtsp P2	199	483	0	0	2	326	0	0	4	373	0	0	4	351	0	0	0	342	0	0
Xtsp P3	3	109	0	450	0	0	0	364	0	0	0	366	0	51	0	317	0	12	0	335

Time period (t)

Figure 1. Optimal product volume that should be purchased from each supplier in the first example

The optimal strategy for this problem is the optimal values of $X_{tsp}, I_{tp}, \forall t \in T, \forall s \in S, \forall p \in P$, i.e. the optimal volume of each product (given in Fig. 1) that should be purchased from each supplier at each time period and the optimal volume of each product (given in Fig. 2) that should be stored in the inventory so that the total cost is minimum and the inventory level is as close as possible to the reference level decided by the decision maker. From Fig. 1, it can be seen that at time period 1, the manufacturer have to purchase 197 unit of product P1, 199 unit of product P2 and 3 unit of product P3 from supplier S1, 2 unit of product P1, 483 unit of product P2 and 109 unit of product P3 from supplier S2, and 298 unit of product P1 from supplier 3.

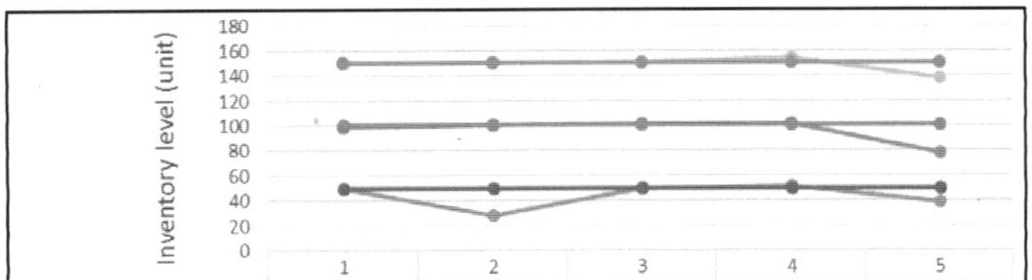

	1	2	3	4	5
INVENTORY LEVEL P1	99	100	101	101	78
INVENTORY LEVEL P2	50	28	50	51	39
INVENTORY LEVEL P3	149	150	150	154	138
Ref level P1	100	100	100	100	100
Ref level P2	50	50	50	50	50
Ref level P3	150	150	150	150	150

Period (t)

Figure 2. Inventory level and reference level in the first example

From Fig 2, it can be seen that at time period 1, the decision maker is deciding that the reference inventory level is 100 unit, 50 unit, and 150 unit of product P1, P2, and P3 respectively. The optimal strategy generated by the model is the manufacturer have to store 99 unit, 50 unit, and 149 unit of product P1, P2, and P3 respectively. The optimal decision for time period 2 to 5 can be derived analogously. The total cost for 5 time periods is 153398. From Fig. 2, it can be concluded that the inventory level is sufficiently closed to the reference level.

4.1.1. Continuous fuzzy demand

For the second example, let the demand $\tilde{D}_{tp}$ is a triangular fuzzy variable defined in Table 1 which can be illustrated by Fig. (3).

Table 1. Triangular membership function of $\tilde{D}_{tp}$

t	p	Triangular fuzzy membership function for $\tilde{D}_{tp}$ i.e. $\mu_{\tilde{D}_{tp}} = (a_{tp}, b_{tp}, c_{tp})$			Expected Value $E[\tilde{D}_{tp}]$
		a_{tp}	b_{tp}	c_{tp}	
1	P1	100	150	200	150
	P2	120	140	340	185
	P3	200	210	240	215
2	P1	180	210	220	205
	P2	160	200	210	192.5
	P3	120	140	160	140
3	P1	140	150	180	155
	P2	210	250	260	242.5
	P3	210	240	280	242.5
4	P1	240	260	285	261.25
	P2	180	220	260	220
	P3	190	210	240	212.5
5	P1	240	250	280	255
	P2	240	250	270	252.5
	P3	310	320	340	322.5

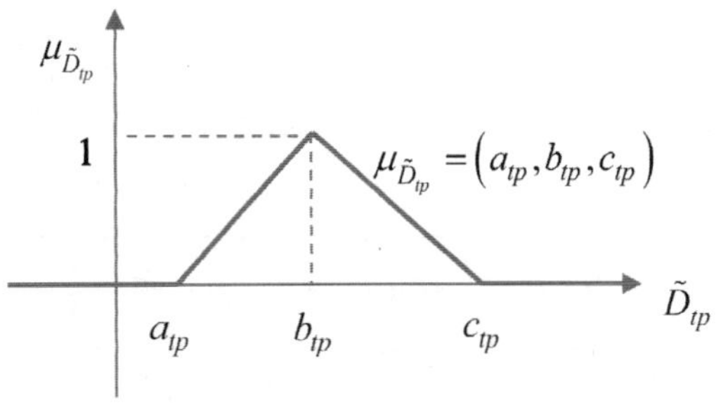

Figure 3. Triangular membership function $\mu_{\tilde{D}_{tp}} = (a_{tp}, b_{tp}, c_{tp})$

We evaluate the model for 5 time periods. The optimal strategy for this example is given in Fig. 4 and Fig. 5. From Fig. 4, it can be seen that at time period 1, 49 unit of product P1 have to be purchased from supplier S1, 210 unit of product P1 and 246 unit of product P2 have to be purchased from supplier S2. From Fig. 5, it can be seen that at time period 1, 100 unit of product P1, 50 unit of product P2, and 149 unit of product P3 have to be stored in the inventory that will be used to meet the demand at time period 2 or afterward. The optimal decision for time period 2 to 5 can be derived analogously. The total cost for 5 time periods is 94027. From Fig. 5, it can be conclude that the inventory level is sufficiently closed to the reference level.

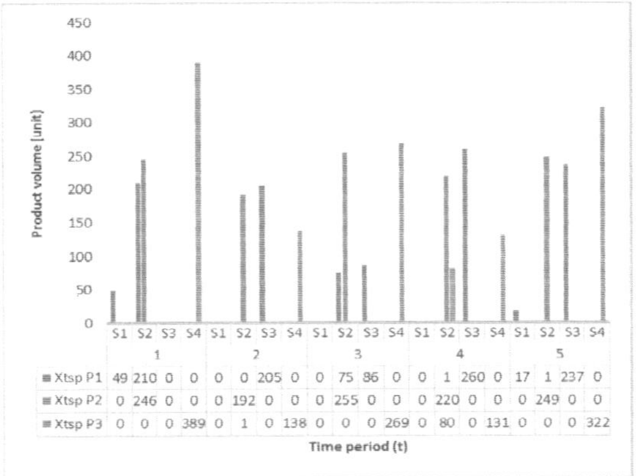

Figure 4. Optimal product volume that should be purchased from each supplier in the second example

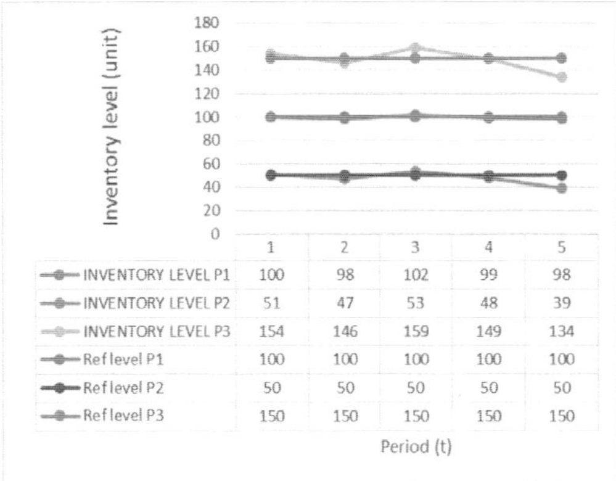

Figure 5. Inventory level and reference level in the second example

5. Conclusion and Future Works

In this paper, the optimal strategy determining for integrated inventory control and supplier selection problem with fuzzy demand was discussed. A mathematical model based on fuzzy expected value was considered to solve the problem. Two numerical examples was performed for model evaluating. From the results, it can be concluded that the problem was solved by the proposed model i.e. the optimal volume of each product that have to be purchased from each supplier for each time period and the optimal volume of each product for each time period that have to be stored were determined with minimum total cost and the inventory level was sufficiently closed to the reference level.

In the future works, we will develop the model so that it also have fuzzy cost parameters. Other methods like chance constrained and goal fuzzy programming, intuitionistic fuzzy programming and optimal control method will also be used to solve the problem and the comparison of them will also be discussed. Application in other problems will also discussed in the future articles.

Acknowledgment
Authors would like to thank DIPA PNBP FSM UNDIP 2017 for financial support to this research.

Appendix. Parameter values for numerical examples

Unit price (UPtsp)

Period	Supplier	P1	P2	P3
All	S1	40	21	20
	S2	41	20	20
	S3	40	22	21
	S4	42	19	18

Transport cost (TCts)

period	S1	S2	S3	S4
all	1	2	1	2

Defect rate (Qtsp)

Period	Supplier	P1	P2	P3
all	S1	0.05	0.02	0.05
	S2	0.02	0.03	0.02
	S3	0.02	0.02	0.05
	S4	0.05	0.75	0.05

Late rate / Delay lead time rate (DLTtsp)

Period	Supplier	P1	P2	P3
all	S1	0.00	0.01	0.01
	S2	0.01	0.01	0.02
	S3	0.02	0.00	0.02

	S4	0.01	0.02	0.00

Storage capacity (Mtp)			
period	P1	P2	P3
all	3000	3500	2500

Defect penalty cost per unit (UPCtsp)				
Period	Supplier	Product		
		P1	P2	P3
all	S1	0.2	0.4	0.25
	S2	0.5	0.6	0.5
	S3	0.75	0.2	0.7
	S4	0.5	1	0.25

Delay penalty cost per unit (UDCtsp)				
Period	Supplier	Product		
		P1	P2	P3
all	S1	0.2	0.2	0.4
	S2	0.5	0.2	0.2
	S3	0.1	0.1	0.1
	S4	0.5	0.5	0.2

Holding cost (HCtp)			
Period	Products		
	P1	P2	P3
all	0.1	0.1	0.2

Period	Total cost budget per period
all	300000

References

[1] Christopher M 2011 *Logistics and supply chain management* 4th ed. (Great Britain: Pearson Education)
[2] Ware N R, Singh S P and Banwet D K 2014 A mixed-integer non-linear program to model dynamic supplier selection problem *Expert Syst. Appl.* **41** 671
[3] Kara S S 2011 Applications supplier selection with an integrated methodology in unknown environment *Expert Syst. Appl.* **38** 2133
[4] Zadeh L A 1978 Fuzzy set as a basis for a theory of possibility *Fuzzy Sets Syst.* **1** 2
[5] Liu B and Liu Y K 2002 Expected value of fuzzy variable and fuzzy expected value models *IEEE Trans. Fuzzy Syst.* **10** 445
[6] Shiraz R K Charles V and Jalalzadeh L 2014 Fuzzy rough DEA model: A possibility and expected value approaches *Expert Syst. Appl.* **41** 434
[7] Wang Y M and Chin K S 2011 Fuzzy data envelopment analysis: A fuzzy expected value approach *Expert Syst. Appl.* **38** 11678
[8] Ghasemi M R, Ignatius J, Lozano S, Emrouznejad A and Hatami-Marbini A 2015 A fuzzy expected value approach under generalized data envelopment analysis *Knowledge-Based Syst.* **89** 148
[9] Virivinti N and Mitra K 2014 Fuzzy Expected Value Analysis of an Industrial Grinding Process *Powder Technol.* **268** 9
[10] Li D 2014 Possibilistic programming approach for fuzzy multidimensional analysis of preference in group decision making *Comput. Ind. Eng.* **73** 1
[11] Wang B, Li Y and Watada J 2017 Multi-period portfolio selection with dynamic risk/expected-return level under fuzzy random uncertainty *Inf. Sci. (Ny).* **385** 1
[12] Gupta P, Mittal G and Mehlawat M K 2013 Expected value multiobjective portfolio rebalancing model with fuzzy parameters *Insur. Math. Econ.* **52** 190
[13] Moussa A M, Kamdem J S and Terraza M 2014 Fuzzy value-at-risk and expected shortfall for portfolios with heavy-tailed returns *Econ. Model.* **39** 247

The Analysis of Forward and Backward Dynamic Programming for Multistage Graph

Anna Angela Sitinjak[1], Elvina Pasaribu[2], Justin E Simarmata[3], Tedy Putra[4], Herman Mawengkang[5]
Department of Mathematics, University of Sumatera Utara, Indonesia

Email :
[1]annaangelasitinjak@yahoo.co.id
[2]elvina.pa1000@yahoo.com
[3]justinesimarmata@yahoo.com
[4]iqrastabat2015@gmail.com
[5]hmawengkang@yahoo.com

Abstract. Dynamic programming is an optimization approach that divides the complex problems into the simple sequences of problems in which they are interrelated leading to decisions. In the dynamic programming, there is no standard formula that can be used to make a certain formulation. In this paper we use forward and backward method to find path which have the minimum cost and to know whether they make the same final decision. Convert the problem into several successive sequential stages starting on from stages 1,2,3 and 4 for forward dynamic programming and the step back from stage 4.3,2,1 for backward dynamic programming and interconnected with a decision rule in each stage. Find the optimal solution with cost principle at next stage. Based on the characteristics of the dynamic programming, the case is divided into several stages and the decision is has to be made (x_k) at each stage. The results obtained at a stage are used for the states in the next stage so that at the forward stage 1, f_1 (s) is obtained and used as a consideration of the decision in the next stage. In the backward, used firstly stage 4, f_4 (s) is obtained and used as a consideration of the decision in the next stage. Cost forward and backward always increase steadily, because the cost in the next stage depends on the cost in the previous stage and formed the decision of each stage by taking the smallest f_k value. Therefore the forward and backward approaches have the same result.

1. Introduction

Mathematical problems cannot be separated from problems with complex or simple categories with various types of solutions. One of complex problem solving is with a dynamic programming. A dynamic programming is an optimization approach that turns complex problems into simple problems and form stages, each stage makes decisions so that form a set of interrelated and consecutive decisions that lead to the final decision. Therefore dynamic programming is known by its characteristic that is multistage on optimization procedure. In the dynamic programming, there is no standard formula that can be used so it needs to make a certain formulation. There are 2 types of problem solving procedures that are forward and backward dynamic programming.

2. Theoretical Framework

Dynamic program is an optimization approach used to solve problems that often occur in daily-life. The problem of dynamic program has characteristics is (Robert Andreani in www.ime.unicamp.br/~andreani/MS515/capitulo7.pdf access in May 30,2017):

1. The problem is divided into several stages by making decisions at each stage.
2. Each stage consists of a number of states and states that correspond to that stage.
3. The effect of the policy decision at each stage is to transform the current state to astate associated with the beginning of the next stage (possibly according to a probability distribution).
4. The solution procedure is designed to find an optimal policy for the overall problem.
5. Given the current state, an optimal policy for the remaining stages is independent ofthe policy decisions adopted in previous stages.
6. The solution procedure begins by finding the optimal policy for the last stage.
7. A recursive relationship that identifies the optimal policy for stage n, given the optimal policy for stage n- 1, is available.

Dynamic program has sequences of decisions so that to form an optimal decision used the optimal policy that is "in an optimal sequence of decisions or choices, each subsequence must also be optimal" (Puntambekar.2010:4-2). This means that we can use the optimal solution of a stage without having to go back to the initial stage for the process from one stage to the next stage.

One of the problems that can be solved with a dynamic programming is the multistage graph problem. Multistage Graph G = (V, E) is a graph with special properties (Puntambekar.2010:4-3):

1. Directed graph
2. All vertices are partitioned into the k stages where k≥ 2
3. Each edge has weight
4. There is only 1 source (let S) and 1 sink (let T)
5. The path from source to sink consists of several stages with each stage has a set of vertices.
6. Its problem is to find the path from S to T (minimum or maximum cost).

Although there is no standard formula for solving a problem with a dynamic programming, it has fixed method namely forward and backward methods.

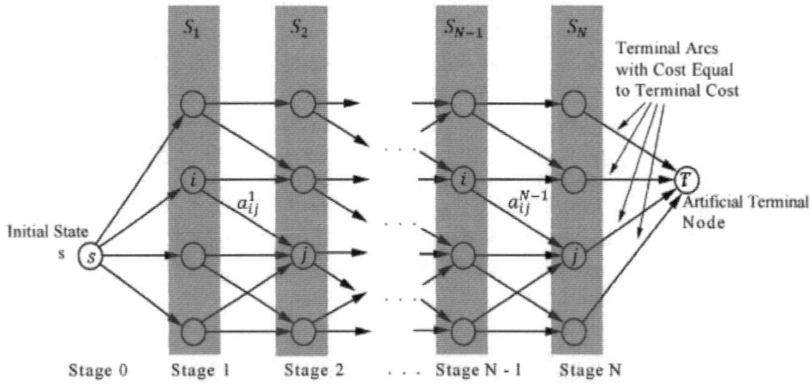

Figure 1.

To solve the multistage graph problem using several stages:

1. Convert the problem to some successive stages which for forward dynamic programming to move from stage 1,2,3, and so on until stage n and move back from stage n, n-1, to stage 1 for backward dynamic programming
2. Complete the last stage by connecting all the previous stages of decisions.
3. Find the optimal solution with the cost principle at k + 1 = (cost generated at k step) + (cost from stage k to stage k + 1) with k = 1,2, ..., n-1 for forward dynamic Programming and at stage k = (cost generated at stage k + 1) + (cost from stage k + 1 to stage k) with k = n, n-1, ..., 1 for backward dynamic programming (Munir in Ika Zulhidayati.2013:26).

3. Discussion.

Find the minimum cost for multistage graph problem by using dynamic programming (web.mit.edu/15.053/www/AMP-Chapter-11.pdf accessed in May 2,2017).

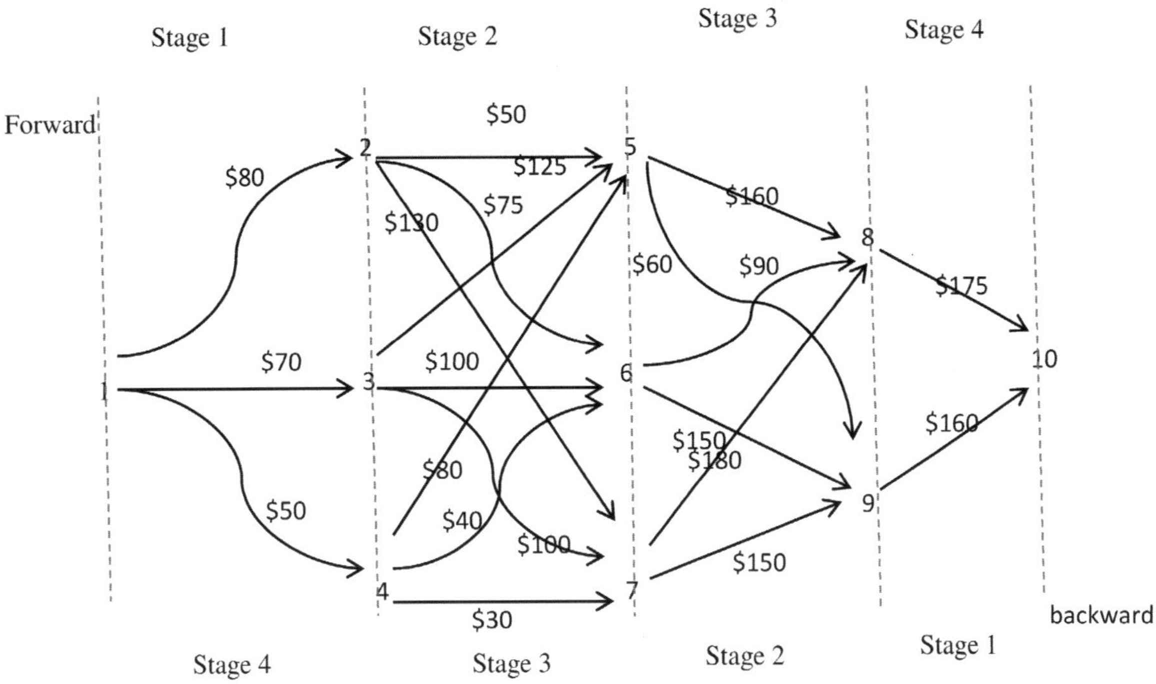

Figure 2.

a. Forward Dynamic Programming

$f_1(s) = c_{x_1 s}$ (initial state)

$f_k(s) = min_{x_k}\{c_{x_k s} + f_{k-1}(x_k)\}$ (recurrence)

with k = 2,3,4

where :

$c_{x_k s}$ = weight

$f_{(s, x_k)}$= total weight

$f_k(s)$ = minimum value

final state: $f_4(10)$ with looking for $f_1(s)$, $f_2(s)$, $f_3(s)$ firstly

stage 1 :

$f_1(s) = c_{x1s}$

Table 1.

s	Optimal Solution	
	$f_1(s)$	x_1^*
2	80	Vertex 1
3	70	Vertex 1
4	50	Vertex 1

Stage 2 :
$f_2(s) = \min_{s2} \{ c_{x2s} + f_1(x_2) \}$

Table 2.

x_2 \ s	$f_2(x_2,s) = c_{x2,s} + f_1(x_2)$			Optimal Solution	
	2	3	4	$f_2(s)$	x_2^*
5	130	195	130	130	2 or 4
6	155	170	90	90	4
7	210	170	80	80	4

Stage 3 :
$f_3(s) = \min_{s3} \{ c_{x3s} + f_2(x_3) \}$

Table 3.

x_3 \ s	$f_3(x_3,s) = c_{x3s} + f_2(x_3)$			Optimal Solution	
	5	6	7	$f_3(s)$	x_3^*
8	290	180	260	180	6
9	190	240	230	190	5

Stage 4 :
$f_4(s) = \min_{s4} \{ c_{x4s} + f_3(x_4) \}$

Table 4.

x_4 \ s	$f_4(x_4,s) = c_{x4,s} + f_3(x_4)$		Optimal Solution	
	8	9	$f_4(s)$	x_4^*
10	355	350	**350**	9

The paths that can be reached with minimum cost are::

a. 1-2-5-9-10 = 350
b. 1-4-5-9-10 = 350

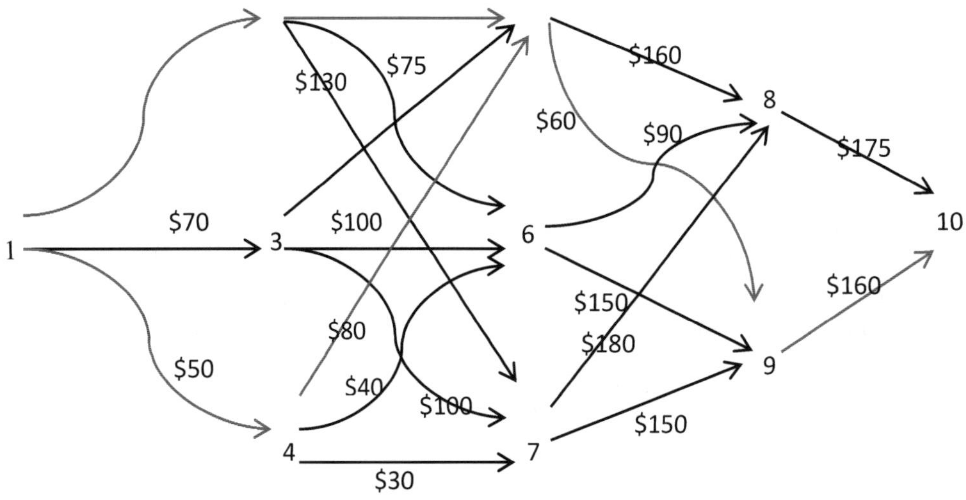

Figure 3.

b. Backward Dynamic Programming

$f_4(s) = c_{x4s}$ (initial state)

$f_k(s) = \min_{xk} \{ c_{sxk} + f_{k+1}(x_k) \}$ (recurrence)

with k = 1,2,3,

where :

x_k = decision variable at the stage k (k=1,2,3)

IOP Conf. Series: Materials Science and Engineering **300** (2018) 012010 doi:10.1088/1757-899X/300/1/012010

c_{sxk} = cost dari s ke x_k
$f_{(s,xk)}$ = total cost of path s to x_k
f_k (s) = minimum value of f_k (s, x_k)
stage 4 :
$f_4(s) = c_{sx4}$

Table 5.

S	Optimal Solution	
	f_4 (s)	x_4
8	175	10
9	160	10

stage 3 :
$f_3(s) = \min_{x3} \{ c_{sx3} + f_4(x_3) \}$

Table 6.

x_3 / s	$f_3(s) = c_{s,x3} + f_4(x_3)$		Optimal Solution	
	8	9	f_3 (s)	x_3
5	335	220	220	9
6	265	310	310	6
7	355	310	310	9

stage 2 :
$f_2(s) = \min_{x2} \{ c_{sx2} + f_3(x_2) \}$

Table 7.

x_2 / s	$f_2(s) = c_{s,x2} + f_3(x_2)$			Optimal Solution	
	5	6	7	f_3 (s)	x_3
2	270	385	440	270	5
3	345	410	410	345	5
4	300	350	340	300	5

stage 1 :
$f_1(s) = \min_{x1} \{ c_{sx1} + f_2(x_1) \}$

Table 8.

x_1 / s	$f_1(s) = c_{s,x1} + f_2(x_1)$			Optimal Solution	
	2	3	4	f_1 (s)	x_1
1	350	415	350	350	2 or 4

Paths with minimum cost are:

a. 1-2-5-9-10 dengan cost 350
b. 1-4-5-9-10 dengan cost 350

IOP Conf. Series: Materials Science and Engineering **300** (2018) 012010 doi:10.1088/1757-899X/300/1/012010

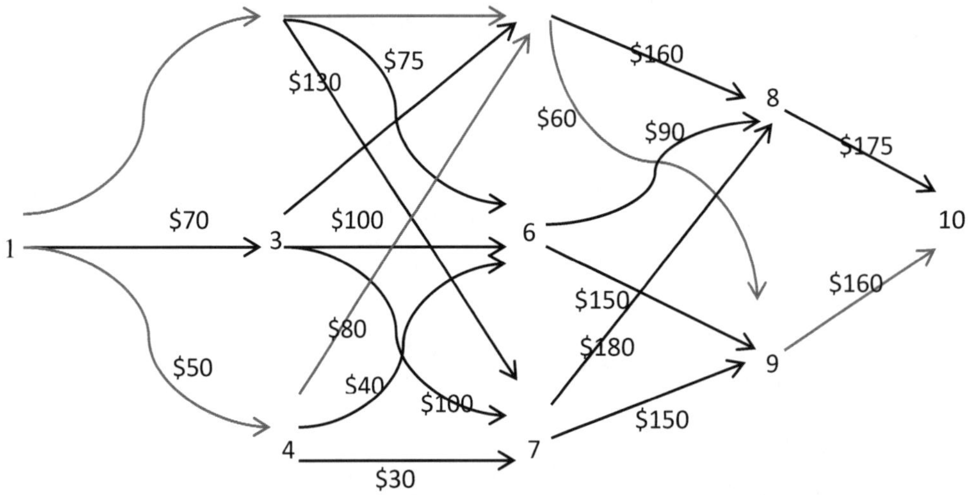

Figure 4.

Based on the optimal solution characteristics of the dynamic programming and the result of the case, it is found that the problem is divided into several stages and taken decision (x_k) at each stage. In this problem, it is made into 4 stages and there are decisions at each stage either forward or backward, where each stage of the problem has a number of states. These decisions are interconnected to form a final decision.

The result obtained at a stage is used for the states in the next stage so that in the forward stage 1, f_1 (s) is obtained and used as a consideration of the decision modifiers in the next stage and at the backward in stage 4, f_4 (s) is obtained and used as consideration of the decision modifier at next stage. This is done until the last stage. In this case only think about cost without thinking of the problem about time and others. Cost in forward and backward always increase steadily, because the cost in the next stage depends on the cost in the previous stage and formed the decision of each stage by taking the smallest value of f_k where optimal decision making follows the principle of optimality. So the recurrence relation is different both the forward and backward but it leads to the same path. Therefore the forward and backward have the same result.

4. Conclusion

Forward dynamic programming has the optimum principle of k + 1 = (cost generated at stage k) + (cost from stage k to stage k + 1) with k = 1,2, ..., N-1. While the principle of backward dynamic programming optimization is cost at the stage k = (cost generated at stage k + 1) + (cost from stage k + 1 to stage k) with k = n, n-1, ..., 1. Although the principle of optimizations both of them are different, but F\forward and backward dynamic programming produce the same path that is 1-2-5-9-10 and 1-4-5-9-10 with cost 350.

References

[1] Nagy G, Salhi S. 2007. Location-routing: Issues, models and methods. European Journal of Operational Research, 177, 649-672.

[2] Andreani, Robert in www.ime.unicamp.br/~andreani/MS515/capitulo7.pdf access in May 30, 2015

[3] Puntambekar,A.A., (2010), *Design and Analysis of Algorithm*, Technical Publication, India.

[4] Zulhidayati, Ika, (2013), *Aplikasi Algoritma Greedy dan Program Dinamis (Dynamic Programming) Pada Permainan Greedy Spiders*, Skripsi,FMIPA, UPI.

[5] web.mit.edu/15.053/www/AMP-Chapter-11.pdf accessed in May 2,2017.

Routing and Scheduling Optimization Model of Sea Transportation

Mika debora br barus[1], Habib asyrafy[2], Esther nababan[3] and Herman mawengkang[4]

Department of Mathematics, University of Sumatera Utara, Medan, Indonesia.

E-mail: [1]Deboramika9@gmail.com, [2]habib.asyrafy@gmail.com, [3]esther@usu.ac.id, [4]hermifa@yahoo.com

Abstract. This paper examines the routing and scheduling optimization model of sea transportation. One of the issues discussed is about the transportation of ships carrying crude oil (tankers) which is distributed to many islands. The consideration is the cost of transportation which consists of travel costs and the cost of layover at the port. Crude oil to be distributed consists of several types. This paper develops routing and scheduling model taking into consideration some objective functions and constraints. The formulation of the mathematical model analyzed is to minimize costs based on the total distance visited by the tanker and minimize the cost of the ports. In order for the model of the problem to be more realistic and the cost calculated to be more appropriate then added a parameter that states the multiplier factor of cost increases as the charge of crude oil is filled.

Keyword: Optimization, Routing and Scheduling, Transportation of crude oil

1. Introduction

The global logistics problem becomes an important issue in increasing the global environments. The integrated optimization of production planning and distribution planning is required from the viewpoint of global supply chain optimization (Nishi, Hiranaka, & Grossmann, 2011; Nishi, Konishi, & Ago, 2007).

This paper examines the routing and scheduling optimization model of sea transportation. One of the issues discussed is about the transportation of ships carrying crude oil (tankers) which is distributed to many islands. The consideration is the cost of transportation which consists of travel costs and layover costs at the port. Crude oil to be distributed consists of several types.

Scheduling of crude oil transportation is a crucial challenge for petrochemical industries. In a petrochemical company, demands of various types of crude oils are given every month. A fleet of very large crude oil carrier (VLCC) tankers is used for international crude oil transportation. Each tanker can load several items of crude oils with finite capacity. The assignment of demands, visiting sequence of ports, and loading volume of multi items of crude oils should be determined simultaneously to minimize the total cost considering the capacity of tankers.

In many practical situations, these decisions are still made manually by the negotiation between the human operators based on the contract with individual supplier. In order to help the decision making of human operators, it is highly required that the practical ship routing and scheduling can be

generated automatically to increase the efficiency and to avoid human errors. It is also easier for new human operators to execute the scheduling system once personnel change occurs.
Based on previous research we have discussed the model of route problem and scheduling for crude oil load planning that is to minimize the total distance visited by tanker and minimize harbor cost imposed from each visited place (Tatsushi Nishi & Tsukasa Izuno, 2014). However, the existing model is not realistic enough to develop a more realistic model. In this research will be discussed about the problem of crude oil delivery more realistic by adding a parameter. Where the parameter states that there is an increase in cost / km when the tanker visited loading place. With the addition of cargo where the greater the burden of crude oil transported by the tanker will affect the additional cost. The load in question is not exceeding the maximum and minimum capacity of the tanker.

2. Literature review

The automating international transportation is demanded rapidly (Waters, 2010). Several works have been addressed related to ship scheduling for maritime transport. An analytical model concerning the international maritime transport market has been studied (Kuruda, Takebayashi, & Tsuji, 2005). An optimization approach to solve the ship scheduling as a generalized set-packing model has been addressed (Kim & Lee, 1997). They generate a feasible ship routing and schedule by a directed acyclic graph. An optimization model as a multi-ship pickup and delivery is proposed in Fagerholt (2001). They use a dynamic programming to solve the shortest path problem. A heuristic search algorithm for ship routing and scheduling with split loads is introduced in Korsvik, Fagerholt, and Laporte (2011) dan Yin, Nishi, and Izuno (2012). In order to solve more practical problems, heuristic approaches are applied (Paessens, 1988). New saving based algoritms are introduced to deal with pick up and delivery of full truckloads problem under time window constraints (Gronalt, Hartl, & Raimann, 2003).

3. Problem description

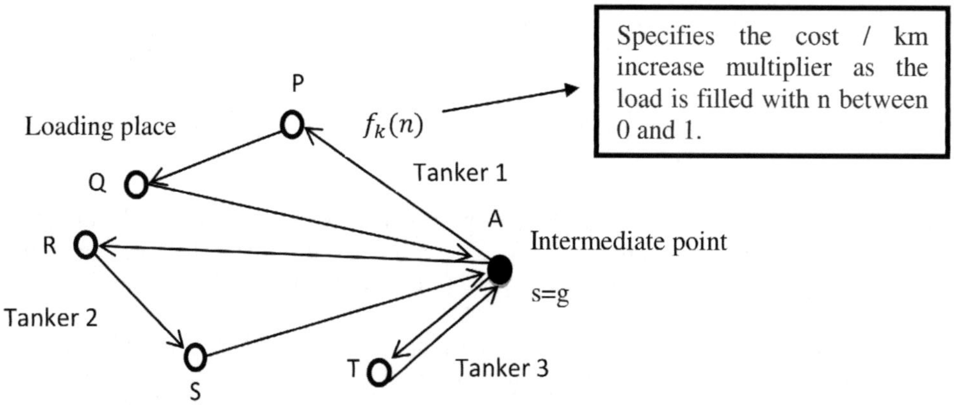

Figure 1. Ship routing and scheduling problem for crude oil transportation

The ship routing and scheduling problem for crude oil transportation mainly consists of two part is the outbound ship loading planning problem that determines the visiting route for a fleet of tankers in various countries. The second part is the inbound ship unloading loaded oil into several domestic factories considering both of inbound and outbound transportation simultaneously . This paper deals with only the first step of the ship routing and scheduling for outbound load planning, whinch is formulated as a vehicle routing problem with split deliveries. L is the set of loading places where crude oil types. Figure 1 shows the model of ship routing and scheduling problem for crude oil load planning. Each loading place has one or more than two type of crude oils. Each tanker has different capacity. All tankers originate and return to a common point after visiting multiple destinations. An

intermediate point (s=g) is the depot which should be started and returned after visiting some of loading places. The different settings : open routes, multiple origins and multiple destinations can be incorporated by setting the graph model of fig. 1. The distances between loading places denoted by dij are not symmetric ($d_{i,j} \neq d_{j,i}$) but satisfy the triangular inequality,($d_{i,j} + d_{j,k} \geq d_{i,k}$). T is the set of alvailable tanklers whose capacity of loading volume is given. The total demand volume Do for each oil type o ∈ O is given by the contract from the suppliers in every month. The problem is a multi product heterogeneous fleet split pickup ship routing problem with finite capacity and loading volume for a fleet of tankers in order to minimize the total cost.
In practical point of views, we assume the following conditions which especially appears in maritime crude oil transportation.

1. The number of loading places visited by a tanker is less than or equal to three. The port charge is imposed if a tanker occupies one port. If one tanker visits one port, visiting time should be sufficient in order to complete loading operations.
2. The loading volume of tanker should be close to be full and the number of tankers should be minimized. This is because the visiting from domestic site to international sites is time consuming and requires lots of costs.
3. The loading condition is different for each loading place.
4. Travel costs increase with increasing cargo

The objective function is the total cost for the operation of all tankers is one month that should be minimized. The total cost consists of the total sum of the bunker costs which is proportion to the distance, and the sum of the port charge which depends on the number of visiting time.

3.1 Problem formulation
The problem is formulated as a mixed integer linear programing problem in this section.

L : set of loading places
O : set of crude oil types
T : set of tankers

Indeks

g : depot, g end point
i,j : loading place
k : tanker,
o :type of oil
s : depot s=L start point s-g the our problem

Decision variables

$q_{k,i,o}$: loading volume of crude oil type o at loading place for tanker k,
$t_{k.i}$: loading time of loading place I for tanker k.
$x_{k,i,j}$: binary variable that takes $1 + f_k(n)$ when tanker (k) visits from loading place i to loading place j, and O other wise.
$\partial_{k.i}$: binary variable that takes 1 when tanker k visits loading place I, and O otherwise.

Parameters

c_i : port charge imposed by visiting loading place i for tanker k,
C_k^{max} : maximum loading volume for tanker k,
$d_{i,j}$: distance between loading place i and loading place j,
D_0 : demand of crude oil type o,
$E_{k,i,j}$: loading date from loading place i to loading place j for tanker k,
M : sufficiently large positive constant
q_i^{min} : minimum loading volume at loading place i.
q_i^{max} : maximum loading volume at loading place i,

w_1 : weighting factor of the total distance traveled by tankers,
w_2 : weighting factor of the total cost imposed by visiting loading place,
τ_m : maximum number of loading places visited by one tanker,
$\mu_{i,o}$: binary constant that takes 1 if crude oil type 0 can be loaded at loading place i, and 0 otherwise.
$f_k(n)$: Specifies the cost / km increase multiplier as the load is filled with n between 0 and 1.

Objective function: The first objective is to minimize the total distances traveled by tankers and the second one is to minimize the port charge imposed by visiting loading places. The weighted sum of the objective function of (1) is minimized.

$$\min w_1 \left(\sum_{k\in T}\sum_{i\in L\cup\{s\}}\sum_{j\in L\cup\{g\}} d_{i,j}x_{k,i,j}\right) + w_2\left(\sum_{k\in T}\sum_{i\in L} c_{i\partial_{k.i}}\right) \quad (1)$$

Constraints : The total demand constraints are denoted by (2)

$$\text{s.t} \sum_{k\in T}\sum_{i\in L} q_{k,i,o} = D_O \quad (\forall_O \in O) \quad (2)$$

The assignment constraints of loading places are denoted by (3) and (4).

$$\sum_{i\in L\cup\{s\}} x_{k,i,j} = \partial_{k,j} \qquad (\forall_k \in T, \forall_j \in L\cup\{g\}) \quad (3)$$

$$\sum_{j\in L\cup\{g\}} x_{k,i,j} = \partial_{k,i} \qquad (\forall_k \in T, \forall_i \in L\cup\{s\}) \quad (4)$$

Eqs. (5) and (6) respectively represent that each tanker starts from the starting point s and stop at the ending point g after visiting loading places.

$$\sum_{i\in L} x_{k_{s.i}} = 1 + f_k(n) \qquad (\forall_k \in T) \quad (5)$$

$$\sum_{i\in L} x_{k_{i.g}} = 1 \qquad (\forall_k \in T) \quad (6)$$

Eq.(7) restricts that the number of visiting time of loading places should be less than or equal to the maximum number of visiting time of loading places τ_m.

$$\sum_{i\in L} \partial_{k.i} \le \tau_m \qquad (\forall_k \in T) \quad (7)$$

Subtour elimination constraints are denoted by (8) where M is a sufficiently large positive number. M is the upper bound of the difference of loading dates for each tanker.

$$t_{k,i} + E_{k,i,j}t_{k,j} - M(1 - x_{k,i,j}) \le 0 \qquad (\forall_k \in T, \forall_i \in L\cup\{s\}, \forall_j \in L\cup\{g\}) \quad (8)$$

The loading volume should be less than the capacity of a tanker where the capacity of tanker is different for each tanker. The constraints are denoted by (9).

$$\sum_{i\in L}\sum_{o\in O} q_{k,i,o} \le C_k^{max} \qquad (\forall k \in T) \quad (9)$$

The minimum and maximum loading volume constraints at each loading place are denoted by (10) and (11).

$$\delta_{k,i}q_i^{min} \sum_{o\in O} \mu_{i,o}\, q_{k,i,o} \qquad (\forall k \in T, \forall_i \in L) \quad (10)$$

$$\sum_{o\in O} \mu_{i,o}\, q_{k,i,o} \le \delta_{k,i}q_i^{max} \qquad (\forall k \in T, \forall_i \in L) \quad (11)$$

where $\mu_{i,o}$ is the given parameter that takes 1 if crude oil type o can be loaded at loading place i, and 0 otherwise. The constraints for the variables are denoted by (12)–(15).

$$x_{k,i,j} \in \{0,1+f_k(n)\} \qquad (\forall k \in T, \forall_i \in L\cup\{s\}, \forall_j \in L\cup\{g\}) \quad (12)$$

$\partial_{k.i} \in \{0,1 + f_k(n)\}$ $(\forall k \in T, \forall_i \in L \cup \{s\} \cup \{g\})$ (13)

$q_{k,i,o} \leq 0$ $(\forall k \in T, \forall i \in L, \forall_O \in O)$ (14)

$t_{k.i} \geq 0$ $(\forall k \in T, \forall i \in L \cup \{s\} \cup \{g\})$ (15)

In order to reduce search space, the valid inequalities of (16) can be imposed. Each loading place must be visited by at least Hi tankers. Hi is obtained from $H_I = \frac{\sum_{o \in O} \mu_{i,O} D_O}{max_k c_k^{max}}$ from the given parameters. We have observed from the computational results that the computation time for general-purpose solvers can be significantly reduced with the constraints of (16)

$\sum_{k \in T} \partial_{k.i} = H_I$ $(\forall i \in L)$ (16)

The original problem (P0) can be regarded as a split delivery vehicle routing problem (SDVRP) that is known to be NP-hard (Salani & Vacca, 2011).

4. Conclusions and future research

The addition of parameters that calculate the current shipload creates these route model and scheduling to be more realistic so that the reckoned ship transportation cost becomes more appropriate. This research only develops model of route and scheduling on tankers that carry crude oil, while transportation problem which is connected with route and scheduling on tanker is difficult problem or Non-Polinomial Hard (NP-Hard Problem). This relates to the issue that is difficult to be solved by the exact method. Therefore it is necessary to apply a heuristic method.

It is expected for next research can develop methods of these problem, route and scheduling in heuristic. The problem of route optimization and scheduling on transport can be developed, constructed and completed using heuristic methods.

References

[1] Andersson, H., Christiansen, M., & Fagerholt, K. (2011). The maritime pickup and delivery Problem with time windows and split loads. INFOR,49(2), 79-91.

[2] Archetti, C., Speranza, M. G., & Hertz, A.(2006). A tabu search algorithm for the split delivery vehicle routing problem. Transportation Science, 40(1), 64-73.

[3] Bach, Lukas, Routing and Scheduling and Problem Optimization Using exact and Heuristic Methods, 2014

[4] Balenguer, J. M., Martinez,M.C., & Mota, E. (2000). A lower bound for the split delivery vehicle routing problem. INFORMS,48(5),801-810.

[5] Banhart, C., Jhonson, E. L., Nemhauser, G. L., Savelbergh, M. W. p,. & Vance, P. H. (1998). Branch-and-price: Column generation for salving huge integer programs. Operations Research, 46(3), 316-329.

[6] Bronmo, G., Nygreen, B., & Lysgaard,J.(2010). Column generation approaches to ship scheduling with flexible cargo sizes. European Journal of operational Research. 200, 139-150.

[7] Cheng, L.., & Duran, M. A. (2004). Logistics for world-wide curde oil transportation sing discrate event simulations and optimal control. Computers and Chemical Engineering, 28, 897-911.

[8] Coccola, M. E., Zamarripa, M., Mendez , C. A., & Espu na, A. (2003). Toward integrated production and distribution management in multi-echelon supply chains. Computers Chemical Engineering (alvailable online).

[9] Chakroborty, partha, *Optimal Routing and Scheduling In Transportation* : Using Genetic Algorithm to Solve Difficult Optimization Problem. 2012

[10] Dongo, R. G., & Cerda,J. (2009). A hybit local improvement algorithm for largescale multi-depot vehicle routing problems with time window. Computers and Chemical Engineering (alvailable online).

[11] Fagerholt, K. (2001). Ship scheduling with soft time window: An optmisation based approach. European J, 131 , 559-571.

[12] Gronalt, M., Hartl, R. F., & Reimann, M.(2003). New Savings based algorithms for time constrained pickup and delivery of full truckloads. European journal of Operation Research.151,502-535.

[13] Hennig, F.,, Nygreen, B., & Lubbecke, M. E. (2012). Nested column generation appelid to the crude oil tanker routing and scheduling problem with split pickup and split delivery. Naval Research Logistics, 59(3-4), 298-310.

[14] Jaumarad, B., Meyer, C., & Vovor, T. (1999). Column/row generation and elimination methods. Les Cahiers du GERAD. G-99-34.

[15] Jin, M., Liu, K., & Bownden, R .O (2007). A two Stage algorithm with valid inequalities for the split delivery vehicle routing problem. International journalof Production Economics, 105,228-242.

[16] Jin, M., Liu, K., & Burak, E (2007) A Column generation approach for the split delivery vehicle routing problem. Operations Research Letters, 36,265-270.

[17] Korsvik,J. E., Fagerholt,K., & Laporte, G.(2011). A Large neighborhood search heuristic for ship routing and scheduling with split loads. Computer and Operation Research, 38,474-483

[18] Karuda, K., Takebayashi, M., & Tsuji, T. (2005). International container transportation network analysis considering post-panamax class container ship. Global Transportation Markets: Analysis and Policy Making Research in Transportation Economics, 13, 369-391.

[19] Li, J., Karimi, I. A., & Srinivasan, R. (2010). Efficeint bulk maritme logistics for the supply and delivery of multiple chemicals. Coputers and Chemical Engineering . 34, 2118-2128.

[20] Marti, Rafael,. Reinelt, Gehard *The Liniear Ordering Problem*: Exact and Heuristic Methods in Combinatorial Optimization. 2010.

[21] Nishi, Tatsushi,. Izuno, Tsukesa,.*Column Generation Heuristic for Ship Routing and Scheduling Problem in Crude Oil Transportation With Split Deliveries*. 2014.

[22] Nishi, T., & Hiranaka, Y. (2009). Lagrangian relaxation and cut generation for sequence dependet setup time flowshop problems. International Journal of Productions Research, 51, 4778-4796.

[23] Nishi, T., Hiranaka, Y ., & inuiguchi, M. (2010). Lagrangian relaxation with cut generation for hybrid flowshop scdulling and problems to minimize the total weighted tardiness. Computers and Operations Research, 37, 189-198.

[24] Nishi, T., Hiranaka ,., & Grossman, I. E. (2011). A bilevel decomposition algorithm for simultaneous production scheduling and conflict-free routing for automated guided vehicles. Competers and Chemical Engineering, 38, 876-888.

[25] Nishi, T., Isoya, Y., & Inuiguchi , M. (2011). An intergeted column generation and largnanian relaxation for salvig flowshop problems to minimize the total weighted tardiness.International Journal of Innovative Computing. Information and Control, 7(11). 6453-6471.

[26] Nishi, T., Konishi, M., & Ago. M.(2007). A distributed decision making system for integratedoptimizationofproductionscheduling anddistributionfor aluminum production line. Computers and Chemical Engineering, 31(10), 1205–1221.

[27] Nishi, T., Muroi, Y., & Inuiguchi, M. (2011). Column generation with dual inequalities for railway crew scheduling problems. Public Transport, 3(1), 25–42.

[28] Nishi, T., Yin, S., & Izuno, T. (2011). Column generation approach to ship scheduling problems for international crude oil transportation. In Proceedings of the 7th IEEE international conference on automation science and engineering (pp. 468–473).

[29] Ohlmann, Jeffrey. W., *A Compressed-annealing Heuristic for the Salesmen Problem With Time Windows.* 2007
[30] Paessens, H.(1988). The savings algorithm for the vehicle routing problem. European Journal of Operational Research, 34, 336–344.
[31] Ropke, S., & Coudeau,J-F (2009). Branch and cut price for the pickup and the delivery problem with time windows. Transportation Science, 43(3), 267-286.
[32] Stalhane, M., Anderson, H., Christiansen, M., Cordeau, J-F.,& Desaulniers, G (2012). A branch-price-and-cut method for a ship routing and scheduling problem with split loads. Computers and Operation Research, 39(12),3361-3375.
[33] Stålhane, M., Anderson, H., Christiansen, M., Cordeau, J.-F., & Desaulniers, G. (2012). A branch-price-and-cut method for a ship routing and scheduling problem with split loads. Computers and Operations Research, 39(12), 3361–3375.
[34] Waters, D. (Ed.). (2010). Global logistics: New directions in supply chain management. UK: The Charted Institute of Logistics and Transport.
[35] Yin, S. Nishi, T., & Izuno,T (2012). A Heuristic Approach for international crude oil transportation scheduling problems. Journal of Advanced Mechanical Design, System, and Manufacturing, 6(5),687-702.

Mathematical model of parking space unit for triangular parking area

Intan Syahrini[1], Teti Sundari[1], Taufiq Iskandar[1,2], Vera Halfiani[2], Said Munzir[1,2,3] and Marwan Ramli[1,2,3]

[1]Department of Mathematics, Syiah Kuala University, Banda Aceh 23111, Indonesia
[2]Dynamic Application and Optimization Research Group, Syiah Kuala University, Banda Aceh 23111, Indonesia
[3]Applied Mathematics Laboratory Syiah Kuala University, Banda Aceh 23111, Indonesia

E-mail: intansyah_89@yahoo.com, tetisundari27@gmail.com, tauis2005@gmail.com, verahalfiani@yahoo.com, smunzir2001@yahoo.com, corresponding author:ramlimarwan@gmail.com or marwan.math@unsyiah.ac.id

Abstract. Parking space unit (PSU) is an effective measure for the area size of a vehicle, including the free space and the width of the door opening of the vehicle (car). This article discusses a mathematical model for parking space of vehicles in triangular shape area. An optimization model for triangular parking lot is developed. Integer Linear Programming (ILP) method is used to determine the maximum number of the PSU. The triangular parking lot is in isosceles and equilateral triangles shape and implements four possible rows and five possible angles for each field. The vehicles which are considered are cars and motorcycles. The results show that the isosceles triangular parking area has 218 units of optimal PSU, which are 84 units of PSU for cars and 134 units of PSU for motorcycles. Equilateral triangular parking area has 688 units of optimal PSU, which are 175 units of PSU for cars and 513 units of PSU for motorcycles.

1. Introduction

Parking Space Unit (PSU) is an effective measure for the area size of a vehicle, including the free space and the width of the door opening of the vehicle. Parking facilities can be on-street or off-street parking [1]. A parking lot design should have a goal to maximize the number of parking space units while considering several things such as parking layout and pedestrian space around the parking area [2]. Parking facilities in public spaces that are easily accessible can also be considered [3]. A wireless-based smart parking system can be one of the solutions to facilitate a parking lot that can allow parking on separate floors, with the result that it will reduce the required land or area for parking space [4]. In other studies, an automated parallel parking system in small mobile vehicles [5] and a simple parking path programming strategy for automatic parking system (APS) [6] were proposed to facilitate the parking system.

AbdelFatah and Taha [7] suggested a design of a parking lot with three possible parking rows and five possible parking angles. Ramli et al. [8] developed the model of AbdelFatah and Taha by establishing four possible parking rows to be applied to the parking lot in Hermes Palace Mall of Banda Aceh. The result shows that by the selecting of efficient parking angles, the parking lots can accommodate the optimal number of vehicles so that it can increase the revenue of parking facility manager. A study on on-street parking at Jl. T. Panglima Polem B.Aceh which is located in the city center was also presented in [9]. The result shows that the parking arrangement on the road is different from the arrangement obtained from the model developed based on the parking demand and requirement. The requirement for a parking area in a public place at the main trading center of Pasar Aceh was studied by Erliana [10]. The result shows that the available parking capacity is unable to meet the actual required parking capacity. Anggraini [11] examined the performance of the street by Cempaka Lima Clinic in Banda Aceh. The result shows that the on-street parking facilities affect the decrease of the street performance level.

Along with the increasing number of vehicles, the provision of parking lots especially outside the road body which can accommodate more vehicles has become indispensable. One solution that can be carried out is to design the parking space unit (PSU) in accordance with the shape of the land so that it can accommodate vehicles in larger quantities. This idea underlies the conduct of this research. Here, an optimization model of PSU will be developed by considering a parking area in triangular shape.

2. The Mathematical Model

The optimization model of parking space unit (PSU) that are discussed in this article can be seen in figure 1. This model is a modification of the model presented by Ramli et al. [6] to the parking lot in square or rectangular shape. The similar approach is applied for the triangular parking lot. The PSU optimization model can be described as four possible parking line. The four possible parking rows are the full X_{ep} exterior row, the X_e exterior, the full X_{ip} interior and the X_i interior. Each possible parking row has possible parking angles of 30°, 45°, 60°, 75° and 90°.

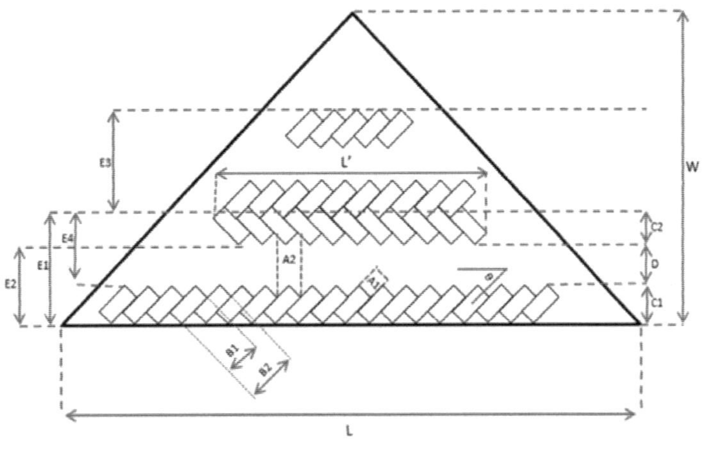

Figure 1. Establishment of triangular parking lot and Parking Space Unit (PSU)

In figure 1, A_1 is the width of the vehicle, A_2 is the width of the vehicle in the row, B_1 is the length of the vehicle, B_2 is the vehicle length in the row, C_1 is the parking row width from the wall to the parking line boundary, C_1 is the width of the parking row in the middle, D is the vehicle alley, E_1 is the full exterior line width, E_2 is the exterior line width, E_3 is the width of the full interior row, E_4 is the width of the interior row, W is the whole parking lot width, L

is the entire parking lot length and L' is the length of the parking lot for the row in the middle without an alley.

Based on figure 1, we can develop the objective function maximizing the PSU based on parking angle with decision variable $X(ep,\theta), X(e,\theta), X(ip,\theta), X(i,\theta)$ and the number of PSU units based on the parking angle $n(ep,\theta), n(e,\theta), n(ip,\theta), n(i,\theta)$ with possible parking corner of the vehicle 30°, 45°, 60°, 75° and 90°. Mathematically, the objective function for the maximum PSU problem can be written as follows,

$$maksZ = \sum_{\theta=30^0,45^0,60^0,75^0,90^0} [n_{(ep,\theta)} + n_{(e,\theta)} + n_{(ip,\theta)} + n_{(i,\theta)}]. \quad (1)$$

The constraints of this problem are the width of the parking lot W, the length of the parking lot L, the length of the parking lot in the middle of L' and the number of vehicles in the form of positive integers.

$$\theta = 30^0, 45^0, 60^0, 75^0, 90^0 [E_{1\theta}X_{(ep,\theta)} + E_{2\theta}X_{(e,\theta)} + E_{3\theta}X_{(ip,\theta)} + E_{4\theta}X_{(i,\theta)}], \quad (2)$$

$$A2n_{(ep,\theta)} - (L + L')X_{(ep,\theta)} \leq 0, \quad (3)$$

$$A2n_{(e,\theta)} - X_{(e,\theta)}L \leq 0, \quad (4)$$

$$A2n_{(ip,\theta)} - (L'_1 + L'_2)X_{(ip,\theta)} \leq 0, \quad (5)$$

$$A2n_{(i,\theta)} - X_{(i,\theta)}L' \leq 0. \quad (6)$$

The optimization coefficients of PSU for the vehicles are given in table 1.

In this article, the triangular parking lot that will be optimized is consists of two types, the isosceles triangular parking area which has two 45° angles and one 90° angle and equilateral triangular parking area. The length and width of each triangular parking area is assumed based on the side comparison of the two types of the triangular shape. The PSU optimization model in the triangular parking area follows the equations (1) to (6). The method applied here is linear integer programming which is calculated by using Lindo software to help find the most optimal results.

3. Results and Discussion

Table 2 shows the simulation results of the PSU optimization model for the isosceles triangular parking lot for cars and motorcycles. It can be seen that the optimal PSU for cars is 82 units consisting of one full exterior row (X_{ep}) with parking angle 75° taking in capacity of 56° PSUs and one full interior row (X_{ip}) with parking angle 60 taking in capacity of 28 PSUs. Meanwhile, the parking lot of the row with the minimum number of PSU will be used as parking space to optimize the number of PSU for motorcycles. For optimum motorcycle PSU, there are 134 PSUs consisting of one full exterior row (X_{ep}) with parking angle 75 accommodating 86 PSUs and two interior lines (X_i) with parking angle 45° and 60° accommodating 23 and 25 PSUs respectively. Based on the results, vehicle in the isosceles triangular parking lot can be illustrated as shown in figure 2.

Table 3 shows the simulation results of the PSU optimization model for equilateral triangular parking area for cars and motorcycles. It can be seen that the optimal PSU for cars is 175 units consisting of one full interior line (X_{ip}) with parking angle 45° accommodating capacity of 175 PSUs. Meanwhile, the parking lot of the row with the minimum number of PSU will be used as parking space for motorcycles. For optimum motorcycle PSU, there are 513 PSUs consisting of a full exterior row (X_{ep}) with parking angle 90° taking in 231 PSUs, a full interior row (X_{ip}) with parking angle 90° taking in 192 PSUs, and one interior line (X_i) with parking angle 90°

Table 1. Coefficient of parking space optimization for both vehicles

Dimension (meter)	Symbol	Angle				
		30°	45°	60°	75°	90°
Car (5 m x 2,5 m)						
Width	A1	2,50	2,50	2,50	2,50	2,50
Width of parallel line	A2	5,00	3,54	2,89	2,59	2,50
Length	B1	5,00	5,00	5,00	5,00	5,00
Length of parallel line	B2	9,33	7,50	6,44	5,67	5,00
Width of the parking line fromwall to border of parking line	C1	4,67	5,30	5,58	5,48	5,00
Width of the middle parking line	C2	3,58	4,42	4,96	5,15	5,00
Circulation alley	D	3,50	3,75	4,50	6,00	7,00
Width of full exterior	E1	11,75	13,47	15,04	16,63	17
Width of exterior	E2	8,17	9,05	10,08	11,48	12,00
Width of full interior	E3	10,66	12,59	14,42	16,30	17,00
Width of interior	E4	7,08	8,17	9,46	11,15	12
Motorcycles (2 m x 0,7 m)						
Width	A1	0,70	0,70	0,70	0,70	0,70
Width of parallel line	A2	1,40	0,99	0,81	0,72	0,70
Length	B1	2,00	2,00	2,00	2,00	2,00
Length of parallel line	B2	3,21	2,70	2,40	2,19	2,00
Width of the parking line fromwall to border of parking line	C1	1,61	1,91	2,08	2,11	2,00
Width of the middle parking line	C2	1,31	1,66	1,91	2,02	2,00
Circulation alley	D	1,60	1,60	1,60	1,60	1,60
Width of full exterior	E1	4,52	5,17	5,59	5,73	5,60
Width of exterior	E2	3,21	3,51	3,68	3,71	3,60
Width of full interior	E3	4,22	4,92	5,42	5,64	5,60
Width of interior	E4	2,91	3,26	3,51	3,62	3,60

taking in 90 PSUs. Based on the results, vehicles in the parking lot in the equilateral triangular shape can be illustrated as shown in figure 3.

Table 2. The simulation results of PSU models for cars and motorcycles in isosceles triangular parking

Angle	30°	45°	60°	75°	90°
Car					
Number of rows $X(ep,\theta)$ / Number of PSU $n(ep,\theta)$	0	0	0	1/56	0
Number of rows $X(e,\theta)$ / Number of PSU $n(e,\theta)$	0	0	0	0	0
Number of rows $X(ip,\theta)$ / Number of PSU $n(ip,]theta)$	0	1/26	1/28	0	0
Number of rows $X(i,\theta)$ / Number of PSU $n(i,\theta)$	0	0	0	0	0
Motorcycles					
Number of rows $X(ep,\theta)$ / Number of PSU $n(ep,\theta)$	0	0	0	1/86	0
Number of rows $X(e,\theta)$ / Number of PSU $n(e,\theta)$	0	0	0	0	0
Number of rows $X(ip,\theta)$ / Number of PSU $n(ip,\theta)$	0	0	0	0	0
Number of rows $X(i,\theta)$ / Number of PSU $n(i,\theta)$	0	1/23	1/25	0	0

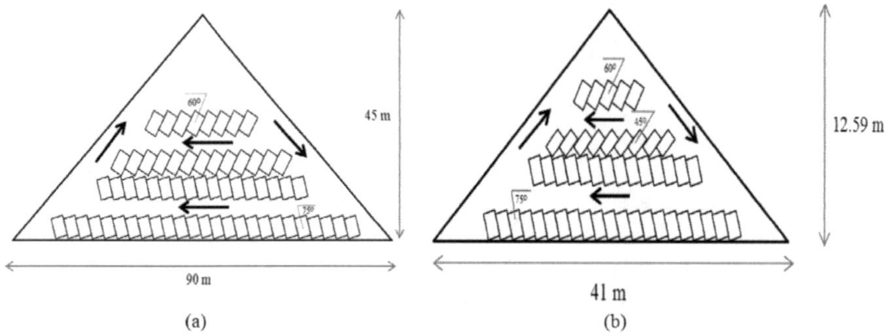

Figure 2. Optimal PSU setting for cars (a) and motorcycles (b) in isosceles triangular parking area

Table 3. Simulation results of PSU model of cars and motorcycles in equilateral triangular parking

Angle	30°	45°	60°	75°	90°
Car					
Number of rows $X(ep,\theta)$ / Number of PSU $n(ep,\theta)$	0	0	1/55	0	0
Number of rows $X(e,\theta)$ / Number of PSU $n(e,\theta)$	0	0	0	0	0
Number of rows $(X(ip,\theta)$ / Number of PSU $n(ip,\theta)$	0	5/175	0	0	0
Number of rows $X(i,\theta)$ / Number of PSU $n(i,\theta)$	0	0	0	0	0
Motorcycles					
Number of rows $X(ep,\theta)$ / Number of PSU $n(ep,\theta)$	0	0	0	0	1/231
Number of rows $X(e,\theta)$ / Number of PSU $(n(e,\theta)$	0	0	0	0	0
Number of rows $X(ip,\theta)$ / Number of PSU $n(ip,\theta)$	0	0	0	0	1/192
Number of rows $X(i,\theta)$ / Number of PSU $n(i,\theta)$	0	0	0	0	1/90

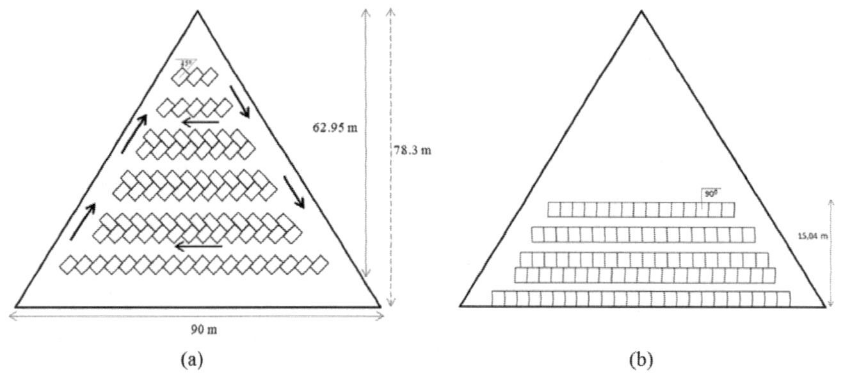

Figure 3. Optimal PSU setting for cars (a) and motorcycles (b) PSU in equilateral triangular parking area

4. Conclusion

The optimization model of parking space unit (PSU) in triangular parking area in this research is examined in two types of shape, namely isosceles and equilateral triangular shapes. The model can be described as four possible lines of parking, which is full exterior line Xep, exterior line Xe, full interior line Xip, and interior line Xi. Each possible parking line has a possible parking angle which are 30°, 45°, 60°, 75° and 90°. The results show that the isosceles triangular parking area has 218 units of optimal PSU, which are 84 units for cars and 134 units for motorcycles. Equilateral triangular parking area has 688 units of optimal PSU, which are 175 units for cars and 513 units for motorcycles.

Acknowledgments

The authors thank the anonymous referees for their valuable suggestions which led to the improvement of the article. This research is funded by Laboratory Grant, Syiah Kuala University, 2017.

References

[1] Directorate General of Land Transportation of Indonesia 1996 *Guide of Technical Organization of Parking Facility* (Jakarta : Directorate General of Land Transportation)
[2] University of Houston 2012 *Parking Lot Design Standards* Section 9.0
[3] Ibrahim D M, Waludjodjati E and Ismail A 2013 Study of public space utilization for parking area at Cikuray Garut street *J. Konstruksi STTGarut*
[4] Aggarwal M, Aggarwal S and Uppal R S 2012 *Int. J. Sci. Res. Pub.* **2(10)** 1-8
[5] Wahab M F, Moel A L, Abu A, Yaacob Z and Legowo A J. 2015 *Eng. Appl. Sci.* **10(16)** 7107-12
[6] Wang J M, Wu S T, Ke C W and Tzeng B K 2013 *Vehicle Engineering* **1(3)** 57-63
[7] Abdelfatah A S and Taha M A 2014 *J. Traff. Logis. Engin.* **2(3)** 176-81
[8] Ramli M, Sary D P and Halfiani V 2016 Optimization model of parking charge and income using Lagrange multiplier method *Proc. AIP Conference Series* pp. 030034-1 – 030034-8
[9] Munzir S, Ihsan M and Amin Z 2010 Linear Programming for Parking Slot Optimization *Proceedings of the 6th IMT-GT Conference on Mathematics, Statistics and Its Applications* (Kuala Lumpur) pp. 462-72
[10] Erliana H, Saleh M S and Anggraini 2014 *J. Teknik Sipil* **3(3)** 188-96
[11] Anggraini R, Burhanuddin and Ilyas M I 2013 Study of public service space availability on the parking facility need and road current performance (case sudy of Cempaka Lima clinic in Banda Aceh) *National Conference of Civil Engineering 7* (Surakarta) T53-T60

IOP Conf. Series: Materials Science and Engineering **300** (2018) 012013 doi:10.1088/1757-899X/300/1/012013

Data-driven Modelling for decision making under uncertainty

Layla Angria S[1], Yunita Dwi Sari[2], Muhammad Zarlis[3], Tulus[4]

Department of Mathematics, University of Sumatera Utara, Medan, Indonesia
Email: [1]lyla_angria@yahoo.com,[2]niitadwisari@gmail.com,[3]m.zarlis@usu.ac.id,
[4] tulusjp@yahoo.com

Abstract. The rise of the issues with the uncertainty of decision making has become a very warm conversation in operation research. Many models have been presented, one of which is with data-driven modelling (DDM). The purpose of this paper is to extract and recognize patterns in data, and find the best model in decision-making problem under uncertainty by using data-driven modeling approach with linear programming, linear and nonlinear differential equation, bayesian approach. Model criteria tested to determine the smallest error, and it will be the best model that can be used.

1. Introduction

One of the important things in a life is the existence of a decision created by the decision maker. Simply put, in everyday life we often make decisions based on the things that support them decision-making is based on available information. Information obtained is usually in the form of data.

Soumendra Mohanty, et. al (2013) In general, data can be defined into 3 important characteristics: composition, context, and conditions**.** Composition refers to the data structure: what is the source, what type of data, what is the nature of the data (most static data or real time stream data) and so on. Context refers to how data is generated, what events are related to the data, how the data sensitivity and others. Conditions refer to the state of the data and whether the data can be used for analysis or need further cleaning and enrichment.

The reality is that according to Dimtris Bertsimas and Aurelie Thiele (2006) the information or data available is incomplete and strategies based on wrong inputs may not be easy to do, or produce poor results when implemented. So that will affect the decision-making.

Based on what is presented above is the necessity of decision making appropriately. Derek W. Bunn (1984) Decision-making is an ubiquitous activity attached to an individual, social group, or group of organizations. In the world of decision-making industry is very important such as in deciding how much consumer demand for a product, transportation problems, water resources and more.

But in decision making there are many parameter uncertainty. Raiffa and Schalaifer (1961) Decision-making with uncertainty is the main arena in the performance of decision-making methodology. Indeed, the main function of decision analysis, as it was originally developed, is to assist in the evaluation of decision-making for options whose ultimate outcome is not known for certain.

Therefore, as is the uncertainty parameters that affecting the decision, must use appropriate models to determine what decisions we will gain and reduce the risk to be generated. The model to be discussed in this issue is the data-driven model. Mosaad Khadr and Mohamed Elshemy (2016) argue that data-driven models is based on analyzing the data characterizing, where a model is built on the basis of finding connections between the system state variables such as input, internal and output variables with only a limited assumption about the system.

2. Essence of data – driven modelling

Solomatine et al. (2008) a data-driven model is based on the analysis of the data about a spesific system. The main concept of data-driven model is to find relationships between the system state variables (input and output) without explicit knowledge of the physical behavior of the system. Task of data-driven modelling is classification: where the task constitutes of assigning a class for an input data point. Association: where association between variables characterising the system is to be identified, which is used in subsequent prediction. Regression: where the task constitutes of predicting a real value associated with an input data point. Clustering: where groups of data points with within group similarity are to be determined.

The process of building a data driven modelling follows general principles adopted in modelling: study the problem – collect data – select model structure – build the model – test the model and (possibly) iterate. In data-driven modelling, not only the model parameters, but also the models structure, is often subject to optimisation.

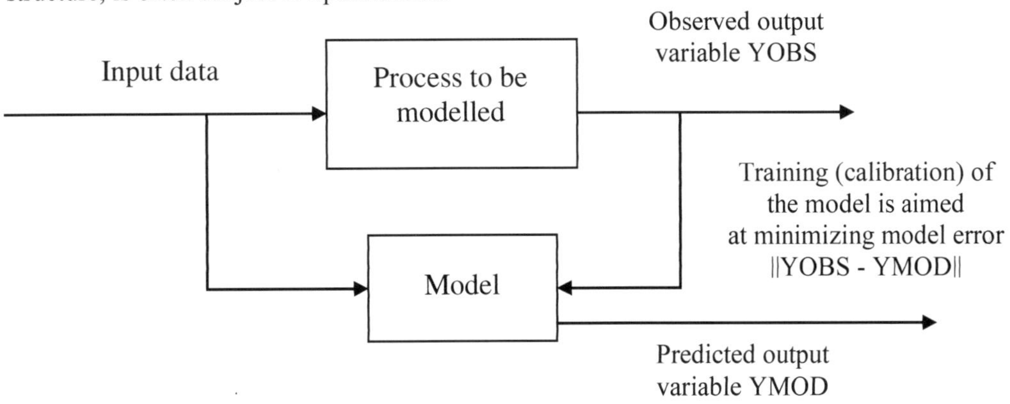

Figure 1. General approach to modelling

a. Regression and Classification

In general two major areas where engineering analysis is required are modeling or simulation of a given occuring phenomena (regression analysis), and identification or categorization of occuring events or pattern (classification analysis). These two major areas come with their unique characteristics in term of objectives to achieve, data sources, and post-calibration implementation, requiring adequate approaches for a succesful model development.

b. Regresion Analysis

The regression analysis includes any technique or algorithm used for mapping or linking several variables among them. The focus is to develop a relationship among one or several dependent variables (outputs) with one or more independent variables (inputs).

c. Clasification Analysis

The classification analysis is related to the categorization or labeling of a given group of variables values into a pre-defined possible list of groups (in some cases the pre-defined groups is not necessary or available). The variables in this case are not necessarily defined as input – outputs groups like in regression analysis.

3. Models and Method

a. Data fitting and model selection

The data that have been obtained are selected according to their respective characters. Data included in linear / nonlinear, stochastic or deterministic. Then modeled with linear / nonlinear equations, arima, ANFIS. Linear equation can be described by the following model:

$$Y = \beta_0 + \beta_1 X + \beta_2 X^2 + \varepsilon \quad (2.1)$$

the curve is represented by the quadratic curve.

$$Y = \beta_0 + \beta_1 X + \beta_2 X^2 + \cdots + \beta_k X^k + \varepsilon \quad (2.2)$$

Where β_0, β_1, ..., β_k are fixed regression parameters and ε is a random error or noise parameter.

Luthkepohl (2005) Time series is basically a measurement data taken chronologically within a certain time. ARMA is a time series forecasting technique. Sadeq (2008) ARMA model can be broken down into AR model *(Autoregressive)* and MA model *(Moving Average).*

$$y_t = b_0 + b_1 y_{t-1} + b_2 y_{t-2} + \cdots + b_n y_{t-n} + \varepsilon_t \quad (2.3)$$

MA Model *(Moving Average)*

$$y_t = a_0 + a_1 \varepsilon_{t-1} + a_2 \varepsilon_{t-2} + \cdots + a_n \varepsilon_{t-n} + \varepsilon_t \quad (2.4)$$

Where:

y_t : time series

$y_{t-1}, y_{t-2}, \dots, y_{t-n}$: previous time series

$\varepsilon_{t-1}, \varepsilon_{t-2}, \dots, \varepsilon_{t-n}, \varepsilon_t$ residual

$a_0, a_1, a_2, \dots, a_n$ constans dan coeffisient MA model

$b_0, b_1, b_2, \dots, b_n$ constants dan coeffisient RA model

Mosaad and Elshemy (2016) An adaptive neuro-fuzzy inference system (ANFIS) is a fuzzy inference system formulated as a feed-forward neural network. Hence, the advantages of a fuzzy system can be combined with a learning algorithm. ANFIS was introduced as an effective tool to represent simple and highly complex functions more powerfully than conventional statistical methods *b. Model and parameter identification*

In the current work, developing model using an appropriate approach e.g with a bayesian approach for both model and parameters of the models were identified.

c. Model selection criteria

1. Normalised root mean squared error (NRMSE) was used to compare the models with different orders for each individual time series. Inferred parameters θ_i were applied back to the system equation to generate time series without stochastic terms, i.e deterministic solution. Note that because parameters were identified in the form of normal distribution, the most probable (mean)value of parameters were plugged into the system equation. Root mean squared error (*RMSE*) between the measurement time series y_t and the inferred deterministic time series $\hat{y}_t$ can be calculated as follows:

$$RMSE = \sqrt{\frac{\sum_{t=1}^{n}(\hat{y}_t - y_t)^2}{n}}$$

2. *Generic form* . As the entire aim was to find a model capable of capturing the common patterns in all time series, a model that could only describe some time series was disregarded.

3. *System stability* . The model has to have a unique stable steady state, which implies that the system's response de- cays with time. This has been checked via calculations of the real parts of corresponding eigenvalues which have to be negative for stability

d. Statistical analysis

The null hypothesis of no difference between the groups of interest was tested at the 5% significance level, and the results are presented as *p* -values. Statistical analysis of the differences in *NRMSE* between two models was performed using the one sample *t* -test, which assesses the normality of data with zero mean and unknown variance at the 5% significance level. The result is presented as *p* – values.

4. Application

4.1 start with the first case where the decision-maker must decide how many items to order at each time period at a single store. (In mathematical terms, the state of the system can be described as a scalar variable, specically, the amount of inventory in the store.) We use the following notation:

$$x_{t+1} = x_t + u_t - w_t \qquad 4.1$$

which yields the closed-form formula:

$$x_{t+1} = x_0 + \sum(u_t - w_t) \qquad 4.2$$

x_t : inventory at the beginning of time period *t*,
u_t : amount ordered at the beginning of time period *t*,
w_t : demand occurring during time period *t*.

Demand is backlogged over time, and orders made at the beginning of a time period arrive at the end of that same period. Therefore, the dynamics of the system can be described as a *linear* equation:

4.2 In the problem of antibody transplant Yan Zhang et. Al (2016) the data-driven model is used to determine the best model. The first model that can considered is a nonlinear differential equation with coefficients in the form of a polinomial function. Model can be described by the following:

$$\frac{d^n}{d^n}x_t + \sum_{i=0}^{n-1} f_{i-1}(x_t)\frac{d^i}{dt^i}x_t + f_0(x_t) = \eta_{\ t} \qquad 4.3$$

$$y_t = x_t + \varepsilon_t \qquad 4.4$$

Linear models with different system orders are considered first. Eq. (1) in Section 3.1.2 transforms into a linear differential equa- tion when $f\, i\, +1\ (x\ t\ )$ are constants, with constant parameters $\theta_i\ (i = 0,1,\ldots,n-1)$ where $f_n(x_t) = \theta_n, \ldots\ ,\ f_2(x_t) = \theta_2,\ f_1(x_t) = \theta_1.\ f_0(x_t)$ is defined as $-\theta_0$ for convenience. A first order linear model was considered first, and it did not show good performance. Then linear models with higher system orders were investigated. In this section, we present the results of system and parameter identification by comparing solutions for linear first, second and third order dynamic equations only:

$$\text{Model 1 } (M_1): \frac{dx_t}{dt} + \theta_1 - \theta_0 = 0 \qquad 4.5$$

$$\text{Model 2 } (M_2): \frac{d^2x_t}{dt^2} + \theta_1\frac{dx_t}{d_t} + \theta_1 x_t - \theta_0 = 0 \qquad 4.6$$

$$\text{Model 3 } (M_3): \frac{d^3x_t}{dt^3} + \theta^3\frac{d^2x_t}{dt^2} + \theta_2\frac{dx_t}{d_t} + \theta_1 x_t - \theta_0 = 0 \qquad 4.7$$

Two nonlinear models were considered: model *NM*1 with nonlinearity in the damping term $(f_1(x_t) = k_1(x_{t)} + \theta_1, f_2(x_{t)} = k_2(x_t) + \theta_2)$ and model *NM*2 with nonlinearity in the *xt* term $(f_1(x_t) = \theta_1, f_2(x_t) = k_2 x_t + \theta_2)$. The corresponding system equations are as follows:

$$NM_1: \frac{d^2x_t}{dt^2} + \theta_2\frac{dx_t}{dt} + (k_1x_t + \theta_1)x_t - \theta_0 = 0 \qquad 4.8$$

$$NM_1: \frac{d^2x_t}{dt^2} + (k_2x_t + \theta_{2)}\frac{dx_t}{dt} + \theta_1 x_t - \theta_0 = 0 \qquad 4.9$$

Table 1

	NRMSE		
	M₁	*M₂*	*M₃*
(a)	*0.272*	*0.053*	*0.014*
(b)	*0.083*	*0.085*	*0.013*
(c)	*0.090*	*0.096*	*0.053*
(d)	*0.088*	*0.071*	*0.073*

4.3 Problem of river flow prediction. In this problem Joko suryanto (2016) Consider 3 models of ARIMA, ANFIS and FFNN. From three models obtained the best model is FFNN (feed forward neural network) with the smallest error value, by follow equation:

$$y_i = f_2\left[\sum_{k=1}^{k} w_{1k} f_2\left(\sum_{j=1}^{j} w_{kj}x_j + b_k\right) + b_1\right] \qquad 4.10$$

$$f_2(p) = \frac{2}{(1+e^{-2p})} - 1 \qquad 4.11$$

5. Conclusion

In making decisions with uncertainty, we should be able to classify the data obtained in order to build a suitable model so as not to cause too much risk. The data obtained must also be in accordance with the objectives to be achieved by the needs maker, so that in making the model does not occur a big error and has biggest corsrelation. E.g in second application problem, the best model is obtained by looking at the smallest error by using a linear model $\frac{d^2x_t}{dt^2} + (k_2x_t + \theta_2)\frac{dx_t}{dt} + \theta_1x_t - \theta_0 = 0$. In the third problem the best model is FFNN by looking at the smallest error and biggest correlation, with error 8,422 and correlation 0,85.

References

[1] Ackoff, R. C. dan F. E. Emery. (1972). *On purposeful system*. Aldine-Atherton. Chicago
[2] Bertsimas, Dimtris. dan Thiele, Aurelie. (2006). *Robust and data-driven optimization: modern decision making under uncertainty*. cambridge
[3] Bunn, W, Derek. (1984). Applied decision analysis. USA
[4] Bertsimas, Dimtris. dan Thiele, Aurelie. (2006). A *Robust Optimization Approach to Inventory Theory. Operation Research*. 54(1):150-168
[5] Bertsimas dan M. Sim. (2004). *The Price of Robustness*. Operation Research
[6] Elshemy, Mohamed. dan Khadr, Mosaad. (2016). *Data-driven modeling for water quality prediction case study: the drains system associated with manzala lake*, egypt. Tanta university
[7] Femandes, Betina. Street Alexandre, dkk. (2016). *An adaptive robust portofolio optimization model with loss constraints based on data-driven polihedral uncertainty sets*. Pontificial catholic university of rio de
[8] J. Daunizeau, KJ. Friston, SJ. Kiebel. (2009). *Variasional bayesian identification and prediction of stochastic nonlinear dynamic causal models*, phys. D nonlinear Phenom. 2089-2118
[9] Meng, Chao. Nageshwaraniyer, Sal Srinivas, dkk. (2013). *Data-driven modeling and simulation frame-work for material handling systems in coal mines*. University of arizona, 766-779
[10] M.J. Beal. (2003). *Varitional Algorithms for Approximate Bayesian inference* (Ph.D. Thesis), The Gatsby Computational Neuroscience Unit. University College London
[11] Mohanty, Soumendra. Harsa Srivatsa, dkk. (2013). *Big data imperatives*. India
[12] Simon, H. A. (1957). *Models of man*. New York
[13] Zhang, Yan. David Briggs, dkk. (2016). *A new data-driven model for post-trasplant antibody dynamics in high risk kidney transplantation*. University of warwick, UK
[14] Solomatine, DP, dan Ostfeld. (2008). *A Data-driven Modelling: Some Past Experience and New Approaches*. J of Hydroinformatics. 3-22
[15] Supranto. (2009). *Teknik Pengambilan Keputusan*. Rineka Cipta. Jakarta
[16] Suryanto, Joko. (2016). *Aplication data-driven technik for monthly river flow prediction*. Unsyiah. Indonesia

Estimating Most Productive Scale Size in Data Envelopment Analysis with Integer Value Data

Yunita Dwi Sari[1], Layla Angria S[2], Syahril Efendi[3], Muhammad Zarlis[4]

Department of Mathematics, Universitas Sumatera Utara, Medan, Indonesia

E-mail: [1]niitadwisari@gmail.com, [2]lyla_angria@yahoo.com, [3]syahnyata1@gmail.com, [4]m.zarlis@usu.ac.id

Abstract. The most productive scale size (MPSS) is a measurement that states how resources should be organized and utilized to achieve optimal results. The most productive scale size (MPSS) can be used as a benchmark for the success of an industry or company in producing goods or services. To estimate the most productive scale size (MPSS), each decision making unit (DMU) should pay attention the level of input-output efficiency, by data envelopment analysis (DEA) method decision making unit (DMU) can identify units used as references that can help to find the cause and solution from inefficiencies can optimize productivity that main advantage in managerial applications. Therefore, data envelopment analysis (DEA) is chosen to estimating most productive scale size (MPSS) that will focus on the input of integer value data with the CCR model and the BCC model. The purpose of this research is to find the best solution for estimating most productive scale size (MPSS) with input of integer value data in data envelopment analysis (DEA) method.
Keywords : Most Productive Scale Size, Data Envelopment Analysis, Integer Value

1. Introduction

The most productive scale size (MPSS) is a measurement that states how resources should be organized and utilized to achieve optimal results. The most productive scale size (MPSS) can be used as a benchmark for the success of an industry or company in producing goods or services. Each decision making unit (DMU) is currently required to increase productivity and high efficiency due to the higher cost required or incurred to run the center of the company. Khodabakhshi (2009) In determining the maximum function of a solution where the value of its objective function is zero, and it is emphasized that the input θ_0^* must be a positive integer, so it is clear that the optimal value of the objective function is the non-negative variable which always have $\phi_0^* \geq \theta_0^*$.

Based on Khodabakhshi's statement that to maximize a solution from its objective function, it is more emphasized on the value of inputs with a positive integer value to obtain an optimal value, so as to estimate the most productive scale size can be more efficient. To estimate the most productive scale size many alternative strategies and improvement methods can be developed, depending on the aspects of output or inputs used as basic aggregates, there are some companies that only have the input of integer value data, for example companies that have a system of distributing goods or services directly to the final customer and not to process the product, but only have the right to carry out business by using the brand, name, system and procedures that have been set before in a certain time and area, so that the basic aggregate to estimate the most productive scale size using only the input of integer value data. Ahn (2012) Data envelopment analysis (DEA) is a method for evaluating performance that allows simultaneous analysis of its performance indicators. Performance appraisals are needed to identify a productivity performance measurement method to be able to determine the efficient productive scale of any decision making unit (DMU).

Ramanathan (2003) Data envelopment analysis (DEA) with CCR model reflects (multiplication) technical efficiency and scale efficiency, while the BCC model reflects only the technical efficiency, so the relative scale efficiency is the ratio on efficiency of CCR model and BCC model. If the result of the ratio is one then the decision making unit (DMU) operates at the most productive scale size. If the ratio of results to less than one means there is still a scale inefficiency in the decision making unit (DMU).

An efficient decision making unit (DMU) with CCR model means technical efficiency and scale efficiency is optimal. While efficient decision making unit (DMU) with BCC model but inefficient with CCR model means experiencing inefficiency to scale, that is because decision making unit (DMU) is only technical efficiency, so that the inefficiency is come from the scale.

Cooper et al. (2011) by data envelopment analysis (DEA), relative efficiency of the decision making unit (DMU) can be defined as the ratio of the total weighted output divided by the total weighted input.

The above statement can be concluded that the data envelopment analysis (DEA) method determines the weights for each input and output of each decision making unit (DMU), where they have non-negative and universal, meaning each decision making unit (DMU) in the existing sample should be able to use the same weighted output / total weighted input and the ratio should be no more than one (total weighted output / total weighted input = 1), because if the value dual is more or less than one then the decision making unit (DMU) is deemed to be inefficient in relative or inefficiency, otherwise if its dual value is equal to one, or its efficiency value is 100%, and then the decision making unit (DMU) is considered efficient.

Based on its performance assessment, the data envelopment analysis (DEA) method can identify units that are used as references that can help to find the cause and solution from inefficiency that optimize productivity, which is a major advantage in managerial applications. Therefore, one possible method to estimate the most productive scale size of an efficient on a decision making unit (DMU) is the data envelopment analysis (DEA) method, since the data envelopment analysis (DEA) method can evaluate the performance of the decision making unit (DMU) and analyzed the evaluation results of the relative efficiency of each comparable decision making unit (DMU). In addition to generating the efficiency value of each decision making unit (DMU), data envelopment analysis (DEA) is also able to show units that are in reference for inefficiency units. So in this research, the data envelopment analysis (DEA) method was chosen to estimate the most productive scale size of the that will focus on the input of integer values data which is expected to find the best solution for estimating most productive scale size of efficiently.

2. Most Productive Scale Size (MPSS)

Khezrimotlagh et al. (2012) DEA is a nonparametric method in proposed research operations to measure the efficiency and productivity of each DMU with multiple inputs and outputs that can only be determined by integer values such as number of employees, books, magazine numbers and total passengers.

Moghaddas and Vaez-Ghasemi (2014) In order to find a point of most productive scale size (MPSS) the analysis will be limited to find the intersection between the CCR border and the combination of the result point after scaling the inefficiency DMU or experience of increasing return to scale or decreasing return to scale using the model equation of proposed model is as follows:

$$\left(\frac{\theta^* X_0 - S^{-*}}{\sum_{j=1}^{n} \lambda_j^*}, \frac{Y_0 - S^{-*}}{\sum_{j=1}^{n} \lambda_j^*}\right) = (\tilde{X}_0, \tilde{Y}_0) \tag{1}$$

Cooper et al. (2011) Figure 1 below shows that the efficiency limits of BCC represented by solid lines and CCR efficiency limits represented by a dotted line from a nonzero technical evaluation point And returns to DMU performance scale simultaneously. It is illustrate with A = (1, 1) as the initial coordinate point in the figure below using the formula (1) to obtain MPSS.

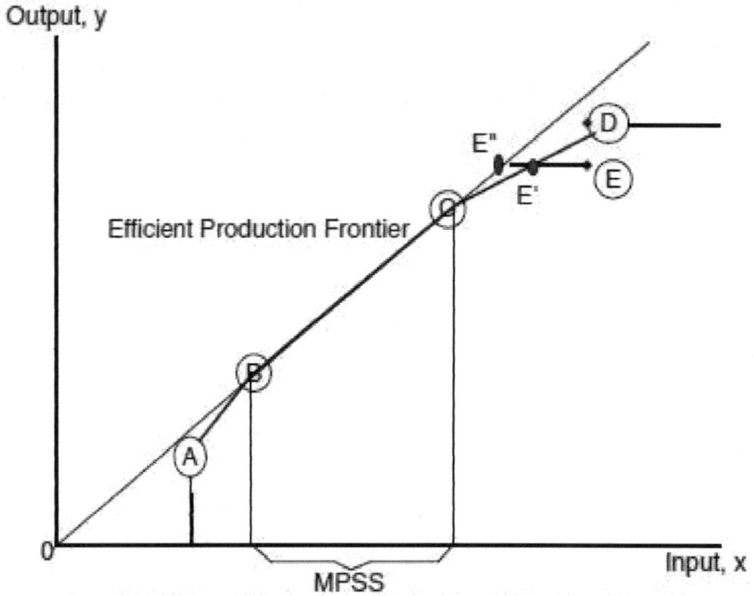

Figure 1. Most Productive Scale Size (MPSS)

Unit B weighs from formula (1) to adjust to unit A, it will be the coordinate points of units B and C. Takes into unit calculations B and C as corresponding to the MPSS point for unit A, and completes with the proposed model where the value of $s^- = 0$ and $s^+ = 0$ which produces the point with coordinates (3/2, 2) and (3, 4) where it is the most productive scale size of efficient measure, while unit E after evaluation the efficiency level turns out that unit E has the potential to be the most productive scale size by scaling. Unit E 'is the result of the first scaling that produces the coordinate point (7 / 2, 9/2), after examining it turns out that unit E' has not reached the most productive scale size, so that unit E 'needs to do the second scaling that produces unit E " Coordinates (27/8, 9/2).

3. Data Envelopment Analysis (DEA)

Khoveyni and Eslami (2013) Fare and Grosskopf stated there is an alternative model to estimate the most productive scale size (MPSS) based on the optimal solution of CCR and BCC is constant return to scale (CRS) also called CCR model and variable return to scale (VRS) called the BCC model.

Cooper et al. (2011) returns to scale (RTS) status in data envelopment analysis (DEA) are as follows:

- If $\theta^*_{CCR} = \theta^*_{BCC}$ or $\sum \lambda^*_j = 1$ This case is called *Constant Return to Scale* (CRS).
- If $\theta^*_{CCR} < \theta^*_{BCC}$ or $\sum \lambda^*_j < 1$ This case is called *Increasing Return to Scale* (IRS).
- If $\theta^*_{CCR} > \theta^*_{BCC}$ or $\sum \lambda^*_j > 1$ This case is called *Decreasing Return to Scale* (DRS).

3.1. Constant Return to Scale (CRS)

Jahanshahloo and Khodabakhshi (2003) This model shows a change in the amount of output that is proportional to the change in the sum of all inputs used. For example if the capital or input is added by x times then the output also increases by x times.

Cooper et al. (2006) the linear form of the CCR model are as follows:

$$\text{Max} \sum_r u_r y_{ro} \tag{2}$$

$$\begin{aligned} s.t \quad & \sum_r u_r y_{ro} - \sum_i v_i x_{io} \leq 0 \\ & \sum_i v_i x_{io} - 1 \\ & u_r, v_i \geq 0 \\ & x_{ij} \in Z^+ \quad \forall i, j \end{aligned}$$

3.2. Variable Return to Scale (VRS)

Jahanshahlo and Khodabakhshi (2003) In this model there are two types of variable return to scale, that are :

- *Decreasing Return to Scale*
  Decreasing return to scale occurs when the number of output changes is not proportional (smaller) than the input change.
- *Increasing Return to Scale*
  Increasing return to scale occurs when changes of all inputs will result in larger output changes than the proportion of input changes.

Wang and Lan (2013) Banker, Charmes, and Cooper express the linear form of the BCC model as follows:

$$\text{Min}\theta - \epsilon(\sum_{i=1}^{m} S_i^- + \sum_{r=1}^{m} S_r^+) \tag{3}$$

$$\begin{aligned} s.t \quad & \sum_{j=1}^{n} \lambda_j x_{ij} + S_i^- = \theta x_{io} \quad i = 1, \ldots, m \\ & \sum_{j=1}^{n} \lambda_j y_{rj} - S_i^+ = y_{ro} \quad r = 1, \ldots, s \\ & \lambda_j \geq 0 \quad j = 1, \ldots, n \\ & x_{ij} \in Z^+ \quad \forall i, j \end{aligned}$$

The efficiency values of BCC is obtained by running the above model for each DMU. The performance efficiency measures of BCC is called pure technical efficiency, which is related to the values obtained from the model allowing the resultant variables, so that the existing scale can be eliminated. In general, the CCR efficiency rating for each DMU will not exceed the efficiency of BCC, which is clearly intuitive as the BCC model analyzes each DMU locally rather than globally.

3.3. Return to Scale (RTS)

Cooper et al. (2011) Linear form of RTS to determine $\sum_{j=1}^{n} \lambda_j^*$ in equation (3) used to scaling process in DMU of increasing return to scale or decreasing return to scale improvement are as follows:

$$\text{Max} \sum_{j=1}^{n} \lambda_j - \epsilon(\sum_{i=1}^{m} S_i^- + \sum_{r=1}^{m} S_r^+) \tag{4}$$

$$\begin{aligned} s.t \quad & \sum_{j=1}^{n} \lambda_j x_{ij} + S_i^- = \theta x_{io} \quad i = 1, \ldots, m \\ & \sum_{j=1}^{n} \lambda_j y_{rj} - S_i^+ = y_{ro} \quad r = 1, \ldots, s \\ & \lambda_j \geq 0 \quad j = 1, \ldots, n \\ & x_{ij} \in Z^+ \quad \forall i, j \end{aligned}$$

4. Application

In this research, the data to be sampled is taken from 10 DMU data that have been research by Moghaddas and Vaez-Ghasemi. Each company has two input value data and two output value data. For input ($I1$) that is data of processed material, ($I2$) is employee data, while output ($O1$) is data from customer satisfaction, and ($O2$) is data from sales. This research focus on the only input of integer value data, so that the data used only inputs derived from employee data. The following will present the efficiency evaluation results of each DMU using DEA with the CCR model and BCC model to estimate the most productive scale size. The 10 DMU data sets presented in Table 1 of the data will be determined which DMUs is efficient and require scaling to be productive. The completion will be done using DEA method with CCR and BCC model.

IOP Conf. Series: Materials Science and Engineering **300** (2018) 012014 doi:10.1088/1757-899X/300/1/012014

Table 1. Evaluation result of efficiency input-output

DMU	X	Y_1	Y_2	CCR	BCC	RTS
1	24	4.1	3.6	0.95	1.00	*Increasing*
2	32	4.0	3.2	0.69	1.00	*Increasing*
3	30	4.4	4.2	0.81	0.67	
4	35	4.4	4.4	0.70	0.57	
5	25	3.6	4.0	0.84	0.80	
6	24	3.9	4.2	0.92	0.83	
7	33	4.5	4.5	0.76	0.61	
8	32	3.6	3.7	0.63	1.00	*Increasing*
9	20	3.6	3.8	1.00	1.00	*Constant*
10	36	4.5	4.6	0.69	0.56	

Table 2. Evaluation result of DMU after scaling

DMU	X	Y_1	Y_2	CCR
1	21.66	3.91	3.42	1.00
2	15.24	2.76	2.21	1.00
8	12.70	2.27	2.33	1.00

Table 3. DMU with input of integer value data

DMU	X	Y_1	Y_2	CCR
1	22	3.97	3.47	1.00
2	15	2.72	2.18	1.00
8	13	2.32	2.37	1.00

From Table 1 it can be seen that DMU 9 has operated on the best scale size efficiency, where $\theta^*_{CCR} = \theta^*_{BCC}$ or $\sum \lambda^*_j = 1$. Where as DMU 1, 2, and 8 are inefficiency to scale, where $\theta^*_{CCR} < \theta^*_{BCC}$ or $\sum \lambda^*_j < 1$, so scaling is required to make DMU 1, 2, and 8 feasible to operate on most productive scale size . For the scaling process on the DMU scale which increases above it is required λ^*_j by using equation (4), so for DMU 1 it is known λ^*_j = 1,05, for DMU 2 it is known λ^*_j =1,45, for DMU 8 is known λ^*_j =1,59 after that substitute λ^*_j in formula (1). After the scaling of the new DMU data is re-evaluated using DEA with CCR model which results can be seen in table 2 which turns out DMU 1, 2, and 8 the objective function is one, which means that the DMU is efficiency and feasible to operate on the best scale size. Because this research focuses on integer value data, the input data which is still in the form of real numbers is changed to form integer and whose results can be seen in table 3.

5. Conclusion

Based on the results of the efficiency evaluation using DEA with CCR model and BCC model to estimate the most productive scale size of the problem samples taken from the 10 DMU data focus on integer value data of Moghaddas research, it appears that DMU 9 is a DMU operating at the best scale size, where $\theta^*_{CCR} = \theta^*_{BCC}$ or $\sum \lambda^*_j = 1$ and $slack = 0$. Where as DMU 1, 2, and 8 are inefficient to scale, where $\theta^*_{CCR} < \theta^*_{BCC}$ or $\sum \lambda^*_j < 1$, so scaling is required to make DMU 1, 2, and 8 feasible to operate on most productive scale size . This research aim to find the best solution to estimate the most productive scale size with the input of integer value data in DEA, with the limitation on the input of integer value data can be more easily to find the most productive scale size and simultaneously affirm the validity of DEA method is as a fairly efficient means to estimate the most productive scale size.

References

[1] Ahn H 2012 Behavioral effects of DEA on performance assessment. *Proceedings of the* 10th *International Conference on DEA*, 48-54.
[2] Cooper W W, Seiford L M, and Tone K 2000 Data Envelopment Analysis Kluwer Academic, Boston, Mass.
[3] Cooper W W, Seiford L M, and Tone K 2011 *Handbook on Data Envelopment Analysis*, Vol. 164 second edition ed F.S. Hillier. Springer.
[4] Cooper W W, Seiford L M, and Tone K 2006 *Introduction to Data Envelopment Analysis and Its Uses : with DEA Solver Software and Reference.* Springer, New York.
[5] Eslami R, Khodabakhshi M, Jahanshahloo G R, Lotfi F H, and Khoveyni M 2012 Estimating most productive scale size with imprecise-chance constrained inputoutput orientation model in data envelopment analysis. *Computers & Industrial Engineering*, 254-261.
[6] Jahanshahloo G R, and Khodabakhshi M 2003 Using inputoutput orientation model for determining most productive scale size in DEA. *Applied Mathematics and Computation*, No. 146. 849-855.
[7] Khezrimotlagh D, Salleh S, and Mohsenpour Z 2012 A Comment on Theory of Integer-Valued Data Envelopment Analysis. *Applied Mathematical Sciences*, Vol. **6**, 2012, No. 116, 5769-5774.
[8] Khodabakhshi M 2009 Estimating most productive scale size with stochastic data in data envelopment analysis. *Economic Modelling*, 968-973.
[9] Khoveyni M, and Eslami R 2013 Estimating right and left returns to scales in data envelopment analysis: A new approach. *International Journal of Data Envelopment Analysis*, Vol. **1**, No. 3, 135-150.
[10] Moghaddas Z, and Vaez-Ghasemi M 2014 Estimating most productive scale size in DEA with real and integer value data. *Int. J. Industrial Mathematics*, Vol. **6**, No. 2, 107-114.
[11] Ramanathan R 2003.*An Introduction to Data Envelopment Analysis: A Tool for Performance Measurement.* Sage Publications India Pvt Ltd.
[12] Wang Y, and Lan Y 2013. Estimating most productive scale size with double frontiers data envelopment analysis. *Economic Modelling*, 182-186.

Mixed Integer Linear Programming model for Crude Palm Oil Supply Chain Planning

Pasukat Sembiring[1], Herman Mawengkang[1], Hendaru Sadyadharma[2], F. Bu'ulolo[1], Fajriana[3]
[1]Department of Mathematics, University of Sumatera Utara
[2]University HKBP Nommensen, Medan, Indonesia
[3]Malaikulsalah University, Lhokseumawe, Aceh, Indonesia

Email: hmawengkang@yahoo.com

Abstract. The production process of crude palm oil (CPO) can be defined as the milling process of raw materials, called fresh fruit bunch (FFB) into end products palm oil. The process usually through a series of steps producing and consuming intermediate products. The CPO milling industry considered in this paper does not have oil palm plantation, therefore the FFB are supplied by several public oil palm plantations. Due to the limited availability of FFB, then it is necessary to choose from which plantations would be appropriate. This paper proposes a mixed integer linear programming model the supply chain integrated problem, which include waste processing. The mathematical programming model is solved using neighborhood search approach.

Keywords: Optimization, Crude Palm Oil, Mixed Integer programming, Supply Chain, Neighborhood Search.

1. Introduction

Jha Crude palm Oil (CPO) industry could belong to a business which needs a network system, as this product is not an end-product. The network system which performs the functions of procurement of materials, processing these materials into intermediate and finished products, and the distribution of these intermediate/ finished products to customers, is called supply chain system. All these facilities are used for fulfilling a customer request. The challenges of supply chain are to produce the right products, in the right quantities, at the right place, at the right time and at appropriate cost.

Due to the complexity of decision making process in supply chains, it is necessary to use modeling techniques that can help identify and implement strategies for designing high performance supply chain networks [10, 13, 14]. The main reasons to use model for the decision making process are large scale nature of the supply chain networks, hierarchical structure of decisions, randomness of various inputs and operations, and dynamic nature of interactions among supply chain elements [10]. Optimization model has become a high technology in supply chain planning and management [4, 5, 7].

Nasiri [6] presented the modelling of an integrated the supply chain system which involve raw material suppliers, several plants, distribution centres, warehouses and customers. Piewthongngam [11] analysed more complicated case for the integrated supply chain. They considered the supply chain case which involved multiple feed mills, multiple plantation, and multiple products. Timpe and Kallrath [2] addressed a mixed integer linear programming model, which integrate production, distribution and marketing and involved plants and sales points. Their paper describes how to cover the relevant features required for the supply chain management along with a multi-site production network.

Dumrongsiri [1] proposes the production planning of palm oil and the capacity planning of palm oil processing plants in Thailand. He creates a mixed integer linear programming model for the integrated production planning. Chungsiriporn [9] use a nonlinear programming model for CPO production planning with minimization of water consumption at the milling process. However they only consider the optimization process to get the CPO. Another CPO production planning which consider only at the milling process is discussed in Amelia [12]. They consider fuzzy logic approach to provide a simpler mechanism in order it would be possible to find the relationship between the processing variables and the amount of CPO and palm kernel losses. Suksa-ard and Reaweewan [3] develop an optimization model in s supply and demand system for a CPO production planning. Their model is to decide which market of CPO should be selected, and also to find out how much demand should be served in the selected market. Sadyadharma [8] present a multi-objective optimization model for a sustainable production planning of CPO considering uncertainty in the reliability of financial risk.

2. Problem Description

The production process of crude palm oil (CPO) can be traced as the milling process of fresh fruit bunch (FFB) into end products palm oil. The process usually through a series of steps producing and consuming intermediate products. These FFB, intermediate and end products (CPO) are necessarily to be inventoried, allowing one to produce and consume them at different moments and rates in time. It should be noted for palm oil the time to keep FFB in inventory is very short.

We consider a CPO milling company who own j milling factories, however the company does not have its own palm plantation. In order the company would be able to produce CPO from its millings, it needs to buy FFB from I public palm plantations. The situation is thus the same as in the two-echelon supply chain system. The management needs to decide from which palm plantation to serve the milling factory and then to put the CPO into inventory (warehouse) afterward to deliver the CPO to the k customers.

3. Mathematical Model

According In this case, the CPO companies is to decide optimally
1. From which palm plantation the FFB should be bought
2. Quantity of FFB to be bought from the chosen palm oil plantation
3. Quantity of CPO to be produced
4. Quantity of CPO to be delivered to each customer
5. Quantity of Waste produced in the milling process such that to minimize the overall operational costs.

There are several assumptions are necessary to be considered..
1. FFB are always available from palm oil plantations,
2. The demands of CPO from customers are deterministic,
3. Out of stock never occur.

Notations to be used.

Decision variables

x_{ij}^{t} Quantity of FFB (ton) to be bought from plantation I for milling j in time period t

y_{j}^{t} Quantity of CPO (ton) to be produced at milling j in time period t

z_{jk}^{t} Quantity of CPO (ton) to be delivered from milling j to customer k in time period t

v_{j}^{t} Quantity of liquid waste (ton) produced after the milling process in milling j in time period t

w_{j}^{t} Quantity of solid waste (ton) produced after the milling process in milling j in time period t

δ_{ij} A binary variable which is 1 if FFB for milling j is bought from plantation I and zero otherwise

Parameters

α_{ij} Transportation cost (Rp.) from plantation I to milling j per kilometer

d_{ij} Distance (km) from plantation I to milling j

β_j Production cost (Rp.) at milling j

γ_{jk} Transportation cost (Rp.) to deliver CPO from milling j to customer k

λ_j Cost (Rp.) for processing liquid waste at milling j

ρ_j Transportation cost (Rp.) for solid waste at milling j

τ_{ij}^t Overall price (Rp.) to buy FFB from plantation i for milling j in time period t

Cm_j^t Milling capacity (ton) at milling $j \in J$ in time period $t \in T$

CP_j^t Production capacity (ton) at milling $j \in J$ in time period $t \in T$

CA_i^t Quantity of FFB (ton) available at plantation $i \in I$ in time period $t \in T$

Sets are defined as follows.

I denotes the set of oil palm plantations

J denotes the set of CPO milling

K denotes the set of customers

T denotes the set of time period

The objective of the decision to be made is to minimize the overall operational costs. Mathematically the objective function can be expressed as follows:

$$\begin{aligned} \textit{Minimize} \quad & \sum_{i\in I}\sum_{j\in J}\sum_{t\in T}(\tau_{ij})(d_{ij})x_{ij}^t + \sum_{j\in J}\sum_{t\in T}\beta_j y_j^t + \sum_{j\in J}\sum_{k\in K}\sum_{t\in T}\gamma_{jk} z_{jk}^t + \\ & \sum_{j\in J}\sum_{t\in T}\lambda_j v_j^t + \sum_{j\in J}\sum_{t\in T}\rho_j w_j^t + \sum_{i\in I}\sum_{j\in J}\sum_{t\in T}\alpha_{ij}\delta_{ij}^t \end{aligned} \quad (1)$$

There are several requirements must be met regarding to limit the decision variables.

$$\sum_{i\in I} x_{ij}^t \le Cm_j^t \qquad \forall j \in J, t \in T \quad (2)$$

$$\sum_{j\in J}\sum_{t\in T} y_j^t \le \sum_{j\in J}\sum_{t\in T} Cp_j^t \quad (3)$$

$$\sum_{i\in I} x_{ij}^t \le \sum_{i\in I} CA_i \delta_{ij}^t \qquad \forall j \in J, t \in T \quad (4)$$

$$\sum_{j\in J} z_{jk}^t \le \sum_{j\in J} y_j^t \qquad \forall k \in K, \mathrm{t} \in \mathrm{T} \quad (5)$$

$$v_j^t + w_j^t \le \sum_{i\in I} x_{ij}^t \qquad \forall j \in J, t \in T \quad (6)$$

$$\sum_{i\in I}\sum_{j\in J}\delta_{ij} = n \quad (7)$$

$$x_{ij}^{t}, y_{j}^{t}, z_{jk}^{t}, v_{j}^{t}, w_{j}^{t} \geq 0, \quad \delta_{ij} \in \{0,1\}, \quad \forall i \in I, j \in J, k \in K, t \in T \tag{8}$$

Expression (2) states that the quantity of FFB to be bought from oil palm plantation $i \in I$ for milling $j \in J$ should be $\leq$ to the capacity of milling $j \in J$ in time period $t \in T$ Constraint (3) describes that the quantity of FFB to be processed in milling $j \in J$ should be $\leq$ to the production capacity of milling $j \in J$ in time period $t \in T$ .Constraint (4) presents that the quantity of FFB to be bought from oil palm plantation $i \in I$ for milling $j \in J$ should be $\leq$ to the availability of FFB in the plantation $i \in I$, if the plantation is chosen, in time period $t \in T$ Equation (5) to state that the quantity of CPO to be delivered to customer $k \in K$ from milling $j \in J$ should be $\leq$ to the quantity produced in milling $j \in J$ in time period $t \in T$ Constraint (6) represents the quantity of solid and liquid waste from milling $j \in J$ should be $\leq$ the quantity of FFB have been processed in time period $t \in T$ .Expression (7) is to ensure that there would be only n plantation available to supply FFB for each milling $j \in J$. Constraints (8) state the type of decision variables.

4. Algorithm

Let

$$x = [x] + f, 0 \leq f \leq 1$$

be the (continuous) solution of the relaxed problem, $[x]$ is the integer component of non-integer variable x and f is the fractional component.

Stage 1.

Step 1. Get row i^* the smallest integer infeasibility, such that $\delta_{i^*} = \min\{f_i, 1-f_i\}$
(This choice is motivated by the desire for minimal deterioration in the objective function, and clearly corresponds to the integer basic with smallest integer infeasibility).

Step 2. Do a pricing operation
$v_{i^*}^{T} = e_{i^*}^{T} B^{-1}$

Step 3. Calculate $\sigma_{ij} = v_{i^*}^{T} \alpha_j$
With j corresponds to

$$\min_{j} \left\{ \left| \frac{d_j}{\alpha_{ij}} \right| \right\}$$

Calculate the maximum movement of nonbasic j at lower bound and upper bound.
Otherwise go to next non-integer nonbasic or superbasic j (if available). Eventually the column j^* is to be increased form LB or decreased from UB. If none go to next i^*.

Step 4. Solve $B\alpha_{j^*} = \alpha_{j^*}$ for α_{j^*}

Step 5. Do ratio test for the basic variables in order to stay feasible due to the releasing of nonbasic j^* from its bounds.

Step 6. Exchange basis

Step 7. If row $i^* = \{\varnothing\}$ go to Stage 2, otherwise
Repeat from step 1.

Stage 2.
Pass 1: adjust integer infeasible superbasics by fractional steps to reach complete integer feasibility.
Pass 2: adjust integer feasible superbasics. The objective of this phase is to conduct a highly localized neighborhood search to verify local optimality.

5. Conclusion
In this paper, we consider an integrated supply chain production – distribution planning faced by decision maker (DM) of a CPO milling industry. This industry needs to buy FFB from several oil palm plantation. Due to the limited availability of FFB then the DM has to decide from which plantations would be appropriate to meet the production capacity of the milling industry. We develop a mixed integer linear programming model for the integrated of production and distribution planning problem of the crude palm oil industry. We solve the model using a direct search approach.

References
[1] A. Dumrongsiri. Mathematical model for production, logistics and plant capacity planning of oil palm bunches. 1st Mae Fah Luang University International Conference, 2012.
[2] C. H. Timpe, J. Kallrat, Optimal planning in large multi-site production networks. Eur. J. Oper. Res. 126, 422–435, 2000.
[3] Ch. Suksa-ard, M. Reweean. Optimization of Supply and demand balance in a palm oil supply chain, Thammasat International Journal of Science and Technology, Vol. 18(2), 2013, 14-31.
[4] D. CARLSSON, M. RONNQVIST, Supply chain management in forestry – Case studies at Sodra Cell AB. *European Journal of Operational Research*, **163** (2005), 589–616.
[5] F. T. S. CHAN, K. VIKAS, Performance optimization of a leagility inspired supply chain model: a CFGTSA algorithm based approach. *International Journal of Production Research*, **47**(3) (2009), 777–799.
[6] G. R. Nasiri, R. Zolfaghari, H. Davoudpour. An Integrated supply chain production-distribution planning with stochastic demands. Computers & Industrial Engineering, Vol. 77, 2014, 35-45.
[7] H. LI, M. AMINI, A Hybrid Optimization Approach to Configure Supply Chain for New Product Diffusion: A Case Study of Multiple-Sourcing Strategy. *International Journal of Production Research*, **50** (2012), 3152–3171.
[8] H. Sadyadharma, Z. Nasution, A. R. Matondang, Herman Mawengkang. Sustainable production planning of crude palm oil industry under uncertainty. International Journal of Engineering Research and Technology, Vol. 2(8), 2013, 1100-1105.
[9] J. Chungsiriporn, S. Prasertsan, C. Bunyakan. Toward cleaner production of palm oil mills: Part 2 Minimization of water consumption and process optimization. Asian Journal on Energy and Environment, Vol. 7(1), 2006, 246-257.
[10] J. S. K. LAU, G. Q. HUANG, K. L. MAK, L. LIANG, Agent-Based Modeling of Supply Chains for Distributed Scheduling. *IEEE Transactions on Systems, Man, and Cybernetics – TSMC*, **36**(5) (2006), 847– 861.
[11] K. Pewthongngam, S. Pathummakul, S. Homkhampad. An ieractive approach to optimize production-distribution planning for an integrated feed Swinecompany. International Journal of Production Economics, Vol. 142(2), 290-301.
[12] L. Amelia, D. A. Wahab, A. Hassan. Modelling palm oil production using fuzzy expert system. Expert System with Applications: An International Journal, Vol. 36(5), 2009, 8735-8749.
[13] S. BISWAS, Y. NARAHARI, Object Oriented Modeling and Decision Support for Supply Chains. *European Journal of Operational Research*, **153**(4) (2004), 704–726.
[14] Y. Gajpal, M. Nourelfath, "Two efficient heuristics to solve integrated load distribution and production planning," Reliability Engineering and System Safety, Vol. 144, pp. 204-214, 2015.

IOP Conf. Series: Materials Science and Engineering **300** (2018) 012016 doi:10.1088/1757-899X/300/1/012016

Optimization Model for Capacity Management and Bed Scheduling for Hospital

Suryati Sitepu[1], Herman Mawengkang[2], Ismail Husein[3]
[1]University HKBP Nomensen Medan, Indonesia
[2,]Mathematics Department, University of Sumatera Utara, Medan, Indonesia
[3]Graduate School of Mathematics, University of Sumatera Utara, Medan, Indonesia

Email: hmawengkang@yahoo.com

Abstract. Hospital is a very important institution to provide health care for people. It is not surprising that nowadays the people's demands for hospital is increasing.. However, due to the rising cost of healthcare services, hospitals need to consider efficiencies in order to overcome these two problems. This paper deals with an integrated strategy of staff capacity management and bed allocation planning to tackle these problems. Mathematically, the strategy can be modeled as an integer linear programming problem. We solve the model using a direct neighborhood search approach, based on the notion of superbasic variables.

Keywords: Capacity management, bed allocation planning, modeling, integer programming, neighborhood search..

1. Introduction

The obvious thing for most people in this planet is to have good health. The legal institution which provide and maintain good health in any country is hospital. Therefore, it is understandable that nowadays the people's demand for hospitals is increasing. One of the the main cause of this rise lies in the ageing populations who are putting heavy demands for health care. The direct impact of this condition is the number of hospitals are increasing. This case also occurs for the people who live in big city, such as, Medan city of North Sumatera province, Indonesia. However, there are still more people from Medan to go to other countries, such as, Malaysia and Singapore, to get better health services. Undoubtedly, the urgent need to tackle this situation is to improve health service performance of hospitals.

All operations related to the health service performance in hospitals are limited in terms of capacity. Therefore, in order to fulfill the patients' demand for health care, the hospitals management should have a planning and controlling the capacity of the operations [1], [2], [3].

Capacity planning decisions are concerned primarily to ensure that the institution has the capabilities to respond to the level of demand experiences, especially to health care industry because not only it relates to the management of highly specialized and costly resources (i.e., nurses, doctors, and advanced medical equipment), but also it makes a difference between life and death in critical conditions [4]. Public hospitals have, in general, more demand for health services than available capacity. Therefore it is important to forecast and manage demand with good precision, in order to adjust capacity or take alternative courses of action for example transfer demand to other facilities. Demand forecasting and management is part of a larger design that intents to provide a systemic solution to global hospital management. Such solution is commonly based on the design of a general process structure for hospitals and which defines the management processes that are needed to optimize the use of resources in doing so and to ensure a predefined service level for patients. The general process structure allowed us to determine the key processes where implementation of new practices would generate most value [5] [6].

For inpatient care facilities in a hospital, this requires information on bed capacity and nursing staff capacity, on a daily as well as annual basis. Quantitative models can be used to calculate capacity needs for different planning purposes and for short, medium and long term planning issues. Although several useful models are described in the international literature [7], [8], [9] many of them are difficult to apply in practice because they require a great deal of data and clerical work [7].

To be able to apply capacity management in practice, models must fulfill different functions: "annual staff planning," "roster scheduling support," and "daily assignment of nurses to wards [10-11]." In addition, "strategic decisions" are sometimes mentioned as a separate planning level [7, 12, 13]. Models based on mathematical optimization techniques from operations research are generally focused on short-term scheduling [14, 15]. Models that do integrate different planning horizons (daily, periodical (1–2 months), and annual) are for example described by [16] and [17]. These models contain connected models for periodical staff planning and daily scheduling. However, models incorporating operational planning issues with tactical and strategic decisions or operational scheduling support with annual staff planning were not found in the literature. Sitepu and Mawengkang [18] use stochastic optimization model to solve hospital nursing – staff management problem.

In general, capacity models are aimed at calculating the number of nurses needed, whereas capacity management models should ideally also give insight into opportunities for improving capacity usage.

Regarding to bed resources capacity planning most approach reported in the literature fall into the use of mathematical programming and simulation. Li et al., [11] present a mathematical programming model approach. They propose a multi-objective decision aiding model for solving allocation of beds problem in a hospital. Their model is based on queuing theory and goal programming (GP). In order to obtain some essential characteristics of access to various departments within a hospital, a waiting line model is explored. The results are used to construct a goal programming frame work, taking account of targets and objectives related to customer service and profits from the hospital manager and all department heads. [19] create a mathematical model based on a dynamic dispatching approach for bed resources allocation considering hospitalization demands, bed capacity and income. The objectives of their model are to maximize income and to minimize the costs to use supplementary beds. Another goal programming model approach for solving re-allocation bed in a hospitals was proposed by [20]. The constraints considered in their model are total number of beds, nursing work hours, waiting time, the definite bed allocation to patient, and the definite bed allocation to ward. An integer linear programming model approach is featured by [14] to solve a hospital bed management problem constrained with budgetary cuts. They implemented their result in French hospitals.

In terms of simulation technique for solving bed allocation problem, the works in the literatures belong to the use of simulation only and a combined simulation-optimization. Due to the inherent uncertainties in the problem, it is reasonable to use simulation technique only. In particular, the use of simulation we should be able to get useful insights on bed allocation. These results can be found in [15], [21], and [22]. Laker et al.[23] use discrete event simulation for solving bed allocation problem in the emergency department of a hospital. An interesting review of using simulation in healthcare is addressed in [24]. However the use of only simulation may not reach the best solution, particularly when the problems involve combinatorial nature. It is not surprising that some literatures propose a combine strategy simulation and optimization. [25] propose a simulation optimization approach for solving resource allocation in an emergency department. The use of multi-objective optimization combined with simulation to tackle the bed allocation problem can be found in [12].

2. Problem Description

Increasing demand for healthcare through hospitals created heavy challenges to their managers and decision makers. The challenges involve high costs, limited budget, and limited resources. Most of hospitals in Medan are encountered with some pressures, such as, shortages of qualified healthcare professionals, limited hospital equipments and facilities, increasing operational costs.

Capacity planning, for hospitals in particular, is concerned with making sure of balancing the quality of health care delivered with the cost of providing that care. Such planning involves predicting the quantity and particular attributes of resources required to deliver health care service at specified levels of cost and quality. The most fundamental measure of hospital capacity planning is the number of inpatient beds accordingly the number of doctors and the number of nurses. Hospital bed capacity decisions have traditionally been made based on target occupancy levels. Certain units in the hospital, such as, intensive care units (ICUs) are often run at much higher utilization levels because of their high costs.

The other important thing that should be considered in order to enhance the service performance of hospitals is waiting time, due to the bed capacity allocation system. Alternative strategy to overcome this situation is to have a well coordinated hospital capacity management along with bed allocation system.

3. Mathematical Model

The decision problem, for our problem, is to maximally coordinate the utilization of multifold resources within the hospital. In our case the multifold resources are doctors, nurses, beds and rooms.. In this case the most appropriate model to be created is a linear integer programming problem.

Notations

Sets

I Set of departments

J Set of doctors type

K Set of nurses type

L Set of available beds

R Set of rooms

E Set of medical equipments

H Set of technicians

Decision variables

XD_{ij} : Number of type $j \in J$ doctors in department $i \in I$

XN_{ik} : Number of type $k \in K$ nurses in department $i \in I$

XDA_{ij} : Number of type $j \in J$ doctors added in department $i \in I$

XNA_{ik} : Number of type $k \in K$ nurses added in department $i \in I$

XT_{ih} : Number of type $h \in H$ technicians in department $i \in I$

XB_{il} : Number of beds $l \in L$ in department $i \in I$

XBA_{il} : Number of beds $l \in L$ added in department $i \in I$

XM_{ie} : Number of medical equipment $h \in H$ in department $i \in I$

Binary variables

Z^{l}_{ir} Equals to 1, if bed $l \in l$ are allocated for room $r \in R$ in department $i \in I$, equals to 0 otherwise.

Y^{i}_{r} Equals to 1 if room $r \in R$ is used in department $i \in I$ Equals to 0 otherwise

Parameters

bd_{ij} : Cost of $j \in J$ type doctors in department $i \in I$

bs_{ik} : Cost of type $k \in K$ nurses in department $i \in I$

bsa_{ik} : Cost of type $k \in K$ nurse-aids in department $i \in I$

bt_{hi} : Cost of type $h \in H$ technicians in department $i \in I$

bt_{il} : Cost of operating beds $l \in L$ in department $i \in I$

bw_{il} : Waiting cost for beds $l \in L$ in department $i \in I$

ba^{i}_{lr} : Cost for allocating bed $l \in L$ for room $r \in R$ if used in department $i \in I$

br^{r}_{i} : Cost to operate room $r \in R$ if used in department $i \in I$

bm^{i}_{er} : Cost to operate medical equipment $e \in E$ for room $r \in R$ if used at department $i \in I$

md_{ij} : Maximum number of doctor for each type can be allocated to each department

mn_{ik} : Maximum number of nurse each type can be allocated to each department

mna_{ik} : Maximum number of nurse-aids each type can be allocated to each department

mb_{il} : Maximum number of beds $l \in L$ can be allocated to each department

ρ_i : Maximum fund available for department $i \in I$

A. *Objective Function*

The objective of this problem is to find all the decision variables such that to minimize total operating cost.

IOP Conf. Series: Materials Science and Engineering **300** (2018) 012016 doi:10.1088/1757-899X/300/1/012016

Mathematically, the objective function can be written as follows.

$$\begin{aligned} Minimize\quad TC = &\sum_{i\in I}\sum_{j\in J} bd_{ij}(XD_{ij} + XDA_{ij}) + \sum_{i\in I}\sum_{k\in K} bs_{ik}(XN_{ik} + XNA_{ik}) + \sum_{i\in I}\sum_{h\in H} bt_{hi}XT_{hi} + \\ &\sum_{i\in I}\sum_{l\in L} bt_{il}(XB_{il} + XBA_{il}) + \sum_{i\in I}\sum_{l\in L} bw_{il}XBA_{il} + \sum_{e\in E}\sum_{r\in R}\sum_{i\in I} bm^{i}_{er}XM^{i}_{er} + \\ &\sum_{l\in L}\sum_{r\in R}\sum_{i\in I} ba^{i}_{lr}Z^{i}_{lr} + \sum_{i\in I}\sum_{r\in R} br^{i}_{r}Y^{i}_{r} \end{aligned} \tag{1}$$

Constraints:

$$XD_{ij} + XDA_{ij} \le md_{ij}, \qquad \forall i \in I, \forall j \in J \tag{2}$$

Constraints (2) state that the number of doctor should not be greater the maximum number of doctors can be allocated to each department.

$$XN_{ik} + XNA_{ik} \le mn_{ik}, \qquad \forall i \in I, \forall k \in K \tag{3}$$

$$XD_{ij} + XDA_{ij} \ge XT_{ih}, \qquad \forall i \in I, \forall j \in J, \forall h \in H \tag{4}$$

$$XB_{il} + XBA_{il} \le mb_{il}, \qquad \forall i \in I, \forall l \in L \tag{5}$$

$$\sum_{l\in L} Z^{i}_{lr} \le Y^{i}_{r}, \qquad \forall i \in I, \forall r \in R \tag{6}$$

Constraints (6) are to make sure that beds are allocated to the available room.

$$\begin{aligned} &\sum_{i\in I}\sum_{j\in J} bd_{ij}XD_{ij} + \sum_{i\in I}\sum_{k\in K} bs_{ik}XN_{ik} + \sum_{i\in I}\sum_{h\in H} bt_{ih}XT_{ih} + \sum_{i\in I}\sum_{l\in L} bt_{il}XB_{il} + \\ &\sum_{i\in I}\sum_{l\in L} bw_{il}XBA_{il} + \sum_{i\in I}\sum_{e\in E}\sum_{r\in R} bm^{i}_{er}XM^{i}_{er} + \sum_{i\in I}\sum_{r\in R}\sum_{l\in L} ba^{i}_{rl}Z^{i}_{rl} + \sum_{i\in I}\sum_{r\in R} cr^{i}_{r}Y^{i}_{r} \le \sum_{i\in I} \rho_i \end{aligned} \tag{7}$$

Constraints (7) state that the budget spent should not be greater than the fund provided.

$$\begin{aligned} &XD_{ij}, \mathrm{XDA}_{ij}, \mathrm{XN}_{ik}, \mathrm{XNA}_{ik}, \mathrm{XT}_{ih}, \mathrm{XB}_{il}, \mathrm{XBA}_{il}, XM^{i}_{er} \ge 0 \ \text{ and integer} \\ &\forall \mathrm{i} \in \mathrm{I}, \forall \mathrm{j} \in \mathrm{J}, \forall \mathrm{k} \in \mathrm{K}, \forall \mathrm{l} \in \mathrm{L}, \forall \mathrm{e} \in \mathrm{E}, \forall \mathrm{h} \in \mathrm{H} \end{aligned} \tag{8}$$

$$Z^{i}_{lr}, \mathrm{Y}^{i}_{r} \in \{0,1\} \qquad \forall i \in I, \forall l \in L, \forall r \in R \tag{9}$$

Expressions (8) and (9) are the nature of decision variables.
It can be seen that the model is in the form of a large scale integer programming problem.

IOP Conf. Series: Materials Science and Engineering **300** (2018) 012016 doi:10.1088/1757-899X/300/1/012016

4. Feasible Neighborhood Heuristic Search

Branch-and-bound approach is a general method for solving linear integer programming problem. However, for large-scale problems such a procedure would be prohibitively expensive in terms of total computing time, and frequently the algorithm terminates without solving the problem. We have adopted the approach of examining a reduced problem in which most of the integer variables are held constant and only a small subset allowed varying in discrete steps.

This may be implemented within the structure of a program by marking all integer variables at their bounds at the continuous solution as nonbasic and solving a reduced problem with these maintained as superbasic..

The procedure may be summarized as follows:

Step 1 : Solve the problem ignoring integrality requirements.

Step 2 : Obtain a (sub-optimal) integer feasible solution, using heuristic rounding of the continuous solution.

Step 3 : Divide the set I of integer variables into the set I_1 at their bounds that were nonbasic at the continuous solution and the set I_2, $I = I_1 + I_2$.

Step 4 : Perform a search on the objective function, maintaining the variables in I_1 nonbasic and allowing only discrete changes in the values of the variables in I_2.

Step 5 : At the solution obtained in step 4, examine the reduced costs of the variables in I_1. If any should be released from their bounds, add them to set I_2 and repeat from step 4, otherwise terminate.

The above summary provides a framework for the development of specific strategies for particular classes of problems. For example, the heuristic rounding in step 2 can be adapted to suit the nature of the constraints, and step 5 may involve adding just one variable at a time to the set I_2.

At a practical level, implementation of the procedure requires the choice of some level of tolerance on the bounds on the variables and also their integer infeasibility. The search in step 4 is affected by such considerations, as a discrete step in a super basic integer variable may only occur if all of the basic integers remain within the specified tolerance of integer feasibility.

In general, unless the structure of the constraints maintains integer feasibility in the integer basic variables for discrete changes in the superbasic, the integers in the set I_2 must be made superbasic. This can always be achieved since it is assumed that a full set of slack variables is included in the problem.

5. Conclusions

This paper presents an optimization model of capacity management problem of a hospital which integrate bed allocation planning.. The optimization model is a large scale mixed integer linear program. We then propose a feasible neigfhbourhood integer search for solving the model.

6. References

[1] Aarabi, M., & Hasanian, S., 2014. *Capacity planning and control: a review. Int. Journal of Scientific & Engineering Research.* Vol. 5, Issue 8: pp. 975-984.

[2] Pierskalla W. P, & Brailer D., 1994. *Application of operations research in health care delivery. In: Pollock S, Barnett A, Rothkopf M (eds) Beyond the profit motive: public sector applications and methodology.* Handbooks in OR&MS, vol 6. North-Holland, New York.

[3] Jack E. P., & Powers T L, 2009. *A review and synthesis of demand and management, capacity management and performance in health-care services.* International Journal of Management Reviews, Vol.11(2), pp. 149-174.

[4] Hans, E. W., van Houdenhoven, M., & Hulshof, P. J., 2012. *A framework for healthcare planning and control.* Handbook of Healthcare System Scheduling (Springer).
[5] Swayne, L. E., Duncan, W. J., & Ginter, P. M., 2012. *Strategic Management of Health Care Organization.* John Wiley and Sons.
[6] Ma, G., & Demeulemeester, E., 2013. *A multi level integrative approach to hospital case mix and capacity planning.* Computers and Operations Research, Vol. 40, 2013, pp. 2198-2207.
[7] Smith-Daniels, V. L, Schweikhart, S. B, & Smith-Danielsn, D. E, 1988. *Capacity management in health care services*. review and future research diresctions *Decision Sci*, Vol. 19, pp. 899-918.
[8] Hurst, K., 2003. *Selecting and Applying Methods for Estimating the Size and Mix of Nursing Teams.* Leeds, Nuffield Institute for Health, pp. 1-19.
[9] Halloran, E. J., & Vermeersch, P. E. H., 1987. *Variability in Nurse Staffing Research data Collection and the Method of Reporting. Journal of Nursing Administration*, pp. 17:26-34.
[10] Ackali E, Cote M J, Lin C. 2006 . *Healthcare Manag. Sci.* **9** 391-404
[11] Li X, Jones D, Tamiz M. 2009 *Journal of the Operational Research Society.* **60** 330-338
[12] Wang Y, Lee L H, Chew E P, Lam S S W, Low S K, Ong M E H, Li H. 2015 *Multi-objective optimization for a hospital inpatient flow process via discrete event simulation*. In Proceedings of the 2015 Winter Simulation Conference, edited by L. Yilmaz, W. K. V. Chan, I. Moon, T. M. K. Roeder, C. Macal, and M. D. Rosseti, 3622-3631. Piscataway, New Jersey: Institute of Electrical and Electronics Engineers, Inc.
[13] Ataollahi F, Bahrami M A, Abesi M, Mobasheri F. 2013 *Middle East Journal of Scientific Research*. **18(11)** 1537-1543.
[14] Bachouch R B, Guinet A, Hajri-Gabouj S. 2012 *Int. J. Production Economics.* **140** 833-843
[15] Harper P R, Shahani A K. 2002 *The Journal of the Operational Research Society.* **53(1)** 11-18.
[16] Abernathy, W. J., Baloff, N., Hershey, J. C., & Wandel, S., 1973. *A Three-Stage Manpower Planning and Schedulling Model-A service sector example. Oper Res*, Vol. 21, pp. 693-711.
[17] Wright, P. D., Bretthauer, K. M., & Cote M. J., 2006. *Reexamining the Nurse Scheduling Problem: Staffing Ratios and Nursing Shortages.* Decision Sciences, Vol. 37, pp. 39-70.
[18] Sitepu S, Mawengkang H. 2015. Int. J. Of Advanced Res. In Computer Eng. & Tech., Vol. 4(1), pp. 44-47.
[19] Wang, T., Guinet, A., Besombes, B,. 2009. *A sizing tool for allocation planning of hospitals bed resources.* S. McClean et al. (Eds.): Intelligent Patient Management, SCI 189, 113-125.
[20] Ataollahi F., Bahrami M. A., Abesi M., & Mobasheri F., 2013. *A goal programming model for reallocation of hospitals' inpatient beds.* Middle east Journal of Scientific Research, Vol. 18(11), pp. 1537-1543.
[21] Akkerman R and Knip M. 2004..*Health Care Management Science*, 7(2):119–126..
[22] Cochran J K, Bharti A. 2006 *Health Care Management Science.* **9(1)** 31-45.
[23] Laker LF, Froehle CM, Lindsell CJ, Ward MJ. The flex track: Flexible partitioning between low and high acuity areas of an emergency depertment. Annals of Emergency Medicine, Vol. 64(6), 2014, 591-603.
[24] Almagooshi, S., 2015. *Simulation modeling in healthcare: Challenges and trends. Prtocedia Manufacturing.* Vol. 3, pp. 301-307.
[25] Keshtkar, L., Salimifard, K., & Faghih, N., 2015. *A simulation optimization approach for resource allocation in an emergency department.* QScience Connect, Vol. 1, pp. 8.

IOP Conf. Series: Materials Science and Engineering **300** (2018) 012017 doi:10.1088/1757-899X/300/1/012017

Periodic Heterogeneous Vehicle Routing Problem With Driver Scheduling

Ellis Mardiana Panggabean[1], Herman Mawengkang[2], Zainal Azis[3], Rina Filia Sari[4]

[1]Graduate School of Mathematics, University of Sumatera Utara

[2] Department of Mathematics, University of Sumatera Utara, Medan, Indonesia

[3]University of Muhammadiyah Sumatera Utara, Medan, Indonesia

[4]Islamic State University, Medan, Indonesia

Email : hmawengkang@yahoo.com

Abstract. The paper develops a model for the optimal management of logistic delivery of a given commodity. The company has different type of vehicles with different capacity to deliver the commodity for customers. The problem is then called Periodic Heterogeneous Vehicle Routing Problem (PHVRP). The goal is to schedule the deliveries according to feasible combinations of delivery days and to determine the scheduling of fleet and driver and routing policies of the vehicles. The objective is to minimize the sum of the costs of all routes over the planning horizon. We propose a combined approach of heuristic algorithm and exact method to solve the problem.

Keywords: Vehicle routing problem, scheduling, combined approach

1. Introduction

The classical Vehicle Routing Problem (VRP) is particularly intended to determine an optimal routing plan for a fleet of homogeneous vehicles such that to satisfy the need of a set of customers. As the root of this logistic problem is derived from the concept of Hamiltonian graph, it is necessarily to impose conditions that each vehicle route starts and ends at the starting point, and each customer is visited once by one vehicle. Since then many researchers have been working in this area to discover new variants and new methodologies. There are a number of survey can be found in literature for VRP, such as, [1], [2], [3], [4], and books [5], [6].

In literature there are some variants of VRP which are grouped according to specific constraints. Some of the well known variants are: Capacitated VRP (CVRP), the vehicles are restricted to carry limited capacity; VRP with time windows (VRPTW), each customer is served within a defined time frame; multiple depots VRP (MDVRP), in this variant goods can be delivered to a customer from a set of depots; VRP with pick-up and delivery (VRPPD), goods not only need to brought from the depot to the customers, but also must be picked-up at a number of customers and brought back to the depot. Taking into account several days of planning for routing problems is another variant of the VRP, known as the priodic VRP (PVRP).

In PVRP, within a given time horizon, there is a set of customers needs to be visited once or several times. There would be a visiting schedules associated with each customer. A fleet of vehicles is available and each vehicle leaves the depot, serves a set of customers, when its work shift or capacity is over, returns to the depot. The problem is to minimize the total length of the routes travelled by the vehicles on the time horizon. This problem is very important in real world applications, such as, distribution for bakery companies [7], blood product distribution [8], or pick-up of raw materials for a manufacture of automobile parts [9].

A survey on PVRP and its extensions can be found in [10]. Due to the complexity of the problem most of the works present heuristic approaches, nevertheless [11] proposed an exact method. [12] addressed a combined of heuristic and exact method for solving PVRP. The extension of the VRP in which one must additionally decide on the fleet composition is known as the

Heterogeneous Vehicle Routing Problem (HVRP) is rooted in the seminal paper of [13] published in 1984 and have recently evolved into a rich research area. There have also been several classifications of the associated literature from different perspectives. [14] provided a general overview of papers with a particular focus on lower bounding techniques and heuristics. The authors also compared the performance of existing heuristics described until 2008 on benchmark instances. [15] presented a review of exact algorithms and a comparison of their computational performance on the capacitated VRP and HVRPs, while [16] reviewed several industrial aspects of combined fleet composition and routing in maritime and road-based transportation. Afshar-Nadjafi and afshar-Nadjafi (2017) propose a constructive heuristic approach for solving HVRP time windows with multi-depot and time-dependent.

Note that all the literature survey of HVRP mentioned in the previous paragraph do not involve PVRP. This paper proposes a model of HVRP in which that the customer can be visited more than once, and there would be a schedule for visiting customers.

This paper is organized as follows. Section 2 describes the problem definition and the mathematical model. Section 3 presents the feasible neighbourhood heuristic search. The algorithm is described in Section 4. Finally Section 5 describes the conclusions

2. Problem Formulation

The decision of the problem is find the feasible transportation tour of a certain type of vehicle such that to minimize the overall operating costs. In order to ease the problem, it is necessary to visualize the problem using graph. Let $\Gamma = (\varsigma , A)$ be a directed graph, where $\varsigma = \{0, 1, \ldots , n\}$ is the vertex set and $A = \{(i, j) : i, j \in \varsigma , i \neq j\}$ is the set of arcs, representing routes between vertex. The depot vertex is indexed by 0. $\varsigma_c = \varsigma \setminus \{0\}$ is the set of customer locations. Each vertex $i \in \varsigma_c$ has a demand $q_i \geq 0$ on each day of the planning horizon, a service time $s_i \geq 0$, a time window $[e_i, l_i]$, where e_i is the earliest time service may begin and l_i is the latest time, and requires a fixed number of visits f_i to be performed according to one of the allowable visit-day patterns. The time window specifying the interval vehicles leave and return to the depot is given by $[e_0, l_0]$. Let $K = \{1,\ldots,k\}$ be the set of available vehicle of all types. A fleet of m vehicles of each type, with capacity Q_k is based at the depot.

For each vehicle $k \in K$, let Q_k denote the capacity in weight. We assume the number of vehicles equals to the number of drivers. Denote the set of n customers (/nodes) by $N = \{1, 2, \ldots, n\}$. Denote the depot by $\{0, n+1\}$. Each vehicle starts from $\{0\}$ and terminates at $\{n+1\}$. Each customer $i \in N$ specifies a set of days to be visited . On each day, customer $i \in N$ requests service with demand of q_i^t in weight, within time window $[a_i, b_i]$. Note that, for the depot $i \in \{0, n+1\}$, we set $q_i^t = 0$. Denote the set of preferable vehicles for visiting customer i by K_i ($K_i \in K$) and the extra service time per pallet by e if a customer is not visited by a preferable vehicle.

This notations used are given as follows :

Set:

DI	The set of internal drivers,
DE	The set of external drivers,
D	The set of drivers $D = DL \cup DE$,
K	The set of vehicles,

N	The set of customers,
N_0	The set of customers and depot $N_0 = \{0, n+1\} \cup N$,
K_i	The set of preferable vehicles for customer $i \in N$,

Parameter:

M_k	The available number of vehicles of type k
Q_k	The weight capacity of vehicle $k \in K$,
$[a_i, b_i]$	The earliest and the latest visit time at node $i \in N_0$,
q_i	The weight demand of node $i \in N_0$
$[g_l, h_l]$	The start time and the latest ending time of driver $l \in D$
α_i	Pick up quantity for customer i
β_i	Delivery quantity for customer i
CV^k	Fixed cost for each type of vehicle $k \in K$ from depot
CVA_{ij}^k	Travelling cost of vehicle $k \in K$ along the edge $(i, j) \in A$
CVN^k	Cost due to the visit of a customer by non preferable vehicle $k \in K$
CDI_l^k	Cost for internal driver $l \in DI$ using vehicle $k \in K$
CDE_l^k	Cost for external driver $l \in DE$ using vehicle $k \in K$
CPQ_j^k	Cost to pick up a number of quantity for customer $j \in N$ using vehicle $k \in K$
CDD_j^k	Cost to deliver a number of quantity for customer $j \in N$ using vehicle $k \in K$

Variables:

x_{ij}^k	Binary variable indicating whether vehicle $k \in K$ travels from node $i \in N_0$ to $j \in N_0$
w_i	Binary variable indicating whether customer $i \in N_0$ is visited by a non-preferable vehicle
v_i^k	The time at which vehicle $k \in K$ starts service at node $i \in N_0$
y_l^k	Binary variable indicating whether vehicle $k \in K$ is assigned to internal driver $l \in DI$
z_l^k	Binary variable indicating whether vehicle $k \in K$ is assigned to external driver $l \in DE$
θ_j^k	Number of pick up demand of customer j served by vehicle $k \in K$
σ_j^k	Number of delivery demands of customer j served by vehicle $k \in K$

The mathematical formulation for this problem is presented as follows:

$$\text{Min} \sum_{k\in K}\sum_{j\in N_0} CV^k x_{oj}^k + \sum_{k\in K}\sum_{(i,j)\in A} CVA_{ij}^k x_{ij}^k + \sum_{i\in N_0}\sum_{k\in K} CVN^k w_i^k + \sum_{l\in DI}\sum_{k\in K} CDI_l^k y_l^k + \sum_{l\in DE}\sum_{k\in K} CDE_l^k z_l^k + \sum_{j\in N}\sum_{k\in K} CPQ_j^k \theta_j^k + \sum_{j\in N}\sum_{k\in K} CDD_j^k \sigma_j^k \tag{1}$$

Subject to

$$\sum_{j \in N_0} x_{0j}^k = 1, \qquad \forall k \in K \tag{2}$$

$$\sum_{k \in K} \sum_{j \in N} x_{ij}^k = 1, \qquad \forall i \in N_0 \tag{3}$$

Constraints (2) and (3)) are to ensure that exactly one vehicle of each type enters and departs from every customer and from the central depot.and comes back to the depot.

$$\sum_{i \in N} x_{ij}^k - \sum_{i \in N} x_{ji}^k = 1; \quad \forall j \in N_0, \forall \mathrm{k} \in \mathrm{K} \tag{4}$$

A flow conservation equation is necessarily needed to maintain the continuity of each vehicle route on each period of time. This equation is presented in Constraints (4).

$$\sum_{j \in N_0} x_{0j}^k \le M_k, \qquad \forall k \in K \tag{5}$$

Constraint (5) represents that each customer is served only by the available and active vehicle of type m.

$$\sum_{j \in N_0} x_{1j}^k \le 1; \qquad \forall k \in K \tag{6}$$

$$\sum_{i \in N, i>1} x_{i1}^k \le 1; \qquad \forall k \in K \tag{7}$$

Constraints (6) and (7) state the availability of vehicles by bounding the number of route, related to vehicle k for each type, directly leaving from and returning to the central depot, not more than one, respectively.

$$\sum_{i \in N} q_i \sum_{j \in N_0} x_{ij}^k \le Q_k; \qquad \forall k \in K_m \tag{8}$$

Constraints (8) ensure that each delivery does not exceed the capacity of each type of vehicle.

$$\sum_{k \in K} \sum_{j \in N_0} x_{ijk}^t = 1 \qquad \forall\, i \in N, t \in T_i \tag{9}$$

$$\sum_{k \in K \setminus K_i} \sum_{j \in N_0} x_{ijk}^t = w_i^t \qquad \forall\, i \in N, t \in T_i \tag{10}$$

$$\sum_{i \in N} \sum_{j \in N_0} q_i^t x_{ijk}^t \le Q_k \qquad \forall\, k \in K, t \in T \tag{11}$$

$$b_i \ge v_{ik}^t \ge a_i \qquad \forall i \in N,\ k \in K, t \in T_i \tag{12}$$

$$v_{0k}^t \geq \sum_{l \in D} (g_l^t \cdot y_{lk}^t) \qquad \forall k \in K, t \in T \qquad (13)$$

$$v_{n+1,k}^t \leq \sum_{l \in D} (h_l^t \cdot y_{ik}^t) \qquad \forall k \in K, t \in T \qquad (14)$$

$$\sum_{k \in K} \theta_{jk}^t = \alpha_j^t \qquad \forall j \in N, t \in T \qquad (15)$$

$$\sum_{k \in K} \sigma_{jk}^t = \beta_j^t \qquad \forall j \in N, t \in T \qquad (16)$$

$$x_{ijk}^t, w_i^t, z_{ik}^t, y_{lk}^t \in \{0,1\} \qquad \forall i, j \in N_0, l \in D, \; k \in K, t \in T \qquad (17)$$

$$v_{ik}^t, u_{ik}^t, r_l^t, s_l^t \geq 0 \qquad \forall i, j \in N_0, l \in D, \; k \in K, t \in T \qquad (18)$$

$$\theta_{jk}^t, \sigma_{jk}^t \in \{0,1,2,\ldots\} \qquad \forall j \in N, k \in K, t \in T \qquad (19)$$

Constraints (11) state that each customer must be visited by one vehicle on each of its delivery days. Constraints (12) define whether each customer is visited by a preferable vehicle. Constraints (13-14) guarantee that the vehicle capacities are respected in both weight and volume.

Constraints (15-16) define the elapsed driving time. More specifically, for the vehicle (k) travelling from customer i to j on day t, the elapsed driving time at j equals the elapsed driving time at i plus the driving time from i to j (i.e., $u_{jk}^t \geq u_{ik}^t + c_{ij}$) if the vehicle does not take a break at customer i (i.e., $z_{ik}^t = 0$); Otherwise, if the vehicle takes a break at customer i (i.e., $z_{ik}^t = 1$), the elapsed driving time at j will be constrained by (10) which make sure it is greater than or equal to the travel time between i and j (i.e., $u_{jk}^t \geq c_{ij}$). Constraints (17) guarantee that the elapsed driving time never exceeds an upper limit F by imposing a break at customer i (i.e., $z_{ik}^t = 1$) if driving from customer i to its successor results in a elapsed driving time greater than F.

Constraints (18) determine the time to start the service at each customer. If j is visited immediately after i, the time v_{jk}^t to start the service at j should be greater than or equal to the service starting time v_{ik}^t at i plus its service duration d_i^t, the extra service time $e \cdot p_i^t$ if i is visited by an inappropriate vehicle (i.e., $w_j^t = 1$), the travel time between the two customers cij, and the break time G if the driver takes a break after serving I (i.e., $z_{ik}^t = 1$). Constraints (19) make sure the services start within the customers' time window.

Constraints (20-21) ensure that the starting time and ending time of each route must lie between the start working time and latest ending time of the assigned driver. Constraints (22) calculate the total travel time for each internal driver. Constraints (23) define the working duration for each driver on every workday, which equals the time the driver returns to the depot minus the time he/she starts work. Constraints (24) make sure that the internal drivers work for no more than a maximum weekly working duration, referred to as 37 week-hour constraints. Constraints (25 – 26) define the pick up and delivery for each customer. Constraints (27-29) define the binary and positive variables used in this formulation.

3. Proposed Method

Constraints (13) – (15) express the nature of the variables used in the model.
Let

$$x = [x] + f, \quad 0 \leq f \leq 1$$

be the (continuous) solution of the relaxed problem, $[x]$ is the integer component of non-integer variable x and f is the fractional component.

Stage 1.

Step 1. Get row i^* the smallest integer infeasibility, such that $\delta_{i*} = \min\{f_i, 1-f_i\}$

(This choice is motivated by the desire for minimal deterioration in the objective function, and clearly corresponds to the integer basic with smallest integer infeasibility).

Step 2. Do a pricing operation

$$v_{i*}^T = e_{i*}^T B^{-1}$$

Step 3. Calculate $\sigma_{ij} = v_{i*}^T \alpha_j$

With j corresponds to

$$\min_j \left\{ \left| \frac{d_j}{\alpha_{ij}} \right| \right\}$$

Calculate the maximum movement of nonbasic j at lower bound and upper bound.

Otherwise go to next non-integer nonbasic or superbasic j (if available). Eventually the column j^* is to be increased form LB or decreased from UB. If none go to next i^*.

Step 4.

Solve $B\alpha_{j*} = \alpha_{j*}$ for α_{j*}

Step 5. Do ratio test for the basic variables in order to stay feasible due to the releasing of nonbasic j^* from its bounds.

Step 6. Exchange basis

Step 7. If row $i^* = \{\varnothing\}$ go to Stage 2, otherwise

Repeat from step 1.

Stage 2. Pass1 : adjust integer infeasible superbasics by fractional steps to reach complete integer feasibility.

Pass2 : adjust integer feasible superbasics. The objective of this phase is to conduct a highly lovalized neighbourhood search to verify local optimality.

4. Conclusions

This paper was intended to develop efficient technique for solving one of the most economic importance problems in optimizing transportation and distribution systems. The aim of this paper was to develop a model of Periodic vehicle Routing with Time Windows, Fleet and Driver Scheduling, Pick-up and Delivery Problem This problem has additional constraint which is the limitation in the number of vehicles. The proposed algorithm employs nearest neighbor heuristic algorithm for solving the model. This algorithm offers appropriate solutions in a very small amount of time.

References

[1] O. Bräysy, & Gendreau, M., 2005. *Vehicle Routing Problem with Times Windows, Part II: Metaheuristics*. Transportation Science, **39(1),** pp. 104-118.

[2] Cordeau, J. F., Laporte, G., Savelsbergh, M.W.F., & Vigo, D., 2007. *Vehivle Routing.* In Barnhart, C. and Laporte, G., edotirs, *Transportation,* handbooks in Operation Research and Management Science, pp. 367-428. North-Holland, Amsterdam.

[3] Golden, B. L., Assad, A. A., & Wasil, E. A., 2002. *Routing Vehicles in the real world: Applications in the solid waste, beverage, food, dairy, and newspaper industries.* In P. Toth and D. Vigo, editors, *The Vehicle Routing Problem,* pp. 245-286. SIAM, Philadelphia, PA.

[4] Laporte G., & Sement, F., 2002. *Classical Heuristics for the Vehicle Routing Problem.* In Toth, P. and Vigo, D., editors, *The vehicle Routing Problem,* SIAM Monographs on Discrete Matheamtics and Applications, pp. 109-128. SIAM, Philadelphia, PA.

[5] Golden, B. L., Raghavan, S., &. Wasil, E. A., 2008. *The vehicle routing problem: latest advances and new challenges.* Springer.

[6] Toth, P., & Vigo, D., 2002. *The Vehicle Routing Problem Society for Industrial and Applied Matheamtics.* Philadelpiha, PA.

[7] Pacheco, J., Alvarez, A., García, I., & Angel-Bello, F., 2011. *Optimizing vehicle routes in a bakeri company allowing flexibility in delivery dates. Journal of Operationsl Research Society* http://dx.doi.org/10.1057/jors.2011.51, 2011.

[8] Hemmelmayr, V., Doerner, K., Hartl, R., & Savelsbergh, M., 2009. *Delivery strategis for blood products supplies. OR Spectrum* **31 (4)** pp. 707-725.

[9] Alegre, J., Laguna, M., & Pacheco, J., 2007. *Optimizing the periodic pick-up for raw materials for a manufacturer of auto parts.* European Journal of Operational Research **179**, pp. 736-746, 2007.

[10] Francis, P., Smilowitz, K., & Tzur, M., 2008. *The period vehicle routing problem and its extensions, in : The Vehicle Routing Problem: Latest Advances and New Challenges.* Vol. 43, Springer, 2008 (Chapter).

[11] Baldacci, R., BArtolini, E., Mingozzi, A., & Valletta, A., 2011. *An Exact algorithm for the period routing problem.* Operations research **59 (1)**, pp. 228-241.

[12] Cacchiani, V., Hemmelmayr, V. C., & Tricoire, F., 2012. *A set-covering based heuristic algorithm for the periodic vehicle routing problem.* Discrete Applied Matehametics , doi:10.1016/j.dam.2012.08.032, 2012.

[13] Golden, B., Assad, A., Levy, L., & Gheysens, F. G., 1984. *The fleet size and mix vehicle* routing problem. Computers and Operations Research, **11,** pp. 49–66.

[14] Baldacci, R., Battarra, M., & Vigo, D., 2008. *Routing a heterogeneous fleet of vehicles.* In B. L. Golden, S.Raghavan, E. A. Wasil, eds. The Vehicle Routing Problem: Latest Advances and New Challenges. pp.1–25. Springer, New York.

[15] Baldacci, R., Toth, P., & Vigo, D. 2010a. Exact algorithms for routing problems under vehicle capacity constraints.Annals of Operations Research 175, pp. 213–245

[16] Hoff, A., Andersson, H., Christiansen,M., Hasle, G., & Løkketangen, A., 2010. *Industrial aspects and literaturesurvey: fleet composition and routing.* Computers & Operations Research 37, pp. 2041–2061.

[17] Afshar-Nadjafi B and Afshar-Nadjafi A. 2017. A constructive heuristic for time-dependent multi depot vehicle routing problem with time windows and heterogeneous fleet. *Journal of King Saud University – Engineering Sciences* 29, 29-34.

Decision Model for Planning and Scheduling of Seafood Product Considering Traceability

Agustin[1], Herman Mawengkang[2], Devy Mathelinea[3]
[1]Graduate School of Environment Management, University of Sumatera Utara
[2,]Mathematics Department, University of Sumatera Utara, Medan, Indonesia
[3]Graduate Studies, University Tun Hussein Onn Malaysia, Johor Darul Takzim, Malaysia

Email: agus.tin58@yahoo.co.id

Abstract. Due to the global challenges, it is necessary for an industrial company to integrate production scheduling and distribution planning, in order to be more efficient and to get more economics advantages. This paper presents seafood production planning and scheduling of a seafood manufacture company which produces simultaneously multi kind of seafood products, located at Aceh Province, Indonesia. The perishability nature of fish highly restricts its storage duration and delivery conditions. Traceability is a tracking requirement to check whether the quality of the product is satisfied. The production and distribution planning problem aims to meet customer demand subject to traceability of the seafood product and other restrictions. The problem is modeled as a mixed integer linear program, and then it is solved using neighborhood search approach.

Keywords: Optimization, modeling, production planning, traceability, neighborhood search.
.

1. Introduction

Indonesia's coastal fisheries can be regarded as an essential source of food, income and cultural heritage to its people. Sixty percent of Indonesian fishers, or 1.6 million, are small-scale fishers, and 85% of their catch is used for human consumption. Additionally, Indonesia has the second longest coastline in the world, and 40% of the population (roughly 100 million people) lives near the coast. Coastal fisheries therefore play an essential role in national food security and nutritional health in Indonesia. In this paper we consider production planning problem which arises in marine fisheries industry in Indonesia. Marine fisheries play an important role in the economic development of Indonesia. This industry could also provide employment to people who live at coastal areas, to increase the financial gain of local government, and to conserve sustainability.

Naturally the fish industry in Indonesia can be found at the coastal area. There are a lot of varieties of fish processed can be produced, depending on the region . Due to the nature conditions , the management of fish industry is still dominated by the local small traditional business, using conventional management strategy. Consequently, they do not have enough knowledge about production and distribution planning and in particular supply chain.

The seafood production is a semi-continuous process and subject to individual characteristics. Seafood is a notably perishable product within the category of food industry . The perishability highly restricts its storage duration and delivery conditions. In order to maintain the quality of the seafood beside perishability, it is necessary to have traceability of the product.

In terms of definition, Hobbs et al. [1] pointed out that there is no agreement on the definition of traceability. However they simply define traceability as the ability to follow the movement of food through specified stages of production, processing, and distribution. In our case we define that traceability is the action to trace and follow the food raw material and finished product through stages of production, inventory and distribution.

Successful traceability processes generally incorporate a carefully devised planning stage to determine when, how, where, and why each traceability link will be created [2]. By aligning traceability tasks with project and organizational goals, and capturing these planning decisions in a trace strategy graph, project stakeholders can ensure that their traceability effort provides effective support for critical software engineering tasks. In prior work, several researchers have proposed the use of strategic traceability graphs [2], [3] to define the traceability goals of a project. In its simplest form, such a graph defines a set of traceable artifacts, the traceability links between them, and the purpose of each of these links. However, such graphs fall short of providing the level of support needed to fully automate the traceability process, especially when artifacts are stored in heterogeneous formats.

The optimal performance of the operations in production-distribution planning is to optimize total cost due to converting the raw fish into final seafood product, keep the products in the inventory and then deliver them to distribution center. In more detail, [4] described the modelling of a more com0-plete integrated supply chain system which composed of raw material suppliers, primary and secondary plants (each one with inventories of raw materials and finished products), distribution centres, warehouses and customer areas. In a situation where uncertainties occur, Bakhrankova et al., [5] used stochastic programming for production planning of fisheries. In the integrated stochastic model, they incorporate both upstream (raw fish quantities) and downstream (price of the finished product) uncertainties, while accounting for fish quality and shelf life requirements.

From the structure of the planning and scheduling of manufacturing problem, it is reasonable that mixed integer linear programming (MILP) models could provide mathematical optimization frameworks to represent the main characteristics of problems. [6] addressed an MILP model, which integrate production, distribution and marketing and involved plants and sales points. Their paper intends to cover the relevant features required for the complete supply chain management with a multi-site production network The MILP is an extensively accepted tool in the dairy industry for well-defined problems (e.g.,[7]; [8], [9]). Bilgen and Celebi [10] present a MILP model addressing the production scheduling and distribution planning problem in a yoghurt production line of multi-product dairy plants. [11] develop a MILP global supply chain model for determining product – plant and customer – distribution center assignments, number of supply chain echelons, and number and locations of distribution centers. [12] formulate a two-stage supply chain model that determines the optimal quantities of products to be produced at each plant, transported from each plant to each distribution center. However, even though they solve a large scale problem, there is no perishability condition involved. [13] present mixed integer linear programming models and computational strategies for the problem of multi-echelon supply chains with inventory under uncertainty. [14] consider a multi-period production system which integrate load distribution and production planning. They use three-phase heuristic approach and Tabu Search for solving the integrated model. Camacho-Vallejo [15] use bilevel mathematical programming model for solving a supply chain's production-distribution planning. An interesting extensive review of the integrated system can be found in ([16], [17], [18]).

It should be noted that all papers that have been mentioned do not consider tracing processes in the production planning.This paper considers a production scheduling and distribution planning problem using tracing process which arises in fisheries industry in Indonesia. The location under investigation for this paper is located at Aceh Province, Indonesia. Marine fisheries play an important role in the economic development of Indonesia, particularly in Aceh province. The main points of this industry are to provide employment to people who live at coastal areas, to increase the financial gain of local government, and to conserve sustainability.

Perishable products and tracing processes, such as fish processed, would impose additional difficulties to the supply chain management due to their limited shelf-life and deterioration of fish quality. These difficulties include the duration of storing the products due to the expired date. Hence, quantities delivered to distribution centers are limited by the shelf-life of goods as well as the distribution center's holding capacity. [19] consider the modelling of the integrated production-distribution planning model for perishable products. The supply chain network considered consists of production facility and multiple distribution centers. [20] discussed an integrated production-distribution planning with scheduling for food catering service industry. Their paper is to find a scheduled production-distribution planning in such a way that the customers service level is satisfied and the total cost is minimized.

The challenge for companies in managing the supply chain of perishable foods is that the value of the product deteriorates significantly over time at rates that are highly dependent on the environment [21]. According to these authors, temperature and humidity are key factors in this process. An additional concern about food production and distribution systems around the world is that they must to be more reliable than manufactory traditional systems. Supply chain of perishables is more susceptible to economic chocks, environmental changes, or even to management errors supported by lack of knowledge [22]. The literature had indicated that in the future food system will have to joint four major characteristics: resilience, sustainability, competitiveness, and ability to manage and meet customer expectations [23]. Due to diseases related to food ingestion and food production globalization [24], costumers had become more conscious of origin and nutritional contents of their own foods. As a consequence of that, an increase of interest on traceability, freshness, and high quality patterns of food is demanded. At the same time, productors have expanded products choice in order to attend costumers' wishes.

This paper concerns with the modeling the integrated production scheduling and distribution planning for several seafood products. There are several plants are setup to produce these seafood products. Due to the shelf life of raw fish and production capacity it is necessarily to conduct traceability and then to schedule to which plant a kind of seafood to be produced. Inventories of raw fish are available at each plant with limited capacity and limited time. The finished qualified sea food products will be delivered to a set of distribution centers, which has limited cooled store rooms. The problem is formulated as a mixed integer linear programming (MILP) model. A direct neighborhood search approach is developed for solving the model.

2. Problem Formulation

This problem is motivated by a fish industry located at the Aceh Province of Indonesia. The industry plans to produce N kinds of local traditional seafood products from J plants. These plants are dispersed in several cities. The finished products will be packed in packaging container and then distributed to a set of L distribution centers. Due to the perishability issue of the raw fish material and the seafood finished products it is necessarily to schedule the storage during the production and distribution operations, and then to conduct traceability for the quality of the products.

The fish industry wishes to plan the production and distribution system with a proper scheduling for the N kind of seafood products within period of time t, t = 1,…, T, in order to fulfill the market demand. There are J plants available. In order to maintain the freshness of fish raw material, each plant has a limited capacity cooled store room with unit holding cost of $\rho_{nt}^{j}, \mathrm{n} \in \mathrm{N}, \mathrm{j} \in \mathrm{J}, \mathrm{t} \in \mathrm{T}$. These raw material have a shelf-life τ_r. If the stored raw materials are not fully used after their shelf-life, they would be discarded. Also there would be tracing processes for the raw fish, these raw fish should be discarded if they do not pass the quality standard. Therefore the level of inventory of raw material at plant j of n kind of fish product before their shelf-life would be denoted as $I_{n,t<\tau_r}^{j}$

There is a set of L distribution centers located around Aceh Province close enough to each plant. Each distribution center l (k = 1, 2, . . . , L) has a nonnegative and deterministic demand D^l_{nt} of n kind fish product in planning period t of the planning horizon. A limited amount of inventory can be stored in cooled room of distribution center l with unit holding cost of ρ^l_{nt} in each planning period t. Note that each finished product has shelf-life τ_f

We assume that the demand for each kind of seafood in the scheduled time period at each distribution center is known. The demands have certain due dates. Unmet demand for a period is not allowed to be transferred to the next period. The unmet demand would be discarded at costs.

The problem for the fish company is to schedule production and distribution, in such a way that will minimize the total cost occurred in the production operations and inventory at each plant and at each distribution center.

For the mathematical description of the model, the following notations are defined.

Sets and indices

- T : Set of time period with index t
- P : Set of fish processed products with index p
- J : Set of plants with index j
- M : Set of raw fish resources with index m
- L : Set of distribution centers with index l

Variables

- x^t_{pj} : Quantity of product p produced to be traced in time period t from plant j (ton)
- z^t_{plj} : Quantity of qualified product p delivered to distribution center l from plant j in time period t (ton)
- u^t_{ij} : Additional amount of raw fish m to purchase in time t for plant j (unit)
- k^t_j : Number of workers required in time period t at plant j (man-period)
- k^{t-}_j : Number of workers laid-off in time period t at plant j (man-period)
- k^{t+}_j : Number of additional workers in time period t at plant j (man-period)
- $I^{t\le\tau_r}_{mj}$: Inventory level of raw fish m at plant j in time period t considering shelf-life τ_r (units)
- $I^{t\le\tau_f}_{pl}$: Quantity of product p to be stored in distribution center l in time period t considering shelf-life τ_f (units)
- B^t_{pj} : Under-fulfillment of product p in period t at plant j (units)
- Q^t_{pl} : Unmet demand of product p in time period t at distribution center l (units)
- y^t_{pj} : Binary variable to state whether a kind of seafood product is set up to be traced at plant j in time period t
- v^t_{pj} : Quantity of product p to be rejected at plant j in time period t

IOP Conf. Series: Materials Science and Engineering **300** (2018) 012018 doi:10.1088/1757-899X/300/1/012018

Parameters

- cp_{pj}^{t} : Cost associated with producing *p* fish products at plant *j* in time period t (Rp.)
- cr_{mj}^{t} : Cost to buy additional raw fish for plant j in time period t (Rp.)
- cwr_{j}^{t} : Cost associated with regular worker at plant j in time period t (Rp.)
- cwa_{j}^{t} : Cost associated with additional worker at plant j in time period t (Rp.)
- cwl_{j}^{t} : Cost associated with laid-off worker at plant j in time period t (Rp.)
- cir_{mj}^{t} : Inventory cost of raw fish m at plant j in time period t (Rp.)
- cuf_{pj}^{t} : Cost associated with under fulfillment of fish product p at plant j in time period t (RP.)
- ct_{plj}^{t} : Transportation cost to deliver fish product p from plant j to DC l in time period t (Rp.)
- $cid_{pl}^{t<\tau_f}$: Inventory cost of fish product p at DC l in time t considering shelf-life τ_f (Rp.)
- cdp_{pl}^{t} : Cost associated with discarding unmet quality of fish product p at DC l in time period t (Rp.)
- ctf_{pj}^{t} : Cost associated with tracing process of fish product p at plant j in time period t (Rp.)
- crf_{pj}^{t} : Cost associated with dumping the rejected fish product p at plant j in time period t (Rp.)
- D_{pj}^{t} : Demand for product p in period t from plant j (units)
- U_{ij}^{t} : Upper bound on additional resources at plant j (units)
- r_{ipj}^{t} : Amount of resource i needed to produce one unit of product p at plant j
- f_{it}^{j} : Amount of resource i available at time t at plant j (units)
- a_p : Number of worker needed to produce one unit of product p (man)
- $UI_{pj}^{t<\tau_r}$: Upper bound on inventory of product p at the plant j in period t before shelf-life (units)
- $UI_{pl}^{t<\tau_f}$: Upper bound on inventory of product p at the distribution center l in period t before shelf-life (units)

3. Mathematical Formulation

The objective of the problem is to minimize the total cost, mathematically can be written as follows.

$$\begin{aligned} Minimize\ z = & \sum_{p\in P}\sum_{j\in J}\sum_{t\in T} cp^t_{pj}x^t_{pj} + \sum_{m\in M}\sum_{j\in J}\sum_{t\in T} cr^t_{mj}u^t_{mj} + \sum_{j\in J}\sum_{t\in T} cwr^t_j k^t_j + \sum_{j\in J}\sum_{t\in T} cwa^t_j k^{t+}_j + \\ & \sum_{j\in J}\sum_{t\in T} cwl^t_j k^{t-}_j + \sum_{m\in M}\sum_{j\in J}\sum_{t\in T} cir^{t<\tau_r}_{mj} I^{t<\tau_r}_{mj} + \sum_{p\in P}\sum_{j\in J}\sum_{t\in T} cuf^t_{pj}B^t_{pj} + \\ & \sum_{p\in P}\sum_{j\in J}\sum_{l\in L}\sum_{t\in T} ct^t_{pjl}z^t_{pjl} + \sum_{p\in P}\sum_{l\in L}\sum_{t\in T} cid^{t<\tau_f}_{pl} I^{t<\tau_f}_{pl} + \sum_{p\in P}\sum_{l\in L}\sum_{t\in T} cdp^t_{pl}Q^t_{pl} + \\ & \sum_{p\in P}\sum_{j\in J}\sum_{t\in T} ctf^t_{pj}x^t_{pj} + \sum_{p\in P}\sum_{j\in J}\sum_{t\in T} crf^t_{pj}v^t_{pj} \end{aligned} \quad (1)$$

Constraints to be met are as follows.

$$\sum_{p\in P} r^t_{ipj}x^t_{pj} \le f^t_{ij} + u^t_{ij} \qquad \forall i \in M,\ \forall t \in T, t < t_r, \forall j \in J \quad (2)$$

Constraint (2) presents the amount of fish resources ι needed to produce fish product p which should have the same amount of available raw fish resources at time t, together with the additional raw resources needed. Note that the raw fish resources are within their shelf life t_r, and have passed the tracing processes.

$$x^t_{pj} \le Cy^t_{pj} \qquad \forall p \in P, j \in J, t \in T \quad (3)$$

Constraint (3) is to make sure that the production of all kind of fish products occur at the scheduled plant

$$u^j_{it} \le U^j_{it} \qquad \forall i \in M,\ \forall t \in T, \forall j \in J \quad (4)$$

Constraint (4) states that the additional raw resources have upper bound.

$$\sum_{p\in P} a_{pj}x^t_{pj} \le k^t_j \qquad \forall t \in T, \forall j \in J \quad (5)$$

The number of regular workers needed is expressed in Constraint (5).

$$I^t_{pj} = I^{t-1}_{pj} + \sum_{l\in L} Z^t_{plj} - D^t_{pj} \quad \forall p \in P, t \in T, l \in L, t < \tau_r \quad (6)$$

$$I^{t<\tau_r}_{pj} \le UI^{t<\tau_r}_{pj} \qquad \forall p \in P, t \in T, j \in J \quad (7)$$

$$I^{t<\tau_f}_{pl} \le UI^l_{nt<\tau_f} \qquad \forall p \in P, l \in L, t \in T \quad (8)$$

Constraints (6) to (8) present the inventories at the plant and distribution center. The shelf life are associated in those expressions.

$$k^j_t = k^j_{t-1} + k^{j+}_t - k^{j-}_t \quad t = 2,\dots T, \forall j \in J \quad (9)$$

Constraints (9) describe that the available workers in any time period equal to the number of the previous period plus any change in the number of worker type during the current period.

$$x^t_{pj} + B^{t-1}_{pj} + I^t_{pj} - B^t_{pj} = D^t_{pj} \quad \forall p \in P, \forall t \in T, \forall j \in J \quad (10)$$

Constraints (10) determine either the produced quantity to be stored in the inventory of the plant or to purchase from other company in order to fulfill the shortfall of meeting the demand.

$$\sum_{p\in P} Z^t_{pjl} \le \sum_{p\in P} UI^t_{pj} \qquad \forall l \in L, \forall j \in J, t \in T \tag{11}$$

Constraints (11) ensure that the amount of all kind seafood product from all plant can be stored in the distribution center inventories.

$$Z^t_{pjl} + Q^t_{pjl} \ge D^t_{pjl} \qquad \forall p \in P, l \in L, j \in J, t \in T \tag{12}$$

Constraints (12) states that the amount of delivery from each plant plus the unmet demand should be at least the same amount of market demand.

$$x^t_{pj}, u^t_{ij}, z^t_{pjl}, v^t_{pj}, I^{t<\tau_r}_{pj}, I^{t<\tau_f}_{pj}, B^t_{pj} \ge 0, \ \forall p \in P, \forall i \in M, \forall t \in T, \forall l \in L, \forall j \in J \tag{13}$$

$$k^j_t, k^{j+}_t, k^{j-}_j \ge 0 \ and \ \text{integer}, \quad \forall t \in \text{T}, \forall \text{j} \in \text{J} \tag{14}$$

$$Z^t_{pjl}, y^t_{pj} = \{0.1\} \qquad \forall p \in P, l \in L, j \in J, t \in T \tag{15}$$

Constraints (13) – (15) express the nature of the variables used in the model.

4. Conclusions

In this paper, we develop a large scale mixed integer linear programming model for the integrated of production scheduling and distribution planning problem of fish industry at Aceh Province, Indonesia with deterministic demand with considering traceability of the products. The model is adequate for solving the planning problem faced by the management of the industry. The model includes the computation of worker which is very useful for the industry in order they will be able to schedule a number of local people, and to conserve the quality of seafood.

5. References

[1] Hobbs J., Bailey, D., Haghiri, M., 2005. Traceability in the Canadian red meat sector: Do consumers care? *Canadian Journal of Agricultural Economics?Revue canadiened'agroeconomie*, 53(1), 47-65.

[2] Ramesh, B., & Jarke, M., 2001. *Towards Reference Models for Requirements Traceability*. IEEE Trans.on Software Engineering, vol. 27, no. 1, Jan, pp. 58-93.

[3] Cleland-Huang, J., Berenbach, B., Clark, S., Settimi, R., & Romanova, E., 2007. *Best Practices for Automated Traceability*. Computer, vol. 40, no. 6, pp. 9.

[4] Cohen, M. A., & Lee, H. L., 1988. *Strategic analysis of integrated production-distribution systems – models and methods.* Oper. Res. 36, 216–228.

[5] K. Bakhrankova, K. T. Midthun, K. T. Uggen, "Stochastic optimization of operational production planning for fisheries," Fisheries Reserarch, vol. 157, 2014, pp. 147-153.

[6] Timpe, C. H., & Kallrat, J., 2000. *Optimal planning in large multi-site production networks.* Eur. J. Oper. Res. 126, 422–435.

[7] Banaszewska, A., Cruijssen, F J. van der Vorst, G. Classen, J. Kampman, "A comprehansive dairy valorization model," J. Dairy Sci., 2012;96:761-779.

[8] Kopanos, G. M., Puigjaner, L., Georgiadis, M C, " Production scheduling in multi-product multistage semicontinuous food process," Ind. Eng. Chem. Res., 2011;50(10):6316-6324.

[9] Kopanos,G M., Puigjaner, L., Georgiadis, M C, "Efficient mathematical frameworks for detailed production scheduling in food processing industries. Computer Chemical Eng., 2012;42:206-21

[10] Bilgen, B., Celebi, Y, " Integrated production scheduling and distribution planning in dairy supply chain by hybrid modeling," Ann. Oper. Res., 2013;211(1):55-82.

[11] Armtzen, B. C., & Brown, G. G., Harrison, T. P., & Trafton, L. L., 1995. *Global supply chain management at digital equipment corporation.* Interfaces, Vol. 25, pp. 69-93.

[12] Jolayemi, J. K., & Olorunniwo, F. O., 2004. *A deterministic model for planning production quantities in a multi-plant, multi-warehouse environment with extensible capacities.* International Journal of Production Economics, Vo. 87, pp. 99-113.

[13] You, F., & Grossman, I. E., 2009. *Multi-echelon supply chain with inventory under uncertainty: MINLP models, computational strategies.* AICHE Journal, Vol. 56, Issue 2, pp. 419-440.

[14] Gajpal, Y., & Nourelfath, M., 2015. *Two efficient heuristics to solve integrated load distribution and production planning.* Reliability Engineering and System Safety, Vol. 144, pp. 204-214.

[15] Jose-Fernando Camacho-Vellejo, R. Munoz-Sanchez, J. L. Gonzalez-Velarde, "A heurstic algorithm for a supply chain's production-distribution planning," Computers & Operations Research, vol. 61, 2015, pp. 110-121

[16] Riemann, M., Neto, R. T., & Bogendorfer, E., 2014. *Joint optimization of production planning and vehicle routing problems: A review of existing strategies.* Pesquisa Operacional 34(2), pp. 189-214.

[17] Chen, Z. L., 2010. *Integrated production and outbound distribution scheduling: review and extensions.* Operations Research, vol. 58, no. 1, pp. 130-148.

[18] Fahimnia, B., Zanjirani Farahani, R., Marian, R., & Luong, L., 2013. *A review and critique on integrated production-distribution planning models and techniques.* Journal of Manufacturing Systems, vol. 32, no. 1, pp. 1-19.

[19] Seyedhosseini, S. M. & Ghoreyshi, S. M., 2014. *An integrated model for production and distribution planning of perishable products with inventory and routing considerations.* Mathematical Problems in Engineering, Vol. 2014, Article ID 475606.

[20] Bouchery, Y., Ghaffari, A., Jemai, Z., & Dallery, Y., 2012. *Including sustainability criteria into inventory models. European Journal of Operations Research,* 222, 229-240.

[21] Blackburn, J. and Scudder, G., 2009. *Supply Chain Strategies for Perishable Products: The Case of Fresh Produce.* Production and Operations Management 18(2): 129–137.

[22] Evans, A. 2009. *The Feeding of the Nine Billion – Global Food Security for the 21st Century.* The Royal Institute of International Affairs, Chatham House, London.

[23] Leat, P. & Revoredo-Giha, C. 2013. *Risk and resilience in agri-food supply chains: the case of the ASDA PorkLink supply chain in Scotland.* Supply Chain Management: An International Journal 18(2): 219-231.

[24] Nepstad, D. C., Stickler, C. M., & Almeida, O. T. 2006. Globalization of the amazon soy and beef industries: Opportunities for conservation. *Conservation Biology* 20(6): 1595-1603.

Optimal Control Inventory Stochastic With Production Deteriorating

Pardi Affandi
Departement of Mathematics FMIPA UNLAM Banjarbaru, Indonesia

p_affandi@unlam.ac.id

Abstract. In this paper, we are using optimal control approach to determine the optimal rate in production. Most of the inventory production models deal with a single item. First build the mathematical models inventory stochastic, in this model we also assume that the items are in the same store. The mathematical model of the problem inventory can be deterministic and stochastic models. In this research will be discussed how to model the stochastic as well as how to solve the inventory model using optimal control techniques. The main tool in the study problems for the necessary optimality conditions in the form of the Pontryagin maximum principle involves the Hamilton function. So we can have the optimal production rate in a production inventory system where items are subject deterioration.
Key words : Inventory production Problem, Stochastic inventory, Control Theory

1. Introduction

Various problems that involve many systems theory, optimal control and some applications. One of them is the inventory problem, the problem is how to manage the change in consumer demand in a finished product. The differential equation model is simplified by using exogenous function, then determined the optimal solution. The next model was developed solution for **t** values towards T. Then use solution with stochastic production deteriorating.

The goal is to determine the average production to minimize some cost function. Related inventory system of optimal control of a deteriorating in the quality of production formed the differential equation, to be determined optimal solutions in the form of the average production so as to minimize some cost function.

2. Theoretical Basic

2.1 The set of convex and convex function

The concept of a convex function underlie some part in the discussion section. The following definitions and theorems related to the set and convex function.

Definition 2.1.1 (Mangasarian) The set $\Gamma \subset \mathbb{R}^n$ called convex set if for to any $x_1, x_2 \in \Gamma$ and for $\lambda \in \mathbb{R}$ such that $0 \leq \lambda \leq 1$ *will apply* $(1-\lambda)(x_1) + \lambda(x_2) \in \Gamma$.

2.2 Nonhomogen Linear Differential Equation and Solution

Definition 2.1.2 (Ross, S.L 1984) Differential equations are equations containing derivatives of one or more dependent variables for one or more independent variables.

Definition 2.1.3 (Ross, S.L 1984) Order linear differential equations-n, with the dependent variable y, and the independent variable x, can be expressed as follows

$$a_0x\frac{d^ny}{dx^n}+a_1x\frac{d^{n-1}y}{dx^{n-1}}+\cdots+a_{n-1}x\frac{dy}{dx}+a_n(x)y=F(x) \qquad (2.1)$$

With a_0 not equal to zero. If F is equal to zero then the equation reduces to

$$a_0x\frac{d^ny}{dx^n}+a_1x\frac{d^{n-1}y}{dx^{n-1}}+\cdots+a_{n-1}x\frac{dy}{dx}+a_nxy=0 \qquad (2.2)$$

called homogeneous differential equation. For $F(x) \neq 0$, referred to as non homogeneous differential equation.

Theorem 2.1.4 (Ross, S.L 1984) Provided y_p a solution to nonhomogeneous linear differential equation (2.1) which does not contain any constants.

2.3 Stochastic Inventory Model

Many phenomena in nature that the occurrence of irregular both in space and time. To model the evolution of a system containing an uncertainty or a system that runs on an unpredictable environment, where the deterministic model is no longer suitable in use for analyze these systems use Stochastic Processes.

Stochastic process is a sequence of events that meet the laws of chance. Stochastic processes are widely used to model the evolution of a system that includes an uncertainty. Or a system that is operating in an unpredictable environment, where the deterministic model is no longer suitable to analyze the system. This model is suitable for the stated model of inventory on a company by the number of average number of requests in which an element of uncertainty. The element of uncertainty is a role model inventory. The company wants sufficient supply to meet customer demand, but production of too many hikes could increase the cost and risk of loss through obsolute or property damage.

2.4 Process Stochastic

The process of stochastic is the set of variables random $\{X(t), t \in T\}$ that describe for $t \in T, X(t)$ dynamic of a process. The process of stochastic is a collection random variable, which is a variable $X(t)$ with distribution limited. Generally, index or parameter shows the time t of the process and the variables randomly $X(t)$ show the status of the process at the T time. The set index time of the process called space time (*time space*) and the set of all values of variables random $X(t)$ which may be called the room status (*state space*). If $T = \{1,2,3,\ldots\}$ can be calculated then the stochastic is discrete and usually expressed with $X(t)$ where $T = \{t \mid t \geq 0\}$ as calculated, then the stochastic is have parameter continuous and is usually expressed by the notation $\{X(t), t \geq 0\}$. Like wise also the room status, room status is called discrete if finite or not finite can calculated and is called continuous if it contains the interval of the line real.

2.5 Optimal Control

In the following discussion, the problems given in the case of optimal control with state and the end time is known. In other words, the target set S shaped $\{S = x_1(t_1)\}$ in the form (x_1, t_1) with x_1 specially element in R^n and t_1 element at (T_1, T_2). Given the state system by the end and the end time unknown $\dot{x}(t) = f[x(t), u(t), t]$ with x(t) vector state sized $nx1$, u(t) input vector sized $mx1$, f a vector valued function. Initially given state is X_0 and initially time is t_0. Target set S form (x_1, t_1) with $t_1 \in (T_1, T_2)$ known value and $t_1 > t_0$. Optimal control problem is to find the admissible control u(t) with the initial value (x_0, t_0) and thefinal value (x_1, t_1) that maximizes the objective function

$J(u) = \int_{t_0}^{t_1} L(x(t), u(t), t)\, dt$. To solve the problems mentioned above optimal control, first determined necessary condition for optimal control are met.

3. Discussion of problems

Many company manufacturing using production inventory system are being made to regulate changes in consumer demand in a finished product. So the consumer and customer demand can be met, but also the production of goods is not in decline. The slump in production inventory may occur due to spoilage or damage to goods inventory of production for a certain period.

For that the company must make a plan so that products finished goods in the warehouse did not fit in decline before the booked by consumers and can manage inventories so as not to delay requests made by consumers. This has caused some problems, in addition to ware housing problems that certainly will add to the cost of storage also makes good planning so that the items are in storage until the decline. So the optimization problem that arises is how to balance between the level of production of goods in order to meet the demands of customers and consumers, but did not issue a production inventory storage costs and huge slump. Then some form of illustrative model used is a constant demand and linear request.

3.1. Matematic modelling

Estimated optimal state for linear systems with noise and noise measurement. Also the white noise process will be defined to obtain Kalman-Bucy filter for continuous time system. Next will be introduced the possibility of controlling the system governed by Ito in stochastic differential equations. In other words, variable controls will be introduced into nonlinear versions of the previous equation. This results in the formulation of stochastic optimal control problems. It should be noted that for such problems, the principle of separation is not generally applicable. Therefore, to simplify the treatment, it is often assumed that state variables can be observed, in the sense that state variables can be directly measured.

Furthermore, much of the literature on this issue uses dynamic programming or Hamilton-Jacobi-Bellman frameworks to determine the maximum stochastic principle. The following will be given the formulation of the stochastic optimal control problem by providing a brief, informal development of the Hamilton-Jacobi-Bellman (HJB) solution form for the solution of the problem equation.

3.2 Optimal Control of Stochastic Production Inventory with deteriorating

The following will introduce the notation and symbols used in the stochastic production inventory model. In this research used generalization of model obtained by Sethi and Thompson (2000). The assumption is used by assuming the plant produces a single homogeneous single and has a finished product in the warehouse. To illustrate the state of the following model in defining parameters and variables in the system, is as follows:

- Xt : inventory level at time t,
- Ut : production rate at time t,
- S (t) : the level of demand at time t,
- T : long period of planning,
- X : factory inventory level k,
- û : factory production objectives,
- x_0 : initial inventory level,
- h : cost coefficient Inventory storage,
- c : coefficient production cost,
- B : residual value per unit of inventory when T,
- z_t : standard Wiener process,
- σ : diffusion coefficient,
- Ø(t) : deteriorating.

The factory inventory level X is the safety stock level that the company wants to keep the amount of inventory safe in storage. Also, U-level factory production level is the most efficient level desired to run the plant. Using the above notation, you can describe the model condition. Equation

$$X(t) = u(t) - S(t) - \emptyset(t),\ x(0) = x_0, \tag{3.1}$$

First is the model flow equation, Which states that inventory levels at time t increase with production and decrease with demand. Then equation (3.1) can be generalized to a stochastic model. By making the inventory level assumption x_t follows the I_{to} Differential Stochastic equation as follows :

$$dX_t = (U_t - S - \emptyset)dt + \sigma dz_t,\quad X_0 = x_0, \tag{3.2}$$

By using process can be expressed as w (t) dt, where w (t) is as white noise process. This process can be interpreted as the sale of random returns or random decomposition in nature. In this model there is no limitation of production rate to be negative. This means that permitted disposal ($u_t < 0$). This states the conditions under which a disposal is not required. Furthermore the inventory level is allowed to become negative.

The solution of this problem will be done by using Hamilton-Jacobi modified The principle generated by the Hamilton-Jacobi-Bellman equation is called that satisfied with a certain value function. Now, the problem is to minimize the total expected cost as determined by the following functional

$$J(u) = E\left\{\int_0^T c(U_t - \hat{u})^2 + h(X_t - \hat{x})^2 dt + BXt\right\} \tag{3.3}$$

Assume that V (x, t) denotes the minimum value of the total expected cost from t to T with $X_t = x$ using the optimal policy from t to T. Then the function is given by:

$$V(x,t) = \min_{U_t} E\left\{\int_t^T c(U_t - \hat{u})^2 + h(X_t - \hat{x})^2 dt + BXt\right\} \tag{3.4}$$

Therefore the function value of $V(x,t)$ must satisfy the Hamilton-Jacobi-Bellman (HJB) equation

$$0 = \max_s\left\{-c(U_t - \hat{u})^2 - h(x - \hat{x})^2 + \frac{\partial V}{\partial t} + \frac{\partial V}{\partial x}(u - S) + \frac{1}{2}\sigma^2\frac{\partial^2 V}{\partial x^2}\right\} \tag{3.5}$$

With boundary conditions $V(x,t) = Bx$ (3.6)

Now to maximize the expression with respect to u by taking its derivative with respect to u and setting it to zero. Procedure of this result

$$\frac{\partial V}{\partial x} - 2c(u - \hat{u}) = 0 \tag{3.7}$$

Therefore, the optimal production rate that minimizes the total cost can be expressed as a function of the function by determining the value of the known equation at the moment in the form of:

$$u(x,t) = \frac{1}{2c}\frac{\partial V}{\partial x} + \hat{u} \tag{3.8}$$

With substitution (3.8) to (3.5) we obtain an equation

$$0 = -\left[\frac{1}{4c}\left(\frac{\partial V}{\partial x}\right)^2 + h(x - \hat{x})^2\right] + \frac{\partial V}{\partial t} + \frac{\partial V}{\partial x}\left[\frac{1}{2c}\left(\frac{\partial V}{\partial x}\right) + \hat{u} - S(t)\right] + \frac{1}{2}\sigma^2\frac{\partial^2 V}{\partial x^2} \tag{3.9}$$

To solve the nonlinear partial differential equation (3.9), let's use the assumption that the solution takes the following form

$$u(x,t) = max\left[0, \frac{\partial V}{\partial x} + \hat{u}\right] \tag{3.10}$$

3.3 The Hamilton-Jacobi-Bellman Equation Model Solution

In this section we will find solutions of stochastic production planning problems with different levels of demand. To solve the nonlinear partial differential equation (3.9), let's use the assumption that the solution takes the following form

$$V(x,t) = Q(t)x^2 + R(t)x + M(t) \tag{3.11}$$

Then $\frac{\partial V}{\partial x} = 2Qx + R$ (3.12)

$$\frac{\partial^2 V}{\partial x^2} = 2Q \tag{3.13}$$

$$\frac{\partial V}{\partial t} = \dot{Q}x^2 + \dot{R}x + \dot{M} \tag{3.14}$$

At which point denotes differentiation with time. Substituting from equation (3.9) to equation (3.11) will give results

$$\frac{1}{4c}[2Qx + R\,]^2 - h[x^2 - 2\hat{x}x + \hat{x}^2] + [\dot{Q}x^2 + \dot{R}x + \dot{M}] + [2Qx + R\ ](\hat{u} - S) + \sigma^2 Q = 0 \tag{3.15}$$

Equation (3.15) should apply to any value of x, we get the following system of ordinary nonlinear differential equations:

$$c\dot{Q} = ch - Q^2 \tag{3.16}$$

$$c\dot{R} + QR = -2ch\hat{x} - 2cQ(\bar{u} - S) \tag{3.17}$$

$$4c\dot{M} = 4ch\hat{x}^2 - R^2 - 4c(\bar{u} - S)R - 4c\sigma^2 Q \tag{3.18}$$

Furthermore, these nonlinear systems will be solved for different cases of demand level with

$$Q(T) = 0,\ R(T) = B,\ M(T) = 0. \tag{3.19}$$

By changing time t into:

$$\tau = e^2\sqrt{h/c}\,(t-T). \tag{3.20}$$

Then, the value will be obtained $\frac{\partial}{\partial t} = 2\sqrt{\frac{h}{c}}\,r\,\frac{\partial}{\partial r}$,

And a certain time will be obtained two different common cases for the level of demand

$\frac{h}{c} - e^{-2} = 0 \Rightarrow r = e^{-2}\sqrt{\frac{h}{c}}T$, and $t = T \Rightarrow r = 1$, Means will be obtained value

$$r \in \left[e^{-2}\sqrt{\frac{h}{c}}T, 1\right]. \tag{3.21}$$

To solve equation (3.16) can be used $\frac{c\dot{Q}}{(ch - Q^2)}$ So it will be obtained

$$Q(\tau) = \frac{\sqrt{ch}\,(\tau-1)}{(\tau+1)} \tag{3.22}$$

Next will be discussed two general cases to determine the average demand. The first case is the case where the constant demand level is S (t) = S_0 = const. In this case the optimal level of production is provided by:

$$u(x,t) = \frac{1}{c}\left[\left(\frac{\sqrt{ch}\,(\tau-1)}{(\tau+1)}\right)x - c(\bar{u} - S_0) + [B + 2c(\bar{u} - S_0)]\frac{\sqrt{\tau}}{(\tau+1)} - \hat{x}\sqrt{ch}\left(\frac{\tau-1}{\tau+1}\right)\right] + \bar{u} \tag{3.23}$$

The functions R (τ) and M (τ) will be obtained by:

$$R(\tau) = -c(\bar{u} - S_0) + 2[B + 2c(\bar{u} - S_0)]\frac{\sqrt{\tau}}{(\tau+1)} - 2\hat{x}\sqrt{ch}\left(\frac{\tau-1}{\tau+1}\right) \tag{3.24}$$

$$M(\tau) = \int\left\{\frac{1}{2\tau}\left[\sqrt{ch}\hat{x}^2 - \frac{R^2}{4\sqrt{ch}} - \sqrt{\frac{c}{h}}(\bar{u} - S_0)R - \sqrt{\frac{c}{h}}\sigma^2 Q\right]\right\}d\tau + M_0 \tag{3.25}$$

In order to find the expected level of inventory, the state equation (3.2) can be solved by taking the mean in relation to the state of the variable x and substituting it with the optimal production rate of (3.23) to obtain:

$$\dot{E}(x) = \frac{Q}{c}E(x) = \frac{1}{c}\left\{[B + 2c(\bar{u} - S_0)]\frac{\sqrt{\tau}}{(\tau+1)} - \hat{x}\sqrt{ch}\left(\frac{\tau-1}{\tau+1}\right)\right\}$$

$$E(x) = \hat{x} - \frac{B + 2c(\bar{u} - S_0)}{2\sqrt{ch}\sqrt{\tau}} + \left[x_0 - \hat{x} + \frac{B + 2\,(u - S_0)}{2\sqrt{ch}\sqrt{r_0}}\right]\frac{\sqrt{r_0}}{1 + r_0}\left(\frac{r+1}{\sqrt{r}}\right) \tag{3.26}$$

Some special cases can be obtained from this, for example when the level of demand. Constant and equal to U level production objectives. In this case the expected supply level will be provided by the following functions:

$$E(x) = \hat{x} + (x_0 - \hat{x})\left[\frac{\sqrt{r_0}\,(r+1)}{(1 + r_0)\sqrt{.}}\right] - \frac{B}{2\sqrt{ch}}\left[1 - \left(\frac{\tau \mp 1}{\tau_0 + 1}\right)\right]\frac{1}{\sqrt{r}} \tag{3.27}$$

So the total expectation cost of $E[V(x,T)]$ will be obtained in the form

$$E[V((x,T),T)] = B\left\{\hat{x} + (x_0 - \hat{x})\left[\frac{\sqrt{r_0}\,(r+1)}{(1 + r_0)\sqrt{r}}\right] - \frac{B}{2\sqrt{ch}}\left[\frac{1}{\sqrt{r}} - \left(\frac{(\tau+1)}{\tau_0 + 1(\sqrt{r})}\right)\right]\right\} \tag{3.28}$$

The second case is the case where the level of time request of the level varies, with the form of the obtained value being S (t) = u-2τ√τ / τ-1. The optimal level of production in this case is given by:

$$u(x,t) = \frac{1}{c}\left[\left(\frac{\sqrt{ch}\,(r-1)}{r+1}\right)x + \frac{(c+b)\sqrt{r}}{1+r} - \sqrt{\frac{‰}{h}}\left(\sqrt{ch}\,\frac{r\sqrt{r}}{1+r} + h\hat{x}\frac{(r-1)}{r+1}\right)\right] + \bar{u} \tag{3.29}$$

In the second case will be obtained, the function R (τ) will be given with

$$R(\tau) = \frac{2(c+b)\sqrt{r}}{\tau+1} - 2\sqrt{\frac{c}{h}}\left(\sqrt{ch}\,\frac{r\sqrt{r}}{1+r}\left(\frac{\tau-1}{\tau+1}\right)\right) \tag{3.30}$$

The following calculation values will be obtained

$$E(x) = \frac{1}{2}\sqrt{\frac{c}{h}}\left[\left(\frac{c+B}{c}\right)\frac{\sqrt{r}}{1+r}\ln\tau - 2\sqrt{\frac{c}{h}}\,\hat{x} + \frac{\sqrt{r}}{1+r}\ln(\tau-1)^2\right] + E_0\frac{\sqrt{r}}{1+r} \tag{3.31}$$

Where $E_0 = (x_0+\hat{x})\left(e^{\sqrt{\frac{h}{c}}T} + e^{-\sqrt{\frac{h}{c}}T}\right) - \left(\frac{c+B}{c}\right)T + \frac{1}{2}\sqrt{\frac{c}{h}}\ln\left(e - 2^{\sqrt{\frac{h}{c}}T} - 1\right)^2$ (3.32)

Then the inventory level of inventory will be provided with

$$E(x) = \frac{1}{2}\sqrt{\frac{c}{h}}\left[\left(\frac{c+B}{c}\right)\frac{\sqrt{r}}{1+r}\ln\tau - 2\sqrt{\frac{c}{h}}\,\hat{x} + \frac{\sqrt{r}}{1+r}\ln(\tau-1)^2 + \frac{\sqrt{r}}{1+r}\right](x_0+\hat{x})\left(e^{\sqrt{\frac{h}{c}}T} + e^{-\sqrt{\frac{h}{c}}T}\right) - \left(\frac{c+B}{c}\right)T + \frac{1}{2}\sqrt{\frac{c}{h}}\ln\left(e - 2^{\sqrt{\frac{h}{c}}T} - 1\right)^2 \tag{3.33}$$

4.Conclusions and Suggestions

From the discussion obtained some conclusions :

Inventory models of both raw material inventory and inventory deterministic and stochastic can be solved by control theory. Proper production arrangements are crucial in order to manage the quantities of the goods produced in a level that balances demand and risk and Optimal control can be used to complete the Inventory model with Stochastic Production Model. For future researchers, it is advisable to research the application of Theory of Control into other areas of research operations.

References

[1] Affandi, P., 2011. *Kendali Optimal system pergudangan dengan produksi yang mengalami kemerosotan*. Tesis, Yogyakarta.

[2] Affandi, P., 2015.*Optimal Inventory Control System With Stochastic Demand*. Ethar , Indonesia.

[3] Astrom, J. K. (1970). *Introduction to stochastic control theory, Academic Press*,New York.

[4] Bensoussan, A., Sethi, S. P., Vickson, R. G. and Derzko, N. (1984). Stochastic production planning with production constraints, SIAM J. Control and Optimization Vol. 22, No.6, 920-935.

[5] Burghes, D.N *Introduction to Control Theory Including Optimal Control* .John Wiley & Sons. New York.

[6] Davis, M. H. (1977). Linear estimation and stochastic control, John Wiley and Sons, New York.

[7] El-Gohary, A. and Bukhari, F. (2003). Optimal control of stochastic preyperdator models, Applied Mathematics and Computations, Vol. 146, 11, 403-415.

[8] El-Gohary, A., Tadj, L. and Al-Rahmah, A. (2006). Optimal control of a stochastic production planning model with different demand rates, International Journal of Applied Mathematics, (Accepted).

[9] Parlar,M. (1985). A stochastic production planning model with a dynamic chance constraint, European Journal of Operational Research, Vol. 20, 255-260.

[10] Perkins, J. R. and Kumar, P. R. (1994). Optimal control of pull manufacturing systems, IEEE Transactions on Automatic Control, 40:12, 2040-2051.

[11] Presman, L. E. Sethi, S. P., Zhang, H. and Bisi, A. (2001). Average cost optimal policy for a

IOP Conf. Series: Materials Science and Engineering **300** (2018) 012019 doi:10.1088/1757-899X/300/1/012019

stochastic two-machine flow shop with limited work-in-process, Nonlinear Analysis, Vol. 47, 5671-5678.

[12] Sagirow, P. (1972). Stochastic methods in the dynamics of satellites, Springer Verlag, New York.

[13] Shepley L.Ross.1984. *Differential Equation.* 3 Editions, John Wiley & Sons. New York.

[14] Shety S.P and Thompson G.L 1985. *Optimal Control Theory : Applied to Management science and economics*. 2 editions.

IOP Conf. Series: Materials Science and Engineering **300** (2018) 012020 doi:10.1088/1757-899X/300/1/012020

Combinatorial Optimization in Project Selection Using Genetic Algorithm

Sari Dewi[1], Sawaluddin[2]
[1,2]Department of Mathematics, Universitas Sumatera Utara, Indonesia

E-mail : [1]dewiq_ui@yahoo.com, [2]sawal@usu.ac.id

Abstract. This paper discusses the problem of project selection in the presence of two objective functions that maximize profit and minimize cost and the existence of some limitations is limited resources availability and time available so that there is need allocation of resources in each project. These resources are human resources, machine resources, raw material resources. This is treated as a consideration to not exceed the budget that has been determined. So that can be formulated mathematics for objective function (multi-objective) with boundaries that fulfilled. To assist the project selection process, a multi-objective combinatorial optimization approach is used to obtain an optimal solution for the selection of the right project. It then described a multi-objective method of genetic algorithm as one method of multi-objective combinatorial optimization approach to simplify the project selection process in a large scope.

1. Introduction

Project selection is the process of assessing each project idea and selecting the project with the highest priority [1]. The objective of the project selection process is to ensure that the selected project benefits optimally for enterprise resources. It also seeks to ensure that relevant investments are made. One of the project management tasks is to choose from a variety of projects that are better adapted to organizational goals [2]. Project management aims as a powerful tool for improving the ability to plan, execute and control activities and how to utilize resources and performance [3].

The uniqueness of the project requires different ways of handling so that the study of project selection activities grows as the development of project selection models to the development of decision-making systems to support such activities. Prior research uses the PMB (Packing-Multiple-Boxer) model or multi-knapsacks model to find solutions in the selection and scheduling of R & D projects [4], based on a simple 0-1 integer programming model, PMB models that can be used to select R & D projects and at the same time can create a selected project schedule. A comprehensive framework is proposed [5] for the project selection problem of uncertainty factors with apparent constraints.

The hybrid approach is used for project selection involving economic procedures and probabilities [6]. Multi objective combinatorial optimization is the process of discovery of one or more optimum solutions as a possible solution for finding the best maximum or minimum value of some conflicting objective (objective) or dual purpose objectives [7]. Techniques for solving multi-objective combinatorial optimization problems can be classified into appropriate estimation algorithms. Some of the research done in solving the problem of project selection with metaheuristic methods such as doing by means of two objective functions that is maximizing profit and maximizing limited resources using multi objective method Ant Colony [15], also using Genetic Algorithm [16] , Multi objective methods of Particel Swarm Optimization [8], and the multi-objective method of Genetic Algorithm [9]. This study considers several objectives of maximizing revenue, minimizing costs and risks that will occur.

2. Problem definition

A project is a complex enterprise with clear objectives, there are schedules, budgets, and consists of interrelated tasks performed by various organizational units [10]. Project selection involves simultaneous comparisons of a number of projects in order to arrive at the project's desired rating [11]. Projects with the lowest criteria are evaluated and then selected with available resources. In some constraints such as models or methods on project selection can be used to increase profits, select investments with limited capital resources, or enhance the competitive position of the organization. The project selection includes a large number of resource allocation methods and techniques [12].

Project management is the application of a set of tools and techniques to direct the use of resources such as time, cost and quality constraints to a unique, complex, and timely outcome [13]. Each job requires a combination of structured tools and techniques to fit the work environment and life cycle from conception to completion. A project is a human activity to achieve clear goals against time scale and achievement [14]. Project management has been developed to plan, coordinate and control complex and diverse project activities with regard to cost, time and quality as the criteria required to measure the project management process [15]. Project management aims as a powerful tool for improving the ability to plan, implement and control activities and how to use resources and performance [3].

3. Mathematical formulation

Suppose that project i $(i = 1, 2, \ldots, H)$will be selected in the period t $(t = 1, 2, \ldots, T)$ required for the selection of housing construction projects. Housing construction is carried out at several locations in the area. So that the housing construction process at each location in period t is not more than one project i will be implemented. This is considered based on the area of land available. Then the selected project i determination is adjusted to the location within period t. The housing development process involves several factors: b_{it} gain (profit), c_{it} cost, t period and limited resources. So the need for an appropriate allocation of resources to meet each project to be selected. The purpose of the allocation of such resources so that the budget or the cost of c_{it} issued in housing construction does not exceed the specified Q_{it} budget and less than the b_{it} gain gained due to limited resources. So the goals that are considered important for housing development can be realized smoothly that is to maximize profits and minimize costs to be incurred on the project. It is proposed to make it easier in determining the project to be selected.

In the project selection process with the planning horizon (period of planning) in period t means that each project has determined the allocation of available resources to meet the needs of each project with a predetermined range of time d_{it}. So it does not exceed the predetermined time limit because if a project i in period t is delayed or exceeds the time limit set then it will impact on all subsequent projects. The allocation of the cost of resources into consideration is the first human resource needs of the project i in which $j = 1, 2, \ldots, e$ both the needs of machine resources m_k on project i where $k = 1, 2, \ldots, s$ and third requirement of raw material ro ro at project i where o = 1, 2, .., v. So that limited resources can be overcome. The gain on project i in period t is denoted by bits.

3.1 Maximize profits

Benefits are a basic form of need that is useful to facilitate the process of activity. Because the benefits in a project have a positive impact on project selection involves the goal of maximizing profits. Then the total profit gained on project i to be implemented is defined in the following equation:

$$\max z_1(x) = \sum_{i=1}^{H}\sum_{t=1}^{T} b_{it}x_{it} \tag{1}$$

3.2. Minimize profits

Cost is the fund used for the project to be implemented. Funds spent on project execution involve many criteria so that the need for handling capable of solving funding problems so that the project implementation process can run in accordance with the plan. In this case the cost used in the form of cost on the resources that must be fulfilled the cost of human resource needs symbolized test, the cost of machine resource needs symbolized and cost of raw materials demand expressed as r_{oi}. Cost on project i in period t is denoted by c_{it}. Then the total cost obtained in the project i to be implemented is defined in the following equation:

$$\min z_2(x) = \sum_{t=1}^{T}\sum_{i=1}^{H} x_{it}\sum_{j=1}^{e} u_{ji}c_{jt} + \sum_{t=1}^{T}\sum_{i=1}^{H} x_{it}\sum_{k=1}^{s} m_{ki}c_{kt} + \sum_{t=1}^{T}\sum_{i=1}^{H} x_{it}\sum_{o=1}^{v} r_{oi}c_{ot} \tag{2}$$

s.t.

$$\sum_{t=1}^{T} x_{it} \le 1 \qquad i = 1,2,\dots H \tag{3}$$

$$\sum_{t=1}^{T} (t + d_{it})x_{it} \le T + 1 \qquad i = 1,2,\dots H \tag{4}$$

$$\sum_{i=1}^{H} u_{ji}x_{it} \le U_{jt} \qquad i = 1,2,\dots e; t = 1,2,\dots,T \tag{5}$$

$$\sum_{i=1}^{H} m_{ki}x_{it} \le M_{kt} \qquad i = 1,2,\dots s; t = 1,2,\dots,T \tag{6}$$

$$\left(\sum_{j=1}^{e} u_{ji} + \sum_{k=1}^{s} m_{ki} + \sum_{o=1}^{v} r_{oi}\right)x_{it} \le Q_{it} \qquad i = 1,2,\dots s; t = 1,2,\dots,T \tag{7}$$

$$\left(\sum_{j=1}^{e} u_{ji} + \sum_{k=1}^{s} m_{ki} + \sum_{o=1}^{v} r_{oi}\right)x_{it} \le b_{it} \qquad i = 1,2,\dots s; t = 1,2,\dots,T \tag{8}$$

$$x_{it} \in \{0,1\} \qquad i = 1,2,\dots s; t = 1,2,\dots,T \tag{9}$$

The objective function (1) is the equation formed for the purpose of maximizing profits by summing the total gains in project i period t. The objective function (2) is the equation formed for the purpose of minimizing the cost by summing all the total cost (total cost of human resource requirements of u_j, total cost of machine needs m_k and total cost of raw materials r_o) on project i period t. Limitation (3) ensures that all selected projects are exercised no later than one in period t. Limitation

(3) guarantees that all projects proposed to be selected have clear planning horizon (horizon planning). Limitation (4) guarantees that all human resource availability of the test on the project i is met according to the allocations established during the project selection process.

Limitation (5) guarantees that all available m_k machine resources on project i are met according to the allocations established during the project selection process. Limitation (6) guarantees that all available raw materials of r_o at project i are met according to the allocation established during the project selection process. Limitation (7) guarantees that all resource allocation of resources in project selection does not exceed the predetermined Qit budget. Limitation (8) ensures that all availability of the required cost does not exceed the profit to be earned 10. The decision variable (9) where 1 is the selected project i and 0 for the unelected project i.

4. Computational Analysis

This section will explain the technicalities of the problem to be solved by discussing a case. The parameters used for this example are presented in Table 1 with the size of the problem that is the problem of project selection on housing construction.

Table 1. Project costs and benefits (amounts in billions of Rupiah)

Alternative Project	Stage 1		Stage 2		Stage 3		Stage 4	
	Cost	Benefit	Cost	Benefit	Cost	Benefit	Cost	Benefit
1	2	3	1	4	7	8	4	6
2	5	6	3	5	8	10	7	10
3	6	8	8	10	10	12	8	12
4	8	10	10	12	0	0	9	14
5	10	13	12	14	0	0	10	15
6	13	15	0	0	0	0	0	0

In Table 1 can be seen about the objective function that becomes the reference that is the cost and profit for each project to be selected (calculated in billions of rupiah) with 4 stages and 6 alternative proposed project. The proposed budget is 20 billion rupiah. In the table presented has assumed the total cost of the total cost of resources to be used (human resources, raw material resources, and engine resources). The experiments were performed in 4 variations of iterations, ie 50, 100, 150 and 200 iterations.

Table 2. Projects selected with Genetic Algorithm

Generation	Vector	Cost	Result
20	1 2 2 2	20	28
100	2 2 2 1	20	27
150	1 1 3 2	20	29
200	1 1 3 2	20	29

For 50 Iteration experiments, 20 billion rupiahs were obtained and 28 billion rupiahs profit with alternative project ie location 1 project 1 (c11 = 2, b11 = 3), location 2 project 2 (c22 = 3, b22 = 5), location 3 project 2 (c23 = 8, b23 = 10), and location 4 project 2 (c24 = 7, b24 = 10). For the 100 Iteration experiments, 20 billion rupiahs and 27 billion rupiahs were earned with alternative project sites (location 1 project 2 (c21 = 5, b21 = 6), location 2 project 2 (c22 = 3, b22 = 5), location 3 project 2 (c23 = 8, b23 = 10), and location 4 project 1 (c14 = 4, b14 = 6).

For the Iteration 150 experiment, the cost of 20 billion rupiahs and profit 29 billion rupiahs with the alternative of location 1 project 1 (c11 = 2, b11 = 3), location 2 project 1 (c12 = 1, b12 = 4), location 3 project 3 (c33 = 10, b33 = 12), and location 4 project 2 (c24 = 7, b24 = 10). For the 200 Iteration experiments, 20 billion rupiahs and 28 billion rupiahs were earned with alternative projects: location 1 project 1 (c11 = 2, b11 = 3), location 2 project 2 (c22 = 3, b22 = 5), location 3 project 2 (c23 = 8, b23 = 10), and location 4 project 2 (c24 = 7, b24 = 10).

Then based on some tests that have been done with 50, 100, 150 and 200 iterations, Harmony search algorithm effectively produces the best solution at 150 and 200 iterations at a cost of 20 billion rupiahs and profit 29 billion rupiahs. Table 2 shows the project to be implemented at location 1 project 1, location 2 project 1, location 3 project 3 and location 4 project 2.

5. Conclusions

Selection of projects to meet the needs of organizations or organizations beyond the limits of resources and time available is difficult in decision making. Thus an organization or company must consider several objectives to meet those needs that are generally objective to the contrary (multi objective). This paper discusses the issue of project selection with the existence of two objective functions that maximize profit and minimize cost and the existence of some limitations are limited resources and available time so that resource allocation in each project is needed. These resources are human resources, machine resources, raw material resources. This is treated as a consideration to not exceed the budget that has been determined. So it can be formulated mathematical form for objective function (multi objective) with boundaries that are fulfilled. To assist the project selection process, the multi-objective combinatorial optimization approach is one way to obtain optimal solutions for selected projects. One of the methods used is Harmony Algorithm Search to simplify the decision-making process on project selection in a large scope. In this paper is given a case example to facilitate the selection of projects with the use of genetic algorithm method.

References:

[1] Momanyi, D. 2013. Effects of capital budgets on cash flow: a case study of Kenya power and lighting company limited. Instutional Repository of KCA University, Kenya.
[2] Meredith, J.R. dan Samuel, S.J. 2009. *Project Management: A managerial Approach, 3rd edm*. Willey & Sons, New York.
[3] Ghasemzadeh, F., Archer, N. dan Iyogun, P. 1999. A Zero-one model for project portfolio selection and scheduling. Operational Research Society. Vol 50, No 7, 745-755.
[4] Sun, H. dan Ma, T. 2005. *A Packing-multiple-boxes model for R&D Project selection and scheduling*. Technovation. Vol 25, No 11, 1355-1361.
[5] Shakhsi, N.M., Torabi, S.H., Iranmaresh. 2011. A Comprehensive framework for project selection problem under uncertainty and real-world Constraints. Computation. Vol 61, No 1, 226-237.
[6] Dutra, C.C., Ribeiro, J.L.D. dan Carvalho, M.M. 2014. An Economic-probabilistic model for project selection and prioritization. Project Management. Vol 32, No 6, 1042-1055.
[7] Bassuer, M., Talbi, E., Nerbo, A. dan Alba E. 2006. *Metaheuristics for Multiobjectif Combinatorial Optimization Problem*. Rapport De Recherche, French
[8] Rabbani, M., Bajestani, M.A. dan Khoshkhou, B.G. 2010. A Multiobjectve Particel Swarm Optimization for Project Selection Problem. Expert System with Application. Vol 37, No 1, 315-321.
[9] Bhattacharyya, R., Kumar, P. dan Kar S. 2011. Fuzzy R&D Portfolio Selection of Interdependent Projects. Computers and Matehmathic with Application. Vol 62, No 10, 3857-3870.
[10] Archibald, R.D. 1992. Managing high technology programs and projects (2nd). Wiley & Sons, NewYork.
[11] Archer, N.P. dan Ghasemzadeh, F. 1999. An integrated framework for project portfolio selection. Project Management. Vol 17, No 4, 207-216.
[12] Dobrovolskiene, N. dan Tamusiuniene, R. 2014. Resource Allocation in project port-folio management :practive in the Baltic states. KSI Transactions on Knowledge Society. Vol 7, No 1, 28-31.
[13] Oisen, P.R. 1971. Can project management be defined?. Project management quarterly. Vol 2, No 1, 12-14.
[14] Reiss, B. 1993. Project Management. E and FN, London.
[15] Atkinson, R.W. 1997. Effective Organisation, Re-Framing the Thinking for information System Project Success. Cassell, London.
[16] Gorbani, S. dan Rabbani, M. 2009. *A New Multiobjective Algorithm For A Project Selection Problem*. Advances in Engineering Software. Vol 40, No 1, 9-14

IOP Conf. Series: Materials Science and Engineering **300** (2018) 012021 doi:10.1088/1757-899X/300/1/012021

Implementation of pattern generation algorithm in forming Gilmore and Gomory model for two dimensional cutting stock problem

Sisca Octarina, Mutia Radiana, Putra B J Bangun

Department of Mathematics, Faculty of Mathematics and Natural Sciences, Sriwijaya University

e-mail address: s.octarina@gmail.com

Abstract. Two dimensional cutting stock problem (CSP) is a problem in determining the cutting pattern from a set of stock with standard length and width to fulfill the demand of items. Cutting patterns were determined in order to minimize the usage of stock. This research implemented pattern generation algorithm to formulate Gilmore and Gomory model of two dimensional CSP. The constraints of Gilmore and Gomory model was performed to assure the strips which cut in the first stage will be used in the second stage. Branch and Cut method was used to obtain the optimal solution. Based on the results, it found many patterns combination, if the optimal cutting patterns which correspond to the first stage were combined with the second stage.

1. Introduction

The printing industry is trying to increase production on a large scale by managing the use of raw materials (stock). The planning aims to determine the minimum amount of raw materials used to fulfill the demand of items. One way that can be done to minimize the use of raw materials is to determine the proper cutting, or in the field of Optimization known as Cutting Stock Problem (CSP).

There are many types of CSP, one-dimensional CSP, two-dimensional CSP, and three-dimensional CSP. Dimension size is determined by the side of the remaining cuts (trim loss).The CSP faced by the paper industry in this study was a two dimensional CSP, which in the stock cutting process focused on the long and wide sides.

Previous research on CSP has been widely practiced. Column generation technique (CGT) can be used in the completion of two-dimensional CSP with guillotine-cutting type and fixed orientation, integer solutions obtained by rounding down on the optimal solution [10]. The completion of two-dimensional 2-stage CSP using the Branch and Price and Gomory Cutting Plane methods has also been done previously [2] by adding Gomory Cutting Plane to linear programming relaxation. Then Mrad [9] has used the Branch and Price method to solve two-dimensional CSP models of type guillotine.

The effective cutting methods can be done by cutting paper from one side to the other side of the line or commonly called guillotine cutting [4-6]. Then Mellouli and Damak [8] implemented an algorithm for determining the cutting pattern called pattern generation (PG) algorithm [12] on multiple two-dimensional CSP using CGT. The application of PG algorithm in arc-flow model can facilitate the search of cutting patterns [1] in one dimensional CSP. For two-dimensional CSP, Octarina et al. [11] created an application for finding the cutting pattern based on the Modified Branch and Bound Algorithm.

Based on this background, this research used PG algorithm in forming Gilmore and Gomory model on two dimensional CSP. The model is extended from the CGT approach proposed by Gilmore and Gomory on a one dimensional CSP. The Gilmore and Gomory model is used to complete two-dimensional CSP on type guillotine and the optimal solution obtained by Branch and Cut method.

2. Research method

Steps conducted in this research are as follows.

1. Determine the data include products name, length and width, and also the demand.
2. Group the data that will be implemented on the PG algorithm in decreasing order.
3. Implement the PG algorithm on two-dimensional CSP to obtain the first cutting pattern and the second cutting pattern.
4. Create a searching tree by using the PG algorithm.
5. Form the cutting pattern table along with cut loss which is depicted from the search tree sequentially by branch.
6. Implement the first and second stage cutting patterns into Gilmore and Gomory models by:
   a. Defines the variables used.
   b. Determine the objective function that produces the minimum stock quantity to meet the demand of each item.
   c. Determine the constraints that ensure strip produced in the first stage is the strips used in the second stage, and will be cut into the requested items. Next define the constraints that ensure all requests item are met.
7. Find the Gilmore and Gomory model solution using the Branch and Cut method and use the LINDO 6.1 program in completing the relaxation LP model.
8. Conclude and obtain the optimal solution.

3. Pattern Generation Algorithm

Let the stock with standard width w'_k and length l'_k, $(k = 1, 2, \dots, h)$ will be cut to n items with the width w_i and length l_i $(i = 1, 2, \dots, n)$. The model of Cutting Stock Problem can be seen in (1).

$$\text{Minimize } z = \sum_{k=1}^{h}\sum_{j=1}^{m_k} c_{jk}x_{jk} + \sum_{i=1}^{n} w_i s_i \tag{1}$$

Subject to:

$$\sum_{k=1}^{h}\sum_{j=1}^{m_k} a_{ijk}x_{jk} - s_i = l_i \quad \text{for all } i, \tag{2}$$

$$x_{jk}, c_{jk}, s_i, a_{ijk} \geq 0 \quad \text{for all } i, j, k, \tag{3}$$

whereas

a_{ijk} is the number of item with width w_i which cut according to j^{th} cutting pattern from k^{th} stock $(i = 1, 2, \dots, n; j = 1, 2, \dots, m; k = 1, 2, \dots, h)$.

x_{jk} is the length of kth stock which cut according to j^{th} cutting pattern.

c_{jk} is the trim loss from k^{th} stock which cut according to j^{th} cutting pattern.

s_i is the remaining length which will produce item with width w_i.

A search tree is used to generate the feasible cutting patterns. The level of the tree represents the required width, which is arranged in decreasing order, the largest size is in the first level while the smallest size is placed at the last level of the tree. The initial vertex from first level is the standard width w'_k which use to generate the cutting pattern.

The branch from i^{th} level of search tree is the multiplication of the number of item and width w_i which represent the number of width item from k^{th} stock to meet w_i width. Vertex from 2nd level to n^{th} is the remaining width after fulfil the certain cutting pattern from the last $i-1$ branch. The last vertex from search tree is the trim loss from different cutting pattern. Search tree was built from up to down and left to right.

The steps for Pattern Generation (PG) algorithms are as follows:

1. Set the width size $w_i (i = 1,2, \dots, n)$ in decreasing order.
2. Use $a_{i1k} = \left\lfloor \dfrac{w'_k - \sum_{z=1}^{i-1} a_{z1k} w_z}{w_i} \right\rfloor$ to fill the first column ($j = 1$) in matrix.
3. Use $c_{jk} = w'_k - \sum_{i=1}^{n} a_{ijk} w_i$ to find the trim loss from each cutting pattern.
4. Set the level index (row index) i to $n-1$.
5. Check the vertex in i^{th} level, e.q. vertex (i,j). If the vertex has the value equals to zero $(a_{ijk} = 0)$, go to Step 7. If not, generate the new column $j = j + 1$ with the following elements:
   a. $a_{zjk} = a_{z(j-1)k}$ $(z = 1, 2, \dots, i-1)$ are elements to fill the vertex which precedes vertex (i,j).
   b. $a_{ijk} = a_{i(j-1)k} - 1$ are elements to fill vertex (i,j).
   c. Fill the remaining vertex from j column. For example, $a_{(i+1)jk}, a_{(i+2)jk}, \dots, a_{njk}$ using
   $$a_{ijk} = \left\lfloor \frac{w'_k - \sum_{z=1}^{i-1} a_{zjk} w_z}{w_i} \right\rfloor$$
6. Use $c_{jk} = w'_k - \sum_{i=1}^{n} a_{ijk} w_i$ to find cut loss from the jth cutting pattern. Go back to Step 4.
7. Set $i = i - 1$. If $i > 0$, repeat Step 5. Otherwise stop the iteration.

4. Result and Analysis

This section discusses the two dimensional CSP by using the same data from [11]. The name, size and number of each product can be seen in Table 1. The stock has the length and width with the size is 1,090 mm and 970 mm respectively.

Table 1. Products size and demand

No	Product Name (Item)	Length (mm)	Width (mm)	Number of order (pieces)
1	Invitation	325	225	300
2	Business Card	90	60	1,000
3	Pamphlets	210	150	3,000
4	Yasin/Book	230	160	500

PG algorithm was applied to find the cutting patterns. The cutting patterns corresponding to the width and length can be seen in Table 2 and 3 respectively.

Then all the cutting patterns are formed to Gilmore and Gomory model that can be seen in (4).

Minimize $z = \lambda_1^0 + \lambda_2^0 + \lambda_3^0 + \lambda_4^0 + \lambda_5^0 + \lambda_6^0 + \lambda_7^0$

with constraints: (4)

$\lambda_2^0 + 2\lambda_3^0 + 7\lambda_5^0 + 5\lambda_6^0 + 12\lambda_7^0 - \lambda_1^1 = 0$

$\lambda_1^0 + \lambda_2^0 + \lambda_3^0 + 3\lambda_4^0 + 3\lambda_6^0 - \lambda_1^2 - \lambda_2^2 - \lambda_3^2 = 0$

$\lambda_1^0 + 2\lambda_2^0 + 3\lambda_3^0 + 2\lambda_4^0 + 2\lambda_5^0 - \lambda_1^3 - \lambda_2^3 - \lambda_3^3 - \lambda_4^3 - \lambda_5^3 - \lambda_6^3 - \lambda_7^3 = 0$

$2\lambda_1^0 + \lambda_2^0 - \lambda_1^4 - \lambda_2^4 - \lambda_3^4 - \lambda_4^4 - \lambda_5^4 - \lambda_6^4 - \lambda_7^4 = 0$

$16\lambda_1^1 + 11\lambda_1^2 + 6\lambda_2^2 + \lambda_3^2 + 11\lambda_1^3 + 6\lambda_2^3 + \lambda_3^3 + 8\lambda_4^3 + 3\lambda_5^3 + 3\lambda_6^3 + 7\lambda_1^4 + 2\lambda_2^4 + 6\lambda_3^4 + \lambda_4^4 + 6\lambda_5^4 +$
$\lambda_6^4 + \lambda_7^4 \geq 1.000$
$2\lambda_1^2 + 4\lambda_2^2 + 6\lambda_3^2 + \lambda_1^3 + 3\lambda_2^3 + 5\lambda_3^3 + 2\lambda_5^3 + \lambda_6^3 + 2\lambda_2^4 + \lambda_3^4 + 3\lambda_4^4 + 2\lambda_6^4 \geq 3.000$
$\lambda_1^3 + \lambda_2^3 + \lambda_3^3 + 3\lambda_4^3 + 3\lambda_5^3 + 4\lambda_6^3 + 6\lambda_7^3 + 2\lambda_1^4 + 2\lambda_2^4 + \lambda_5^4 + \lambda_6^4 \geq 500$
$\lambda_1^4 + \lambda_2^4 + 2\lambda_3^4 + 2\lambda_4^4 + 2\lambda_5^4 + 2\lambda_6^4 + 4\lambda_7^4 \geq 300$
$\bar{\lambda} \geq 0$ and *integer*
with $\bar{\lambda} = (\lambda_1^0, \ldots, \lambda_j^0, \lambda_1^1, \ldots, \lambda_j^1, \lambda_1^2, \ldots, \lambda_j^2, \ldots, \lambda_1^{m'}, \ldots, \lambda_j^{m'})^T$

Table 2. Cutting patterns corresponding to the width

The	Number of items				*Cut Loss*
j^{th}pattern	225 mm	160 mm	150 mm	60 mm	(mm)
1	4	0	0	1	10
2	2	1	2	1	0
3	2	1	0	6	0
4	2	0	3	1	10
5	2	0	1	6	10
6	1	2	2	2	5
7	1	2	0	7	5
8	0	6	0	0	10
9	0	4	1	3	0
10	0	3	2	3	10
11	0	3	0	8	10
12	0	1	5	1	0
13	0	1	3	6	0
14	0	1	1	11	0
15	0	0	6	1	10
16	0	0	4	6	10
17	0	0	2	11	10
18	0	0	0	16	10

Table 3. Cutting patterns corresponding to the length

The	Number of items				*Cut Loss*
j^{th}pattern	325 mm	230 mm	210 mm	90 mm	(mm)
1	2	1	1	0	0
2	1	2	1	1	5
3	0	3	1	2	10
4	0	2	3	0	0
5	0	2	0	7	0
6	0	0	3	5	10
7	0	0	0	12	10

Next the model in (4) can be rewritten in Table 4. Optimal solution of Gilmore and Gomory model showed the value of $z = 142$ with $\lambda_1^0 = 38$, $\lambda_4^0 = 99$, $\lambda_6^0 = 5$, $\lambda_1^1 = 25$, $\lambda_1^2 = 3$, $\lambda_3^2 = 347$, $\lambda_3^3 = 182$, $\lambda_4^3 = 1$ $\lambda_7^3 = 53$, $\lambda_3^4 = 2$, and $\lambda_7^4 = 74$. It means the 1st, 4th, and 6th cutting patterns can be used in the first stage, whereas the cutting patterns that can be used in the second stage are as follows.

1. The 1st cutting pattern will be used in the strip with length of 90 mm.
2. The 1st and 3rd cutting patterns will be used in the strip with length of 210 mm.
3. The 3rd , 4th, and 7th cutting patterns will be used in the strip with length of 230 mm, and
4. The 3rd and 7th cutting patterns will be used in the strip with length of 325 mm.

IOP Conf. Series: Materials Science and Engineering **300** (2018) 012021 doi:10.1088/1757-899X/300/1/012021

Table 4. Gilmore and Gomory Model

(l_i, w_i)	λ_1^0	λ_2^0	λ_3^0	λ_4^0	λ_5^0	λ_6^0	λ_7^0	λ_1^1	λ_1^2	λ_2^2	λ_3^2	λ_1^3	λ_2^3	λ_3^3	λ_4^3	λ_5^3	λ_6^3	λ_7^3	λ_1^4	λ_2^4	λ_3^4	λ_4^4	λ_5^4	λ_6^4	λ_7^4	
(90, 60)	0	1	2	0	7	5	12	−1																		= 0
(210, 150)	1	1	1	3	0	3	0		−1	−1	−1															= 0
(230, 160)	1	2	3	2	2	0	0					−1	−1	−1	−1	−1	−1	−1								= 0
(325, 225)	2	1	0	0	0	0	0												−1	−1	−1	−1	−1	−1	−1	= 0
(90, 60)								16	11	6	1	11	6	1	8	3	3	0	7	2	6	1	6	1	1	≥ 1.000
(210, 150)								0	2	4	6	1	3	5	0	2	1	0	0	2	1	3	0	2	0	≥ 3.000
(230, 160)								0	0	0	0	1	1	1	3	3	4	6	2	2	0	0	1	1	0	≥ 500
(325, 225)								0	0	0	0	0	0	0	0	0	0	0	1	1	2	2	2	2	4	≥ 300
Objective function	1	1	1	1	1	1	1																			

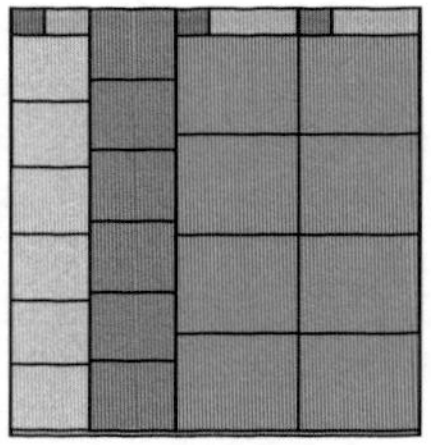
Figure 1. First Combination

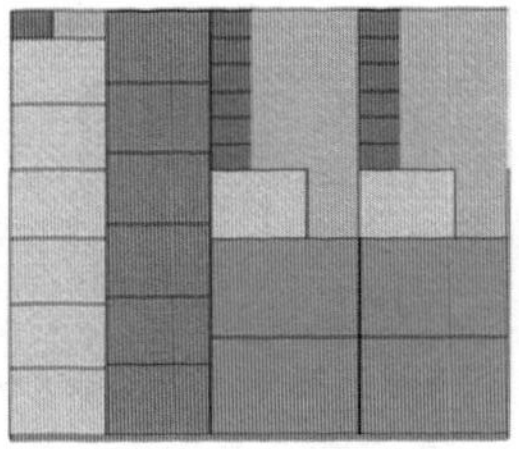
Figure 2. Second Combination

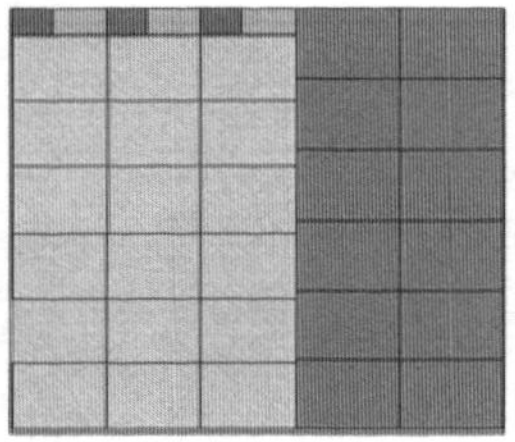
Figure 3. Third Combination

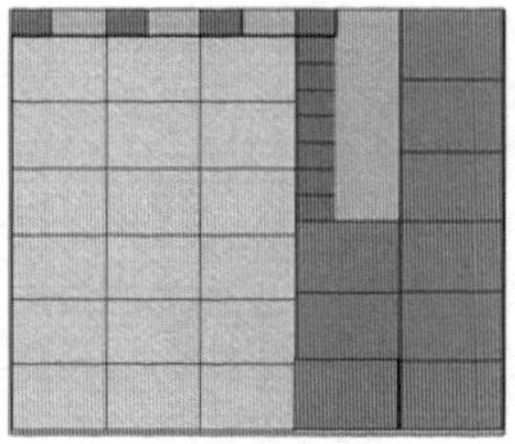
Figure 4. Fourth Combination

Based on the implementation of pattern generation algorithm on cutting pattern determination, it found 57 cutting patterns corresponding to the width and 34 cutting patterns corresponding to the length. The cutting pattern chosen is the pattern that has the minimum cut loss and at the same time can meet the availability of items ordered, so the selected pattern is a pattern with a cut loss of no more than 10 millimeters as many as 18 cutting patterns corresponding to the width and 7 cutting patterns according to the length.

The selected cutting patterns, formed into the Gilmore and Gomory model and completed using Branch and Cut method and LINDO 6.1 program. The addition of Gomory restrictions in this study did not adequately affect the solution obtained due to the large number of variables. The number of items resulting from stock cuts are 142 sheets based on the optimal cutting pattern. The optimal solution obtained is substituted into Gilmore and Gomory model constraints, resulting in 1,056 sheets of business cards, 3,000 pamphlets, 503 pieces of Yasin / books, and 300 pieces of invitations. Each item produced is not less than the request, the resulting excess amount can be used for the next order.

The optimal solution from the first and second stage can be combined to the new cutting patterns, as shown in Figures 1-7.

Figure 5. Fifth Combination

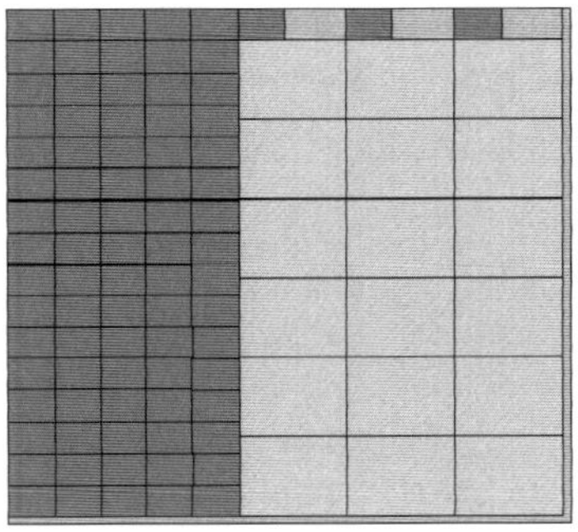

Figure 6. Sixth Combination

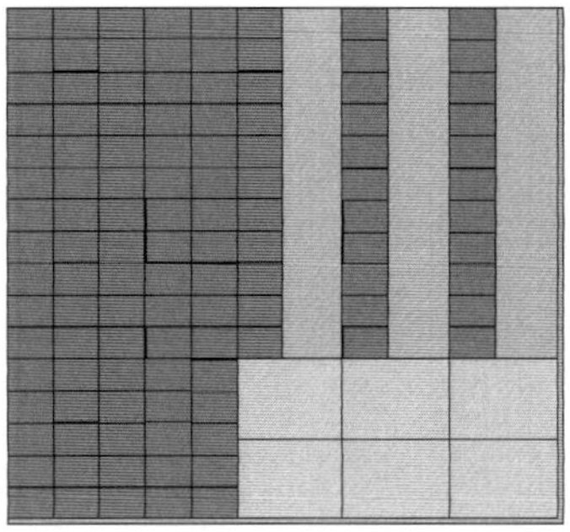

Figure 7. Seventh Combination

Note: means trim loss; means pamphlets; means invitations; means business card; means Yasin/books

Figures 1-7, shows that there are 7 combinations of cutting patterns that can be used. Trim loss from each cutting patterns can be used again to produce the other items that fixed the size.

5. Conclusion

From the result, it can be concluded that implementation of PG algorithm helps the searching of cutting patterns become easier. Then, the patterns can be modelled as Gilmore and Gomory model whereas the objective function is to minimize the number of cutting stock and the constraints assure that the strips from first stage can be reused in the second stage in order to meet all the demands.

Acknowledgement

The research leading to this study was financially supported by Directorate of Higher Education Indonesia (DIKTI) for support through "Penelitian Produk Terapan", 2017.

References

[1] Bangun P B J, Octarina S, and Apriani R 2016 Penyelesaian algoritma *pattern generation* dengan model *arc-flow* pada *cutting stock problem* (CSP) satu dimensi *Proceeding Annual Research Seminar* Fakultas Ilmu Komputer Universitas Sriwijaya

[2] Belov G and Scheithauer G 2004 A branch and cut and price algorithm for one dimensional stock cutting and two dimensional two stage cutting *European Journal of Operational Research* **171**(1) 85-106

[3] Chen D S, Batson R G, and Dang Y 2010 *Applied Integer Programming Modeling and Solution* (New Jersey: John Wiley & Sons)

[4] Gilmore P and Gomory R 1961 A linear programming approach to the cutting stock problem *Operations Research* **9**(6) 849 - 859

[5] Gilmore P and Gomory R 1963 A linear programming approach to the cutting stock problem II *Operations Research* **11**(6) 863 - 888

[6] Gilmore P and Gomory R 1965 Multistage cutting stock problems of two and more dimension *Operations Research* **13**(1): 94 – 120

[7] Macedo R A S G 2011 Models and algorithms for hard optimization problems *Thesis* Universidade do Minho Braga

[8] Mellouli A and Dammak A 2008 An algorithm for the two dimensional cutting stock problem based on a pattern generation procedure *Information and Management Science* **19**(2): 201 – 218

[9] Mrad M, Meftahi I, and Haouari M 2012 A branch-and-price algorithm for the two stage guillotine cutting stock problem *Journal of the Operational Research Society*, 1-9

[10] Novianingsih K, Hadianti R, and Uttunggadewa S 2007 Column generation technique for solving two dimensional cutting stock problems: method of stripe approach *Journal of The Indonesian Mathematical Society* (MIHMI) **13**(2), 161-172

[11] Octarina S, Bangun P B J, and Hutapea S 2016 The application to find cutting pattern in two dimensional cutting stock problem *Accepted and to be published in Journal of Informatics and Mathematical Science*

[12] Suliman S M A 2001 Pattern generating procedure for the cutting stock problem *International Journal of Production Economics*, **74** : 293-301

An Improved Search Approach for Solving Non-Convex Mixed-Integer Non Linear Programming Problems

Joni Wilson Sitopu[1], Herman Mawengkang[2], Riri Syafitri Lubis[1]
[1]Graduate School of Mathematics, University of Sumatera Utara, Medan, Indonesia
[2] Department of Mathematics, University of Sumatera Utara, Medan, Indonesia

Email: hmawengkang@yahoo.com

Abstract. The nonlinear mathematical programming problem addressed in this paper has a structure characterized by a subset of variables restricted to assume discrete values, which are linear and separable from the continuous variables. The strategy of releasing nonbasic variables from their bounds, combined with the "active constraint" method, has been developed. This strategy is used to force the appropriate non-integer basic variables to move to their neighbourhood integer points. Successful implementation of these algorithms was achieved on various test problems.

Keywords: Nonlinear programming, active constraints, direct search, global optimization

1. Introduction

There are a lot of optimization problem involved integer or discrete variables which occur naturally in a nonlinear function. Mixed Integer Nonlinear Programming (MINLP) refers to these kind of optimization problems with continuous and discrete variables and nonlinearities in the objective function and constraints. This paper concerns with a special class of MINLP problem in which the discrete variables are linear and separable from the continuous variables.

This problem is defined by the following model.

$$\textit{Minimize} \quad f(x, y) \tag{1}$$

$$\text{s.t.} \quad g_i(x, y) \le b_i, \quad i = 1, \ldots, m \tag{2}$$

$$x \in X, \ \ y \in Y \tag{3}$$

where f and g_i , $i = 1, \ldots, m$, are real-valued function on $\mathbb{R}^{n+q}$, $X \subseteq \mathbb{R}^n$, and Y is a continuous subset in $\mathbb{R}^q$.

There are many diverse and important applications of MINLPs. A small subset of these applications includes portfolio optimization [1], [2], block layout design in the manufacturing and service sectors [3], network design with queuing delay constraints [4], integrated design and control of chemical processes [5], drinking water distribution systems security [6]. Energy system [7], [8], and multiperiod supply chain problems [9].

The needs in such diverse areas have motivated research and development in MINLP solver technology, particularly in algorithms for handling large-scale, highly combinatorial and highly nonlinear problems.

Methods for solving MINLPs include innovative approaches and related techniques taken and extended from MIP, such as, Outer Approximation (OA) methods [10, 11], Branch-and-Bound (B&B) [12, 13, 24], Integrating Successive Quadratic Programming (SQP) and B&B [15], Generalized Bender's Decomposition (GBD) [16], and Branch and Cut [17] for solving MINLPs have been discussed in the literature since the early 1980's. These approaches generally rely on the successive solutions of closely related NLP problems. For example, B&B starts out forming a pure continuous NLP problem by dropping the integrality requirements of the discrete variables (often called the relaxed MINLP or RMINLP). Moreover, each node of the emerging B&B tree represents a solution of the RMINLP with adjusted bounds on the discrete variables. A survey regarding to solving a nonconvex MINLP can be found in [18].

Whenever the MINLP problem become too large to be solved by exact method, most of researchers move to heuristic method. Heuristic approaches to solving MINLPs include Variable Neighbourhood Search [19], automatically tuned variable fixing strategies [20], feasible neighbourhood search [21], Feasibility Pump [23, 27], heuristics based on Iterative Rounding [25]. [19] propose a MINLP heuristic called the Relaxed-Exact-Continuous-Integer Problem Exploration (RECIPE) algorithm. The algorithm puts together a global search phase based on Variable Neighbourhood Search [19] and a local search phase based on a MINLP heuristic. Masmoudi et al. (017) propose a hybrid of genetic algorithm for solving MINLP. In heuristic approaches, however, one of the main algorithmic difficulties connected to MINLPs is to find a feasible solution.

Due to the fact that the functions in MINLPs are not smooth, therefore in this paper we use a direct search method, known as unconstrained optimization techniques that do not explicitly use derivatives. More information regarding to direct search method in optimization can be found in (Vayasis, 1991).

In this paper we address a strategy of releasing nonbasic variables from their bounds, combined with the "active constrained" method and the notion of superbasic for efficiently tackling a particular class of MINLP problems.

The rest of this paper is organized as follows. In Section 2 we give a brief notion of neighbourhood search. The basic approach of the proposed method is presented in Section 3. How to derive the proposed method is given in Section 4. The algorithm is presented in Section 5. Section 6 addresses a computational experience. The conclusions can be found in Section 7.

2. Neighbourhood Search

It should be noted that, generally, in integer programming the reduced gradient vector, which is normally used to detect an optimality condition, is not available, even though the problems are convex. Thus we need to impose a certain condition for the local testing search procedure in order to assure that we have obtained the "best" suboptimal integer feasible solution.

Scarf [18] has proposed a quantity test to replace the pricing test for optimality in the integer programming problem. The test is conducted by a search through the neighbors of a proposed feasible point to see whether a nearby point is also feasible and yields an improvement to the objective function.

Let $[\beta]_k$ be an integer point belongs to a finite set of neighborhood $N([\beta]_{k})$. We define a neighbourhood system associated with $[\beta]_k$, that is, if such an integer point satisfies the following two requirements

1. If$[\beta]_j \in N([\beta]_k)$ then $[\beta]_k \in [\beta]_j), j \neq k$.
2. $N([\beta]_k) = [\beta]_k + N(0)$

With respect to the neighborhood system mentioned above, the proposed integerizing strategy can be described as follows.

Given a non-integer component, x_k, of an optimal vector , x_B. The adjacent points of x_k, being considered are $[x_k]$ and $[x_k] + 1$. If one of these points satisfies the constraints and yields a minimum deterioration of the optimal objective value we move to another component, if not we have integer-feasible solution.

Let $[x_k]$ be the integer feasible point which satisfies the above conditions. We could then say if $[x_k] + 1 \in N([x_k])$ implies that the point $[x_k] + 1$ is either infeasible or yields an inferior value to the objective function obtained with respect to $[x_k]$. In this case $[x_k]$ is said to be an "optimal" integer feasible solution to the integer programming problem. Obviously, in our case, a neigbourhood search is conducted through proposed feasible points such that the integer feasible solution would be at the least distance from the optimal continuous solution.

3. Solving Nonlinear Programming Problem

It has been our experience that many nonlinear programming problems are inordinately large because they are attempting to approximate, by piecewise linearization, what is essentially a nonlinear problem. It also appears that many real-life problems are such that only a small percentage of the variables are involved nonlinearly in the objective function. Therefore we should be able to write a class of nonlinear programming problems with the following form:

$$\text{minimize} \quad F(x) = f(x^N) + c^T x, \tag{4}$$
$$\text{subject to} \quad Ax = b, \tag{5}$$
$$\boldsymbol{l} \leq \boldsymbol{x} \leq \boldsymbol{u} \tag{6}$$

where A is $m \times n$, $m \leq n$. Due to the separable property, w could partition $\boldsymbol{x}$ into a linear portion $\boldsymbol{x}^L$ and a nonlinear portion $\boldsymbol{x}^N$ **.** The partition can be expressed as

$$\boldsymbol{x} = \begin{bmatrix} \boldsymbol{x}^N \\ \boldsymbol{x}^L \end{bmatrix}$$

where $\boldsymbol{x}^N$ are called the *nonlinear variables.* Note that A and $\boldsymbol{c}$ operate on all variables $\boldsymbol{x}$. In some cases the part of $\boldsymbol{c}^T\boldsymbol{x}$ involving $\boldsymbol{x}^N$ may be incorporated into $f(\boldsymbol{x}^N)$; in other cases **c** may be zero. We assume that the function $f(\boldsymbol{x}^N)$ is continuously differentiable in the feasible region, with gradient

$$\nabla f(\boldsymbol{x}^N) = \boldsymbol{g}(\boldsymbol{x}^N),$$

and we assume that both $\boldsymbol{f}$ and $\boldsymbol{g}$ can be computed at any feasible point $\boldsymbol{x}^N$.

In essence the algorithm is an extension of the revised simplex method . To use some of the associated terminology, it might be described as an extension which permits more than m variables to be basic.

Partitioning $\boldsymbol{x}$ and $\boldsymbol{F}(\boldsymbol{x})$ into linear and nonlinear terms is of considerable practical importance; for descriptive purposes, however, it is convenient to denote $\boldsymbol{F}(\boldsymbol{x})$ and $\nabla \boldsymbol{F}(\boldsymbol{x})$ simply by $f(\boldsymbol{x})$ and $\boldsymbol{g}(\boldsymbol{x})$.

With a few conventional exceptions, we use upper-case letters for matrices, lower-case for vectors and Greek lower-case for scalars. The quantity $\varepsilon > 0$ represents the precision of floating-point arithmetic.

4. Basic of The Method

Basic solutions; justification for standard form

Before proceeding to the nonlinear problem, we need to introduce some linear programming background. In particular, equations (1) – (3) with $f(\boldsymbol{x}^N) = 0$ are the standard form for stating linear programs and for solving them in practical implementations of the simplex method. ***A*** *basic solution* is characterized by having at most m "basic" variables lying between their bounds while the remaining $n - m$ "nonbasic" variables are equal to one bound or other. An associated square basis matrix $\boldsymbol{B}$ is drawn from the columns of the constraint matrix $\boldsymbol{A}$, and as the simplex method proceeds the columns of $\boldsymbol{B}$ are replaced one at a time.

Note that it is assumed in the standard form that A and x contain a full identity matrix and a full set of slack variables, respectively. (General equality and inequality constraints are accommodated by placing suitable upper and lower bounds on the slacks.) There are many practical reasons for retaining the standard form here. Full justification would require much implementation background, but very briefly, a sparse triangular factorization of $\boldsymbol{B}$ can be maintained more easily if columns (but not rows) of $\boldsymbol{B}$ are altered. Further, although the total number of variables is apparently greater, it is very easy to take advantage of the unit vectors associated with slacks, whenever $\boldsymbol{B}$ is re-factorized.

A. *Superbasic variables*

One virtue of the concept of basic solutions is the emphasis thereby given to the upper and lower bounds, $\boldsymbol{l} \leq \boldsymbol{x} \leq \boldsymbol{u}$. It is misleading to regard these as "sparse constraints"; more importantly, they serve directly to eliminate a large percentage of the variables. The simplex method is therefore free to focus its attention on transforming (factorizing) just B, rather than the whole of A.

With nonlinear problems we cannot expect an optimal point to be a basic solution. However, if the number of nonlinear variables is small, it seems reasonable to suppose that an optimal solution will be "nearly" basic. Thus, as a simple generalization we introduce the notion of *superbasic* variables and partition the set of general constraints (2) as follows:

$$\boldsymbol{Ax} = \begin{array}{c} \begin{array}{ccc} m & s & n-m-s \end{array} \\ \left[\begin{array}{c|c|c} \boldsymbol{B} & \boldsymbol{S} & \boldsymbol{N} \end{array}\right] \\ \begin{array}{ccc} \text{basics} & \text{super-basics} & \text{nonbasics} \end{array} \end{array} \begin{bmatrix} \boldsymbol{x}_B \\ \boldsymbol{x}_S \\ \boldsymbol{x}_N \end{bmatrix} = \boldsymbol{b} \tag{7}$$

The matrix B is square and nonsingular as in the simplex method, S is $m \times s$ with $0 \leq s \leq n-m$, and N is the remaining columns of A. The associated variables $\boldsymbol{x}_B$, $\boldsymbol{x}_s$, $\boldsymbol{x}_N$ are called the *basics, superbasics* and *nonbasics* respectively. Both basics and superbasics are free to vary between their bounds. The name is chosen to highlight the superbasics' role as "driving force"; they may be moved in any direction at all (preferably one that improves the objective value), and the basics are then obliged to change in a definite way to maintain feasibility with respect to the constraints $\boldsymbol{Ax} = \boldsymbol{b}$.

Our expectation that solutions will often be "nearly" basic is confirmed by the following theorem:

Theorem 1. *Suppose a nonlinear program has t variables occurring nonlinearly (in either the objective or the constraints). An optimal solution exists in which the number of superbasic variables s satisfies* $s \leq t$.

Derivation of method

We will assume that $f(x)$ can be expanded in a Taylor's series with remainder of second order:

$$f(\boldsymbol{x} + \Delta\boldsymbol{x}) = f(x) + g(\boldsymbol{x})^T \Delta\boldsymbol{x} + \tfrac{1}{2} \Delta\boldsymbol{x}^T G(\boldsymbol{x} + \gamma \Delta x)\, \Delta\boldsymbol{x} \tag{8}$$

where $0 \leq \gamma \leq 1$, and $G(\boldsymbol{x} + \gamma\Delta\boldsymbol{x})$ is the Hessian matrix of second partial derivatives evaluated at some point between $\boldsymbol{x}$ and $\boldsymbol{x} + \Delta\boldsymbol{x}$. Note that G is a constant matrix if $f(x)$ is a quadratic function.

Let us partition $\Delta\boldsymbol{x}$ and $\boldsymbol{g}(\boldsymbol{x})$ corresponding to the partitioning of A.

If $f(x)$ were truly quadratic, we could obtain a constrained stationary point at $\boldsymbol{x} + \Delta\boldsymbol{x}$ by requiring two properties of the step $\Delta\boldsymbol{x}$:

Property 1.

$$\begin{bmatrix} B & S & N \\ 0 & 0 & I \end{bmatrix} \begin{bmatrix} \Delta\boldsymbol{x}_B \\ \Delta\boldsymbol{x}_S \\ \Delta\boldsymbol{x}_N \end{bmatrix} = 0 \tag{9}$$

i.e., the step remains on the surface given by the intersection of the active constraints.

Property 2.

$$\begin{bmatrix} g_B \\ g_S \\ g_N \end{bmatrix} + G \begin{bmatrix} \Delta \boldsymbol{x}_B \\ \Delta \boldsymbol{x}_S \\ \Delta \boldsymbol{x}_N \end{bmatrix} = \begin{bmatrix} B^T & 0 \\ S^T & 0 \\ N^T & I \end{bmatrix} \begin{bmatrix} \mu \\ \lambda \end{bmatrix} \tag{10}$$

i.e., the gradient at $\boldsymbol{x} + \Delta\boldsymbol{x}$ (given by the left hand-side of (10)) is orthogonal to the surface of active constraints and is therefore expressible as a linear combination of the active constraint normals.

For a more general function $f(x)$, the step $\Delta\boldsymbol{x}$ may not lead directly to a stationary point, but we shall use Properties 1 and 2 to determine a feasible descent direction.

From (9) we have:

$$\Delta\boldsymbol{x}^{\mathrm{N}} = 0, \tag{11}$$

and

$$\Delta\boldsymbol{x}_{\mathbf{B}} = -W\Delta\boldsymbol{x}_s, \tag{12}$$

where

$$W = B^{-1}s. \tag{13}$$

Thus,

$$\Delta\boldsymbol{x} = \begin{bmatrix} -W \\ I \\ 0 \\ \end{bmatrix} \Delta\boldsymbol{x}_s$$

The following matrix can be used to simplify Eq. (10)

$$\begin{bmatrix} I & 0 & 0 \\ -W^T & I & 0 \\ 0 & 0 & I \end{bmatrix} \tag{14}$$

First it provides an expression for estimates of the Lagrange multipliers for the general constraints:

$$B^T \mu = g_B + \begin{bmatrix} I & 0 & 0 \end{bmatrix} G \begin{bmatrix} -W \\ I \\ 0 \end{bmatrix} \Delta\boldsymbol{x}_s \tag{15}$$

Note that when $\|\Delta\boldsymbol{x}_s\| = 0$ (which will mean $\boldsymbol{x}$ is stationary) we have

$$B^T \mu = g_B \tag{16}$$

As we observed that μ is analogous to the pricing vector π in the revised simplex method. (From now on we shall denote the solution of (16) by π.) Next we have from (10) that

$$\boldsymbol{\lambda} = g_N - N^T \mu + \begin{bmatrix} 0 & 0 & I \end{bmatrix} G \begin{bmatrix} -W \\ I \\ 0 \end{bmatrix} \Delta\boldsymbol{x}_s \tag{17}$$

and again when $\|\Delta\boldsymbol{x}_s\| = 0$ this equation reduces to

$$\boldsymbol{\lambda} = g_N - N^T \pi \tag{18}$$

which is analogous to the vector of reduced costs in linear programming.

The third result from equation (10), following pre-multiplication by the matrix (11), is an expression for the appropriate step:

$$\begin{bmatrix} -W^T & I & 0 \end{bmatrix} G \begin{bmatrix} -W \\ I \\ 0 \end{bmatrix} \Delta \boldsymbol{x}_s = -\boldsymbol{h} \tag{19}$$

where

$$\boldsymbol{h} = \begin{bmatrix} -W^T & I & 0 \end{bmatrix} \boldsymbol{g} = \boldsymbol{g}_s - W^T \boldsymbol{g}_B = \boldsymbol{g}_s - S^T \pi \tag{20}$$

The form of equation (16) suggests that

$$\begin{bmatrix} -W^T & I & 0 \end{bmatrix} G \begin{bmatrix} -W \\ I \\ 0 \end{bmatrix} \tag{21}$$

can be regarded as a "reduced" Hessian and $\boldsymbol{h} = \begin{bmatrix} -W^T & I & 0 \end{bmatrix} \boldsymbol{g}$ a reduced gradient, with (19) giving a Newton step in the independent variables $\Delta \boldsymbol{x}_s$. Note that $\|\boldsymbol{h}\| = 0$ becomes a necessary condition for a stationary point on the current set of active constraints, which, if the reduced Hessian is nonsingular, implies that $\|\Delta \boldsymbol{x}_s\| = 0$.

Summary

We considered a class of algorithms in which the search direction along the surface of active constraints is characterized as being in the range of a matrix **Z** which is orthogonal to the matrix of constraint normals. Thus, if $\hat{A}\boldsymbol{x} = \hat{\boldsymbol{b}}$ is the current set of $n - s$ active constraints, Z is an $n \times s$ matrix such that

$$\hat{A}Z = 0. \tag{22}$$

The main steps to be performed at each iteration are as follows.

(A) Compute the reduced gradient $\boldsymbol{g}_A = Z^T\boldsymbol{g}$

(B) Form some approximation to the reduced Hessian, viz.

$$G_A \doteqdot Z^T G Z$$

(C) Obtain an approximate solution to the system of equations

$$Z^T G Z \boldsymbol{p}_A = -Z^T \boldsymbol{g} \tag{23}$$

by solving the system

$$G_A \boldsymbol{p}_A = -g_A.$$

(D) Compute the search direction $\boldsymbol{p} = Z\boldsymbol{p}_A$.

(E) Perform a iinesearch to find an approximation to a^*, where

$$f(\boldsymbol{x} + \alpha^* \boldsymbol{p}) = \min_{\substack{\alpha \\ \{x+\alpha p \text{ feasibel}\}}} f(\boldsymbol{x} + \alpha \boldsymbol{p})$$

Apart from having full column rank, eq. (22) is (algebraically) the only constraint on Z and thus Z may take several forms. The particular Z corresponding to our own procedure is of the form

$$Z = \begin{bmatrix} -W \\ I \\ 0 \end{bmatrix} = \begin{bmatrix} -b^{-1}S \\ I \\ 0 \end{bmatrix} \begin{matrix} \}m \\ \}s \\ \}n-m-s \end{matrix} \tag{24}$$

This is a convenient representation which we will refer to for exposition purposes in later sections, but we emphasize that computationally we work only with S and a triangular (LU) factorization of B. The matrix Z itself is never computed.

5. The Algorithm

After solving the relaxed problem, the procedure for searching a suboptimal but integer-feasible solution from an optimal continuous solution can be described as follows.

Let

$$x = [x] + f, \quad 0 \le f \le 1$$

be the (continuous) solution of the relaxed problem, $[x]$ is the integer component of non-integer variable x and f is the fractional component.

Stage 1.

Step 1. Get row i^* the smallest integer infeasibility, such that $\delta_{i^*} = \min\{f_i, 1 - f_i\}$

Step 2. Calculate

$v_{i^*}^T = e_{i^*}^T B^{-1}$

this is a pricing operation

Step 3. Calculate $\sigma_{ij} = v_{i^*}^T a_j$

With j corresponds to

$$\min_j \left\{ \left| \frac{d_j}{\sigma_{ij}} \right| \right\}$$

I. *For nonbasic j at lower bound*

If $\sigma_{ij} < 0$ and $\delta_{i^*} = f_i$ calculate $\Delta = \frac{(1-\delta_{i^*})}{-\sigma_{ij}}$

If $\sigma_{ij} > 0$ and $\delta_{i^*} = 1 - f_i$ calculate $\Delta = \frac{(1-\delta_{i^*})}{\sigma_{ij}}$

If $\sigma_{ij} < 0$ and $\delta_{i^*} = 1 - f_i$ calculate $\Delta = \frac{\delta_{i^*}}{-\sigma_{ij}}$

If $\sigma_{ij} > 0$ and $\delta_{i^*} = f_i$ calculate $\Delta = \frac{\delta_{i^*}}{\sigma_{ij}}$

II. *For nonbasic j at upper bound*

If $\sigma_{ij} < 0$ and $\delta_{i^*} = 1 - f_i$ calculate $\Delta = \frac{(1-\delta_{i^*})}{-\sigma_{ij}}$

If $\sigma_{ij} > 0$ and $\delta_{i^*} = f_i$ calculate $\Delta = \frac{(1-\delta_{i^*})}{\sigma_{ij}}$

If $\sigma_{ij} > 0$ and $\delta_{i^*} = 1 - f_i$ calculate $\Delta = \frac{\delta_{i^*}}{\sigma_{ij}}$

If $\sigma_{ij} < 0$ and $\delta_{i^*} = f_i$ calculate $\Delta = \frac{\delta_{i^*}}{-\sigma_{ij}}$

Otherwise go to next non-integer nonbasic or superbasic j (if available). Eventually the column j^* is to be increased form LB or decreased from UB. If none go to next i^*.

Step 4. Calculate

$\alpha_{j^*} = B^{-1} \alpha_{j^*}$

i.e. solve $B\alpha_{j^*} = \alpha_{j^*}$ for α_{j^*}

Step 5. Ratio test; there would be three possibilities for the basic variables in order to stay feasible due to the releasing of nonbasic j^* from its bounds.

If j^* lower bound

Let

$$A = \min_{i' \neq i^* | \alpha_{ij^*} > 0} \left\{ \frac{x_{B_{i'}} - l_{i'}}{\alpha_{ij^*}} \right\}$$

$$B = \min_{i' \neq i^* | \alpha_{ij^*} < 0} \left\{ \frac{u_{i'} - x_{B_{i'}}}{-\alpha_{ij^*}} \right\}$$

$$C = \Delta$$

the maximum movement of j^* depends on: $\theta^* = \min(A, B, C)$

If j^* upper bound

Let

$$A' = \min_{i' \neq i^* | \alpha_{ij^*} < 0} \left\{ \frac{x_{B_{i'}} - l_{i'}}{\alpha_{ij^*}} \right\}$$
$$B' = \min_{i' \neq i^* | \alpha_{ij^*} > 0} \left\{ \frac{u_{i'} - x_{B_{i'}}}{-\alpha_{ij^*}} \right\}$$
$$C' = \Delta$$

The maximum movement of j^* depends on: $\theta^* = \min(A', B', C')$

Step 6. Exchanging basis for the three possibilities

1. If *A* or *A'*
   - $x_{B_{i'}}$ becomes nonbasic at lower bound $l_{i'}$
   - x_{j^*} becomes basic (replaces $x_{B_{i'}}$)
   - x_{i^*} stays basic (non-integer)
2. If *B* or *B'*
   - $x_{B_{i'}}$ becomes nonbasic at upper bound $u_{i'}$
   - x_{j^*} becomes basic (replaces $x_{B_{i'}}$)
   - x_{i^*} stays basic (non-integer)
3. If *C* or *C'*
   - x_{j^*} becomes basic (replaces x_{i^*})
   - x_{i^*} becomes superbasic at integer-valued

Step 7. If row $i^* = \{\square\}$ go to Stage 2, otherwise
Repeat from step 1.

Stage 2. Do integer lines search to improve the integer feasible solution

6. Computational Experience: A Plant Upgrading Problem

A plant upgrading problem, cited in Daellenbach et al. (1983), describes a decision problem faced by a manufacturer of an electrical wire in order to upgrade its machinery. At present some of the machines are nearly worn out or obsolete, so that running costs are becoming unacceptably high. A shift in technology has caused an expansion in demand for the type of wire this manufacturer produces. However, the present plant cannot meet the demand projected for the near future.

There are two types of wire produced by this manufacturer: bare wire and covered wire. Each of these two products has two different sizes. In terms of the machines availability, the manufacturer has various options, including retaining and modifying some existing machines and purchasing new machines. Therefore it is desirable to draw a recommended configuration of the plant to meet the projected demand as well as to schedule which machine should produce each type of product such that the total running cost is minimized.

7. Mathematical Statement of the Problem

Let y_i be the number of machines of options i, and x_{ij} be the proportion of annual machines time on machines of options i to produce wire size *j*, $(i = 1,\ldots,5 \text{ and } j = 1,2)$.

The problem can be formulated mathematically as follows.

The objective of the problem is to minimize the total running cost (in \$1000's)

$$\text{Minimize } P = 30y_1 + 33.6x_{11} + 32.88x_{12} + 50y_2 + 47.4x_{21}$$
$$+47.04x_{22} + 80y_3 + 58.8x_{31} + 57x_{32} + 100y_4$$
$$+62.4x_{41} + 54.7x_{42} + 140y_5 + 86.4x_{51} + 82.8x_{52}$$

There are several constraints have to be satisfied. Firstly is to meet the projected sales demand for each type and size of wire.

For bare wire of size 1 and 2

$$5880x_{11} + 8820x_{21} - 7200x_{31} - 9600x_{41} \geq 3000 \quad \text{(BARE1)}$$

$$4704x_{12} + 8232x_{22} - 6000x_{32} - 7800x_{42} \geq 2000 \text{ (BARE2)}$$

For covered wire of size 1 and 2

$$6984x_{31} + 9312x_{41} + 9312x_{51} \geq 14000 \text{ (COVRD1)}$$

$$5820x_{32} + 7566x_{42} + 6984x_{52} \geq 10000 \text{ (COVRD2)}$$

Next is to make certain that the number of machines available for each option provides a production capacity that covers the production needs. Hence, the number of machine of type i is required for the production of all wire sizes.

$$y_i \geq x_{i1} + x_{i2},\ i = 1,\ldots,5 \text{ (MATCH}i\text{)}$$

The restrictions of the machine options

$$y_1 \leq 2 \text{ (REST1)}$$

$$y_3 + y_4 \leq 1 \text{ (REST2)}$$

8. Solution to the Problem

The plant upgrading problem mentioned in the previous section is a MILP problem. It has 11 constraints, 10 continuous variables and 5 integer variables.

The relaxed problem is solved on PC optimization software. We then integerized the continuous result using our proposed algorithm. The continuous results as well as the integer results of the plant upgrading problem are presented in Table 1. The total computational time for both continuous and integer results is 2.5 seconds.

Table 1. Results for Plant Upgrading Problem

Variable	Activity in Cont. Soln.	Activity After Integ. Process
x_{11}	0.5102	0.51020
x_{12}	0.0	0.42517
x_{21}	0.0	0.0
x_{22}	0.24295	0.0
x_{31}	0.0	0.0
x_{32}	0.0	0.0
x_{41}	0.0	0.0
x_{42}	0.0	0.0
x_{51}	1.50344	1.50344
x_{52}	1.43184	1.43184
y_1	0.5102	1.0
y_2	0.24295	0.0
y_3	0.0	0.0
y_4	0.0	0.0
y_5	2.93528	3.0
Obj. Value (P)	715.41817	729.57606

9. Conclusions

This paper has presented a direct search method for achieving integer-feasibility for a class of mixed-integer nonlinear programming problems in a relatively short time. The direct search approach used the strategy of releasing nonbasic variable from their bounds, combined with the "active constraint" method and the notion of superbasic. After solving a problem by ignoring the integrality requirements, this strategy is used to force the appropriate non-integer basic variables to move to their neighborhoods integer points.

A study of the criteria for choosing a nonbasic variable to work with in the integerizing strategy has also been made. The number of integerizing steps would be finite if the number of integer variables contained in the problem is finite. However, it should be noted that the computational time for the integerizing process does not necessarily depend on the number of integer variables, since many of the integer variables may have an integer value at the continuous optimal solution.

The new direct search method has been shown to be successful on a range of problems, while not always able to achieve global optimality. In a number of cases to obtain the suboptimal point is acceptable, since the exponential complexity of the combinatorial problems in general precludes branch-and-bound, except on small to medium problems.

Computational testing of the procedure presented this paper has demonstrated that it is a promised approach for large problems.

10. References

[1] Bienstock, D., 1996. *Computational study of a family of mixed-integer quadraticprogramming problems, Mathematical Programming.* **74**, pp. 121-140.

[2] Jobst, N. J., Horniman, M. D., Lucas, C. A., & Mitra, G., 2001. *Computational aspects of alternative portfolio selection models in the presence of discrete asset choice constraints, Quantitative Finance* **1**, pp. 489-501.

[3] Danna, E., Rothberg, E., & LePape, C., 2005. *Exploring relaxation induced neighborhoods to improve MIP solutions, Mathematical Programming.* **102**, pp. 71-90.

[4] Sawaya, N., 2006. R*eformulations, relaxations and cutting planes for generalized disjunctive programming.* PhD thesis, Chemical Engineering Department, Carnegie Mellon University.

[5] Flores-Tlacuahuac, A., & Biegler, L. T., 2007. *Simultaneous mixed-integer dynamic optimization for integrated design and control, Computers and Chemical Engineering.* **31**, pp. 648–656.

[6] Laird, C. D., Biegler, L. T., & Van Bloemen Waanders, B., 2006. *A mixed integer approach for obtaining unique solutions in source inversion of drinking water networks, Journal of Water Resources Planning and Management, Special Issue on Drinking Water Distribution Systems Security.* **132**, pp. 242-251.

[7] Kim, J. S., & Edgar, T. F., 2014. *Optimal scheduling of combined heat and powerplants using mixed-integer nonlinear programming.* Energy, **77,** pp. 675-690.

[8] Yang, Y., Zhang, S., & Xiao, Y., 2015. *An MILP (mixed integer linear programming) model for optimal design of district scale distributed energy resource system.* Energy, 90, pp. 1901-1915.

[9] Kaya, O., Urek, B., 2016. *A mixed integer nonlinear model and heuristic solutions for location, inventory and pricing decisions in a closed loop supply chain.* Competers & Operations Research, **65,** pp. 93-103.

[10] Duran, M. A., & Grossmann, I. E., 1986. *An Outer-Approximation Algorithm For a class of Mixed-Integer Nonlinear Programs.* Mathematical Programming **36,** pp. 307.

[11] Roger Fletcher & Sven Leyffer, 1994. *Solving Mixed Integer Nonlinear Programming by Outer Approximation.* Mathematical Programming, **66,** pp. 327.

[12] Belotti, P., Lee, J., Liberti, L., & Margot, F., 2009. *A. Wachter, Branching and Bounds Tightening Techniques for non-convex MINLP.* Optimization Methods and Software, **24(4)**, pp. 597-634.

[13] Gupta, O. K., & Ravindran, A., 1985. *Branch and Bound Experiments in Convex Nonlinear Integer Programming.* Manage Sci., **31 (12)**, pp. 1533-1546 (1985).

[14] Nannicini, G., & Belotti, P., 2009. *Local Branching for MINPs.* Technical Report workingpaper, CMU.

[15] Leyffer, S., 2001. *Integrating SQP and Branch and Bound for mixed integer nonlinear programming.* Computational Optimization and Applications, **18**, pp. 295-309.

[16] Geoffrion, A. M., 1972. *A Generalized Benders Decomposition.* J. Op-tim. Theory Appl., **10 (4),** pp. 237-260.

[17] Vigerske, S., & Gleixner, A., 2017. *SCIP: Global optimization of mixed integer nonlinear programs in a branch and cut framework.* ZIB Report, pp. 16-24 (Revised version May 2017).

[18] Burer, S., & Letchford, A., 2012. *Non-convex mixed-integer nonlinear programming: A survey, Surv. Oper. Res.* Manag. Sci., **17**, pp. 97-106.

[19] Liberti, L., Nannicini, G., & Mladenovic, N., 2009. *A good recipe for solving MINLPs.* In: V. Maniezzo, T. Stutzle, S.Voss(eds.) Metaheuristics: Hybridizing metaheuristics and Mathematical Programming. Annals of Information Systems, **10**, pp. 231-245. Springer.

[20] Berthold, T., & Gleixner, A., 2010. *Undercover-Primal MINLP Heuristic.* In P. Bonami, L. Liberti, A. Miller and A. Sartenaer (eds.) Proceedings of the European Workshop on MINLP, pp 103-113, Marseille.

[21] Mawengkang, H., & Murtagh, B. A., 1986. *Solving Nonlinear Integer Programs with Large-Scale Optimization Software.* Annals of Operations Research, pp. 5425-437.

[22] D'Ambrosio, C., Frangioni, A., Liberti, L., & Lodi, A., 2010. *A storm of Feasibility Pump for non-convex MINLP.* Tech. Rep. OR-10-13, DEIS, Universita di Bologna.

[23] Fischetti, M., Glover, F., & Lodi, A., 2005. *The Feasibility Pump.* Mathematical Programming A **104(1),** 91-104.

[24] Nannicini, G., Belotti, P., 2009. *Rounding-based heuristics for non-convex MINLPs.* In: P. Bonami, L. Liberti, A. Miller, A. Sartenear (eds). Proceedings of the European Workshop on MINLP. CIRM, Marseille, France.

[25] Liberti, L., Miladenovie, N., & Nannicini, G., 2011. *A Recipe for Finding Good Solutions to MINLPs.* Mathematical Programming Computation, August 2011.

[26] Vavasis, S. A., 1991. *Nonlinear Optimization: Complexity Issues.* Oxford University Press, Oxford.

[27] D'Ambrosio, C., Frangioni, A., Liberti, L., & Lodi, A., 2010. *Experiments with a Feasibility Pump approachfor non-convex MINLPs.* In: P. Festa(ed) Proceedings of the 9th Synposiumon Experimental Algorithms (SEA 2010), Lecture Notes in Computer Science, **6049.** Springer, Berlin.

[28] Scarf, H. E., 1986. *Testing for Optimality in the Absence of Convexity.* In: W. P. Heller, R. M Starr and D. A. Starett (Eds) Cambridge University Press, pp. 117-134

[29] Tamara G. Kolda., Robert M. Lewis, Virginia Torezon, 2003. *Optimization by Direct Search: New Perspective on Some Classical and Modern Methods.* SIAM Review, **45 (1)**, pp. 385-482.

[30] Westerlund, T., & Petersson, F., 1995. *A Cutting Plane Method for Solving Convex MINLP Problems.* Computers Chem. Eng., **19,** pp. 131-136.

IOP Conf. Series: Materials Science and Engineering **300** (2018) 012023 doi:10.1088/1757-899X/300/1/012023

An Integer Programming Model For Solving Heterogeneous Vehicle Routing Problem With Hard Time Window considering Service Choice

Enny Susilawati[1], Herman Mawengkang[2], Syahril Efendi[3]
[1]Graduate School of Mathematics, University of Sumatera Utara, Medan, Indonesia
[2]Department of Mathematics, University of Sumatera Utara, Medan, Indonesia
[3]Faculty of Computer Science and Information Technology, University of Sumatera Utara, Medan, Indonesia

Email: hmawengkang@yahoo.com

Abstract. Generally a Vehicle Routing Problem with time windows (VRPTW) can be defined as a problem to determine the optimal set of routes used by a fleet of vehicles to serve a given set of customers with service time restrictions; the objective is to minimize the total travel cost (related to the travel times or distances) and operational cost (related to the number of vehicles used). In this paper we address a variant of the VRPTW in which the fleet of vehicle is heterogenic due to the different size of demand from customers. The problem, called Heterogeneous VRP (HVRP) also includes service levels. We use integer programming model to describe the problem. A feasible neighbourhood approach is proposed to solve the model.

Keywords: Vehicle routing problem, Heterogenic fleet, time restricted, integer programming, feasible neighbourhood search

1. Introduction

Vehicle Routing Problem (VRP) is a key element of many distribution systems which involve routing and scheduling of vehicles through a set of nodes. This is a well known combinatorial optimization problem which requires the determination of an optimal set of routes used by a fleet of vehicles to serve a set of customers, taking into account various operational constraints.VRP was first introduced by Dantzig and Ramser (1959). Since then many researchers have been working in this area to discover new methodologies in selecting the best routes in order to find the better solutions. Studies on this NP-hard problem have resulted in several exact and heuristic techniques of general applicability (Cordeau et al., 2007; El-Sherbeny, 2010; Yeun et al., 2008; Cordeau et al., 2001). Drexl (2011) presents a compact review of vehicle routing literature.

The VRPTW is a generalisation of VRP. It can be reviewed as a combined vehicle routing and scheduling problem which often arises in many real-world applications. It is to optimise the use of a fleet of vehicles that must make a number of stops to serve a set of customers, and to specify which customers should be served by each vehicle and in what order to minimise the cost, subject to vehicle capacity and service time restrictions (Ellabib et al. 2002). The problem involves assignment of vehicles to trips such that the assignment cost and the corresponding routing cost are minimal.

In terms of graph the VRPTW can be defined as follows: Let $G = (V, E)$ be a connected digraph consisting of a set of n + 1 nodes, each of which can be reached only within a specified time interval or time window, and a set E of arcs with non-negative weights representing travel distances and associated travel times. Let one of the nodes be designated as the depot. Each node i, except the depot, requests a service of size q_i.

The VRPTW has been the subject of intensive research efforts for both heuristic and exact optimisation approaches. Early surveys of solution techniques for the VRPTW can be found in Golden and Assad (1986), Desrochers et al. (1990). The main focus in Desrosiers et al. (1995) and Cordeau et al. (2001) were exact solution techniques. Further details on these exact methods can be found in Larsen (1999) and Cook and Rich (1999). Because of the high complexity level of the VRPTW and its wide applicability to real-life situations, solution techniques capable of producing highquality solutions in limited time, i.e. heuristics, are the most popular. Shi et al. (2012) propose a modified artificial bee colony algorithm to solve VRPTW. A hybrid metaheuristic based on Large Neighbourhood Search approach and Variable Neighbourhood Search technique was proposed by Rincon-Garcia et al. (2017) to tackle the VRPTW.

The company focused in this paper has various type of vehicle in its operation to deliver goods for customers. Each type of vehicle has different capacities. The variant of VRP which considers mixed fleet of vehicles is called Heterogeneous VRP (HVRP), introduced firstly by Golden et al. (1984). This generalization is important in practical terms, for most of customers demand are served by several type of vehicles (Hoff et al., 2010; Koc, 2016).The objective of the HVRP is to find fleet composition and a corresponding routing plan that minimizes the total cost.

As this is a combinatorial problem it is not surprising that the most approach addressed for solving the HVRP is heuristics. Liu and Shen (1999) were the first to tackle the HVRPTW and developed a number of parallel insertions heuristics based on the insertion scheme of (Solomon, 1987) and embedding in the calculations of the relevant criteria the acquisition costs of (Golden et al. (1984). Braysy et al. (2008) presented a deterministic annealing meta heuristic for the HVRPTW, outperforming the results of (Liu and Shen (1999), and then Braysy et al. (2009) developed a linearly scalable hybrid threshold-accepting and guided local search meta heuristic for solving large scale HVRPTW instances. Coelho et al. (2016) designed a trajectory search heuristic to solve a large-scale HVRPTW considering multi-trips. Repousis and Tarantillis (2010) presented an Adaptive Memory Programming solution approach for the HVRPTW that provides very good results in the majority of the benchmark instances examined. Subramaniam et al. (2012) proposed a hybrid algorithm for the problem. Their algorithm is composed by an Iterrated Local Search (ILS) based heuristic and a Set Partitioning (SP) formulation. Yousefikhoshbakht et al. (2016) present a modified column generation to solve HVRPTW.

This paper concerns with HVRP with hard time windows and service choice. The basic framework of the vehicle routing part can be viewed as a Vehicle Routing Problem with Time Windows (VRPTW) in which there are a limited number of vehicles, characterized by different capacities are available and the customers have a specified time windows for services. We address a large-scale mixed integer programming formulation to model the problem. A feasible neighbourhood heuristic search is proposed to get the sub-optimal integer feasible solution

2. Problem Description of The Hvrptw-SC

This paper considers a problem faced by a service company located in Medan city, North Sumatra Province, Indonesia. The basic frame work of HVRPTW can be defined graphically as follows. Let $G=(V,A)$ be a complete directed graph, where V = $\{0, 1, \ldots, n\}$ is the vertex set and $A = \{(i,j)$: $i,j \in$ V , $i \neq j\}$ is the set of route. For each route $(i,j) \in A$ a distance (or travel) cost c_{ij} is defined. Vertex 0 (i = 0) is the depot vertex, center of service , where the vehicle fleet is located. Define V_c is the set of customers' vertex. Each vertex $i \in V_c$ has a known fixed daily demand $w_i \geq 0$, a service time $s_i \geq 0$, and a service time windows $[a_i,b_i]$ In particular, at depot the demand $w = 0$ and service time t = 0.

As mentioned before this is a heterogeneous problem, therefore the fleet of K vehicles is composed by m different type of vehicles, each with capacity Q_m . The number of vehicles available to be used for vehicle type m is n_m. Let K_m is the set of vehicle type m then each customer is served accordingly by exactly one vehicle. At the depot ($i=0$), define a time window for vehicles to leave and to return to depot with [a_0, b_0]. We then can say that the arrival time of a vehicle at customer i is a_i and its departure time is b_i . Each type of vehicle is imposed with a fixed cost, f_m . Further more ,a fixed acquisition cost f_k is incurred for each of vehicle k in the routes. Each route originates and terminates at the central depot and must satisfy the time window constraints, i.e., a vehicle cannot start servicing customer i before a_i and after b_i however, the vehicle can arrive before a_i and wait for service.

Each customer node $i \in V_c$ has a known daily demand, W_i, service frequency, σ^t and a minimum service frequency, F_i, measured in days, t, per period. The demand accumulated between visits, w_i , is a function of the daily demand of the node. The stopping cost at a node i, τ_i , is a function of the frequency of the schedule since more items accumulate with less frequent service and, therefore, require more time to load/unload. Associated with each arc $(i, j) \in A$ is a known travel cost, c_{ij}. There is a set K of vehicles, each with capacity C, and T is the set of workdays in the planning horizon.

To associate with service choice, it is reasonable to define a service gain (in terms finance), δ^t, as an incentive to offer more frequent service to customers. Therefore we can say that service gain is a function demand's rate w_i .

We define the decision variables as follows.
Firstly we define the decision variables as follows.

- $x_{0j}^k = \begin{cases} 1 & \text{if vehicle type k} \in \text{K to deliver from depot to customer j} \in V_c; \\ 0 & \text{otherwise;} \end{cases}$
- $x_{ij}^m = \begin{cases} 1 & \text{if vehicle type m} \in K_m \text{ to deliver for } (i, j) \in V_c \text{ ,i} \neq \text{j;} \\ 0 & \text{otherwise;} \end{cases}$
- $z_0^m = \begin{cases} 1 & \text{if vehicle type m} \in K \text{ is available and active at depot ;} \\ 0 & \text{otherwise;} \end{cases}$
- l_i^m Arrival time for vehicle type $m \in K_m$ at customer $i \in V_c$ (non-negative continuous variable)
- u_i^m Duration of service of vehicle type $m \in K_m$ at customer $i \in V_c$ (non-negative continuous variable)

3. The mathematical model

First we formulate the objective function. The decision to be made is to choose the route of vehicles to serve customers demand, such that to minimize cost. Expression (1) is the objective function which describes the minimum of the travel costs.

In this basic framework the manager of the catering company wants to use the available vehicle for each type efficiently, such that the total cost is minimized. The total cost consists of traveling cost of all vehicle used and the cost for the availability of vehicle in the planning horizon time of a day.

$$Minimize \sum_{j \in V_c} c_{oj} \sum_{k \in K} x_{0j}^k + \sum_{(i,j) \in V_c} \sum_{m \in K_m} \sum_{t \in T} (\tau_{ij}^t \sigma_{ij}^t - w_{ij}^t \delta_{ij}^t) x_{ijm}^t + \sum_{m \in K_m} f_m z_o^m \quad (1)$$

Subject to

$$\sum_{k\in K} x_{0j}^{k} = 1, \qquad \forall j \in V_c \tag{2}$$

$$\sum_{k\in K}\sum_{j\in V} x_{ij}^{k} = 1, \qquad \forall i \in V_c \tag{3}$$

$$\sum_{i\in V} x_{ij}^{k} - \sum_{i\in V} x_{ji}^{k} = 1; \quad \forall j \in V_c, \forall \mathrm{k} \in \mathrm{K} \tag{4}$$

$$x_{ij}^{m} \le z_0^{m}, \quad (i,j) \in V_c, \forall m \in K_m \tag{5}$$

$$\sum_{j\in V_c} x_{1j}^{k} \le 1; \qquad \forall k \in K \tag{6}$$

$$\sum_{i\in V_c, i>1} x_{i1}^{k} \le 1; \qquad \forall k \in K \tag{7}$$

$$\sum_{i\in V_c} d_i \sum_{j\in V_c} x_{ij}^{m} \le Q_m; \quad \forall m \in K_m \tag{8}$$

$$x_{ij}^{m}(l_i^{m} + u_i^{m} + s_i + t_{ij} - l_j^{m}) = 0; \quad \forall m \in K_m, (i,j) \in A \tag{9}$$

$$l_i^{m} \le a_i \sum_{j\in V_c} x_{ij}^{m}; \qquad \forall m \in K_m, i \in V_c \tag{10}$$

$$a_i \sum_{j\in V_c} x_{ij}^{m} \le l_i^{m} + u_i^{m} \le b_i \sum_{j\in V_c} x_{ij}^{m}; \quad \forall m \in K_m, i \in V_c \tag{11}$$

$$\sum_{j\in V_c} w_j x_{oj}^{m} \le n_m; \qquad \forall m \in K_m \tag{12}$$

$$x_{0j}^{k}, x_{ij}^{m}, z_0^{m} \in \{0,1\}; \qquad \forall i \in V, \forall j \in V_c, \forall k \in K, \forall m \in K_m \tag{13}$$

$$l_i^{m}, u_i^{m} \ge 0; \qquad \forall i \in V_c, \forall m \in K_m \tag{14}$$

Constraints (2) and (3) are to guarantee that only one vehicle regardless their type are allowed to enter and depart from every customer node and comes back to the depot. Constraint (4) describes a flow conservation equation to maintain the continuity of each vehicle route on each period of time. Constraint (5) states that each customer is served only by the available and active vehicle of the corresponding type. Constraints (6) and (7) are to check the availability of vehicles by bounding the number of route, related to vehicle k for each type, directly leaving from and returning to the central depot, not more than one, respectively. Constraints (8) ensure that each delivery does not exceed the capacity of each type of vehicle. Constraints (9) establishes the equilibrium among the arrival time, duration of service, service time and travel time at customers in the routes assigned. Constraints (10) and (11) present strict time window for each customer. Constraint (12) guarantees that the number availability of active vehicle does not exceed the number of vehicle available at the central depot. Constraint (13) to state the binary variables, and Constraint (14) to define the continuous variables.

4. The Method Proposed

Stage 1.

Step 1. Solve the relaxed problem. If the result is a feasible integer solution. Stop.
The original problem has been solved.
Otherwise go to step 2.

Step 2. Get row i^* such that $\delta_{i*} = \min\{f_i, 1-f_i\}$

Step 3. Perform a pricing operation

$$v_{i*}^T = e_{i*}^T B^{-1}$$

Step 4. Calculate the maximum movement of nonbasic j, $\sigma_{ij} = v_{i*}^T \alpha_j$

With j corresponds to

$$\min_j \left\{ \left| \frac{d_j}{\alpha_{ij}} \right| \right\}$$

Eventually the column j^* is to be increased form LB or decreased from UB. If none go to next i^*.

Step 5. Solve $B\alpha_{j*} = \alpha_{j*}$ for α_{j*}

Step 6. Do ratio test for the basic variables in order to stay feasible due to the releasing of nonbasic j^* from its bounds.

Step 7. Exchange basis

Step 8. If row $i^* = \{\emptyset\}$ go to Stage 2, otherwise
Repeat from step 2.

Stage 2.

Step 1 adjust integer infeasible superbasics by fractional steps to reach complete integer feasibility.

Step 2 adjust integer feasible superbasics. The objective of this phase is to conduct a neighbourhood search to verify local optimality.

5. Conclusion

This paper was intended to present a solution for one of the most important problems in Supply Chain Management and Distribution problems. The aim of this paper was to develop a model of heterogeneous vehicle routing problem (HVRP) with time windows considering service-choice. This problem has imposed a benefit that could be gained as incentive for offering serviced to customers. The proposed algorithm employs nearest neighbor heuristic algorithm for solving the model.

References

[1] Cordeau J.-F., Laporte G., Savelsbergh M.W.P., Vigo D.: *Vehicle Routing*. In *Handbook in OR & MS, Vol. 14*, C. Barnhart and G. Laporte (eds.), Elsevier B.V. 2007, 367–428.

[2] El-Sherbeny N.A.: *Vehicle routing with time windows: An overview of exact, heuristic and metaheuristic methods*, Journal of King Saud University (Science), 22(2010), 123–131.

[3] Yeun L.C., Ismail W.R., Omar K., Zirour M.: *Vehicle Routing Problem: Models and Solutions*, Journal of Quality Measurement and Analysis, 4/1(2008), 205–218.

[4] Cordeau J.-F., Laporte G, Mecier A.: *A unified tabu search heuristic for vehicle routing problems with time windows*, Journal of Operational Research Society, 52(2001), 928–936.

[5] Drexl M. 2011. Applications of the vehicle routing problem with trailers and transshipment. Technical Report LM-2011-06, Johannes Gutenberg University Mainz.

[6] I. Ellabib , Otman A.B., & Calamai P. 2002. An Experimental Study of a Simple Ant Colony System for the Vehicle Routing Problem with Time Windows. ANTS, LNCS 2463: 53-64.

[7] B. L. Golden. & Assad A.A. 1986. Perspectives on Vehicle Routing: Exciting New Developments. *Operations Research* **34**: 803–809.

[8] Desrochers J K, Lenstra J K, Savelsbergh M W P. 1990. European J. of Operational Research,

46(3), 322-332.
[9] Desrosiers J, Dumas Y, Solomon M M, Soumis F. 1995. Network Routing, Handbooks in Operations Research and Management Science, vol.8. North-Holland, Amsterdam, 35-139.
[10] Cordeau J F , Laporte G and Mercier A. 2001. A unified tabu search heuristic for vehicle routing problems with time windows. Journal of Operational Research Society 52: 928–936.
[11] J. Larsen . 1999. Parallelization of the Vehicle Routing Problem with TimeWindows. Ph.D. thesis, Institute of Mathematical Modelling, Technical University of Denmark, Lyngby, Denmark
[12] W. Cook & Rich J.L. 1999. A Parallel Cutting-Plane Algorithm for the Vehicle Routing Problems with Time Windows. Working Paper, Department of Computational and Applied Mathematics, Rice University, Houston, U.S.A.
[13] Shi, Y-J, Meng, F-W, Shen G-J. 2012. A modified artificial bee colony algorithm for vehicle routing problems with time windows. Information Technology Journal, 1-6.
[14] Yousefikoshbakht M, Dolatnejad A, Didehvar F, Rahmati F. 2016. A modified column generation to solve the heterogeneous fixed fleet open vehicle routing problem. Journal of Engineering, Vol. 2016, pp. 1-12.
[15] Rincon-Garcia N, Waterson B J, Cherrett T J. 2017. A hybrid metaheuristic for the time-dependent vehicle routing problem with hard time windows. International Journal of Industrial Engineering Computations, 8, 141-160.
[16] Coelho V N, Grasas A, Ramalhinho H, Coelho I M, Souza M J F, Cruz R C. 2016. An ILS-based algorithm to solve a large-scale real heterogeneous fleet VRP with multi-trips and docking constraints. European Journal of Operational Research, 250, 367-376.
[17] Golden, B., Assad, A., Levy, L., & Gheysens, F. G. 1984. Computers & Oper. Res., 11, 49-66.
[18] Hoff, A., H. Andersson, M. Christiansen, G. Hasle and A. Lokketangen, 2010. Computers & Operations Research, 37.(12), 2041-2061.
[19] Koc, C.,T. Bektas, O. Jabali, G. Laporte., 2016. European Journal of Operational Research, 249, 1-21
[20] Liu, F. H., Shen, S. Y. 1999. Journal of the Operational Research Society, (50)7, 721-732.
[21] Solomon, M.M. 1987. Operations Research., 35(2), 254-265
[22] Braysy, O., Dullaert, W., Hasle, G., Mester, D., Gendreau, M., 2008. Transportation Science 42 (3), 371– 386.
[23] Braysy, O., Porkka, P.P., Dullaert, W., Repoussis, P.P., Tarantilis, C.D., 2009. Expert System and Applications 36 (4), 8460–8475.
[24] Repoussis, P.P., Tarantilis, C.D., 2010. Transportation Research Part C 18 (5), 695–712.
[25] Subramanian, A., P. H. V. Penna, E. Uchoa, L. S. Ochi., 2012. European Journal of Operational Research, 221, pp. 285-295.

Discrete Optimization Model for Vehicle Routing Problem with Scheduling Side Cosntraints

Dedy Juliandri[1], Herman Mawengkang[2], F. Bu'ulolo[2]
[1]Graduate School of Mathematics, University of Sumatera Utara, Medan, Indonesia
[2]Department of Mathematics, University of Sumatera Utara, Medan, Indonesia

Email: hmawengkang@yahoo.com

Abstract. Vehicle Routing Problem (VRP) is an important element of many logistic systems which involve routing and scheduling of vehicles from a depot to a set of customers node. This is a hard combinatorial optimization problem with the objective to find an optimal set of routes used by a fleet of vehicles to serve the demands a set of customers It is required that these vehicles return to the depot after serving customers' demand. The problem incorporates time windows, fleet and driver scheduling, pick-up and delivery in the planning horizon. The goal is to determine the scheduling of fleet and driver and routing policies of the vehicles. The objective is to minimize the overall costs of all routes over the planning horizon. We model the problem as a linear mixed integer program. We develop a combination of heuristics and exact method for solving the model.
Keywords: Integer programming, Transportation, Scheduling, Combined method

1. Introduction

Vehicle Routing Problem (VRP) is one of the important issues that exist in transportation system. This is a well known combinatorial optimization problem which consists of a customer population with deterministic demands, and a central depot which acts as the base of a homogeneous fleet of vehicles. The objective is to design a set of vehicle routes starting and terminating at the central depot, such that the demand of customers is totally satisfied, each customer is visited once by a single vehicle, the total demand of the customers assigned to a route does not exceed vehicle capacity, and to minimize the overall travel cost, taking into account various operational constraints.

VRP was first introduced by Dantzig and Ramser (1959). Due to the model has a lot applications, particularly in logistic system, and the combinatorial nature contained in the structure of the problem, it is not surprising that many researchers have been working in this area to discover new methodologies in order to solve the problems efficiently. There are a number of survey can be found in literature for VRP, such as, Caceres-Cruz et al (2014), Braysy and Gendreau (2005), Cordeau et al. (2007), Golden et al. (2002), Laporte and Sement (2002), and books (Golden et al. (2008), Toth and Vigo (2002)).

Mathematically, *VRP* can be defined as follows: vehicles with a fixed capacity Q must deliver order quantities nonnegative q_i ($i = 1,\ldots,n$) of goods to n customers from a single depot ($i = 0$). These vehicles must come back to the depot after serving customers. Knowing the distance d_{ij} between customers i and j ($i, j = 1,\ldots,n$), the objective of the problem is to minimize the total distance traveled by the vehicles in a way that only one vehicle handles the deliveries for a given customer and the total quantity of goods that a single vehicle delivers is not larger than Q.

There are three types of approach have been proposed for solving VRP, viz., exact, heuristics, and hybrid (a combined between heuristics and exact method).
Bettinelli et al. (2011) propose an exact method called branch-and-cut-and-price for solving VRP. Another exact method called column generation was used by Ceselli et al. (2009) and Goel (2010). Kok et al. (2010) propose dynamic programming as an exact method for solving VRP. More exact solution for VRP can be found in Baldacci et al (2010), and Baldacci et al. (2012).

Due to NP-hardness of VRP, some meta-heuristics are developed to solve it such as genetic algorithms (Cheng and Wang, 2009; Ursani et al., 2011; Vidal et al., 2013), ant colony (Ding et al., 2012; Yu and Yang, 2011), Tabu search (Belhaiza et al., 2013; Ho and Haugland, 2004), simulated annealing (Ban˜ os et al., 2013; Deng et al., 2009; Kuo, 2010; Tavakkoli-Moghaddam et al., 2007, 2011). Hybrid approach for solving VRP can be found in Guimarans (2012).

Pickup and delivery problems (PDPs) are a class of vehicle routing problems in which objects or people have to be transported between an origin and a destination. They can be classified into three different groups. The first group consists of many-to-many problems, in which any vertex can serve as a source or as a destination for any commodity. An example of a many-to-many problem is the Swapping Problem (Anily and Hassin, 1992). In this problem, every vertex may initially contain an object of a known type of commodity as well as a desired type of commodity. The problem consists of constructing a route performing the pickups and deliveries of the objects in such a way that at the end of the route, every vertex possesses an object of the desired type of commodity. Problems in the second group are called one-to-many-to-one problems. In these problems commodities are initially available at the depot and are destined to the customer vertices; in addition, commodities available at the customers are destined to the depot. Finally, in one-to-one problems, each commodity (which can be seen as a request) has a given origin and a given destination. Problems of this type arise, for example, in courier operations and door-to-door transportation services.

The existing literature on the VRP time windows with pick-up and delivery mainly (VRPTWPD) focuses on heuristic algorithms (see e.g., Nanry and Barnes (2000), Røpke and Pisinger (2006)). However, there are also a few studies that have introduced exact solution algorithms (see, e.g., Dumas et al. (1991), Savelsbergh and Sol (1995), Røpke and Cordeau (2009)).

In this paper we add the scheduling of fleet and driver in the VRPTWPD. The problem can be modeled as a mixed integer linear program. We use a direct search method for solving the problem.

2. VRP Problem formulation

Before we model the VRPTWPD, it is necessary to model the VRP as a framework to model the advanced problem. Let a graph $G(V, A)$ is given with nodes $V = C \cup \{0\}$ and arcs A, in VRP C is a representation of the set of customers, which 0 is the depot. Moreover, we have a set R of resources which e.g. can be load and/or time. Each resource $r \in R$ has a resource window $[a_i^r, b_i^r]$ that must be met upon arrival to node $i \in V$, and a consumption $_{ij}^{r} \geq 0$ for using arc $(i, j) \in A$. A resource consumption at a node $i \in C$ is modeled by a resource consumption at edge (i, j), and hence usually $_{0j}^{r} \geq 0$ for all $j \in C$. A global capacity limit Q can be modeled by imposing a resource window $[0, Q]$ for the depot node 0.

The VRP can now be stated as: Find a set of routes starting and ending at the depot node 0 satisfying all resource windows, such that the cost is minimized and all customers C are visited. A solution to the VRP will consist of a number of routes which starts from depot and at the end will be back to the depot.

In the following let c_{ij} be the cost of arc $(i, j) \in A$, x_{ij} be the binary variable indicating the use of arc $(i, j) \in A$, and $_{ij}^{r}$ (the resource stamp) be the consumption of resource $r \in R$ at the beginning of arc $(i, j) \in A$. Let $\delta^+(i)$ and $\delta^-(i)$ be the set of outgoing respectively ingoing arcs of node $i \in V$. Combining the two index model from Bard et al. [3] with the constraints ensuring the time windows for the asymmetric travelling salesman by Ascheuer et al. [1]. A mathematical model can be formulated as follows:

$$\min \quad \sum_{(i,j)\in A} c_{ij} x_{ij} \tag{1}$$

$$\text{s.t.} \quad \sum_{(s,j)\in\delta^+(i)} x_{ij} = 1 \qquad \forall\, i \in C \tag{2}$$

$$\sum_{(s,j)\in\delta^-(i)} x_{ij} = \sum_{(s,j)\in\delta^+(i)} x_{ij} \qquad \forall\, i \in V \tag{3}$$

$$\sum_{(s,j)\in\delta^-(i)} (T^r_{ji} + {}^{r}_{ji} x_{ij}) \le \sum_{(s,j)\in\delta^+(i)} T^r_{ji} \qquad \forall\, r \in R, \forall\, i \in C \tag{4}$$

$$a^r_i x_{ij} \le T^r_{ji} \le b^r_i x_{ij} \qquad \forall\, r \in R,\ \forall\, (i,j) \in A \tag{5}$$

$$T^r_{ij} \ge 0 \qquad \forall\, r \in R,\ \forall\, (i,j) \in A \tag{6}$$

$$x_{ij} \in \{0,1\} \qquad \forall\, (i,j) \in A \tag{7}$$

The objective (1) sums up the cost of the used arcs. Constraints (2) ensure that each customer is visited exactly once, and (3) are the flow conservation constraints. Constraints (4) and (5) ensure the resource windows are satisfied. It is assumed that the bounds on the depot are always satisfied. Note, that no sub-tours can be present since only one resource stamp per arc exists and the arc weights are positive for all $(i, j) \in A : i \in C$.

For a one dimensional resource such as *load* a stronger lower bound of the LP relaxation can be obtained by replacing (4) to (6) with $\sum_{(i,j)\in\delta^+(S)} x_{ij} \ge r(S)$, where $r(S)$ is a minimum number of vehicles needed to service the set S. All though this model can not be directly solved it is possible to overcome this problem by only including the constraints that are violated.

3. Mathematical Formulation of VRPFDPDTW

To formulate the model, firstly we should refer to the formulation of VRP given in Section 2. Now we add another formulation regarding to the fleet and driver scheduling and also pick-up and delivery. Let T is the planning horizon and D is the set of drivers. The set of workdays for driver $l \in D$ is denoted by $T_l \subseteq T$. The start working time and latest ending time for driver $l \in D$ on day $t \in T$ are given by g^t_l and h^t_l, respectively. For each driver $l \in D$, let H denote the maximum weekly working duration. We denote the maximum elapsed driving time without break by F and the duration of a break by G.

Let K denote the set of vehicles. For each vehicle $k \in K$, let Q_k and P_k denote the capacity in weight and in volume, respectively. We assume the number of vehicles equals to the number of drivers. Denote the set of n customers (nodes) by $N = \{1, 2, \ldots, n\}$. Denote the depot by $\{0, n+1\}$. Each vehicle starts from depot, $\{0\}$, and terminates at depot, $\{0\}$. Each customer $i \in N$ specifies a set of days to be visited, denoted by $T_i \subseteq T$. On each day $t \in T_i$, customer $i \in N$ requests service with demand of q^t_i in weight and p^t_i in volume, service duration d^t_i and time window $[a_i, b_i]$. Note that, for the depot on day t, we set $q^t_i = p^t_i = d^t_i = 0$.

We define binary variable x^t_{ijk} to be 1 if vehicle k travels from node i to j on day t, binary variable w^t_i to be 1 if customer i is not visited by a preferred vehicle on day t. Variable v^t_{ik} is the time that vehicle

k visits node i on day t. Binary variable z_{ik}^t indicates whether vehicle k takes a break after serving customer i on day t. Variable u_{ik}^t is the elapsed driving time for vehicle k at customer i after the previous break on day t. Binary variable y_{lk}^t is set to 1 if vehicle k is assigned to driver l on day t. Variables r_l^t and s_l^t are the total working duration and the total travel time for driver l on day t, respectively.

We define the notations to be used as follows.

Set:

T	The set of workdays in the planning horizon,
D	The set of drivers $D = D_I \cup D_E$,
T_l	The set of workdays for driver $l \in D$,
K	The set of vehicles,
N	The set of customers,
N_0	The set of customers and depot $N_0 = \{0\} \cup N$,
K_i	The set of preferable vehicles for customer $i \in N$,
T_i	The set of days on which customer $i \in N$ orders,

Parameter:

Q_k	The weight capacity of vehicle $k \in K$,
P_k	The volume capacity of vehicle $k \in K$,
c_{ij}	The travel cost (time) from node $i \in N_0$ to node $j \in N_0$,
cd_{lk}	The cost for driver $l \in D$ for vehicle $k \in K$,
$[a_i, b_i]$	The earliest and the latest visit time at node $i \in N_0$,
d_i^t	The service time of node $i \in N_0$ on day $t \in T_i$,
q_i^t	The weight demand of node $i \in N_0$ on day $t \in T_i$,
p_i^t	The volume demand of node $i \in N_0$ on day $t \in T_i$,
$[g_l^t, h_l^t]$	The start time and the latest ending time of driver $l \in D$ on day $t \in T$,
α_i^t	Pick up quantity for customer i on day $t \in T_i$,
β_i^t	Delivery quantity for customer i on day $t \in T_i$,
F	The maximum elapsed driving time without break,
G	The duration of break for drivers,

Variables:

x_{ilk}^t	Binary variable indicating whether vehicle $k \in K$ travels from node $i \in N_0$ to $j \in N_0$ on day $t \in T$,
v_{ik}^t	The time at which vehicle $k \in K$ starts service at node $i \in N_0$ on day $t \in T$,
z_{ik}^t	Binary variable indicating whether vehicle $k \in K$ takes break after serving node $i \in N_0$ on day $t \in T$,
u_{ik}^t	The elapsed driving time of vehicle $k \in K$ at node $i \in N_0$ after the previous break on day $t \in T$,
y_{lk}^t	Binary variable indicating whether vehicle $k \in K$ is assigned to driver $l \in D$ on day $t \in T$,

r_l^t The total working duration of driver $l \in D$ on day $t \in T$,

s_l^t The total travel distance of driver $l \in D$ on day $t \in T$,

θ_{jk}^t Number of pick up demand of customer j served by vehicle $k \in K$ on day $t \in T$

σ_{jk}^t Number of delivery demands of customer j served by vehicle $k \in K$ on day $t \in T$

The problem can be presented as a mixed integer linear programming model.

The objective of the problem is to minimize costs, which consists of the travel cost from depot to customers, travel cost from customer to customer, and the cost for drivers. Mathematically the objective can be written as

$$Minimize \sum_{j\in N}\sum_{k\in K}\sum_{t\in T} c_{0jk}^t x_{ojk}^t + \sum_{i\in N}\sum_{j\in N, j>i}\sum_{k\in K}\sum_{t\in T} c_{ijk}^t + \sum_{l\in D}\sum_{k\in K}\sum_{t\in T} cd_{lk}^t y_{lk}^t \quad (8)$$

Subject to:

$$\sum_{j\in N} x_{ji}^t = 1 \qquad \forall i \in N_0, t \in T \quad (9)$$

$$\sum_{i\in N_0} x_{ij}^t = 1 \qquad \forall j \in N, t \in T \quad (10)$$

$$\sum_{i\in N}\sum_{j\in N_0} q_i^t x_{ijk}^t \le Q_k \qquad \forall k \in K, t \in T \quad (11)$$

$$\sum_{i\in N}\sum_{j\in N_0} p_i^t x_{ijk}^t \le P_k \qquad \forall k \in K, t \in T \quad (12)$$

$$u_{jk}^t \ge u_{ik}^t + c_{ij} - M(1 - x_{ijk}^t) - Mz_{ik}^t \qquad \forall i, j \in N_0,\ k \in K, t \in T \quad (13)$$

$$u_{jk}^t \ge c_{ij} - M(1 - x_{ijk}^t) \qquad \forall i, j \in N,\ k \in K, t \in T \quad (14)$$

$$u_{ik}^t + \sum_{j\in N_0} c_{ij} x_{ijk}^t - F \le Mz_{ik}^t \qquad \forall i \in N_0,\ k \in K, t \in T \quad (15)$$

$$b_i \ge v_{ik}^t \ge a_i \qquad \forall i \in N,\ k \in K, t \in T_i \quad (16)$$

$$v_{0k}^t \ge \sum_{l\in D}(g_l^t \cdot y_{lk}^t) \qquad \forall k \in K, t \in T \quad (17)$$

$$v_{n+1,k}^t \le \sum_{l\in D}(h_l^t \cdot y_{ik}^t) \qquad \forall k \in K, t \in T \quad (18)$$

$$s_l^t \ge \sum_{i\in N_0}\sum_{j\in N_0} c_{ij} x_{ijk}^t - M(1 - y_{ik}^t) \qquad \forall l \in D, k \in K, t \in T_i \quad (19)$$

$$r_l^t \ge v_{n+1,k}^t - g_l^t - M(1 - y_{lk}^t) \qquad \forall l \in D, k \in K, t \in T_i \quad (20)$$

$$\sum_{k\in K} \theta_{jk}^t = \alpha_j^t \qquad \forall j \in N, t \in T \quad (21)$$

$$\sum_{k\in K} \sigma_{jk}^t = \beta_j^t \qquad \forall j \in N, t \in T \quad (22)$$

$$x_{ijk}^t, w_i^t, z_{ik}^t, y_{lk}^t \in \{0,1\} \qquad \forall i, j \in N_0, l \in D,\ k \in K, t \in T \quad (23)$$

$$v_{ik}^t, u_{ik}^t, r_l^t, s_l^t \ge 0 \qquad \forall i, j \in N_0, l \in D,\ k \in K, t \in T \quad (24)$$

$$\theta_{jk}^t, \sigma_{jk}^t \in \{0,1,2,\ldots\} \qquad \forall j \in N, k \in K, t \in T \quad (25)$$

Constraints (9-10) are to ensure that only one customer is to be visited. Constraints (11-12) guarantee that the vehicle capacities are respected in both weight and volume. Constraints (13-14) define the elapsed driving time. More specifically, for the vehicle (k) travelling from customer i to j on day t, the elapsed driving time at j equals the elapsed driving time at i plus the driving time from i to j (i.e., $u^t_{jk} \geq u^t_{ik} + c_{ij}$) if the vehicle does not take a break at customer i (i.e., $z^t_{ik} = 0$); Otherwise, if the vehicle takes a break at customer i (i.e., $z^t_{ik} = 1$), the elapsed driving time at j will be constrained by (1) which make sure it is greater than or equal to the travel time between i and j (i.e., $u^t_{jk} \geq c_{ij}$). Constraints (15) guarantee that the elapsed driving time never exceeds an upper limit F by imposing a break at customer i (i.e., $z^t_{ik} = 1$) if driving from customer i to its successor results in a elapsed driving time greater than F. Constraints (16) make sure the services start within the customers' time window.

Constraints (17-18) describe that the starting time and ending time of each route must be between the start working time and latest ending time of the assigned driver. Constraint (19) is to calculate the total travel time for each assigned driver. Constraints (20) are necessary to define the working duration for each assigned driver on every workday. Constraints (21 – 22) define the pick up and delivery for each customer. Constraints (23-25) define the binary and other variables used in this formulation.

4. The Algorithm

Let

$$x = [x] + f, \quad 0 \leq f \leq 1$$

be the (continuous) solution of the relaxed linear programming problem, $[x]$ is the integer component of non-integer basic variable x and f is the fractional component.

There are three stages.

The first stage.

Step 1. Get row i^* such that $\delta_{i^*} = \min\{f_i, 1 - f_i\}$

(This is to minimize the deterioration of the objective function, and clearly corresponds to the integer basic infeasibility).

Step 2. Obtain

$$v^T_{i^*} = e^T_{i^*} B^{-1}$$

Step 3. Calculate the maximum movement of nonbasic j, using $\sigma_{ij} = v^T_{i^*} \alpha_j$

With j corresponds to

$$\min_j \left\{ \left| \frac{d_j}{\alpha_{ij}} \right| \right\}$$

Otherwise go to next non-integer nonbasic or superbasic j (if available). Eventually the column j^* is to be increased form LB or decreased from UB. If none go to next i^*.

Step 4.

Solve $B\alpha_{j^*} = \alpha_{j^*}$ for α_{j^*}

Step 5. Do ratio test for the basic variables due to the releasing of nonbasic j^* from its bounds.

Step 6. Perform the exchange basis

Step 7. If row $i^* = \{\varnothing\}$ go to Stage 2, otherwise

Repeat from step 1.

Stage 2. Pass1 : perform a movement of integer infeasible superbasics by fractional steps to obtain the complete integer feasibility.

Pass2 : do a heuristic for the integer feasible superbasics. The objective of this phase is to conduct a highly localized neighbourhood search to verify local optimality.

5. Conclusions

This paper is to develop efficient technique for solving one of the most economic important problems in optimizing transportation and distribution systems. The aim of this paper is to develop a model of vehicle routing with Time Windows, Fleet and Driver Scheduling, Pick-up and Delivery Problem This problem has additional constraints which are the limitation in the number of vehicles. The proposed algorithm employs nearest neighbor heuristic algorithm for solving the model.

References

[1] B. L. Golden, A. A. Assad, and E. A. Wasil, "Routing Vehicles in the real world: Applications in the solid waste, beverage, food, dairy, and newspaper industries". In P. Toth and D. Vigo, editors, *The Vehicle Routing Problem,* pages 245-286. SIAM, Philadelphia, PA, 2002.

[2] B. L. Golden, S. Raghavan, and E. A. Wasil, "The vehicle routing problem: latest advances and new challenges". Springer, 2008.

[3] G. Laporte and F. Sement. Classical Heuristics for the Vehicle Routing Problem. In Toth, P. and Vigo, D., editors, *The vehicle Routing Problem,* SIAM Monographs on Discrete Matheamtics and Applications, pages 109-128. SIAM, Philadelphia, PA, 2002.

[4] J.-F. Cordeau, G. Laporte, M.W.F. Savelsbergh, and D. Vigo. Vehivle Routing. In Barnhart, C. and Laporte, G., edotirs, *Transportation,* handbooks in Operation Research and Management Science, Pages 367-428. North-Holland, Amsterdam, 2007.

[5] O. Bräysy and M. Gendreau. "Vehicle Routing Problem with Times Windows, Part II: Metaheuristics". *Transportation Science,* 39(1): 104-118, 2005.

[6] P. Toth and D. Vigo. "The Vehicle Routing Problem Society for Industrial and Applied Matheamtics", Philadelpiha, PA, 2002.

[7] G.B. Dantzig and J.H. Ramser, "The truck dispatching problem", Management Science, Vol. 6, 1959, 80.

[8] A. Bettinelli, A. Ceselli, and G. Righini. 2011. A branch-and-cut-and-price algorithm for the multi-depot heterogeneous vehicle routing problem with time windows. Transportation Research Part C: Emerging Technologies 19, 5 (2011), 723–740.

[9] Ceselli, G. Righini, and M. Salani. 2009. A column generation algorithm for a rich vehicle-routing problem. Transportation Science 43, 1 (2009), 56–69.

[10] A. Goel. 2010. A column generation heuristic for the general vehicle routing problem. In Learning and Intelligent Optimization, C. Blum and R. Battiti (Eds.). Lecture Notes in Computer Science, Vol. 6073. Springer, 1–9.

[11] A.L. Kok, C.M. Meyer, H. Kopfer, and J.M.J. Schutten. 2010. A dynamic programming heuristic for the vehicle routing problem with time windows and European community social legislation. Transportation Science 44, 4 (2010), 442–454.

[12] R. Baldacci, E. Bartolini, A. Mingozzi, and R. Roberti. 2010. An exact solution framework for a broad class of vehicle routing problems. Computational Management Science 7, 3 (2010), 229–268.

[13] R. Baldacci, A. Mingozzi, and R. Roberti. 2012. Recent exact algorithms for solving the vehicle routing problem under capacity and time window constraints. European Journal of Operational Research 218, 1 (2012), 1–6.

[14] Cheng, C.B., Wang, K.P., 2009. Solving a vehicle routing problem with time windows by a decomposition technique and a genetic algorithm. Expert Syst. Appl. 36 (4), 7758–7763.

[15] Ursani, Z., Essam, D., Cornforth, D., Stocker, R., 2011. Localized genetic algorithm for vehicle routing problem with time windows. Appl. Soft Comput. 11 (8), 5375–5390.

[16] Vidal, T., Crainic, T.G., Gendreau, M., Prins, C., 2013. A hybrid genetic algorithm with adaptive diversity management for a large class of vehicle routing problems with time windows. Comput. Oper. Res. 40 (1), 475–489.

[17] Ding, Q., Hu, X., Sun, L., Wang, Y., 2012. An improved ant colony optimization and its application to vehicle routing problem with time windows. Neurocomputing 98 (3), 101–107.

[18] Yu, B., Yang, Z.Z., 2011. An ant colony optimization model: the period vehicle routing problem with time windows. Transp. Res. Part E: Logist. Transp. Rev. 47 (2), 166–181.

[19] Belhaiza, S., Hansen, P., Laporte, G., 2013. A hybrid variable neighborhood Tabu search heuristic for the vehicle routing problem with multiple time windows. Comput. Oper. Res. (in press), http://dx.doi.org/10.1016/j.cor.2013.08.010.

[20] Ho, S.C., Haugland, D., 2004. At Tabu search heuristic for the vehicle routing problems with time windows and split deliveries. Comput. Oper. Res. 31 (12), 1947–1964.

[21] Ban˜ os, R., Ortega, J., Gil, C., Ferna′ ndez, A., Toro, F., 2013. A simulated annealing based parallel multi-objective approach to vehicle routing problems with time windows. Expert Syst. Appl. 40 (5), 1696–1707.

[22] Deng, A.M., Mau, C., Zhou, Y.T., 2009. Optimizing research of an improved simulated annealing algorithm to soft time windows vehicle routing problem with pick-up and delivery. Syst. Eng. Theory Pract. 29 (5), 186–192.

[23] Kuo, Y., 2010. Using simulated annealing to minimize fuel consumption for the time-dependent vehicle routing problem. Comput. Ind. Eng. 59 (1), 157–165.

[24] Tavakkoli-Moghaddam, R., Gazanfari, M., Alinaghian, M., Salamatbakhsh, A., Norouzi, N., 2011. A new mathematical model for a competitive vehicle routing problems with time windows solved by simulated annealing. J. Manuf. Syst. 30 (2), 83–92.

[25] D. Guimarans. 2012. Hybrid algorithms for solving routing problems. Ph.D. Dissertation. Universitat Aut`onoma Barcelona.

[26] Nanry, W, W Barnes. 2000. Solving the pickup and delivery problem with time windows using reactive tabu search. Transportation Research Part B: Methodological 34(2) 107–121.

[27] Røpke, S, D Pisinger. 2006. An adaptive large neighborhood search heuristic for the pickup and delivery problem with time windows. Transportation Science 40(4) 455–472.

[28] Røpke, S, J-F Cordeau. 2009. Branch and cut and price for the pickup and delivery problem with time windows. Transportation Science 43(3) 267–286.

[29] Dumas, Y, J Desrosiers, F Soumis. 1991. The pickup and delivery problem with time windows. European Journal of Operational Research 54(1) 7–22

[30] Savelsbergh, M, M Sol. 1995. The general pickup and delivery problem. Transportation Science 29(1) 17–29.

IOP Conf. Series: Materials Science and Engineering **300** (2018) 012025 doi:10.1088/1757-899X/300/1/012025

Model Evaluation To Measuring Efficiencies of ICT Development In Indonesia Region Using DEA

Syahril Efendi[1], M. Fadly Syahputra[2], M. Anggia Muchtar[3]
[1,2,3] Department of Computer Science, Faculty of Computer Science and Information Technology, Universitas Sumatera Utara, Medan 2015 Indonesia

Abstract. ICT Pura or digital city is a program designed by the Indonesian government with the main objective is to determine the level of readiness of each district and city in each province in the era of digital economy. It is necessarily to evaluate whether a city or a region that was successfully managing ICT better than other city and significantly contributes to the communities and living systems. Data envelopment analysis (DEA) is a well known technique to estimate efficiency and returns to scale through the construction of a best practice frontier, based on non-parametric mathematical programming approach. This paper addresses DEA BCC method to get index of efficiencies for all region in Indonesia covered by ICT Pura. Numerical result is given.

1. Introduction

During these last few years, the Indonesian Government had made an effort to develop nationally the application of information-communication in public sectors, called ICT-Pura. ICT Pura could mean a "ICT's City" or "Digital City". It is given to a city that was successfully managing ICT better than other city and thus significantly contributes to the communities of environment and living systems .

ICT Pura program is designed by the Indonesian government to meet several main objectives, which is to determine the level of readiness of each district and city in each province in the era of digital economy which will be begin in 2015, to measure the size of the real gap between the target and actual conditions on every district and city to be prepared in a national strategy. It is also to provide motivation, support, incentives, and appreciation for the District and the City, for their hard effort and their preparation to face the era of the digital society through a variety of program development and implementation of ICT in the respective areas.

ICT Pura is expected to show the real situation of the city's readiness for the challenge of ICT development, including measuring the amount of technology gap between regions and city. So from the real world conditions, ICT Pura will represent the strengths, weaknesses, challenges, and opportunities of the city, and it can be a step to plan the next strategy for ICT problem and optimization from the previous ICT Pura result.

Performance measurement and benchmarking are important tools for improvement in highly competitive and rapidly changing business environment of our era. Clearly, performance measurement concept is related with the measurement and improvement of efficiency to a great extent. Investigation and further analysis of worst and best performers in a business environment play a key role in deriving useful information to understand the current state of the processes and to identify the opportunities for improvement in efficiency. Several modelling methods can be found in Economics and Operational Research (OR) literature attempting to assess efficiency of different types of business operations, units or processes, called Data Envelopment Analysis (DEA).

DEA evaluates the relative performances of comparable units having the authority to make decisions. These units are called decision making units (DMUs). In DEA model it is necessarily to have multiple input and output quantities of DMUs. The model can be formulated as a linear program. Each DMU is considered one by one and a linear program is solved for each DMU. The efficiency of a DMU is defined as the ratio of weighted outputs to weighted inputs. These weights are positive decision variable of the linear program. When a DMU is under consideration, DEA maximizes the efficiency ratio of that DMU by changing the weights of inputs and outputs. There is only one constraint type in standard DEA, none of the DMUs can be more efficient than 100%. Weighted outputs of a DMU cannot be more than weighted inputs of that DMU. DEA approach is based on the original work [1]. and became popular with the work [2]. Standard DEA models only use the input and output quantities of DMUs. The efficiency of a DMU is measured relative to all other DMUs with the simple restriction that all DMUs lie on or below an efficient frontier [3].

An interactive DEA procedure (IDEA) was developed by [4]. IDEA incorporates the DM preference information to DEA in an iterative manner by setting minimum (maximum) acceptable levels of outputs (inputs). The authors illustrated IDEA model in the assessment of physics departments of UK Universities. [5] introduced the value efficiency analysis approach. By this approach, DEA uses the preference information of DM and assigns efficiency scores by using the estimated value function of DM. [6] introduced a systematic approach for the performance analysis of academic research in R&D institutions and universities using value efficiency analysis.

Referance [7] analyzed the policy effectiveness of player teams in a business game. Firstly, the efficient DMUs are determined by DEA. Afterwards, input weights are restricted since the author knows the policies of teams, i.e. the importance of inputs. Restrictions are made with constraints such as the weight of first input is not less than that of second input or the weight of first input is not less than 20% of the sum of all input weights. By introducing weight restrictions in DEA, some efficient DMUs become inefficient. The author classifies inefficiency as technical inefficiency and policy inefficiency. Technical inefficiency is the inefficiency of DMU in DEA model. Policy inefficiency is the difference of efficiency values of DMU in DEA model and weight restricted DEA model. Since this difference is due to the inefficient policy of the DMU.

A general analysis of DEA models is made by [8]. In this study DEA models are classified and new DEA approaches are shown with the use of multi criteria decision making approaches. [9] analyzed and compared data envelopment analysis models and multiple objective linear programming. Both articles have represented different reviews of DEA models and approaches.

DEA is used in efficiency, effectiveness and/or performance analysis of banks and bank branches by [10][11][12]. Measuring efficiency for public service with DEA is proposed by [13]. The application of DEA in transport sector is wide-spread, particularly in the evaluation of airports, ports, railways and urban transport companies [14].

Referance [15] use DEA to evaluate suppliers performance and segment suppliers. They develop DEA with chance constrained due to restriction imposed to fullfil on-time delivery . They show that there are statistically significant relationship between the dimensions of information sharing by the buying firm and the classification of supplier firms at varying level of on-time delivery performance risk. In evaluating hotel performance using DEA was proposed by [16]. They use as the input units are operations, marketing and environmental capabilities, the outputs are diversification strategy on performance of hotel industry. Another application of DEA in hotel efficiency is proposed by [17]. Referance [18] investigate to evaluate a problem in supply chain management called Supplier Total Cost of Ownership (TCO). In order to analyse the supplier performance (or efficiency), they use DEA. The result they get efficiency indexes and ranking of suppliers, whilst requiring substantially less effort to perform the analysis. Referance [19] propose DEA to analyse comparative performance of airline strategic alliances.

2. Data Envelopment Analysis

In general, methods of efficiency measurement are based upon the estimation of a production frontier and can be classified into two basic groups as parametric and non-parametric approaches. Parametric frontiers are based on specific functional forms and can be either deterministic or stochastic. On the non-parametric efficiency evaluation side, Data Envelopment Analysis (DEA) is a well established method aiming to identify relative efficiency of Decision Making Units (DMUs) producing multiple outputs through the use of multiple inputs. It does not require any assumptions about the functional form. The efficiency of a DMU is measured relative to all other DMUs with the simple restriction that all DMUs are members of a production possibility set and they lie on or below an efficient frontier. Efficiency can be simply defined as the ratio of output to input. More output per unit of input reflects relatively greater efficiency. If the greatest possible output per unit of input is achieved, a state of *absolute* or *optimum efficiency* has been achieved and it is not possible to become more efficient without new technology or other changes in the production process.

Let there are n DMU to be evaluated, each DMU has m input and s output. We can write $x_{ij}(i = 1,\cdots,m)$ and $y_{rj}(r = 1,\cdots,s)$, respectively, as input data and output data from $DMU_j,(j=1,...,n)$. The efficiency of DMU_j can be formulated as

$$max\, h_0 = \frac{\sum_{r=1}^{s} u_r y_{ro}}{\sum_{i=1}^{m} v_i x_{io}} \quad (1)$$

where u_r and v_i are the weight of, respectively, r^{th} output and i^{th} input .
The efficiency of DMU can now be found from the following solution of a linear programming problem

$$max\ h_0 = \frac{\sum_{r=1}^{s} u_r y_{ro}}{\sum_{i=1}^{m} v_i x_{io}} \quad (2)$$

$$s.t.\ \frac{\sum_{r=1}^{s} u_r y_{rj}}{\sum_{i=1}^{m} v_i x_{ij}} \le 1,\ j = 1,\cdots,n \quad (3)$$

$$u_r, v_i \ge 0,\ r = 1,\cdots,s;\ i = 1,\cdots,m \quad (4)$$

where subscript 0 is assigned for the DMU being evaluated, x_{ij} is the ith input of DMU_j and $x_{ij} > 0$, $i = 1,2,\cdots,m$ and $j = 1,2,\cdots,n$. Coupled with that, y_{rj} is the rth output being observed of DMU_j and $y_{rj} > 0$ for $i = 1,2,\cdots,m$ and $j = 1,2,\cdots,n$. The decision variables for the problem are u_r and v_i reprsenting weight values as the solution of problem (2, 3, 4). However the problem (2, 3, 4) has unbounded solution, as if u^* and v^* are the optimal solution, then for every $\alpha > 0$, (αu^* and αv^*) are optimal. Using transformation of Charnes-Cooper, then the representative solution of (u,v) can be chosen with condition :

$$\sum_{i=1}^{m} v_i x_{io} = 1 \quad (5)$$

Therefore we can get an equivalent linear programming of Problem (2, 3, 4) as a fractional programming. If we take the numerator of the problem equals to one then we will get an equivalent linear program, which can be written as:

$$\max\ h_0 = \sum_{r=1}^{s} u_r y_{r0} \quad (6)$$

$$\text{s.t.}\ \sum_{r=1}^{s} u_r y_{rj} - \sum_{i=1}^{m} v_i x_{ij} \le 0,\ \ j = 1,2,\cdots,n \quad (7)$$

$$\sum_{i=1}^{m} v_i x_{io} = 1 \quad (8)$$

$$u_r, v_i \ge 0,\ r = 1,\cdots,s;\ i = 1,\cdots,m \quad (9)$$

Linear programming stated as Problem (6, 7, 8, 9) is called as CCR, from Charnes-Cooper-Rhodes (*Multiplier* model) with *input-output* orientation. Maximizing is carried out by choosing between *"virtual" multiplie* (weight values) *u* and *v* which can give a result of the maximum rate of ratio *"virtual" output* and *"virtual" input*. Such a problem can be formulated as a linear program for each DMU_0, written mathematically with the following model.

$$\theta_0^* = \min\ \theta_0 \quad (10)$$

$$\text{s.t.}\ \sum_{j=1}^{n} \lambda_j y_{rj} \ge y_{r0},\ \ r = 1,2,\cdots,s \quad (11)$$

$$\sum_{j=1}^{n} \lambda_j x_{ij} \le \theta_0 x_{i0},\ \ i = 1,2,\cdots,m \quad (12)$$

IOP Conf. Series: Materials Science and Engineering **300** (2018) 012025 doi:10.1088/1757-899X/300/1/012025

$$\lambda_j \geq 0,\ j = 1,2,\cdots,n \quad (13)$$

Problem (10, 11, 12, 13) can give an optimal solution θ_0^*, as the value of efficient. It is called technical efficient or CCR efficient for a particular DMU_0 (*envelopment model*). In order to get the overall efficience of DMU, it is neccessarily to repeat the process for each DMU_j , j = 1,2,...,n . The value of θ is always less than or equal to one.

Now, we need to get the result of DEA model in terms of variable return to scale value. This result can be achieved if Problem (10 – 13) can be modified as convex form by inserting additional constraint $\sum_{j=1}^{n} \lambda_j = 1$.

Now, the problem can be expressed as:

$$\theta_0^* = \min\ \theta_0 \quad (14)$$

$$\text{s.t. } \sum_{j=1}^{n} \lambda_j y_{rj} \geq y_{r0},\quad r = 1,2,\cdots,s \quad (15)$$

$$\sum_{j=1}^{n} \lambda_j x_{ij} \leq \theta_0 x_{i0},\quad i = 1,2,\cdots,m \quad (16)$$

$$\sum_{j=1}^{n} \lambda_j = 1 \quad (17)$$

$$\lambda_j \geq 0,\ j = 1,2,\cdots,n \quad (18)$$

The BCC input-oriented model can be expressed in the following equations; where x_{ij} and y_{rj} are inputs and outputs, and s_i^- and s_r^+ are the corresponding slacks:

$$\theta_0^* = \min\ \theta_0 - \varepsilon\left(\sum_{i=1}^{m} s_i^- + \sum_{r=1}^{s} s_r^+\right) \quad (19)$$

$$\text{s.t. } \sum_{j=1}^{n} \lambda_j y_{rj} - s_r^+ = y_{r0},\quad r = 1,2,\cdots,s \quad (20)$$

$$\sum_{j=1}^{n} \lambda_j x_{ij} + s_i^- = \theta_0 x_{i0},\quad i = 1,2,\cdots,m \quad (21)$$

$$\sum_{j=1}^{n} \lambda_j = 1 \quad (22)$$

$$\lambda_j \geq 0,\ j = 1,2,\cdots,n \quad (23)$$

The title is set 17 point Times Bold, flush left, unjustified. The first letter of the title should be capitalized with the rest in lower case. It should not be indented. Leave 28 mm of space above the title and 10 mm after the title.

3. Problem Formulation

The research is carried out using the flow of mathematical modeling based on mapping process and the calculation of ICT Pura index of the whole Districts/Cities in Indonesia. The process had been done by the Indonesian Government in year 2011 up to 2013. The results are used as a based to present "ICT Pura Award" to the Districts and Cities, for their efforts to develop communication and information technology in their regions. The index results can be formulated as a mathematical model. From the model we would be able to obtain four dimension treated as variables of the efficiency measurement of creating and developing ICT at the District/Cities, using DEA. These four dimensions are: ICT Use (Intensity), ICT Readinees (Infrastructure), ICT Capability (Skills) dan ICT Impact (Outcomes).

4. Calculating Index

ICT Pura index can be said as an indicator to describe the level of readiness of a District/City in Indonesia in facing or adapting with the digital community environment. Referring to ITU (*International Telecommunication Union*), WSIS (*World Summit of Information Society*) and *The World Bank Institute* the paradigm for calculating index can be shown in Figure 1.

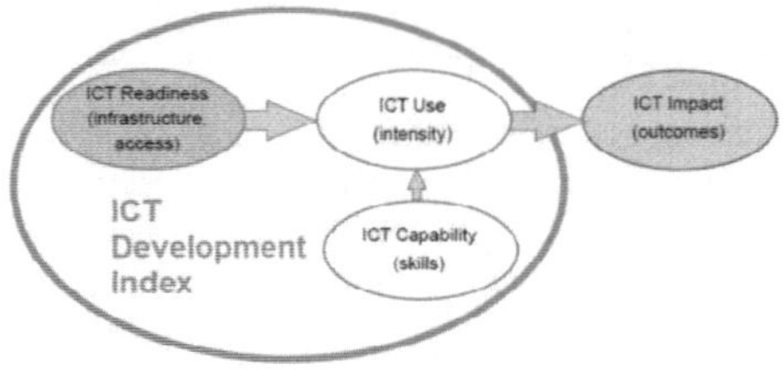

Figure 1. Paradigm in Calculating Index

The Index of ICT Pura can be found from Eq. (14)

$$\text{Index}_{\text{ICT Pura}} = NR_{IC}*25\% + NR_{IU}*40\% + NR_{IR}*20\% + NR_{II}*15\% \quad (24)$$

where

NR_{IC} = the average value of component ICT Capability
NR_{IU} = the average value of component ICT Usage
NR_{IR} = the average value of component ICT Readiness
NR_{II} = the average value of component ICT Impact

According to the scoring system used in the mapping questionnaire, the value of ICT Pura Index is in the range of 0 to 5. The meaning of each index is described in Table 1.

Table1. Description of Index

0	= City/District not ready at all	(I-mula)
1	= City/District just started	(I-pratama)
2	= City/District almost ready	(I-muda)
3	= City/District ready	(I-madya)
4	= City/District ready, and able to compete	(I-utama)
5	=City/District ready, and being fart ahead	(I-paripurna)

The results of ICT Pura Index for the year 2011, 2012 and 2013 for Districts/Cities around can be found in Table 2.

Table 2. The Results of ICT Pura Index of Each Province in Indonesia.

No.	**DMU Province**	**Capability 25%**	**Usability 40%**	**Readiness 20%**	**Impact 15%**	**Ix_ICTPura**
1	Aceh	2.69	1.93	2.29	2.99	2.34
2	Sumater Utara	2.80	2.38	2.74	2.92	2.62
3	Sumatera Barat	2.51	1.87	2.31	2.51	2.22
4	Riau	2.52	1.81	2.08	2.20	2.12
5	Kep. Riau	2.57	1.92	2.42	2.57	2.29
6	Jambi	1.74	1.45	1.53	1.92	1.63
7	Sumatera Selatan	2.27	1.77	2.09	2.40	2.06
8	Bengkulu	2.46	1.78	2.13	2.46	2.12
9	Lampung	2.21	1.83	1.93	2.57	2.06
10	Kep. Bangka Belitung	2.55	1.88	2.26	3.00	2.30
11	DKI Jakarta	3.02	2.41	2.07	3.77	2.73
12	Jawa Barat	2.92	2.46	2.86	3.05	2.73
13	Jawa Tengah	3.10	2.43	2.86	3.13	2.79
14	Banten	2.74	1.86	2.40	2.62	2.29
15	Jawa Timur	3.06	2.49	2.79	3.28	2.66

16	DI Yogyakarta	3.32	2.68	3.13	3.32	3.03
17	Bali	3.04	2.17	2.87	2.90	2.62
18	Nusa T. Barat	2.64	1.89	2.29	2.46	2.25
19	Nusa T. Timur	1.82	1.44	1.41	1.97	1.64
20	Kalimantan Barat	1.99	1.42	1.74	1.91	1.71
21	Kalimantan Tengah	2.16	1.60	1.92	2.39	1.92
22	Kalimantan Selatan	2.12	2.02	2.11	2.50	2.13
23	Kalimantan Timur	2.65	2.04	2.19	3.37	2.45
24	Sulawesi Utara	2.00	1.42	1.58	2.71	1.80
25	Sulawesi Tengah	2.43	1.59	1.92	2.56	2.04
26	Sulawesi Selatan	2.86	2.07	2.46	2.90	2.47
27	Sulawesi Tenggara	2.11	1.42	1.62	2.69	1.86
28	Gorontalo	2.42	1.98	1.79	2.77	2.20
29	Sulawesi Barat	2.01	1.58	1.76	2.63	1.89
30	Maluku	1.54	1.40	1.21	2.70	1.59
31	Maluku Utara	2.19	1.50	1.75	2.37	1.87
32	Papua	2.36	1.85	1.98	2.70	2.15
33	Papua Barat	1.75	1.34	1.39	2.19	1.59

In this paper, we propose the calculation to get the level of efficiency of developing ICT in each province (DMU), using BCC DEA of Problem (14, 15, 16, 17, 18).

$$\theta_0^* = \min \theta_0 \quad (25)$$

$$\text{s.t. } \sum_{j=1}^{n} \lambda_j y_{rj} \geq y_{r0}, \quad r = 1,2,\cdots,s \quad (26)$$

$$\sum_{j=1}^{n} \lambda_j x_{ij} \leq \theta_0 x_{i0}, \quad i = 1,2,\cdots,m \quad (27)$$

$$\sum_{j=1}^{n} \lambda_j = 1 \quad (28)$$

$$\lambda_j \geq 0, \ j = 1,2,\cdots,n \quad (29)$$

The result of the efficiency level for all Province in Indonesia and the Category for each Province as shown is shown in Table 3.

Table 3. Level of Efficiency and Category for Each DMU (Province)

No.	DMU Province	Capability X1	Usability X2	Readiness X3	Impact Y	Efficiency level	Category
1	Aceh	2.69	1.93	2.29	2.99	86.72%	I-Madya
2	Sumater Utara	2.80	2.38	2.74	2.92	67.55%	I-Madya
3	Sumatera Barat	2.51	1.87	2.31	2.51	73.67%	I-Madya
4	Riau	2.52	1.81	2.08	2.20	74.10%	I-Madya
5	Kep. Riau	2.57	1.92	2.42	2.57	72.12%	I-Madya
6	Jambi	1.74	1.45	1.53	1.92	94.50%	I-Muda
7	Sumatera Selatan	2.27	1.77	2.09	2.40	77.10%	I-Madya
8	Bengkulu	2.46	1.78	2.13	2.46	77.07%	I-Madya
9	Lampung	2.21	1.83	1.93	2.57	75.67%	I-Madya
10	Kep. Bangka Belitung	2.55	1.88	2.26	3.00	89.53%	I-Madya

11	DKI Jakarta	3.02	2.41	2.07	3.77	100.00%	I-Madya
12	Jawa Barat	2.92	2.46	2.86	3.05	70.34%	I-Madya
13	Jawa Tengah	3.10	2.43	2.86	3.13	74.32%	I-Madya
14	Banten	2.74	1.86	2.40	2.62	74.76%	I-Madya
15	Jawa Timur	3.06	2.49	2.79	3.28	78.21%	I-Madya
16	DI Yogyakarta	3.32	2.68	3.13	3.32	74.08%	I-Utama
17	Bali	3.04	2.17	2.87	2.90	73.22%	I-Madya
18	Nusa T. Barat	2.64	1.89	2.29	2.46	72.58%	I-Madya
19	Nusa T. Timur	1.82	1.44	1.41	1.97	94.42%	I-Muda
20	Kalimantan Barat	1.99	1.42	1.74	1.91	94.37%	I-Muda
21	Kalimantan Tengah	2.16	1.60	1.92	2.39	85.22%	I-Muda
22	Kalimantan Selatan	2.12	2.02	2.11	2.50	72.64%	I-Madya
23	Kalimantan Timur	2.65	2.04	2.19	3.37	99.63%	I-Madya
24	Sulawesi Utara	2.00	1.42	1.58	2.71	99.26%	I-Muda
25	Sulawesi Tengah	2.43	1.59	1.92	2.56	87.01%	I-Madya
26	Sulawesi Selatan	2.86	2.07	2.46	2.90	76.75%	I-Madya
27	Sulawesi Tenggara	2.11	1.42	1.62	2.69	98.51%	I-Muda
28	Gorontalo	2.42	1.98	1.79	2.77	74.04%	I-Madya
29	Sulawesi Barat	2.01	1.58	1.76	2.63	88.09%	I-Muda
30	Maluku	1.54	1.40	1.21	2.70	100.00%	I-Muda
31	Maluku Utara	2.19	1.50	1.75	2.37	90.75%	I-Muda
32	Papua	2.36	1.85	1.98	2.70	75.68%	I-Madya
33	Papua Barat	1.75	1.34	1.39	2.19	100.00%	I-Muda

From Table 3, we should be able to observe that there is a discrepancy between Level of Efficiency and the Category of a Province obtained. For example, Aceh Province has the ICT Pura Index 2.34 and Category I-Madya, meaning that this province is in category ready to develop, although the efficiency level is only 86.72%. Yogyakarta obtains the highest Index, 3.03. Therefore this province belongs to Category I-Utama. Unfortunately, the level of efficiency is only 74,08%, worse than Aceh. However, it is different with Maluku province, from the Table 5 we can see that this province has the least ICT-Pura index, which is 1.59 with Category I-Muda. However, the level of efficiency to develop its ICT is 100%.

Figure 2 describes graphically the ICT-Pura index and the level of efficiencies for each provinces.

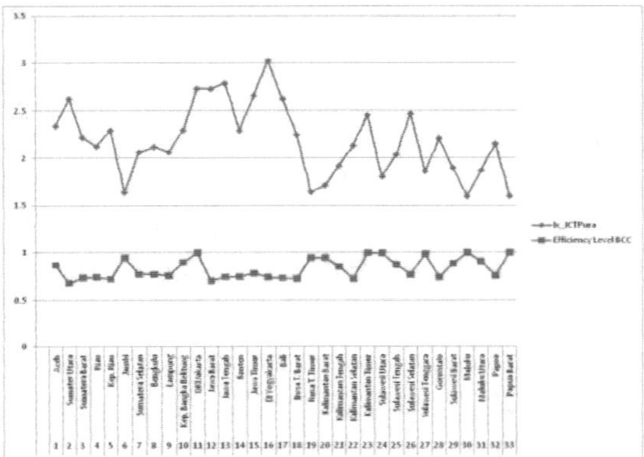

Figure 2. Results for ICT-Pura index and Level of Efficiency of each Provinces.

5. Conclusion

In order to be able to compete in the "Digital Economic" era, each province in Indonesia has been urged to develop its potential in the field of Information-Communication-Technology. Indonesia has given motivations to fulfill this needs through what is called ICT Pura. This paper presents a way to see the readiness of each province in Indonesia for the Digital Economy era by measuring the level of efficiency via DEA model. Each province is called DMU. The results obtained show that it is not enough just to use Category to see the readiness of each province, but we need level of efficiency, in such a way the local Government should able to detect the weakness of his/her Province.

Acknowledgment

Special thanks to the Ministry of Higher Education and Research Technology for supporting this research under a scheme of Hibah Bersaing Research with contract no.78/UN5.2.3.1/PPM/SP/2015.

References

[1] Farrell, M. J.(1957). The Measurement of Productive Efficiency. *Journal of the Royal Statistical Society A 120,* 253-281.

[2] Charnes, A., Cooper, W. W., & Rhodes, E.(1978). Measuring the efficiency of decision-making units. *European Journal of Operational Research 2,* 429-444.

[3] Seiford, L. M., & Thrall, R. M.(1990). Recent developments in DEA: the mathematical programming approach to frontier analysis. *Journal of Econometrics 46*, 7-38.

[4] Post, T., & Spronk, J.(1999). Performance benchmarking using interactive data envelopment analysis. *European Journal of Operational Research 115,* 472-487.

[5] Halme, M., Joro, T., Korhonen, P., Salo, S., & Wallenius, J.(1999). A value efficiency approach to incorporating preference information in data envelopment analysis. *Management Science 45,* 103-115.

[6] Korhonen, P., Tainio, R., & Wallenius, J. (2001). Value efficiency analysis of academic research. *European Journal of Operational Research 130,* 121-132.

[7] Kornbluth, J. S. H.(1991). Policy effectiveness using cone restricted data envelopment analysis. *The Journal of the Operational Research Society 42,* 1097-1104.

[8] Kleine, A.(2004). A general model framework for DEA. *Omega 32,* 17-23.

[9] Joro, T., Korhonen, P., & Wallenius, J.(1998). Structural comparison data envelopment analysis and multiple objective linear programming. *Management Science 44,* 962-970.

[10] Fukuyama, H.(1995). Measuring efficiency and productivity growth in Japanese banking: A non-parametric frontier approach. *Applied Financial Economics, 5*, 95-117.

[11] Favero, C., & Papi, L.(1995). Technical efficiency and scale efficiency in the Italian banking system: A non-parametric approach. *Applied Economics*, *27,* 385-395.

[12] Drake, L.(2001). Efficiency and productivity change in UK banking. *Applied Financial Economics, 11,* 557-571

[13] Smith, P. C., & Street, A.(2004). *Measuring efficiency of public services: the limit analysis. Working paper,* Centre for Health Economics, University of York.

[14] Markovits-Somogyi, R.(2011). Measuring Efficiency in Transport: The State of The Art of Applying Data Envelopment Analysis, *Transport 26*(1), 11-19.

[15] Ross, A. D., Kuzu, K., & Li, W.(2016). Exploring supplier performance risk and the buyer's role using chance-constrained data envelopment analysis. *European Journal of Operational Research, 250,* 966-978.

[16] Ramanathan, R., Ramanathan, U., & Zhang, Y.(2016). Linking oprations, marketing and environmental capabilities and diversification to hotel performance: A data envelopment analysis approach. *Int. J. Production Economics, 176,* 111-122.

[17] Jacobs, R.(2001). Alternative Methods to Examine Hospital Efficiency: Data Envelopment Analysis and Stochastic Frontier Analysis. *Health Care Management Science 4*, 103-116.

[18] Visani, F., Barbieri, P., Mart, F., Di Lascio, L., Raffoni, A., & Vigo, D.(2016). Supplier's total

cost ownership evaluation: a data envelopment analysis approach. *Omega, 61,* 141-154.

[19] Min, H., & Joo, S.(2016). A comparative performance analysis of airline strategic alliances using data envelopment analysis. *Journal of Air Transport Management, 52,* 99-110.

An Integer Programming Model for Multi-Echelon Supply Chain Decision Problem Considering Inventories

Amin Harahap[1], Herman Mawengkang[2], Siswadi[1], Syahril Effendi[3]
[1]Graduate school of Mathematics, University of Sumatera Utara, Medan, Indonesia
[2]Department of Mathematics, University Sumatera Utara, Medan, Indonesia
[3]Faculty of Computer Science and Information Technology, University of Sumatera Utara, Medan, Indonesia

Email: hmawengkang@yahoo.com

Abstract. In this paper we address a problem that is of significance to the industry, namely the optimal decision of a multi-echelon supply chain and the associated inventory systems. By using the guaranteed service approach to model the multi-echelon inventory system, we develop a mixed integer ;programming model to simultaneously optimize the transportation, inventory and network structure of a multi-echelon supply chain. To solve the model we develop a direct search approach using a strategy of releasing nonbasic variables from their bounds, combined with the "active constraint" method. This strategy is used to force the appropriate non-integer basic variables to move to their neighbourhood integer points.

Keywords: Mixed Integer Linear Programming, Supply chain, Multi-Echelon, Inventory.

1. Introduction

Most companies nowadays are organized into networks of manufacturing and distribution sites that procure raw materials, process them into finished goods, and distribute the finished goods to customers. The goal is to deliver the right product to the right place at the right time for the right price. These production-distribution networks are what we call "supply chains."

Supply Chain Management has attracted companies' attention as they realize the potential cost benefits of integrating decisions with other members of their supply chain. The primary cost factors within a supply chain can be put into the categories of production, transportation and inventory. The signature of supply chain management is the integration of activities. Effective supply chain members invariably integrate the wishes and concerns of their downstream members into their operations while simultaneously ensuring integration with their upstream members.

It is widely accepted that supply chain and its management are the logical progression of developments in logistics and logistics management respectively (Cooper and Ellram, 1993;Bowersox and Closs, 1997; Kent and Flint, 1997; Ganeshan et al., 1998).

Inventory optimization remains one of the key challenges in supply chain management. Typically, large amounts of working capital are tied up in today's supply chains, restricting the opportunities for growth that are essential for a company's success in competitive markets. However, researches have shown that inventories have a high opportunity to reduce them within supply chain, and hence increasing competiveness factors and reduce production costs.

From industrial aspects inventory management is a very important problem, however most of the models in the literature consider inventory management and supply chain network design separately, even though they are closely interrelated (Viau et al., 2007). This is due to the fact that minimizing transportation and inventory costs are considered as being two contradictory objectives. Nevertheless , there are some related works on supply chain optimization that take into account the inventory costs, but they do not mention any detailed inventory management policy (Daskin et al., 2002; Shen et al., 2003; sourirajan et al., 2007; You and Grossman, 2008; You and Grossman, 2010). In these models, the safety stock level is treated as a parameter, and can be considered as a lower bound of the total inventory level (Rodriguez et al., 2013; Bok et al., (2000); Verderame et al., (2008); Varma et al., (2007); Schulz et al., (2006), or it is considered as an inventory target that would lead to some penalty costs if violated (Jackson and Grossmann, 2003).

Most of the literature put focuses on inventory structure single stage integrated with supply chain design. Daskin et al (2002), Shen et al (2003) and Darnius and Mawengkang (2014) present a joint location-inventory model, which extends the classical uncapacitated facility location model to include nonlinear working inventory and safety stock costs for a two-stage supply chain network, so that decisions on the installation of distribution centers (DCs) and the detailed inventory replenishment decisions are jointly optimized. To simplify the problem, inventories in the retailers are neglected, and they also assume that all the DCs have the same constant replenishment lead time, and the demand at each customer has the same variance-to-mean ratio. You and Grossmann (2008) proposed a mixed-integer nonlinear programming (MINLP) approach to study a more general model based on the one developed by Daskin et al (2002) and Shen et al (2003), relaxing the assumption on identical variance-to-mean ratio for customer demands.

Supply chain management in the process industries has long been used as a tool to define production and distribution policies, as well as product allocation. This is the case of Cohen and Lee (1988) who described the modelling of a supply chain composed of raw material vendors, primary and secondary plants (each one with inventories of raw materials and finished products), distribution centres, warehouses and customer areas. Later, Cohen and Moon (1991) used supply chain optimisation to analyse the impact of scale, complexity (the operating costs are a function of the utilisation rates and number of products being processed in each facility) and weight of each cost factor (e.g. production, transportation and allocation costs) on the optimal design and utilisation patterns of the supply chain systems.

Timpe and Kallrath (2000) described an MILP model, which combined production, distribution and marketing and involved plants and sales points, to cover the relevant features required for the complete supply chain management of a multi-site production network. Jayaraman and Pirkul (2001) developed a Capacitated Plant Location Problem (CPLP) type model for planning and coordination of production and distribution facilities for multiple commodities, comprising raw materials suppliers, production sites, warehouses and customer areas. The authors followed a holistic approach to the supply chain, resulting in a deterministic, steady-state, multiechelon problem. Park (2005) considered both integrated and decoupled production and distribution planning problem, consisting of multiple plants, retailers, items over multiple periods. The author proposed mixed integer optimisation models and a two-phase heuristic solution to maximise the total net profit.

The objective of our model is to construct an integrating production, inventory and routing decisions within two-echelon context that minimizes total costs while meeting the following requirements: decide how many warehouses are needed, where to locate the opened warehouses, and how to allocate products from plants to warehouse and then to customers ; find out how much to keep in inventory at any time period; and create how to build the vehicles' routes starting from opened warehouse to customers and back to that warehouse. In general distribution problems can be considered as a combinatorial optimization model. Therefore the model is formulated as a Mixed Integer Program (MIP). A direct feasible search approach is developed for solving the model.

2. Model formulation

This model deals with the distribution of several products N from a single manufacturer to a set of candidate warehouse (W) and then to a set of retailers, I.

To build a model of the problem some assumptions are needed.

(i) The retailers deal with deterministic demand.

(ii) The fleet of vehicles are homogen with the same capacity.

(iii) Out-of-stock situations never occur.
(iv) The route considered only from warehouse to retailers..
(v) The inventory holding cost for warehouses is the same for all candidate warehouses.
(vi) Inventory level at retailers is limited by constraint, namely: physical capacity at retailers site.

Let G = (V, E) be an undirected network where V is a set of nodes comprised of a subset I of m potential warehouse sites and a subset J = V/I of n customers. E is a set of route connecting each pair of nodes in V. We define a feasible route as a route in which a vehicle starts from a candidate warehouse, visits a number of retailers, and comes back to the same warehouse. As such, a feasible route does not pass by more than one warehouse. As a result, the number of possible feasible routes will be $W \times 2^{I}$, where W and I are the number of warehouses and retailers, respectively. Feasible routes and the associated parameters (α_{ir} and β_{rw}, as explained later) are required to define the model and, therefore, generated before solving the model. We assume vehicle capacity to be larger than the maximum customer demand at any time period. Moreover, in any time period, each vehicle travels at most on one route, and customers are visited at most once.

The objective is to minimize total cost incurred in the inventory-location problem. The total cost consists of three components. They are as follows:
(i) warehouse fixed-location cost: the cost to establish and operate a warehouse;
(ii) retailer unit-inventory holding cost: the cost to store products at retailer; and
(iii) routing cost: the cost associated with transportation the goods from plants to warehouse and the cost from warehouse to retailers.

Sets

O = Set plants, O = 1,…, |O|
N = Set of products, N = 1,…n
W = Set of candidate warehouses $W = 0, \ldots, |W|$
I = Set of retailers, $I = 0, \ldots, |I|$
V = Set of nodes, $V = O \cup W \cup I$
T = Set of time periods $T = 1, \ldots, |T|$
R = Set of all feasible routes
K = Set of homogeneous vehicles $K = 1, \ldots, |K|$

Parameters

F_w = Fixed cost of opening and operating warehouse $w \in W$
C = Vehicle capacity
d_{ijt} = Demand of product $j \in N$ at customer $i \in I$ in time period $t = 1, \ldots, T, \ldots, T + \tau_{max} - 1$
u_{ijt} = Upper bound inventory level of product $j \in N$ at customer $i \in I$ in time period $t \in T$, $u_{it} = \left(\Sigma_{>t}^{<t+\ max} d_i\right)$
h_{ijt} = Inventory holding cost of product $j \in N$ at customer $i \in I$ in time period $t \in T$
I_{ij0} = Inventory level of product $j \in N$ at customer $i \in I$ at the beginning of time period $t = 1$

$$\beta_{rw} = \begin{cases} 1 & \text{if route } r \in R \text{ visits warehouse } w \in W; \\ 0 & \text{otherwise} \end{cases}$$

$c1_{ow}$ = Transportation cost from node plant $o \in O$ to warehouse $w \in W$
$c2_{wi}$ = Transportation cost from node warehouse $w \in W$ to retailer $i \in I$
c_r = Transportation cost of route $r \in R$

Decision Variables

$$X_{ij} = \begin{cases} 1 & \text{if warehouse } i \in W \text{ is served by plant } j \in O; \\ 0 & \text{otherwise} \end{cases}$$

$$Y_{ij} = \begin{cases} 1 & \text{if demand of retailer } i \in I \text{ is served by warehouse } j \in W \text{ ;} \\ 0 & \text{otherwise} \end{cases}$$

I_{it} = Inventory level of product $j \in N$ at customer $i \in I$ at the end of time period $t \in T$

$$\theta_{rt} = \begin{cases} 1 & \text{if route } r \in R \text{ is selected in time period } t \in T; \\ 0 & \text{otherwise} \end{cases}$$

$$m_w = \begin{cases} 1 & \text{if warehouse is opened at location } w \in W; \\ 0 & \text{otherwise} \end{cases}$$

a_{jirt} = Quantity of product $j \in N$ delivered to customer $i \in I$ by route $r \in R$ in time period $t \in T$

3. The Model

The problem can be formulated as a MIP problem which has a mathematical form as follows.
The objective function is to minimize installation cost of warehouses, and all transportation costs.

$$\min \sum_{w \in W} f_w m_w + \sum_{t \in T} \left(\sum_{r \in R} c_r \theta_{rt} + \sum_{j \in N} \sum_{i \in I} h_{jit} I_{jit} \right) + \sum_{i \in O} \sum_{j \in W} c1_{ij} X_{ij} + \sum_{i \in O} \sum_{j \in I} c2_{ij} Y_{ij} \quad (1)$$

Subject to
$$\sum_{r \in R} \theta_{rt} \le 1 \qquad \forall t \in T \quad (2)$$

$$\sum_{i \in O} X_{ij} = m_j; \qquad \forall j \in W \quad (3)$$

$$\sum_{j \in W} Y_{ji} = 1; \qquad \forall i \in I \quad (4)$$

$$\sum_{i \in I} a_{jirt} \le C\theta_{rt} \qquad \forall j \in N, r \in R, t \in T \quad (5)$$

$$I_{ijt-1} + \sum_{r \in R} \alpha_{jir} a_{jirt} = d_{jit} + I_{jit} \qquad \forall j \in N, i \in I, t \in T \quad (6)$$

$$I_{ijt} \le u_{ijt} \qquad \forall i \in I, t \in T, j \in N$$

(7)
$$\theta_{rt} \le \sum_{w \in W} \beta_{rw} m_w \qquad \forall r \in R, t \in T$$

(8)
$$\sum_{r \in R} \theta_{rt} \le |K| \qquad \forall t \in T$$

(9)

$$X_{ij} = \{0,1\}; \qquad \forall i \in O, \forall j \in W \quad (10)$$

$$Y_{ij} = \{0,1\}; \qquad \forall i \in I, \forall j \in W \quad (11)$$

$$\theta_{rt} \in \{0,1\} \qquad \forall r \in R, t \in T \quad (12)$$

IOP Conf. Series: Materials Science and Engineering **300** (2018) 012026 doi:10.1088/1757-899X/300/1/012026

$$m_w \in \{0,1\} \quad \forall w \in W \tag{13}$$

$$\alpha_{jirt}, I_{jit} \geq 0 \quad \forall j \in N, i \in I, \; r \in R, r \in T \tag{14}$$

Constraints (2) guarantee that a customer is visited once at most in any time period. Constraints (3) state that if warehouse j is installed then it should be served by only one plant i. Constraints (4) describe that each demand from retailer j should be served only by warehouse i. Constraints (5) account for the vehicle capacities. Inventory balance equations are represented in Constraints (6). Constraints (7) is to ensure that the inventory level at a customer never exceeds the total demand in the next consecutive time periods. Constraints (7) guarantee that routes start and end with open warehouses only. Constraint (9) limit the maximum number of routes at any time period to be no greater than the number of vehicles. Finally, Constraints (10) to (13) for the binary variables, and constraints (14) ensure that quantities to be shipped to customers and inventory levels are non-negative.

4. The Algorithm

Stage 1.

Step 1. Get row i^* the smallest integer infeasibility, such that $\delta_{i^*} = \min\{f_i, 1 - f_i\}$

Step 2. Calculate

$$v_{i^*}^T = e_{i^*}^T B^{-1}$$

this is a pricing operation

Step 3. Calculate $\sigma_{ij} = v_{i^*}^T a_j$

With j corresponds to

$$\min_j \left\{ \left| \frac{d_j}{\sigma_{ij}} \right| \right\}$$

I. *For nonbasic j at lower bound*

If $\sigma_{ij} < 0$ and $\delta_{i^*} = f_i$ calculate $\Delta = \frac{(1-\delta_{i^*})}{-\sigma_{ij}}$

If $\sigma_{ij} > 0$ and $\delta_{i^*} = 1 - f_i$ calculate $\Delta = \frac{(1-\delta_{i^*})}{\sigma_{ij}}$

If $\sigma_{ij} < 0$ and $\delta_{i^*} = 1 - f_i$ calculate $\Delta = \frac{\delta_{i^*}}{-\sigma_{ij}}$

If $\sigma_{ij} > 0$ and $\delta_{i^*} = f_i$ calculate $\Delta = \frac{\delta_{i^*}}{\sigma_{ij}}$

II. *For nonbasic j at upper bound*

If $\sigma_{ij} < 0$ and $\delta_{i^*} = 1 - f_i$ calculate $\Delta = \frac{(1-\delta_{i^*})}{-\sigma_{ij}}$

If $\sigma_{ij} > 0$ and $\delta_{i^*} = f_i$ calculate $\Delta = \frac{(1-\delta_{i^*})}{\sigma_{ij}}$

If $\sigma_{ij} > 0$ and $\delta_{i^*} = 1 - f_i$ calculate $\Delta = \frac{\delta_{i^*}}{\sigma_{ij}}$

If $\sigma_{ij} < 0$ and $\delta_{i^*} = f_i$ calculate $\Delta = \frac{\delta_{i^*}}{-\sigma_{ij}}$

Otherwise go to next non-integer nonbasic or superbasic j (if available). Eventually the column j^* is to be increased form LB or decreased from UB. If none go to next i^*.

Step 4. Calculate

$\alpha_{j^*} = B^{-1}\alpha_{j^*}$

i.e. solve $B\alpha_{j^*} = \alpha_{j^*}$ for α_{j^*}

Step 5. Ratio test; there would be three possibilities for the basic variables in order to stay feasible due to the releasing of nonbasic j^* from its bounds.

If j^* lower bound

Let

$$A = \min_{i' \neq i^* | \alpha_{ij^*} > 0} \left\{ \frac{x_{B_{i'}} - l_{i'}}{\alpha_{ij^*}} \right\}$$

$$B = \min_{i' \neq i^* | \alpha_{ij^*} < 0} \left\{ \frac{u_{i'} - x_{B_{i'}}}{-\alpha_{ij^*}} \right\}$$

$$C = \Delta$$

the maximum movement of j^* depends on: $\theta^* = \min(A, B, C)$

If j^* upper bound

Let

$$A' = \min_{i' \neq i^* | \alpha_{ij^*} < 0} \left\{ \frac{x_{B_{i'}} - l_{i'}}{\alpha_{ij^*}} \right\}$$

$$B' = \min_{i' \neq i^* | \alpha_{ij^*} > 0} \left\{ \frac{u_{i'} - x_{B_{i'}}}{-\alpha_{ij^*}} \right\}$$

$$C' = \Delta$$

The maximum movement of j^* depends on: $\theta^* = \min(A', B', C')$

Step 6. Exchanging basis for the three possibilities

1. If A or A'
   - $x_{B_{i'}}$ becomes nonbasic at lower bound $l_{i'}$
   - x_{j^*} becomes basic (replaces $x_{B_{i'}}$)
   - x_{i^*} stays basic (non-integer)
2. If B or B'
   - $x_{B_{i'}}$ becomes nonbasic at upper bound $u_{i'}$
   - x_{j^*} becomes basic (replaces $x_{B_{i'}}$)
   - x_{i^*} stays basic (non-integer)
3. If C or C'
   - x_{j^*} becomes basic (replaces x_{i^*})
   - x_{i^*} becomes superbasic at integer-valued

Step 7. If row $i^* = \{\emptyset\}$ go to Stage 2, otherwise

Repeat from step 1.

Stage 2. Do integer lines search to improve the integer feasible solution

5. Conclusion

In this paper, we have presented a mixed integer programming (MILP) model that determines the optimal network design, transportation and inventory levels of a multi-echelon supply chain with the presence of customer demand to be fulfilled from the chosen warehouse. To solve the resulting MILP problem efficiently for large scale instances, a direct search algorithm was proposed.

Reference

[1] Bok, J.-K.; Grossmann, I. E.; Park, S., "Supply Chain Optimization in Continuous Flexible Process Networks". *Industrial & Engineering Chemistry Research* **39, (5),** 1279-1290, (2000).

[2] Verderame, P. M.; Floudas, C. A., "Integrated Operational Planning and Medium-Term Scheduling of a Large-Scale Industrial Batch Plants". *Industrial and Engineering Chemistry Research,* **47,** 4845-4860, (2008).

[3] Varma, V. A.; Reklaitis, G. V.; Blau, G. E.; Pekny, J. F., "Enterprise-wide modeling & optimization – An overview of emerging research challenges and opportunities". *Computers & Chemical Engineering,* **31**, 692-711, (2007).

[4] Schulz, E. P.; Diaz, M. S.; Bandoni, J. A., "Supply chain optimization of large-scale continuous process". *Computers & Chemical Engineering* **29,** 1305-1316, (2006).

[5] Jackson, J. R.; Grossmann, I. E., "Temporal decomposition scheme for nonlinear multisite production planning and distribution models". *Industrial & Engineering Chemistry Research,* **42, (13),** 3045-3055, (2003).

[6] Daskin, M. S.; Coullard, C.; Shen, Z.-J. M., "An inventory-location model: formulation, solution algorithm and computational results". *Annals of Operations Research,* **110,** 83-106, (2002).

[7] Shen, Z.-J. M.; Coullard, C.; Daskin, M. S., "A joint location-inventory model". *Transportation Science,* **37,** 40-55,(2003).

[8] Sourirajan, K.; Ozsen, L.; Uzsoy, R., "A single-product network design model with lead time and safety stock considerations". *IIE Transactions,* **39,** 411-424, (2007).

[9] You, F.; Grossmann, I. E., "Mixed-Integer Nonlinear Programming Models and Algorithms for Large-Scale Supply Chain Design with Stochastic Inventory Management". *Industrial & Engineering Chemistry Research,* **47, (20),** 7802, (2008).

[10] Cohen, M.A., Lee, H.L., Strategic analysis of integrated production-distribution systems – models and methods. Oper. Res. 36, 216–228, 1988.

[11] Cohen, M.A., Moon, S., An integrated plant loading model with economies of scale and scope. Eur. J. Oper. Res. 50, 266–279, 1991.

[12] Jayaraman, V., Pirkul, H., Planning and coordination of production and distribution facilities for multiple commodities. Eur. J. Oper. Res. 133, 394–408, 2001.

[13] Park, Y.B., An integrated approach for production and distribution planning in supply chain management. Int. J. Prod. Res. 43, 1205–1224, 2005.

[14] Timpe, C.H., Kallrath, J., Optimal planning in large multi-site production networks. Eur. J. Oper. Res. 126, 422–435, 2000.

[15] COOPER, M. C. & ELLRAM, L. M. (1993) Characteristics of Supply Chain Management

[16] BOWERSOX, D. J. & CLOSS, D. J. (1997) *Logistics Management - the integrated supply chain process*, McGraw-Hill.

[17] KENT, J. L. & FLINT, D. J. (1997) Perspectives on the evolution of logistics thought. *Journal of Business Logistics,* 18.

[18] GANESHAN, R., JACK, E., MAGAZINE, M. J. & STEPHENS, P. (1998) A Taxonomic

[19] Review of Supply Chain Management Research. *Quantitative Models for Supply Chain Management.* Kluwer Academic Publishers You, F., Grossmann, I.E., 2010, Integrated Multi-Echelon Supply Chain Design with Inventories Under Uncertainty: MINLP Models, Computational Strategies. AIChE Journal, 56, 419–440.

IOP Conf. Series: Materials Science and Engineering **300** (2018) 012026 doi:10.1088/1757-899X/300/1/012026

[20] Viau M A, Trepanier M, Baptiste P. 2007. Integration of inventory control in spread supply chains, CIRRELT-2007-43
[21] Rodriguez M A, Vecchietti A R, Grossmann I E, Harjunkoski L. 2013. Chemical Engineering Transactions, Vol. 32, 1171-1176.
[22] Darnius O, Mawengkang H. 2014. International J. of Advanced Res. In Comp. Sc. And Software Eng., 3(4), 1-6.

Model estimation of claim risk and premium for motor vehicle insurance by using Bayesian method

Sukono[1], Riaman[2], E. Lesmana[3*], R. Wulandari[4], H. Napitupulu[5*], S. Supian[6]

[1,2,3,4,5,6]Department of Mathematics, FMIPA, Universitas Padjadjaran, INDONESIA
Email: sukono@unpad.ac.id; riaman_02@yahoo.com; man.msie@gmail.com; ratnawulandari624@gmail.com; napitupuluherlina@gmail.com; sudradjat@unpad.ac.id.

[*]corresponding author: man.msie@gmail.com; napitupuluherlina@gmail.com

Abstract. Risk models need to be estimated by the insurance company in order to predict the magnitude of the claim and determine the premiums charged to the insured. This is intended to prevent losses in the future. In this paper, we discuss the estimation of risk model claims and motor vehicle insurance premiums using Bayesian methods approach. It is assumed that the frequency of claims follow a Poisson distribution, while a number of claims assumed to follow a Gamma distribution. The estimation of parameters of the distribution of the frequency and amount of claims are made by using Bayesian methods. Furthermore, the estimator distribution of frequency and amount of claims are used to estimate the aggregate risk models as well as the value of the mean and variance. The mean and variance estimator that aggregate risk, was used to predict the premium eligible to be charged to the insured. Based on the analysis results, it is shown that the frequency of claims follow a Poisson distribution with parameter values λ is 5.827. While a number of claims follow the Gamma distribution with parameter values p is 7.922 and θ is 1.414. Therefore, the obtained values of the mean and variance of the aggregate claims respectively are IDR 32,667,489.88 and IDR 38,453,900,000,000.00. In this paper the prediction of the pure premium eligible charged to the insured is obtained, which amounting to IDR 2,722,290.82. The prediction of the claims and premiums aggregate can be used as a reference for the insurance company's decision-making in management of reserves and premiums of motor vehicle insurance.

Keywords: Poisson distribution, Gamma distribution, Bayesian method, aggregate claims, premium calculation

1. Introduction

Motor vehicle insurance is one of the important branches of non-life insurance type. Even in many countries, motor vehicle insurance is the largest total premium revenue earner. Indonesia is included in the country that gets the largest total insurance premium, from motor vehicle insurance. One of the driving factors causing the auto insurance industry to grow rapidly is the increasing number of motor vehicles in Indonesia every year. As a risk taker and recipient institution, insurance companies must be able to anticipate the risks if there are many claims. Because otherwise, would cause losses that could make the insurer bankrupty [11]. In risk management, insurance companies must know the character of the risk. The purpose is to predict the losses event in the future [4].

In modeling the claim loss there are two important measures to be considered, namely the frequency of claims and the amount or severity of claims. The frequency of claims, as usual is properly modelled by using discrete distributions, including binomial, geometric, negative binomial and Poisson. Whereas, claim severity represent a large loss of an insurance claim, which is generally modelled by using a non-negative continuous distribution. For example the exponential and Pareto distributions, as well as the attached traits such as tail and quartile properties [4,1]. To carry out the distribution parameter estimation of claims frequency and claims severity, according Jaroengeratikun [10] and Bolstad [3], the Bayesian method can be applied to estimate the parameters of a model of the distribution of losses. The advantage of the Bayesian method is that the prior information of the parameters involved in the risk model needs to be determined first, then the sample data is observed to produce a posterior distribution. Similar studies ever undertaken include Migon et al. [13] and Sukono et al. [16], conducting a risk analysis of health insurance claims, where the model parameter estimation uses the Bayesian method approach. Meanwhile, Eliasson [6] applied the Bayesian method for the estimation of the credibility parameters for non-life insurance pricing on historical data of individual claims. In the analysis of non-life insurance, the risk distribution model of loss is an important concern for insurance companies. The risk distribution model is very useful in determining the premium to be paid by the insured to the insurer.

Based on the previous explanation, this paper intends to apply how to assess a collective risk models in non-life insurance using Bayesian methods. The Bayesian method is used to estimate the parameters of the claim frequency model and the amount of the claim, in order to be used in the calculation of risk, and also to determine the premiums the insured must pay to the insurer.

2. Methology

In this section the aims is to discuss the methodology, which includes the discussion of: collective risk model, parameter estimation, and principle of premiums calculation. We begin with a discussion of the collective risk model as follows.

2.1 Collective risk model

This section aims to discuss the collective risk model. If we let the total frequency of claims for an insurance portfolio in the time period t is denoted by N_t, then a large number of aggregate claims S_t is as follows [9]:

$$S_t = \sum_{i=1}^{N_t} X_{t,i} \,. \tag{1}$$

where $X_{t,i}$ is the *i*-th amount of claim which occurred during the time period t, or so-called amount of individual claims [4,5].

The assumptions used in this model are:

- Amount of claim $X_{t,i}$ is a non-negative random variables that are independent and identically distributed.
- Claim frequency N_t is a random variable and independent toward the *i*-th amount of claim $X_{t,i}$.

An important thing to do is to evaluate the frequency distribution model of claims, as well as the appraisal of the large distribution model of claims. After the estimator of the calim frequency distribution model and the estimator of the amount of claim distribution model have been obtained, it can then be used to determine the expected value $E(S_t)$ and variance $Var(S_t)$. The expected value and the variance are the measures to estimate the total loss in the collective risk model. In the collective risk model, the expected and variance magnitude of total loss can be calculated using the following equation [12,14]:

$$E(S_t) = E[E(S_t \mid N_t)] = E[E(X_1 + X_2 + ... + X_N)] = E[N.E(X)]$$

$$= E(X)E(N). \tag{2}$$

$$Var(S_t) = E[Var(S \mid N)] + Var[E(S \mid N)] = E[N.Var(X)] + Var[N.E(X)]$$
$$= E(N)Var(X) + [E(X)]^2 Var(N). \tag{3}$$

2.2 Maximum Likelihood and Bayesian methods

This section aims to discuss the maximum likelihood and Bayesian estimation methods. Maximum Likelihood is one of the most commonly methods used for performing an estimation of parameters. Assumed that $X_1, X_2, ..., X_n$ are random sample of the population with a probability density function $f(x \mid \theta_1, \theta_2, ...\theta_k)$, likelihood function of the random sample is expressed by [4]:

$$l(\theta \mid x) = l(\theta_1, \theta_2, ..., \theta_k \mid x) = \prod_{i=1}^{n} f(x_i \mid \theta_1, \theta_2, ...\theta_k). \tag{4}$$

Furthermore, estimation of parameter θ is performed which gives the maximum value of (4). The estimation of these parameters in this paper is done by using Bayesian method. According to Boldstad [3] and Gill [8], the Bayesian method is one of the parameter estimation methods that uses preliminary information on parameters θ, or so-called as the prior distribution, and information from the observed data that has been obtained. After the sample information is taken, and the prior has been determined, the posterior distribution is determined by multiplying the prior with the sample information obtained from the likelihood. Where this prior is independent of its likelihood, and the posterior distribution is given by:

$$f(\theta \mid x) = \frac{f(\theta, x)}{f(x)} = \frac{f(\theta)l(x \mid \theta)}{\int_0^\infty f(\theta)l(x \mid \theta)d\theta} \propto f(\theta)l(x \mid \theta),$$

with posterior distribution $f(\theta \mid x)$, prior distribution $f(\theta)$, and Likelihood function $l(x \mid \theta)$.

2.3 Parameters estimation of claim frequency model

This section aims to discuss the parameters estimation of the claim frequency model. Assumed that frequency of claim N is a discrete random sample that Poisson distributed with mean λ. The probability density function of the Poisson distribution is expressed as follows [2,4]:

$$p(n) = \frac{e^{-\lambda}\lambda^n}{n!}, \; n = 1,2,3,...$$

Therefore, the likelihood function of the sample data which following the Poisson distribution is expressed by:

$$l(n_i \mid \lambda) = \frac{e^{-t\lambda}\lambda^{\sum_{i=1}^{t} n_i}}{\prod_{i=1}^{t} n_i!}.$$

According to Ntzoufras [15], the frequency of claims N which Poisson distributed, has a prior conjugate that Gamma-distributed with a probability density function expressed by:

$$f(\lambda) = \frac{\beta_\lambda^{\alpha_\lambda} \lambda^{\alpha_\lambda - 1} e^{-\lambda\beta_\lambda}}{\Gamma(\alpha_\lambda)}, 0 \le \lambda < \infty.$$

Posterior of parameter λ can be expressed by multiplication between likelihood $l(n_i \mid \lambda)$ with prior $f(\lambda)$. If the equation which not related to the parameter λ is neglected, the proportional posterior equation is obtained as follows:

$$f(\lambda \mid n_1, n_2, ..., n_t) \propto \lambda^{\alpha_\lambda + \sum_{i=1}^{t} n_i - 1} e^{-\lambda(\beta_\lambda + t)}.$$

So it can be said that posterior of the parameter λ also Gamma distributed as expressed as [3,7]:

$$\lambda \mid n_i \sim Gamma(\alpha_\lambda + \sum_{i=1}^{t} n_i, \beta_\lambda + t).$$

Therefore, if the posterior of the parameter λ is Gamma Distributed, then the mean or expected value is obtained, expressed by:

$$E(\lambda \mid n_i) = \hat{\mu}_\lambda = \frac{\alpha_\lambda + \sum_{i=1}^{t} n_i}{\beta_\lambda + t},$$

while the variance of the posterior is expressed by:

$$Var(\lambda \mid n_i) = \hat{\sigma}_\lambda{}^2 = \frac{\alpha_\lambda + \sum_{i=1}^{t} n_i}{(\beta_\lambda + t)^2}.$$

2.4 Parameters estimation of claim amount model

In this section it is intended to discuss about the parameters estimation of the claim amount model. In this paper the amount of claim X is assumed to be a continuous random sample that Gamma distributed with parameter p and θ. The probability density function that follows the Gamma distribution is expressed by [2,4]:

$$f(x) = \frac{\theta^p x^{p-1} e^{-x\theta}}{\Gamma(p)}.$$

Furthermore, the likelihood function of sample data of claim amount which follows Gamma distribution is expressed by:

$$l(x_i \mid p, \theta) = \frac{\theta^{pn}}{\Gamma(p)^n} \left(\prod_{i=1}^{n} x_i{}^{p-1} \right) e^{-\theta \sum_{i=1}^{n} x_i}.$$

According to Ntzoufras [15], amount of claims which is Gamma distributed has a prior conjugate that Gamma distributed with a hyperparameter α_θ and β_θ. The probability density function is expressed as follows:

$$f(\theta) = \frac{\beta_\theta{}^{\alpha_\theta} \theta^{\alpha_\theta - 1} e^{-\theta\beta_\theta}}{\Gamma(\alpha_\theta)}, 0 \le \theta < \infty.$$

Posterior to parameter θ can be expressed by multiplication between likelihood $l(x_i \mid p, \theta)$ to prior $f(\theta)$. If the equation which not related to the parameter θ is neglected, the proportional posterior equations are obtained as [4], [7]:

$$f(\theta \mid p, x) \propto \theta^{pn + \alpha_\theta - 1} e^{-\theta(\sum_{i=1}^{n} x_i + \beta_\theta)}.$$

So it can be said that posterior $f(\theta \mid p, x)$ is Gamma distributed with parameter $\alpha_{\theta,1} = pn + \alpha_\theta$ and $\beta_{\theta,1} = \sum_{i=1}^{n} x_i + \beta_\theta$.

Because of the posterior parameter θ is Gamma distributed then the mean can be obtained, expressed by:

$$E(\theta \mid x_i) = \hat{\mu}_\theta = \frac{pn + \alpha_\theta}{\sum_{i=1}^{n} x_i + \beta_\theta},$$

while the variance of the posterior is expressed by:

$$Var(\theta \mid x_i) = \hat{\sigma}_\theta{}^2 = \frac{pn + \alpha_\theta}{\left(\sum_{i=1}^{n} x_i + \beta_\theta\right)^2}.$$

2.5 Premium calculation models

This section aims to discuss the principles of premium calculation. Premiums are calculated on the basis of the principle of equality, or the principle of premium calculation. There are several principles used for the calculation of premiums, namely [4,16]:

a. Pure Premium Principle
The calculation of pure premium principle is done by using following equation:

$$p(t) = E(S), \tag{5}$$

where $E(S) = \dfrac{E(S_t)}{t}$.

b. Expectation Value Principle
The premium calculation of Expectation Value Principle is done by using following equation:

$$p(t) = (1+\theta)E(S_t), \tag{6}$$

where $\theta > 0$ represents a premium loading factor.

c. Variance Principle
The premium calculation of Variance Value principle is done by using following equation:

$$p(t) = E(S) + \alpha Var(S), \tag{7}$$

where $Var(S) = \dfrac{Var(S_t)}{t}$ and $\alpha > 0$ represents a premium loading factor

d. Standard Deviation Principle
The premium calculation of Standard Deviation Principle is done by using equations:

$$p(t) = E(S) + \alpha\sqrt{Var(S)}, \tag{8}$$

where $\alpha > 0$.

3. Results and discussion

In this section the aims is to study the results and discussions that include: analyzed data, an estimation of the claims frequency distribution model, the estimation of the claims amount distribution model, and the calculation of insurance premiums. It starts with a discussion of the claims frequency distribution, as follows.

3.1 Analyzed data

The data used in this study, is the data of motor vehicle insurance claims on non-life insurance company PT. "X" with periods of 2015 to 2016. Data are grouped into claims frequency and claim amount. A summary of claims data is presented in Table 1.

Table 1. Data of claim amount and frequency of claim

No	Interval	Frequency
1	1,978,435 - 3,121,754	6
2	3,121,755 - 4,265,074	13
3	4,265,075 - 5,408,393	20
4	5,408,395 - 6,551,712	12
5	6,551,713 - 7,695,031	8
6	7,695,031 - 8,838,351	4
7	8,838,352 - 9,981,670	7
Total	392,307,367	70

Furthermore, based on claims frequency data and the claims amount in Table 1, the most suitable or fit distribution is determined by using *Easyfit 5.6* software, and it is tested statistically by using a goodness of fit test.

3.2 Estimating of the claim frequency distribution model

In this section it aims to estimate the claim frequency distribution model. The steps include: identification of the distribution model, the estimation of the distribution model, and testing the significance of the distribution estimator.

- ***Identification of the claim frequency data distribution model***

Identification of the claim frequency data distribution model is done by using statistical *software EasyFit 5.6.* The result of curve matching can be seen in Figure 1.

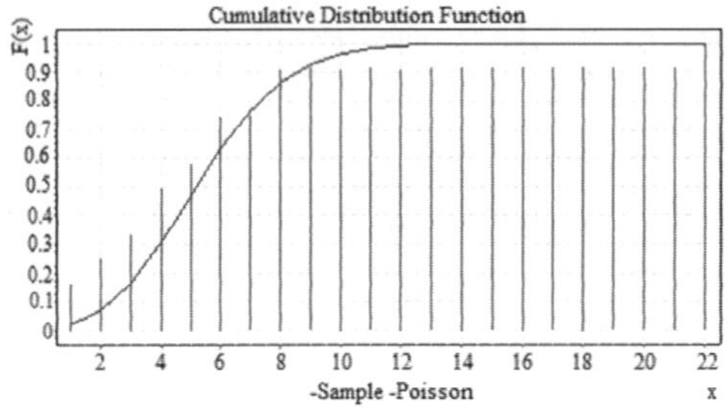

Figure 1 Histogram and cumulative distribution function curve of claims frequency data

Based on the results of curve fitting using *Easyfit 5.6* software, the suitable model for the frequency of claims is obtained, that is the Poisson distribution. Estimated parameters using the maximum likelihood method $\lambda = 5{,}8333$. To determine the most suitable distribution, the *Kolmogorov-Smirnov* test was used on the Poisson distribution by using *EasyFit 5.6* software. Based on the *Kolmogorov-Smirnov* test, the most suitable model for the claims frequency data is the Poisson distribution.

- ***Parameters estimating of the claim frequency distribution model***

Based on Figure 2, the selected prior value is $\alpha_\lambda = 0.001$ and $\beta_\lambda = 0.001$, because the Bayesian estimation approximates the likelihood estimation of the parameter λ.

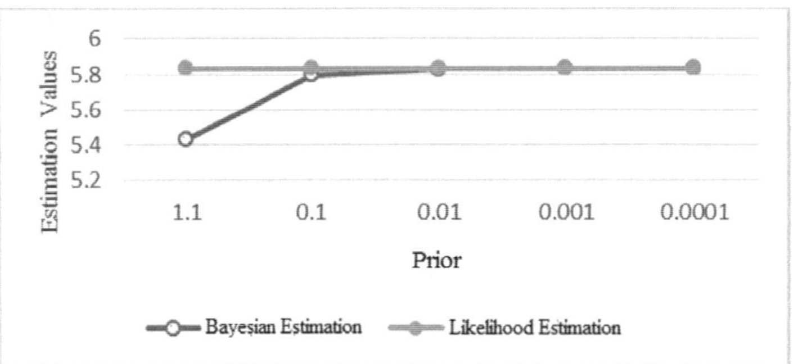

Figure 2 Comparison of parameter λ by estimation of Likelihood and Bayesian

Because the claim frequency data is known, and the values of α_λ and β_λ have been determined, then the posterior distribution of the parameters λ can be expressed by: $\lambda \mid n_i \sim Gamma(70.001,\ 12.001)$. The statistical summary of the mean and variance of parameter λ which was observed over the twelve months expressed in Table 2.

Table 2. Statistical summary of Bayesian estimation of parameter λ which is obtained manually

Parameter	Posterior α_λ	Posterior β_λ	Mean	Variance	Standard Deviation
λ	70.001	12.001	5.832931	0.486037	0.697163

The OpenBUGS program is used to obtain a statistical summary of parameters λ, by simulating the sample data from the posterior distribution in several iterations. In this study, the three chains are used to simulate the posterior distribution of samples, wherein each chain was observed to test the convergence of the parameters visually by using trace plot. The iteration performed on each chain is 10,000 iterations. The statistical summary with the OpenBUGS program yields Bayesian parameter estimation λ, which comprises the mean and standard deviations expressed in Table 3.

Table 3. Statistical summary of Bayesian estimation of parameter λ which is obtained by OpenBUGS program

Parameter	Mean	Standard Deviation	2.5% Percentil	97.5 % Percentil
λ	5.827	0.6985	4.544	7.276

3.3 Estimating of the claims amount distribution model

In this section the purpose is to evaluate the distribution model of the severity of the claim. The steps include: identification of the distribution model, the estimation of the distribution model, and testing the significance of the distribution estimator.

- ***Identification of the claim amount data distribution model***

Identification of the distribution model of the claims amount made by using statistical *software Easyfit 5.6.* The result of curve fitting can be seen in Figure 3.

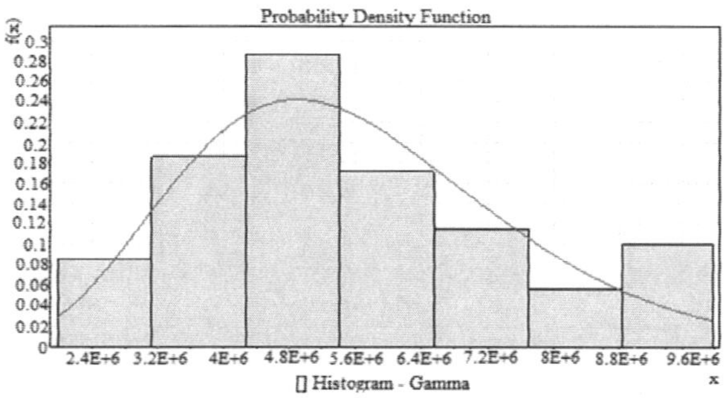

Figure 3. Histogram and density function of claims severity data

Based on the results of curve fitting using *Easyfit 5.6* software, the obtained distribution model which suitable for amount of claims that Gamma distributed with parameters $p = 7.9216$ and $\theta = 1.41437\times10^{-6}$. To determine the most suitable distribution, *Kolmogorov-Smirnov* test was used on the Gamma distribution by the help of *EasyFit 5.6* software. Based on the *Kolmogorov-Smirnov* test, the most suitable model for amount of claim data is Gamma distribution.

- ***Parameter estimating of the claim amount data distribution model***

Based on Figure 3, the selected prior value is $\alpha_\theta = 0.001$ and $\beta_\theta = 0.001$, since the Bayesian estimated value is close to the likelihood estimate of the parameter θ.

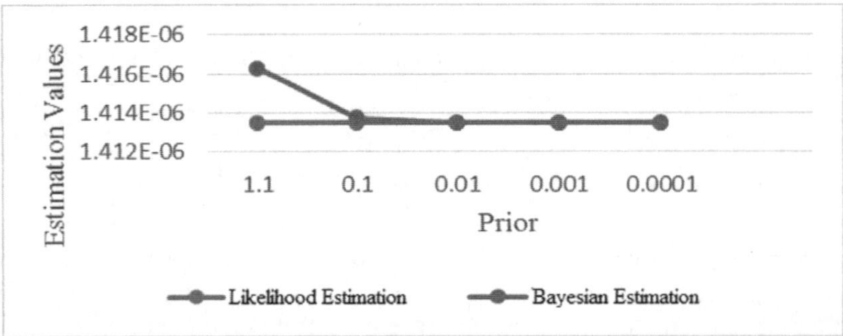

Figure 4. Comparison of parameter estimation of θ by using likelihood and Bayesian

Based on Equation (14), the known value of each hyperparameter α_θ and β_θ are:

$$\alpha_{\theta,1} = pn + \alpha_\theta = (7.9216)(70) + 0.001 = 554.513$$

$$\beta_{\theta,1} = \sum_{i=1}^{n} x_i + \beta_\theta = 392{,}307{,}367 + 0.001 = 392{,}307{,}367.001$$

A statistical summary of the mean and variance of parameter θ, which was observed over the twelve months expressed in Table 4.

Table 4. Statistical summary of Bayesian estimation parameter θ which is obtained manually

Parameter	Posterior α_θ	Posterior β_θ	Mean	Variance	Amount of Claim
θ	554.513	392,307,367.001	1.41×10^{-6}	3.60×10^{-15}	5,604,380.85

The posterior distribution explains the confidence level of the parameters contained in the sample data. Statistical summary of posterior parameter θ shown in Table 5, which consists of the mean and standard deviation. These results are obtained by using statistical software OpenBUGS of three parallel chains and the iterations are done of 10,000 times.

Three parallel chains with an iteration of 10,000 times in each chain are used to test the convergence of parameter θ, which is checked visually by using *trace* plot. Convergence test is done by observe whether the three chain are overlapping each other or not in the *trace* plot.

Table 5. Statistical summary of Bayesian estimation parameter θ which is obtained by OpenBUGS program

Parameter	$E(\theta)$	Standard Deviation	2.5% Percentil	97.5 % Percentil	*E(X)*
θ	1.413×10^{-6}	6.028×10^{-8}	1.298×10^{-6}	1.534×10^{-6}	5,606,227.884

Based on Table 5, it is obtained that the mean of claims approved by the insurance company to be paid to the insured amounting to IDR5,606,227,884.

3.4 Estimating of the collective risk model

This section the purpose is to estimate the collective risk model. Collective risk model is used for the calculation of premiums. Based on equation (1) the obtained number of aggregate claims is $S_t = 392{,}307{,}367$. Based on equation (2) the obtained expectation of collective risk model is $E(S_t) = (5.827)(5{,}606{,}227.884) = 32{,}667{,}489.88$.

While based on equation (3) the obtained variance of collective risk model is:

$Var(S_t) = (5.827)(3,967,610,000,000) + (5,606,227.884)^2(0.4879) = 38,453,900,000,000.$

The expected value and the variance of the previous collective risk model, can be used to predict the premium to be paid by the insured.

3.5 The Calculating of Premium

In this section the intention is to make a calculation of premiums amount to be paid by the insured to the insurance company. The amount of premium depends on the value of the claim frequency and amount of the aggregate claim (collective). In this study, the assumed charge factor at a premium for θ is 0.1 and α is 0.1.

- By using equation (17), the obtained calculation of pure premium is: $p(t) = 2,722,290.823$.
- By using equation (6), the obtained calculation of the expected value premium principles is: $p(t) = 2,994,519.906$.
- By using equation (7), the obtained calculation of variance premium principles is: $p(t) = 320,452,000,000$.
- By using equation (8), the obtained calculation of standard deviation premium principles is: $p(t) = 2,901,301.783$.

4. Conclusions

In this paper, we discussed the estimation of claims risk model and motor vehicle insurance premiums by using Bayesian methods approach. Based on the data processing, the frequency of claims is Poisson distributed with the value of estimated parameter $\lambda = 5,827$, and total claim amount of Gamma distributed with estimated parameter $p = 7,9216$ and $\theta = 1,41347 \times 10^{-6}$. By using both distribution estimators the aggregate claim distribution was formed, then the aggregate claim is obtained on non-life insurance companies amounting to IDR32,667,489.88 with a variance of IDR38,453,900,000,000.00. The value of these expectations and variance is used by the insurance company as a reference in determining the premium value. In this paper, the prediction of the pure premium to be paid by the insured to the insurance company is IDR2,722,290.82, the prediction of the expectation premium is IDR2,994,519.91, the prediction of the variance premium is IDR320,450,000,000.00, and the prediction of standard deviation premium is IDR2,901,301.78. Based on some predictions of such premiums amount, the insurance company can determine the yearly reasonable and affordable premium amount for the insured.

Acknowledgment

Further thanks to the Rector, Director of DRPMI, and Dean of FMIPA, Universitas Padjadjaran, which has a grant program of the Academic Leadership Grant (ALG) under the coordination of Prof. Dr. Sudradjat, and a grant program of Competence Research of Lecturer of Unpad (Riset Kompetensi Dosen Unpad/RKDU) under the coordination of Dr. Sukono, which is a means to increase research activities and publications to researchers at Universitas Padjadjaran.

References

[1] Al-Noor N H and Bawi S F 2015 Bayes Estimators for the Parameters of the Inverted Exponential Distribution under Symmetric and Asymmetric Loss Function. *J. of Natural Sciences Resarch*, 5(4) pp 45-52
[2] Azevedo F C D, Oliveira, T A, and Oliveira, A 2016 Modeling Non-Life Insurance Price For Risk Without Historical Information *REVSTAT – Stat. J.* **14** 2 April 2016 pp 171–192
[3] Bolstad W M 2007 *Introduction to Bayesian Statistics Second Edition* (America: A John Wiley & Sons. Inc)
[4] Dickson D C M 2005 *Insurance Risk and Ruin* (Cambridge: Cambridge University Press)

[5] Djuric Z 2013 Collective Risk Model In Non-Life Insurance *Economic Horizons* **15** 2 May-Agustus 2013: pp 167-175

[6] Eliasson D 2015 Bayesian Credibility Methods for Pricing Non-life Insurance on Individual Claims History *Working Paper* Postal address: Mathematical Statistics Stockholm University, SE-106 91, Sweden. E-mail: daniel@danieleliasson.com.

[7] Gamerman, et al. 2006 *Markov Chain Monte Carlo: Stochastic Simulation for Bayesian Inference* (2nd ed) London : Chapman & Hall/CRC

[8] Gill J 2002 *Bayesian Methods A Social and Behavioral Science approach* (London : Chapman & Hall/CRC)

[9] Inanoglu H and Jacobs H 2009 Models for Risk Aggregation and Sensitivity Analysis: an Application to Bank Economic Capital *J. of Risk and Financial Management* 2(2009), pp 118-189

[10] Jaroengeratikun U, Bodhisuwan W, and Thongteeraparp A 2012 A Bayesian Inference of Non-Life Insurance Based on Claim Counting Process with Periodic Claim Intensity *Open J. of Statistics* No 2 April 2012: pp 177-183

[11] Lumbanbatu R M 2015 Modeling Insurance Claims using a Compound Distribution. *International J. of Science and Research* (IJSR) pp 1505-06

[12] Manimaran R, Balakrishnan V, and Narayanan V 2014 A Collective Risk Theory in Reinsurance *Int. J. of Innovation in Science and Math* 2 Issue 1, ISSN (Online): 2347–9051 Copyright © 2014 IJISM, All right reserved pp 151- 153

[13] Migon et al. 2006 Bayesian Analysis Of A Helath Insurance Model. *J. of Actuarial Practice* **13** 62 pp 61-80

[14] Nino S and Paolo C G 2010 A Collective Risk Model for Claims Reserve Distribution *29th Int. Congress of Actuaries* - ICA 2010, Cape Town – March 7-12th 2010 pp 1-22

[15] Ntzoufras I 2009 *Bayesian Modeling Using WinBUGS: An Introduction* USA:Wikey. http://as.wiley.com/WileyCDA/WileyTitle/productCd-047014114X.html

[16] Sukono, Suyudi M., Islamiyati F, and Supian S 2017 Estimation Model of Life Insurance Claims Risk for Cancer Patients by Using Bayesian Method. *IOP Conf. Series: Materials Science and Engineering* 166 (2017) 012022 doi:10.1088/1757-899X/166/1/012022. pp. 1-9.

Econometric Model of Rice Policy Based On Presidential Instruction

Surya Abadi Sembiring and Julia Hutauruk
Faculty of Agriculture, Catholic University of St. Thomas SU

Email: sabadi@ymail.com, juliahutauruk@ymail.com

Abstract. The objective of research is to build an econometric model based on Presidential Instruction rice policy. The data was monthly time series from March 2005 to September 2009. Rice policy model specification using simultaneous equation, consisting of 14 structural equations and four identity equation, which was estimated using Two Stages Least Squares (2SLS) method. The results show that: (1) an increase of government purchasing price of dried harvest paddy has a positive impact on to increase in total rice production and community rice stock, (2) an increase community rice stock lead to decrease the rice imports, (3) an increase of the realization of the distribution of subsidized ZA fertilizers and the realization of the distribution of subsidized NPK fertilizers has a positive impact on to increase in total rice production and community rice stock and to reduce rice imports, (4) the price of the dried harvest paddy is highly responsive to the water content of dried harvest paddy both the short run and long run, (5) the quantity of rice imported is highly responsive to the imported rice price, both short run and long run.

1. Introduction

According to Timmer [19], most government intervention in their economies to raise resources, at least an indirect effect on food prices facing consumers or crops prices facing farmers. Streeten[17] argued that policymakers are confront with a fundamental dilemma, between high prices for food or lower price food. Tobias et al, [18] described many Asian countries through policy instrument to achieve rice self-sufficiency and stabilize domestic prices. They discussed production policy in Indonesia such as fertilizers subsidy farmers, price supports and food subsidy for rice price.

Based on presidential instruction (*Inpres*), policymaker sets the objectives of rice policy are farmer's income, rural economic development, stabilization of national economy and food security Sembiring. [14]. Sembiring [13] constructed model of rice policy based on Inpres to describe the impact of government policies on farmers income, food price and food security. Dawe [2] discussed that the benefits of stabilizing food prices. Ellis [6] argued that in the cases Indonesian state, Bulog achieves an ostensibly difficult price stabilizations role.

Jha and Srinivasan [7] concluded that price stabilization is the most important element of food policy in India and many other countries in both developing and developed countries. Cumming et al, [1] argued that India, Indonesia, Pakistan and the Philippines are implementing the stabilization of grain prices, Poultan et al, [12] described the impact food price instability on household. The result show that poor consumers, high prices reduce real income, especially during low season

Before the 2010, research on rice policy based on Presidential Instruction (Inpres) was limited in Indonesia. With the development of behavioral equation, Sembiring [13,15] constructed economic analysis on rice policy based on Presidential Instruction. His derived from Timbergen framework policy analysis constructed simultaneous equation. The Timbergen framework is also known as the objectives-constraints-instruments approach to policy analysis Ellis [5]. Sembiring [15] suggested that policy instrument to achieve food security an increase the government purchased price of dried harvest paddy, an increase distribution of fertilizers, an increase of the number household receive Raskin and to develop irrigation system.

The objective of this study is to constructed econometric model based on Timbergen policy analysis. This study is expected to describe the correlation policy objectives – constrains and policy instrument based on rice policy on presidential instruction.

2. Data and Methods

Data type is monthly time series data Sembiring [13]. The econometric model of rice policy uses a simultaneous equation, consisting of 14 structural equations and four identity equations. The identity equations in this model based on Sembiring [13]. According to Hallam [8], the process of econometric modelling consists of specification, estimation, validation, and finally application Rice policy model specification using simultaneous equation, consisting of 14 structural equations and 4 identity equation. System contains 18 equations (G) with 50 variabels (K). An equation retailer rice price containing 7 variables (M) and the counting rule (K-M) > (G-1, Koutsiannis [11] gives (50 -7) > (18-1), retailer rice price equation is over identified. Therefore other equation is must over identified. When the equation is over identified it can be proved by 2SLS (Two Stage Least Squares).

3. Result and Discuss

According to Table 2, all the equation has the coefficient of determination R2 than those for quantity of rice import and rate of inflation more than 0.3366. The coefficient of determination R2 close to zero is 0.0704, it means that 7.04 per cent of the variation in quantity of rice import can be attributed to variations in price of import rice, community rice stock and lagged quantity of rice import, Thomas [20]. Definition of variables, as shown in Table 1.

According to Table 2, each lagged variable has a positive sign and significant at the 1 per cent confidence level whereas lagged quantity of rice import significant at the 15 per cent confidence level. According to the estimated equation, all the equation has statistic F than those for quantity of rice import and rate of inflation are significant at the 1 per cent. It means that the explanatory variables of each equation together can explain the endogenous variables well. According to the "(1)" paddy harvested area depends positively and significantly on HPGP, CHIT and LLAPT but for the HJRT variable is negative and significant. Based on the "(2)" yield of paddy depend negatively and significantly HPZA and HPNP, but for T and LYPIT is negative and significant. This is consistent with Sembiring et al, [16], their found that the government policies to increase NPK fertilizers by 15 per cent lead on decrease paddy harvested area 0,288 per cent

Table 1. Definition of the Variables

Variable	Definition	Variable	Definition	Variable	Definition
LAPT	Paddy Harveste Area	QPIT	Paddy Production	QBIT	Rice Production
HJRT	Price of Producen Corn	QCBB	Bulog Rice Stock	QCBG	Government Rice Stock
QMBT	Quantity of Rice Import	QCBD	Community Rice Stock	YPIT	Yield of Paddy
HGKP	Price of Dried Harvest Paddy	HBRT	Price of Retailer Price	HMBT	Price of Import Rice
INFT	Rate of Inflation	HPBB	Bulog Purchases Rice	T	Trend of Technology
CHIT	Average Rainfall	INFB	Rice Contribution to Inflation	STBF	Bulog Rice Final Stock
STOB	Bulog Operational Stock	EXRT	Rupiah Exchange Rate	STGF	Government Rice Final Stock
RAST	Rice for Poor People	SDBI	Indonesia Rice Surplus	D	Dummy Market Operation
HPGP	Government Purchase Price of Dried Harvest Paddy	HPZA	Price of Subsidized ZA Fertilizers	HPNP	Price of Subsidized NPK Fertilizers
RGHZ	Rasio HPGP to Ceiling Retailer Price of ZA Fertilizers	RGHN	Ratio HPGP to Ceiling Retailer Price of NPK Fertilizers	RTRA	Number of Household Receive Raskin
DCBP	Distribution of Government Rice	QBLD	Rice for Seed, Shrinkage and Others	PBPT	PBPT
RPZA	Realization of the Distribution of Subsidized ZA Fertiizer	RPNP	Realization of the Distribution of Subsidized NPK Fertilzer	KAGP	Dried Harvest Paddy with Water Content
LLAPT	Lagged Paddy Harvested Area	LYPIT	Lagged Yield of Paddy	LHBRT	Lagged Price of Retailer Price
LQMBT	Lagged Quantity of Rice Import	LINFT	Lagged Rate of Inflation	LQCBG	Lagged Government Rice Stock

LRAST	Lagged Rice for Poor People	LHPZA	Lagged Price od Subsidized ZA Fertilizer	LHPNP	Lagged Price of Subsidized NPK Fertilizer
LHGKP	Lagged Price of Dried Harvest Paddy	LDCBP	Lagged Distribution of Government Rice	LHMBT	Lagged Price of Import Rice
LSTBF	Lagged Bulog Rice Final Stock	LSTGF	Lagged Government Rice Final Stock		

All coefficients sign as in "(7)" than STGF agree with prior expectations and all variable than RTRA are statistically significant. The negative sign of STGF indicates that it leads to reduce RAST and significant at the 1 per cent confidence level. The coefficient of RTRA variable is positive and insignificant. All coefficients sign as in "(8)" and in "(9)" agree with prior expectations. In "(8)" all variable than QCBD are statistically significant. In "(9)" all variable than those for HBRT and INFB are statistically significant. The positive sign of QMBT indicates that it leads to increase HMBT and insignificant. This result is similar to Dawe finding [3] found that an increase of Indonesian rice import by one million ton increased price of word rice by US $ 8,56 per ton, while Dawe [4] found that price of word rice increase by US $ 12,38 per ton.

As in "(5)" all variable except QCBB are statistically significant and all coefficients agree with prior expectations. Price of dried harvest paddy depend negatively and significantly on the KAGP. The coefficient of HPGP, and LHGKP are positive and significant but for QCBB is insignificant. As in "(6)" show all variable except QCBB are statistically significant and all coefficients than DCBP agree with prior expectations. The coefficients of HMBT, HGKP, STBF, and LHBRT are positive but the coefficient of DCBP is negative, tending to contradict the hypothesis. Price of retailer rice depends negatively and significantly on RAST. The negative coefficient of RAST means the increase in RAST reduce price retailer rice.

All coefficients sign as in "(7)" than STGF agree with prior expectations and all variable than RTRA are statistically significant. The negative sign of STGF indicates that it leads to reduce RAST and significant at the 1 per cent confidence level. The coefficient of RTRA variable is positive and insignificant. All coefficients sign as in "(8)" and in "(9)" agree with prior expectations. In "(8)" all variable than QCBD are statistically significant. In "(9)" all variable than those for HBRT and INFB are statistically significant. The positive sign of QMBT indicates that it leads to increase HMBT and insignificant. This result is similar to Dawe finding [3] found that an increase of Indonesian rice import by one million ton increased price of word rice by US $ 8,56 per ton, while Dawe [4] found that price of word rice increase by US $ 12,38 per ton.

Table 2. The estimated econometric model of rice policy based on presidential instruction

EN# The structural and identity equation

(1) LAPT = - 82,7540 + 0,2726 HPGP***** - 0,0583 HJRT + 0,8625 CHIT***** + 0,5614 LLAPT*,
(0,488/*1,113*) (-0,310/*-0,297*) (0,147/*0,334*)
R^2 = 0.33665, DW = 1.7571, F = 6.22*

(2) YPIT = 17,6361 - 0,0007 HPZA***** - 0,0005 HPNP** + 0,0259 T* + 0,6816 LYPIT*,
(-0,281/*-0,882*) (-0,377/*-1,185*) (0,155/*0,486*)
R^2 = 0.89721, DW = 1.7670, F =106.92*

(3) HPZA = 314,4367 + 19,7054 RGHZ - 0,8266 RPZA***** + 0,8333 LHPZA*,
(0,020/*0,121*) (-0,026/*-0,156*)
R^2 = 0.71429, DW = 1.8594, F = 41.67*

(4) HPNP = 686,0720 + 71,7761 RGHN - 1,9149 RPNP **** + 0,8241 LHPNP*,
(0,022/*0,122*) (-0,029/*-0,168*)
R^2 = 0.69947, DW = 1.8740, F = 38.79*

(5) HGKP = 2590,6480 + 0,1527 HPGP*** - 131,7160 KAGP* + 0,0076 QCBB + 0,8422 LHGKP*
(0,127/*0,806*) (-1,096/*-6,951*) (0,005/*0,034*)
R^2 = 0.94945, DW = 1.6793, F = 230.08*

*(6) HBRT = 83,8785+ 0,0139 HMBT**** + 0,5390 HGKP* + 0,3464 DCBP**** - 0,1781 RAST*** + 0,0288 STBF****

(0,010/*0,036*) (0,250/*0,919*) (0,001/*0,003*) (-0,007/*-0,026*) (0,008/*0,031*)

+ 0,7280 LHBRT*, R^2 = 0.99559, DW = 1.7138 F = 1769.69*

(7) RAST = -15,1637 + 0,8692 STOB* + 0,0005 RTRA - 0,7067 QCBG* + 0,6998 STGF* + 0,1365 LRAST*
(0,955/*1,106*) (0,039/*0,045*) (-1,509/*-1,747*) (1,458/*1,689*)
R^2 = 0.96650, DW = 2.4375, F = 277.00*

(8) QMBT = 134,9944 - 0,0203 HMBT**** - 0,0008 QCBD + 0,1536 LQMBT****
(-1.440/*-1,701*) (-0,046/*-0,055*)
R^2 = 0.07043 DW = 2.0606, F = 1.26

(9) HMBT = -118,9950 + 0,0453HBRT + 0,0120 QMBT + 0,0050 SDBI** + 0,9392 LHMBT*
(0,064/*1,056*) (0,0001/*0,003*) (0,004/*0,060*)
R^2 = 0.94732, DW = 1.4393, F = 220.28*

(10) INFT = 1,7776 - 0,0002 HBRT***** + 2,3971 INFB* + 0,00003 EXRT* + 0,0822 LINFT
(-1.286/*-1,401*) (0,164/*0,178*) (-0,397/*-0,432*)
R^2 = 0.15795, DW = 1.9606, F = 2.30

(11) DCBP = - 11,5089 - 0,0027HBRT + 17,8076 D* + 0,0526 QCBG*** + 1,8137INFT + 0,4888 LDCB*
(-1,356/*-2,652*) (0,684/*1,338*) (2,189/*4,283*) (0,133/*0,260*)
R^2 = 0.44954, DW = 1.8251, F = 7.84*

(12) STBF = - 31,2846 - 0,4063HBRT + 0,6202 HPBB* + 0,6276 STOB*** + 0,7046 LSTBF
(-1,380/*-4,671*) (1,606/*5,439*) (0,092/*0,311*)
R^2 = 0.71080, DW = 1.9574 F = 30.11*

(13) STGF = 43,7619 - 0,4436 DCBP*** + 0,9147 LSTGF* , R^2 = 0.80437, DW = 2.0353, F =104.85*
(-0,011/*-0,128*)

(14) QCBG = -8,6144 + 0,9318 PBPT* + 0,0212 STBF* + 0,9276 LQCBG* R^2 = 0.95181, DW = 1.4655 F = 329.20*
(0,030/*0,415*) (0,075/*1,030*)

(15) QPIT = LAPT x YPIT

(16) QBIT: Fk x QPIT

(17) QBLD = Fp x QBIT

(18) QCBD = QBIT - QBLD

Note: * significant 1%, ** significant 5%, *** significant 10%, **** significant 15%, and *****significant 20%.
In this table the number in parentheses are short run and long run elasticity , and italic number is long run elasticity, Intriligator [9] ; the coefficient of determination R^2, Durbin-Watson statistics are given as DW, and F statictic value. # equations numbering.

Based in "(10)" and "(11)", the effect of retailer rice price on INFT is found to significant and negative as was not theoretically expected, in addition, the coefficient HBRT on DCBP is negative and is not significant as was not theoretically expected. In "(10)" all variable than those for EXRT and LINFT are statistically significant and in "(11)" all variable than those for HBRT and INFT are statistically significant.

All coefficients sign in "(13)" and "(14)" all variables agree with prior expectations and are statistically significant. Whereas in "(12)" the coefficient of the HBRT and STOB are wrong sign and all variable than those for HBRT and LSTBF are statistically significant. The remaining fourth equations are identities. In "(15)" is total paddy production, in "(16)" total rice production, in "(17)" rice for seed, shrinkage and other, and in "(18)" Community rice stock is the difference between total rice production and rice for seed and others.

The elasticity estimates are shown in Table 2. According to the study on elasticity both short run and long run, there are 6 structural equation have highly responsiveness between endogenous variable and explanatory variables, both short run and long run elasticity: (i) the estimates of the elasticity of HGKP with respect to the KAGP are -1,0969 and -6,951, respectively in equation 5. It means, increasing KAGP by 1 per cent reduce HGKP by more than 1,0969 per cent in short run and 6,951 per cent in long run. (ii) the short run and long run elasticity RAST are -1,509 and -1,747 with respect to QCBG, and 1,458 and 1,689 with respect to STGF in "(7)", (iii) the estimates of the elasticity of QMBT with respect to the HMBT are -1,440 and -1,701, respectively in equation 8. This indicates that, increasing HMBT by 1 per cent reduce QMBT by more than 1,440 per cent in short run and 1,701 per cent in long run (iv) the short run and long run elasticity INFT are -1,286 and -1,401 with respect to SDBI in "(10)",

(v) the short run and long run elasticity DCBP are -1,356 and -2,652 with respect to HBRT, and 2,189 and 4,283 with respect to QCBG in "(11)", and (vi) in "(12)", the short run and long run elasticity STBF are -1,380 and -4,671 with respect to HBRT, and 1,6068 and 5,4399 with respect to HPBB.

An increase of HPGP has a positive impact on the LAPT, and an increase of the LAPT affects an increase of YPIT. The increase in YPIT was followed by an increase in QBIT, although QBLD increase, QCBD also increased. An increase in QCBD leads to the QMBT to decrease. The increase of HPGP has a positive impact on HGKP. An increase HGKP pushed up the HBRT. In terms of input usage, an increase in RGHZ and RGHN to increase HPZA and HPNP, respectively. Whereas, an increase RPZA and RPNP to reduce HPZA and PHNP, respectively, followed an increase YPIT. An increase YPIT, to increase QPIT and QBIT and QCBD. An increase QCBD to reduce QMBT.

4. Conclusion

An increase of government purchasing price of dried harvest paddy has a positive impact on the increase of total rice production and community rice stock. An increase community rice stock leads to decrease the rice imports. An increase of the realization of the distribution of subsidized ZA fertilizers and the realization of the distribution of subsidized NPK fertilizers have a positive impact on the increase in total rice production and community rice stock and to reduce rice imports. The price of the dried harvest paddy is highly responsive to the water content of dried harvest paddy both in the short run and long run.

5. Another Section of Your Paper

Policy instrument to achieve the objective policy are increasing the government purchased price of dried harvest paddy, distribution of fertilizers, and improve technology.

6. Acknowledgments

The authors would like to thank Higher Education (Dikti) for financial support on this research project.

References

[1] Cumming J, Rashid RS and Gulati A 2006 Grain price stabilization experiences in Asia: What have we learned? Food Policy 31 302

[2] Dawe D 2001 How far down the path to free trade ? The importance of rice price stabilization in developing Asia. *Food Policy* 26 163

[3] Dawe D 2002 The Changing structure of the world rice market, 1950-2000. *Food Policy 27 355.*

[4] Dawe D 2008 Can Indonesia trust the world rice market? *Bulletin of Indonesian Economic Studies*, 34 (1) 115.

[5] Ellis F 1992 *Agricultural Policies in Developing Countries*. (Cambridge: Cambridge University Press).

[6] Ellis F 1993 Private trade and public role in staple food marketing. The case of rice in Indonesia. Food Policy 18 428.

[7] Jha S, Srinivasan PV 1999 Grain price stabilizations in India: Evaluation of policy alternatives. *Agricultural Economics* 21 93.

[8] Hallam D 1990 *Econometric Modelling of Agricultural Commodity Markets*. (London :Routledge).

[9] Intriligator, MD 1980. *Econometric Models, Techniques, and Applications*. (New Dehli: Prentice –Hall of India) p 466-499

[10] Kako T, Gemma M and Ito S. 1997 Implications of the minumum access rice Import on supply and demand balance of rice in Japan. *Agricultural Economics* 16 (3) 193

[11] Koutsoyiannis, A 1977. Theory of Econometrics: An Introductory Exposition of Econometric Methods. (New York :Harper and Row Publishers Inc).p 352-353

[12] Poultan C, et al. 2006. State intervention for food price stabilization in Africa: Can it work? Food

Policy 31 342.

[13] Sembiring S A 2011 Economic Analysis of Presidential Instruction on National Rice Policy Year 2005-2008. Doctoral Dissertation. Graduate School, Bogor Agricultural University, Bogor.

[14] Sembiring SA 2014 Proc. Nat. Editor Simbolon A. (Medan: Bina Media Perintis). p 20-32

[15] Sembiring SA 2011 Towards Green Technology for Better Future. Symp.,Editor: Halimmatudahliana., Suryanto. D. (Medan: USU Press)

[16] Sembiring SA, Harianto, Siregar H, dan Saragih B 2012 Dampak kebijakan pemerintah melalui Instruksi Presiden tahun 2005-2008 tentang kebijakan perberasan terhadap ketahanan pangan. Forum Pascasarjana. Vol 35 15.

[17] Streeten P 1987 Transitional Measures and Political Support for Food Price Policy in Food Policy: Integrating Supply, Distribution and Consumption. Edited by Gittinger. J.P, J.Leslie., C. Hoisington. (Baltomore: The Johns Hopkins University Press). p 277-281

[18] Tobias A, Molina I, Valera HG, Mottaleb KA, Mohanty S 2012 Handbook on rice policy for Asia. (Los Banos: International Rice Research Institute). p 22.

[19] Timmer, C P 1987 Price Policy and the Political Economy of Markets in Food Policy: Integrating Supply, Distribution and Consumption. Edited by Gittinger.J.P, J.Leslie., C. Hoisington. (Baltomore: The Johns Hopkins University Press). P 264-276

[20] Thomas, R L 1997 Modern Econometrics An Introduction. (Harlow: Addison Wesley Longman) p 89.

IOP Conf. Series: Materials Science and Engineering **300** (2018) 012029 doi:10.1088/1757-899X/300/1/012029

An Improved Version of the Cervical Vertebral Maturation (CVM) Method for the Assessment of Mandibular Growth in Deutero-Malay Sub Race

Fahmi Oscandar[1], Yuti Malinda[2], H. Azhari[3], Nani Murniati[4], Sing Yeh Ong[5], Subiyanto[6], Sudradjat Supian[7*]

[1,3]Department of Dentomaxillofacial Radiology-Forensic Odontology, Faculty of Dentistry, Universitas Padjadjaran, Indonesia

[2]Department of Oral Biology-Microbiology-Forensic Odontology, Faculty of Dentistry, Universitas Padjadjaran, Indonesia

[4]Depertment of Oral Biology-Anatomy-Forensic Odontology, Faculty of Dentistry, Universitas Padjadjaran, Indonesia

[5]Undergraduate of Faculty of Dentistry, Universitas Padjadjaran, Indonesia

[6]Department of Marine Science, Faculty of Fishery and Marine Science, Universitas Padjadjaran, Indonesia

[7]Department of Mathematics, Faculty of Mathematics and Natural Science, Universitas Padjadjaran, Indonesia

*Corresponding author: sudradjat@unpad.ac.id

Abstract. In this paper, Cervical Vertebral Maturation method was used to assess the mandibular growth in Deutero-Malay sub race. Twenty eight laterals Cephalometric radiographs of Deutero-Malay sub race aged 9-15 were observed. The observation used stratified random sampling by measuring the quantitative and qualitative assessment of the 2nd through 4th cervical vertebra of the subjects. It produced the diagram of developmental stages of cervical vertebrae for Deutero-Malay sub race. The diagram can be used to determine mandibular growth in term of qualitative by matching the shape of cervical vertebrae. It was obtained that the Cervical Vertebral Maturation method can be used to assess mandibular growth in Deutero-Malay sub race by matching the shape of cervical vertebrae to the diagram of developmental stages of cervical vertebrae. In addition, Cervical Vertebral Maturation method can be used to identification person's age.

1. Introduction

Recent studies have shown that cervical vertebral growth is linked intimately to the maturation and growth of mandibular. To determine the normal growth rate, some clinical studies identified that during the circumpubertal growth period is when it shows greatest response. With this lead in research, the subject target for this thesis are male and female subjects aged 9 to 15 [1].

In the study of forensic odontology, determination of normal growth rate is essential to aid in identifying a person's credibility to optimal growth rate. By determining the normal growth of a person, it can aid other researches to estimate the age of a certain subject using various types of age estimation methods.

An improved version of Cervical Vertebral Maturation (CVM) method, a maturational index that have been proposed to evaluate skeletal maturity in patient has proven to be effective in determining growth rates in terms of changes in mandibular size and body height. The CVM method comprises of 6 stages (Cvs 1 - Cvs 6), each stage shows the different stages of cervical vertebral growth throughout the pubertal years provided by an analysis which will be carried out in two ways that is visual analysis and cephalometric analysis [2].

The motive of choosing this tittle is to understand and to provide an understanding of how the cervical vertebral maturation method can be used to determine mandibular growth for deutero-Malay sub race ethnics in Indonesia by using the CVM method.

The main reason of choosing deutero-Malay sub race as the subject of this thesis is because there are no data or researches that have been carried out on this race before in Indonesia. Since the research is to be done in Indonesia, it's vastest ethnics currently populating this country is the deutero-Malay sub race and it is only wise to choose this race as my thesis subject.

2. Materials and Methods

The morphology of the bodies of second (odontoid process, C2), third (C3), and fourth (C4) cervical vertebrae were analysed in 6 consecutive observations. The analysis comprised of visual and cephalometric appraisals of morphological characteristics of the cervical vertebrae.

Visual analysis will be the assessment through shape evaluated by visual inspection. Two sets of variables were analysed: (1) presence of a concavity at the lower border of the body of C2, C3, and C4; and (2) shape of the body of C3 and C4. The following explains the assessment of shape [3]:

1. Trapezoid (the superior border is tapered from posterior to anterior);
2. Rectangular horizontal (the heights of the posterior and anterior borders are equal; the superior and inferior borders are longer than the anterior and posterior borders);
3. Squared (the posterior, superior, anterior and inferior borders are equal);
4. Rectangular vertical (the posterior and anterior borders are longer than the superior and inferior borders).

Cephalometric analysis is by measuring the percentage of increase and decrease in concavity of C2 through C4. The cephalometric analysis is carried out by plotting several points on the digitalized cervical vertebral bodies from the cephalogram and is traced and digitized for the description of morphological characteristics. From the 13 points that were plotted and measured from the same lateral cephalogram, three liner and four ratio measurements which is adopted partially from Hellsing [4, 5] are made.

The subjects of this research are 28 deutero-Malay sub race patients age 9 to 15 who come to the Dentomaxillo facial Radiology Installation of Dental Hospital, Faculty of Dentistry, Padjadjaran University. The sampling method of this research is stratified random sampling and the research method is observational analytical study.

3. Results and Discussion

Table 1 shows the visual analysis of the morphologic characteristic of cervical vertebrae (C2, C3, C4). The features of examined vertebrae at the seven observations can be summarized as follows:

T9. A concavity is evident at the lower border of C2 in 50% of the subjects. Concavity is absent in C3 and C4 in 100% of subjects. The bodies of C3 and C4 are trapezoid in shape for all subjects.
T10. A concavity is not observed in all the lower borders of C2, C3, and C4 in all subjects. Both C3 and C4 are still trapezoid in shape in 100% of subjects.
T11. A concavity is present at the lower border of C2 in 75% of subjects and only 25% of subjects on C3. No concavity is present at the lower border of C4. 75% of subjects have trapezoid shaped body in C3 and C4 and only 25% of subjects have rectangular horizontal shaped body in C3 and C4.
T12. A concavity is present at the lower border of C2 and C3 in 50% of the subjects, but only 25% of subjects have concavity at the lower border of C4. The shape of both C3 and C4 can either be trapezoid or rectangular horizontal.
T13. This stage is characterized by the presence of concavity in all the lower border of C2, C3, and C4. 50% of subjects have concavity at the lower border of C2 and C4, and an astonishing 75% of subjects with concavity at lower border of C3. There is also a presence of squared shaped C3 and C4 body with 50% of subjects carrying squared shaped body in C4 and the remaining with the shape of rectangular horizontal.
T14. By this stage, a concavity can be observed in all the lower borders of C2, C3, and C4 in all the subjects except for a non-significant 25% of subject with no concavity at the lower border of C2. Evidence of rectangular vertical shaped body indicates the starting of maturation peak even though there is only 25% of subjects with rectangular vertical shaped body in C3 and C4 with the remaining body shaped in square.
T15. A concavity is present in all the lower borders of C3, and C4 in 100% of the subjects and 75% of subjects in C2 with an exception of a non-significant 25% of subject with no concavity at the lower border of C2. The body of C3 is rectangular vertical in 50% of subjects and the remaining with trapezoid and square shaped body. The body of C4 is also rectangular vertical in 50% of subjects and squared in the remaining 50%. This indicates the happening of mandibular growth peak.

Table 2 shows the descriptive statistics for cephalometric measurements of vertebral morphologic characteristics, together with statistical comparisons between observations.

There are no significant differences for any of the measurements assessed between T9 and T10. The depth of concavities at the lower border of the second (C2Conc) and third (C3Conc) cervical vertebra was significantly greater at T13 when compared to T12.

In the transition from T10 to T11, the height of anterior border of both C3 and C4 increases significantly, leading to significant decrements in the ratio between the heights of the posterior and anterior borders of the vertebral bodies (C3PAR and C4PAR).

At T12, the depth of concavity at the lower border of C4 (C4Conc) becomes significantly greater than at T11. In the transition from T11 to T12, the height of the anterior borders of both C3 and C4 have increase significantly again, thus leading to significant decrements both in the ratio between the heights of the posterior and anterior borders of the vertebral bodies (C3PAR and C4PAR) and in the ratio between the length of the base and the anterior height of the vertebral bodies (C3BAR and C4BAR). On average, C3PAR and C4PAR now have a ratio of approximately 1:1, an indication that both C3 and C4 vertebral bodies are rectangular horizontal in shape.

At T13, the depth of concavity at the lower border of C2 and C3 has the greatest and most significant increase when compared to T12. This indicates the peak of mandibular growth is greatest during this transition. Decrements of the ratio between the length of the base and the anterior height of the vertebral bodies (C3BAR and (C4BAR). The mean values for these measurements indicate that the vertebral bodies become progressively more squared in shape.

At T14 and T15, the depth of concavity at the lower border of C2, C3, and C4 remain greatly significant than the earlier stages. One-third of the cases show a rectangular vertical shape of one or both C3 and C4 vertebral bodies.

4. Conclusion

The objective of the present investigation was to aid in determining the condition of a patient if the growth rate is normal, and more specifically, a direct appraisal of the skeletal maturity of the mandible in relation to the changes in morphological features of the cervical vertebrae.

Secondly, an evaluation of the morphological features of the cervical vertebral bodies restricted to those that are visible on the lateral cephalogram even when a protective collar is worn, as originally proposed by Hassel and Farman [3]. Third, a description of the cervical vertebral morphology at each developmental stage which will allow the clinician to apply the method based on the information derived from a single cephalogram, and lastly, an evidence to support and prove if a subject has a normal growth rate by comparison of the morphological changes of the cervical vertebrae with the length of the mandibular growth.

The weakness of the research is that the subjects used are not consecutive as there are no available consecutive radiographs in the radiology installation. In addition to that, the research is conducted by merging the measurements from male and female subjects which may have reduced the accuracy by a little. Third and lastly, the capturing of the cephalometric radiographs is performed by different operators which may have produced a difference in radiograph in terms of technique of capturing and positioning of the patient.

In Figure 1, the diagram of developmental stages of cervical vertebrae guidance is as such. Stage DS1 is accounted for T9 and T10 showing no presence of concavity on all lower borders of cervical vertebrae. Shape of C3 and C4 is trapezoid. T11 is categorised into stage DS2 where C2 shows concavity at the lower border. The length of C3ua-C3la on the anterior border of C3 and C4 shows increase in length however, still not long enough to change the shape of C3 and C4 therefor remain as trapezoid shape.

DS3 is marked with concavity at the lower border of C2 and C3. The shape change in C3 from trapezoid shape to rectangular horizontal in shape which in this case is seen in stage T12. Stage T12 is included in the DS4 because of the presence of shape change in C4 from trapezoid to rectangular horizontal in shape while C2, C3 and C4 remain to have concavity present at the lower borders. DS5 includes the change in both C3 and C4 in terms of shape. Transitioning from rectangular horizontal to square shaped, T13 shows the distinctive changes thus is grouped in this category.

Stage DS6 can be said as the starting of a static shape change stage as it marks the starting of shape transition from square to rectangular vertical in C3 and C4 in which will remain unchanged in the upcoming stages and this characteristic is observed in T14. T15 has similar characteristics as seen in T14 but with a higher prevalence showing rectangular vertical shape in C3 and C4 vertebras.

By applying the knowledge and referring to the Figure 1 and Table 3 of developmental stages of cervical vertebral growth, a physician will be able to determine if a deutero-Malay sub race individual has a normal growth rate by comparing his or her cephalometric radiograph to the references above.

Acknowledgment

We would like to thank the financial support from Universitas Padjadjaran through the grant program of the Academic Leadership Grant (ALG) under the coordination of Prof. Dr. Sudradjat.

References

[1] O'Reilly M T, Yanniello G J. 1988. Mandibular growth changes and maturation of cervical vertebrae--a longitudinal cephalometric study. *Angle Orthod.*, 58(2):179-184. doi:10.1043/0003- 3219(1988)058<0179:MGCAMO>2.0.CO;2.

[2] Franchi L, Baccetti T, McNamara J A. 2000. Mandibular growth as related to cervical vertebral maturation and body height. *Am J Orthod Dentofac Orthop.*, 118(3):335-340.

[3] Hassel B, Farman A G. 1995. Skeletal maturation evaluation using cervical vertebrae. *Am J Orthod Dentofac Orthop.*,107(1):58-66. doi:10.1016/S0889- 5406(95)70157-5.

[4] Hellsing E. 1991. Cervical vertebral dimensions in 8-, 11-, and 15-year-old children. *Acta Odontol Scand.*, 49(4):207-213. doi:10.3109/00016359109005909.

[5] Flores-Mir C, Burgess C A, Champney M, Jensen R J, Pitcher M R, Major P W. 2006. Correlation of skeletal maturation stages determined by cervical vertebrae and hand-wrist evaluations. *Angle Orthod.*, 76(1):1-5. doi:10.1043/0003- 3219(2006)076[0001:COSMSD] 2.0.CO;2.

Table 1 Results of Qualitative Analysis of Cervical Vertebral (C2-C4) Characteristics at the Six Consecutive Observations (T_9-T_{15})

	T9		T10		T11		T12	
	Yes	No	Yes	No	Yes	No	Yes	No
Concavity at the lower border of C2	2 (50%)	2 (50%)	0 (0%)	4 (100%)	3 (75%)	1 (25%)	2 (50%)	2 (50%)
Concavity at the lower border of C3	0 (0%)	4 (100%)	0 (0%)	4 (100%)	1 (25%)	3 (75%)	2 (50%)	2 (50%)
Concavity at the lower border of C4	0 (0%)	4 (100%)	0 (0%)	4 (100%)	0 (0%)	4 (100%)	1 (25%)	3 (75%)
C3 shape: trapezoid	4 (100%)	0 (0%)	4 (100%)	0 (0%)	3 (75%)	0 (0%)	2 (50%)	0 (0%)
C4 shape: trapezoid	4 (100%)	0 (0%)	4 (100%)	0 (0%)	1 (25%)	0 (0%)	1 (25%)	0 (0%)
C3 shape: rectangular horizontal	0 (0%)	0 (0%)	0 (0%)	0 (0%)	1 (25%)	0 (0%)	2 (50%)	0 (0%)
C4 shape: rectangular horizontal	0 (0%)	0 (0%)	0 (0%)	0 (0%)	0 (0%)	0 (0%)	3 (75%)	0 (0%)
C3 shape: squared	0 (0%)	0 (0%)	0 (0%)	0 (0%)	0 (0%)	0 (0%)	0 (0%)	0 (0%)
C4 shape: squared	0 (0%)	0 (0%)	0 (0%)	0 (0%)	0 (0%)	0 (0%)	0 (0%)	0 (0%)
C3 shape: rectangular vertical	0 (0%)	0 (0%)	0 (0%)	0 (0%)	0 (0%)	0 (0%)	0 (0%)	0 (0%)
C4 shape: rectangular vertical	0 (0%)	0 (0%)	0 (0%)	0 (0%)	0 (0%)	0 (0%)	0 (0%)	0 (0%)

Table 1 Extended

	T13		T14		T15	
	Yes	No	Yes	No	Yes	No
Concavity at the lower border of C2	2 (50%)	2 (50%)	3 (75%)	1 (25%)	3 (75%)	1 (25%)
Concavity at the lower border of C3	3 (75%)	1 (25%)	4 (100%)	0 (0%)	4 (100%)	0 (0%)
Concavity at the lower border of C4	2 (50%)	2 (50%)	4 (100%)	0 (0%)	4 (100%)	0 (0%)
C3 shape: trapezoid	1 (25%)	0 (0%)	2 (50%)	0 (0%)	1 (25%)	0 (0%)
C4 shape: trapezoid	1 (25%)	0 (0%)	2 (50%)	0 (0%)	0 (0%)	0 (0%)
C3 shape: rectangular horizontal	2 (50%)	0 (0%)	0 (0%)	0 (0%)	0 (0%)	0 (0%)
C4 shape: rectangular horizontal	1 (25%)	0 (0%)	0 (0%)	0 (0%)	0 (0%)	0 (0%)
C3 shape: squared	1 (25%)	0 (0%)	1 (25%)	0 (0%)	1 (25%)	0 (0%)
C4 shape: squared	2 (50%)	0 (0%)	1 (25%)	0 (0%)	2 (50%)	0 (0%)
C3 shape: rectangular vertical	0 (0%)	0 (0%)	1 (25%)	0 (0%)	2 (50%)	0 (0%)
C4 shape: rectangular vertical	0 (0%)	0 (0%)	1 (25%)	0 (0%)	2 (50%)	0 (0%)

Table 2 Results of Quantitative Analysis: Descriptive Statistics and Statistical Comparisons (ANOVA for Repeated Measurements with post-hoc Scheffe's test) on the Measurements at the Six Consecutive Cephalometric Observations (T_9-T_{15}) *

	T9			T10			T11			T12		
	M	SD	SEM	M	SD	SEM	M	SD	SEM	M	SD	SEM
Age (months)	101.75	3.30	0.16	111.25	3.86	0.18	126.50	1.73	0.07	138.50	1.29	0.05
Co-Gn (mm)	195.79	17.72	0.63	193.90	16.08	0.57	198.46	9.26	0.32	193.91	12.58	0.55
C2Conc (mm)	1.655	0.063	0.035	0.00	0.00	0.00	1.53	0.18	0.08	1.37	0.81	0.49
C3Conc (mm)	0.00	0.00	0.00	0.00	0.00	0.00	0.60	0.00	0.00	1.25	0.92	0.58
C4Conc (mm)	0.00	0.00	0.00	0.00	0.00	0.00	0.00	0.00	0.00	1.34	0.00	0.00
C3PAR (ratio)	1.12	0.88	0.83	1.11	0.60	0.57	1.22	0.60	0.54	1.06	0.30	0.29
C3BAR (ratio)	1.37	0.91	0.78	1.53	0.25	0.20	1.94	0.32	0.46	1.67	0.84	0.65
C4PAR (ratio)	1.20	0.60	0.55	1.31	1.18	1.04	1.42	1.49	1.24	1.26	1.47	1.37
C4BAR (ratio)	1.62	0.45	0.35	1.47	1.00	0.82	1.93	0.21	0.15	1.67	1.41	1.08

	T13			T14			T15		
	M	SD	SEM	M	SD	SEM	M	SD	SEM
Age (months)	149.75	1.70	0.06	162.50	4.20	0.16	173.75	3.30	0.12
Co-Gn (mm)	205.36	14.25	0.49	209.27	8.37	0.28	221.43	7.36	0.24
C2Conc (mm)	**2.54**†	0.21	0.09	1.90	0.63	0.26	2.49	0.67	0.24
C3Conc (mm)	**2.92**†	0.92	0.31	1.60	0.87	0.34	2.54	0.93	0.29

C4Conc (mm)	1.76	0.47	0.25	1.93	0.67	0.28	**2.52**†	0.51	0.16
C3PAR (ratio)	**1.04**†	0.71	0.70	1.12	0.28	0.26	1.04	0.70	0.70
C3BAR (ratio)	**1.24**†	0.66	0.59	1.18	0.44	0.40	0.94	0.48	0.50
C4PAR (ratio)	2.09	0.88	0.85	**1.06**†	0.48	0.47	1.02	0.97	0.97
C4BAR (ratio)	1.28	0.45	0.40	1.18†	0.40	0.37	1.01	0.27	0.26

*M indicates mean; SD indicates standard deviation; and SEM indicates standard error of mean.

† Statistically significant with respect to preceding observation.

Table 3 Table of Developmental Stages of Cervical Vertebral Growth

Developmental Stages	Explanation	Observation Stages
DS1	No concavity on lower border of C2, C3 and C4. C3 and C4 trapezoid shape.	T9 and T10
DS2	Concavity at lower border of C2. C3 and C4 trapezoid shape.	T11
DS3	Concavity at lower border of C2 and C3. C3 rectangular horizontal shape, C4 trapezoid shape.	T12
DS4	Concavity at lower border of C2, C3 and C4. C3 and C4 rectangular horizontal shape.	T12
DS5	Concavity at lower border of C2, C3 and C4. C3 and C4 square shape.	T13
DS6	Concavity at lower border of C2, C3 and C4. C3 and C4 rectangular vertical shape.	T14
DS7	Concavity at lower border of C2, C3 and C4. C3 and C4 remain rectangular vertical shape.	T15

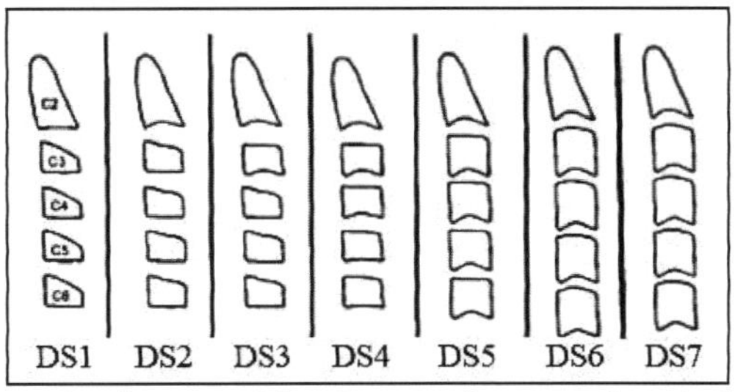

Figure 1. The diagram of developmental stages of cervical vert

IOP Conf. Series: Materials Science and Engineering **300** (2018) 012030 doi:10.1088/1757-899X/300/1/012030

The Robustness of Two Independent Samples t Test Using Monte Carlo Simulation with R

[1]Prana Ugiana Gio, [2]Elly Rosmaini

[1,2] Department of Mathematics, University of Sumatera Utara

Email: gioprana89@gmail.com, ellyrosmaini@gmail.com

Abstract. Two independent samples t test is a parametric statistical method that have several assumptions underlying this method. The assumption applied in the use of two independent samples t test is normal population and equal variance. These assumptions will affect accuracy of the result if the underlying assumptions satisfied or not. Many people use the t statistic to compare two samples even when the underlying assumptions are in doubt. This paper is intended as an introductory article that gives brief overview about the robustness or sensitivity of this popular test statistic with respect to changes in the assumptions. One way to investigate the robustness or sensitivity of two independent samples t test using Monte Carlo simulation with R.

1. Introduction

Suppose given two independent samples, $x_1, x_2, x_3, \ldots, x_m$ and $y_1, y_2, y_3, \ldots, y_n$. Note that m and n are respectively the number of element in x and y sample. In this case wishes to test the hypothesis that mean of the x population are equal with mean of the y population that mathematically can be stated

$$H_0: \mu_x = \mu_y. \tag{1}$$

Let $\bar{x}$ and $\bar{y}$ denote the mean of x and y sample, with standard deviation for each sample s_x and s_y. The standard test to test (1) is based on the t statistic

$$T = \frac{\bar{x} - \bar{y}}{s_p\sqrt{\frac{1}{m} + \frac{1}{n}}}, \tag{2}$$

with s_p denotes pooled standard deviation that can be calculated

$$s_p = \sqrt{\frac{(m-1)s_x^2 + (n-1)s_y^2}{m+n-2}}. \tag{3}$$

Under the hypothesis H_0 (Walpole et al., 2007; Albert, 2009; Mann, 2013), the test statistic T has t distribution with $m + n - 2$ degree of freedom, when:

- Both the x and y samples are independent random samples from normal distributions.
- The standard deviation of x and y populations are equals, $\sigma_x = \sigma_y$

Suppose the significance level of the test is stated at α. The hypothesis H_0 will be rejected when

$$|T| \geq t_{n+m-2,\frac{\alpha}{2}}, \qquad (4)$$

where $t_{df,\alpha}$ is the $(1-\alpha)$ quantile of a t random variable with df degrees of freedom (Albert, 2009).

If the underlying assumptions are satisfied, namely x population and y population are normal distribution and have equal variances, then the level of significance of t test will be stated at α (Albert, 2009). But, in practice, many people use t statistic to compare two samples even when the underlying assumption are in doubt (Albert, 2009). Therefore, one of the interesting problem is to investigate sensitivity and robustness of this popular statistic test respect to the change of the underlying assumptions using Monte Carlo simulation with R.

2. Probability Distribution

2.1. Normal Distribution

A random variable X with probability density function

$$f(x) = \frac{1}{\sqrt{2\pi\sigma}} e^{-\frac{(x-\mu)^2}{2\sigma^2}} \qquad (5)$$

is normal random variable with parameters μ where $-\infty < \mu < \infty$ and $\sigma > 0$ (Montgomery and Runger, 2014). Note that μ is location parameter and σ is scale parameter.

2.2. t Distribution

Let $x_1, x_2, \ldots, x_n$ be a random sample from a normal distribution with unknown mean μ and unknown variance σ^2. The random variable

$$T = \frac{\bar{x} - \mu}{\frac{s}{\sqrt{n}}} \qquad (6)$$

has a t distribution with $n-1$ degrees of freedom (Montgomery and Runger, 2014). The t probability density function is

$$f(t) = \frac{\Gamma\left[\frac{(k+1)}{2}\right]}{\sqrt{\pi k}\,\Gamma\left(\frac{k}{2}\right)} \times \frac{1}{\left[\left(\frac{t^2}{k}\right)+1\right]^{\frac{(k+1)}{2}}} \quad ; -\infty < t < \infty \quad (7)$$

where k is the number of degrees of freedom. The mean and variance of t distribution are zero and $\frac{k}{k-2}$ (for $k > 2$), respectively (Montgomery and Runger, 2014).

2.3. Exponential Distribution

L Montgomery and Runger (2014) a continuous random variable X with probability density function

$$f(x) = \lambda e^{-\lambda x} \quad ; 0 \leq x < \infty \qquad (8)$$

is exponential random variable with parameter $\lambda > 0$. Note that λ is also called rate parameter. If the random variable X has an exponential distribution with parameter λ,

$$\mu = \frac{1}{\lambda} \ and \ \sigma^2 = \frac{1}{\lambda^2}. \qquad (9)$$

3. Monte Carlo Simulation

To apply the Monte Carol method, the analyst constructs a mathematical model that simulates a real system. A large number of random sampling of the model is applied yielding a large number of random samples of output results from the model. The origin began in the 1940s by three scientists, John von Neumann, Stanislaw Ulam and Nicholas Metropolis who were employed on a secret assignment in the Los Alamos National Laboratory, while working on a nuclear weapon project called the Manhattan Project. The Manhattan team formulated a model of a system they were studying that included input variables, and a series of algorithms that were too complicated to analytically solve (Thomopoulos, 2013).

The method is based on running the model many times as in random sampling. For each sample, random variates are generated on each input variable, computations are run through the model yielding random outcomes on each output variable. Since each input is random, the outcomes are random. In the same way, they generated thousands of such samples and achieved thousands of outcomes for each output variable. In order to carry out this method, a large stream of random numbers were needed (Thomopoulos, 2013).

The Monte Carlo method proved to be successful and was an important instrument in the Manhattan Project. After the War, during the 1940s, the method was continually in use and became a prominent tool in the development of the hydrogen bomb. The Rand Corporation and the U.S. Air Force were two of the top organizations that were funding and circulating information on the use of the Monte Carlo method. Soon, applications started popping up in all sorts of situations in business, engineering, science and finance (Thomopoulos, 2013).

4. R Programming Language

R is a scripting language for statistical data manipulation and analysis. A lot of new functions are contributed by users, many of whom are prominent statisticians (Matloff, 2011). R is a very suitable platform for writing a simulation algorithm. One can generate random samples from a wide variety of probability distributions, and R has an extensive set of data analysis capabilities for summarizing and graphing the simulation output (Albert, 2009). In this paper, simple R function has been designed to investigate the robustness or sensitivity of two independent samples t test (t statistic) using Monte Carlo simulation with R. with respect to changes in the assumptions.

5. Programming a Monte Carlo Simulation

One way to investigate the robustness or sensitivity of two independent samples t test (t statistic) using Monte Carlo simulation (Albert, 2009). In this paper use R for writing a simulation algorithm. The following is algorithm to do the Monte Carlo simulation to investigate the robustness or sensitivity of two independent samples t test (t statistic).

- First of all is generated random sample, namely $x_1, x_2, x_3, \dots, x_m$ for x sample and $y_1, y_2, y_3, \dots, y_n$ for y sample. The random sample is generated from various of probability distribution. In this paper, the size for each sample is 10 ($m = n = 10$).
- Next, compute t statistic from x and y sample. The formula to compute t statistic can be seen in (2).
- Determine t critical value. The t critical value is depended on degrees of freedom and stated significance level α.
- The t statistic is compared with t critical value to determine whether H_0 is rejected or no.

In this paper, step 1–4 is repeated 10000 times. Albert (2009) the estimates the true significance level can be computed

$$\hat{\alpha}^T = \frac{Number\ of\ rejections\ of\ H_0}{N} \tag{10}$$

6. Sensitivity of t Statistic of Two Independent Samples t Test Respect to The Change of Assumption Using Monte Carlo Simulation

Table 1 is presented true significance level of the t-test computed by Monte Carlo experiment. Based on Table 1, the random sample is generated from various kind of probability distribution, namely normal, t, and exponential distribution. Based on the result in Table 1, when first population and second population are normal distribution with the equal mean and spread (N(0,1) and N(0,1)), the inaccuracy of the estimated true significance level respect to the stated significance level is 0,51%. But there is an increase the innacuracy, namely 17,09%, when the spread is significantly different (N(0,1) and N(0,10)). When first population and second population are t distribution with the equal spread ($t_{df=3}$ and $t_{df=3}$), the inaccuracy of the estimated true significance level respect to the stated significance level is 5,12%. But there is an decrease the innacuracy, namely 3,34%, when the spread is different ($t_{df=3}$ and $t_{df=3}$). The inaccuracy of the estimated true significance level respect to the stated significance level is 4,68%, when first population and second population are exponential distribution with the same rate (exp(5) and exp(5)).

Table 1. True Significance Levels of the T-Test Computed by Monte Carlo Experiments

Simulation	N(0,1) & N(0,1)	N(0,1) & N(0,10)	$t_{df=3}$ & $t_{df=3}$	$t_{df=3}$ & $t_{df=25}$	Exp(5) and Exp(5)	N(10,2) and Exp(1/10)	N(1,1) and Exp(1)
1	0.0986	0.114	0.0952	0.0939	0.0948	0.1568	0.0998
2	0.1001	0.1194	0.0896	0.0952	0.0923	0.1473	0.1096
3	0.1021	0.1182	0.093	0.0956	0.0938	0.1514	0.1061
4	0.1044	0.1174	0.0974	0.0995	0.0957	0.151	0.1088
5	0.0986	0.1178	0.0929	0.0946	0.0995	0.1506	0.1043
6	0.1015	0.1131	0.0972	0.093	0.0929	0.1577	0.1047
7	0.1014	0.1178	0.0963	0.1014	0.0983	0.1489	0.1096
8	0.098	0.112	0.0954	0.1014	0.0948	0.1536	0.0997
9	0.099	0.1254	0.0971	0.0977	0.0949	0.1546	0.1083
10	0.1	0.1109	0.0945	0.0975	0.095	0.1518	0.1049
11	0.1039	0.1137	0.1023	0.0926	0.0957	0.1499	0.1013
12	0.103	0.1188	0.092	0.0953	0.0912	0.1585	0.1070
13	0.0981	0.1205	0.0994	0.1023	0.0936	0.1489	0.1083
14	0.093	0.1162	0.0858	0.0948	0.0967	0.1514	0.1066
15	0.1012	0.1148	0.0945	0.095	0.098	0.1532	0.1083
16	0.0998	0.1178	0.0937	0.0985	0.0949	0.1501	0.0997
17	0.1036	0.1109	0.096	0.0945	0.1041	0.1579	0.1087
18	0.0955	0.113	0.0953	0.0978	0.0995	0.1543	0.1094
19	0.0995	0.1145	0.1	0.0943	0.0966	0.1591	0.1069
20	0.1022	0.1207	0.0957	0.0962	0.0942	0.1507	0.1076
21	0.1001	0.1135	0.0952	0.1018	0.0955	0.144	0.1071
22	0.1036	0.1166	0.0966	0.0954	0.0949	0.1605	0.1094
23	0.1021	0.1214	0.0917	0.0966	0.0931	0.1522	0.1032
24	0.1055	0.1189	0.0943	0.1006	0.0937	0.1542	0.1075

25	0.0976	0.1225	0.094	0.0954	0.0954	0.1536	0.1053
26	0.1	0.1172	0.0945	0.0953	0.093	0.1587	0.1068
27	0.0997	0.1214	0.0923	0.092	0.0936	0.1588	0.1102
28	0.0975	0.122	0.0963	0.0952	0.0933	0.1527	0.1055
29	0.1029	0.1167	0.094	0.0975	0.0944	0.1541	0.1037
30	0.1028	0.1156	0.0942	0.0988	0.0961	0.1576	0.1037
Average	0.10051	0.11709	0.09488	0.09665667	0.0953166	0.15347	0.1060
Inaccuracy (%)	0.51	17.09	5.12	3.34333333	4.6833333	53.47	6.066666

The last, first population is normal distribution and second population is exponential population, with equal mean but different spread, the inaccuracy of the estimated true significance level respect to the stated significance level is 53,47%. Note that, normal and exponential distribution substantially have different shape. When the first population is normal distribution and second population is exponential population, with equal mean and spread, the inaccuracy of the estimated true significance level respect to the stated significance level is 6,06%.

7. Conclusion

Based on the result of the investigation, the inaccuracy of the estimated true significance level respect to the stated significance level is lower than 10% when first population and second population have equal mean and spread, eventhough substantially have different shape of distribution. In this case, the stated significant level is 10%. The inaccuracy of the estimated true significance level respect to the stated significance level is higher than 10% when first population and second population have equal mean but significantly different spread. But it does not hold for t population. It means that, when first population and second population are t distribution, the inaccuracy of the estimated true significance level respect to the stated significance level is lower than 10%, eventhough have big difference of degrees of freedom.

References

[1] Albert, J. 2009. *Bayesian Computation with R, 2nd Edition*. New York: Springer.
[2] Mann, P. S. 2013. *Introductory Statistics, 8th Edition*. United States of America: John Wiley & Sons.
[3] Walpole, R. E., Myers, R. H., Myers, S. L., and Ye, K. 2007. *Probabilty and Statistics for Engineers and Scientist, 8th Edition*. Prentice Hall.
[4] Montgomery, D. C. and Runger, G. C. 2014. *Applied Statistics and Probability for Engineers, 6th Edition*. John Wiley and Sons.
[5] Thomopoulos, N. T. 2013. *Essentials of Monte Carlo Simulation, Statistical Methods for Building Simulation Models*. New York: Springer.
[6] Matloff, N. 2011. *The Art of R Programming, A Tour of Statistical Software Design*. San Fransisco: No Starch Press.

Spatial hierarchical Bayes estimation of mean years of schooling

Dwi A S Wahyuni[1], **Sutarman Wage**[2] **and Open Darnius**[3]

Department of Mathematics, Universitas Sumatera Utara, Medan 20155 Indonesia

E-mail: [1]dwiasihseptiwahyuni@gmail.com; [2]sutarman@usu.ac.id; [3]open@usu.ac.id

Abstract. A spatial hierarchical bayes for estimating mean years of schooling district level is proposed. We developed spatial hierarchical bayes within a Monte Carlo simulation study with *R* software. The simulation generated posterior distribution invers gamma. The spatial correlation used rook contiguity for each district. Hierarchical bayes method with spatial weighted provides smaller relative bias and relative root mean square.

1. Introduction

Mean years of schooling is one of the key indicators of social performance. Statistics Indonesia annually publishes mean years of schooling to regency level. Regency level divided into some small area. The small area such as district level, village level, etc. Local government are needed mean years of schooling in small area level for regional planning and fund allocation. Mean years of schooling collected by national socio economic survey. The sample size of national socio economic survey is not sufficient to provide mean years of schooling to smaller level. The bayes method is applied in small area estimations by borrowing information from other small areas that have similar characteristics or values in the past and the values of variables that are related to the variables being observed [5].

The application of bayes methods in small area parameters consists of four kinds methods of Best Linear Unbiased Prediction (BLUP), Empirical Best Linear Unbiased Prediction (EBLUP), Empirical Bayes (EB), and Hierarchical Bayes (HB). Hierarchical bayes method can solve very complex models for the data based on very simple models as building blocks [3]. HB methods also can give solution for complex small area model by markov chain monte carlo methods [4]. Model parameters in hierarchical bayes are treated as random variables and assigned a prior distribution [2]. The use of sampling weights in bayesian hierarchical models for small area estimation is a review of Chen et al [1] presenting bayesian spatial smoothing models that result in less Mean Square Error (MSE) values than the general bayesian method approach without spatial weights.

This study aims to obtain the estimation of the mean years of schooling parameter in district level in Tapanuli Tengah Regency using spatial hierarchical bayes method. The hierarchical bayes method has several advantages, such as: can be applied to estimate general indicators defined as functions of the response variable model, hierarchical bayes estimators are unbiased and optimal models when minimizing posterior variance, the bootstrap method for MSE estimation is not required so that the total timing less than empirical bayes methods [5]. The estimation parameter of mean years of schooling district level is done by utilizing National socio economic survey 2015 and Village potential 2014 data as the auxiliary variables.

2. Spatial Hierarchical Bayes Model

Let y_i be the value of the modeling estimation in the i-th region where i is the number of small areas in-sample ($i = 1, ..., m$). Auxiliary variables $x_{ij} = (x_{i1}, ..., x_{in})^T$ available for a small area $i = 1, ..., m$ and variable $j = 1, ..., n$. Sampling model is the first stage in hierarchical bayes written as follows:

$$y_i|\theta_i, \sigma_v^2\, N_p(\theta_i, \psi_i)$$

where ψ_i is a sampling variance that is assumed to be equal for all areas to form a model $\theta_i = \alpha + x_i\beta + e_i + v_i$ where θ_i is the actual parameter value of the target variable in the i-th area, v_i describes the spatial correlation of random effects [5].

Secondstage model or population model spatial hierarchical bayes is:

$$\theta_i|\beta, \sigma_v^2 \sim N(x_i^T\beta, \sigma_v^2 D^{-1})$$

where $D = \lambda R + (1-\lambda) I$ and R is a spatial weighted matrix with members r_{ii} is the number of neighboring area in the i-th and $r_{il} = 1$ if l is neighboring area i and $r_{il} = 0$ if others , $i \neq l$. Value of λ explained spatial autocorrelation parameters whose value ranges from 0 to 1. The spatial weighted matrix (R) used in this study uses a rook contiguity type defined as starting 1 for the adjacent areas in the north, south, west and east or called the side of the common side while 0 (zero) for others. This method will give a value of 1 if the i-region coincides with the j-region and 0 (zero) if the i-region does not coincide with j-region.

Both stages of spatial hierarchical bayes are then explained into the conditional distribution of Gibbs Sampling to be processed through Markov Chain Monte Carlo (MCMC). The established markov chain is $\eta = (\mu^T, \lambda^T)$ where $\mu = (\theta_1, \ldots, \theta_m)^T$. θ is small area parameter and $\lambda = \left(\beta^T, \sigma_v^2\right)^T$ is model parameters. Partition Gibbs Sampling are:

(i) $[\theta_i|\beta, \lambda, \sigma_v^2, y_i] \sim MVN[\Lambda y, (1-\Lambda)X\beta, \Lambda E]$ dimana $\Lambda = (E^{-1} + D/\sigma_V^2)^{-1}E^{-1}$ dengan $E = \{\tilde{\sigma}_1^2, ..., \tilde{\sigma}_1^2\}$ dan $X = (x_1, ..., x_m)^T$

(ii) $[\beta|\theta, \lambda, \sigma_v^2, y_i] \sim MVN[(X'DX)^{-1}X'D\theta, \sigma_v^2(X'DX)^{-1}]$

(iii) $[\lambda|\theta, \beta, \sigma_v^2] \sim |[\lambda R + (1-\lambda)I]^{-1}|^{\frac{1}{2}} x \exp\{-\frac{1}{2\sigma_v^2}(\theta - X\beta)'[\lambda R + (1-\lambda)I](\theta - X\beta)\}$

(iv) $[\sigma_v^2|\theta, \beta, \lambda] \sim IG[\frac{m}{2} + a, \frac{1}{2}[(\theta - X\beta)'D(\theta - X\beta) + b]$

Based on the form of Gibbs Sampling, parameter estimation θ,β,σ_v^2 can be generated directly from (i), (ii), and (iv) via the Gibbs Sampling algorithm because these three parameters have a clear, normal and Inverse Gamma (IG) distribution.

3. Simulated Models

Methods based on simulated models using sample data proposed for small area estimation [4]. Data analysis was performed to obtain the best method of spatial hierarchical bayes through simulation. Simulation is done by R software. Simulation is generating data from the posterior distribution $\sigma_v^2 \sim IG(a, b)$ where averages and variance are assumed to be equal ($a = b$) raised with values $0.0001; 0.001; 0.01; 0.1; 1$ on hierarchical bayes and spatial hierarchical bayes models. Simulation which gives the Relative Bias (RB) and the smallest Relative Root Mean Square Error ($RRMSE$) values used to simulate the sampling variance. Here is the formula for calculating RB and $RRMSE$. $RB_i = \frac{1}{K}\frac{\sum_{k=1}^{K}(\widehat{\overline{Y}}_i^{(k)} - \overline{Y}_i)}{\overline{Y}_i}$ and $RRMSE_i = \frac{\sqrt{\frac{1}{k}\sum_{k=1}^{K}(\widehat{\overline{Y}}_i^{(k)} - \overline{Y}_i)^2}}{\overline{Y}_i}$

The value of sampling variance ψ_i raised with value 0,001;0,01;1.The simulation steps are as follows:

(i) Building a population consist of 18 small areas in this study is the district area. The independent variable is generated from the normal distribution . The population is a random component derived from the normal distribution $x_{ij} \overset{iid}{\sim} N(0,1)$ so that the linear mixed model built in this scenario will satisfy the normalized assumption. In the spatial hierarchical bayes model is given an additional treatment of spatial weights. Spatial weights are formed based on rook contiguity type which is 1 if area i is squashed side with region j and 0 (zero) if region i does not coincide with j.
(ii) Generating independent variables x_{ij} as many as N units, value x_{ij} used to spatial hierarchical bayes model in the simulation process.
(iii) Generating data with inverse gamma distribution parameters (a, b) set to $a = b = 0.0001$, $a = b = 0.001$, $a = b = 0.01$, $a = b = 0.1$, $a = b = 1$.
(iv) Calculating RB and RRMSE from the simulation $IG(0.0001; 0.0001)$, $IG \sim (0.001; 0.001)$, $IG \sim (0.01; 0.01)$, $IG \sim (0.1; 01)$, $IG \sim (1; 1)$
(v) Generating data with ψ_i amount 0.001; 0.01; and 1 and value θ from direct estimate.
(vi) Calculating RB dan RRMSE from simulation $\psi_i = 0.001$, $\psi_i = 0.01$, and $\psi_i = 1$.
(vii) Calculating the response variable value y_{ij} depend on coefficient value β which has been specified for each small area.

In simulation, every iteration process is done, the new k-algorithm is obtained for the parameter β,σ^2, Y from district in the first until eighteenth. Algorithm is done as much as 5.000 iterations. The value of the parameter estimation is derived from the stationary markov mean value. The following is the result of iteration of parameter estimation of spatial hierarchical bayes beta parameter and response variable.

In the process of spatial hierarchical bayes modelling using spatial weights rook contiguity. Figure 1 simulation process stops when the burn in process has been completed on the trace plot. So that the posterior distribution has converged. Data comes from the target distribution and produces a fairly stable value because it has not formed a certain pattern. The density plot image shows the distribution pattern of the observed parameter observers tending to be symmetrical. The autocorrelation plot shows that the autocorrelation values in the first lag are close to one and then the values continue to decrease to zero, so it can be said that in the chain there is a correlation of estimation of parameters between iterations. The correlation indicates that the algorithm is already within the target distribution area.

Table 1 and 2 show that the mean and distribution variance the inverse gamma that gives the smallest RB and RRMSE value is the value of $a = b = 1$ where a is the average and b is the variance of inverse gamma distribution. Table 1 shows that 12 of the 18 district have the smallest RB values on $a = b = 1$. Table 2 shows that the smallest $RRMSE$ values spread almost evenly distributed on the average value and variance of inverse gamma distributions. However, the smallest $RRMSE$ values are mostly on average and variance 1. By therefore it can be concluded that the posterior distribution with the mean value and variance 1 gives the smallest RB and $RRMSE$ values.

Table 3 shows RB and RRMSE values of spatial hierarchical bayes with generated distribution data posterior inverse gamma (1,1) and variance sampling 1. The simulation result show that the RB reaches a sufficiently variable minimum value on the spatial HB model. However, the minimum RRMSE value is maximized on the model spatial hierarchical bayes. The simulation results show that hierarchical spatial bayes gives the least RRMSE value when variance sampling 1 and generated by inverse gamma distribution with average and variance 1.

Figure 1. *Trace Plot, Density Plot, Autocorrelation Plot* Beta 1

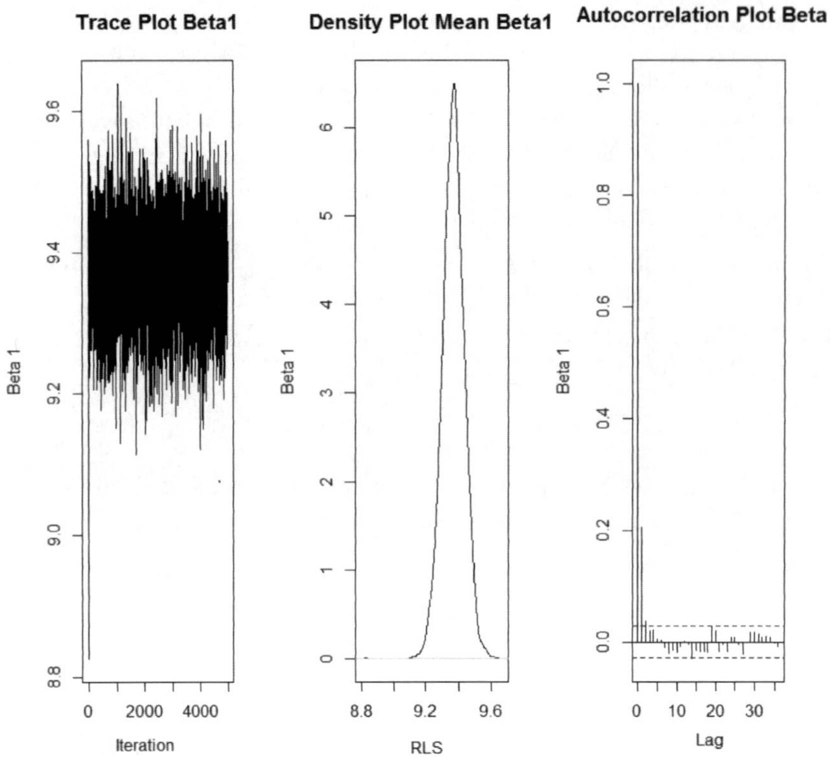

Table 1. Comparison of relative biased of spatial hierarchical bayes methods with a simulated posterior inverse gamma distribution

District	$a = b = 0.0001$	$a = b = 0.001$	$a = b = 0.01$	$a = b = 0.1$	$a = b = 1$
Pinang Sori	-0.062	0.058	-0.066	-0.089	-0.107
Badiri	-0.020	-0.020	-0.018	-0.019	-0.022
Sibabangun	-0.039	-0.029	-0.047	-0.085	-0.101
Lumut	-0.229	-0.211	-0.259	-0.354	-0.401
Sukabangun	-0.049	-0.048	-0.058	-0.079	-0.091
Pandan	0.241	0.253	0.237	0.199	0.184
Tukka	-0.148	-0.138	-0.163	-0.209	-0.235
Sarudik	-0.038	-0.049	-0.030	-0.026	-0.051
Tapian Nauli	0.034	0.041	0.031	0.006	-0.007
Kolang	0.041	0.043	0.038	0.044	0.047
Sorkam	0.070	0.067	0.066	0.073	0.061
Sorkam Barat	0.129	0.128	0.136	0.168	0.186
Barus	0.160	0.159	0.164	0.188	0.194
Sosorgadong	0.093	0.096	0.087	0.081	0.067
Andam Dewi	-0.052	-0.068	-0.022	-0.091	-0.142
Barus Utara	0.130	0.132	0.141	0.182	0.206
Manduamas	-0.228	-0.218	-0.237	-0.248	-0.263
Sirandorung	0.084	0.081	0.095	0.137	0.151

Table 2. Comparison of relative root mean square error of spatial hierarchical bayes method with a simulated posterior inverse gamma distribution

District	$a = b = 0.0001$	$a = b = 0.001$	$a = b = 0.01$	$a = b = 0.1$	$a = b = 1$
Pinang Sori	0.209	0.216	0.204	0.206	0.212
Badiri	0.235	0.217	0.220	0.190	0.184
Sibabangun	0.172	0.192	0.173	0.185	0.191
Lumut	0.322	0.353	0.329	0.372	0.397
Sukabangun	0.210	0.211	0.201	0.202	0.203
Pandan	0.227	0.240	0.216	0.185	0.172
Tukka	0.250	0.273	0.258	0.284	0.299
Sarudik	0.174	0.193	0.162	0.155	0.155
Tapian Nauli	0.196	0.195	0.177	0.165	0.160
Kolang	0.184	0.179	0.172	0.160	0.156
Sorkam	0.203	0.193	0.188	0.167	0.161
Sorkam Barat	0.230	0.213	0.216	0.210	0.217
Barus	0.217	0.206	0.209	0.209	0.210
Sosorgadong	0.198	0.203	0.181	0.163	0.154
Andam Dewi	0.650	0.557	0.561	0.347	0.350
Barus Utara	0.254	0.236	0.232	0.227	0.238
Manduamas	0.387	0.391	0.376	0.377	0.385
Sirandorung	0.251	0.221	0.234	0.214	0.215

Table 3. Relative bias and relative root mean square error of spatial hierarchical bayes modelling of Tapanuli Tengah Regency 2015

District	RB	RRMSE
Pinang Sori	-0.062	0.2177
Badiri	-0.021	0.2302
Sibabangun	-0.035	0.183
Lumut	-0.224	0.341
Sukabangun	-0.051	0.210
Pandan	0.243	0.230
Tukka	-0.146	0.262
Sarudik	0.045	0.183
Tapian Nauli	0.034	0.195
Kolang	0.037	0.183
Sorkam	0.067	0.202
Sorkam Barat	0.127	0.222
Barus	0.157	0.210
Sosorgadong	0.089	0.199
Andam Dewi	-0.062	0.603
Barus Utara	0.126	0.244
Manduamas	-0.225	0.388
Sirandorung	0.080	0.235

The modeling results show that the simulation process of the spatial hierarchical bayes method is constructed by generating a posterior distribution of inverse gamma with a mean value and variance of 0.0001; 0.001; 0.01; 0.1; 1 then the best RB and RRMSE value raised the value of the sampling variance of 0.001; 0.01; 1. The spatial hierarchical bayes model which gives the smallest RB and RRMSE values is the posterior inverse gamma distribution with averages and variance 1 and the sampling variance 1.

4. Conclusion

Small area estimation modelling use spatial hierarchical bayes with iteration process as much as 5,000 times. Some conclusions as follows:

(i) The process of generating data can be done by varying the mean value and variance of a significant posterior gamma distribution so that the convergence of the markov chain is achieved more quickly and results in smaller RRMSE values.

(ii) The spatial hierarchical bayes model which gives the smallest RB and RRMSE values is the posterior inverse gamma distribution with averages and variance 1 and the sampling variance 1.

5. References

[1] Chen C, Wakefield J, and Lumely T. (2014). The Use of Sampling Weights in Bayesian Hierarchical Models For Small Area Estimation. *Spatial and Spatio-temporal Epidemiology*, Vol. 11, 33–43.

[2] Guadarrama M, Molina I, dan Rao J.N.K. (2015). A Comparison of Small Area Estimation Methods for Poverty Mapping. *Statistics and Econometrics*, Vol. 15, No. 5, 1–24.

[3] Molina, Isabel, Nandram B, and Rao J.N.K. (2014). Small Area Estimation of General Parameters With Application to Poverty Indicator: A Hierarchical Bayes Approach. *The Annuals of Applied Statistics*, Vol. 8, No. 2, 852–885.

[4] Rao, J.N.K (2008). Some Methods For Small Area Estimation. *Rivista Internazionale in Science Sociali*, No.4, 387–406.

[5] Rao, J.N.K and Molina I (2015). *Small Area Estimation* . (New Jersey: John Wiley & Sons)

IOP Conf. Series: Materials Science and Engineering **300** (2018) 012032 doi:10.1088/1757-899X/300/1/012032

Forecasting dengue hemorrhagic fever cases using ARIMA model: a case study in Asahan district

Fazidah A Siregar[1], Tri Makmur[2], Saprin S[3]
[1]Faculty of Public Health, University of Sumatera Utara, Medan, 20155, North Sumatera, Indonesia
[2]Faculty of Medicine, Islamic University of Sumatera Utara,Medan, North Sumatera, Indonesia
[3]Asahan district health department, North Sumatera Province

*Email : fazidah@usu.ac.id

Abstract. Time series analysis had been increasingly used to forecast the number of dengue hemorrhagic fever in many studies. Since no vaccine exist and poor public health infrastructure, predicting the occurrence of dengue hemorrhagic fever (DHF) is crucial. This study was conducted to determine trend and forecasting the occurrence of DHF in Asahan district, North Sumatera Province. Monthly reported dengue cases for the years 2012-2016 were obtained from the district health offices. A time series analysis was conducted by Autoregressive integrated moving average (ARIMA) modeling to forecast the occurrence of DHF. The results demonstrated that the reported DHF cases showed a seasonal variation. The SARIMA $(1,0,0)(0,1,1)_{12}$ model was the best model and adequate for the data. The SARIMA model for DHF is necessary and could applied to predict the incidence of DHF in Asahan district and assist with design public health maesures to prevent and control the diseases.

1. Introduction

Dengue hemorrhagic fever (DHF) still health problem worldwide and currently more than 100 countries are endemic for dengue virus infection with an estimated 50 million dengue infections occur annually [1,2]. According to *World Health Organization* (WHO), Indonesia ranked second for DHF cases in Asia. The Ministry of Health of Indonesia reported that North Sumatera province ranked third highest for DHF cases with 16 dengue endemic districts [3]. Asahan district is one of them with the incidence of DHF increased annually, from 31/100,000 population in 2012 to 102/100,000 population in 2016 [4].

The DHF is a complex problem. The dynamic of dengue infection are driven by complex interaction between host, vectors and viruses that are influenced by environmental. On the other hand, many factors contribute to DHF incidence, including socio-demographic, behavioural, cultural and environmental factors. Moreover, *Aedes* mosquito as the main vector for DHF play role on transmission also influenced by climate and environmental [5,6].

Currently, no specific treatment and no vaccines are available for dengue hemorrhagic fever. In addition, lack of effective vector control measures and poor disease surveillance, made the dengue prevention measures may often be implemented late, thereby reducing the opportunities for preventing transmission and controlling the epidemic. DHF outbreaks can be predicted by epidemiological modeling thus enabling the health systems to be prepare to manage outbreaks. Therefore, enhancing the preparedness by predicting the occurence of DHF made crucial. Time series methodology as forecasting model has been increasingly used in the field of epidemiological research of infectious disease, such as malaria [7,8] dan dengue hemorrhagic fever [9-13].

Detailed information about when and where DHF occurred in the past can be useful to describe the magnitude problem and predict the disease in the future time. This could assist public health intervention to prevent and control the disease. This study was undertaken to develop a prediction model for DHF using time series data over the past decade in Asahan district and to forecast the monthly DHF incidence for the year 2017.

2. Methods

This study was conducted in Asahan district, an endemic area for DHF cases in North Sumatera Province. Asahan is situated south of North Sumatera Povince, situated between 2.300 – 3.260 North Latitude and 99.01- 100.000. It covers a surface area of 4,580,75 km2 with total population (2010) was 1,024,369 (Figure1).

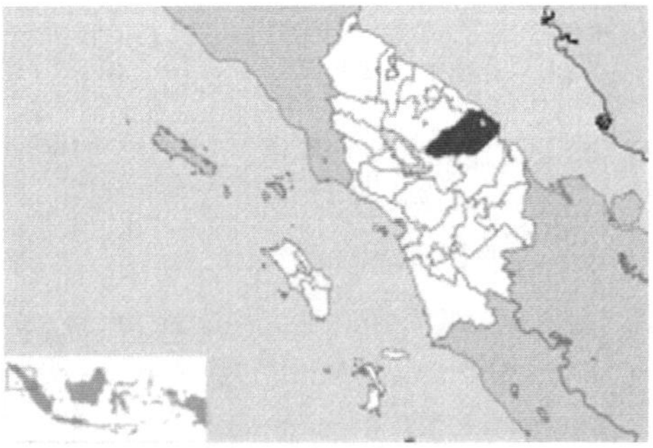

Figure 1. Asahan district in North Sumatera Province

2.1. Data collection

Reported monthly dengue hemorrhagic cases for January 2012 through December 2016 were obtained from Asahan district health offices.

2.2. Statistical analysis

Autoregressive integrated moving average (ARIMA) models have been used for analyzing time series data containing seasonal trends to develop a forecasting model that was first popularized by Box and Jenkins [14]. This method for selecting an appropriate ARIMA model for estimating and forecasting an univariate time-series consisted of identification, estimation, diagnostic checking and forecasting. ARIMA model may include autoregressive (p) terms, differencing (d) terms and moving average (q) operations and is represented by ARIMA (p, d, q). A stationary time series is one whose statistical properties such as mean and variance are constant over time. Seasonality usually causes the series to be nonstationary. Therefore, Seasonal ARIMA (SARIMA) is an extension of the ARIMA method to a series in which a pattern repeats seasonally over time and is represented as SARIMA (p, d, q) (P, D, Q)s, which that the Seasonal autoregressive (P), seasonal differencing (D), and seasonal moving average parameters (Q); s defines the number of time periods until the pattern repeats again for a monthly data it is.

Minitab version 16.0 was used to analysis and determine the best-fitting model. First, check for stationary was made with the aid of a control chart. After verifying that the series was stationary, an ARIMA model was developed. Before using the model for forecasting, it was checked for adequacy. This was done through examining the ACF and PACF of the residuals. Further, Ljung-Box test was used to provide an indication of whether the model was correctly specified. A significant value less than 0.05 was considered to acknowledge the presence of structure in the observed series, which was not accounted for by the model; therefore, we ignored the model if it had significant value. After the best model was identified, forecast for monthly values of the year 2017 were made.

3. Results and Discussions

The reported monthly DHF cases during 2012-2016 exhibited a seasonal pattern (Figure 2) and increased significantly during the study period (Figure 3). The number of monthly DHF cases began to increase in June, peaked in October and then gradually decreased

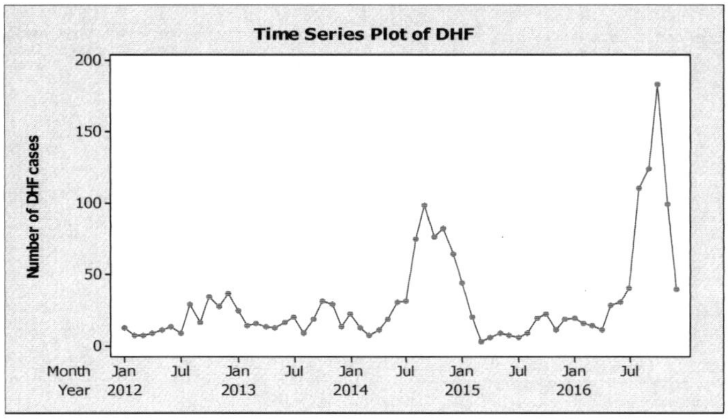

Figure 2. Monthly number of DHF cases in Asahan district from 2012 to 2016

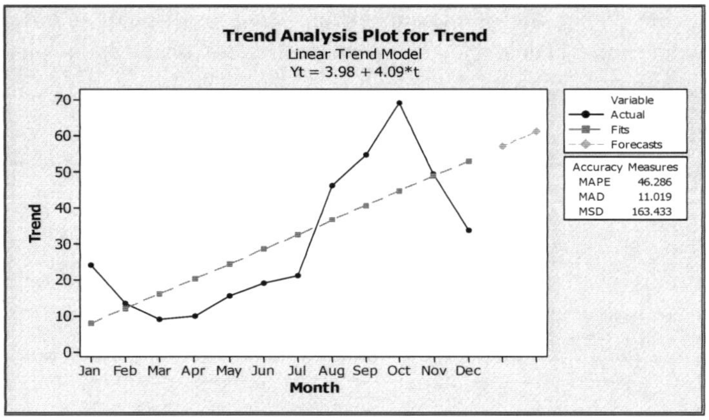

Figure 3. Trend analysis of the number of DHF cases in Asahan district from 2012 to 2016

The time series plot of the reported DHF cases displayed seasonal fluctuations and therefore deemed non stationary. The ACF of DHF time-series and the PACF of DHF time series are expressed in Figure 4. Decrease in autocorrelation values after lag 2 indicated non stasionary data; consequently, there was need to include a first-lag difference term in the SARIMA model structure (d = 1) as presented in Figure 5.

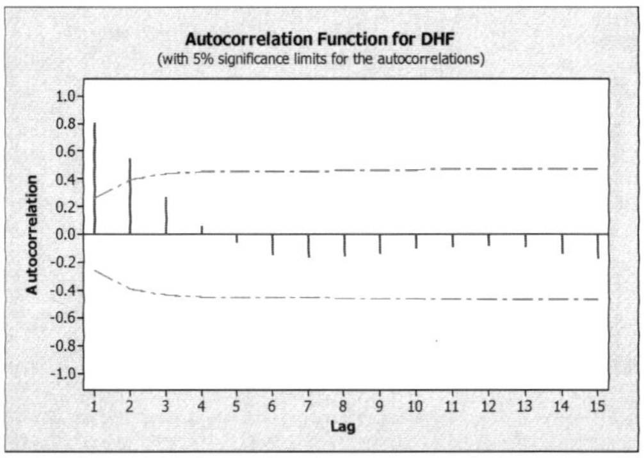

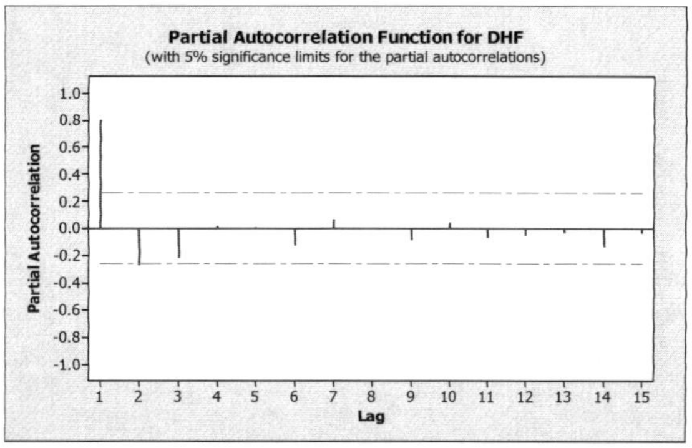

Figure 4. ACF and PACF of the reported DHF cases

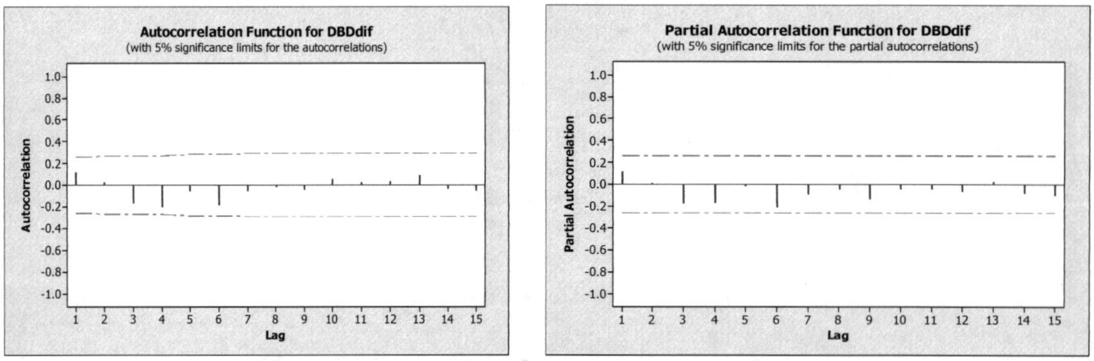

Figure 5. ACF and PACF of DHF difference first lag

Among the statistical models, SARIMA $(1,0,0)(0,1,1)_{12}$ was selected as the best model, with the lowest mean square error of 442 and p value (L Jung-Box) 0.602 . L jung-Box test > 0.05 suggested that there were no significant autocorrelation between residuals at different lag times and the residuals were white noise [Table 1]. This was further corroborated by plotting the ACF and PACF of the residuals as presented in Figure 6.

Table 1. Characteristics of SARIMA models

No	Model	*p value*				**MS**	p value
		AR	SAR	MA	SMA		(LJung Box)
1	$(1,0,0)(1,0,0)_{12}$	0.000	0.684	-	-	424	0.451
2	$(1,0,0)(0,1,1)_{12}$	0.000	-	-	0.013	442	0.602
3	$(0,1,1)(0,1,1)_{12}$	-	-	0.326	0.005	478	0.575
4	$(1,1,0)(0,1,1)_{12}$	0.756	-	-	0.010	507	0.779

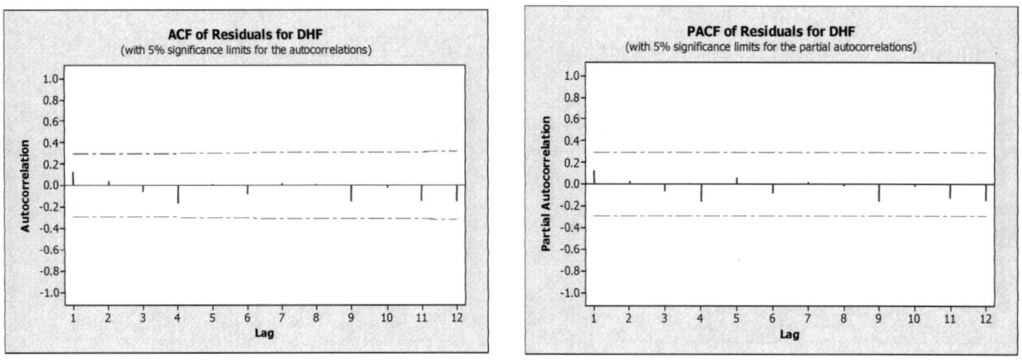

Figure 6. ACF and PACF of the residual of SARIMA $(1,0,0)(0,1,1)_{12}$

The SARIMA model was constructed using data collected between Januray 2012 until December 2016 and was verified using data collected from January to June 2017. The validity of model was tested and used to forecast monthly DHF cases in 2017. The result show that the actual value and the predicted value matched reasonably well as presented in Figure 7.

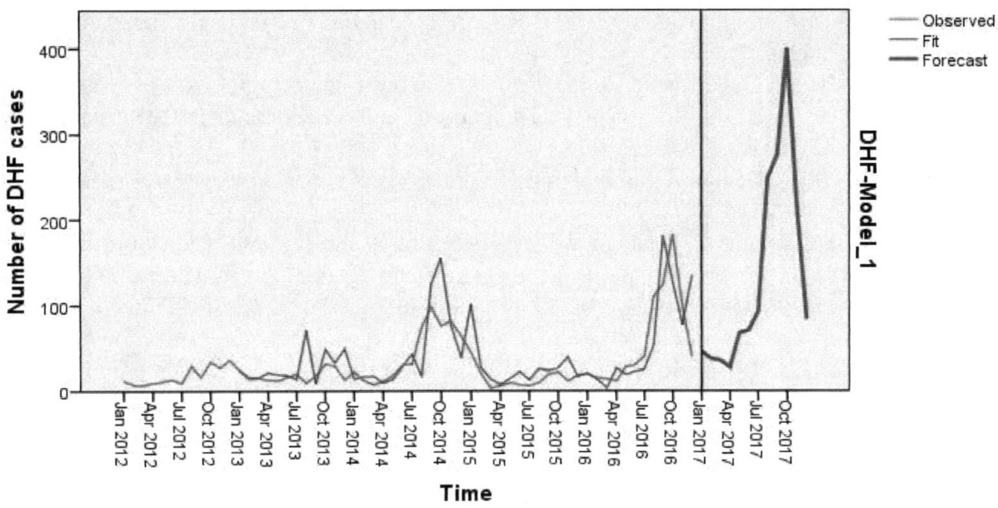

Figure 7. The actual monthly DHF cases from 2012 - 2016 and the monthly DHF cases predicted for 2017 by using SARIMA model in Asahan district

ARIMA models useful in modelling dependence of a time series between observation [15] and to be adequate tools for use in epidemiological surveillance [16]. Our study provides an example of applying a SARIMA model to forecast incidence of DHF in Asahan district. These models have been utilized to forecast DHF incidence in several countries [9,10,12,16,17].

Among all models, SARIMA (1,0,0) $(0,1,1)_{12}$ was the most suitable predictive model in this study, which showed the lowest mean square error. A study in Brazil found SARIMA $(2,1,3)(1,1,1)_{12}$ model was best fit for the dengue incidence data [10]. While a study by Luz *et al* in Rio de Janeiro, Brazil found that SARIMA (2,0,0) $(1,0,0)_{12}$ model was deemed best fit and no seasonal differencing [9]. Choudhury *et al.*, reported SARIMA (1,0,0) $(1, 1, 1)_{12}$ as the most suitable model for forecasting dengue incidence in Dhaka, Bangladesh [18]. On another study in Thailand found SARIMA (2,0,1) $(0, 2, 0)_{12}$, ARIMA (1,0,1), and SARIMA (2, 1, 0) $(0, 1, 1)_{12}$ models as most suitable for DHF cases [17,19,20].

ARIMA modeling is a useful tool for interpreting surveillance data and forecast of the cases to help guide timely prevention and control measures. In addition, may providing decision-makers a clear idea in designing an effective prevention and control strategy for DHF. Further research is recommended to integrate the forecasting model into the existing disease control program in terms in reducing the disease occurrence.

4. Conclusions

SARIMA $(1,0,0)(0,1,1)_{12}$ model provides the predicted incidence of DHF in Asahan district. Application this model could be used to predict the incidence of DHF in the future time and assist with design public health maesures to prevent and control the disease.

Acknowledgments

We are grateful to the head of the Asahan district health office and staff for their assistance in this study. We are also thankful to all of the participants and people who were involved in this study.

References

[1]. Guzman, A. and Istúriz, R. E. (2010. Update on the global spread of dengue. *International journal of antimicrobial agents* **36** S40
[2]. Guzman, M. G et al (2011). Dengue: a continuing global threat. *Nat Rev Microbiol.* **8** S7

[3]. Ministry of Health of Indonesia. (2011). Report of dengue hemorrhagic cases in Indonesia, Jakarta, Indonesia.
[4]. Asahan district Health Department.(2016). Report of dengue hemorragic cases in Asahan district.
[5]. Rodriguez, H. & de la Hoz R, F. (2005). Dengue and vector behaviour in Ca'queza, Colombia, 2004. *Revista de Salud Publica.* **7(1)** 1-15.
[6]. Halstead, S.B.(2008). Dengue virus-mosquito interactions. *Annual Review Entomology.* **53** 273-291.
[7]. Wangdi K, Singhasivanon P, Silawan T, Lawpoolsri S, White NJ, and Kaewkungwal J. (2010). Development of temporal modelling forecasting and prediction of malaria infections using time-series and ARIMAX analyses: A case study in endemic districts of Bhutan. *Malar J* . **9** 251.
[8]. Loha E, Lindtjørn B.(2010).Model variations in predicting incidence of Plasmodium falciparum malaria using 1998-2007 morbidity and meteorological data from south Ethiopia. *Malar J.* **9** 166.
[9]. Luz, P. M., Mendes, B. V., Codeço, C. T., Struchiner, C. J and Galvani, A. P. (2008). Time series analysis of dengue incidence in Rio de Janeiro, Brazil. *The American Journal of Tropical Medicine and Hygiene.* **79(6)** 933-939.
[10]. Martinez, E., Soares da Silva, E. & Dal Fabbro, A. (2011). A SARIMA forecasting model to predict the number of cases of dengue in Campinas, State of São Paulo, Brazil. *Revista da Sociedade Brasileira de Medicina Tropical* **44(4)** 436-440.
[11]. Lu, L., Lin, H., Tian, L., Yang, W., Sun, J. Liu.Q (2009). Time series analysis of dengue fever and weather in Guangzhou,China*BioMed Central Public Health* **9(395)** 1-5.
[12]. S. Wongkoon, M. Pollar, M. Jaroensutasinee, and K. Jaroensutasinee (2007). Predicting DHF Incidence in Northern Thailand using Time Series Analysis Technique *International Scholarly and Scientific Research & Innovation* **1(8)** 484-488.
[13]. Bhatnagar S, Lai V, Gupta SD, Gupta OP.(2012). Forecasting incidence of dengue in Rajasthan, using time series analyses. *Indian Journal of Public Health.* **56(4)** 281-285.
[14]. Box GE, Jenkins GM.Time series analysis: Forecasting and control. San Francisco:Holden-Day; 1976
[15]. Gharbi, M., Quenel, P., Gustave, J., Cassadou, S., Ruche, G.L., Girdary, L. and Maramma, L. (2011). Time series analysis of dengue incidence in Guadeloupe, French West Indies: forecasting models using climate variables as predictors. *BioMed Central Infectious Diseases,* **11(1)** 1-13
[16] Nobre FF, Monteiro AB, Telles PR, Williamson GD. Dynamic linear model and SARIMA(2001): A comparison of their forecasting performance in epidemiology. Stat Med **20** 3051-69.
[17]. Promprou S, Jaroensutasinee M, Jaroensutasinee K. Forecasting dengue haemorrhagic fever cases in Southern Thailand using ARIMA models.(2006). *Dengue Bull* **30** 99-106.
[18]. Choudhury MA, Banu S, Islam MA. Forecasting dengue incidence in Dhaka, Bangladesh (2008) : A time series analysis. *Dengue Bull* **32** 29-37.
[19]. Silawan T, Singhasivanon P, Kaewkungwal J, Nimmanitya S, Suwonkerd W.(2008) Temporal patterns and forecast of dengue infection in Northeastern Thailand. *Southeast Asian J Trop Med Med Public Health* **39** 90-8.
[20]. Wongkoon S, Pollar M, Jaroensutasinee M, Jaroensutasinee K. (2008). Predicting DHF incidence in Northern Thailand using time series analysis technique. *International Journal of Biological and Life Sciences* **4** 117-121.

Volatility in GARCH Models of Business Tendency Index

Dwi A S Wahyuni[1], Sutarman Wage[2] and Ateng Hartono[3]

[1] [2]Department of Mathematics, Universitas Sumatera Utara, Medan 20155 Indonesia;
[3]Statistics Indonesia, Medan, Indonesia

E-mail: [1]dwiasihseptiwahyuni@gmail.com; [2]sutarman@usu.ac.id; [3]ateng@bps.go.id

Abstract. This paper aims to obtain a model of business tendency index by considering volatility factor. Volatility factor detected by ARCH (Autoregressive Conditional Heteroscedasticity). The ARCH checking was performed using the Lagrange multiplier test. The modeling is Generalized Autoregressive Conditional Heteroscedasticity (GARCH) are able to overcome volatility problems by incorporating past residual elements and residual variants.

1. Introduction

The business condition and economy in Indonesia experienced a sharp change. This is in line with the growth of Indonesia's economic growth every quarter from 2000 to 2017 which has fluctuated and tends to decline by 2017 compared to 2000. Statistics Indonesia has measured the Business Tendency Index (BTI) that is able to provide information on the state of business and the economy in the short term, which is in the current quarter and the prediction of the next quarter. The BTI is one of financial time series. Modeling financial time series is a complex problem that is difficult to reproduce artificially using stochastic models [2]. However, long-term economic development information is needed by businessman to take action in business decision-making. Therefore accurate measurements are needed to determine the state of the business and the long-term economy.

The movement of BTI figures for second quarter 2000 until first quarter 2017 experienced very sharp fluctuations. The BTI in the first quarter of 2017 amounted to 103.42, this value decreased from fourth quarter of 2016 amounted to 106,7. The value of BTI in third quarter of 2016 amounted to 107,89. This indicates that business conditions very fluctuated compared to the previous quarter. On the other hand, Indonesia's economic growth in the first quarter compared to the fourth quarter contracted by 0.34 percent where the components of exports of goods and services is the largest component that supports economic growth. The business condition in the first quarter 2017 increased was caused by three components, such as the use of production/business capacity with the achievement index of 104.60, operating income with the achievement of the index value of 104.54 and the average number of working hours with the achievement of the index value of 101.13.

This fluctuation makes it difficult for businessmen to predict the movement of BTI in the coming quarter. The change of BTI figure that is very drastic every quarter is supposed to contain volatility elements so that in statistical modeling it is necessary to consider the heteroscedasticity element in the model. The simplest measurement of volatility can see from typically non independent and identically distributed with leptocurtic and skewed of distributions [3]. GARCH model calculate volatility by demeanded squared return [1]. GARCH model work simply if volatility is changing gradually over time because The Business Tendency Index Indonesia figures in this study were modeled using ARCH and GARCH.

2. ARCH and GARCH Models

ARCH models were introduced by Engle (1982) and their GARCH extension is due to Bollerslev (1986). The key concept in these models are the conditional variance, that is, the variance conditional on the past. In the classical GARCH models, the conditional variance is expresses as a linear function of squared past values of the series [2]. GARCH models is hardly compatible with the assumption of constant conditional variance, that is called conditional heteroscedasticity: $\mathrm{Var}\left(\varepsilon_t|\varepsilon_{t-1}, \varepsilon_{t-2}, \ldots\right) \neq$ const. Conditional heteroscedasticity is perfectly compatible with stationary.

A process (ε_t) is called a GARCH (p,q) process if its first two conditional moments exist and satisfy: [2]

(i) $E\left(\varepsilon_t|\varepsilon_u, u < t\right) = 0, t \in \mathrm{Z}$

(ii) There exist constant $\omega, \alpha_i, i = 1, ..., q$ and $\beta_j, j = 1, ..., p$, such that $\sigma_t^2 = Var(\varepsilon_t|\varepsilon_u, u < t) = \omega + \sum_{i=1}^{q} \alpha_i \varepsilon_{t-1}^2 + \sum_{j=1}^{p} \beta_j \sigma_{t-j}^2, t \in \mathrm{Z}$ or $\sigma_t^2 = Var(\varepsilon_t|\varepsilon_u, u < t) = \omega + \alpha(B)\varepsilon_{t-1}^2 + \beta(B)\sigma_{t-j}^2, t \in \mathrm{Z}$

The GARCH (1,1) model when $p = q$, has the form: $\varepsilon_t = \sigma_t \eta_t$ $\sigma_t^2 = \omega + \alpha\varepsilon_{t-1}^2 + \beta\sigma_{t-1}^2$ with $\omega \geq 0, \alpha \geq 0, \beta \geq 0$. The conditional variant σ_t^2 (as it depends on the previous period) has three parts, namely the mean (ω), the volatility of the previous period σ_{t-1}^2 (called ARCH), previous period variant σ_{t-1}^2 (called GARCH).

Testing the existence of heteroscedasticity in time series data uses Lagrange Multiplier. Lagrange Multiplier works by linear regression equation $h_t = \alpha_0 + \alpha_1 u_{t-1}^2 + \alpha_2 u_{t-2}^2 + ... + \alpha_m u_{t-m}^2$ so that the coefficient of determination (R^2). The Lagrange Multiplier (LM) test statistic is $T \times R^2$ where T is the number of observations. The decision maker of LM test statistics is when the value of LM larger than chisquare table. Then reject the initial hypothesis (no heteroskedasticity). The value of α is the value of signifinance and m is number of parameters.

3. GARCH Model of BTI

BTI in this research is BTI from second quarter of 2000 to first quarter of 2017 or as much as 68 observation. Table 1 shows BTI was highest in the second quarter of 2000 is 122.5. The lowest BTI occurred in the first quarter of 2006 is 95.12.

Table 1. Descriptive Statistics

Mean	Maximum	Minimum	Std. Dev	Skewness	Kurtosis	Observations
106.5866	122.5000	95.1200	5.3520	0.2199	3.2759	68

BTI numbers fluctuate during the observation period. The following graph of the development of the Indonesian business tendency index in the second quarter of 2000 to first quarter of 2017.

Figure 1 shows that BTI Indonesia has high volatility. In other words, drastic increase and decrease can occur at any time. BTI in first quarter of 2006 then jumped to 108.5 in second quarter of 2006. The right econometric model to estimate high volatility data is ARCH and GARCH analysis.

Figure 1. Indonesia Business Tendency Index second quarter of 2000 to first quarter of 2017

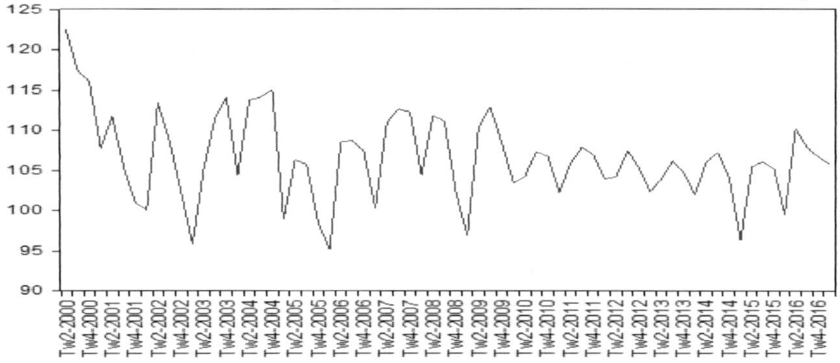

Source: Statistics Indonesia

Table 2. Test Stationary

		t-statistics	Prob*
Augmented Dickey-Fuller test statistic		-5.091501	0.0001
Test critical values:	1 percent level	-3.53862	
	5 percent level	-2.908420	
	10 percent level	-2.591799	

Table 2 shows testing stationarity of data BTI using unit root test (unit root test). Table 2 shows that the probability value of 0.0001 means the data is not stationary. Because of the data not stationary, we must checked heteroscedasticity using Lagrange Multiplier (LM).

The existence of ARCH from box jenkins model can be checked using Lagrange Multiplier test. The result of LM test shows that the data of Indonesian business tendency index from second quarter of 2000 to first quarter of 2017 contains heteroscedasticity. It is characterized by a probability value of F - statistic less than 5 percent is 0.0099 and value of Obs*R-squared more than chi-square table.

Table 3. LM Test

Breusch-Godfrey Serial Correlation LM Test:			
F-statistic	7.057053	Prob.F(1,66)	0.0099
Obs*R-squared	6.568560	Prob. Chi-Square(1)	0.0104

Figure 2. Testing Normality Error

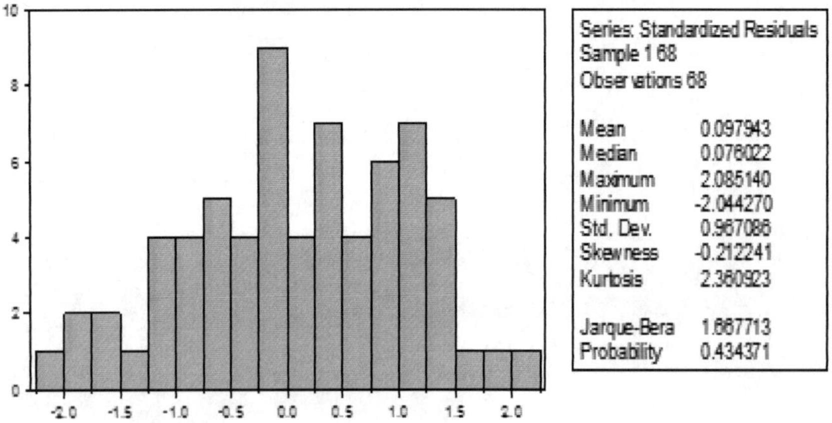

GARCH Estimation

In the BTI modeling, several model simulations were performed to obtain the best model. The measurements to obtain the best model comparison use Akaike Info Criterion (AIC) and Schwarz Information Criterion (SIC) values. Models that provide the smallest AIC and SIC values are selected as the best model for the BTI Indonesia.

Table 4. GARCH Model Comparison for Business Tendency Index

Model	AIC	SIC
GARCH(1,0)	6,126480	6,224399
GARCH(0,1)	6,023624	6,12544
GARCH(1,1)	5,859331	5,989890
GARCH(1,1) Regresorvarian	5,994182	6,158711

Selection of GARCH model with different order to find the most appropriate model. The best model is GARCH (1,1). The GARCH model has been obtained, evaluated to ensure that the model is feasible. The GARCH model is evaluated in several ways as follows: Normal error checking is used to determine whether or not there is heteroscedasticity in residual. Figure 2 shows testing normality error shows that the value of Jarque-Bera and probability is more than 5 percent each of 1.667713 and 0.434371. With a 5 percent error rate it can be concluded that the residual is normally distributed. The result of residual randomness test using Correlogram Q-Statistics shows that the probability value is greater than = 5 percent so that the annoying variable has no autocorrelation element.

Table 5 shows heteroscedasticity test that the probability value of F statistic and probability value of chi-square is greater than 0,05. The ARCH effect test using Lagrange Multiplier is performed to ensure that the GARCH (1,1) model does not contain heteroscedasticity. Table 6 shows results of model estimated. GARCH Model (1,1) on the BTI is

$\sigma^2 = 0,7361 - 0,2011\varepsilon^2_{t-1} + 1,1074\sigma^2_{t-1}$

$GARCH = 0,73610,2011 * RESID(-1)^2 - 1,1074 * GARCH(-1)$

Table 5. Heteroskedasticity Test

Heteroskedasticity Test: ARCH			
F-statistic	3.158376	Prob.F(1,65)	0.0802
Obs*R-squared	3.104698	Prob. Chi-Square(1)	0.0781

Table 6. Results of Model Estimated

Coefficient	ARCH(1,0)	GARCH(0,1)	GARCH(1,1)	GARCH(1,1) Regressor variant
C	106.0179	105.7457	105.8258	105.5697
AR(1)				0.2674
Variance equation				
C	15.4864	0.5786	0.7361	1.6385
Resid(-1)2	0.42692	0.9291	-0.2011	0.1549
GARCH(-1)			1.1074	0.7541

4. Conclusion

GARCH model can know the volatility that occurs in movement of business tendency index. Based on model evaluation results, it can be concluded that GARCH (1,1) model is suitable to predict business tendency index with 5 percent error rate.

References

[1] Molnar, Peter. (2016). High Low Range in GARCH Models of Stock Return Volatility. *Applied Economics*, Vol. 48, No. 51, 4977–4991.
[2] Francq, Christian and Jean Michel Zakoian. (2010). *GARCH Models*. Willey, United Kingdom.
[3] Mills, Terence S and Raphael N. Markellos. (2008). *The Econometric Modelling of Financial Time Series, Third Edition*. Cambridge University Press, New York.

Model of Decision Making through Consensus in Ranking Case

Gim Tarigan, Open Darnius

Departement of Mathematics, University of Sumatera Utara

Email: gintarigansibero@gmail.com, open@usu.ac.id

Abstract. The basic problem to determine ranking consensus is a problem to combine some rankings those are decided by two or more Decision Maker (DM) into ranking consensus. DM is frequently asked to present their preferences over a group of objects in terms of ranks, for example to determine a new project, new product, a candidate in a election, and so on. The problem in ranking can be classified into two major categories; namely, cardinal and ordinal rankings. The objective of the study is to obtin the ranking consensus by appying some algorithms and methods. The algorithms and methods used in this study were partial algorithm, optimal ranking consensus, BAK (Borde-Kendal)Model. A method proposed as an alternative in ranking conssensus is a Weighted Distance Forward-Backward (WDFB) method, which gave a little difference i ranking consensus result compare to the result oethe example solved by Cook, et.al (2005).

1. Introduction

In making decision, sometimes one has to select one among some alternative options, meaning that at least there are two or more alternative options to be chosen. In pratice there might be two or more Decision Maker (DM) involved in making a decision based on one or more considerations or certain criteria. The Consensus Ranking (CR) Method computes an overall ranking that most closely represents the ranking of thr majority of judges. Some advantages of the CR method are the it is intuitive for understanding, easy to verify, and difficult to manipulate. The main idea in ranking consensus how to decide a rank that satisfied the preference of all DM. A decision making in the problem of ranking consensus simply lies in formulating the right choice among a variety of DM's preference after an evaluation or assessment they made based on the given criteria.

In this life, humans are almost at all times involved in decision making, even in their daily lives which contexts may not be too complicated. When one makes a decision, there is a process running in his brain that determines the decision made. If the decision is made for simple things such as which clothes to choose, one does not need to formulate according to a regulation or model that needs to be analyzed mathematically. However, if what needs to be decided is complicated and highly risky, it requires a further study. The determination of ranking consensus based on some subjective preferences of Dm is a quite rapidly growing problem and is studied by some mathematics experts. Determining ranking among some objects with quantitative criteria is probably easy, but it is not easy for the objects with ordinal qualitative criteria. In order to be more objective to determine the ranking of the objects with qualitative criteria, it usually requires more than one adjudicator as the decision making team. This team, in the process, gives their individual ranking based on each preference before formulating final result of the ranking. A decision maker frequently needs an aid equipment such as scientific, logical and structured analysis. The means of analysis is a decision making model implemented to make a decision of complicated problem by combining a group of ranking in order to obtain a ranking consensus or compromise rank which has been proposed individual preference have been proposed by many researches such as Kemeny and Snell (1962), Kendall (1962), and Inada (1969).

This ranking consensus problem can be categorized into two major primary groups; namely, Cardinal and Ordianal problems. The formulation of Cardinal ranking occurs when a decision maker is able to express the preference level of an alternative over the other alternatives with one utility function. On the other hand, Ordinal ranking does not require a certain preference level. A complete ordinal making form n alternatives is set to be integers (1, 2, ..., n). This study focused on ordinal ranking. The interest of ordinal ranking representation and formulation requires minimum amount of information, where each decision maker only decides one preference over the other preference. The simplest consensus development of ordinal ranking is majority method.

Borda, J.C (1781) proposed "sign method" to develop consensus on opinions by determining the mean of the ranking allocated by the decision maker to each alternative and the winning alternative is the one with the lowest mean. The same version of this model is represented by Kendall (1962). Kendall is the first person to study ordinal ranking in statistical framework with problem evaluation approach. Kendall's solution is to arrange the alternative rankings according to the number of equivalent rankings. Kendall organizes ranking by means of sign method. Borda-Kendall (BK) technique is the most widely used ranking consensus method in the pratice due to its simple calculation. Cook and Seifor (1982) further studied BK technique and proposed "minimum variation" method to determine ranking consensus. The popular method to develop consensus is to define the distance function on the group of all rankings and later determine the closest ordinal ranking which may mean a minimum distance.

Some other researchers use integer programming and goal programming to solve ranking consensus problem. Ali et.al (1986) presented integer distance function. Cook et.al (2006) used an extensive experiment simulation to compare integer programming by means of heuristic procedure. Iz and Jellassi (1991) used goal programming to measure the individual preference of the group members through Ordinal ranking scheme.

Multiple Criteria is also commonly used it formulate and solve the problem of ranking consensus. Cook and Krees (1991) propose a weighted ordinal ranking model in which each set of a alternatives is ordianlly ranked over a set of criteria. The criteria are:

1. The urgency of weigth allocated to each criteria.
2. The urgency of various positions at which an alternative can be placed, and
3. The precision in which a decision maker is able to differentiate the alternatives over a certain criteria. the clonclusion is that the optimal ranking consensus method cannot be emphasized on the decision maker. It is better to understand the strength and weakness of each method and allow the decision maker to select a certain method. The main problem in his study emerged when a k group of decision makers are asked to organize the ranking of n objects as the alternatives, and how to combine the decision makers' making into are one ranking consensus, or the so called ranking compromise.

2. Method

2.1. Consensus Ranking Concept

Consensus ranking concept in the ordinal qualitative criteria is meant basic framework which leads to problem approach that can mathematically be analogized in order to facilitate understanding in searching for the alternative of solution.

This concept began with an assumption that there was a group of N object which would be ranked by the K persons as the team of decision makers (DM) in the ranking, and a ranking method was done partially in which each DM was given as many ($n \leq N$) objects randomly, uaing replacement method. After that, each DM freely ranked n object they owned based on individual preferences to be recommended for a consensus ranking with skenario which called Weighted Distance Forward Backward (WDFB).

2.1.1. Weighted Distance Forward Backward (WDFB) Method

Weighted Distance Forward Backward (WDFB) method suggested as a method for determining ranking consensus method consists in three scenarios. Defining distance in ranking was made by giving value of one for a distance of two adjascent object in a single ranking; for example, in the ranking arrangement of three object, a,b, and c was two. The first scenario in this method was begin by determining the weight toward the distance in ranking with the following formula:

$$\frac{n+1-i}{\sum_{i=1}^{n} i} \tag{1}$$

The weigth is used to calculate a Object Ranking Value (ORV). ORV was calculated for each DM by multiplying the distance between a pair of object in ranking forward and the weight of the object ranking distance. ORV was then added for each dominating object of the entire Dm and is donated as Foeward Object Ranking Value (FORV). The second scenario was also begun by determining ranking weight according to the weight formula in the first scenario, but with minus mark as follows:

$$-\frac{n+1-i}{\sum_{i=1}^{n} i} \tag{2}$$

ORV was the calculated by multiplying the distance between a pair of object in a rank backward and the weight of the object rank distance. ORV was then added for each dominating object of the entire DM and is donated as Backward Object Ranking Value (BORV). In the last scenario, FORV was added with BORV as the basic for ranking consensus object by putting the right order called Basic Value of Consensus Ranking Object (BVCRO) from maximum to minimum. This was what it called Consensus Ranking.

3. Result and Discussion

In order to implement the WDFB method as it was suggested to be method of ranking consensus, an example completed by Cook, et.all (2005) was taken, and the usage of the main algorithm of optimal ranking consensus its compromised ranking was 2 > 1 > 4 >6 > 3 > 5. The WDFB method proposed in this paper solvet the example, and it is purposely done compare their result.

The problem can be regarded as the chioice of the best n=4 proposals out of N=6 research proposals to be financed, and there are five DMs involved as assesor to obtain the ranking of the six proposal were ranked 1,2,3,4,5,and 6, the result of the ranking of each DM for the four ranked proposals would be as follows:

DMProposal	*Rangking*
1{1,2,3,5}	$1 \succ 3 \succ 2 \succ 5$
2{1,2,4,6}	$2 \succ 1 \succ 4 \succ 6$
3{3,4,5,6}	$4 \succ 3 \succ 5 \succ 6$
4{1,4,5,6}	$6 \succ 1 \succ 4 \succ 5$
5{1,2,3,6}	$6 \succ 2 \succ 3 \succ 1$

3.1. Implementation of WDFB Method

The settlement as the ranking consensus by using WDFB method was done according to the stages which had been designed in the scenario as follows:
Scenario I. Forward Weighted Ranking Distance: for ranking 1, 2, 3, and 4 consecutively it was found $\frac{4}{10}, \frac{3}{10}, \frac{2}{10}, \frac{1}{10},$ while the distance among the objects in each Dm was determined in the example made by DM I as follows: the distance from object 1 to object 2 was 1, annotated with d (1,2) = 1, the distance form object 1 to object 3 was 2, annotated with d (1,3) = 2, and the distance from object 1 to object 5 was 3, annotated with d (1,5). Furthermore, the value of each ranking object is calculated by multiplying the distance with weights with in the forward ranking. In this example for the ranking made by DM1 of the object 1 was

$ORB1=d(1,2)* \frac{4}{10}= \frac{4}{10}$

$ORB1=d(1,3)* \frac{4}{10}, = \frac{8}{10}$

$ORB1=d(1,5)* \frac{4}{10}= \frac{12}{10}.$

The ranking value of object 2 and object 3 could be obtained as follows:

$ORB2=d(2,3)* \frac{3}{10} =\frac{3}{10}$

$ORB2=d(2,5)*\frac{3}{10} = \frac{6}{10},$ and

$ORB3=d(3,5)*\frac{2}{10} = \frac{2}{10},$ so that it could be calculated for each DM with a similiar way.

Scenario II. Backward Weighted Ranking Distance: for ranking 1, 2, 3, and 4 consecutively, it was found $\frac{-4}{10},\frac{-3}{10},\frac{-2}{10},\frac{-1}{10},$ while the distance among the objects in the ranking for each Dm could be determined. In the example made by DM 1, it was found that the distance from object 5 to object 3 was 1, annotated with d(5,3)=1, the distance form object 5 to object 2 was 2, annotated with d(5,2)=2, and the distance from object 5 to object 1 was 3, annotated with d(1,5). Furthermore, the value of each ranking object is calculated by multiplying the disatance with weights within the backward ranking. In this example for the ranking made by
DM1 of the object 5 are:

$ORB5=d(5,3)*\frac{-4}{10}= \frac{-4}{10}$

$ORB5=d(5,2)*\frac{-4}{10}=\frac{-8}{10}$

$ORB5=d(5,1)*\frac{-4}{10}=\frac{-12}{10}.$ The ranking value of object 3 and object 2 could be calculated with the same method and it was found that

$ORB3=d(3,2)*\frac{-3}{10}=\frac{-3}{10}$

$ORB3=d(3,1)*\frac{-3}{10}=\frac{-6}{10}$

$ORB2=d(2,1)*\frac{-2}{10}=\frac{-2}{10},$ so that it could be calculated for each DM with a similiar way.

Scenario III: the result of the calculation of each DM was added up in order to obtain FORV and BORV. The total of FORB and BORV for each object was made as the making consensus ranking. Form the calculation of each DM, it could be found the Basic Value of Consensus Ranking (BVCR) as follows:

BVCRO1=1,4	BVCRO2=2,4
BVCRO3=0,7	BVCRO4=1,0
BVCRO5=-5,5	BVCRO6=0

So that the Consensus Ranking Proposal was P2, P1, P4, P3, P6, P5, and thus the proposal which was financed, based on the consensus, was Proposal Number 2,1,4, and 3. In terma of Cook, et.al example's notation, the result is 2 › 1 › 4 › 3. The difference lies on proposal3 instead of proposal 6.

4. Conclusion

WDFB Method with in the three proposed scenarios have slight differences in completion of consensus ranking problem in the investigated example of Cook, at.al (2005) using the main algorithm model optmal ranking consensus. This requires more in-depth studies that can be done in future studies. Can be suggested that this method can be modeled, and further can be done in a simulation study comparing it with existing methods.

Acknowledgments
The authors gratefully acknowledges support of Reseach Institute of University Sumatera Utara under BPPTN program with the cantract number 6049/UN5.1.R/PPM2016 tanggal 19 Juli 2016.

References

[1] Ali, I., Cook, W.D. and Kress, M. (1986)" Ordinal ranking and intensity of preference: a linear propgraming approach ", Management Science, Vol. 32, no. 12, pp.1642-1647.
[2] Borda. J.C (1781)"Memoire sur les elections au scrutiny,"in Histoire de 1'Academie Royale des sciences, Paris.
[3] Borgart. K. P (1975), "Preference structures-II: distances between asymmetric relations," SIAM Journal on Applied Mathematics, vol. 29, pp. 254-262.
[4] Cook W.D. (2006) 'Distance-based and adhoc consensus models in ordinal preference ranking. European Journal of Operational Research 172, pp. 369-385
[5] Cook W.D, B. Golany, M. Penn and T. Raviv (2006). Creating a Consensus Ranking of Proposals from Reviewers' Partial Ordinal Ranking. Working Paper, York Uversit.
[6] Cook W.D, B. Golany, M. Penn, T. Raviv (2005). Optimal Allocation of Proposals to Reviewers to Facilitate Effective Ranking, Management Science forthcoming.
[7] Cook W.D, Kress. M (1991). Ordinal in formation and preference structures: decision models and applications, Prentice-Hall, Nedw Jersey.
[8] Cook W.D, M. Kress (1990). An math generation model for weak ranking of players in tournament, Journal of the Operational Research society, 41(12),1111-1119.
[9] Cook W.D, L.M. Seiford (1982). The Borda-Kendall Consensus methods for priority ranking problems, Management Science, 28(6), 621-637.
[10] Cook W.D, L.M. Seiford (1978). Priority ranking and consensus formation, Management Science, 24(16), 1721-1732.
[11] Cook W.D, and Saipe.A.L (1976), "Committe approach to priority planning: the median ranking method, "Cahiers du Centre d'Etudes de Recherche Operationnelle, vol.18, no.3, pp.337-351.
[12] Foroughi AA and Tamiz M (2005). An effective total ranking model for a ranked voting system Omega 33:491496
[13] Golany, B., M. Kress. (1993). A multi-criteria evaluation of methods for obtaining weights from ratio-scale matrices, European Journal of Operational Research, 69(2), 210-220.
[14] Inada. K (1969)," The simple majority decision rule,"Econometria, vol.37, no.3, pp. 490-560
[15] Iz. P and Jelassi. M. T (1991) "an interactive group decision aid for multi-objective problems: an empirical assessment, "Omega, vol. 18, no.6, pp. 191-198.
[16] Kemeny. J. Gand and Snell. J. L. (1962). "Preference ranking: an axiomatic approach, "in Mtahematical Models in the Social Science, chapter 2, pp.9-23, Ginn,Boston,Mas,USA.
[17] Kendall. M (1962), Rank Correlation Methods, Hafner, New York, NY, USA,3rd edition.
[18] Keesey. R.E (1974) Modren Parliamentary Procedur, Houghton Mifflin, Boston, Mass, USA.
[19] Kirwood.C.W, Sarin.R.K (1985). Ranking with partial information –a method and an application Operations Research, 33(1), 38-48.
[20] Kruger'H.A, Kearney'.W.D (2008) Consensus ranking – An ICT security awareness case styudy. Journal Science Direct,North-West University.

[21] Majdid Tavana, Frank Lo Pinto, and Jmaes W. Smither (2007) A Hybrid Distance – Based Ideal – Seeking Consensus Ranking Model. Journal of applied Mathematics and Decision Science, Article Id 20489, 18 pages.

[22] Muralidharan. C, Anantharaman. N, and Deshmukh S.G, (2002) “A multi-criteria group decision – making model for supplier rating, “The Journal of Supply Chain Mangement, vol. 38, no.4, pp.22-33

[23] Wang. YM, Chin. KS and Yang. JB (2007) Three new models for preference voting and aggregation. Journal of the Operational Research Society 58,1389-1393.

IOP Conf. Series: Materials Science and Engineering **300** (2018) 012035 doi:10.1088/1757-899X/300/1/012035

Analysis of forecasting malaria case with climatic factors as predictor in Mandailing Natal Regency: a time series study

D Aulia[1], S F Ayu[2], and A Matondang[3]
[1,3]*Faculty of Public Health, University of Sumatera Utara*
[2]*Faculty of Agribusiness, University of Sumatera Utara*

Email : [1]aulia_destanul@yahoo.com, [2]srifajar.ayu@gmail.com, [3]apriyantinamatondang9@gmail.com

Abstract. Malaria is the most contagious global concern. As a public health problem with outbreaks, affect the quality of life and economy, also could lead to death. Therefore, this research is to forecast malaria cases with climatic factors as predictors in Mandailing Natal Regency. The total number of positive malaria cases on January 2008 to December 2016 were taken from health department of Mandailing Natal Regency. Climates data such as rainfall, humidity, and temperature were taken from Center of Statistic Department of Mandailing Natal Regency. E-views ver. 9 is used to analyze this study. Autoregressive integrated average, ARIMA (0,1,1) $(1,0,0)^{12}$ is the best model to explain the 67,2% variability data in time series study. Rainfall (*P* value = 0.0005), temperature (*P* value = 0,0029) and humidity (*P* value = 0.0001) are significant predictors for malaria transmission. Seasonal adjusted factor (SAF) in November and March shows peak for malaria cases.

1. Introduction

Malaria cases have decreased to 18% in 2015, from 262 million cases in 2000 to 214 million cases in 2015. Most cases in 2015 occurred in Africa (88%), Southeast Asia (10%), and the Eastern Mediterranean (2%). It was estimated that there has decreased to 37% cases in 2000-2015. Malaria transmission decreased from 106 countries in 2000 to 57 countries in 2015, resulting in transmission of malaria more than 75%. Furthermore, 18 countries ere able to reduce the case of malaria to 50-75% [1].

North Sumatra Province was not categorized as the highest and lowest of Annual Paracitic Incidence (API). The province was decreased in API and has reached the target of health ministry plan's for malaria morbidity (API) less than 1 per 1,000 risk population in 2015. Nevertheless, there are 3 regencies / cities in North Sumatera Province that have not reached national targets and have high API number. The highest prevalence of malaria with API are Mandailing Natal with 6.88%, GunungSitoli is 3.38%, and Batu Bara Regency is 1.40% [2].

The case of malaria in Mandailing Natal has decreased from 4029 in 2008 to 3465 patients in 2011. While in 2012 there was an increased of 7544 cases, then decreased significantly from 754 to 905 patients in 2016. This is due to the fluctuation of malaria cases every month as malaria is still problem.

Studies of malaria epidemics shows the relationship among climatic conditions: rainfall, temperature, and humidity. The transmission is seasonal, with the peak happens in raining season when the temperature is between 20^0C and 30^0C and the mean relative humidity is at least to 60% because of the increasing breeding sites [4].

2. Methods

The study was conducted in Health Departement of Mandailing Natal. This regency occupied in 662,070 km^2 area, which is divided into 23 districts and 407 villages (definitive sub-district). According to Departement of Centre Statistic Mandailing Natal Regency, the total population of Mandailing Natal was 211.506 in 2015. The altitude area 2.145 metres and influenced by monsoon, dry and rainy season temperature, and precipitation [5]. Rainy season in Mandailing Natal starts in December to March, but dry season starts in June to September. This situation changed each year after the transition period in April to May and in October to November [6].

Data of positives malaria cases occur in January 2008 to December 2016 takes from Health Departement of Mandailing Natal. It represented the malaria cases in this research. The data for rainfall, humidity, and temperature data were taken from Departement of Centre Statistic of Mandailing Natal. E-views ver. 9 software was used for analysed this data. Stationarity data was used with auto correlation function (ACF) and partial autocorrelation function (PACF). Seasonal adjusted factor (SAF) was determined the peak of seasonal variation. The Ljung-Box was used to determining if the model was good to acted the confounding factors, forecasting of the incidence malaria cases

3. Result and discussion

3.1 Result

The total number of laboratory confirmed about malaria case in 2008 to 2016 is showed in (Table 1). It was increased in 2012 and 2013 as shown in Table 1. The exploration of malaria infections : rainfall, temperature, and humidity in 2008 to 2016 shown that no trend and suggestion in seasonal dependency. The autocorrelation function (ACF) was significant in peak with a lag of 12 months (autocorrelation=0.137; Box-Ljung statistics (P = 0.027), partial autocorrelation function (PACF) was significant in peak with a lag of 12.

Table 1. Malaria cases in Mandailing Natal during the research period (*n*=108)

Month	2008	2009	2010	2011	2012	2013	2014	2015	2016
January	429	237	238	260	660	809	500	441	174
February	251	216	375	223	711	638	494	360	180
March	512	298	293	387	698	764	356	346	187
April	262	329	315	329	699	778	469	304	105
May	103	278	227	250	640	904	413	217	87
June	278	218	241	201	553	627	389	283	27
July	312	109	214	97	566	500	260	213	29
Agustus	465	389	108	210	327	365	439	321	32
September	273	317	396	256	546	683	637	329	23
October	258	349	366	305	643	602	536	259	22
November	498	371	356	598	857	401	455	183	20
December	388	286	354	349	796	473	350	165	19
Total	4029	3397	3483	3465	7696	7544	5298	3421	905

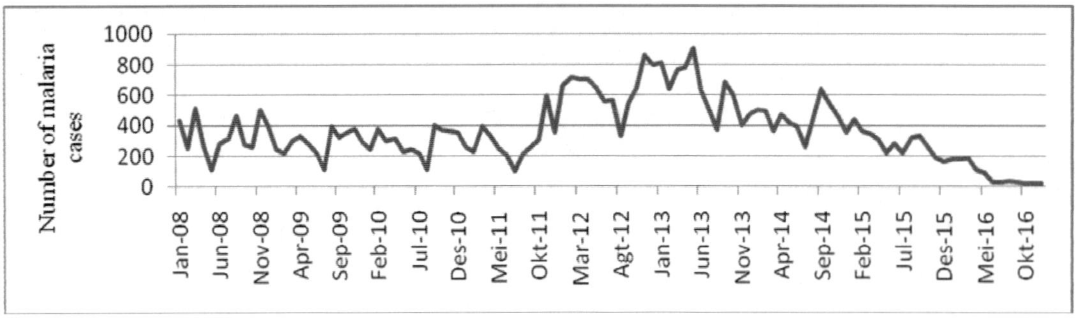
Number of malaria cases
1000
800
600
400
200
0
Jan-08 Jun-08 Nov-08 Apr-09 Sep-09 Feb-10 Jul-10 Des-10 Mei-11 Okt-11 Mar-12 Agt-12 Jan-13 Jun-13 Nov-13 Apr-14 Sep-14 Feb-15 Jul-15 Des-15 Mei-16 Okt-16

(a)

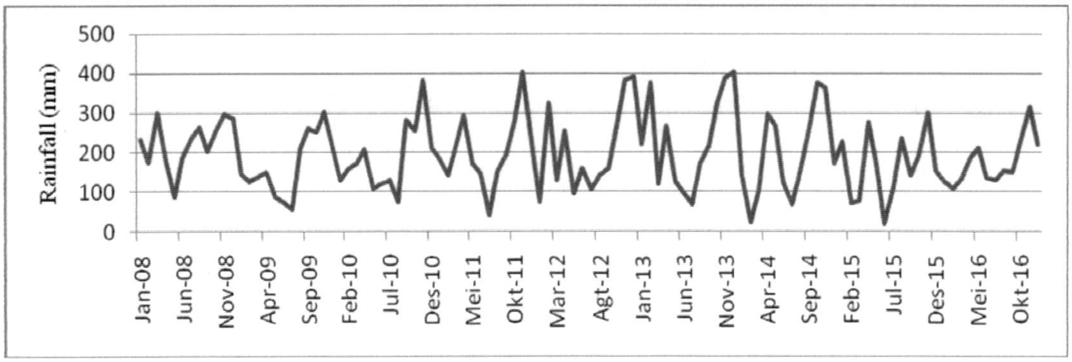
Rainfall (mm)
500
400
300
200
100
0
Jan-08 Jun-08 Nov-08 Apr-09 Sep-09 Feb-10 Jul-10 Des-10 Mei-11 Okt-11 Mar-12 Agt-12 Jan-13 Jun-13 Nov-13 Apr-14 Sep-14 Feb-15 Jul-15 Des-15 Mei-16 Okt-16

(b)

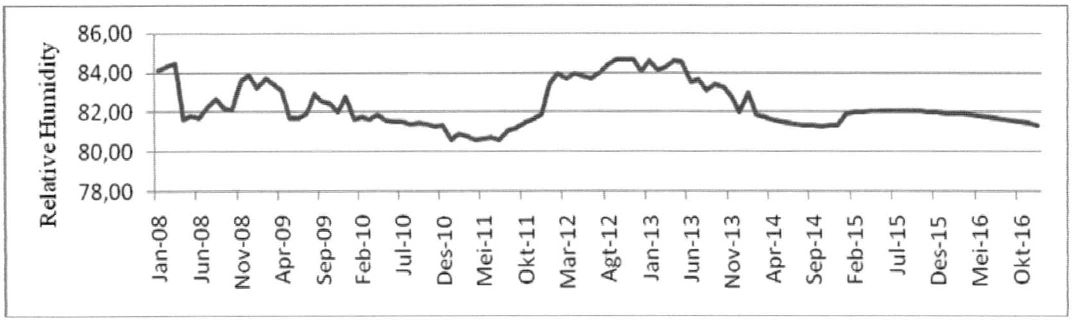
Relative Humidity
86,00
84,00
82,00
80,00
78,00
Jan-08 Jun-08 Nov-08 Apr-09 Sep-09 Feb-10 Jul-10 Des-10 Mei-11 Okt-11 Mar-12 Agt-12 Jan-13 Jun-13 Nov-13 Apr-14 Sep-14 Feb-15 Jul-15 Des-15 Mei-16 Okt-16

(c)

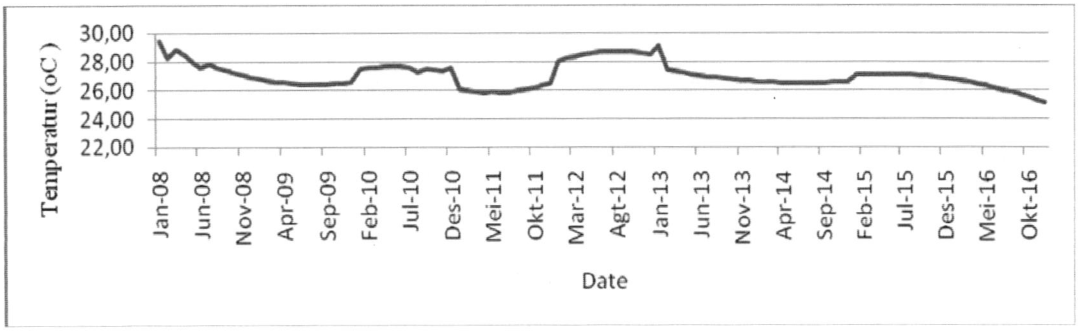

(d)

Figure 1. Rainfall, relative humidity, and mean maximum temperature of malaria cases with in January 2008 to December 2016 (a-d)

Observations have component with not seasonal variation. Table 2 shown that in November and March, the seasonal adjustment factor (SAF) was greater than other month. This period coincides with the rainy season in Mandailing Natal Regency. In November had the highest with 10,72 SAF, followed in March which is 10,12 SAF. So that the peak of malaria transmission which risk is 10,72 times and 10,12 more than typical month.

Table 2. Seasonal Adjustment Factor (SAF) for Malaria Cases

Month	Observated Cases	SAF
January	3748	9,22
February	3448	8,45
March	3841	10,12
April	3590	9,51
May	3119	8,58
June	2817	7,65
July	2300	6,32
August	2656	8,46
September	3460	9,8
October	3340	9,68
November	3739	10,72
December	3180	9,23

Eviews ver. 9 was the best model for forecasted malaria cases with rainfall, temperature, and humidity as predictor. ARIMA $(0,1,1)(1,0,0)^{12}$ model was the best statistical for this time series study. There were three climatic factors, mean rainfall (*P* value = 0.005), temperature (*P* value = 0.0029) and relative humidity (*P* value = 0.001), the lagged was significant.

The Ljung-Box (modified Box-Pierce) statistics shows in (Table 2). ARIMA$(0,1,1)(1,0,0)^{12}$. There was a difference of statistics with stationary *R*-squared used can estimate estimate of the variation and proportion. *R*-squared was better with trend or seasonal pattern.The stationary squared (up to a maximum value of 1) indicated was good, and 0,672 of A value have shown in Table 3.

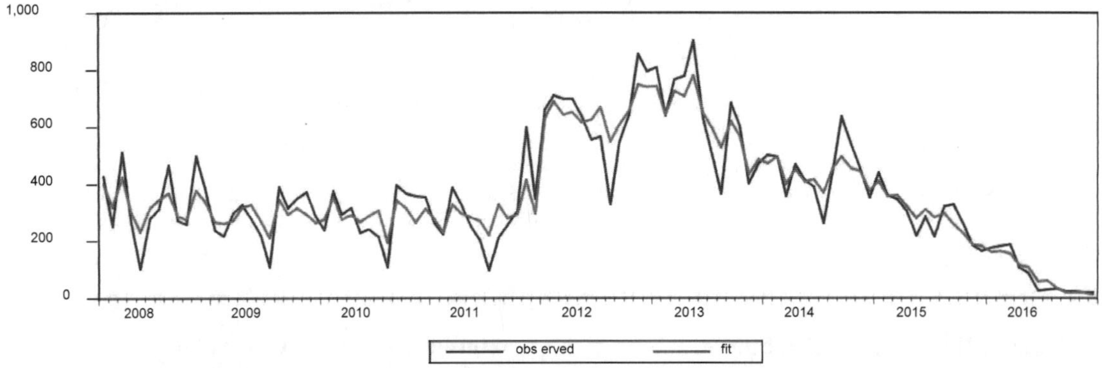

Figure 2. The actual and predicted of malaria

(ARIMA) model was using in this research for identificated, estimate, and forecastied. ARIMA model for seasonal was $(p, d, q)(P,D,Q)^s$, where p and P- are the autoregressive andseasonal autoregressive, respectively, d and D- are the nonseasonal differences and seasonal differencing, respectively,q and Q- are the moving average parameters and seasonal moving average parameters, respectively, and s represents thelength of the seasonal period

ARIMA (0,1,1) $(1,0,0)^{12}$ model used for forecasting the monthly malaria for next period which in January 2017 to December 2018 that have significant predictors in rainfall, temperature, and humidity which were lagged for one month. Model of forcasted shows 701 cases in 2017 and 1890 cases in 2018 with significant peaks during rainy season (Figure 3).

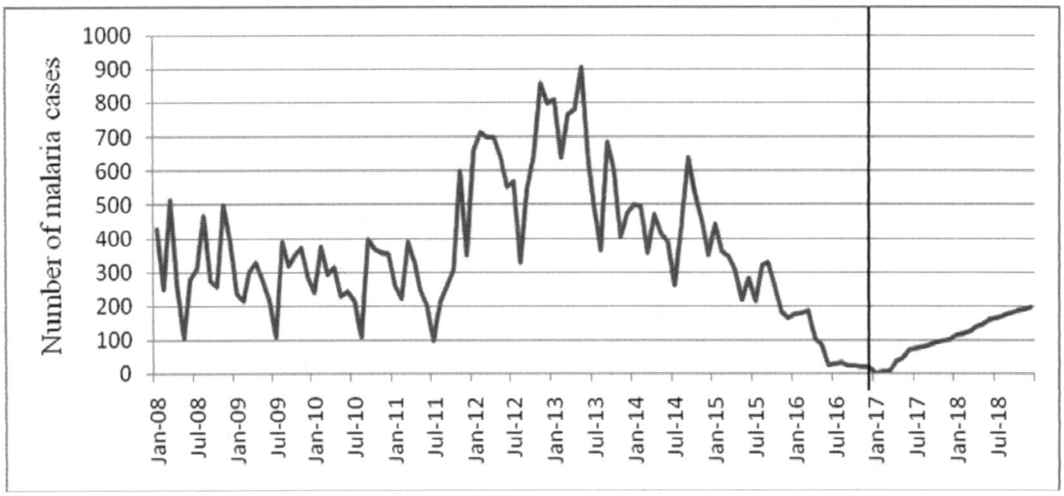

Figure 3. Fluctuation of malaria cases in Mandailing Natal Regency

Table 3. Stationary of malaria case monthly

Model of malaria parameter	Stationary R^2	P value	Model type
Malaria infections	0.672	0,000000	ARIMA (0,1,1) $(1,0,0)^{12}$

3.2 Discussion

ARIMA model was the best of a time series in epidemiological surveillance for this research. ARIMA (0,1,1) $(1,0,0)^{12}$ was a useful software for estimated the total malaria case for monthly in future with using the climatic factors like rainfall, temperatur, and relative humidity as predictors. These climatic was contributed because anopheles vector took two weeks to completing their life cycle and two more weeks for the generation of parasites to be a new host. Wangdi et al. Research shown that ARIMA (2,1,1) $(0,1,1)^{12}$ was the best model for predicted malaria in Bhutan [9]. ARIMA model was also used for forecasting malaria cases in Sri Lanka [10] and Ethiopia [11]. Kumar et al research also taken ARIMA (0,1,1) (0,1,0) for predicted malaria in New Delhi [7].

Malaria transmission was found between in November and March. Rainy season in Mandailing Natal in December to last March. From this we can take that there is a strong correlation malaria with rainfall in Mandailing Natal with peak was occuring in March.

It was related with Widiastuti research, that rainfall have be increasing the malaria incidence in near beach of Kalianda with 48 cases every month, with highest proportion is 89% after rainfall in 7 months regularly. In Kotabumi, the malaria incidance 45 cases with rainfall 1 per month and highest proportion of 93 % after rainfall 3 months regularly. The two district with limit rainfall was 25 mm, it was explain the malaria incidence [12].

Temperature also acts significant role in malaria transmission. Temperature is expected to increase transmission and prevalence of malaria with shortening the incubation period of the parasite in the mosquitoes. Sporogonic cycles take about in 9 to 10 days at temperatures of 28 °C but higher than 30 °C and below 16 °C have negative impact on parasite development [13]. Malaria transmission also depends on relative humidity since mosquitoes have a limited range of tolerable relative humidity. The high surface area to ratio volume of mosquitoes makes them especially sensitive to desiccation in low humidity levels [14].

Transmission of malaria will happen in the relative humidity is at least 60% as one of predictor. The research of Mofu showed that there was relationship between Anopheles density and humidity, where the correlation results $p = 0.002$ with average humidity at the time of the study was 82.6%. The highest density of Anopheles (4.1 head / person / hour) in humidity was 85.3% and the lowest was 1.0 / person / hr at 78.5% and 76.0% air moisture with average humidity when the study was 82.6% [15].

There were many factors such as the type of parasite, movement, immunity for malaria, resistance of insecticide that can contributed for malaria case.

4. Conclusion and recommendation

4.1 Conclusion

Based on the research, rainfall, temperature and relative humidity are predictors and have strong correlations with malaria cases in Mandailing Natal. ARIMA in time series study is the best model for this research to predict malaria case monthly, annually and will happen in next year.

4.2 Recommendation

This data can be used to make a plan and policy about malaria prevention and inform the best time for people to travel into this area.

References

[1] WHO 2015 *World Malaria Report 2015* (Prancis : World Health Organization)

[2] Health Office of North Sumatera 2016 *Prevention of Malaria Diseases Report 2015* (Medan Health Office of North Sumatera)

[3] Gomez E, Otero, Van H, and Aguirre J 2007 *Malaria Journal* 6 (1) 129

[4] Gillies T 1988 *Principles and Practice of Malariology* 453–485 (Scotland)

[5] Prevention Center office of Mandailing Natal Regency 2016 *Malaria Profile of Mandailing Natal Regency 2015*

[6] Goverment of Mandailing Natal Regency 2014 *Medium Term Development Plan of Mandailing Natal Regency 2011-2016*

[7] Kumar, Mangal A, and Singh S 2014 *Malaria Research and Treatment in Delhi* 482851 (6)

[8] Helfenstein U 1991 *International Journal of Epidemiology* 20 (3) 808

[9] Wangdi, Singhasivanon P, and Kaewkungwal J 2010 *Malaria Journal* 9 251

[10] Vounatsou P, Galappaththy G, and Amerasinghe P, 2008 *Malaria Journal* 7 (76)

[11] Abeku A, De Vlas S, and G. Borsboom 2002 *The American Journal of* Tropical Medicine *and Hygiene, 7, (10) 851*

[12] Widiatusti 2009. *IPB Journal* Institution of Agricultur Bogor.

[13] Craig M, Snow R, and Sueur D 1999 *Parasitology Today*, 15 (3) 105

[14] Yamana T and Eltahir E 2013 *Parasites & Vectors* 9 235

[15] Mofu, R 2013 *Journal of Environment Health* 12 (2)

IOP Conf. Series: Materials Science and Engineering **300** (2018) 012036 doi:10.1088/1757-899X/300/1/012036

Estimating parameter of Rayleigh distribution by using Maximum Likelihood method and Bayes method

Fitri Ardianti[1] and Sutarman[2]

Department of Mathematics, University of Sumatera Utara, Medan, Indonesia

E-mail: [1]`fitriardianti96@gmail.com`; [2]`sutarman@usu.ac.id`

Abstract. In this paper, we use Maximum Likelihood estimation and Bayes method under some risk function to estimate parameter of Rayleigh distribution to know the best method. The prior knowledge which used in Bayes method is Jeffrey's non-informative prior. Maximum likelihood estimation and Bayes method under precautionary loss function, entropy loss function, loss function-L_1 will be compared. We compare these methods by bias and MSE value using R program. After that, the result will be displayed in tables to facilitate the comparisons.

1. Introduction

Rayleigh distribution is a continuous probability distribution family which is used as a model of life time. This distribution has long been considered to have significant applications in many fields such as survival analysis, reliability theory, and especially communication engineering. This distribution is a special case of two parameter Weibull distribution with the shape parameter equal to 2. The Rayleigh distribution was originally derived by Lord Rayleigh [1].

Siddiqui [2] showed that the Rayleigh amplitude distribution (the distribution of the power or amplitude of electronic waves received through a scattering medium) is the asymptotic distribution of a two-dimensional random walk. Several authors have contributed to this model. Howlader and Hossain [3] obtained Bayes estimators for the scale parameter and the reliability function in the case of Type-II censored sampling. Abd Elfattah, et al. [4] studied the efficiency of the maximum likelihood estimates of the parameter under three cases, namely, Type-I, Type-II and progressive Type-II censored sampling schemes. Hendi, et al. [5] obtained Bayes estimators of the scale parameter, reliability function and failure rate by using non-informative prior anda Hartigan prior based on upper record values. Dey and Das [6] obtained Bayesian predictive intervals of the parameter of Rayleigh distribution. Dey [7] also obtained Bayes estimators for the parameter and reliability function of the Rayleigh distribution under different loss function. Dey [7] also studied the Bayes estimators for the parameter and reliability function of Rayleigh distribution based on complete as well as Type-II consored samples, also compared relative risk functions.

Soliman [8], Mostert [9] compared the Bayesian estimators under the linear exponential (LINEX) loss function and squared-error loss (SEL) function. Dey and Dey [10] did similiar work for the complete model by applying Jeffreys prior and a loss function as proposed by Al-Bayyati [11]. Ferreira, et al. [12] extend concepts in the literature for the censored Rayleigh model by considering new loss function, namely the Al-Bayyati loss (ABL) and comparing it to other known result.

Al-Mayali and Al-Shaibani [13] compare some estimators of Rayleigh distribution with simulation. The method which was used in their research were maximum likelihood estimation, Bayes estimation, shrinkage estimation, and Bayesian shrinkage estimation and after that applied the result in simulation study to compare the result, and obtain which method was the best [13]. Singh and Srivastava [14] use Bayesian estimation of Inverse Maxwell distribution under squared error, precautionary, entropy, and another two loss function for using quasi-prior have been obtained.

2. Rayleigh Distribution

The Rayleigh distribution has a wide range of applications including life testing experiments and clinical studies. This distribution has probability density function (pdf):

$$f(x|\theta) = \frac{x}{\theta^2} e^{-\frac{x^2}{2\theta^2}}, x, \theta > 0 \tag{1}$$

3. Maximum Likelihood Estimation

Let x_1, x_2, $\ldots$, x_n be random sample of the population with density $f(x, \theta)$ where $\theta(\theta_1, \theta_2, \ldots, \theta_k)$ are unknown parameters, the likelihood function is written:

$$L(\theta_1, \theta_2, ..., \theta_3) = \pi_{i=1}^{n} f(x_i; \theta) \tag{2}$$

The likelihood function is a function of an unknown parameter. In the application $L(\theta)$ denotes the joint probability density function of the random sample. If S parameter space is an open interval and $L(\theta)$ is a function that can be derived and assumed to be maximum at S. Then the equation of maximum likelihood is:

$$\frac{d}{d\theta} L(\theta) = 0 \tag{3}$$

The parameters are estimated with solving the equation (3)

4. Bayes Method

Let $f(y|\theta); \theta \in \Theta$ be the probability density function of lifetime distribution of a component or an animate, where the parameter space Θ is known but the true value of θ is unknown. Let $g(\theta)$ be the prior density function of the random variable θ. Let $\underline{y} = (y_1, ..., y_n)$ be an independent observations from $f(y; \theta)$. Then using Bayes's theorem (1763) the posterior distribution $f(\theta|\underline{y})$ of θ is given by:

$$f(\theta|\underline{y}) = \frac{f(\underline{y}|\theta) g(\theta)}{\int_\theta f(y|\theta) g(\theta) d\theta} \tag{4}$$

where $f(\underline{y}|\theta)$ is the joint probability density function of $\underline{y} = (y_1, ..., y_n)$. For given sample $\underline{y}$, the posterior p.d.f $f(\theta|\underline{y})$ is the basis for most types of Bayesian inference [14].

5. Risk Function

Let $L(\hat{\theta}, \theta)$ be the loss function associated with an estimate of the parameter θ. Let $g_\theta(\theta|W = w)$ be the posterior distribution of the random variable Θ. Then the *risk* associated with $\hat{\theta}$ is the expected value of the loss function with the respect to the posterior distribution of θ [14]. The risk function is:

$$risk = \int_\theta L(\hat{\theta}, \theta) g_\Theta(\theta|W = w) d\theta \tag{5}$$

The fundamental problems in Bayesian analysis is that of the choice of prior distribution and loss function $L(\hat{\theta}, \theta)$ which may be appropriate for the situation at hand. The Bayes estimation $\hat{\theta}$ of θ is the course optimal relative to the loss function chosen. A commonly used loss function is the squared error loss function (SELF)
$L(\hat{\theta}, \theta) = (\hat{\theta} - \theta)$

(a) Squarred error loss function (SELF)

$$L(\hat{\theta}, \theta) = (\hat{\theta} - \theta)^2 \tag{6}$$

The Bayes estimator under the above loss function, say $\hat{\theta}$ the posterior mean, i.e.

$$\hat{\theta} = E(\theta) \tag{7}$$

The risk function is given by:

$$R(\hat{\theta}) = E(\hat{\theta})^2 - 2\theta E(\hat{\theta}) + \theta^2 \tag{8}$$

(b) Precautionary Loss Function

The posterior expectation of loss function is:

$$E[L(\hat{\theta} - \theta)] = E(\frac{\theta^2}{\hat{\theta}}) + E(\hat{\theta}) - E(\theta) \tag{9}$$

The value of $\hat{\theta}$ that minimize (9), denoted by $\hat{\theta}_p$ is obtained by solving the following equation

$$\frac{d}{d\theta} E[L(\hat{\theta} - \theta)] = 0 \tag{10}$$

$$\bar{\theta}_p = [E(\theta^2)]^{\frac{1}{2}} \tag{11}$$

(c) Entropy Loss Function

The Bayes estimator of entropy loss function is:

$$\bar{\theta}_e = [E(\frac{1}{\theta})]^{-1} \tag{12}$$

(d) Loss function-L_1

Consider the loss function given by:

$$L_1(\hat{\theta}, \theta) = (\frac{\hat{\theta}}{\theta} - 1)^2 \tag{13}$$

The Bayes estimator of under loss function-L_1 is:

$$\hat{\theta}_1 = \frac{E(\frac{1}{\theta})}{E(\frac{1}{\theta})^2} \tag{14}$$

(e) Loss function-L_2
Consider the loss function given by:

$$L_2(\hat{\theta}, \theta) = (\frac{\theta}{\hat{\theta}} - 1)^2 \tag{15}$$

The Bayes estimator of under loss function-L_2 is:

$$\bar{\theta}_2 = \frac{E(\theta^2)}{E(\theta)} \tag{16}$$

6. Result

(a) Estimating parameter by using Maximum Likelihood method
In this paper, we will estimate one parameter of Rayleigh distribution by using maximum likelihood estimation. The probability of density function (pdf) of this distribution is:

$$f(x|\sigma) = \frac{x}{\sigma^2} e^{-\frac{x^2}{2\sigma^2}}, x, \sigma > 0 \tag{17}$$

and From the equation (14), we will find the maximum likelihood function:

$$L(x, \sigma) = \Pi_{i=1}^{n} x_i (\frac{1}{\sigma^2})^n exp(-\frac{1}{2} \sum_{i=1}^{n} (\frac{x_i}{\sigma})^2) \tag{18}$$

After that, we will log the equation (18) and the last differential partial of the equation, so we will find the estimator of rayliegh distribution by using maximum likelihood function. It will showed:

$$ln[L(x, \sigma)] = ln[\Pi_{i=1}^{n} x_i (\frac{1}{\sigma^2})^n exp(-\frac{1}{2} \sum_{i=1}^{n} (\frac{x_i}{\sigma})^2)] \tag{19}$$

$$\hat{\sigma}_{MLE} = \sqrt{\frac{1}{2n} \sum_{i=1}^{n} {x_i}^2} \tag{20}$$

(b) Estimating parameter by using Bayes method
The fisrt step to estimate the parameter by using bayes method is we have to find the prior distribution, because bayes method is a method which calculate the role of prior. In this paper, the prior distribution which used is non-informatif prior. The non-informative prior which is used in this paper is Jeffrey's method. The non-infomatif prior of the rayleigh distribution is:

$$f(\sigma) = \sqrt{I(\sigma^2)} \propto \frac{1}{\sigma} \tag{21}$$

by the equation (21), we will find the posterior distribution:

$$f(\sigma|x) \propto \frac{c}{\sigma^{2n+1}} \Pi_{i=1}^{n} x_i exp(-\frac{1}{2} \sum_{i=1}^{n} (\frac{x_i}{\sigma})^2) \tag{22}$$

from the equation (22), we will find the posterior density function:

$$\pi * (\sigma|x_i) = \frac{(\sum_{i=1}^{n} {x_i}^2)^n exp[-\frac{1}{2}\sum_{i=1}^{n}(\frac{x_i}{\sigma})^2]}{\sigma^{2n+1}2^n\Gamma(n+\frac{3}{2})}. \tag{23}$$

The Bayes estimator is an estimator that minimizes the risk function by representing the expected of loss function. There are some loss function that can be applied to the bayes metod such as squarred error loss function, precautionary loss function, entropy loss function-L_1 and loss function-L_2 [14].
In this paper, the author use three loss functions to estimate rayleigh distribution in bayesian approach. Namely, precautionary loss function, entropy loss function and loss function-L_1.

i. Precautionary loss function
The Bayes estimator under precautionary loss functon:

$$\sigma_p = [E(\sigma^2)]^{\frac{1}{2}} \tag{24}$$

$$= \sqrt{\int_0^\infty \sigma^2 \pi * (\sigma|x_i)d\sigma} \tag{25}$$

$$= \sqrt{\frac{\Gamma(n+\frac{1}{2})\sum_{i=1}^{n} x_i^2)}{2\Gamma(n+\frac{3}{2})}} \tag{26}$$

ii. Entropy loss function
The Bayes estimator under Entropy loss functon:

$$\sigma_e = [E(\frac{1}{\sigma})]^{-1} \tag{27}$$

$$= [\int_0^\infty \frac{1}{\sigma}\pi * (\sigma|x_i)d\sigma]^{-1} \tag{28}$$

$$= \frac{\Gamma(n+\frac{3}{2})\sqrt{\sum_{i=1}^{n} x_i^2}}{\sqrt{2}\Gamma(n+2)} \tag{29}$$

iii. Loss function-$\mathbf{L_1}$
The Bayes estimator under Loss function-L_1:

$$\sigma_1 = \frac{E(\frac{1}{\sigma})}{E(\frac{1}{\sigma^2})} \tag{30}$$

$$= \frac{\sqrt{2}\Gamma(n+2)}{2\Gamma(n+\frac{5}{2})}\sqrt{\sum_{i=1}^{n} x_i^2} \tag{31}$$

7. Simulation

Estimate by using Maximum likelihood and Bayes method will be compared using simulation. Data simulations were performed by generating various types of data conditions involving four different sample sizes n=10, n=25, n=50, and n=100. The data obtained were analyzed to estimate the parameters using Maximum Likelihood and Bayes methods. The next step is calculate the bias and MSE (Mean Square Error) values of both methods. The simulation in this paper is done with R program.

From Table 7.1 it can be seen that the bias value of both methods shows the smaller the bias value with the larger the sample size. The bias value of the Bayes method with loss function-L_1 shows smaller numbers than Maximum Likelihood method and Bayes method with precautionary loss function loss function, entropy loss function, and loss function-L_1.

For the MSE value, from Table 7.2 shows the smaller error value with the condition of the larger the sample size. The MSE value in the Maximum Likelihood method shows smaller numbers than the Bayes method with the loss function of precautionary loss function, entropy loss function, and loss function-L_1. From the data simulation, results in the table above shows that Bayes method is not always better in estimating the parameters compared to Maximum Likelihood method.

References

[1] Johnson N L, Kotz S, and Balakrishman S 1994 *Continuos Univariate Distributions*, **456**
[2] Siddiqui M M (1962) *Some problems connected with Rayleigh distributions*, **60D**, 167-174.
[3] Howlader H A and Hossain A 1995 *On Bayesian estimation and prediction from Rayleigh distribution based on type-II censored data*, **24**(9), 2249-2259.
[4] Abd Elfattah A M, Hassan A S and Ziedan D M (2006) *Efficiency of Maximum Likelihood Estimators under Different Censored Sampling Schemes for Rayleigh Distribution*
[5] Hendi M L, Abu S E and Alraddadi A A 2007 *A Bayesian Analysis of Record Statistics from Rayleigh Model*, International Mathematical Forum, **2** (13), 619-631.
[6] Dey S and Das M K (2007) *A Note on Prediction Inerval for a Rayleigh Distribution: Bayesian Approach*, American Journal of Mathematical and Management Science, **1&2**, 43-28.
[7] Dey S 2009 *Comparison of Bayes Estimators of the parameter and reliability function for Rayleigh distribution under different loss function*, Malaysian Journal of Mathematical Science, **3**, 217-261.
[8] Soliman A A 2000 *Comparison of LINEX and quadratic Bayes estimators for the Rayleigh distribution*, Communications in Statistics - Theory and Methods, **29**(1), 95-107.
[9] Mostert P J *A Bayesian method to analyse cancer lifetimes using Rayleigh models*, Unpublished PhD thesis, University of South Africa.
[10] Dey S and Dey T 2011 *Rayleigh distribution revisited via an extension of Jeffreys prior information and a new loss function*, Revstat, **9**(3), 213-226.
[11] Al-Bayyati H N *Comparing methods of estimating Weibull failure models using simulation*, Unpublished PhD thesis, College of Administration and Economics, Baghdad University, Iraq.
[12] Ferreira J T, Bekker A, and Arashi M 2015 *Objective bayesian estimators for the right censored rayliegh distribution: evaluating The Al-Bayyati loss function*, Revstat, **14** 433-454.
[13] Al-Mayali and Al-Shaibani 2013 *A Comparison for some of the estimators of rayleigh distribution with simulation*, Revstat, **11**(4).
[14] Singh K L and Srivastava R S (2012) *Bayesian estimation of parameter of inverse maxwell distribution via size-biased sampling*, International Journal of Science and Research(IJSR), **3.358**, 2319-7064.

Table 7.1 Bias Value of Estimation Rayleigh Distribution

N	σ	$\hat{\sigma}_{MLE}$	$\hat{\sigma}_{BSp}$	$\hat{\sigma}_{BSe}$	$\hat{\sigma}_{BS1}$	$Bias_{MLE}$	$Bias_{BSp}$	$Bias_{BSe}$	$Bias_{BS1}$
10	0.5	0.7496453	0.7315789	0.7066852	0.691493	-0.2846832	-0.2898723	-0.2970224	-0.3013859
	1	0.7496453	0.7315789	0.7066852	0.691493	-0.7846832	-0.7898723	-0.7970224	-0.8013859
	1.5	0.7496453	0.7315789	0.7066852	0.691493	-1.284683	-1.289872	-1.297022	-1.301386
	2	0.7496453	0.7315789	0.7066852	0.691493	-1.784683	-1.789872	-1.797022	-1.801386
	2.5	0.7496453	0.7315789	0.7066852	0.691493	-2.284683	-2.289872	-2.297022	-2.301386
25	0.5	0.6645003	0.6579534	0.6484712	0.6423827	-0.3881442	-0.3892462	-0.3908424	-0.3918672
	1	0.6645003	0.6579534	0.6484712	0.6423827	-0.8881442	-0.8892462	-0.8908424	-0.8918672
	1.5	0.6645003	0.6579534	0.6484712	0.6423827	-1.388144	-1.389246	-1.390842	-1.391867
	2	0.6645003	0.6579534	0.6484712	0.6423827	-1.888144	-1.889246	-1.890842	-1.891867
	2.5	0.6645003	0.6579534	0.6484712	0.6423827	-2.388144	-2.389246	-2.390842	-2.391867
50	0,5	0.7435811	0.7398909	0.7344527	0.7308961	-0.4095031	-0.4099522	-0.4106141	-0.4110469
	1	0.7435811	0.7398909	0.7344527	0.7308961	-0.9095031	-0.9099522	-0.9106141	-0.9110469
	1,5	0.7435811	0.7398909	0.7344527	0.7308961	-1.409503	-1.409952	-1.410614	-1.411047
	2	0.7435811	0.7398909	0.7344527	0.7308961	-1.909503	-1.909952	-1.910614	-1.911047
	2.5	0.7435811	0.7398909	0.7344527	0.7308961	-2.409503	-2.409952	-2.410614	-2.411047
100	0.5	0.7096232	0.7078558	0.7052282	0.7034933	-0.4382613	-0.438415	-0.4386436	-0.4387946
	1	0.7096232	0.7078558	0.7052282	0.7034933	-0.9382613	-0.938415	-0.9386436	-0.9387946
	1.5	0.7096232	0.7078558	0.7052282	0.7034933	-1.438261	-1.438415	-1.438644	-1.438795
	2	0.7096232	0.7078558	0.7052282	0.7034933	-1.938261	-1.938415	-1.938644	-1.938795
	2.5	0.7096232	0.7078558	0.7052282	0.7034933	-2.438261	-2.438415	-2.438644	-2.438795

Table 7.2 MSE (*Mean Square Error*) Value of Estimation Rayleigh Distribution

N	σ	$\hat{\sigma}_{MLE}$	$\hat{\sigma}_{BSp}$	$\hat{\sigma}_{BSe}$	$\hat{\sigma}_{BS1}$	MSE_{MLE}	MSE_{BSp}	MSE_{BSe}	MSE_{BS1}
10	0.5	0.7496453	0.7315789	0.7066852	0.691493	0.1299173	0.1317209	0.1342944	0.1359151
	1	0.7496453	0.7315789	0.7066852	0.691493	0.6646005	0.6715932	0.6813168	0.687301
	1.5	0.7496453	0.7315789	0.7066852	0.691493	1.699283	1.711465	1.728338	1.738687
	2	0.7496453	0.7315789	0.7066852	0.691493	3.233966	3.251337	3.27536	3.290073
	2.5	0.7496453	0.7315789	0.7066852	0.691493	5.268649	5.291209	5.322382	5.341459
25	0.5	0.6645003	0.6579534	0.6484712	0.6423827	0.1885944	0.1890773	0.1897811	0.1902356
	1	0.6645003	0.6579534	0.6484712	0.6423827	0.8267386	0.8283235	0.8306235	0.8321028
	1.5	0.6645003	0.6579534	0.6484712	0.6423827	1.964882	1.967569	1.971465	1.973969
	2	0.6645003	0.6579534	0.6484712	0.6423827	3.603026	3.606815	3.612307	3.615836
	2.5	0.6645003	0.6579534	0.6484712	0.6423827	5.74117	5.746061	5.753149	5.757703
50	0.5	0.7435811	0.7398909	0.7344527	0.7308961	0.196964	0.1971867	0.1975158	0.1977314
	1	0.7435811	0.7398909	0.7344527	0.7308961	0.8564671	0.8571389	0.8581299	0.8587783
	1.5	0.7435811	0.7398909	0.7344527	0.7308961	2.01597	2.017091	2.018744	2.019825
	2	0.7435811	0.7398909	0.7344527	0.7308961	3.675473	3.677043	3.679358	3.680872
	2.5	0.7435811	0.7398909	0.7344527	0.7308961	5.834976	5.836995	5.839972	5.841919
100	0.5	0.7096232	0.7078558	0.7052282	0.7034933	0.2095575	0.2096487	0.2097844	0.2098742
	1	0.7096232	0.7078558	0.7052282	0.7034933	0.8978188	0.8980637	0.898428	0.8986688
	1.5	0.7096232	0.7078558	0.7052282	0.7034933	2.086079	2.086479	2.087073	2.087465
	2	0.7096232	0.7078558	0.7052282	0.7034933	3.77434	3.774894	3.775717	3.77626
	2.5	0.7096232	0.7078558	0.7052282	0.7034933	5.962601	5.963309	5.964361	5.965055

IOP Conf. Series: Materials Science and Engineering **300** (2018) 012037 doi:10.1088/1757-899X/300/1/012037

Research of Face Recognition with Fisher Linear Discriminant

R Rahim[1*], T Afriliansyah[2], H Winata[3], D Nofriansyah[4], Ratnadewi[5] and S Aryza[6]

[1]Department of Informatics Engineering, Institut Teknologi Medan, Medan, Indonesia
[2]STKIP Bumi Persada Lhokseumawe, Jalan Medan Banda Aceh No. 59, Lhokseumawe, Indonesia
[3]Department of System Computer, STMIK Triguna Dharma, Jl. Jenderal Abdul Haris Nasution No.73, Medan 20219, Indonesia
[4]Department of Information System, STMIK Triguna Dharma, Jl. Jenderal Abdul Haris Nasution No.73, Medan 20219, Indonesia
[5]Electrical Engineering, Maranatha Christian University, Jalan Prof. Drg. Surya Sumantri No.65, Bandung 40164, Indonesia
[6]Faculty Engineering, Universitas Pembangunan Panca Budi, Jl. Jend Gatot Subroto KM. 4.5, Medan, 20122, Indonesia
[1,6]School of Computer and Communication Engineering, Universiti Malaysia Perlis, Perlis, Malaysia

*usurobbi85@zoho.com

Abstract. Face identification systems are developing rapidly, and these developments drive the advancement of biometric-based identification systems that have high accuracy. However, to develop a good face recognition system and to have high accuracy is something that's hard to find. Human faces have diverse expressions and attribute changes such as eyeglasses, mustache, beard and others. Fisher Linear Discriminant (FLD) is a class-specific method that distinguishes facial image images into classes and also creates distance between classes and intra classes so as to produce better classification.

1. Introduction

Technological developments are growing rapidly today especially in the field of information technology [1], various applications have been designed to seek information about one's self-data, one of which is the application of self-knowledge [2]. Facial recognition is one of the pattern recognition approaches for personal identification purposes in addition to other biometric approaches such as fingerprint recognition, signature, and retina of the eye and so on [2]. The recognition of the face image is related to an object that is never the same, because of the parts that can change [2] [3] [4]. These changes can be caused by facial expressions, shooting angles, or changes in accessories on the face. In this experiment, the same object with some of the differences must be identified as a single object [5].

Generally, facial recognition has several stages to identify the face of face detection, face alignment, feature extraction and also face matching [6]. Many methods are used in facial recognition applications, one of which is the Fisher Linear Discriminant method which is the basis of the Fisher face and the development of Principal Component Analysis (PCA) [2] [5] [7]. The Fisher Linear Discriminant (FLD) method is used to find the average face value of each class and to calculate the distribution matrix in the class and the spreading matrix between the classes and to find the optimal projection value of the image to obtain the face weights [5] [8]. The Fisher Linear Discriminant Method (FLD) is the most widely used method for pattern recognition. Fisher Linear Discriminant (FLD) is an example of a class-specific method, as it seeks to form the scatter between classes and intra classes so as to produce a better classification [9].

This paper conducted experiments to test the resemblance of the face with Fisher Linear Discriminant algorithm, the testing process is done gradually by showing the workings of Fisher Linear Discriminant in examining the similarity of the face and compare the image in the test so that the percentage of similarity of the face is known.

2. Fisher Linear Discriminant

Fisher Linear Discriminant (FLD), also known as Linear Discriminant Analysis (LDA) was discovered by Robert Fisher in 1936 for taxonomic classification and became one of the most widely used techniques in pattern recognition [3] [4] [6] [9]. FLD is an example of a class-specific method, and it is a well-known method for classifying class into class-and-class matrices of classes that could be taken for classification, as it seeks to form inter-class and intra-class scatterings so as to produce a ranking better [6] [9] [10]. Following steps do the calculation of Fisher Linear Discriminant (FLD) method:

a. Specify the number of classes and members of each class.

b. Establish the average value of the *training set (m)* and the average face value of each class (average class face, m_i), The formula as below

$$m = \frac{1}{N}\sum_{i=1}^{N} X_i \tag{1}$$

The formula calculation looks for the average value of each class:

$$m_i = \frac{1}{N_i}\sum_{j=1,x'j\in X_i}^{N} x_j' \tag{2}$$

c. Perform an inter-class matrix of inter-class (between-class scatter matrix, S_b) and in-class scatter-matrix, S_W matrices. By calculating the formula:

$$S_W = \sum_{i=1}^{C}\sum_{j=1,x_j'\in x_i}^{N_i}(x_j' - m_i)(x_j' - m_i)^T \tag{3}$$

$$S_B = \sum_{i=1}^{C} N_i\,(m_i - m)\,(m_i - m)^T \tag{4}$$

d. Find eigenvalue or W_{FLD} projection matrix with formula calculation:

$$W_{FLD} = \lambda \text{ and } W_{FLD} = \frac{S_B}{S_W} = S_B.S_W^{-1} \tag{5}$$

e. Calculate the value of *eigenvector (w)* by the formula:

$$w = S_W^{-1}\,(m_1 - m_2) \tag{6}$$

f. Calculate the total scatter matrix to obtain the PCA projection matrix

$$S_T = \sum_{i=1}^{N}(m_i - m)(m_i - m)^T \tag{7}$$

g. Calculate the optimal projection matrix W_{opt}^T

$$W_{opt}^T = W_{FLD}^T\, W_{PCA}^T \tag{8}$$

h. Calculate the weight value of each of the face images with the optimal projection matrix W_{opt}^T with the calculation of the formula.

$$F_i = X_i\, W_{opt}^T \tag{9}$$

Before performing face recognition with Euclidean Distance similarity calculation [11] [12], the facial image to be tested with the average value of training set with the formula calculation:

$$\phi_{input} = X_{input} - m \tag{10}$$

After the normalization value of the test face image is obtained, then the further project is examined into the optimal projection matrix.

$$F_{test} = W_{opt}^{T} \, \Phi_{input} \tag{11}$$

3. Result and Discussion

The experiments were conducted by examining 3 (three) face images that have similarities, of the three face images compared and know which image has the highest similarity, the testing process is done by using Fisher Linear Discriminant algorithm with the stages that have been determined, for more details can be considered in the following process:

a. Grayscale Process and Threshold Image Training

The face image is changed from the RGB shape into the grayscale form and then turn into a black and white image by doing the threshold, and the result is as follows:

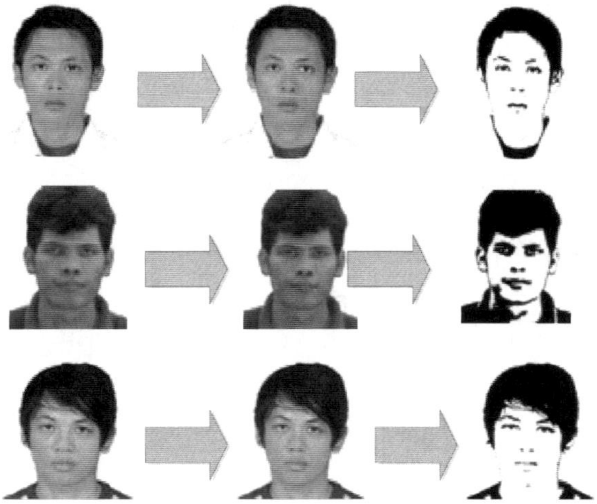

Figure 1. RGB Image Change to Grayscale and Threshold

Once the training image is formed in the threshold, three images can be obtained with a size of 3 x 3 pixels as can be seen in Figure below

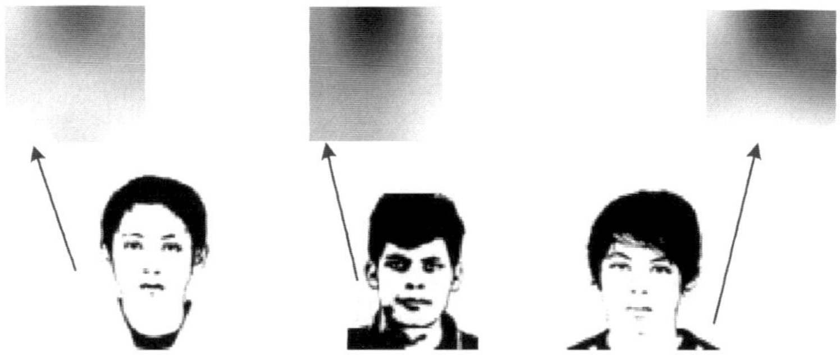

$$\begin{bmatrix} 214 & 106 & 190 \\ 236 & 234 & 223 \\ 253 & 212 & 250 \end{bmatrix} \quad \begin{bmatrix} 139 & 0 & 127 \\ 196 & 174 & 209 \\ 117 & 153 & 178 \end{bmatrix} \quad \begin{bmatrix} 180 & 23 & 151 \\ 204 & 173 & 173 \\ 229 & 249 & 225 \end{bmatrix}$$

Figure 2. Image Training Results Threshold

b. Fisher Linear Discriminant Calculation

The use of the formula from steps 1-8 will be tested by the formula already described previously with the pixel value contained in FIG. 2, the result of the testing process using the Fisher Linear Discriminant method and the Euclidean distance to recognize the face can be seen in the following tables:

Table 1. Weight Training Image Value

Face of	j	$F_{j,1}$	$F_{j,2}$	$F_{j,3}$	$F_{j,4}$	$F_{j,5}$
1	1	-185.072	-189.196	-172.310	-238.204	-259.480
	2	-104.195	-96.621	-83.608	-201.076	-214.892
2	3	-135.977	-130.039	-114.449	-191.628	-206.433
		$F_{j,6}$	$F_{j,7}$	$F_{j,8}$	$F_{j,1}$	$F_{j,6}$
1		-242.412	-249.451	-266.149	-247.350	-243.412
		-199.871	-151.034	-164.574	-154.433	-199.871
2		-192.679	-239.666	-263.224	-247.824	-192.679

Table 2. Different Squares of Each Image Training of Imagery Test

j	$(F_{j,1} - F_{uji,1})^2$	$(F_{j,2} - F_{uji,2})^2$	$(F_{j,3} - F_{uji,3})^2$	$(F_{j,4} - F_{uji,4})^2$	$(F_{j,5} - F_{uji,5})^2$
1	$((-185.072)-(20.32))^2$	$((-189.196)-(35.52))^2$	$((-172.310)-(13.98))^2$	$((-238.204)-(13.79))^2$	$((-259.480)-(34.76))^2$
2	$((-104.195)-(20.32))^2$	$((-96.621)-(35.52))^2$	$((-83.608)-(13.98))^2$	$((-201.076)-(13.79))^2$	$((-214.892)-(34.76))^2$
3	$((-135.997)-(20.32))^2$	$((-130.039)-(35.52))^2$	$((-114.449)-(13.98))^2$	$((-191.628)-(13.79))^2$	$((-206.433)-(34.76))^2$
j	$(F_{j,6} - F_{uji,6})^2$	$(F_{j,7} - F_{uji,7})^2$	$(F_{j,8} - F_{uji,8})^2$	$(F_{j,9} - F_{uji,9})^2$	$(F_{j,6} - F_{uji,6})^2$
1	$((-243.412)-(21.86))^2$	$((-249.451)-(16.39))^2$	$((-266.149)-(27.43))^2$	$((-248.350)-(9.86))^2$	$((-243.412)-(21.86))^2$
2	$((-199.871)-(21.86))^2$	$((-151.034)-(16.39))^2$	$((-164.574)-(27.43))^2$	$((-154.433)-(9.86))^2$	$((-199.871)-(21.86))^2$
3	$((-192.679)-(21.86))^2$	$((-239.666)-(16.39))^2$	$((-263.224)-(27.43))^2$	$((-247.824)-(9.86))^2$	$((-192.679)-(21.86))^2$

Table 3. Sum of Squares of Each Image of Training on Test Image

j	$(F_{j,1} - F_{uji,1})^2$	$(F_{j,2} - F_{uji,2})^2$	$(F_{j,3} - F_{uji,3})^2$	$(F_{j,4} - F_{uji,4})^2$	$(F_{j,5} - F_{uji,5})^2$
1	42187.51	49161.26	34705.86	63503.49	86852.47
2	15505.57	17463.19	9524.41	46169.54	62330.6
3	24430.75	27412.23	16495.31	42198.60	58178.40
j	$(F_{j,6} - F_{uji,6})^2$	$(F_{j,7} - F_{uji,7})^2$	$(F_{j,8} - F_{uji,8})^2$	$(F_{j,9} - F_{uji,9})^2$	Sum Result $(F_{j,i} - F_{uji,i})^2$
1	65677.84	70674.14	86191.62	66158.21	565112.4
2	49166.80	28032.50	36867.49	26992.97	292053.08
	49166,80	28032,50	36867,49	26992,97	292053,08
3	46029.08	65567.28	84482.71	66402.28	431196.64

Table 4. Final Result Calculation of Euclidean Distance Similarity

Face -	j	Sum Results $(F_{j,i} - F_{test,i})^2$	$d_v = \sum_{j=1}^{m} \lVert F_{\text{train of j,w}} - F_{\text{test,w}} \rVert$	Result
1	1	565112.4	751.7396	
	2	292053.08	540.4193	**Minimal**
2	3	431196.64	656.6556	

Table 4 generates a minimum value of 540.41935. The value lies on the 2nd face so that the face recognition tests have similarities on the 2nd object. The output of this facial recognition process is a recognizable facial image.

4. Conclusion

Based on the discussions and evaluations that have been done on facial recognition application using Fisher Linear Discriminant (FLD) method, it is concluded that the application of Fisher Linear Discriminant (FLD) method in face recognition has high accuracy because this method divides class specifically and face recognition application with Fisher Linear Discriminant (FLD) method is able to recognize faces with various expressions are also able to recognize the face with attribute changes such as glasses, whiskers, and others.

IOP Conf. Series: Materials Science and Engineering **300** (2018) 012037 doi:10.1088/1757-899X/300/1/012037

References

[1] R. Rahim, A. S. Ahmar, A. P. Ardyanti and D. Nofriansyah, "Visual Approach of Searching Process using Boyer-Moore Algorithm," in *International Conference on Information and Communication Technology 2017*, Medan, 2017.

[2] S. Thakur, J. K. Sing, D. K. Basu and M. Nasipuri, "Face Recognition Using Fisher Linear Discriminant Analysis and Support Vector Machine," *Springer : International Conference on Contemporary Computing,* pp. 318-326, 2009.

[3] J. Pan, X. S. Wang and Y. H. Cheng, "Single-Sample Face Recognition Based on LPP Feature Transfer," *IEEE Access,* vol. 4, pp. 2873 - 2884, 2016.

[4] H. Shin, "An extension of Fisher's discriminant analysis for stochastic processes," *Elsevier : Journal of Multivariate Analysis,* vol. 99, no. 6, pp. 1191-1216, 2008.

[5] B. Zou, L. Li, Z. Xu, T. Luo and Y. Y. Tang, "Generalization Performance of Fisher Linear Discriminant Based on Markov Sampling," *IEEE Transactions on Neural Networks and Learning Systems,* vol. 24, no. 2, pp. 288 - 300, 2013.

[6] Y. Li, W. Yuan, H. Sang and X. Li, "Combination recognition of face and ear based on two-dimensional fisher linear discriminant," in *4th IEEE International Conference on Software Engineering and Service Science (ICSESS)*, Beijing, China , 2013.

[7] Y. Xu, Z. Li, J. Yang and D. Zhang, "A Survey of Dictionary Learning Algorithms for Face Recognition," *IEEE Access ,* pp. 1-12, 2017.

[8] Y. Zhang and Y. Wo, "A fusion iris feature extraction method based on fisher linear discriminant," in *International Conference on Machine Learning and Cybernetics (ICMLC)*, Tianjin, China , 2014.

[9] F. Z. Okwonu and A. R. Othman, "A Model Classification Technique for Linear Discriminant Analysis for Two Groups," *International Journal of Computer Science ,* vol. 9, no. 3, pp. 125-128, 2012.

[10] B. Sunitha and M. Seetha, "ENHANCING THE PERFORMANCE OF SVM CLASSIFICATION BASED ON FISHER LINEAR DISCRIMINANT," *International Journal of Research in Engineering and Technology,* vol. 3, no. 5, pp. 48-54, 2014.

[11] R. Rahim, S. Nurarif, M. Ramadhan, S. Aisyah and W. Purba, "Comparison Searching Process of Linear, Binary and Interpolation Algorithm," in *International Conference on Information and Communication Technology 2017*, Medan, 2017.

[12] R. Rahim, M. Dahria, M. Syahril and B. Anwar, "Combination of the Blowfishand Lempel-Ziv-Welch algorithms for text compression," *World Transactions on Engineering and Technology Education,* vol. 15, no. 3, pp. 292-297, 2017.

IOP Conf. Series: Materials Science and Engineering **300** (2018) 012038 doi:10.1088/1757-899X/300/1/012038

Station Blackout Analysis of HTGR-Type Experimental Power Reactor

Syarip[1,3], Aliq Zuhdi[2], Sabilul Falah[2]
[1] Centre for Accelerator Science & Technology, Yogyakarta, Indonesia.
[2] Polytechnic Institute of Nuclear Technology, Yogyakarta, Indonesia.

[3] E-mail: syarip@batan.go.id

Abstract. The National Nuclear Energy Agency of Indonesia has decided to build an experimental power reactor of high-temperature gas-cooled reactor (HTGR) type located at Puspiptek Complex. The purpose of this project is to demonstrate a small modular nuclear power plant that can be operated safely. One of the reactor safety characteristics is the reliability of the reactor to the station blackout (SBO) event. The event was observed due to relatively high disturbance frequency of electricity network in Indonesia. The PCTRAN-HTR functional simulator code was used to observe fuel and coolant temperature, and coolant pressure during the SBO event. The reactor simulated at 10 MW for 7200 s then the SBO occurred for 1-3 minutes. The analysis result shows that the reactor power decreases automatically as the temperature increase during SBO accident without operator's active action. The fuel temperature increased by 36.57 °C every minute during SBO and the power decreased by 0.069 MW every °C fuel temperature rise at the condition of anticipated transient without reactor scram. Whilst, the maximum coolant (helium) temperature and pressure are 1004 °C and 9.2 MPa respectively. The maximum fuel temperature is 1282 °C, this value still far below the fuel temperature limiting condition i.e. 1600 °C, its mean that the HTGR has a very good inherent safety system.

1. Introduction

Energy statistics show that electricity demand in Indonesia is increasing every year. The growth rate of electricity demand in 2000 to 2014 is 5.12% per year, and the growth should be balanced with adequate electricity production. The more electricity needed, the more electricity production increase. Power generation in Indonesia mostly uses energy from fossil fuels such as petroleum and coal. The percentage of self-produced electricity energy (including rent) per primary energy type is: natural gas 49,312.48 GWh (28.13%), coal 84,076.12 GWh (47.96%), and oil 26,433.18 GWh (15.08%) [1,2]. Continuous use of these resources causes the supply to run low. In addition, the resulting gas emissions can trigger global warming.

One potential alternative energy source is nuclear energy that can generate enormous energy. In addition, nuclear power plants do not produce air pollution so it is more environmentally friendly. The negative impact of nuclear energy is the presence of radiation produced in the fission process of fuel. However, the radiation can be contained well by a layered defense system in the reactor during normal operation of the reactor.

One of the fourth generation power reactors that have higher inherent safety, more economical, and more efficient power generation is HTGR (high temperature gas cooled reactor). This type of reactor can also be used to produce hydrogen which is needed by the fertilizer industry. HTGR uses helium cooling, developed from GCR (gas cooled reactor) and AGCR (advanced gas cooled reactor)

CO_2-cooled [3,4]. The HTGR reactor is a type of reactor that will be built as an experimental power reactor (EPR) in Serpong. EPR uses uranium fuel and is capable of generating power up to 10 MW thermal (MWt) or equivalent to 3 MW of electricity to supply electricity needs for Puspiptek, Serpong, South Tangerang [5]. Prior to the 10 MW EPR development carried out in the densely populated areas of Serpong, research or testing needs to be done. Lesson learnd from the major accidents that have occurred in previous reactors such as Three Mile Island, Chernobyl, and Fukushima, safety is a major aspect that needs to be considered and studied more deeply so that in the development and operation of EPR can be run as expected.

According to PLN's 2014 statistics, electricity distribution in the Greater Jakarta and Tangerang areas has a significant disturbance index of 2.45 times/customer with a duration of 3.87 hours/customer [1]. This data indicates that in the Puspiptek region has a high probability of losing electricity from the network that could trigger the station blackout. The station blackout is one of the most common trigger events causing damage to the reactor core. The station blackout (SBO) occurs when the reactor has no electricity supply at all. Electrical energy is needed even if the reactor is off to circulate heat on the reactor core. The primary objective of accident management in nuclear power plants is to minimize fission product release into the environment by cooling the core and containment at the same time. However, under extreme condition of extended station black-out (ESBO), as in Fukushima, electric power is totally unavailable to cool the core and the containment. The incident was happened to the Fukushima reactor due to the tsunami which cut off all connections to the source of the power grid and also caused flooding of the emergency generator and made it not functioning. As a result, the pump loses power to circulate water into the reactor so that the reactor core temperature is increased due to the decay heat not being well circulated. Ultimately, the heat causes melting of the reactor core units 1, 2, and 3.

Therefore, the in-depth study on the effect of SBO condition on the EPR or HTGR reactor that will be built in Serpong needs to be done. As a research tool, PCTRAN-HTR is used to simulate reactor behavior and analyze value changes in important parameters when SBO occurs. PCTRAN is a reactor simulation software developed by Micros-Simulation Technology that has validated by FSAR (Final Safety Analysis Report) and uses actual data [6]. The purpose of the analyses presented in this paper is to identify the behavior of important EPR or HTGR parameters in case of total SBO. The PCTRAN-HTR simulator code has also been used to simulate and to analyze the optimum control rods positions in achieving reactor criticality for the continuation of reactor operation [7]. This functional simulator provides a good tool for specialists working in the field HTGR operation and safety.

2. Description of HTGR

The HTGR is a fourth generation power plant developed from GCR (gas cooled reactor) and AGCR (advanced gas cooled reactor). HTGR has many advantages over previous generation reactors such as higher inherent safety, more economical, and more efficient power generation. Heat generated from HTGR can be used for hydrogen production as well as for applications in other fields. HTGR has the potential to become one of the advanced reactors that will continue to be developed. The primary cooling system uses helium as a reactor core coolant. The helium gas flows through the fuel and carries the heat generated by the ^{235}U fission reaction and is supplied to the steam generator. Then, the water in the secondary cooling system absorbs the heat energy to drive the turbine. This type of reactor is capable of producing temperatures up to 1000 ° C that can be used for hydrogen production [4,8,9].

2.1. HTGR reactor fuel

The fuel used in HTGR is in the form of ceramic coated fuel particles (CFP) of tri structural isotropic (TRISO) type, having a diameter of ~ 1 mm. The TRISO fuel consists of a UO_2 fuel kernel enclosed by a porous carbon layer (to release a fission gas fuel), an inner carbon pyrolytic layer (IPyC), silicon carbide (SiC), and an outer pyrolytic carbon layer (OPyC). The TRISO fuel structure is shown in Figure 1 [5]. The TRISO coating can prevent the release of fission products in the fuel because it has a structure that is resistant to very high temperatures. This type of fuel is not damaged when it reaches 2000 °C. Normal HTGR operation is not more than 1250 °C and at the worst accident condition, the fuel temperature is maintained not to exceed 1600 °C, for the HTGR design with helium output temperature of 850 °C [9]. The fraction of fuel damage as a function of temperature is shown in Figure 2.

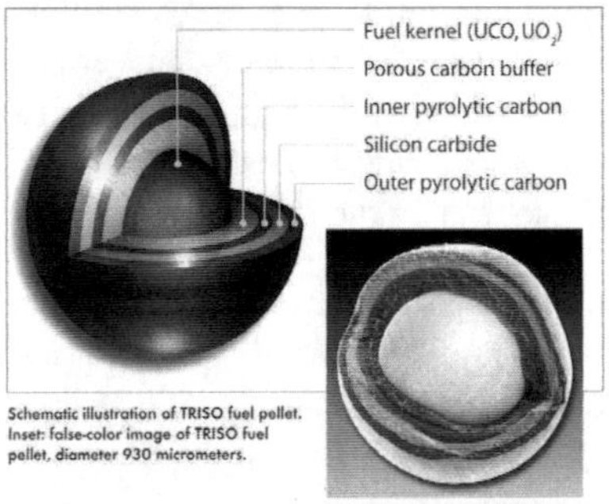

Figure 1. TRISO fuel layer [8]

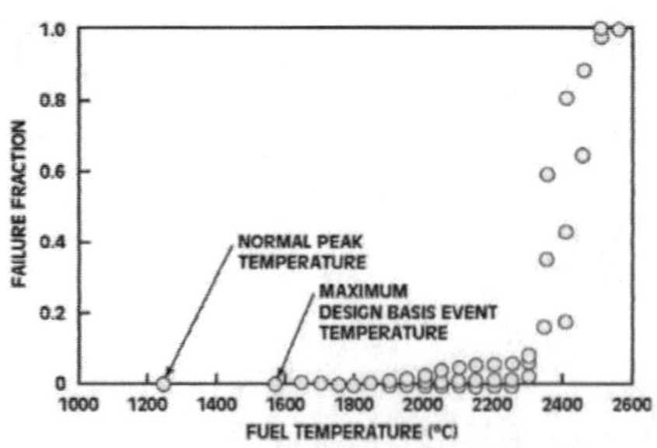

Figure 2. Fraction of fuel damage [9]

HTGR uses helium gas with a high pressure as a heat transfer medium in the primary system. Helium includes noble gases that are difficult to react and have no heat transfer limit or phase changes. Helium also has a density that has very little impact on reactivity. In Loss of Cooling Accidents (LOCA) the leaking helium in containment does not undergo condensation as in light water reactor (LWR), so the containment can last longer [10]. The HTGR has a relatively low power density so that the changes in temperature occurred in a relatively long time. Damage to the cooling system causes a very slow rise of transient temperature in the reactor core.

2.2. Experimental power reactor (EPR)

The development of nuclear reactor for electricity generation (NPP) in Indonesia until now has not yet been realized. Various issues as constraints of the program, among them are that many people still worried about the safety and security, and not yet convinced that the country has the capability to build and operate the reactor safely. The government through the National Nuclear Energy Agency (BATAN) is initiating for the development of experimental power reactor (EPR) of HTGR type with the main objective is to demonstrate a small NPP which can be operated safely [5].

It is expected by demonstrating the EPR therefore, the trust of the public towards the safety and security in operation of the NPP power reactor and the capability of the country for operating the nuclear reactor would increase. The trust of the public is very important for the success of the plan in the development of a large NPP. The EPR scheme is shown in Figure 3.

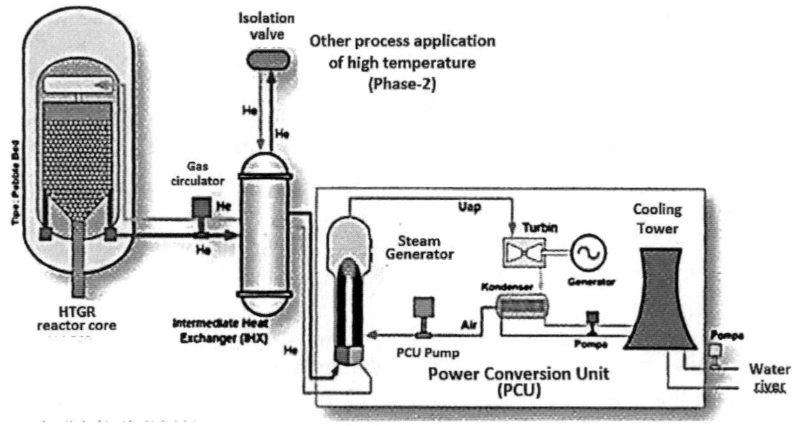

Figure 3. The EPR scheme [5]

3. Material and methods

The loss of off-site power (LOOP) event occurs when all electrical power to the nuclear power plant from the power grid is lost. A station blackout (SBO) is a complete failure of both off-site and on-site alternating current (AC) power sources [10,11]. The blackout station is a condition when the reactor does not get a power supply at all. The blackout station begins with the LOOP and is followed by the failure of the emergency diesel generator which is the onsite source of electricity [12,13,14]. When a blackout station occurs all systems requiring electrical energy fail to function, including the most important is the malfunction of the reactor coolant circulation pump. This condition may cause the residual heat of the decay to accumulate on the reactor core and cause the fuel temperature to increase.

In 1975, the "Reactor Safety Study" (NUREG-75/140) showed that the SBO could be an important contributor to the total nuclear accident. In addition, the experience gained during operating a nuclear reactor shows that the reliability of emergency AC power source systems both inside and outside the site may still below expectations [15]. The total SBO events are analyzed using PCTRAN-HTR, a computer software created by Micro-Simulation Technology to simulate HTGR operation. The PCTRAN-HTR simulator can be used for calculation of reactor operating parameters until the maximum power of 200 MW. (10 MWe ≤ P ≤ ~60 MWe) such as neutron flux and reactor power distributions, plant efficiency, electric power, load factor, core power density, heat flux, mean burn-up, and excess reactivity. The cooling system parameter can also be calculated such as: flow rate, pressure, gas temperatures at core inlet-outlet, inlet-outlet steam temperatures, steam pressure & temperature at turbine inlet, feed-water temperature, condenser pressure, steam mass flow, gas coolant radioactivity and fuel temperature for various reactor power levels [6]. The reactor parameters can be displayed graphically and numerically and can be saved in output files.

3.1. Mathematical model for helium coolant system

The basic thermal hydraulics of the mathematical model used in PCTRAN-HTR simulator is first-principle in mass and energy balance. This ensures credible and realistic simulations [6]. For a gas-cooled reactor, the primary and secondary helium systems are modelled by two control volumes described by the ideal gas law:

$$PV = N\,R\,T \tag{1}$$

where : P = pressure, V = volume, a constant for either the primary or secondary gas system, N = mole number, R = gas constant, and T = absolute temperature. The pressure and temperature transient during a transient are then governed by the simple derivative of the above equation and mass and energy input to the system as follows (8):

$$\begin{aligned} V\frac{dP}{dt} &= \left(\frac{dN}{dt}\right)RT + NR\frac{dT}{dt} \\ \frac{dN}{dt} &= \frac{1}{M_a}\left(W_{in} - W_{out}\right) \\ MC_p\frac{dT}{dt} &= Q \end{aligned} \tag{2}$$

where, M_a = molecular weight of the gas, W_{in} = gas flow influx, W_{out} = gas flow out-flux, C_p = specific heat of the gas at constant pressure, and M = total mass of the gas in the volume.

PCTRAN-HTTR uses a semi-empirical method for reactor coolant flow that encompasses both forced and natural circulation conditions. For forced circulation when the reactor coolant pumps are on, full (volumetric) rated flow is assumed. In the course of pump trip, the flow coasts down exponentially until stable natural circulation is established (3).

$$W(t) = W_1 + (W_0 - W_1)\, e^{-t/\tau} \tag{3}$$

where, W_0 is the nominal rated flow and W_1 is the eventual natural circulation flow, which is usually a few percent of the nominal. τ is the characteristic time. This equation is also used for pump run-out to exceed 100% of its nominal value, i.e., $W_1 > W_0$. Whiles, the reactor hot and cold leg temperatures for a HTR are calculated by the heat balance from the reactor core and removal rate by the steam generators at a given loop flow rate as follows:

$$\Delta T = \frac{Q_{SG}}{W_{RC}C_p}, \quad T_H = T_{Avg} + \frac{\Delta T}{2}, \quad T_C = T_{Avg} - \frac{\Delta T}{2} \tag{4}$$

where T_{avg} is the reactor coolant (RC) average temperature.

3.2. Fuel temperature of the reactor core

A simplified model accounting for the temperatures of fuel and cladding has been constructed in PCTRAN-HTTR. This model can simulate: thermal power transmitted into the coolant in contrast to nuclear power generated by the core during normal operation; fuel and cladding heatup during accident conditions. Core thermal power Q_{MWT} is represented by (5)

$$Q_{MWT} = UF * (T_F - T_{avg}) \tag{5}$$

Where T_F is the known average fuel temperature at 100% rated power. The heat transfer coefficient UF is then calculated such that the core thermal power is equal to the neutron power from the kinetics equation at steady state.

Transient fuel temperature will be calculated by the imbalance between nuclear power and thermal power. The heat transfer coefficient UF varies with the core flow to the 0.8^{th} power according to Dittus-Boetller correlation during forced circulation and flow coast down. It will settle at a small value for natural circulation. The graphite sleeve and reflector's temperature is calculated similarly.

The simulated reactor is a 10 MW HTGR, the duration of SBO is varied from 0.5, 1, 1.5, 2, 2.5, 3 minutes on reactor failed and successfully scram. The basic data parameter used in the simulation is presented in Table 1.

Table 1. Input basic data

Parameter	Value
Rated thermal power per core (MW)	10
Rated neutron flux at full power (n/cm^{2}s)	3.10^{14}
Charging flow (kg/s)	5
Main feed water enthalpy (kW/kJ)	600
Total fuel mass (kg)	13333.33
Nominal core helium flow rate (kg/s)	6.66
Initial average fuel temperature (°C)	1300
Initial gas temperature (°C)	878.6
Initial gas pressure (MPa)	7.10^{8}
Power range high flux trip (n/cm^2/sec)	4.10^{15}
Pressure of steam generator (SG) at zero power (MPa)	1
Pressure of SG at 100% power (MPa)	3.5
Volume of SG (m^3)	10

4. Results and Discussion.

The simulation of the 10 MW HTGR reactor is carried out by operating the reactor for 2 hours (7200 seconds) until the reactor in steady state condition so that all parameters are stable. The SBO condition is simulated from 1 minute up to 3 minutes.

4.1. Reactor fail to scram

Reactor fails to scram is the condition that the reactor still in operating condition when the incident occurred and can be shut down after a while. Simulation of blackout station on a reactor condition fail to scram was done to know how reliable the reactor when the accident occurred. The blackout station causes a loss of helium flow and a secondary coolant so that the reactor experiences a shortage of coolant circulation. The phenomenon that occurs when a station blackout and reactor fails to scram is the rise in fuel temperature and decrease inlet helium temperature. When a blackout station occurs, the helium flow falls sharply until the power source is recovered. During 1 minute of the blackout station, the helium flow decrease reached 3.3 kg/s from the initial of 6.5 kg/s to 3.2 kg/s. After the blackout station ends, the fuel temperature, helium outlet, and helium inlet return to normal within 5 minutes. Changes in fuel temperature, helium outlet, and helium inlet are shown in Figure 4. At station blackout condition of 1 minute, power decreased 25% or 2.5 MW. The power of the reactor is initially 10 MW turned to 7.5 MW. The decrease is caused by the fuel temperature increase by 61 °C. The power and fuel temperature relation is shown in Figure 5.

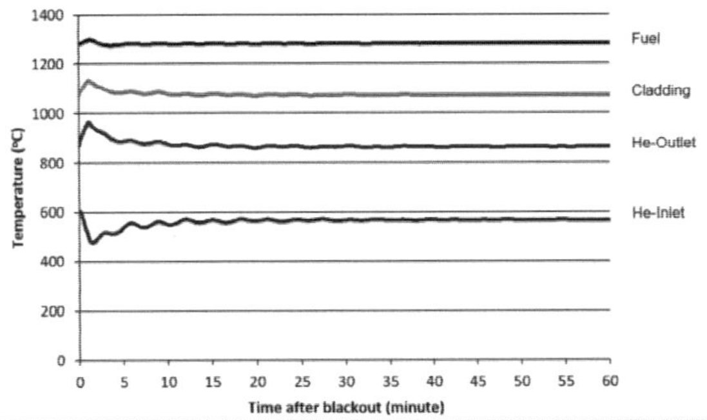

Figure 4. Temperature changes for 1 minute SBO and reactor fail to scram

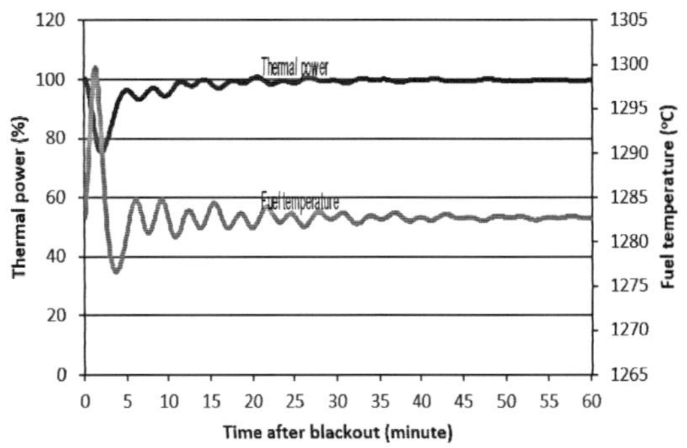

Figure 5. The reactor power and fuel temperature at 1 minute SBO

The temperature and power profiles for the duration of SBO of 1.5m, 2m, and 3m are similar with the curve profile of Figure 6 and Figure 7. The temperature and power changes for time variation of station blackout are shown in Table 1 and Table 2. The rate of change in fuel temperature is 36.57 °C/min, this is obtained from (Table 2) the average temperature different (ΔT) of fuel divided by the time duration of SBO. The rate of decrease in power of each temperature increase is obtained from (Table 3) the average different (Δ) power divided by ΔT of fuel i.e. 0.069 MW/°C. The increase in fuel temperature causes the reactor power to decrease due to the inherent safety of HTGR i.e. with negative reactivity coefficient of the fuel temperature (Doppler effects) this phenomenon is shown in Figure 7.

Table 2. Temperature change

No	SBO (m)	ΔT fuel (°C)	ΔT he-out (°C)	ΔT he-in (°C)
1	0.5	34	57	-75
2	1	22	106	-128
3	1.5	88	165	-171
4	2	47	203	-214
5	2.5	59	250	-259
6	3	71	289	-323

Table 3. Reactor power change

No	SBO (m)	ΔT fuel (°C)	Δρ Doppler (% dk/k)	Δ power (MW)
1	0.5	34	-0.05	-0.05
2	1	22	-0.03	-2,5
3	1.5	88	-0.12	-1.8
4	2	47	-0.07	-4,6
5	2.5	59	-0.08	-5,5
6	3	71	-0.1	-6,3

4.2. Reactor successfully scram

Reactor successfully scram is the condition that the reactor automatically shut down when the incident occurred. In this condition, the station blackout does not cause a rise in reactor fuel temperature. However, the temperature of the inlet helium increases and the pressure also increases. Although the reactor has undergone a scram, a circulation pump is still required to cool the reactor from the fission products decay heat or residual heat. The decay heat can cause an increase in fuel temperature. The trend of fuel and helium inlet-outlet temperatures for the 3 minutes of SBO incident is shown in Figure 6 and the trend of its pressure change is shown in Figure 7.

The temperature and power profiles for SBO incident duration of 1.5 m, 2 m, and 3 m are similar, such as shown in Figure 6 and Figure 7. It can be seen from Figure 6 and Figure 7 that at a 3 minutes SBO incident the maximum fuel temperature is 1282 °C and the maximum helium gas coolant inlet is 1004 °C. The fuel maximum fuel temperature at this SBO incident is still far bellows fuel temperature design limit i.e. 1600 °C such as described in Figure 2 [9]. The results are in accordance with the similar analysis results such as reported by several researchers [16,17,18,19].

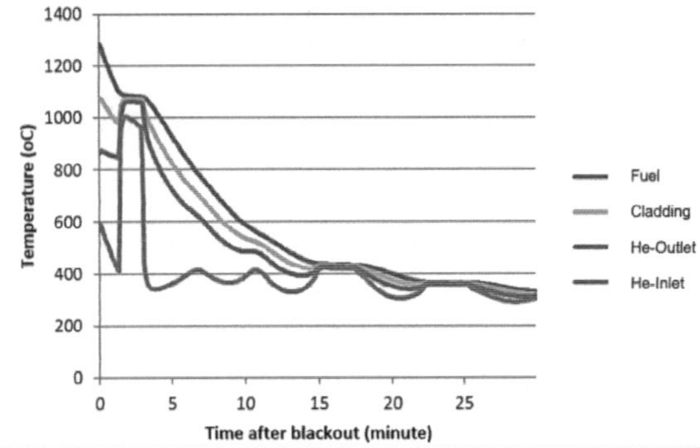

Figure 6. Fuel and helium coolant temperature changes for 3 minutes SBO and reactor successfully scram

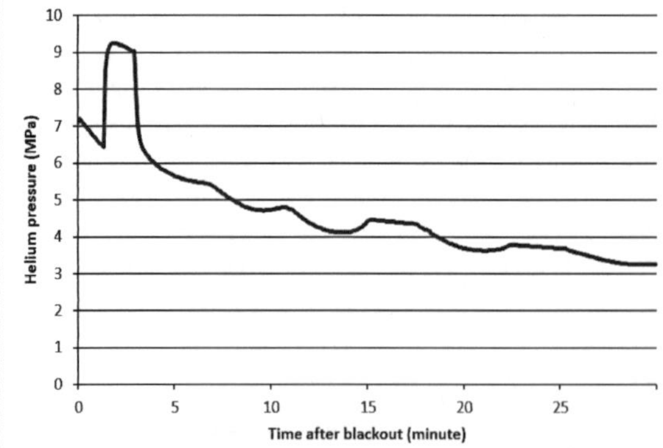

Figure 7. The helium pressure change at SBO 3 minutes with reactor scram

5. Conclusion

The station blackout incident on EPR 10 MW of HTGR type which fails to scram causes an increase in fuel temperature with the rate of 36.57 °C/minute. The increase in fuel temperature causes a decrease in reactor power with the rate of 0.069 MW/°C. The maximum coolant temperature and pressure reach 1004 °C and 9.24 MPa respectively, whiles the maximum fuel temperature is 1282 °C. The fuel maximum fuel temperature at this SBO incident is still far bellows fuel temperature design limit of 1600 °C. This result proves that the HTGR has a good inherent safety and the reactor in a safe condition without operator intervention during the SBO event.

6. Acknowledgments

The authors would like to thank the Director of Center for Accelerator Science and Technology Yogyakarta, to the Rector of Polytechnic Institute of Nuclear Technology, as well as to the Head of Reactor Division and all staffs for their supports.

References

[1] PLN. 2015 *Statistik PLN* (in Indonesian). ISSN: 0852-8179. Jakarta: Sekretariat Perusahaan PT PLN

[2] National Energy Council (DEN). 2016 *Indonesia energy outlook 2016*. Jakarta: National Energy Council, Secretariat General.

[3] IAEA. 2003 *Evaluation of high temperature gas cooled reactor performance: Benchmark analysis related to initial testing of the HTTR and HTR-10*, IAEA-TECDOC-1382.

[4] IAEA. 2008 Accident analysis for nuclear power plants with modular high temperature gas cooled reactors, Safety Report Series No.54. Vienna.

[5] BATAN. 2016 *Rencana pembangunan RDE*, http://www.batan.go.id/index.php/id/reaktor-daya-eksperimental-rde

[6] Micro-Simulation Technology. 2015 *PCTRAN-HTR simulator software for PC*. New Jersey: Micro-Simulation Technology.

[7] Syarip, Khoirul Anam, Dwi Priyantoro. 2016 Post reactor scram control rods position adjustment analysis for the Indonesian experimental power reactor concept, Ganendra Journal of Nuclear Science and Technology, (Jurnal Iptek Nuklir Ganendra), **19** 2 83-93

[8] Hu, S., dkk. 2006 *Commissioning and operation experience and safety experiments on HTR-10.* Proceedings of the 3rd International Topical Meeting on High Temperature Reactor Technology, D00000052, Johanesburg.

[9] Reitsma, F. 2012 *HTGR Safety Design*. IAEA Course on High temperature Gas Cooled Reactor Technology. IAEA.

[10] Kadak A.C. 2016 The status of the US high-temperature gas reactors. Engineering **2** 119–123.

[11] X. Yan, H. Noguchi, H. Sato, Y. Tachibana, K. Kunitomi, R. Hino. 2014 *A hybrid HTGR system producing electricity, hydrogen and such other products as water demanded in the Middle East*, Nuclear Engineering and Design **271** 20–29.

[12] Andrija Volkanovski and Andrej Prošek. 2015 *Delayed station blackout event and nuclear safety*, Science and Technology of Nuclear Installations, Volume **2015**, Article ID 192601, 9 pages.

[13] Feng, B. 2014 *Characteristics of helium gas with high temperature and high pressure flowing through a 90-degree elbow*. Journal Power Engineering Volume 2014, Article ID 764283. http://dx.doi.org/10.1155/2014/764283.

[14] Volkanovski, A., Prosek, A. 2011 *Station blackout and nuclear safety*. Proceedings International Conference Nuclear Energy for New Europe.

[15] Baranowsky, P. W. 1988 *Evaluation of station blackout accidents,* Report NUREG-1032. Washington: U.S.A. Nuclear Regulatory Commission.

[16] Jong Woon Park, Wook-Cheol Seol. 2016 *Considerations for severe accident management under extended station blackout conditions in nuclear power plants*, Progress in Nuclear Energy **88** 245-256

[17] X. Yan, H. Noguchi, H. Sato, Y. Tachibana, K. Kunitomi, R. Hino. 2014 *A hybrid HTGR system producing electricity, hydrogen and such other products as water demanded in the Middle East*, Nuclear Engineering and Design **271** 20–29.

[18] Syarip, Subki, I.R, and Canton, M.H., 1996 *Availability analysis of the AP600 passive core cooling system,* IAEA-TECDOC No-920.

[19] Lommers L.J., Mays B.E., Shahrokhi F. 2014 *Passive heat removal impact on AREVA HTR design*. Nucl. Eng. **271**:569–77.

IOP Conf. Series: Materials Science and Engineering **300** (2018) 012039 doi:10.1088/1757-899X/300/1/012039

Directed Diffusion Modelling for Tesso Nilo National Parks Case Study

Indra Yasri, Ery Safrianti

Electrical Engineering Department - Universitas Riau
Kampus Bina Widya, Jl. HR Subrantas KM 12.5, Pekanbaru 28293

indyasri@yahoo.com

Abstract. — Directed Diffusion (DD has ability to achieve energy efficiency in Wireless Sensor Network (WSN). This paper proposes Directed Diffusion (DD) modelling for Tesso Nilo National Parks (TNNP) case study. There are 4 stages of scenarios involved in this modelling. It's started by appointing of sampling area through GPS coordinate. The sampling area is determined by optimization processes from 500m x 500m up to 1000m x 1000m with 100m increment in between. The next stage is sensor node placement. Sensor node is distributed in sampling area with three different quantities i.e. 20 nodes, 30 nodes and 40 nodes. One of those quantities is choose as an optimized sensor node placement. The third stage is to implement all scenarios in stages 1 and stages 2 on DD modelling. In the last stage, the evaluation process to achieve most energy efficient in the combination of optimized sampling area and optimized sensor node placement on Direct Diffusion (DD) routing protocol. The result shows combination between sampling area 500m x 500m and 20 nodes able to achieve energy efficient to support a forest preventive fire system at Tesso Nilo National Parks.

1. Introduction

The long dry season has triggered several forest fire cases in the last decade in the Indonesia generally and especially in Riau Province. In this province mostly forest fire cases exposed to peat land and this become problematic to handle it [11]. Base on record of Riau Province Natural Resource Conservation Agency, Pelalawan Residency with Tesso Nilo National Park (TNNP) has contributed the most hotspot among others Residency in Riau Province. Also according to this agency, almost every year there are several forest fire cases at TNNP. According to these facts, it is necessary to provide a preventive forest fire system instead of handling a fire case.

The recent technology of preventive forest fire system using of Wireless Sensor Network (WSN) [1,2]. A Wireless Sensor Network (WSN) consists of many sensor nodes. Those sensors are equipped by communication system that make them able to communicate among them directly or indirectly through a base-station (BS). A huge number of sensors able to perform sensing across a wider geographical zone with higher accuracy. Individually, every sensor node equipped by sensing instrument, processing instrument, transmission instrument, moveable system, position tracking system, and power system.

Regardless of the numerous applications of WSNs, the networks in WSNs has several limitations, e.g., inadequate energy supply, restricted computing power, and restricted communication channel of the wireless connections among sensor nodes. One of the primary challenge in the design of WSNs is to deliver the data communication with a little bit longer network life time and try to avoid connectivity

loss by deploying some tricky energy management approaches. These approaches are known as routing protocol [3].
There are progressive works on routing protocol [4,5,6,7,8,9]. Data Aggregation (DA) approach [3,12] is one of them. The basic idea of DA is to gather the incoming data from individual sensor node through avoiding severance, optimizing the transmissions number; hence lead to network energy saving and it will extend its lifetime. Unlike old-style node to node routing, DA manage to traces tracks from many sources to a sole destination that gathers a solid in-network from reappearance data.
Directed Diffusion is one of popular Data Aggregation methods which proposed by [10]. This method proposes a data- centric and application aware pattern in the perception that all data produced by sensor nodes is labeled by attribute-value pairs. Practically, sensors quantify events and generate gradients of evidence in their surrounding regions. Data is requested by base station through spreading interests. A task required to be performed by the network is described by interest. The network will absorb the interest hop-by-hop, and will be distributed by each node to its regions. Gradients will prepare to retrieve data matching the enquiry towards the demanding node while the interest is distributed all over the network, i.e., a Base Station may look for data thru broadcasting interests and these interests are propagated by intermediate nodes. Once a sensor receives the interest from a particular sensor node, this sensor prepares a gradient to that sensor nodes. This procedure remains until gradients are built between sources and Base Station. Basically, a gradient composed by a direction and attribute value. Each individual neighbor has different gradient strength and also reflects to vary of data flow capacity. Generally, the procedure of Directed Diffusion involves the following stages: interest delivery, gradient developing, and data broadcasting. The best track among multi tracks will be reinforced when the interests is matched to the gradients to anticipate flooding condition. Aggregation of data is performed in the middle of process to minimize cost of communication.

2. Methodology

There are 4 stages of procedure involved in this modelling i.e. appointing of sampling area, sensor node placement, implementation of all scenarios in stages 1 and stages 2 on directed diffusion processes and enclosed by optimized sampling area and optimized sensor node placement on cDirect Diffusion (DD) routing protocol. The first stage begins to determine GPS coordinate for TNTN area. The initial sampling area is started from 500m*500m to 1000m*1000m with 100 m increment in between. The overall procedure is described in figure 1.

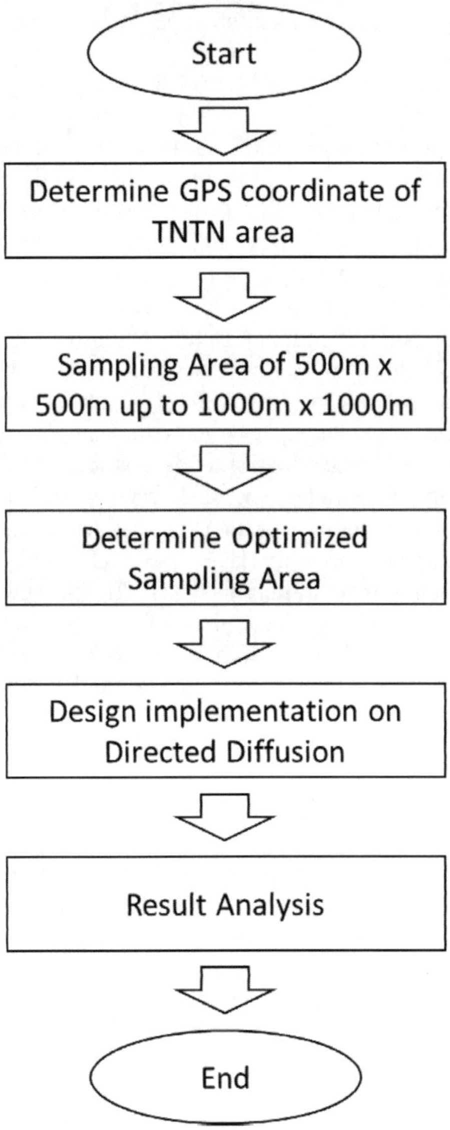

Figure 1. Modelling Procedure.

Initially, the sample area of 500m x 500m is determined. This area based on a certain number of hotspot which occurs frequently at Tesso Nilo National Park on the years of 2014 and 2015. The next procedure is program set up on Java Simulator (J-Sim). The aims of program are sensor node density arrangement which are set with various numbers i.e. 20 nodes, 30 nodes and 40 nodes. These various numbers of sensor node arrangement are deployed on various size of sampling area i.e. 500m x 500m, 600m x 600m, 700m x 700m, 800m x800m, 900m x 900m and 1000m x 1000m. The most energy efficient will be achieved from those various combinations between sampling area and sensor node density arrangement. The result is extracted based on output linearity. In the next procedure the result is implemented on Directed Diffusion (DD) routing protocol. This implementation is done through Java Simulator (J-Sim).

Directed Diffusion is initiated by sampling area of 500m x 500 with various sensor node density arrangement i.e. 20 nodes, 30 nodes and 40 nodes. The sink node is set on the position as follow: x=250 and y=250. The target node and sensor node is assigned randomly. Lastly, the simulation time is set to 1800 Seconds. Once the simulation is completed, the plotting sensor nodes are arranged in the cluster formation.

3. Result and Discussions

The result is presented according to the described scenarios in section II. Firstly, DD implementation for the sampling area of 500m x500m and density sensor nodes of 20 is shown in figure 2.

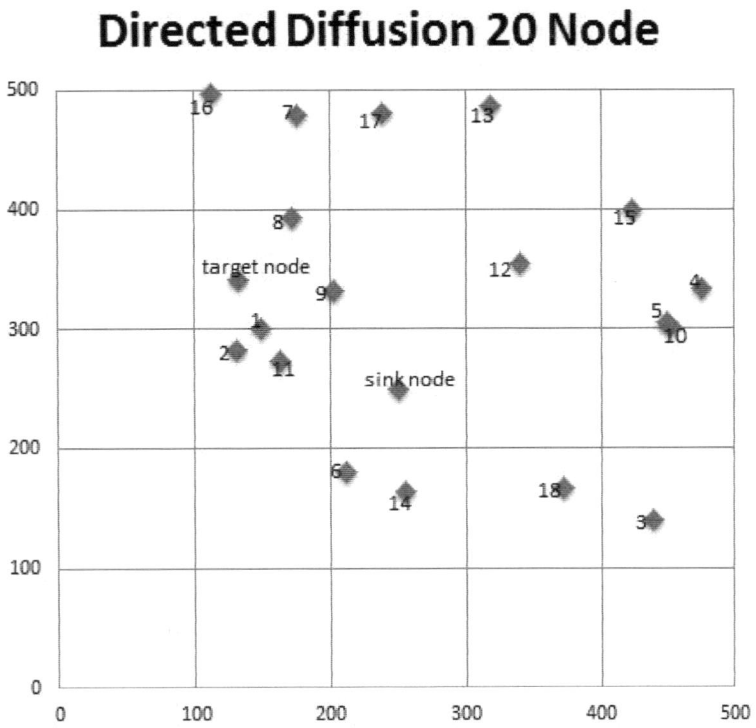

Figure 2. Directed Diffusion for 20 Node.

Figure 2 shows DD implementation with 20 nodes. These nodes are distributed within the area 500m x 500m. Two of them are target node and sink node. The target node is located at the area of 400m x 200m and sink node is located at 300m x 300m.
The simulation also examines the energy used in various of sampling area and various of sensor node density as shown in figure 3.

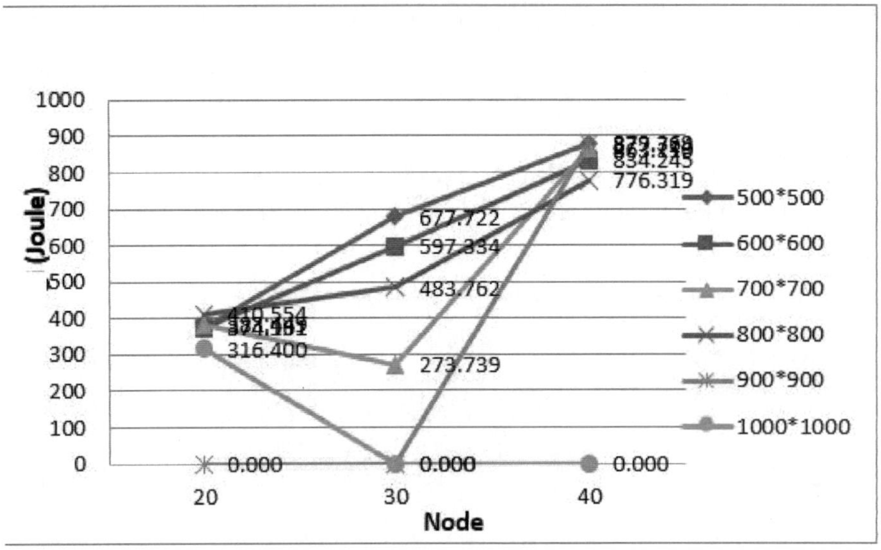

Figure 3. Energy used in various area and nodes density.

The area 500m x 500m to area 600m x 600m shows linearity in energy used with 0,83 Joule increment. Nonetheless for the area 700m x 700m to 800m x 800m shows non-linearity and less-accurate energy used.

Each of above scenarios is executed with 5 iterations. The table 1, table 2 and table 3 are shown different energy used after 5 iterations.

Table 1. The result of 5 iterations for 20 nodes sections.

Iteration	The energy used (Joule)
1	208.377
2	305.882
3	97.575
4	191.262
5	365.496

Table 2. The result of 5 iterations for 30 nodes sections.

Iteration	The energy used (Joule)
1	340.799
2	313.790
3	343.173
4	193.783
5	214.605

Table 3. The result of 5 iterations for 40 nodes sections.

Iteration	The energy used (Joule)
1	429.574
2	661.487
3	347.524
4	560.951
5	432.082

The intention to performs execution of five iterations for each particular density sensor nodes is to get the range of energy used. Then the overall of energy used for different sensor density is shown in figure 4.

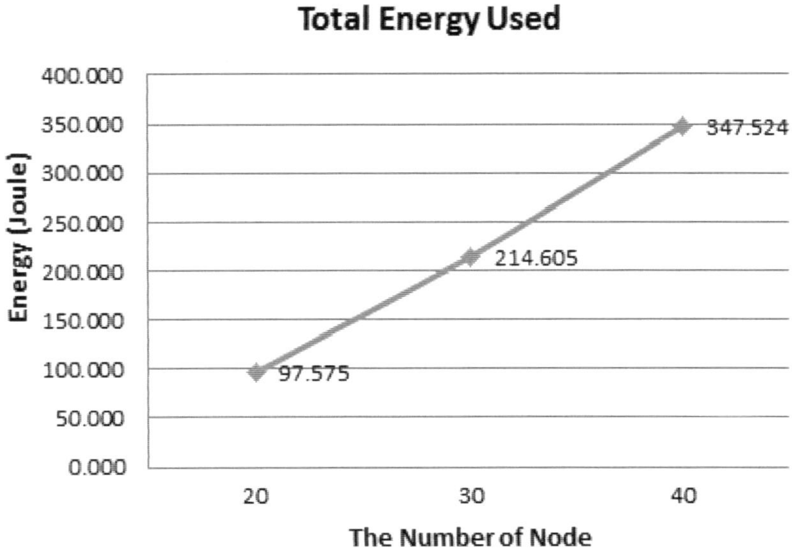

Figure 4. Directed Diffusion for 3 different numbers of Nodes.

The total energy used has shown the linearity on energy used as the growing of the sensor nodes density.

4. Conclusions

The Directed Diffusion (DD) implementation for Sampling area of 500m x 500m able to achieve energy efficiency when applied to different sensor node density i.e. 20 nodes, 30 nodes and 40 nodes. The energy efficiency is indicated by linearity on energy used. The biggest energy used is occurred at 40 nodes with 347.524 Joule.

Acknowledgment

This research is funded by DRPM Kemenristekdikti 2017. Thanks to DRPM Kemenristekdikti for this funding.

IOP Conf. Series: Materials Science and Engineering **300** (2018) 012039 doi:10.1088/1757-899X/300/1/012039

References

[1] Teguh R, Honma T, Usop A, Shin H, Igarashi H. Detection and Verification of Potential Peat Fire Using Wireless Sensor Network and UAV. InProceedings of the International Conference Information Technology and Electrical Engineering, Yogyakarta, Indonesia 2012 Jul 12 (pp. 6-10).

[2] Putra EH, Hariyawan MY, Gunawan A. Wireless Sensor Network for Forest Fire Detection. TELKOMNIKA (Telecommunication Computing Electronics and Control). 2013 Sep 1;11(3):563-74.

[3] Al-Karaki JN, Kamal AE. Routing techniques in wireless sensor networks: a survey. IEEE wireless communications. 2004 Dec;11(6):6-28.

[4] Teguh R, Igarashi H. Optimization of sensor network topology in deployed in inhomogeneous lossy media. TELKOMNIKA (Telecommunication Computing Electronics and Control). 2015 Jun 1;13(2):469-77.

[5] Goyal, Deepak, and Malay Ranjan Tripathy. "Routing protocols in wireless sensor networks: a survey." 2012 Second International Conference on Advanced Computing & Communication Technologies. IEEE, 2012.

[6] Ben-Othman J, Yahya B. Energy efficient and QoS based routing protocol for wireless sensor networks. Journal of Parallel and Distributed Computing. 2010 Aug 31;70(8):849-57.

[7] Manzoor B, Javaid N, Rehman O, Akbar M, Nadeem Q, Iqbal A, Ishfaq M. Q-LEACH: A new routing protocol for WSNs. Procedia Computer Science. 2013 Dec 31;19:926-31.

[8] Ming Lu Y, WS Wong V. An energy-efficient multipath routing protocol for wireless sensor networks. International Journal of Communication Systems. 2007 Jul 1;20(7):747-66.

[9] Ren F, Zhang J, He T, Lin C, Ren SK. EBRP: energy-balanced routing protocol for data gathering in wireless sensor networks. IEEE Transactions on Parallel and Distributed Systems. 2011 Dec;22(12):2108-25.

[10] Intanagonwiwat, C., Govindan, R., Estrin, D., Heidemann, J., & Silva, F. (2003). Directed diffusion for wireless sensor networking. IEEE/ACM Transactions on Networking, 11(1), 2–16. http://doi.org/10.1109/TNET.2002.808417

[11] Nigel Sizer et al. “Preventing Forest Fires in Indonesia: Focus on Riau Province, Peatland, and Illegal Burning”. Available: http://www.wri.org/blog/2014/04/preventing-forest-fires-indonesia-focus-riau-province-peatland-and-illegal-burning

[12] I.Yasri, Y.Rahayu, Febrizal, Indah Kurniati, Santi R.Sipayung: Clustered Data Aggregation Based Routing Protocol Wireless Sensor Networks for A Forest Preventive Fire System at Tesso Nilo National Park. ICON EEI 2016, ISBN: 978-979-792-724-0

An Implementation Of Elias Delta Code And ElGamal Algorithm In Image Compression And Security

Dian Rachmawati[1*], Mohammad Andri Budiman[1*], and Cut Amalia Saffiera[1*]

[1]Departemen Ilmu Komputer, Fakultas Ilmu Komputer dan Teknologi Informasi, Universitas Sumatera Utara, Jl. Universitas No. 9-A, Medan 20155, Indonesia

*Email: dian.rachmawati@usu.ac.id, mandrib@usu.ac.id, amalia.ssafiera@gmail.com

Abstract. In data transmission such as transferring an image, confidentiality, integrity, and efficiency of data storage aspects are highly needed. To maintain the confidentiality and integrity of data, one of the techniques used is ElGamal. The strength of this algorithm is found on the difficulty of calculating discrete logs in a large prime modulus. ElGamal belongs to the class of Asymmetric Key Algorithm and resulted in enlargement of the file size, therefore data compression is required. Elias Delta Code is one of the compression algorithms that use delta code table. The image was first compressed using Elias Delta Code Algorithm, then the result of the compression was encrypted by using ElGamal algorithm. Prime test was implemented using Agrawal Biswas Algorithm. The result showed that ElGamal method could maintain the confidentiality and integrity of data with MSE and PSNR values 0 and infinity. The Elias Delta Code method generated compression ratio and space-saving each with average values of 62.49%, and 37.51%.

1. Introduction

Cryptography is the science of using mathematics to transform the contents of information or message in secure mode and also immuned to attack [10]. The main objectives of cryptography are confidentiality, integrity, authentication, and non-repudiation [5]. The process of converting the original message (plaintext) into a secured message (cipher text) is called encryption, while on the other hand, the process of converting ciphertext to plaintext is called decryption [2]. The security process is performed on every bit of the message to allow data size enlargement [4]. ElGamal is one of the algorithms of cryptography that includes an asymmetric algorithm. El-Gamal is based on the difficulty of calculating discrete logs in a large prime modulus [9].

Data compression is a technique that commonly used to reduce the size of data in order to store it much more compactly and also to decrease its transfer time [3]. Data compression has been widely applied to various types of data, text, images, and video [11].

The images may cover a huge amount of the memory both in the RAM and in the storage. Image compression is for reducing the redundancy and irrelevance of image to allow them to either store or transmit the data in a better way[8]. Image compression can be classified as lossy and lossless. Lossy compression is a method that bases on the assumption that the current data files save more information than human beings can perceive. Thus the irrelevant data can be removed, so there is information lost during the compression process [7]. Lossless compression is a data compression method that allows the original data to be completely reorganized, without losing any information [6]. Elias Delta Code belongs to the lossless method. In this code, each character is represented by mapping the source code to a number of bit variables [11].

In Singh and Kumar [9], the weakness of the ElGamal algorithm is caused by the magnification of size. Therefore, it appears that data compression technique can reduce or decrease the size of the data. Elias Delta code is one of universal code that encodes the positive integers. An advantage of universal code is not mandatory to find probabilities of the source string exactly[1]. So, this research will combine both Elias Data Code to compress and ElGamal Algorithm to encrypt the image for getting the better result.

2. Method

In this implementation there are 3 main menus namely: Key Generator to generate public and private key, Sender Menu to compress image and encryption, and Recipient Menu to decrypt and decompress files.

2.1 Steps for generating key:

1. Randomize a large prime number p with Agrawal Biswas test prime Algorithm. The test as follows:
   a. Choose the number p that will be tested, with p >100.
   b. Generate random Z such that $3 < Z < p - 2$.
   c. If $(1 + Z)^p \ mod \ p = (1 + Z^p \ mod \ p) mod \ p$, then p is prime.
   d. If not, p is not a prime number.
2. Randomize α as primitive root of p.
3. Randomize an integer d such that $1 < d < p - 2$.
4. Compute $\beta = \alpha^d \ mod \ p$.
5. Get the public key (p, α, β) and private key (p, d).

Then on the Sender menu, image was first compressed using Elias Delta Code Algorithm, then the result of the compression is encrypted by using ElGamal algorithm.

2.2 Steps for compressing the image:

1. Get all the frequency of all color value on each pixel in the selected image.
2. Create a compression table based on frequency (descending order).
3. Create Elias Delta Code Table
   a. Write n in binary. The left most bit will be 1.
   b. Calculate the number of bits, remove the leftmost bit from.
   c. Add the calculation in the binary on the left of n, after the leftmost bit of n is removed.
   d. Subtract 1 from the calculation in step 2 and add the number of zeros to the code.
4. Replace File Content According To Elias Delta Code Table. Add padding bits and flag.
5. Generate Elias Delta Code To ASCII Code.
6. Calculate compression ratio and space saving.

2.3 Steps for encrypting:

1. Get the public key (p, α, β).
2. Encryption the compression value (P) on each block, with calculate $c_1 = \alpha^r mod \ p$ and $c_2 = P \ x \ \beta^r mod \ p$
3. Save the result value of encryption.

Then on the Recipient menu, get the result of compressed and encrypted file. Then, the file is decrypted using Elgamal and decompressed using Elias Delta Algorithm to get the original image.

2.4 Steps for decrypting the file:

1. Get the private key (p, d).
2. Decryption on each block using $P = c_2 \times {c_1}^{p-1-d} \; mod \; p$.

2.5 Steps for decompressing:

1. Convert P from string to binary digit.
2. Read bit flag (8-bit from right) to know how many padding bits.
3. Check starts from the first bit (from left), n = 1
   a. If the bit is contained in the Elias Delta Code table, change it to the appropriate string.
   b. If bit is not in the table, read the string in front of it, bit n + 1 until the bits are contained in the table.
4. Repeat steps a and b until finished bit before padding.
5. Return the string value on each pixel to get the original image

Figure 1 shows that pixel coordinate (300,1) of the image with intensity value (210,210,190).

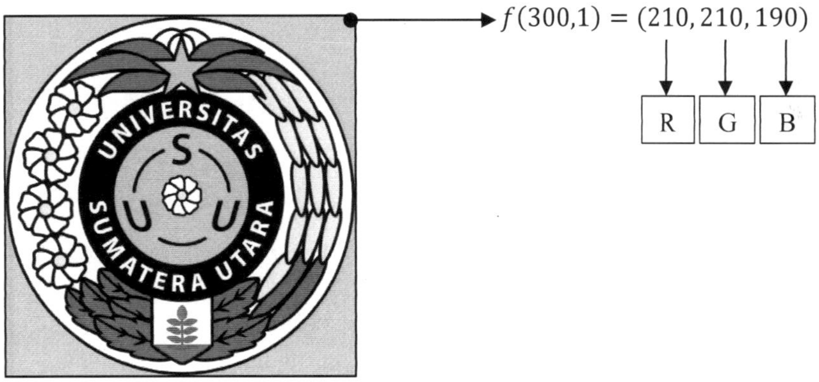

Figure 1. Sample of RGB value for true color image

From Figure 1, the steps for compression and encryption of the pixel (300,1) are as follows:

1. Compression Process
   a. Create a compression table based on frequency (descending order) as showed in table 1.

Table 1. The Compression Table with Elias Delta Code

Value	Frequency	Elias Delta Code
210	2	1
190	1	0100

   b. Replace value of pixel according to elias delta code table.
   The pixel values of 210,210,190 changes to string bit = "110100".
   c. Add padding bits and flag.
   For the number of string bits is 6, it has to take addition two extra bit 0. Then the padding bits is "00", and the flag is the binary of 2 such as "00000010". Finally the result of compression string is "110100**00***00000010*".
   d. Generate Elias Delta Code To ASCII Code equal to 208, 2.

2. Generate Key
   a. Randomize a prime number $p = 271$. Test with Agrawal Biswas Algorithm as follows:
      - Generate random Z such that $3 < Z < p - 2$. Example $Z = 200$.
      - Do the first calculation
        $$(1+Z)^p \; mod \; p = (1+200)^{271} \; mod \; 271 = 201$$
      - Do the second calculaltion
        $$(1+Z^p \; mod \; p) \; mod \; p = (1+200^{271} \; mod \; 271) \; mod \; 271 = (1+200) \; mod \; 271 = 201$$

      For the result of both calculation is equal, then 271 is a prime number.
   b. Randomize primitive root $\alpha = 107$. The test as follows.
      - Calculate $q = \frac{p-1}{2} = \frac{271-1}{2} = 135$
      - Calculate α^2 mod p and α^q mod p.
        $$\alpha^2 \; mod \; p = 107^2 \; mod \; 271 = 67$$
        $$\alpha^q \; mod \; p = 107^{135} \; mod \; 271 = 270$$

      107^2 mod 271 $\neq$ 1 and 107^{135} mod 271$\neq$ 1, then α is primitive root.
   c. Randomize $d = 96$.
      $$\beta = \alpha^d \; mod \; p. = 107^{96} \; mod \; 271 = 39$$
   d. Then get the public key $(p,\alpha,\beta) = (271,107, 39)$ and private key $(p, d) = (271, 96)$.

3. Encryption Process
   a. Get the public key $(p,\alpha,\beta) = (271,107,39)$.
   b. Randomize $r < p - 1$. r = 50.
   c. Calculate c_1 and c_2 on each block
      - $$c_1[1] = \alpha^r mod \; p = 107^{50} mod \; 271 = 238$$
        $$c_2[1] = P_1 \; x \; \beta^r mod \; p = 208 \; x \; 39^{50} \; mod \; 271 = 193$$
      - $$c_1[2] = \alpha^r mod \; p = 107^{50} mod \; 271 = 238$$
        $$c_2[2] = P_1 \; x \; \beta^r mod \; p$$

$= 2\ x\ 39^{50}\ mod\ 271$
$= 67$

4. Decryption Process
   a. Get the private key $(p, d) = (271, 96)$.
   b. Decryption on each block with calculate $P = c_2 \times c_1^{p-1-d} \mod p$.

$P_1 = c_2[1] \times c_1[1]^{p-1-d} \mod p$
$= 193 \times 2381^{271-1-96} \mod 271$
$= 208$

$P_2 = c_2[2] \times c_1[2]^{p-1-d} \mod p$
$= 67 \times 2381^{271-1-96} \mod 271$
$= 2$

5. Decompression Process
   a. Convert P from string to binary digit.
      String bit = "1101000000000010".
   b. Read bit flag (8-bit from right) = "00000010", then padding bits = 2.
   c. Check starts from the first bit, the value is 1. Because bit "1" is contained in the Elias Delta Code table, change it to the appropriate string of value 210.
   d. Check the next bit, the value is 1, change to 210.
   e. Check the next bit, the value is 0. Because bit is not available in the table, read the next string or bit n + 1, until those bits are available in the table. That is "0100", then change it to appropriate string of the value 190.
   f. Input those value to get the intensity value of pixel of (210,210,190).

3. Results and Discussions

The experiments were performed on Windows 10 Notebook with Intel Core i3 processor, 64-bit architecture, and 2048MB RAM. The development environment used for coding C# scripts is SharpDevelop. The results of the experiments of each set are presented in Tables 2 and 3 as follows.

Table 2. The Compression result for 5 sample images.

Resolution (pixel)	Original Size (kb)	Compressed Size (kb)	Compressed Ratio (%)	Space Saving (%)
200 x 200	156.30	99.59	63.72	36.28
400 x 400	625.05	392.88	62.86	37.14
600 x 600	1406.30	875.85	62.28	37.72
800 x 800	2500.05	1546.66	61.87	38.13
1000 x 1000	3906.30	2410.73	62.49	38.29

It can be seen in tabel 1 that the average compressed ratio is 62.49% and the average space saving is 37.51 %.

Table 3. Measure of image quality for 5 sample images.

Resolution (pixel)	MSE	PSNR
200 x 200	0	Infinity
400 x 400	0	Infinity
600 x 600	0	Infinity
800 x 800	0	Infinity
1000 x 1000	0	Infinity

It can be showed in table 3 that input and output image are exactly the same.

4. Conclusions

In conclusion, the integrity of the image is maintained, with the value of MSE and PSNR 0 and infinity. Also, the average of compression ratio and space saving is 62.49% and 37.51% respectively.

5. Acknowledgments

The authors gratefully acknowledge that the present research is supported by Fund Dissemination IPTEKS Research Results for Lecturers / Researchers Universitas Sumatera Utara.

References

[1] Aarthi, R.S, Muralidharan, D, and Swaminathan, P 2012 *Double Compression Of Test Data Using Huffman Code* Journal of Theoretical and Applied Information Technology **39** (2) 104-113

[2] Ayushi. 2010. *A Symmetric Key Cryptographic Algorithm*. International Journal of Computer Applications **1** (15) 1-2

[3] Budiman, M.A and Rachmawati, D. 2017. *On Using Goldbach GO Codes and Even-Rodeh Codes for Text Compression.* Material Science and Engineering 180.

[4] Cai, Jin-Yi and Cusick, Thomas W. 1999. *A Lattice-Based Public-Key Cryptosystem*. Information and Computation **151** 17-31.

[5] Garg et al.2014. *Comparison of Asymmetric Algorithms in Cryptography*. International Journal of Computer Science and Mobile Computing **3** (4) 1190-1196.

[6] Kavitha, P. 2016. *A Survey on Lossless and Lossy Data Compression Methods*. International Journal of Computer Science & Engineering Technology **7** (03) 110-114.

[7] Mamta Sharma. 2012. *Compression Using Huffman Coding.* International Journal of Computer Science And Network Security **5** (5)

[8] Singh, Manjari. 2016. *Various Image Compression Techniques: Lossy and Lossless*. International Journal of Computer Applications **142** (6) 23-26

[9] Singh, Rashmi dan Shiv Kumar. 2012. *Elgamal's Algorithm in Cryptography.* International Journal of Scientific & Engineering Research **3** (12) 1-4.

[10] Thakkar, Jaydip. 2015. *An Encryption and Decryption More Secure Elgamal Cryptosystem.* International Journal of Science Technology & Engineering **1** (12)

[11] Walder et al. 2011. *Fast Decoding Algorithms For Variabl-Lengths Codes*. 2012. Information Sciences **183** 66–91.

IOP Conf. Series: Materials Science and Engineering **300** (2018) 012041 doi:10.1088/1757-899X/300/1/012041

Optimal Path Determination for Flying Vehicle to Search an Object

R. Heru Tjahjana, R. Heri Soelistyo U, L. Ratnasari, B. Irawanto

Department of Mathematics, Diponegoro University, Jl. Prof. Sudarto Tembalang Semarang Indonesia, 50275

heru.math.undip@ gmail.com

Abstract. In this paper, a method to determine optimal path for flying vehicle to search an object is proposed. Background of the paper is controlling air vehicle to search an object. Optimal path determination is one of the most popular problem in optimization. This paper describe model of control design for a flying vehicle to search an object, and focus on the optimal path that used to search an object. In this paper, optimal control model is used to control flying vehicle to make the vehicle move in optimal path. If the vehicle move in optimal path, then the path to reach the searched object also optimal. The cost Functional is one of the most important things in optimal control design, in this paper the cost functional make the air vehicle can move as soon as possible to reach the object. The axis reference of flying vehicle uses N-E-D (North-East-Down) coordinate system. The result of this paper are the theorems which say that the cost functional make the control optimal and make the vehicle move in optimal path are proved analytically. The other result of this paper also shows the cost functional which used is convex. The convexity of the cost functional is use for guarantee the existence of optimal control. This paper also expose some simulations to show an optimal path for flying vehicle to search an object. The optimization method which used to find the optimal control and optimal path vehicle in this paper is Pontryagin Minimum Principle.

1. Introduction

The researchs about the flying vehicle attractt the attention of the researchers. Many researches has conducted the research about flying vehicle. The examples of recent research related with the flying vehicle topic are [1]-[8], these papers are different from this this paper, these papers do not use optimal control approach. The results of the researchs which related with the flying vehicle research conducted by the main authors expose about fuel saving can be read in [9]-[10]. This paper describe model of one flying vehicle that move to search an object use optimal control approach. The exposition of this paper inspired by sad event like an aircraft fall or miss in the midle of its flight. After fall or after performs emergency landed prosess in the land or in the water an aircraft produce special signal. This special signal use to help the search and rescue team to find the loss aircraft. The searcher aircraft or the other aircraft will catch the signal from the loss aircraft and the seracher aircraft will move reach it. The detail of the main problem of this paper will expose in the next section.

2. Problem Statement

Searching an object in wide area is an interesting problem. This exposition describe the object with special signal that loss in the field and the flying vehicle has a task to find it. This paper will expose how to find the object with the flying vehicle. The problem of this paper is how to determine optimal path which describe a flying vehicles to search an object. The flying vehicle after catch the signal from the object then fly to the object. The main problem is how to make the vehicle flying along the optimal path and as soon as possible reach the lost object.

3. Mathematical Model

This paper use optimal control to solve the problem. In optimal control approach, the mathematical model is needed. The assumption is need to expose before the model is chosen.

3.1 Assumption

Beside the object which searched completed by a signal producer during searching time, this paper make some assumptions as follows: the flying vehicle move in the same height or mathematically move in two dimentions, no disturbance between object and flying vehicle, and the vehicle has fuel enough along the motion to reach the object.

3.2 The model of flying vehicle

The model in this paper divides into two parts. The first model is model of the flying vehicle and the second one is model for describe the vehicle's task. In this paper the model of flying vehicle use Dubin's model as follows

$$\begin{aligned} \dot{r}_n &= V\cos\psi\cos\gamma^c \\ \dot{r}_e &= V\sin\psi\cos\gamma^c \\ \dot{r}_d &= V\sin\gamma \\ \dot{\psi} &= \frac{g}{V}\tan\phi^c \end{aligned} \tag{1}$$

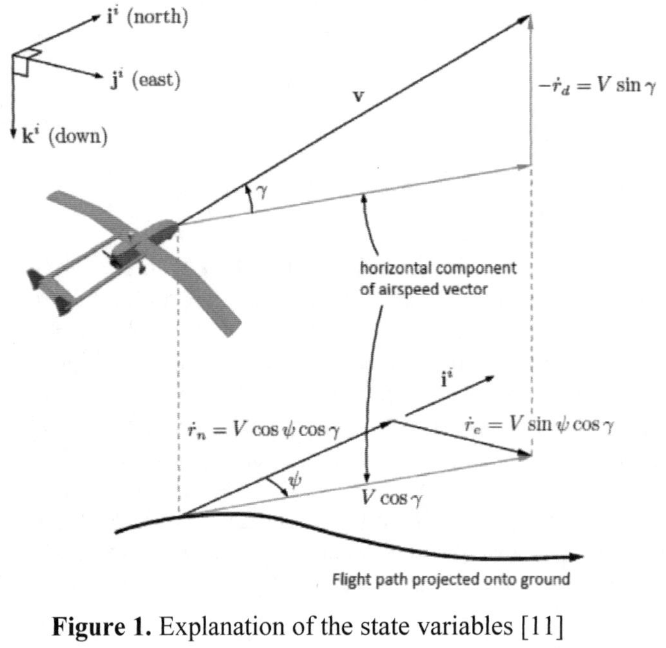

Figure 1. Explanation of the state variables [11]

Under assumption that the vehicle move in the same height along the motion then model in equation(1) can be simplified as follows

$$\begin{aligned} \dot{r}_n &= u_1 \,\mathrm{V} \cos\psi \\ \dot{r}_e &= u_1 \,\mathrm{V} \sin\psi \\ \dot{r}_d &= 0 \\ \dot{\psi} &= u_2. \end{aligned} \tag{2}$$

In this paper, model (2) can make simpler as follows

$$\begin{aligned} \dot{r}_n &= u_1 \cos\psi \\ \dot{r}_e &= u_1 \sin\psi \\ \dot{\psi} &= u_2. \end{aligned} \tag{3}$$

3.3 Model of Cost Functional

Cost functional is the most important thing in optimal control. The cost functional or index performance make the system works satisfy the goal to solve the problem. Since the goal of the paper writing is how can to take position an serched object as soon as possible then we use cost funtional that must be minimized as follows

$$J = \int_0^T dt. \tag{4}$$

4. Mathematical Analysis

4.1 Controllability and Observability

Since the solution of this paper through optimal control, the mathematical analysis is needed. The first Analysis is Controllability. The controllability model (3) is needed to checked.

Theorem 4.1.1

The model of vflying vehicle is controllable

Proof:

If model (3) linearized, the result is not controllable. The controllability can show via non linear system. The process to show that the model controllable can start from to write the model to Affine form. The Affine form of the model (3) can be presented as follows

$$\dot{\mathbf{x}} = \mathbf{f}(\mathbf{x}) + \boldsymbol{g}_1 u_1 + \boldsymbol{g}_2 u_{2,} \tag{5}$$

with $\mathbf{x} = \begin{bmatrix} r_n \\ r_e \\ \psi \end{bmatrix}$ $\mathbf{f}(\mathbf{x}) = \mathbf{0}, \boldsymbol{g}_1 = \begin{bmatrix} \sin\Psi \\ \cos\psi \\ 0 \end{bmatrix}$ and $\boldsymbol{g}_2 = \begin{bmatrix} 0 \\ 0 \\ 1 \end{bmatrix}$. Then the process is determination of controllability matrix. In this process, need Lie Bracket is needed. Finally, the controllability matrix has rank 3. From this fact, can be concluded that the system is locally accessible and since $\mathbf{f}(\mathbf{x}) = \mathbf{0}$ then the system is controllable. ■

The next analysis is observability. The observability of the model given in the next theorem as follows.

Theorem 4.1.2

The model of vflying vehicle is observable

Proof:

To check the observability, the observability matrix is formed. After that, the rank of obesrvability matrix is computed. In this case, the observabilty matrix is 3, and can be conclude that the model (3) is observable. ■

4.2 Existence of Optimal Control

Optimal control is used in this paper, so the existence of optimal control is needed to showed. The exixtence theorem is given as follows.

Theorem 4.2.1

The optimal control for the peoblem of this paper is exist

Proof:

The convexity of the cost functional is guareantee that the optimal control is exist. Consider the cost functional which given in equation number (4). The integrad is constant function. The constant function is convex, so the cost fountional is convex. If the cost functional is convex then the optimal control is exist. The other things that guarantee the existence of optimal control is restriction of the control. Consider from the declaration of the Dubin's aircraft model that the controls are bounded, from these facts can be conclude that the exixtence of optimal control is guaranteed. ■

4.3 Optimal Path

After the existence of optimal control is guareanted, if the optimal control is used for drive the flying vehicles from the origin to reach the searched object then the vehicle's path that obtained is optimal path. The path that produced with the optimal control is optimal path.The optimal path also can be seen from the value of Hamiltonian Function. Hamiltonian function is connect between model of flying vehicle or dynamical system and cost functional. The value of Hamiltonian function is constant as long as in optimal path. In other words, the derivative of Hamiltonian function with respect to time as long as optimal trajectory is zero.

4.4 Method to determination optimal path

For determination optimal path this paper use optimal control approach. The optimal path begin from the initial condition of the searcher flying vehicle until the position of searched object. The proposed method to determine optimal path can be seen as optimal control problem as follows:

4.4.1 The dynamical system use the dynamical model of flying vehicle.

4.4.2 The intitial condition use initial position and orientation the searcher vehicle and the boundary condition use the initial position and orientation from the serched object which detected from the signal that produced by the searched object.

4.4.3 The task of the flying searcher vehicle is move as soon as posible to reach the searche object. This task described in cost functional (4). The existence of optimal control is guaranteed from section 4.4.4. From the Hamiltonial function can be derivated into Hamiltonian system.

4.4.4 The optimal path equation is the solution of diferential equation system which contain Hamiltonian system, optimal control, initial and boundary condition.

5. Simulation

5.1 Simulation Scenario

The simulation begin from the lost object in the wide field. Like in the assumption the lost object produce special signal. The signal can detected by a flying vehicles, as a position that must be reached by the flying vehicles as soon as possible. The initial position of the flying vehicle and the position that must be reached or position of searced object can be seen as Two Points Boundary Value Problem. If the initial position and orientation of flying vehicle is given as X_0 and the position and orintationof searced object is detected as X_f, the flying vehicle, forced fly from the initial position to the the reached object with minimized the time. This simulation scenario approaching the real situation when the object lost in wide area and as long as in searching time the object produce special signal.

5.2 Simulation Result

The first simulation is to show that optimal control can ce used to solve the main problem in relative short distance and the result given in the Figure 2. The unit in this simulalation sectionas follows $(km,\quad km,\quad radian)$. The axis in simulation result is described as the erect and the horizontal axis. The erect axis is North and the horizontal axis is East. In the first simulation the initial condition of flying vehicle is (0,0,1) and the position of the searched object is (1,1,1). The initial conditions of the second, the third and the fourth simulations are as same as the first simulation (0,0,1) but the bondary condition of the the second, the third and the fourth simulations respectively are given by (20,40,1),(30,20,1) and (5,5,1) also the simulation result respectively are given in Figure 3, Figure 4 and Figure 5. From these figures can be maked conclusion that optimal control approach can work well to solve the problem.

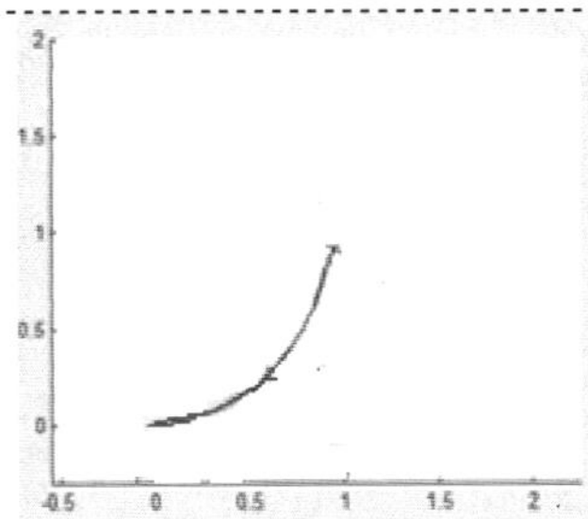

Figure 2. The simulation result for short distance

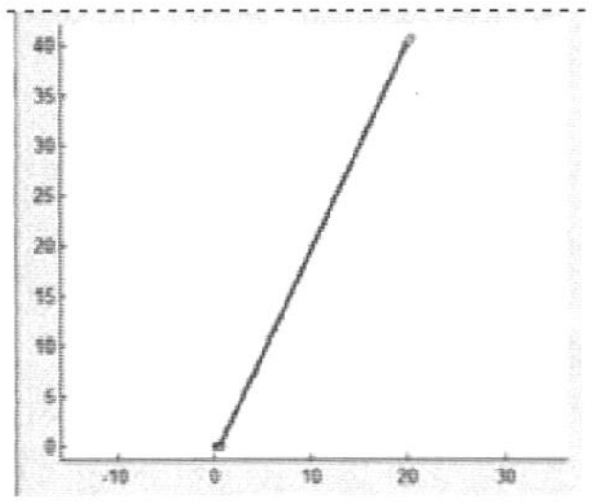

Figure 3. Optimal path of the second simulation

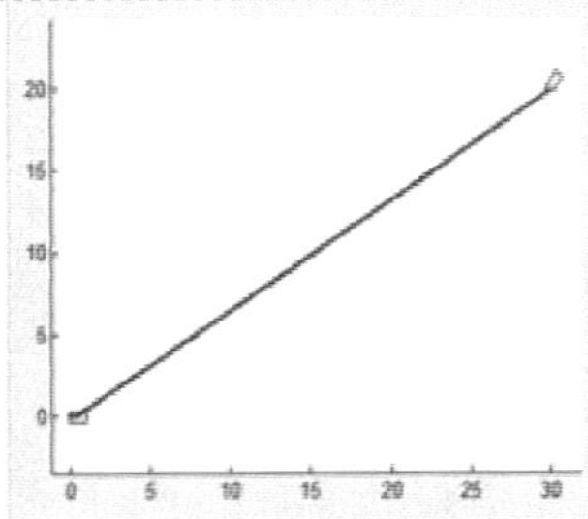

Figure 4. Optimal path of the third simulation

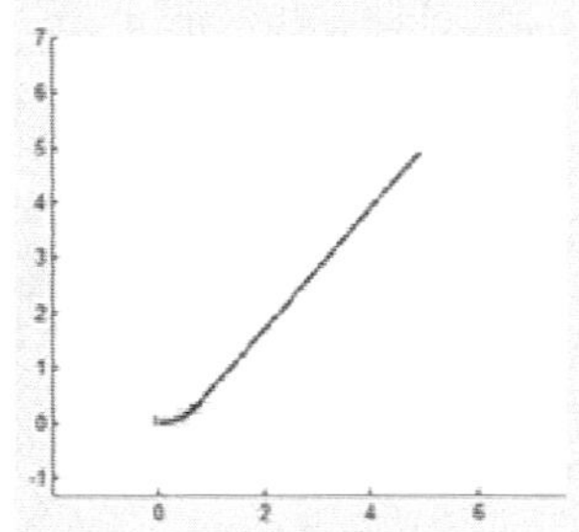

Figure 5. Optimal path of the fourth simulation

6. Conclusion

Problem to search an object and during the searching time the lost object produce special signal which catched by the flying vehicle as a searcher vehicles can be solved by optimal control approach. From the mathematical analysis, the model (3) is feasible to solve the problem throguh optimal control method. The flying vehicle forced as soon as to reach the lost object by the cost functional. From the simualtion results obtained conclusion that optimal control can be used to solve the main problem.

7. References

[1] Faessler M , Fontana F, Forster C, Mueggler E, Pizzoli M and Scaramuzza D 2016 *Journal of Field Robotic* **33** 431

[2] Huang A S, Bachrach A, Henry P, Krainin M, Maturana D, Fox D and Roy N 2017 *Springer Tracts in Advanced Robotics* **100** 235
[3] Bhandari, S, Brown, Nc, Guzman, I, Sherman,T, Tellez J and Gomez E 2017 *Proc. AIAA SciTech Forum* AIAA 2017-1153
[4] Zhang B and Duan H 2017 *IEEE/ACM Transactions on Computational Biology and Bioinformatics* **14** 97
[5] Conrad J, Wallach D, Kalweit F, Lindel P and Templin S 2017 *Lecture Notes in Computer Science* **10272** 424
[6] Siddall R, Ancel OA and Kovač M 2017 *Journal of the Royal Society Interface Focus* **7**, 20160085
[7] Andert F, Ammann N, Krause S, Lorenz S, Bratanov, D and Mejias L 2017 *Journal of Intelligent & Robotic Systems* **87** 1
[8] Aguilar W G, Luna M A, Moya J F, Abad V, Parra H and Ruiz H 2017 *Proc. Int. Conf. on Semantic Computing (ICSC)* San Diego CA p 509
[9] Tjahjana, R H and Sasongko P S 2014 *Applied Mathematical Sciences* **8** 733
[10] Tjahjana, R H and Sasongko P S 2014 *Applied Mathematical Sciences* **8** 5689
[11] Owen M, Beard RW and Mc Lain T W 2015 *Handbook of Unmanned Aerial Vehicles* ed. Valavanis, KP and Vachtsevanos, GJ (Dordrecht: Springer) p 1677

Acknowledgments
The authors thank to Faculty of Science and Mathematics Diponegoro University Indonesia for fund this research under research grant with contract number 1645a/UN7.5.8/PP/2017.

IOP Conf. Series: Materials Science and Engineering **300** (2018) 012042 doi:10.1088/1757-899X/300/1/012042

Hybrid Cryptosystem Using Tiny Encryption Algorithm and LUC Algorithm

Dian Rachmawati [1*], Amer Sharif [1*], Jaysilen[1*], and Mohammad Andri Budiman[1*]

[1]Program Studi S1 Ilmu Komputer, Fakultas Ilmu Komputer dan Teknologi Informasi, Universitas Sumatera Utara, Jl. Alumni No 9-A, Medan 20155, Indonesia

*Email: dian.rachmawati@usu.ac.id, amersharifdjamin@gmail.com, jaysilen07@gmail.com, mandrib@usu.ac.id

Abstract. Security becomes a very important issue in data transmission and there are so many methods to make files more secure. One of that method is cryptography. Cryptography is a method to secure file by writing the hidden code to cover the original file. Therefore, if the people do not involve in cryptography, they cannot decrypt the hidden code to read the original file. There are many methods are used in cryptography, one of that method is hybrid cryptosystem. A hybrid cryptosystem is a method that uses a symmetric algorithm to secure the file and use an asymmetric algorithm to secure the symmetric algorithm key. In this research, TEA algorithm is used as symmetric algorithm and LUC algorithm is used as an asymmetric algorithm. The system is tested by encrypting and decrypting the file by using TEA algorithm and using LUC algorithm to encrypt and decrypt the TEA key. The result of this research is by using TEA Algorithm to encrypt the file, the cipher text form is the character from ASCII (American Standard for Information Interchange) table in the form of hexadecimal numbers and the cipher text size increase by sixteen bytes as the plaintext length is increased by eight characters.

1. Introduction

Cryptography is a method to send the file in secret, which is mean only the recipient can read the file [1]. Cryptography has its etymology, kryptos means hidden, and graphein means to write [2]. Based on its etymology, cryptography is a technique to secure file by encoding the original file to become some scramble code which is difficult to understand by the public [2]. There is two main process in cryptography, the process of transform file into scramble code is called encryption and the process to reverse the scramble code back to the original file is called decryption [3].

In this research, three main components in cryptography are used, which is cipher text, plaintext, and the keys. Cipher text is the scramble code to secure the original file [2] and plaintext is a file that needs to be secured [2]

Tiny Encryption Algorithm (TEA) is a symmetric algorithm which is found by David Wheeler and Roger Needham from Cambridge University. TEA is designed to be fastest and most efficient cryptographic algorithm compared to other symmetric algorithms like IDEA, DES, and AES [4]. The plaintext is divided into two halves and the key is divided into four 32-bit blocks [5]. XOR, ADD, and SHIFT operator are used in encryption and decryption process, which means the cipher text size is small and easy to send to recipient [6]. The same key is used by TEA for encryption and decryption process, it mean that the security of this algorithm depends on its key. Once the key is divulged, everyone can encrypt and decrypt the file.[2]

LUC is an asymmetric algorithm developed by Peter Smith and Michael Lennon based on Lucas Function. LUC algorithm has two different keys which mean the security of this algorithm does not depend on its key. LUC public keys are used to encrypt plaintext and LUC private key is used to decrypt the cipher text [7].

LUC Algorithm has key generation phase in order to produce LUC public key to encrypt plaintext and LUC private key to decrypt the cipher text. Even the LUC public key can be made public [2]. Although LUC public key is so secure, it produces huge size cipher text and sender will be difficult to send the cipher text to the recipient.

Hybrid Cryptosystem is a method to secure file by combining both of symmetric and asymmetric algorithm in order to secure plaintext [3]. In the hybrid cryptosystem, a file is secured by using the symmetric algorithm and symmetric key is secured by using asymmetric algorithm [3]. By using this method, the sender has small-size cipher text because plaintext is secured by the symmetric algorithm and secure key because a symmetric key is secured by asymmetric algorithm. In this research, TEA is used as symmetric algorithm and LUC algorithm is used as asymmetric algorithm.

2. Method

The scheme of securing file by using a hybrid cryptosystem method is showed on Figure 1.

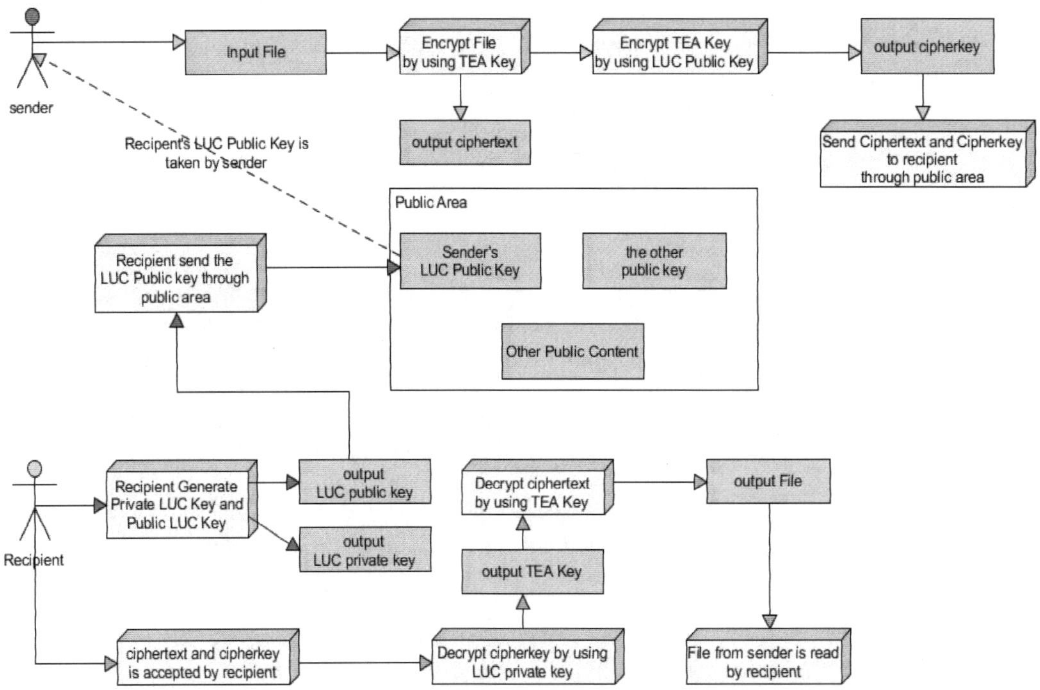

Figure 1 Hybrid Cryptosystem Method Scheme

Based on figure 1,the first step to using hybrid cryptosystem method is generate LUC public key and the LUC private key by following this step [8].

- Select two large prime number(P,Q) and P must be co-prime to Q or GCD (Greatest Common Division) between P and Q is 1
- Calculate the value of N by using this equation :
$$N = P * Q \quad (1)$$
- Calculate the euler function (T) by using this equation :
$$T = (P - 1) * (P + 1) * (Q - 1) * (Q + 1) \quad (2)$$
- Select the random number (E) and E is smaller than T but larger than 1
- check if E is co-prime to T or not.
- If E is co-prime to T, then E is the LUC public key
- Calculate the R(N) value by using this equation :
$$R(N) = LCM\{(P - 1) * (P + 1) * (Q - 1) * (Q + 1)\} \quad (3)$$
- Calculate the D value iteratively until D reach this equation :
$$D * E \ Mod \ R(N) = 1 \quad (4)$$
- After D value is generated, then D is the LUC private key.

After both of keys are generated, recipient send LUC public key to sender, so sender can use the LUC public key to encrypt the TEA key. Before encrypt the TEA key, sender must encrypt the file first using TEA key in order to produce ciphertext by following this step: [6]

- Convert plaintext and key into encoding table. In this research, the ASCII (American Standard for Information Interchange) encoding table is used,
- Divide the plaintext into two blocks and each block have 32-bit data (block Y [0], Block Z [0])
- split the key into four blocks and each block have 32-bit data (block K [0], block K [1], block K [2], block K [3]
- Using this Equation to start encryption:
$$Y[i+1] = Y[i] + \{(Z[i] << 4) + K[0]) \wedge (Z[i] + delta) \wedge ((Z[i] >> 5) + K[1]))\} \quad (5)$$
$$Z[i+1] = Z[i] + \{(Y[i] << 4) + K[2]) \wedge (Y[i] + delta) \wedge ((Y[i] >> 5) + K[3]))\} \quad (6)$$
$0 \leq i \leq 31$" ; i is Natural Number;
Delta is the golden number or 9E3779B9
The cipher text will be available in block Y [32] and in block Z [32].

Sender encrypt the TEA key by using LUC public key which is already accepted from recipient earlier by following this step [7] :

- Convert The TEA key into encoding table
- Input LUC Encryption Key (e)
- Using this pattern:
$$V[0] = 2 \quad (7)$$
$$V[1] = TEA\ Key = P \quad (8)$$
$$V[i] = (P * V[I - 1] - V[I - 2]) mod\ N; i \geq 2 \quad (9)$$
- Repeat the (8) process until I = E
$$V[E] = (P * V[e - 1] - V[e - 2]) \ mod\ N. \quad (10)$$
- V[E] is the cipherkey or V[E] = C.

After recipient accept both of cipher text and cipher key, recipient uses LUC Private Key to decrypt the cipher key in order to produce the TEA key by following this step [8] :

- Select the cipher key
- Input LUC Decryption Key (d)
- Using this pattern:
$$V[0] = 2 \quad (11)$$
$$V[1] = Cipherkey = C \quad (12)$$
$$V[I] = (C * V[i - 1] - V[i - 2]) \ mod\ N\ i \geq 2. \quad (13)$$

- When I = d, use this equation :

$$V[d] = (C * V[d-1] - V[d-2]) \bmod N. \quad (14)$$

- V[d] is the TEA Key

After TEA key is produced, recipient can use it to decrypt the ciphertext by following this step [4] :

- Divide the cipher text into two halves, and each halve have 32-bit data (block Y[32], Block Z[32])
- Divide the key into four blocks, and each block have 32-bit data (block K[0], block K[1], block K[2], block K[3]
- Using this Equation to start Encryption.

  Y[i-1] = Y[i] - {(Z[i] <<4) +K[0]) ^ (Z[i] + delta) ^ ((Z[i] >> 5) + K[1]))} (15)

  Z [i-1] = Z[i] - {(Y[i] <<4) + K[2]) ^ (Y[i] + delta) ^ ((Y[i] >> 5) + K [3]))} (16)

  i start from 32 down to 1; i is a Natural Number

  Delta is the golden number or 9E3779B9
- Plaintext will come out when it reach block Y[0] and block Z[0]

After recipient decrypts the ciphertext, the hybrid cryptosystem method to secure file already done. Because sender use a symmetric algorithm (TEA) to encrypt the file and an asymmetric algorithm (LUC Public Key) to encrypt the TEA Key. Meanwhile, recipients use the asymmetric algorithm (LUC Private Key) to decrypt the cipher key in order to produce the TEA key which is used to decrypt the ciphertext.

3. Results and Discussions

These experiments are conducted on the Windows 10 Pro Creator Update Notebook which has 64-bit architecture Intel Core i3 processor and 4096 MB RAM. The result of this research is presented in Table 1 as follows:

Table 1. The Comparison between plaintext size and ciphertext size

Sample Number	Plaintext size (byte)	Ciphertext size (byte)
1	1	16
2	2	16
3	8	16
4	9	32
5	10	32
6	17	48
7	18	48
8	25	64
9	33	80
10	41	96
11	56	112
12	95	192
13	96	208
14	106	224
15	112	224

Table 1 shows when the plaintext size is increased by eight characters, the ciphertext size will increase by sixteen bytes.

The integrity is the main purpose in cryptography, that means the file was sent by the sender, the same file must be accepted by a recipient and there is the result of securing file by using hybrid cryptosystem method as shown in Figure 2.

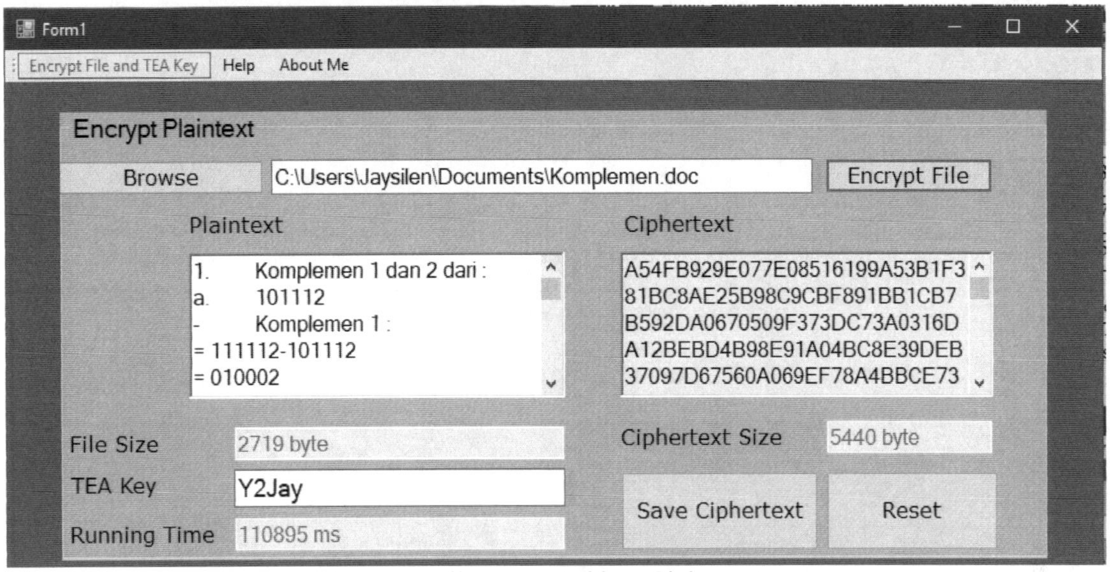

Figure 2 : File Encrypted into Ciphertext

Based on figure 2, the plaintext has been encrypted by using TEA key and produce ciphertext. Because hybrid cryptosystem method is used to secure file, the sender must encrypt the TEA key use LUC Public Key, which is available on Figure 3.

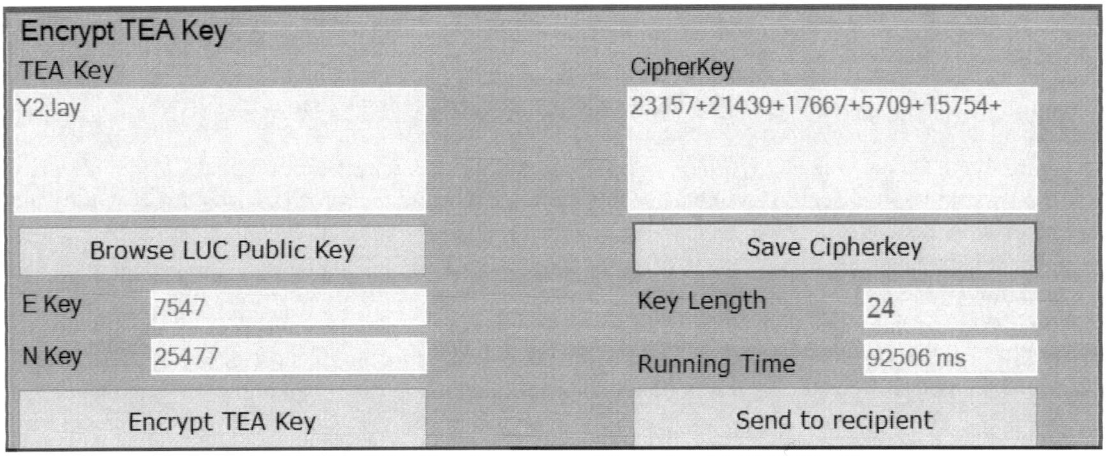

Figure 3 The TEA Key is encrypted by using LUC Public Key

After the TEA key is encrypted and produce the cipher key, both of cipher key and ciphertext is sent to a recipient. Then, recipient use LUC private key to decrypt the cipher key on Figure 4.

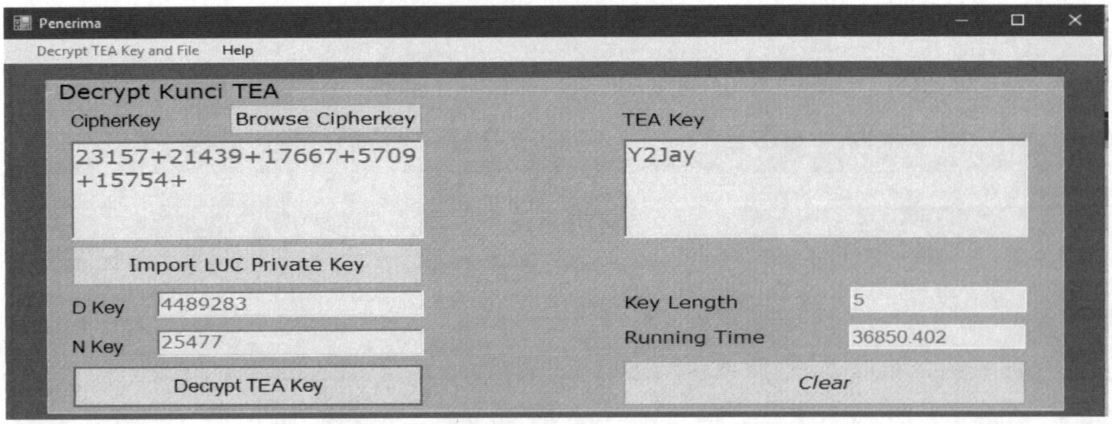

Figure 4 : Cipherkey is decrypted into TEA key

The recipient already have the TEA key to decrypt the ciphertext which is showed on Figure 5.

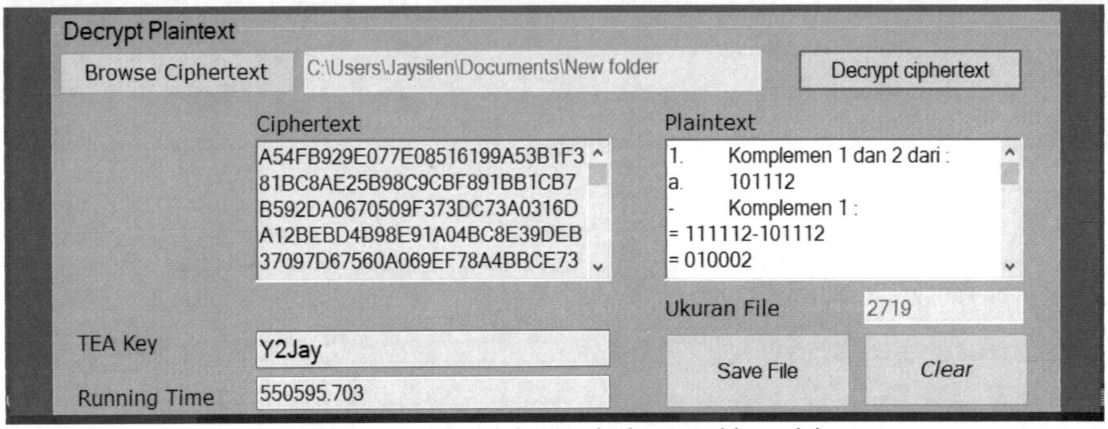

Figure 5. The Ciphertext is decrypted into plaintext

Figure 4 and Figure 5 show hybrid cryptosystem method has met the requirement of the integrity aspect on cryptography because recipient accepts the same file and the same key which was sent by the user before.

4. Conclusions

The conclusion of this research is TEA key is used to encrypt the plaintext and decrypt the ciphertext, meanwhile, LUC Algorithm is used to encrypt the TEA key and decrypt the cipher key. The hybrid cryptosystem method by using TEA and LUC Algorithm met the requirement of the integrity aspect on cryptography. The ciphertext size is increased by sixteen bytes as the plaintext length is increased by eight characters. This system can secure the file which has this extension: *.pdf, *.txt, *.rtf, *.doc, *.docx, and *.otd.

Acknowledgments

The authors gratefully acknowledge that the present research is supported by Fund Dissemination IPTEKS Research Results for Lecturers / Researchers Universitas Sumatera Utara.

References

[1] R. Mollin, An Introduction to Cryptography, 2nd Ed., Florida: Chapman & Hall/CRC, 2007.

IOP Conf. Series: Materials Science and Engineering **300** (2018) 012042 doi:10.1088/1757-899X/300/1/012042

[2] B. Schneier, Applied Cryptography, 2nd Ed., John Wiley & Sons, Inc, 1996.

[3] N. Smart, Cryptography: An Introduction, 3rd Ed., Bristol City, 1999.

[4] S. A. Y. Hunn, S. Z. Naziri and N. Idris, "The Development of Tiny Encryption Algorithm(TEA) Crypto-Core for Mobile Systems," *IEEE Internatioal Conference on Electronics Design, Systems and Applications,* pp. 45-49, 2012.

[5] J.Balakrishna, P. Daniel, and R. Chandel "Design & Implementation of VLSI Architecture for XTEA," *International Conference on Advanced Computing, Communication and Network*, Hamirpur, 2011.

[6] D. Wheeler, R. Needham "TEA, a Tiny Encryption Algorithm", *First International Conference on Security in Pervasive Computing*, Boppard, 2003.

[7] Z. M. Ali, "Parallel Computation Algorithm for LUC Cryptosystem Based on Binary Number," *Journal of Theoretical And Applied Information Technology,* vol. 44, no. 1, pp. 12-21, 2012.

[8] P. J. Smith and M. J. J. Lennon, *LUC: A New Public Key System,* Auckland: The University of Auckland, 1993.

Nonlinear evolution of Benjamin-Feir wave group based on third order solution of Benjamin-Bona-Mahony equation

Zahnur[1,2], Vera Halfiani[3], Salmawaty[4], Tulus[5] and Marwan Ramli[3,4]

[1] PhD Students of University of Sumatera Utara, Indonesia
[2] Department of Informatics, Syiah Kuala University, Banda Aceh 23111, Indonesia
[3] Dynamic Application and Optimization Research Group, Syiah Kuala University, Banda Aceh 23111, Indonesia
[4] Department of Mathematics, Syiah Kuala University, Banda Aceh 23111, Indonesia
[5] Department of Mathematics, University of Sumatra Utara, Medan 20155, Indonesia

E-mail: zahnur@gmail.com, verahalfiani@yahoo.com, salma_waty@yahoo.com, tulus@usu.ac.id, corresponding author : ramlimarwan@gmail.com; marwan.math@unsyiah.ac.id

Abstract. This study concerns on the evolution of trichromatic wave group . It has been known that the trichromatic wave group undergoes an instability during its propagation, which results wave deformation and amplification on the waves amplitude. The previous results on the KdV wave group showed that the nonlinear effect will deform the wave and lead to large wave whose amplitude is higher than the initial input. In this study we consider the Benjamin-Bona-Mahony equation and the theory of third order side band approximation to investigate the peaking and splitting phenomena of the wave groups which is initially in trichromatic signal. The wave amplitude amplification and the maximum position will be observed through a quantity called Maximal Temporal Amplitude (MTA) which measures the maximum amplitude of the waves over time.

1. Introduction

The evolution of nonlinear wave can by portrayed as the occurrence of spontaneous high amplitude waves. When the wave height exceeds the significant wave height of measured wave train by factor more than 2.2, the wave is known as an extreme or rouge wave [1]. The extreme wave can appear from nowhere and then disappear without a trace [2]. There are some mechanisms of extreme wave occurrence; wave-current interaction, geometrical focusing, soliton collision, focusing due to dispersion, and modulational instability [2–4]. Modulational instability happens as the effect of nonlinearity. The instability causes growing modulation of the wave's envelope due to the energy focusing. This extreme wave behavior has been actively studied and appears not only in hydrodynamic context but also in optical waves, quantum electronics, and plasma physics [5–9].

Extreme wave phenomenon may occur to the bichromatic wave group which is a superposition of two monochromatic waves with the same amplitudes and slightly different frequencies. The wave which is initially in bichromatic form experiences instability and focusing during its propagation. New side bands occur and modulate the wave causing an amplification to the amplitude [10]. This mechanism had been also investigated numerically and experimentally in [11] and [12]. Marwan [13] studied the deformation of this wave based on the third order asymptotic series solution of Korteweg de Vries (KdV) equation. The wave propagates and undergoes amplitude amplification until it reaches the maximum peaking and then followed by the decrease at its amplitude. The similar approach up to the fifth order solution of KdV equation was also inspected in [14]. The result suggests that the higher order of the solution gives higher maximum amplification. However, these wave's maximum amplifications are still far below the ones obtained from numerical software HUBRIS which is governed based on Laplace equation [12] and Boussinesq equation [15]. The evolution of bichromatic wave group was also investigated in [16] on the basis of Benjamin-Bona-Mahony (BBM) equation. The study suggests that the initial amplitude and frequency difference of the bichromatic signal may affects the amplitude amplification factor and the maximum position.

An instability also happens to the trichromatic wave group. Trichromatic wave is formed when a monochromatic wave is perturbed by a pair of side bands with smaller amplitude and slightly different frequencies. This wave is also known as Benjamin-Feir type wave. The study of the deformation of this wave was studied in [17] by employing the KdV equation as the wave model. The amplification and extreme position was inspected by utilizing a quantity called maximal temporal amplitude (MTA). This quantity measures the maximum wave's amplitude at each spatial position during the observation time [13–15]. The evolution of rouge wave has been also investigated through the Nonlinear Schrodinger (NLS) equation. NLS also yields solutions which experience instability during the wave propagation [18–20]. This equation appears as the envelope of wave groups of many nonlinear equations such as KdV [6, 21–23], BBM [24], and KP [9].

In this study, we are interested in investigating the evolution of a wave group which is initially in the trichromatic wave form. The wave will be developed based on the third order asymptotic solution of BBM equation. The amplitude amplification and the maximum position will be inspected by generating the MTA of the wave.

2. Third Order Side Band Solution of BBM Model

Benjamin-Bona-Mahony (BBM) equation was developed as a revised model of KdV equation. Benjamin et al [25] explained that the classic KdV equation failed the unidirectional assumption when the wave number value is high. The regularized long wave model they proposed can explain the wave physical behavior better. BBM equation is mathematically stated as [25],

$$\eta_t + \eta_x + \eta\eta_x - \eta_{xxt} = 0, \tag{1}$$

where η is elevation, x and t are respectively spatial and time variables.

We develop the solution of (1) through asymptotic expansion up to the third order. Suppose that the solution of (1) is expressed in the asymptotic series below,

$$\eta(x,t) = \varepsilon\eta^{(1)} + \varepsilon^2\eta^{(2)} + \varepsilon^3\eta^{(3)} + O(\varepsilon^4), \tag{2}$$

where ε is real number. The solution for $\eta^{(i)}$ will be determined at each order. Furthermore, we consider trichromatic wave as the initial wave form:

$$\eta(x,t) = \sum_{p=1}^{3} a_p e^{i\theta_p} + c.c. \tag{3}$$

where a_p represents amplitude, $\theta_p = \omega_p t - k_p x + \psi_p$, ω_p is frequency, k_p is wavenumber, ψ_p is phase shift, and $c.c.$ is the complex conjugate. To investigate the deformation of the wave as the peaking phenomena, we define the trichromatic wave as a monochromatic wave being perturbed by a pair of side bands with smaller amplitude. Hence we assign $a_2 > a_1$, $a_1 = a_3$, and $\omega_1 = \omega_2 - \nu$, $\omega_3 = \omega_2 + \nu$, where ν is small real number. Also, to resolve the secular terms which may occur at higher order, we applied Linstedt-Poincare method and expand the wave number in the series $k_p = k_p^{(0)} + \varepsilon k_p^{(1)} + \varepsilon^2 k_p^{(2)} + O(\varepsilon^3)$, for each $p = 1, 2, 3$.

We substitute the wave number series and (2) into (1), collect the terms at the same order of ε, and solve the equation separately at every order. For the first order $O(\varepsilon)$, we obtain an equation equivalent to the linear form of equation (1). Assuming the solution is in the trichromatic wave form (3) yields the dispersion relation $\Omega(k_p^{(0)}) = \omega_p = k_p^{(0)} \Big/ \left(1 + (k_p^{(0)})^2\right)$, for $p = 1, 2, 3$.

On the second order $O(\varepsilon^2)$, there appears secular terms which will resonance with the homogeneous solution of first order equation and cause the solution of higher order have higher amplitude. To eliminate these terms, we set $k_p^{(1)} = 0$. Furthermore, on the equation of $O(\varepsilon^3)$, the value of $k_p^{(2)}$ needs to be assigned to treat the resonant terms that are appeared. The value of the correction wave number reads

$$k_p^{(2)} = \frac{-(K_1 + K_2)}{a_p - 2a_p k_p^{(0)} \omega_p},$$

where $K_1 = a_p k_p^{(0)} \Big(a_1^2 (A_{+,1p} + 2A_{-,11}) + a_2^2 (A_{+,2p} + 2A_{+,22}) + a_3^2 (A_{+,3p} + 2A_{-,33}) + a_t^2 (A_{+,pt} + A_{-,pt} + A_{-,tp}) + a_s^2 (A_{+,ps} + A_{-,ps} + A_{-,sp}) \Big)$,

$$K_2 = \begin{cases} a_t^2 a_s \left(A_{+,tt} + A_{-,ts} + A_{-,st}\right) \left(2k_t^{(0)} - k_s^{(0)}\right), & p = 1 \\ a_p a_t a_s \left((A_{+,st} + A_{+,ts}) + (A_{-,sp} + A_{-,ps}) + (A_{-,tp} + A_{-,pt})\right) \left(k_t^{(0)} + k_s^{(0)} - k_p^{(0)}\right), & p = 2 \\ a_s^2 a_t (A_{+,ss} + A_{-,st} + A_{-,ts})(2k_s^{(0)} - k_t^{(0)}), & p = 3 \end{cases}$$

$t = (p \bmod 3) + 1$, $s = (p - 2 \bmod 3) + 1$.

We will investigate the evolution of the trichromtic wave modulated by the side bands. Hence, assuming $a_2 = a, a_1 = a_3 = \delta a$, δ is small real number. Equation (3) as the solution of first order equation can be rewritten as

$$\eta^{(1)} = 2(\delta a \cos\theta_1 + a\cos\theta_2 + \delta a \cos\theta_3) = 2a(1 + 2\delta\cos\Delta\theta)\cos\theta_2, \tag{4}$$

where $\Delta\theta = (\theta_3 - \theta_1)/2 = (\nu t - \kappa t + \Delta\psi)$. We obtain that the solution of $O(\varepsilon^2)$ is waves consisting of frequencies which are far from the frequencies of trichromatic waves (4). These waves may not significantly affect the wave modulation and the extreme position. The waves with higher frequency also comes up in third order solution. Omitting these terms, we only consider the waves with near frequencies as the side bands. Hence, the third order side band waves and the free waves are stated as follow,

$$\begin{aligned} \eta_{sb}^{(3)} &= \delta^2 a^3 \Big((B_{-,123} + B_{-,213}) \cos(\theta_1 - \Delta\theta_r) + B_{-,112} \cos(\theta_1 - \Delta\theta_l) + \\ &(B_{-,1231} + B_{-,321}) \cos(\theta_3 + \Delta\theta_l) + B_{-,332} \cos(\theta_3 + \Delta\theta_r) \Big) + \\ &\delta^3 a^3 \Big(B_{-,113} cos(\theta_1 - 2\Delta\theta) + B_{-,331} cos(\theta_3 + 2\Delta\theta) \Big), \\ \eta_{sb,fw}^{(3)} &= \delta^2 a^3 \Big((B_{-,123} + B_{-,213}) \cos(\varphi(\theta_1 - \Delta\theta_r)) + B_{-,112} \cos(\varphi(\theta_1 - \Delta\theta_l)) + \\ &(B_{-,1231} + B_{-,321}) \cos(\varphi(\theta_3 + \Delta\theta_l)) + B_{-,332} \cos(\varphi(\theta_3 + \Delta\theta_r)) \Big) + \\ &\delta^3 a^3 \Big(B_{-,113} cos(\varphi(\theta_1 - 2\Delta\theta)) + B_{-,331} cos(\varphi(\theta_3 + 2\Delta\theta)) \Big), \end{aligned}$$

where $\Delta\theta_l = \theta_2 - \theta_1$, $\Delta\theta_r = \theta_3 - \theta_2$, $\varphi(\theta_p) = \omega_p t - \Omega^{-1}(\omega_p)x + \psi_p$,

$$B_{+,pqr} = \frac{C_+ A_{+,pq}}{D_+}, \qquad B_{-,pqr} = \frac{C_-(A_{+,pq} + A_{+,pr} + A_{+,rq})}{D_-}, \qquad C_\pm = \left(k_p^{(0)} + k_q^{(0)} \pm k_r^{(0)}\right),$$

$$D_\pm = (\omega_p + \omega_q \pm \omega_r) - (k_p^{(0)} + k_q^{(0)} \pm k_r^{(0)}) + (k_p^{(0)} + k_q^{(0)} \pm k_r^{(0)})^2 (\omega_p + \omega_q \pm \omega_r),$$

$$A_{\pm,pq} = \pm k_q^{(0)} \Big/ \left((\omega_p \pm \omega_q) - (k_p^{(0)} \pm k_q^{(0)}) + (k_p^{(0)} \pm k_q^{(0)})^2 (\omega_p \pm \omega_q)\right), \text{ and } A_{-,pq} = 0 \text{ for } p = q.$$

Finally, the side band solution is $\eta(x,t) \approx \eta^{(1)} + \eta_{sb}^{(3)} - \eta_{sb,fw}^{(3)}$.

3. Extreme Position and Amplitude Amplification

The extreme position of the side band modulated waves will be inspected by employing Maximal Temporal Amplitude (MTA). MTA can be mathematically stated as [13].

$$\text{MTA}(x) = \max_t \eta(x,t)$$

The result will be presented in the laboratory measure. Thus, nondimensional BBM equation stated in (1) needs to be converted to its physical variables (see [16]). Here, we consider a water tank with water depth 5 m and gravitational acceleration 9.8 m/s^2. Taking $a = 0.2$ m and $\delta = 0.5$ m gives the initial amplitude of the trichromatic signal at 0.8 m, while setting $\omega_{2,lab} = 0.5$ rad/s, $\nu = 0.02$ rad/s yields wave numbers $k_{1,lab}^{(0)} = 7.00$ rad/m, $k_{2,lab}^{(0)} = 6.71$ rad/m, $k_{3,lab}^{(0)} = 6.44$ rad/m. We ignore the contribution of the phase shift, thus $\psi_1 = \psi_2 = \psi_3 = 0$.

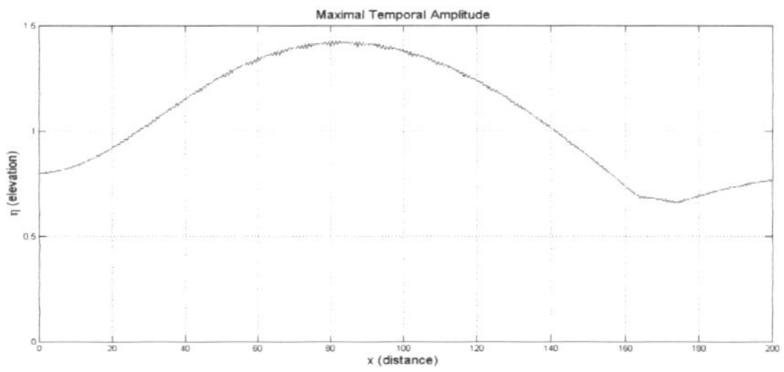

Figure 1. MTA of trichromatic wave signal for $a_{lab} = 0.2$ m, $\delta = 0.5$, $\omega_{2,lab} = 0.5$ rad/s, $\nu = 0.02$ rad/s.

Figure 1 exhibits the maximum amplitude of the signal along the spatial axis. The wave is initially at the amplitude 0.8 m at $x = 0$ comes through an amplitude increase until it reaches the maximum position at $x = 81.5$ m with maximum amplitude $\eta_{max} = 1.42$ m. Thus, the wave is amplified as much as 1.78 times its initial amplitude. The signal deformation at some positions is presented in figure 2. Figure 2(a) is the trichromatic signal at $x = 0$. New groups of waves appear at $x = 40$ m (figure 2(b)), modulate the wave and cause the peaking on the amplitude. The signal at maximum position is shown in figure 2(c). After reaching the maximum peaking, the wave amplitude starts to decrease as presented in figure 2(d). The peaking phenomenon is affected by the appearance of new side band waves during the propagation of the wave. The spectra of the waves forming the propagating wave are illustrated in figure 3. It can be observed that at the initial position $x = 0$, there is trichromatic wave form and during the propagation there grows new side bands modulating the trichromatic signal.

Some parameters affect the maximum position and amplitude amplification factor (AAF). AAF is calculated by comparing the highest value of MTA with the value of MTA at the initial position [14]. Table 1 presents the values of maximum elevation and position of the wave for some values of the monochromatic wave's frequencies $\omega_{2,lab}$ and the frequency differences ν. The results indicate that higher frequency yields higher amplitude amplification while extending

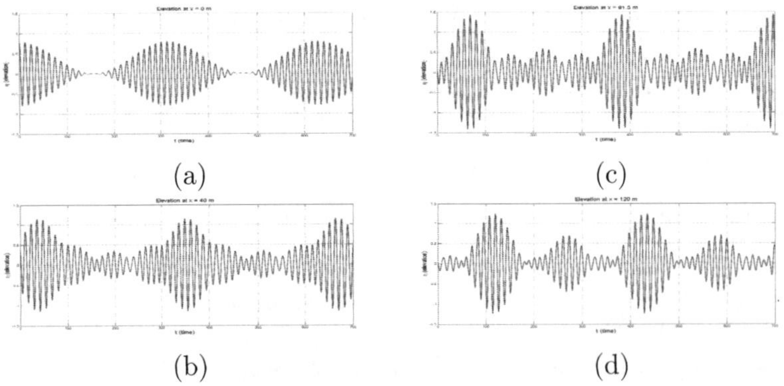

Figure 2. Wave signals on time domain at (a) $x = 0\,\mathrm{m}$ (b) $x = 40\,\mathrm{m}$ (c) $x = 81.5\,\mathrm{m}$ (d) $x = 120\,\mathrm{m}$ for parameter values $a_{lab} = 0.2\,\mathrm{m}, \delta = 0.5, \omega_{2,lab} = 0.5\,\mathrm{rad/s}, \nu = 0.02\,\mathrm{rad/s}$.

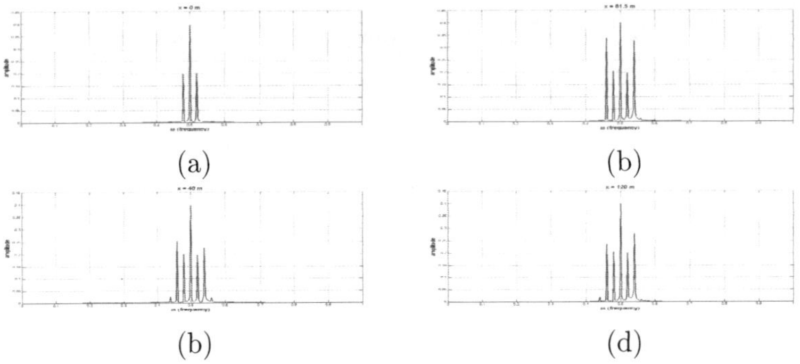

Figure 3. Amplitude spectra of the wave signals at (a) $x = 0\,\mathrm{m}$ (b) $x = 40\,\mathrm{m}$ (c) $x = 81.5\,\mathrm{m}$ (d) $x = 120\,\mathrm{m}$ for parameter values $a_{lab} = 0.2\,\mathrm{m}, \delta = 0.5, \omega_{2,lab} = 0.5\,\mathrm{rad/s}, \nu = 0.02\,\mathrm{rad/s}$.

the distance of the maximum peaking from the initial position. On the contrary, the higher frequency difference will decrease the amplification but makes the maximum peaking position closer. Meanwhile, the change in amplitude a will also influence the amplification and maximum position. While the higher initial amplitude gives higher amplification, it shortens the distance of wave traveling to reach the maximum peaking position. These results is presented in table 2.

Table 1. Maximum position and AAF for some values of frequencies and frequency differences with $a_{lab} = 0.2\,\mathrm{m}, \delta = 0.5$.

$\omega_{2,lab}$(rad/s)	ν(rad/s)	x_{max}(m)	η_{max}(m)	AAF
0.4	0.02	55.5	1.25	1.56
0.4	0.03	31	1.00	1.24
0.45	0.02	69.5	1.34	1.68
0.45	0.03	42	1.06	1.33
0.5	0.02	81.5	1.42	1.78
0.5	0.03	57	1.13	1.41

Table 2. Maximum position and AAF for some values of amplitudes a_{lab} with $\delta = 0.5, \omega_{2,lab} = 0.5\,\text{rad/s}, \nu = 0.02\,\text{rad/s}$.

a_{lab}(m)	x_{max}(m)	η_{max}(m)	AAF
0.16	108.5	1.00	1.57
0.18	94.5	1.21	1.68
0.23	66.5	1.79	1.94
0.25	59	2.05	2.05

4. Conclusion

We have derived a model of nonlinear wave propagation which is initially in trichromatic wave form based on BBM model. The wave deforms during its propagation and experiences a peaking and splitting phenomena. The amplitude amplification can be observed through the MTA. The higher wave frequency for the initial signal will produce higher AAF but farther maximum peaking position. Conversely, the closer frequency difference among the supporting wave of the trichromatic signal decreases the amplitude amplification but makes the maximum position nearer to the initial position. Meanwhile, the higher initial amplitude gives higher AAF and closer maximum MTA.

Acknowledgments

The authors would like to thank the anonymous referees for their valuable suggestions which led to the improvement of this article. This research is funded by Professor Research Grant and H-Index Research Grant, Syiah Kuala University, 2017.

References

[1] Dean R G 1990 *Freak Waves: A Possible Explanation, Water Wave Kinetics* (Amsterdam: Kluwer) pp. 609-621
[2] Akhmediev N, Soto-Crespo J M and Ankiewicz A 2009 *Phys. Lett. A* **373** 2137-2145
[3] Kharif C, Pelinovsky E and Slunyaev A 2009 *Rogue Waves in the Ocean* (Berlin: Springer)
[4] Onorato M, Residori S, Bortolozzo U, Montinad A and Arecchi F T 2013 *Phys. Rep.* **528** 47-89
[5] Akhmediev N N and Ankiewicz A 1997 *Solitons: Nonlinear Pulses and Beams* (London: Chapman and Hall)
[6] El-Labany S K, Moslem W M, El-Bedwehy N A, Sabry R and Abd El-Razek H N 2012 *Astrophys. Space Sci.* **338** 3-8
[7] Bacha M, Boukhalfa S and Tribeche M 2012 *Astrophys. Space Sci.* **341** 591-595
[8] Rahman A U and Ali S 2014 *Astrophys. Space Sci.* **351** 165-172
[9] El-Wakil S A, Abulwafa E M, Elhanbaly A and El-Shewy E K 2014 *Astrophys. Space Sci.* **353** 501-506
[10] Marwan and Andonowati 2003 *J. Math. Sci.* **8** 81-87
[11] Westhuis J, van Groesen E and Huijsmans R H M 2001 *J. Waterway Port Coast. Ocean Eng.* **127** 334-342
[12] Stansberg C 1997 *Proc. WAVES '97* **2** 1227-1241
[13] Marwan 2010 *Tamkang J. Math.* **41 (1)** 51-69
[14] Ramli M, Munzir S, Khairuman T and Halfiani V 2014 *Far East J. Math Sci.* **90 (1)** 97-117
[15] Ramli M 2016 *IAENG Int. J. App. Math.* **46 (1)** 29-34
[16] Halfiani V and Ramli M 2017 *J. Phys. : Conf. Ser.* **890** 012011
[17] Ramli M 2015 *Far East J. Math. Sci.* **97 (8)** 925-937
[18] Ramli M 2009 *Int. J. Engineer.* **22 (3)** 243-249
[19] Karjanto N, van Groesen E, Peterson P 2002 *J. Indones. Math. Soc.* **8 (4)** 39-47
[20] Slunyaev A, Sergeeva A and Pelinovsky E 2015 *Phys. D* **303** 18-27
[21] Boyd J P and Guan Y C 2001 *Math. Comput. Sim.* **55 (4-6)** 317-328
[22] Bruckner M C, Düll W P and Schneider G 2014 *J. Math. Anal. Appl.* **414** 166-175
[23] Schneider G 2011 *Advances in Mathematical Physics* Article ID 854719 4 pages
[24] Halfiani V, Salmawaty and Ramli M 2017 *Far East J. Math Sci.* **102 (6)** 1077-1090
[25] Benjamin T B, Bona J L and Mahony J J 1972 *Philos. Trans. Roy. Soc. London, Ser. A* **272** 47-78

IOP Conf. Series: Materials Science and Engineering **300** (2018) 012044 doi:10.1088/1757-899X/300/1/012044

Analysis Local Convergence of Gauss-Newton Method

Rahmi Wahidah Siregar[1], Tulus[2] and Marwan Ramli[3]
Department of Mathematics, Universitas Sumatera Utara, Medan, Indonesia

E-mail: [1]rahmisiregar51@yahoo.com, [2]tulus_jp@yahoo.com and [3]marwan301@yahoo.com

Abstract. The Gauss-Newton method is a very efficient, simple method used to solve nonlinear least-squares problems. This can be seen as a modification of the newton method to find the minimum value of a function. In solving nonlinear problems, the Gauss Newton Algorithm is used to minimize the sum of quadratic function values, which in its completion does not require the calculation or estimate of the derivatives of the two functions f (x) hence numerically more efficient with direct or iterative processes. The Gauss Newton method studied in this study is restricted to functions of one or two variables. The results of Gauss Newton's method analysis consisted of convergence at simple roots and multiple roots. Newton's method often converges quickly, especially when the iteration begins to be close enough to the desired root. However, if iteration begins far from the searched root, this method can be missed without warning. Implementation of this method usually detects and overcomes the convergence failures.

1. Introduction

The development of an increasingly advanced world makes the problem more complex, no exception problems involving mathematical problems. One of the problems that is often encountered in mathematics and science and engineering is to find the root of the equation, ie values that satisfy f (x) = 0 (Proinov, 2009). Numerical methods are techniques in which mathematical problems are formulated in such a way that they can be solved by arithmetic operations. Numerical methods are used to solve problems where analytic calculations can not be used. This numerical method is presented in the form of algorithms that can be calculated quickly and easily. Given that the algorithm developed in the numerical method is the Gauss Newton algorithm then in the algorithm it will appear the term iteration is the repetition of the calculation process.
View the least-squares nonlinear problem:

$$\min_{x \varepsilon \Omega} \|F(x)\|^{2} \qquad (1)$$

Where $\Omega \subseteq \mathbb{X}$ is the open set and $F \rightarrow Y$ is a non-linearly deferred function, with X and Y being real or complex. The problem arises when the data is good, when $\mathbb{X} = \mathbb{R}^n\ dan\ \mathbb{Y} = \mathbb{R}^m$, where m is the number of observations and n is the number of parameters that Dennis and Schnabel (1996) and Nocedal (1999) have indicated.

$F'(x)$ is denoted by a derivative of F at point $x \in \Omega$. When $F'(x)$ is an invertive derivative of the problem of stationary point found from the problem (1.1) that is the solution of the nonlinear equation $F'(x) * F(x) = 0$, Where A^* denotes the adjoint of operator A, Find the least squares solution of nonlinear equations

$$F(x) = 0 \qquad (2)$$

This problem has been extensively studied by Dedieu and Kim (2002) and Dedieu and Shub (2000) for functional analysis and in the Riemannian context by Adler et al. (2002).

Where $F^{'}(x)$ is injektif and has a closed image for all $\in \Omega$, Gauss Newton method finds the stationary point of the above problem, described as follows where point $x_0 \in \Omega$ defines

$$x_{k+1} = x_k - [F^{'}(x_k) * F^{'}(x_k)]^{-1} F^{'}(x_k) * F(x_k), \quad k = 0,1,...$$

If the above method converges to $x_* \in \Omega$, then x_* is the stationary point of the problem (1), but we can not conclude that x_* the solution of the problem (1) or $F(x_*) = 0$. To ensure that the stationary point x_* is the solution of the problem (1), we must apply the optimal conditions. It should be pointed out that if $F^{'}(x)$ can be reversed for all $x \in \Omega$, then Gauss Newton's method becomes Newton's method. Literatures discussing the convergence of Newton and Gauss Newton methods include Adler et al. (2002), Alvarez et al. (2008), Argyros and Hilout (2010), Chen (2008), Proinon (2010).

It is also known that the Gauss Newton method may fail or even fail to be properly defined by Dennis (1996) and Dedieu (2000). To ensure that the method is well defined and convergence to the stationary point of the problem some conditions to be worn such as the classical convergence analysis shown by Dennis (1996) and Nocedal (1999) require that Lipschitz comply that the initial iteration is close enough for the solution but can not make us see clearly how big the convergence of spherical radius for analytic functions, Dedineu and Shub (2000) have given an approximation of the convergence radius and criteria for the convergence of the Newtonian Gauss method. One of the obstacles in using Gauss Newton's method is to compute the value of the function derivative. This is not always easy if done manually, especially for certain functions, even if calculations are done by calculator or computer. Therefore it is necessary to find the appropriate software to implement the newton method which does not require manual derivative calculation, matlab can be used for this purpose. The Gauss-Newton algorithm is a very efficient, simple method used to solve nonlinear least-squares problems (Cox et al., 2004). This can be seen as a modification of the newton method to find the minimum value of a function. In solving non-linear problems, the Gauss Newton Algorithm is used to minimize the sum of quadratic function values, which in its completion does not require the calculation or estimate of the derivatives of the two functions f (x) hence numerically more efficient with direct or iterative processes. The Gauss Newton method studied in this study is restricted to functions of one or two variables. The analysis of Gauss Newton's method involves convergence of simple roots and multiple roots. Newton's method often converges quickly, especially when the iteration begins to be close enough to the desired root. However, if iteration begins far from the searched root, this method can be missed without warning. Implementation of this method usually detects and overcomes the convergence failures.

The speed of convergence is basically a value that measures how quickly the solution of iterated results leads to the actual solution. Let $f(x) = 0$ be the equation to be solved with the exact solution x^*dan $x_k(k = 1,2,...)$ is the numerical solution of the ke-k iteration result. Difference $|x_k - x^*|$ Measure how far the numerical solution of the actual solution is to the k-iteration or usually referred to as the error (in the k-iteration) and of course the desired property is $|x_{k+1} - x^*| < |x_k - x^*|$. This study will analyze the local convergence of the Gauss Newton method and solve some examples of the application of the Gauss Newton method.

2. Literature Review

2.1 Non-linear least-squares problems

In mathematical terms, optimization usually involves maximizing or minimizing. In the least nonlinear least squares problem is a nonlinear quadratic minimization problem. This is called the least squares because it minimizes the sum of squares of the function. This type of problem occurs just as the model functions into data: if $\phi(x;t)$ represents the function model with t as the independent variable, then every $r_j\,(x) = \phi(x;t_j) - y_j$, where $d(t_j, y_j)$ is the set of data points. Two common algorithms to solve the quadratic problem are the Gauss-Newton method and the Levenberg Marquardt Algorithm.

The least squares problem is to minimize the difference between a set of data and model functions that approach the data. Considering a set of data $d(t_j, y_j)$ and the function model $\phi(x; t_j)$, we obtain the function difference with the equation $r_j(x) = \phi(x; t_j) - y_j$ where y_j is the y component at the data point on t_j. Therefore the purpose function f (x) is the sum of the squares of the nonlinear functions:

$$f(x) = \frac{1}{2} \sum_{j=1}^{m} r_j^2(x) \quad (3)$$

This equation includes the sum of all r_j components of the vector r the remaining components of m given by the vector

$$r(x) = (r_1(x), r_2(x), \dots, r_m(x))^T$$

When calculating the gradient of f (x), it is necessary to find the residual gradient vector. Jacobian J (x) is the matrix of all $\nabla r_j(x)$:

$$J(x) = \left[\frac{\partial r_j}{\partial x_i}\right] j = 1, \dots, m; i = 1, \dots n = \begin{bmatrix} \nabla r_1(x)^T \\ \nabla r_2(x)^T \\ \vdots \\ \nabla r_m(x)^T \end{bmatrix}$$

Gradient and Hessian can be expressed in terms of Jacobian:

$$\nabla f(x) = \sum_{j=1}^{m} r_j(x) \nabla r_j(x) = J(x)^T r(x)$$

$$\nabla^2 f(x) = \sum_{j=1}^{m} r_j(x) \nabla r_j(x)^T + \sum_{j=1}^{m} r_j(x) \nabla^2 r_j(x)$$

$$= J(x)^T J(x) + \sum_{j=1}^{m} r_j(x) \nabla^2 r_j(x)$$

The Gauss-Newton method is a very efficient, simple method used to solve nonlinear least-squares problems (Cox et al., 2004). This can be seen as a modification of the newton method to find the minimum value of a function. In solving non-linear problems, the Gauss Newton Algorithm is used to minimize the sum of quadratic function values, which in its completion does not require the calculation or estimate of the derivatives of the two functions f (x) hence numerically more efficient with direct or iterative processes. The Gauss Newton method studied in this study is restricted to functions of one or two variables. The analysis of Gauss Newton's method involves convergence of simple roots and multiple roots. Newton's method often converges quickly, especially when the iteration begins to be close enough to the desired root. However, if iteration begins far from the searched root, this method can be missed without warning. Implementation of this method usually detects and overcomes the convergence failures.

It is also known that the Gauss Newton method may fail or even fail to be properly defined by Dennis (1996) and Dedieu (2000). To ensure that this method is well defined and convergence to the stationary point of the problem of some conditions to be worn eg a classical convergence analysis is

indicated by see Dennis (1996) and Nocedal (1999) necessitates meeting F 'Lipschiatz that the initial iteration is close enough for the solution But it can not make us see clearly how big the convergence of the spherical radius for analytic functions, Dedineu and Shub (2000) have provided an approximation of the convergence radius and criteria for the convergence of the Newtonian Gauss method.
Several literatures discussing convergence are presented by Jinhai Chen (2006) discussing the convergence of the newtonian gauss method by solving the nonlinear least squares problem, Li (2011) discusses the theory of convex processes with the convergence of repetitive sequences generated by the Gauss-Newton method for problems Convex inclusions, Ferreira and Oliveira (2011) discuss about the local convergence of the proximal Gauss-Newton method to solve nonlinear least squares problems in Hilbert space settings.

2.2 Newton Method

The newton method is one of the open methods for determining root solutions of nonlinear equations or determining one of the roots of a nonlinear equation with f (x) = 0, with the following key principles :

1. This method approaches the curve f (x) with the tangent line (gradient) at a starting point.
2. The next assessment value is the point of intersection between the gradient of the curve with the x axis.

Let's say : $f(x) = e^{x} - 5x^2 = 0$, with $x_0 = 1$

Table 1. Iterations of the Newton Method for $\boldsymbol{f(x) = e^{x} - 5x^2 = 0}$

iterasi	x(k)	f(x)	f'(x)
0	1	-2,2822	-7,2812
1	0,6866	-0,3701	-4,8791
2	0,6107	-0,0228	-4,2653
3	0,6053	0,0003	-4,2212
4	0,6051	0	-4,2210

2.3 Gauss Newton Method

The Gauss-Newton method is a very efficient, simple method used to solve nonlinear least-squares problems (Cox et al., 2004). This can be seen as a modification of the newton method to find the minimum value of a function. In solving non-linear problems, the Gauss Newton Algorithm is used to minimize the sum of quadratic function values, which in its completion does not require the calculation or estimate of the derivatives of the two functions f (x) hence numerically more efficient with direct or iterative processes.

Gauss Newton's Algorithm

Step 0. Take $x_0 \in \mathbb{R}^n, k := 0$.
Step 1: If $\|g_k\| \leq \epsilon$, stop.
Step 2: Perform until convergent

Step 2.1: Complete $(J(x_k)^T J(x_k) s_k = -J^T(x_k) f(x_k)$, for s_k.

Step 2.2: Set $x_{k+1} = x_k + s_k, k := k + 1$. kembali ke langkah 1. Go back to step 1.

Description: that at each iteration, step 2.1 of this method is equivalent to solving the linearisation of least squares problem.

$$\min_{s} \frac{1}{2} ||J(x_k)s + f(x_k)||_2^2$$

The Gauss-Newton method can also be written as a fixed point iteration as follows

$$x_{k+1} = G(x_k)$$

Where $G(x) \equiv x - J^{+}(x)f(x)$ and $J(x)^{+} \equiv (J^{T}(x)J(x))^{-1}J^{T}(x)$ is defined in Moore -Penrose Pseudo-Invers on J (x). In solving non-linear problems, the Gauss-Newton Algorithm is a modification of the newton method to find the minimum value of a function. Gauss-Newton is important for rapid convergence close to the solution, but like Newton's method, efficiency depends on having an accurate early guess.

2.4 Convergence of the Gauss-Newton Method

An efficient condition for the convergence of the Gauss-Newton method is known in this case where the normal equations for linearisation of least squares problems in (3.16) are completed in step 2.1 at each iteration. Given some of the results. The following assumptions are made in order to build the theory:

A.1 There is $x^{*} \in \mathbb{R}^{n}$ such that $J^{T}(x^{*})f(x^{*}) = 0$

A.2 Jacobian $J(x^{*})$ matrix at x^{*} full rank n.

With the notation ρ (A) to show the spectral radius of the matrix A at n x n, and defined:

$$\varrho = \rho((J(x^{*})^{T}J(x^{*}))^{-1}Q(x^{*}))$$

Here's the theorem on the local convergence of the Gauss-Newton method.

Theorem 1: Suppose assuming A.1 and A.2 if $\varrho < 1$, so that iteration of gauss newton converges locally to x^{*} for that ε > 0 such that $\{x_k\}$ the generation of gauss newton algorithms converges at x^{*} for all $x_0 \in D \equiv \{x| ||x - x^{*}||_2 < \varepsilon\}$.

Proof: Theorem 1 has a geometric interpretation as described in previous research by Wedin (1974) and Ake Bjorck (1996) in his research showing that the S surface in $\mathbb{R}^{m}$ is given by the parametric representation $y = f(x), x \in \mathbb{R}^{n}$, and let M be the point at S with the coordinates $f(x^{*})$, by taking O as the origin of the coornidic system. The OM vector is orthogonal to the plane of contact with the surface S to M.

Theorem 2: In the study of Wedin (1974) suppose assume theorem 1 and that $f(x^{*})$ is not zero, so

$$\varrho = ||f(x^{*})||_2\, \chi$$

Where χ is the maximum maximum curvature of surface S at point M corresponding to the normal direction $w^{*} = f(x^{*})/||f(x^{*})||_2$.

In residuals with a zero value, where $f(x^{*}) = 0$, the relation in equation continues. In this case the origin of O lies on the surface of S and χ denotes the maximum principle curvature S toward normal to the tangent surface at O. When we have $Q(x^{*}) = 0$ dan karena $\varrho = 0$, the result is still valid.

For Gauss Newton's method to converge, therefore, it is sufficient to maximize the principle of curvature χ of the surface S at the point $f(x^{*})$ to satisfy $\frac{1}{\chi} > ||f(x^{*})||_2$. This condition applies if and only if $\nabla^{2}\phi(x^{*})$ is positive definite at $x^{*}$and ensures that $x^{*}$is a local minimum of the objective function φ indicated by Ake Bjorck (1996). The relation in equation 3.19 shows that the convergence condition in theorem 3.3 is invariant in the transformation of the problem of nonlinear least squares by local diffeomorphism because of the sum $||f(x^{*})||_2\, \chi$ Properties in the study by Wedin (1974). The evidence of this result depends on the iterative theory for the iterative process with continuous stationary point iterations shown by Ortega and Rheinboldt (1970) studies. This theory ensures local convergence at a linear level. However, in Dennis and Schanbel's (1983) study, the condition is more rigorous for local convergence. Conditions that provide a higher degree of order of convergence can be inferred from this theory.

3. Conclusion

The Gauss-Newton method is a highly efficient iterative method used to solve nonlinear least-squares problems. This can be seen as a modification of the newton method to find the minimum value of a function. The Gauss-Newton method studied in this study is restricted to functions of one or two variables. The analysis of Gauss Newton's method involves convergence of simple roots and multiple roots. Gauss Newton's method often converges quickly, especially when the iteration begins to be close enough to the desired root. But if iteration begins far from the searched root, this method can slip off without warning, its convergence may be slow, it probably will not converge at all. Implementation of this method usually detects and overcomes the convergence failures.

References:

[1]. O.P. Ferreira, Local convergence of Newton's method in Banach space from the viewpoint of the majorant principle, IMA J. Numer. Anal. 29 (3) (2009) 746–759.

[2]. R.L. Adler, J.P. Dedieu, J.Y. Margulies, M. Martens, M. Shub, Newton's method on Riemannian manifolds and a geometric model for the human spine, IMA J. Numer. Anal. 22 (3) (2002) 359–390.

[3]. F. Alvarez, J. Bolte, J. Munier, A unifying local convergence result for Newton's method in Riemannian manifolds, Found. Comput. Math. 8 (2) (2008) 197–226.

[4]. I.K. Argyros, S. Hilout, Improved generalized differentiability conditions for Newton-like methods, J. Complexity 22 (3) (2010) 316–333.

[5]. L. Blum, F. Cucker, M. Shub, S. Smale, Complexity and Real Computation, Springer-Verlag, New York, 1998. With a foreword by Richard M. Karp.

[6]. Chapra, Canale (1996). Metode Numerik Jilid 1. . PT Gelora Aksara Pratama. Erlangga.

[7]. J. Chen, The convergence analysis of inexact Gauss–Newton methods for nonlinear problems, Comput. Optim. Appl. 40 (1) (2008) 97–118.

[8]. J. Chen, W. Li, Convergence of Gauss–Newton's method and uniqueness of the solution, Appl. Math. Comput. 170 (1) (2005) 686–705.

[9]. J. Chen, W. Li, Convergence behaviour of inexact Newton methods under weak Lipschitz condition, J. Comput. Appl. Math. 191 (1) (2006) 143–164.

IOP Conf. Series: Materials Science and Engineering **300** (2018) 012045 doi:10.1088/1757-899X/300/1/012045

Optimal control for Malaria disease through vaccination

Said Munzir[1], Muhammad Nasir[2] and Marwan Ramli[1]

[1] Department of Mathematics, Syiah Kuala University, Banda Aceh 23111, Indonesia
[2] Master's Program of Mathematics, Syiah Kuala University, Banda Aceh 23111, Indonesia

E-mail: smunzir2001@yahoo.com, sir_math69@yahoo.com, corresponding author : ramlimarwan@gmail.com; marwan.math@unsyiah.ac.id

Abstract. Malaria is a disease caused by an amoeba (single-celled animal) type of plasmodium where anopheles mosquito serves as the carrier. This study examines the optimal control problem of malaria disease spread based on Aron and May (1982) SIR type models and seeks the optimal solution by minimizing the prevention of the spreading of malaria by vaccine. The aim is to investigate optimal control strategies on preventing the spread of malaria by vaccination. The problem in this research is solved using analytical approach. The analytical method uses the Pontryagin Minimum Principle with the symbolic help of MATLAB software to obtain optimal control result and to analyse the spread of malaria with vaccination control.

1. Introduction

Malaria is a disease caused by an amoeba (single-celled animal) type of plasmodium where anopheles mosquito serves as the carrier. There are four malaria parasites which can infect human body, *plasmodium falciparum*, *plasmodium vivax*, *plasmodium ovale*, and *plasmodium malariae*. Malaria disease at one place can be different form another place, depending on the existence of the infected people, the malaria vector, the parasites, and the environmental factors [1]. Malaria disease is a contagious disease which generally suffered by people living in tropical or subtropical area. The disease may also affects people who live in the swamp area or unhealthy environment. Fever, cold, and cold sweat may appear as the symptoms. The malaria symptoms typically begin a few days after the infected mosquito bite the patient. The infection period depends on some factors such as infection rate, method and time of treatment, and the immune due to drugs, and other biological factors.

The spread of malaria disease which can be out of control requires an effort to study the spread pattern of the disease. Mathematical model can be applied to help predict and control the epidemic in the future. The malaria transmission can be approximated by mathematical model and simulated by applying the real problem case. The result is expected to be beneficial to resolve the malaria spread. Epidemic model of malaria in a population can consists of three groups of subpopulation, which are susceptible, infected, and recovered groups. Susceptible is a group of healthy people with a possibility to be infected by malaria disease. Infected is a group of infected people with a possibility to be susceptible again. Recovered represents a group of people who is recovered from the malaria disease and become immune to the disease. This system is known as SIR type epidemic model [2, 3].

The spread of the disease can be controlled through administering vaccination. Vaccination is conducted by giving an antigen compound which functions to improve the body's immune to viruses or disease. To determine the best strategy in administering vaccine to a population of malaria epidemic, the optimal control theory can be applied. The objective of this control through vaccination is to minimize the number of infected individuals as well as the cost [4].

This study concerns on investigating the malaria disease spread by employing mathematical model proposed by Aron and May [2]. Here, the model will be further developed by adding a control to the recovered population. This control aims to decrease the number of infected population through administering vaccine while minimizing the amount and cost of the vaccination. The model will be solve analytically with the symbolic help of MATLAB software.

2. Optimal Control Model Formulation

The dynamic system which will be discussed here is formulated in the system of ordinary differential equations consisting of three dependent state variables with respect to time, $x(t)$, $y(t)$, and $z(t)$. Here, $x(t)$ represents the number of people in susceptible group, $y(t)$ represents the number of people in infected group, and $z(t)$ represents the number of people in recovered group. The recovered population is assume to developed a partial immune as the result of their body's reaction when they were infected by the disease and recovered. The diagram of the population transfer among these group is presented in figure 1. In figure 1, h and ρ respectively

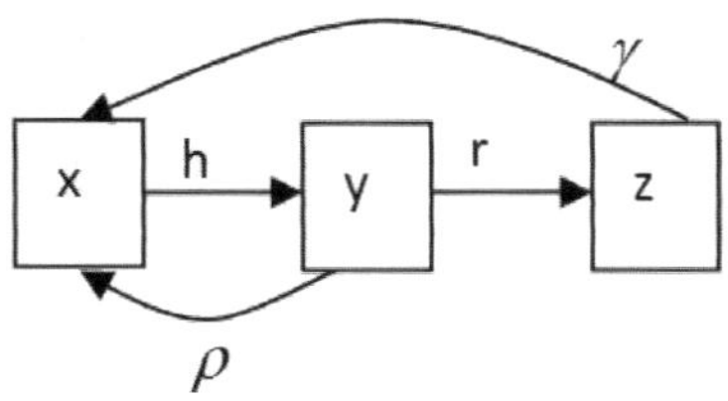

Figure 1. Malaria transmission diagram [2]

denote the transfer rates from susceptible to infected and from infected to susceptible. The transfer rates from infected to partially immune and from partially immune to the susceptible are respectively denoted by r and γ.

Ronald [5] suggested that the immune is achieved and developed over time, then there is the rate of immune development $r(t)$ with its initial value is defined as $r(0) = 0$. When a person is exposed to the disease, there is a delay for a person to achieve an immune. The immune tends to asymptotically decrease to a certain value r_∞. Therefore, the rate of change of the immune system is presented as follow

$$\frac{dr}{dt} = \sigma t(r_\infty - r), \quad r(0) = 0. \tag{1}$$

If σ denotes twice of the rate, the equation (1) becomes $\frac{dr}{dt} = 2\sigma t(r_\infty - r)$. Hence, we have

$$r(t) = r_\infty(1 - e^{-\sigma t^2}).$$

Also, we define a control variable representing the amount of vaccine given to the recovered population. Hence, the SIR model with the addition of control variable $u(t)$ is given as

$$\begin{aligned}\frac{dx}{dt} &= -hx + \rho y + \gamma z - ux,\\ \frac{dy}{dt} &= hx - \rho y - r_\infty(1 - e^{-\sigma t^2})y,\\ \frac{dz}{dt} &= r_\infty(1 - e^{-\sigma t^2})y - \gamma(h)z + ux.\end{aligned} \tag{2}$$

The parameter h represents the transfer rate from susceptible to infected, ρ is the transfer rate from infected to susceptible, γ is the transfer rate from recovered to susceptible, r_{inf} is the transfer rate from infected to recovered, and u is the amount of vaccine which is being administered where $u = \{u(t) : 0 \leq u \leq 1\}$.

Furthermore, we define an objective equation which minimizes the vaccination cost:

$$\min J(u) = \int_{t_0}^{t_f} y(t)^2 + u(t)^2\, dt, \tag{3}$$

where t_0 is the initial time and t_f is the final time. The optimal control system will be solved to determine an optimal value of $u(t)$ such that $J(u(t)) = \min\{J(u(t) : u \in U\}$ where $U = \{u(t) : 0 \leq u \leq 1, 0 \leq t \leq t_f\}$.

3. Results and Simulations

3.1. Analytical solution and simulation of malaria disease spread

The system (2) and (3) will be solved analytically. Here, we assign the parameters' values $h = 1.99, \rho = 0.074$, and $\gamma = 0.05$. Forming the Hamiltonian and applying the Pontryagin Minimum Principle yields the following optimal state and costate equations,

$$\begin{aligned}\frac{dx}{dt} &= 0.074y - 1.99x + 0.05z - 1.0\lambda_1 x^2 + \lambda_3 x^2,\\ \frac{dy}{dt} &= 1.99x - 0.187y,\\ \frac{dz}{dt} &= 0.113y - 0.05z + \lambda_1 x^2 - 1.0\lambda_3 x^2,\\ \frac{d\lambda_1}{dt} &= \lambda_1(x(\lambda_1 - 1.0\lambda_3) + \lambda_1 x - 1.0\lambda_3 x + 1.99) - 1.99\lambda_2 - 1.0(\lambda_1 - 1.0 * \lambda_3)(\lambda_1 x - 1.0\lambda_3 x) -\\ &\quad 1.0\lambda_3(x(\lambda_1 - 1.0\lambda_3) + \lambda_1 x - 1.0\lambda_3 x),\\ \frac{d\lambda_2}{dt} &= 0.187\lambda_2 - 0.074\lambda_1 - 0.113\lambda_3 - 1.0,\\ \frac{d\lambda_3}{dt} &= 0.05\lambda_3 - 0.05\lambda_1.\end{aligned} \tag{4}$$

Based on the equation (4), the optimal control solution will be determined to investigate the optimal strategy to control the transmission of the malaria disease.

Figure 2 displays that the number of the susceptible population decreases while the number of infected and recovered population increases over time. The optimal control also increases over time. The optimal control strategy obtained from the Hamiltonian's result for the population in the system is $x = 0.0001$, $y = 0.6958$ dan $z = 0.3041$. Therefore, the strategy solution for optimally control the system is by giving the amount of vaccination $u = 3173$ to the population in the system.

IOP Conf. Series: Materials Science and Engineering **300** (2018) 012045 doi:10.1088/1757-899X/300/1/012045

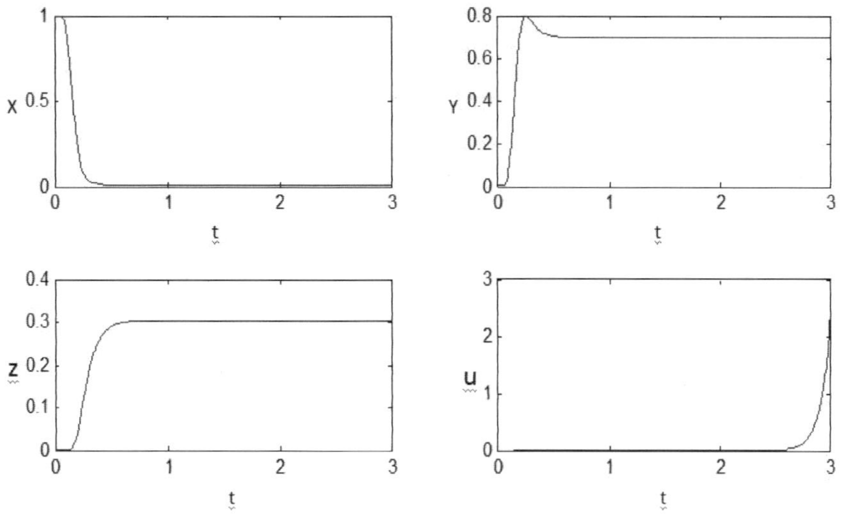

Figure 2. The graph of state and control variables for $h = 1.99, \rho = 0.074, \gamma = 0.05$.

3.2. Analitical solution and simulation for modified objective function with penalty
In this section, we modified the objective function by adding penalties to the infected and control variable in the performance index. The objective function with penalties can be written as

$$\min J(u) = \int_{t_0}^{t_f} \eta y(t)^2 + \frac{\alpha}{2} u(t)^2 \, dt, \tag{5}$$

where η is the infected state penalty and α is the vaccine administering control penalty. Here, the penalties' values should be chosen in the interval $(0, 1)$.

Some combinations of penalties' values are chosen to find the values which yield the best solution for the optimal control system. Figure 3 presents the optimal solution for $\eta = 0.9$ and $\alpha = 0.2$.

The rate of change of the population obtained from the model with penalties indicate a change with respect to time. Figure 3 displays that there are increase and decrease in the number of population over time in each group. The optimal control strategy obtained from this simulation is $x = 0.1109$, $y = 0,6864$, $z = 0.3136$. Therefore, the strategy solution for optimally control the system is by giving the amount of vaccination $u = 1198$.

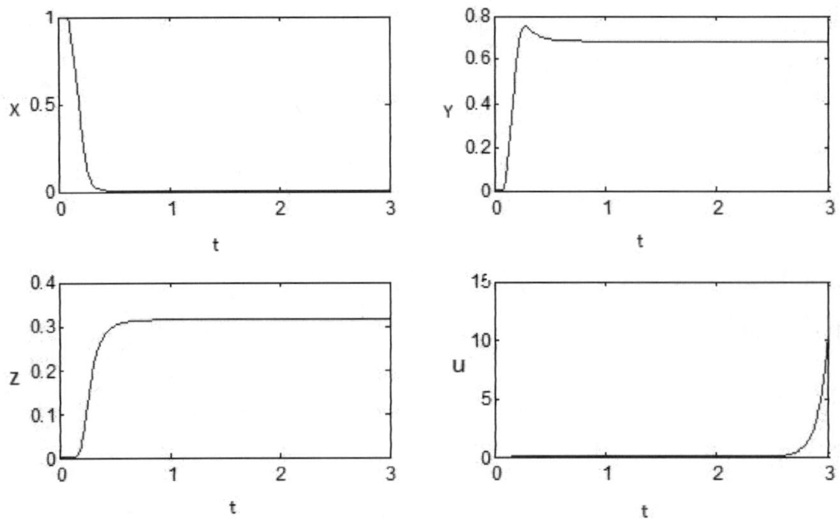

Figure 3. The graph of state and control variables for $h = 1.99, \rho = 0.074, \gamma = 0.05$ and penaties' values $\eta = 0.9, \alpha = 0.2$.

4. Conclusion
The transmission of malaria disease can be investigated through a mathematical model. To control the disease spread in a population, vaccine can be administered to population. We administered the vaccination variable to the recovered population and applied optimal control theory to minimize the amount of vaccine being administered and the cost. The addition of penalties to the objective function gives better value to the outcome. In this case the system does not only produces smaller amount of the vaccination but also gives the optimal solution for the system. The results of the simulations suggest that control strategy through the optimal control system with penalty is more effective than the one without penalty.

Acknowledgments
The authors thank the anonymous referees for their valuable suggestions which led to the improvement of the article. This research is funded by Laboratory Grant, Syiah Kuala University, 2017.

References
[1] Sudjadi B and Laila S 2006 *Biology Sciences in Life* (Jakarta: Yudistira)
[2] Aron J L and May R M 1982 *The Population Dynamics of Infectious Disease* (London: Chapman and Hall)
[3] Koela J C and Antia R 2003 Epidemiological Model for the Spread of Anti-Malarial Resistance *Malaria J.* **2 (3)**
[4] Elhia M, Balatif O, Bouyagghroumni J, Labriji E and Rachik M 2012 Optimal control applied to the spread of influenza A (HINI) *Appl. Math. Sci.* **6** 4057-65
[5] Yeargers E K, Herod J V, Shonkweiler R W. 1996 *An Introduction to the Mathematics of Biology: with Computer Algebra Models* (New York: Springer)

IOP Conf. Series: Materials Science and Engineering **300** (2018) 012046 doi:10.1088/1757-899X/300/1/012046

Dynamic Models Increase Understanding of Geometry Through Proof

Sari Adeliza[1], Tulus[2], Marwan Ramli[3]
[1,2,3]Departement of Mathematics, Universitas Sumatera Utara, Indonesia
[1]sariadeliza09@gmail.com, [2]tulus_jp@yahoo.com, [3]marwan301@yahoo.com

Abstract. This research paper is intended to explain and development dynamic models in enhancing the understanding of geometry in understanding the concepts that relates to geometry.This research is done by making steps or structure of workmanship that facilitate students in understanding and working on problems on geometry through proof. Students perform four proofs namely two-column proof, paragraph proof, proof trees, flow proof. The study on relevant literature search and the previous research results show theoretically that dynamic models are highly relevant models for understanding geometric concepts through proof. So that students can understand and have the ability to prove.

1. Introduction

Mathematical proof is a convincing demonstration of the formula, the theorem is true, with the aid of logic and mathematics. Mathematical proof is one of the aspects that must be considered in the mathematics learning in schools recommended from NCTM (2000). The few or many experiences of students in preparing a first junior high school proof will have an impact on the ability to prove when attending college at the first level, as stated by Moore (1994) that one of the reasons why students encounter difficulties in proof is the experience in constructing Evidence is limited to proof of school geometry. Accordingly, based on the results of a study conducted by Sabri (2003) on the concept of mathematical proof of prospective teachers it is suggested that the junior secondary school curriculum should prepare better students in mathematical proof learning. This suggests that the concept of proof of junior high students is very weak.

Based on the above description, the researcher wants to improve the understanding of the students' geometry through mathematical proof and proofing by students with dynamic model. So as to improve the competence of students' ability in solving problems on geometry material through proof.

Dynamic model is an abstraction and simplification of a complex system, but attempted to represent the system properly. Furthermore, based on the dynamic model obtained, simulations of policy scenarios are based on logically developed assumptions (Sterman, 2000). Dynamic models describe the behavior of the system over time.

This study aims to develop a learning model that can overcome difficulties and improve the thinking stage of junior secondary students. So as to improve the competence of the ability of junior high students to develop optimally and to increase students' interest to familiarize themselves with using proof in solving geometry problems with dynamic models.

2. Understanding Geometry

These are all the required expertise required for all other branches of mathematics, including science.

2.1 Troubleshooting Steps in Mathematics

According to Poyla (1973) there are 4 steps that need to be implemented to solve the math problem that is:

1). Understand and deepen the problem, 2). Finding a strategy, 3). Using strategies to solve problems and 4). Review it again and reflect on the solution or initial state.

There is no chance of solving a problem if we cannot understand it first. This process requires not only the introduction of what to look for but also the different types of important information combined together to get answers.

2.2 Dynamic Models

Dynamic model is an abstraction and simplification of a complex system, but attempted to represent the system properly. Furthermore, based on the dynamic model obtained, simulation of policy scenarios based on logically developed assumptions (Sterman, 2000). The use of system dynamics methodology is more emphasized to the objectives of enhancing the notion of how the behavior of the system arises from its structure over time.

3. Research Methods

This research has a purpose to know how dynamic model of increasing understanding of geometry through proof hence this research is literature research in field of education about mathematics especially related to dynamic model, improvement of geometry understanding and proof. Relevant research findings will be presented in the discussion section. The steps taken in this research are

a. Identify dynamic models of increasing understanding of geometry through verification.
b. Demonstrate dynamic models in increasing understanding of geometry through proof.

3.1 Proof Representations that Support Developing and Writing Proofs

The purpose of this section is to highlight four different ways that proofs can be represented in geometry and discuss how these various representations have the potential to facilitate proving.

As pointed out by Anderson (1983), successful attempts at proof generation can be divided into two major episodes " an episode in which a student attempts to find a plan for the proof and an episode in which the student translates that plan into an actual proof". The proof forms that we highlight include proof tree, two-column proof, flow proof, and paragraph proof.

3.1.1 Two-Column Proof

A two-column proof lists the numbered statements in the left column and a reason for each statement in the right column (Larson, 2001). The two-column form requires that students record the claims that make up their argument (in the statements column) as well as their justification for these claims (in the reasons column). However, the consecutively numbered steps of the proof may lead students to believe that the deductive process is more linear than it actually is.

3.1.2 Paragragh Proof

This form is more conversational than the other proof forms (Larson, 2001). Paragraph proofs are more like ordinary writing and can be less intimidating (EDC,2009). For this reason, they look more like an explanation than a structured mathematical device (EDC,2009). However the lack of structure could also be a detriment. In particular, some teachers have concluded that the paragraph form was not appropriate for high school students because students tended to leave out the reasons that justified their statement. As a result, students would often come to invalid conclusions (Cirillo,2008).

3.1.3 Proof Trees

The proof tree is an outline for action, where the action is *writing* the proof. Anderson (1983) described the proof tree as follows :

> The students must either try to search forward from the givens trying to find some set of paths that converge satisfactorily on the statement to be proven, or must try to search backward from the statement to be proven, trying to find some set of dependencies that lead back to the givens (p.194)

The proof tree could also be a useful tool to scaffold the work of determining what the given premises are or what conclusion can be proved.

3.1.4 Flow Proof

A flow proof uses the same statements and reasons as a two-column proof, but the logical flow connecting the statements is indicated by Arrows (Larson, 2001) and separated into different "branches". The flow proof helps students to brainstorm , working through the most difficult parts of solving a proof : (1) understanding the working information-analyzing the given and the diagram- and (2) knowing what additional information is needed to solve the proof-analyzing what is being proved (Brandell, 1994). A disadvantage to this proof form might be that students are not required to supply reasons that justify their statements in the way that the "Reason" column of the two-column proof forces them to do.

3. Discussion of research results

Nine different problems (presented in no particular order) that illustrate how students may be provided with opportunities to expand their role in the process of proving. Here's the discussion of 9 student problems in proving:

1. In problem 1, the student is provided with a conjecture (the diagonals of a rectangle are congruen) and a corresponding diagram and asked to write the "Given" and the "Prove" statements.
2. In Problems 2, the student is provided with the "Given" and the "Prove" statements but is asked to draw the diagram.
3. In Problem 3, when provided with a particular theorem, the student is asked to do all three of these tasks (write the "Given", the "Proven" and draw the diagram).
4. Problem 4 is similar to the first three in that students are invited to determine the "Given", but this time they also provided with the statement to be proved as well as the proof of that statement. Problem 4 is similar to the "fill in" type proofs that we have seen in some textbooks (Serra : 2008) except that rather than having students fill in the statements or reasons, they are filling in the premises.
5. Next, in Problem 5, students are asked to draw a conclusion or determine what could be proved when provided with particular "Given" conditions and a corresponding diagram.
6. In Problem 6, students are asked to determine what auxiliary line might be drawn in order to construct the proof that two angles are congruent. This is not a common problem posed to students because, typically, teachers either construct the auxiliary lines for their students or a hint is provided in the textbook that helps students determine where this line should be drawn (Herbs & Brach : 2006).
7. Problem 7 is unique in the sense that the student is asked what could be proved, but the givens are ambiguous. Leaving the problem more open-ended affords students opportunities to write conjectures. It is expected that the student will consider two different cases corresponding to whether the quadrilateral is concave or convex. In both cases the student could argue that the remaining pair of sides are congruent to each other.
8. Finally, in Problems 8 and 9, students have the opportunity to take part in analyzing proofs. In Problem 8, a paragraph proof is provided, and students are asked to find the error.
9. Problem 9, students are provided with a proof and asked to determine what theorem was proved.

In this section, we proposed nine problems that illustrate how teachers could increase their student's involment in proving by having them make reasoned mathematical conjectures, use conjectures to set up a proof, and evaluate mathematical proofs by looking for errors and determining what was proved. In the next, we address the issue of supporting students in proving by commenting on multiple proof representations.

4. Conducting Proof

IOP Conf. Series: Materials Science and Engineering **300** (2018) 012046 doi:10.1088/1757-899X/300/1/012046

In the study of mathematics, especially about the learning of geometry where in solving the problem one on mathematics.

The four steps above are recognized by different names: recognize, plan, do, check again (*see, plan, do, check*). In this way the student will have a more comprehensive understanding of the student's answers to the problem and make the student more confident in the truth.

The following is an example of a question in which the student performs a proof with the empirical experience of the researcher on some students.

1. Problem
   Show two right triangles are congruent if the hypotenuse and a leg of one are congruent to the corresponding parts of the other. Given : Right ΔABC with are right angle at C and Right ΔDEF with are right angle at F.
2. How is the problem solved?
   For this matter you can also use a theory that you know. For this problem is the use of congruence in the triangle. You can check with an image size. This means that you first need to think about it.
3. Purpose of achievement
   After understanding the problem then draw the triangle in accordance with what is known in the matter, then see the properties of triangle angles and the size of each side of both triangles.
4. Settlement process
   Plan for problem solving to verify and Learning Outcomes is students are able to solve problems and verify mathematical problems
5. Stages of Teaching
   - Remind again how to solve problems in the first mathematics and ask students to write something they know when they encounter a problem (draw two triangles by naming each corner from the known in the matter, then look at the size on each side and angle On both triangles). And develop a plan that can be used or that have been d sorted to solve the problem.
   - Run a plan and discuss what they think is more useful and appropriate in solving the problem.
   - Discussing the skills and abilities they have to figure out an unknown size or explanation. This might help students mention what they already know and get from their settlement.
   - While students solve problems ask and ask students to explain their reasoning. If students are unable to provide further answers. Then you may want to suggest more directly by using the nature of congruence and give solution.
6. Expansion of the problem
   Can you find a side that has the same length?
   If a side is the same length can you determine that each corner also has the same magnitude on the two triangles?
7. Solution
   On this issue everything has to be done by proving that the two triangles are congruent. Then there are four ways to prove that the two triangles are congruent by using dynamic model stages.

To Prove : $\Delta ABC \cong \Delta DEF$.

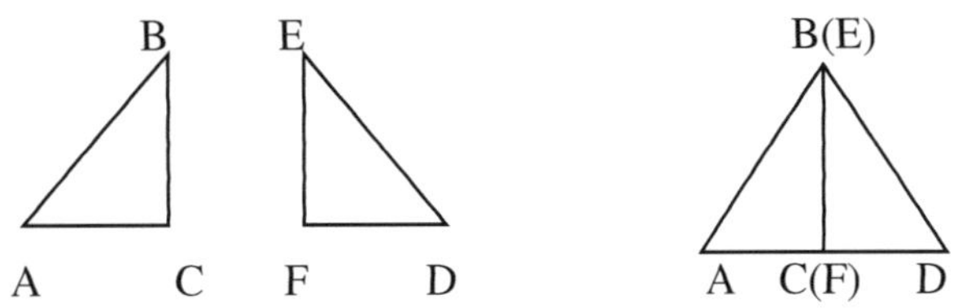

Figure 1. Two Triangles

Plan : Move the two given triangles together so that $\overline{BC}$ coincides with $\overline{EF}$, forming an isosceles triangle. The given triangles are proved congruent by using Theorem 1 and SAA.

A. A two-column proof lists the numbered statements in the left column and a reason for each statement in the right column.

Table 1. Two-Column Proof

Statements	Reasons
1. $\overline{BC} \cong \overline{EF}$	1. Given
2. Move triangles ABC and DEF together so that $\overline{BC}$ coincides with $\overline{EF}$, A and D are on opposite sides of $\overline{BC}$.	2. A geometric figure may be moved without changing its size or shape. Equal lines may be made to coincide.
3. $\angle C$ and $\angle F$ are right angles	3. Given
4. $\angle ACD$ is a straight angle	4. The whole equals the sum of its parts
5. $\overline{AD}$ is a straight line segment.	5. The sides of a straight angle lie in a straight line.
6. $\overline{AB} \cong \overline{DE}$	6. Given
7. $\angle A \cong \angle D$	7. If two sides of a triangle are congruent the angles opposite these sides are congruent.
8. $\Delta ADC \cong \Delta BCD$	8. SAA.

B. A paragraph proof describes the logical argument with sentences. It is more conversational than a two-column proof.

Since ΔABC dan ΔDEF is two isosceles triangles with $\overline{AB} \cong \overline{DE}$ and $\overline{BC} \cong \overline{EF}$ Then move ΔABC and ΔDEF so that $\overline{BC}$ coincides with $\overline{EF}$, A and D are on opposite sides of $\overline{BC}$. If $\angle C$ and $\angle F$ are right angles then $\angle ACD$ is a straight angle and $\overline{AD}$ is a straight line segment. If two sides of a triangle are congruent then $\angle A \cong \angle D$ the angles opposite these sides are congruent. Thus, $\angle ADC \cong \angle BCD$. Because SAA, Thus $\Delta ABC \cong \Delta DEF$.

C. A Proof Tree

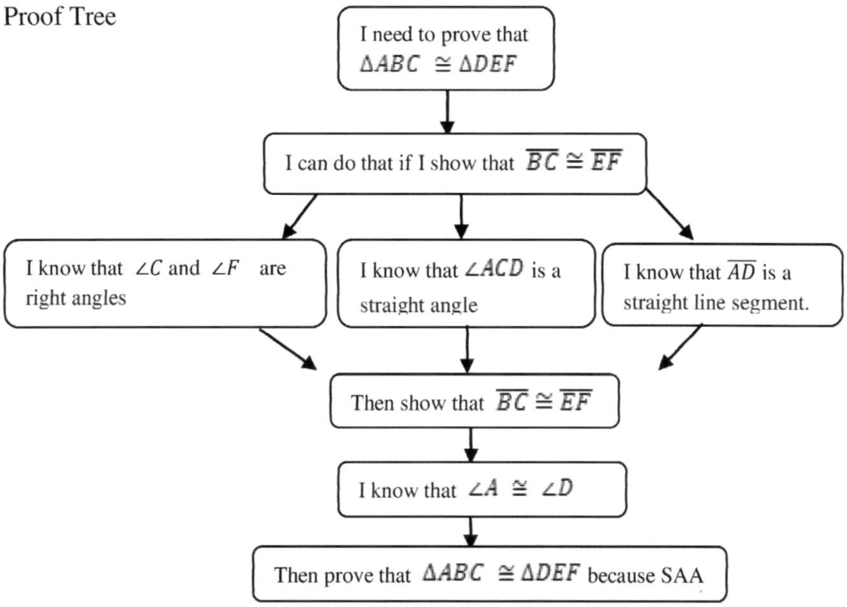

Figure 2. A proof tree

A proof tree is an outline or plan of action that specifies a set of geometric rules that allows students to get from the givens of the problem through intermediate levels of statemen, to the to-be-proven statement.

D. A Flow Proof

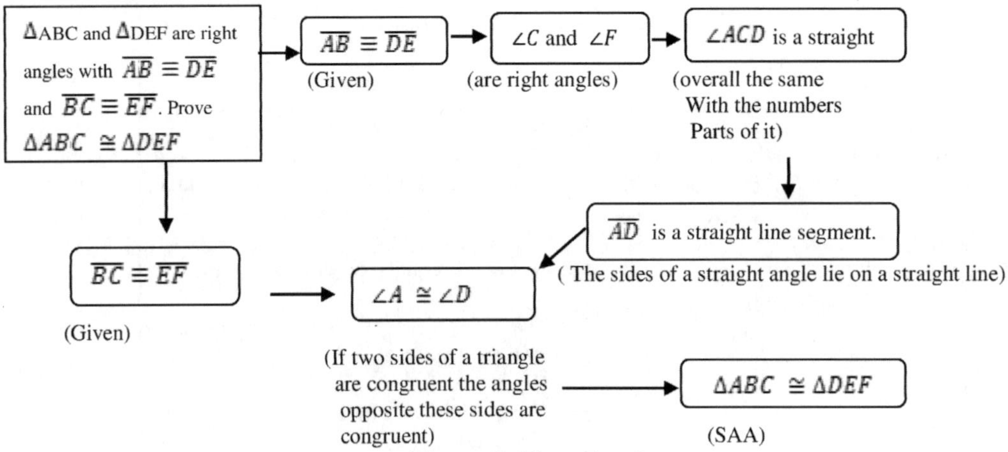

Figure 3. Flow Proof

A flow proof uses the same statements and reasons as a two-column proof, but the logical flow connecting the statements is indicated by arrows. Depending on whether it is the plan or the proof itself, students may or may not choose to write the reasons beneath the statements.

5. Conclusion

After the researchers conducted a discussion of the data obtained from the results of the study, the researchers took the conclusion, namely:

1. Based on the empirical experience undertaken on the students, each of the evidences has a contribution to influence the increase in understanding of geometry in students. Of the four proofs performed with the dynamic model indicates that two-column proof is more understood by the students. Seen from the results some students do after learning the problem solving on geometry through the four proofs.
2. Teachers are fully aware of the purpose of this, teachers should pose problems to the students in order to give a greater role in proving. Teachers introduce problems for students to write down reasons, write statements to be proven, as well as ideas on diagrams. Students should be provided with opportunities to make excuses and evaluate mathematical and authentic opinions. Finally, teachers advance and provide different types of reasoning and methods on proof. Teaching the exercises so that students engage in solving problems, guessing, writing conditional statements to prove and then explanations and prove their allegations can assign students more opportunities to engage in mathematics.

References

[1] Anderson, J. R. (1983). Acquisition of proof skills in geometry. In R, S. Michalski, J. G. Carbonell, & T. M. Mitchell (Eds.), *Machine learning: Anarticificial intelligence approach* (pp. 191−219). Palo alto, CA: Tioga Publishing.

[2] Brandell, J. L. (1994). Helping students write paragraph proofs in geometry, *The Mathematics Teacher*,87, 498−502.

[3] Cirillo, Michelle and Herbst, G.Patricio.(2012). *Moving Toward More Authentic Proof Practices in Geometry* Vol. 21, No. 2, 11–33. The Mathematics Educator

[4] National Council of Teachers of Mathematics (NCTM). (2000). *Principles and Standards for SchoolMathematics.*

IOP Conf. Series: Materials Science and Engineering **300** (2018) 012046 doi:10.1088/1757-899X/300/1/012046

Reston, VA: NCTM.
[5] Hanna, G. (1983). *Rigorous Proof in Mathematics Education* . OISE Press,xsToronto.
[6] Herbst, P. G. (2002). Establishing a custom of proving in American school geometry: Evolution of the two-column proof in the early twentieth century. *Educational Studies in Mathematics, 49,*283—312.
[7] Herbst, P. G., & Brach, C. (2006). Proving and doing proofs in high school geometry classes: What is it that is going on for students? *Cognition and instruction,24,* 73—122.
[8] Kilpatrick, J., Swafford, J., & Findell, B (Eds). (2001). *Adding it up: Helping children learn mathematics.* Washington, DC : National academy Press.
[9] Lakatos, I. (1976). *Proofs and refutations.* New York, NY: Cambridge University Press.
[10] Lampert, M. (1992). Practices and problems in teaching authentic mathematics. In F. K.Oser,A.
[11] Larson, R., Boswell, L., & Stiff, L. (2001). *Geometry.* Boston, MA: McDougal Littell.
[12] Lewis, H. (1978). *Geometry: A contemporary course.* New York, NY: Random House/McCormickMathers.
[13] Polya, G. (1981). *Mathematical discovery:On understanding, learning, and teaching problem solving (combined Ed).* New York, NY: Wiley.
[14] Rich, B., Thomas, C. (2001). *Geometry.* Department of Mathematics Brooklyn Technical High School, New York.
[15] Stylianides, A. J. (2007). Proof and proving in school mathematics. *Journal for Research in Mathematics Education,* 38, 289—321.

Mathematical model of snake-type multi-directional wave generation

Muarif[1], Vera Halfiani[2], Siti Rusdiana[3], Said Munzir[2,3] and Marwan Ramli[2,3]

[1]Mathematics Graduate Study Program, Syiah Kuala University, Banda Aceh 23111, Indonesia
[2]Dynamic Application and Optimization Groups, Syiah Kuala University, Banda Aceh 23111, Indonesia
[3]Department of Mathematics, Syiah Kuala University, Banda Aceh 23111, Indonesia

Email: muarif89@gmail.com, verahalfiani@yahoo.com, ninanizarli@yahoo.com, smunzir2001@yahoo.com, corresponding author : ramlimarwan@gmail.com; marwan.math@unsyiah.ac.id

Abstract. Research on extreme wave generation is one intensive research on water wave study because the fact that the occurrence of this wave in the ocean can cause serious damage to the ships and offshore structures. One method to be used to generate the wave is self-correcting. This method controls the signal on the wavemakers in a wave tank. Some studies also consider the nonlinear wave generation in a wave tank by using numerical approach. Study on wave generation is essential in the effectiveness and efficiency of offshore structure model testing before it can be operated in the ocean. Generally, there are two types of wavemakers implemented in the hydrodynamic laboratory, piston-type and flap-type. The flap-type is preferred to conduct a testing to a ship in deep water. Single flap wavemaker has been explained in many studies yet snake-type wavemaker (has more than one flap) is still a case needed to be examined. Hence, the formulation in controlling the wavemaker need to be precisely analyzed such that the given input can generate the desired wave in the space-limited wave tank. By applying the same analogy and methodhology as the previous study, this article represents multi-directional wave generation by implementing snake-type wavemakers.

1. Introduction

In ocean, an extreme wave can occur unpredictibly, but its effect can cause serious damage to the ships and offshore structures (see [1–8]). Therefore any information about waves is very important for the design of ships and offshore structures. This information can be accuired by studying wave generation. The wave can be generated by a wavemaker. The wavemaker has been used a hundred years in a wave tank for ship testing [9]. It is really important to do testing to a model in constructing any maritime structures [10]. A wave tank in a hydrodynamic laboratory is a facility where maritime structures and ships can be tested on a model scale.

Study on extreme wave itself has been actively conducted. Marwan [11] studied the deformation of propagating wave which is initially bichromatic signal based on the third order solution of Korteweg-de Vries (KdV) equation. The wave is reported experiencing amplitude amplification. The extention of the study was presented in [12] where the fifth order solution of KdV with bichromatic signal gave higher amplification. Ramli [13] applied the same signal to the Boussinesq equation which yielded even higher amplitude amplification compared to the both previous results. An extreme wave generation also studied on the basis of Benjamin-Bona-Mahony (BBM) equation in [14] and the results suggested that the amplitude amplification is affected by by the initial amplitude and frequency difference between the supports of the bichromatic signal. Signal in trichromatic form also trigger an extreme wave. This phenomenon stated in [15]. Extreme wave can also be studied via the wave group envelope. It has been studied that the wave groups of some nonlinear wave equations satisfying Nonlinear Schrodinger equation, such as KdV [16, 17] and BBM [18].

Signaling problem is an important factor in generating wave. A signal is given to the wavemaker such that the desired wave profile will appear in the wave tank. The wave tank has a wavemaker at one side and a wave absorber that act as the beach at the other side [19]. According to O'Dea and Newman [9], wavemakers installed in testing tanks today are typically either hinged flaps (rotating about a pivot point below the still water level), or translating pistons. Furthermore they said that hinged flaps are more commonly found in deep-water tanks (water depth large compared to typical wave lengths), while pistons are more commonly found in shallow tanks used to study coastal waves, Tsunamis, etc.

Generally,there are two types of wavemaker which are widely used in hydrodynamic laboratories, piston and flap type wavemakers [19], see figure 1. Beside that there is also snake type wavemaker. This wavemaker has many part that can move like a flap or piston. Single flap wavemaker has been explained in many studies (see [19–21]). In this study, we will consider specifically the snake type of wavemaker (three dimensionals) which moves like a flap with extended boundary condition to the system of the water motion in the tank, see figure 2.

Figure 1. Wavemaker type : (a) Flap, (b) Piston [9].

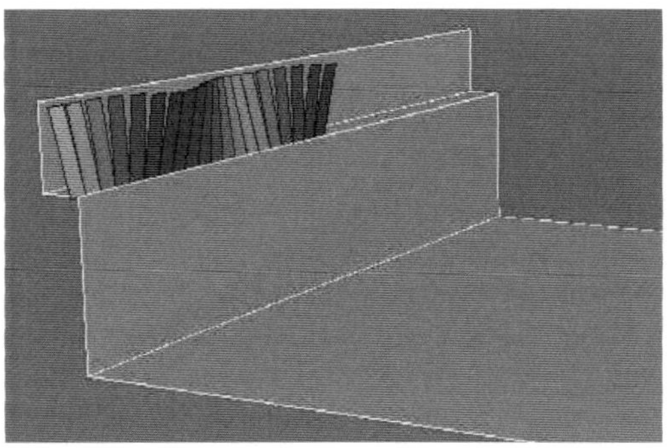

Figure 2. Snake type wavemaker [9].

2. Wavemaker Formulation

Assume that the wavemaker in this case looks like figure 3.

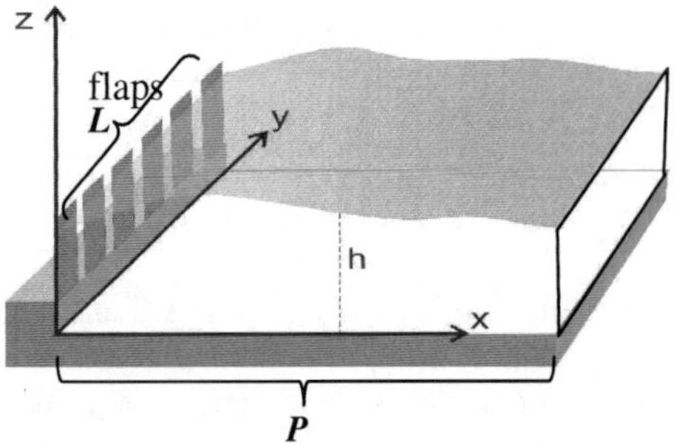

Figure 3. Scheme of snake type wavemaker

In this case, we consider the governing equation for the fluid motion under a periodic water wave is the Laplace equation in linier form for a single flap.

$$\frac{\partial^2 X}{\partial x^2}+\frac{\partial^2 Y}{\partial y^2}+\frac{\partial^2 Z}{\partial z^2}=0-h\leq z\ \leq\eta(x,t),0\leq y\ \leq L,0\leq x\ \leq P. \tag{1}$$

Dean and Dalrymple [13] stated that the boundary conditions for this case are:

- Bottom boundary condition for horizontal bottom (no flow condition/assumed no leak),

$$-\frac{\partial\phi}{\partial z}=0 \text{ at } z=-h. \tag{2}$$

- Dynamic free surface boundary condition,

$$\eta-\frac{1}{g}\frac{\partial\phi}{\partial t}=0, \text{ at } z=0. \tag{3}$$

- Kinematic free surface boundary condition,

$$-\frac{\partial\phi}{\partial z}=\frac{\partial\eta}{\partial t}, \text{ at } z=0. \tag{4}$$

- Lateral boundary condition at the wavemaker,

$$U(z)\cos(\lambda y-\sigma t)=-\frac{\partial\phi}{\partial x}, \text{ at } x=0.$$

where ϕ is the velocity potensial of the propagating wave,η is the elevation above the still water level, x, y and z are spaces dimension, t is time dimension, h is the water depth, λ is a wave number and g is gravitational acceleration.

Now in this study we will consider one more condition. Assumed that, there is no leak at the wall of the wave tank and neglect reflective wave of the wall, we can write:

- Wall boundary conditions

$$\frac{\partial\phi}{\partial y}=0 \text{ at } y=L\frac{\partial\phi}{\partial x}=0 \text{ at } x=P, \tag{5}$$

while L and P represents the wide and the lenght of the wave tank respectively. Assumed that the solution to equation (1) take form,

$$\phi(x,y,z,t) = X(x).Y(y).Z(z).T(t), \tag{6}$$

where $X(x)$ is some function that depends only on x, $Y(y)$ depends only on y, $Z(z)$ depends only on z, and $T(t)$ varies only with time.

Using separation variable methode we got,

$$Z = Ae^{-k_p z} + Be^{k_p z}, \tag{7}$$
$$Y = Ee^{i\lambda y} + Fe^{-i\lambda y}, \tag{8}$$
$$X = Ce^{-i\sqrt{k_p^2-\lambda^2}x} + De^{i\sqrt{k_p^2-\lambda^2}x}, \tag{9}$$

where A, B, C, D, E, F, k_p, and λ are constants. Now substitute equations (7), (8), and (9) to equation (6) we get,

$$\phi = \left(Ae^{-k_p z} + Be^{k_p z}\right)\left(Ce^{-i\sqrt{k_p^2-\lambda^2}x} + De^{i\sqrt{k_p^2-\lambda^2}x}\right)\left(Ee^{i\lambda y} + Fe^{-i\lambda y}\right)T(t), \tag{10}$$

ϕ must be periodic in time by the lateral boundary conditions, so $T(t)$ specify as $T(t) = \sin\sigma t$. Equation (10) can be written as

$$\phi = \left(Ae^{-k_p z} + Be^{k_p z}\right)\left(Ce^{-i\sqrt{k_p^2-\lambda^2}x} + De^{i\sqrt{k_p^2-\lambda^2}x}\right)\left(Ee^{i\lambda y} + Fe^{-i\lambda y}\right)\sin\sigma t. \tag{11}$$

Now equation (11) by using the bottom boundary condition (equation (2)) takes form,

$$\phi = G\cosh k_p(h+z)\left(Ce^{-i\sqrt{k_p^2-\lambda^2}x} + De^{i\sqrt{k_p^2-\lambda^2}x}\right)\left(Ee^{i\lambda y} + Fe^{-i\lambda y}\right)\sin\sigma t,$$

with $G = 2Ae^{k_p h}$ a new constant. Now by applying the wall boundary conditions (equation (5)), ϕ takes final form,

$$\phi = A_p\cosh k_p(h+z)\cos\left(\sqrt{k_p^2-\lambda^2}(P-x)\right)\cos\lambda(y-L)\sin\sigma t. \tag{12}$$

Here we got a different equation from Dean and Dalrymple [13], because in this study we use the wall boundary conditions (equation (5)).

Elevation calculated by using dynamic free surface boundary condition (equation (3)),

$$\eta = \frac{1}{g}\frac{\partial\phi}{\partial t}\bigg|_{z=0},$$

$$\eta = \left[\frac{A_p\sigma}{g}\cosh k_p h\right]\cos\left(\sqrt{k_p^2-\lambda^2}(P-x)\right)\cos\lambda(y-L)\cos\sigma t.$$

IOP Conf. Series: Materials Science and Engineering **300** (2018) 012047 doi:10.1088/1757-899X/300/1/012047

Now by using kinematic free surface boundary condition (equation (4)) σ will be determined,

$$-\left.\frac{\partial \phi}{\partial z}\right|_{z=0} = \frac{\partial \eta}{\partial t},$$

$$-A_p\, k_p \sinh k_p h \cos\left(\sqrt{k_p^2-\lambda^2}(P-x)\right)\cos\lambda(y-L)\sin\sigma t$$
$$= -\frac{A_p\sigma^2}{g}\cosh k_p h \cos\left(\sqrt{k_p^2-\lambda^2}(P-x)\right)\cos\lambda(y-L)\sin\sigma t$$
$$\sigma^2 = g k_p \tanh k_p h.$$

For a complete solution, A_p need to be determined. This is evaluated by the lateral boundary condition at the wavemaker.

$$U(z)\cos(\lambda y-\sigma t) = -\left.\frac{\partial\phi}{\partial x}\right|_{x=0},$$

$$U(z)\cos(\lambda y-\sigma t) = -A_p\sqrt{k_p^2-\lambda^2}\cosh k_p(h+z)\sin\left(\sqrt{k_p^2-\lambda^2}P\right)\cos\lambda(y-L)\sin\sigma t. \quad (13)$$

To find A_p, equation (13) is multiplied by $\cosh k_p(h+z)$ and integrated from $-h$ to 0. Therefore we get,

$$A_p = -\frac{4k_p\cos(hy-\sigma t)\int_{-h}^{0}U(z)\cosh(k_p(h+z)dz}{\sqrt{k_p^2-\lambda^2}\left(\cosh(k_p h)-1-h\right)\sin\left(\sqrt{k_p^2-\lambda^2}\,p\right)\cos\lambda(y-L)\sin\sigma t}. \quad (14)$$

Now if we introduce θ as a directional angle made by the wave orthogonal to x axis (figure 4), where λ is the wave number in the y direction and $\sqrt{k_p^2-\lambda^2}$ is the wave number in the x direction. k_p represents the wave number in the propagation direction.

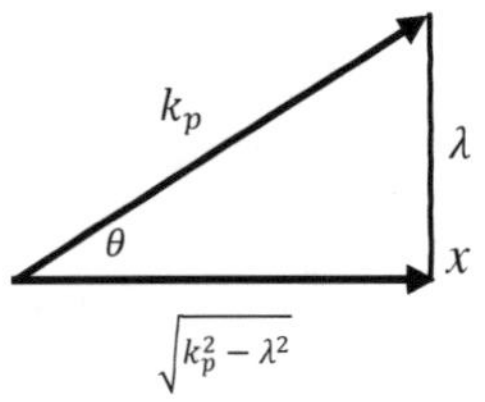

Figure 4. Definition for θ [13].

According to figure 4, $\sqrt{k_p^2-\lambda^2} = k_p\cos\theta$ and $\lambda = k_p\sin\theta$, substituted to equation (12), now velocity potential can be written,

$$\phi(x,y,z,t) = A_p\cosh k_p(h+z)\cos\left(k_p\cos\theta\,(P-x)\right)\cos k_p\sin\theta\,(y-L)\sin\sigma t, \quad (15)$$

where A_p is given by equation (14). Equation (15) shows the solution for a single flap wavemaker. Now due to linearity of the problem we can superimposed numerous single flap wavemaker motions and become a snake type wavemaker. So velocity potential can be written,

$$\sum_{1}^{n} \phi_n(x,y,z,t) = \sum_{1}^{n} A_p \cosh k_p(h+z) \cos\left(k_p \cos\theta\,(P-x)\right)$$
$$\cos\left(k_p \sin\theta\,(y-L)\right) \sin\sigma(t+\gamma_n),$$

with $n=1,2,3,...$ is the number of the flap and γ_n is phase of flaps. Equation above is proposed as model for snake-type multi-directional wave generation.

3. Conclusion

We have considered snake type wavemaker by linear theory approximation. It has been shown that the wave produced by snake type wavemaker can be considered as linear superposition of single flap wavemaker motion. Mathematical model of snake-type multi-directional wave generation can be determined reasonably well from linear wave theory. This study takes into account only the mathematical model of snake type wavemaker.

Acknowledgment

The authors thank the anonymous referees for their valuable suggestions which led to the improvement of the article. This research is funded by Professor and H-index Research Grant, Syiah Kuala University, 2017.

References

[1] Earle M D 1975 *J. Geophys* **80** 377-79
[2] Soares C G, Cherneva Z and Antao E M 2004 *Journal of Geophysical Research* **109** 7 pages
[3] Mori N, Liu P C and Yasuda T 2002 *Ocean Engineering* **29** 1399-414.
[4] Divinsky B V, Lopatikin L L, Pelinovsky E N and Slyungaev A V 2004 *Doklady Earth Science* **395** 438-43
[5] Trulsen K and Dysthe K 1997 *Proc. of the 21st Symposium on Naval Hydrodynamics* 550-558
[6] Smith R 1976 *J. Fluid Mech.* **77** 417-31
[7] Tooli E M and Gregersen B 2001 *The Open Ocean Engineering J.* **4** 24-33.
[8] Waseda T, Tamura H and Kinoshita T 2012 *J. Marine Sci. Tech.* **17** 305-314
[9] O'Dea J F and Newman J N 2007 Numerical Studies of Directional Wavemaker Performance *28th American Towing Tank Conference* (Michigan: Ann Arbor)
[10] Finnegan W and Goggins J 2012 *Ocean Engineering* **43** 23-31
[11] Marwan 2010 *Tamkang J. Math.* **41 (1)** 51-69
[12] Ramli M, Munzir S, Khairuman T and Halfiani V 2014 *Far East J. Math Sci.* **90 (1)** 97-117
[13] Ramli M 2016 *IAENG Int. J. App. Math.* **46 (1)** 29-34
[14] Halfiani V and Ramli M 2017 *J. Phys : Conf. Ser.* **890** 012011
[15] Ramli M 2015 *Far East J. Math Sci.* **97 (8)** 925-937
[16] Boyd J P and Guan Y C 2001 *Math. Comp. Sim.* **55 (4-6)** 317-28
[17] Bruckner M C, Dull W P and Schneider G 2014 *J. Math. Anal. Appl.* **414** 166-75
[18] Halfani V, Salmawaty and Ramli M 2017 *Far East J. Math Sci.* **102 (6)** 1077-1090
[19] Kusumawinahyu W M, Karjanto N and Klopman G 2006 *J. Indones. Math. Soc. (MIHMI)* **12 (1)** 41-57
[20] Halfiani V, Muarif and Ramli M 2015 *Bulletin of Mathematics* **7(1)** 51-63
[21] Dean R G and Dalrymple R A 1991 *Water Wave Mechanics for Engineers and Scientists, Volume 2 of Advanced Series on Ocean Engineering* (Singapore:World Scientific)

Description of Student's Metacognitive Ability in Understanding and Solving Mathematics Problem

Herlina Ahmad[1], Febryanti[2], Fatimah[3], Muthmainnah[4]

[1,2,3]Department of Mathematics Education, Universitas Al Asyariah Mandar, Polewali Mandar, Indonesia
[4]Department of Language Education, Universitas Al Asyariah Mandar, Polewali Mandar, Indonesia

E-mail: herlinaahmad39@gmail.com

Abstract. This research was conducted qualitative which was aim to describe metacognitive ability to understand and solve the problems of mathematics. The subject of the research was the first year students at computer and networking department of SMK Mega Link Majene. The sample was taken by purposive sampling technique. The data obtained used the research instrument based on the form of students achievements were collected by using test of student's achievement and interview guidance. The technique of collecting data researcher had observation to ascertain the model that used by teacher was teaching model of developing metacognitive. The technique of data analysis in this research was reduction data, presentation and conclusion. Based on the whole findings in this study it was shown that student's metacognitive ability generally not develops optimally. It was because of limited scope of the materials, and cognitive teaching strategy handled by verbal presentation and trained continuously in facing cognitive tasks, such as understanding and solving problem.

1. Introduction

One of the things that distinguish school mathematics with one as a science is the presentation of the object study. The Mathematics materials at school, the presentation of objects of mathematical study should not start with a definition or theorem, but should be adjusted to the intellectual development of students. Another thing that distinguishes school mathematics with mathematics as a science is its degree of abstraction. Unlike the object of direct study that becomes main target in learning mathematics, the object is not directly regarded as the impact of accompaniment, so it does not receive more serious attention from teachers and students. This is reflected in measurement of the achievement of mathematics learning results that have not involved these indirect objects. This is reinforced by [2] describes that the mathematical studies is one of the object of a difficulty of a teacher to teach mathematics school. Consequently, a teacher should try to "reduce" the abstract nature of mathematical objects (facts, concepts, operations or principles) to make it easier for students to grasp the subject matter of mathematics at school. In other words, mathematics teacher should teach mathematics concretely.

The development of cognitive psychology in this recent years, there are also ways to evaluate the achievement of learning outcomes, especially for the cognitive domain. One of the interesting efforts is the development of educational experts in revising "Bloom's Taxonomy" about the cognitive dimension. The domain of knowledge in the 2013 curriculum using [1] Bloom Bloomfield taxonomy revises Bloom's Taxonomy of cognitive aspects into two dimensions: (1) the dimensions of cognitive processes and (2) the dimensions of knowledge. The relationship between cognitive process and knowledge dimension of students' thinking development which is known by cognitive process dimension based on the formulation of Basic Competence of knowledge (KD-3) has relationship with the knowledge form (knowledge dimensions) [4]. Remembering process (C1), understanding (C2), and applying (C3). The analysis thinking development process (C4) to creativity one (C6) has a relationship with metacognitive knowledge form will be described in table 1.

Learning strategy (cognitive strategy) is one of the cognitive components that is closely related to the student's activity in capturing and understanding material. The [6] classifies the cognitive strategies as learning outcomes, so they need to teach and train to students. According to [2] Metacognition is concerned with students thinking about their own thinking and their ability to use certain learning strategies appropriately. The relation with this line to understanding an idea of metacognition concept as the process of "thinking about own thinking" in order to build a strategy to solve the problem.

The [7] explain that self-knowledge is an important component of metacognition. In [7] self-knowledge model includes the strengths and weaknesses of a person about cognitive and learning. For example, students who know that they generally work easier on multiple-choice tests than essay tests have self-knowledge to face the test. While [11] suggests the notion of metacognition as one of the thinking process about own thinking processes in order to build a strategy to solve the problem.

The theoretical descriptions above, supported by [8] finds that: (a) metacognition plays an important role in problem solving, (b) students are more skilled at solving problems if they have metacognitive knowledge, (c) teachers emphasize specific strategies for solving problems and pay less attention to important features of other problem-solving activities; The result finds more achievements at the secondary level in elementary school on the important of mathematical reasoning and problem-solving strategies.

The metacognitive ability is enough potential to improve the meaningfulness of students' learning mathematics in classroom, so one of the dimension aspects of interesting knowledge to study more deeply, both theoretically and empirically through the mathematics learning study is the metacognitive knowledge because it is the most complex and the highest aspect in the taxonomy and the metacognitive aspect is more related to the object of indirect study of mathematics which has been less attention from teachers and students. Based on the considerations above, the researchers are interested in analyzing and describing the students' metacognitive ability to understand the material and solve mathematics problems.

2. Models

The research was conducted field research, the researchers obtained by collecting data based on empirical experience in the field using qualitative research design. This research intended to conduct an investigation by describing the state of the object/subject of research at the present based on the facts [5].

The subject of the research was 20 students of the first year at computer and networking department at SMK Mega Link Majene.

The flow diagram in the study of students' metacognitive ability description to understand and solve mathematics problems can be seen in Figure 1.

The main instruments in this research were self-researcher and supporting instrument:

a. Test

The test used to obtain data about students' metacognitive abilities. Types of the tests in this study was a description test consisting of two types namely (1) test to understand the materials and (2) test to solve mathematics problems.

The metacognitive ability Indicators to understand the material were (a) Students can highlight important formulas from the matrix subject, (b) Students can make marginal notes on concepts and principles on the matrix subject, (c) Students can make a summary of the matrix discussion, (d) Students can create a concept mapping from the subject matter of the matrix. Besides that, The indicators of metacognitive ability to solve mathematical problems were (a) Students can use heuristic strategy, (b) Students can apply reverse thinking strategy, (c) Students can apply forward thinking procedure, (d) Students can apply inductive thinking procedure, (e) Students can apply deductive thinking procedure.

b. Interviews

The Interviews aim at obtaining information about the students' metacognitive ability to understand the material and solve mathematics problems. The participants were taken 9 participants as a sampling technique. The ability of students in the classroom was divided into 5 categories namely students with very high ability, high ability, middle level ability, low level ability and very low ability.

The validity of a data based on triangulation data. Triangulation data involves validity of the data by utilizing to compare others data [10]. In this research the triangulation data was obtained on students answers' analysis result.

Data analysis technique that used in this research was qualitative descriptive data analysis technique. [9] mentions that there are three steps in processing qualitative data such as data reduction, data presentation, and conclusion or verification.

3. Results

3.1 The students metacognition ability in understanding the material

Based on the analysis of metacognitive ability type in understanding the materials, the research finding 2 students who reached the category in very high ability category ($3{,}5 \leq x \leq 4$) based on the four indicators of understanding the material, it turned out the fourth indicator was making concept mapping of the matrix as subject matter was relatively more difficult to understand by the students, when compared to the other indicators that highlight the important formulas were make marginal notes about concepts and principles, and make material summaries. Here is an example of students work with high category. The following table analysis of student's metacognitive ability in understanding the material can be seen in table 2.

The student example of work (high category) can be seen in figure 3. Based on the students work at number 4 (make a map of the concept of matrix material you have learned) find one student is able to create concept maps correctly).

3.2 Metacognitive ability of students in solving problems

The result of metacognitive ability test in solving problems based on the indicators of retrograde thinking in solving mathematical problems were relatively more difficult, when compared to other strategies such as heuristics, forward thinking, inductive thinking, and deductive thinking. Here in example of student work in the high category. The following table analysis of student's metacognitive ability of students in solving problems can be seen in table 3. Example of student work (high category) can be seen in figure 3.2, in question number 2 students were asked to solve the problem by using a backward thinking strategy. The subject appeared at the first step was correct by answering the question from the beginning, but the subject made an error in operation 4 (-2) - 3 (-1) the result should be -8- (-3) = -8 + 3 = -5. The subjects did not solve the mathematics problems carefully.

4. Conclusion and discussion

The researcher collected data in this research based on the data analysis types of understanding test as well as in-depth interviews, the researcher findings one of the indicator is making concept mapping considered more complicated by the students, while for the test in solving mathematical problems type and in-depth interviews finding that the indicators using backward thinking were considered more complicated by the students.

From the overall findings in this research illustrated that in general metacognitive ability of students have not grown / developed optimally. This was because the scope of the material is limited, i.e. only in one subject and teaching cognitive strategy was not enough just by verbal delivery, but should be constantly trained in dealing with cognitive tasks, such as understanding and solving a problem, these findings were correlate to [6] explanation, he said that cognitive strategies (metacognitive) by which a person regulates his own behavior in attention, learning, remembering and thinking, is not studied in the once-so, but through improvement over a long period of time.

5. References

[1] Anderson, O.W. and Krathwohl, D.R. 2001. *A Taxonomy for Learning, Teaching, and Assessing (A Revision of Blom's Taxonomy of Educational Objectives)*. New York: Addision Wesley Longman, Inc.

[2] Barbara K. Hofer. 2004. Epistemological Understanding as a Metacognitive Process: Thinking Aloud During Online Searching. Educational Psychologist. Vol. 39 , Iss. 1, 2004, Pages 43-55

[3] Barker, L and Brown, A.L. 1984. *Metacognitive Skill and Reading*. Hand Book of Reading Research. 1 (353). P.v. 394.

[4] Bimbingan Teknis Implementasi Kurikulum 2013 SMK. 2017. *Analisis Dokumen SKL, KI-KD*. Direktorat Pembinaan Sekolah Menengah Kejuruan. Direktorat Jenderal Pendidikan Menengah. Kementerian Pendidikan Dan Kebudayaan.

[5] Bogdan. R.S. 1982. *Qualitative Research for Education an Introduction to Theory and Methods*. (Boston: Allyn and Bacon).

[6] Gagne, R.M. 1985. *The Condicions of Learning and Theory of Instruction.* (Holt: Rinehart and Winston)

[7] Flavell, J. 1979. *Theories of Learning in Educational Psychology: Metacognition Theory*.

[8] Keiichi, Shigematsu. 2000. *Metacognition in Mathematics Education*. Mathematics Education in Japan. Japan: JSME, Juli 2000.

[9] Milles, M.B dan Huberman, A.M. 2007. *Analisis Data Kualitatif. Buku Sumber Tentang Metode-metode Baru*. Jakarta: UI Press.

[10] Moleong, L.J. 2013. *Metodologi Penelitian Kualitatif (Edisi Revisi)*. Bandung: PT. Remaja Rosda Karya.

[11] O'Neil Jr, H.F. dan Brown, R.S. 1997. *Differensial Effects of Question Formats in Math Assesment on Metacognition and Affect*. Los Angeles: CRESST-CSE University of California.

Table 1. The Relationship between Dimensions of Cognitive Process and Dimensions of Knowledge

No	Thinking Development of Bloom's Taxonomy Revised by Anderson (*Cognitive Process Dimension)*	Knowledge Dimension	Evidence
1	Remembering (C1)	Factual Knowladge	Lower Order Thinking Skills (LOTS)
2	Understanding /Interpreting concept (C2)	Conceptual Knowladge	
3	Applying (C3)	Procedural Knowladge	
4	Analyzing (C4)	Metacognitif Knowledge	Higher Order Thinking Skills (HOTS)
5	Evaluating (C5)		
6	Creating (C6)		

IOP Conf. Series: Materials Science and Engineering **300** (2018) 012048 doi:10.1088/1757-899X/300/1/012048

Table 2. Analysis of student metacognitive ability in understanding the material.

Number subject	Indicator understanding the material				Final score	Category
	Underline	Make marginal note	Make a material summary	Create a mapping concept		
1	4	4	4	4	4	Very high ability
2	4	4	4	3	3,75	Very high ability
3	4	3	3	2	3	High ability
4	4	3	3	1	2,75	High ability
5	3	3	3	2	2,75	High ability
6	3	3	2	1	2,25	Middle level ability
7	3	3	1	1	2	Middle level ability
8	2	2	1	0	1,25	Low level ability
9	1	1	0	0	0,5	Very low level ability

Table 3. Analysis of student metacognitive ability of students solving problems.

Number subject	Indicator metacognitive ability of students in solving problems					Final score	Category
	Heuristic strategy	Reverse thinking strategy	Forward thinking procedure	Inductive thinking procedure	Deductive thinking procedure		
1	4	2	4	3	3	3,2	High ability
2	4	2	3	2	2	2,6	High ability
3	3	1	2	2	2	2	Middle level ability
4	3	1	2	2	2	2	Middle level ability
5	2	0	2	1	1	1,2	Low level ability
6	3	0	1	1	1	1,2	Low level ability
7	2	0	1	1	1	1	Low level ability
8	1	0	1	0	0	0,4	Very low level ability
9	2	0	0	0	0	0,4	Very low level ability

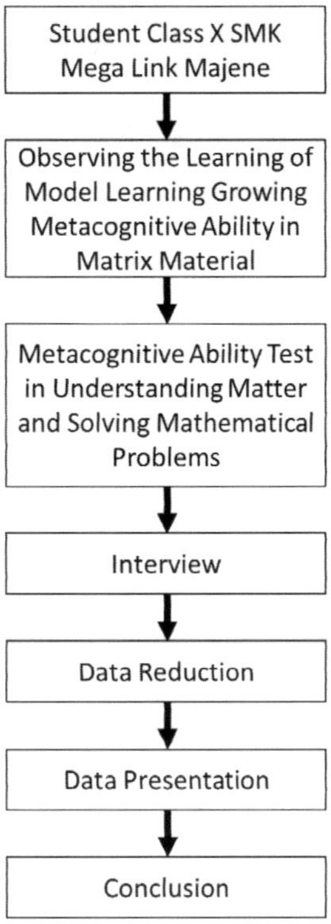

Figure 1. Flow diagram in the study of students' metacognitive ability description in understanding and solving mathematics problems.

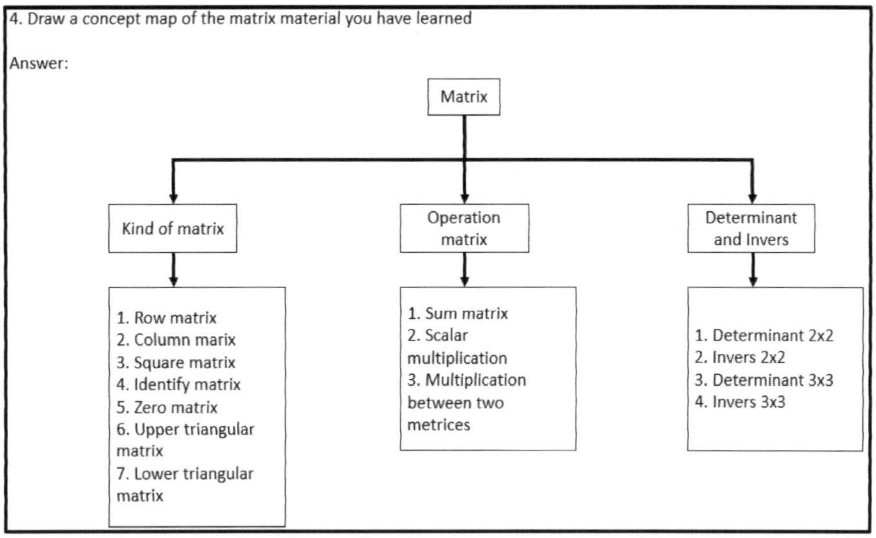

Figure 2. Answer about the number 4, draw a concept map of the matrix material.

IOP Conf. Series: Materials Science and Engineering **300** (2018) 012048 doi:10.1088/1757-899X/300/1/012048

2. Given two matrix $A = \begin{bmatrix} 4 & -1 \\ 3 & -2 \end{bmatrix}$ dan $B = \begin{bmatrix} a & x \\ -b & c \end{bmatrix}$ $if\ A^{-1} = B^T$, then the value b.

Answer:

$A^{-1} = B^T$

$B^T = A^{-1}$

$$\begin{bmatrix} a & -b \\ x & c \end{bmatrix} = \frac{1}{det.A}\begin{bmatrix} -2 & 1 \\ -3 & 4 \end{bmatrix}$$

$$\begin{bmatrix} a & -b \\ x & c \end{bmatrix} = \frac{1}{(4.(-2)) - (3.(-1))}\begin{bmatrix} -2 & 1 \\ -3 & 4 \end{bmatrix}$$

$$\begin{bmatrix} a & -b \\ x & c \end{bmatrix} = \frac{1}{-8-3}\begin{bmatrix} -2 & 1 \\ -3 & 4 \end{bmatrix}$$

$$\begin{bmatrix} a & -b \\ x & c \end{bmatrix} = \frac{1}{-11}\begin{bmatrix} -2 & 1 \\ -3 & 4 \end{bmatrix}$$

$$\begin{bmatrix} a & -b \\ x & c \end{bmatrix} = \begin{bmatrix} \frac{2}{11} & -\frac{1}{11} \\ \frac{3}{11} & -\frac{4}{11} \end{bmatrix}$$

$$-b = -\frac{1}{11}$$

$$b = \frac{1}{11}$$

Figure 3. Answer about the number 2, students were asked to solve the problem by using a backward thinking strategy. The subject appeared at the first step was correct by answering the question from the beginning, but the subject made an error in operation 4 (-2) - 3 (-1) the result should be -8- (-3) = -8 + 3 = -5. The subjects did not solve the mathematics problems carefully.

IOP Conf. Series: Materials Science and Engineering **300** (2018) 012049 doi:10.1088/1757-899X/300/1/012049

Modification of 2-D Time-Domain Shallow Water Wave Equation using Asymptotic Expansion Method

Teuku Khairuman[1,4], Nasruddin MN[1], Tulus[2] and Marwan Ramli[3]

[1] Department of Physics, Faculty of Science, University of Sumatera Utara, Medan, INA
[2] Department of Mathematics, Faculty of Science, University of Sumatera Utara, Medan, INA
[3] Department of Mathematics, Faculty of Science, Syiah Kuala University, Aceh, INA

E-mail: teuku.khairuman@students.usu.ac.id

Abstract. Generally, research on the tsunami wave propagation model can be conducted by using a linear model of shallow water theory, where a non-linear side on high order is ignored. In line with research on the investigation of the tsunami waves, the Boussinesq equation model underwent a change aimed to obtain an improved quality of the dispersion relation and non-linearity by increasing the order to be higher. To solve non-linear sides at high order is used a asymptotic expansion method. This method can be used to solve non linear partial differential equations. In the present work, we found that this method needs much computational time and memory with the increase of the number of elements.

1. Introduction

Wave propagation, like tsunami waves, can be caused by undersea movement. The problem of tsunami waves that caused coastal damage has been widely studied by experts. The damage caused by the waves relates to how far, high and rapid the spreading of water when entering the coastal area. Studies conducted to understand the evolution of tsunami wave propagation have been done by using shallow water theories, Boussinesq equations, non-hydrostatic wave equations, and Navier-Stokes equations [1, 2]. The use of the Boussinesq equation for tsunami wave propagation has also been used by Groesen and Klopman [3] and Adytia [4]. The Boussinesq equation is an equation that modeled wave propagation on surfaces propagating in two opposite directions [5].

One form of the Boussinesq equation, especially for shallow waters known as the shallow water wave equation [6–8]. In line with the study of tsunami wave investigation, the Boussinesq equation model undergoes a change aimed at improving the quality of dispersion relation and inclination by increasing the order to be higher. The complexity in the numerical implementation arising from the application of the high order Boussinesq model would be modified and solved by asymptotic expansion methods.

2. Theory

Tsunami waves have an initial wavelength of several tens or hundreds of km. In governing equations, nonlinear convective tribes and ocean floor friction are relatively small and negligible,

[4] Permanent Address: Department of Physics, Faculty of Science, Syiah Kuala University, Aceh, INA

while the influence of frequency dispersion depends on wavelength and can be included in linear model governing equations. Non linear effect is not significant, while the effect of frequency dispersion can be important for shorter waves. Therefore, deep ocean tsunami propagation can be modeled by linear Boussinesq equations. Furthermore, tsunamis move across the ocean from the initial location of a tsunami to a coastal area where the water depth is relatively shallower.

A number of governing equations are used to simulate the propagation of tsunami waves, such as linear Boussinesq equations. After ignoring the nonlinear terms, the linear Boussinesq equation can be written as follows [9–11]:

$$\partial_t u + g\partial_x \eta + u\partial_x u - \frac{H^2}{3}\partial_{xxt} u = 0 \tag{1}$$

$$\partial_t \eta + \partial_x (Hu) + \partial_x (\eta u) = 0 \tag{2}$$

3. Methods

In this article, asymptotic expansion method was used to approximate a analytic solution of shallow water wave equation. To apply this method, the first step had to do is to make an assumption related to the form of asymptotic method that will be used. Here, the asymptotic expansion forms of elevation $\eta(x,t)$ and velocity $u(x,t)$ are written as the following

$$\begin{aligned} \eta(x,t) &\approx \varepsilon\eta^{(1)}(x,t) + \varepsilon^2\eta^{(2)}(x,t) + \varepsilon^3\eta^{(3)}(x,t) \\ u(x,t) &\approx \varepsilon u^{(1)}(x,t) + \varepsilon^2 u^{(2)}(x,t) + \varepsilon^3 u^{(3)}(x,t) \end{aligned} \tag{3}$$

where ε is a small parameter in the first order that states the ratio between wave amplitude and depth. η^j and u^j respectively represent solutions for elevation and velocity at the $j - order$, $j = 1, 2, \ldots$ [5].

4. Results and Discussion

The form of asymptotic expansion of equation(3) is substituted into equations (1) dan (2). After each side is grouped according to the order of each ε and assigns the value of each part to zero, it is obtained

$$O(\varepsilon): \qquad \partial_t u^{(1)} + g\partial_x \eta^{(1)} - \frac{H^2}{3}\partial_{xxt} u^{(1)} = 0 \tag{4}$$

$$\partial_t \eta^{(1)} + H\partial_x u^{(1)} = 0 \tag{5}$$

$$O(\varepsilon^2): \qquad \partial_t u^{(2)} + g\partial_x \eta^{(2)} - \frac{H^2}{3}\partial_{xxt} u^{(2)} = -u^{(1)}\partial_x u^{(1)} \tag{6}$$

$$\partial_t \eta^{(2)} + H\partial_x u^{(2)} = -(\eta^{(1)}\partial_x u^{(1)} + u^{(1)}\partial_x \eta^{(1)}) \tag{7}$$

$$O(\varepsilon^3): \qquad \partial_t u^{(3)} + g\partial_x \eta^{(3)} - \frac{H^2}{3}\partial_{xxt} u^{(3)} = -u^{(1)}\partial_x u^{(2)} - u^{(2)}\partial_x u^{(1)} \tag{8}$$

$$\partial_t \eta^{(3)} + H\partial_x u^{(3)} = -\partial_x(\eta^{(1)}u^{(2)} + \eta^{(2)}u^{(1)}) \tag{9}$$

4.1. First-order solution

Eliminating $u^{(1)}$ from equations (4 and 5), the following equation can be obtained in term of η.

$$\partial_{tt}\eta^{(1)} - gH\partial_{xx}\eta^{(1)} = \frac{gH^3}{3}\partial_{xxxx}\eta^{(1)} \tag{10}$$

Solution for first-order equations (10) can be written in the form of cosine equations as follows

$$\eta^{(1)}(x,t) = A_1 \cos(kx - \omega t) \tag{11}$$

where A_1, k, and ω, respectively, are wave amplitude, the wave number, and dispersion relation of wave. Hereafter, the form of this first-order solution is derived respectively to x and t, so that it is obtained

$$\frac{\omega^2}{k^2} = gH\left(1 + \frac{(kH)^2}{3}\right) \tag{12}$$

4.2. Second-order solution

In the second order we obtain the non homogeneous linear differential equation as follows

$$\partial_{tt}\eta^{(2)} - gH\partial_{xx}\eta^{(2)} = \frac{gH^3}{3}\partial_{xxxx}\eta^{(2)} - \frac{H}{2}\partial_{xx}(u^{(1)})^2 - \partial_{xt}(u^{(1)}\eta^{(1)})) \tag{13}$$

The following two solutions are taken as solutions to the equation (13).

$$\begin{aligned} \eta^{(2)}(x,t) &= A_2 \cos(2(kx - \omega t)) \\ u^{(1)}(x,t) &= B_1 \cos(kx - \omega t) \end{aligned} \tag{14}$$

Equation 14 is derived respectively to x and t so we get

$$\begin{aligned} \partial_{xx}(u^{(1)})^2 &= -2B_1^2k^2 \cos(2(kx - \omega t)) \\ \partial_{xt}(u^{(1)}\eta^{(1)}) &= 2A_1B_1k\omega \cos(2(kx - \omega t)) \end{aligned} \tag{15}$$

Then equations (12, 14 and 15) are substituted to equation (13), so that the value of A_2 can be written as the following

$$A_2 = \frac{3}{5}\left(\frac{HB_1^2k^2 + 2A_1B_1k\omega}{gH^3k^4}\right) \tag{16}$$

4.3. Third-order solution

As the second-order, in the third order we obtain the non homogeneous linear differential equation as follows

$$\partial_{tt}\eta^{(3)} - gH\partial_{xx}\eta^{(3)} = \frac{gH^3}{3}\partial_{xxxx}\eta^{(3)} + H\partial_{xx}(u^{(1)}u^{(2)}) - \partial_{xt}(\eta^{(1)}u^{(2)} + \eta^{(2)}u^{(1)})) \tag{17}$$

By applying the same working principle as in the previous order, for the third order equation, it is known that the multiplication of the inhomogeneous parts yields

$$\begin{aligned} \partial_{xx}(u^{(1)}u^{(2)}) &= -B_1 B_2 k^2 \left(9\cos(3(kx-\omega t)) + \tfrac{1}{9}\cos(kx-\omega t)\right) \\ \partial_{xt}(\eta^{(1)}u^{(2)} + \eta^{(2)}u^{(1)}) &= (A_1 B_2 + A_2 B_1)k\omega \left(9\cos(3(kx-\omega t)) + \tfrac{1}{9}\cos(kx-\omega t)\right) \end{aligned} \tag{18}$$

where $u^{(2)}(x,t) = B_2\cos(kx-\omega t)$. Therefore, the solution form for equation (17) can be written as

$$\eta^{(3)}(x,t) = C_1\cos(3(kx-\omega t)) \tag{19}$$

The substitution of equations (18 and 19) into equation (17) gives the following results

$$C_1 = \frac{1}{60}\left(\frac{B_1B_2k + (A_1B_2 + A_2B_1)\omega}{H^2k^3\cos(3(kx-\omega t))}\right)\left(9\cos(3(kx-\omega t)) + \tfrac{1}{9}\cos(kx-\omega t)\right) \tag{20}$$

Based on the results obtained as written in equation (20), then equation (19) can be rewritten to

$$\eta^{(3)}(x,t) = \frac{1}{60}\left(\frac{B_1B_2k + (A_1B_2 + A_2B_1)\omega}{H^2k^3}\right)\left(9\cos(3(kx-\omega t)) + \tfrac{1}{9}\cos(kx-\omega t)\right) \tag{21}$$

4.4. First solution

Three solutions for each order have been completed. Thus, the Boussinesq equation solution for tsunami wave propagation obtained by asymptotic expansion method until the third order can be rewritten as

$$\begin{aligned} \eta(x,t) = \quad & \varepsilon A_1\cos(kx-\omega t) \\ & + \varepsilon^2 A_2\cos(2(kx-\omega t)) \\ & + \varepsilon^3 C_1\left(9\cos(3(kx-\omega t)) + \tfrac{1}{9}\cos(kx-\omega t)\right) \end{aligned} \tag{22}$$

with

$$C_1 = \frac{1}{60}\left(\frac{B_1B_2k + (A_1B_2 + A_2B_1)\omega}{H^2k^3}\right) \tag{23}$$

and

$$B_2 = \frac{A_2\omega - \frac{1}{2}A_1B_1k}{kH} \tag{24}$$

4.5. Correction of Wave Numbers

Regarding the essence of applying the asymptotic expansion method is to approach a value, which is an analytical solution of the shallow water equation. That is, the approximate solutions must be convergent. Therefore, to obtain a convergent solution, further correction of other parameters of the wave equation is needed, in this case, that is the wave numbers (k). To do this, the procedure applied is the same as of for obtaining the solution of each order, as described in the previous section.

Assuming that the wave numbers (k) has the following expansion shape,

$$k \approx k^{(0)} + \varepsilon k^{(1)} + \varepsilon^2 k^{(2)} \tag{25}$$

Now, equation (25) is substituted to equation (22) to replace value of (k), so that equation (22) is rewritten as the following

$$\begin{aligned} \eta(x,t) = \quad & \varepsilon A_1 \cos\left((k^{(0)} + \varepsilon k^{(1)} + \varepsilon^2 k^{(2)})x - \omega t\right) \\ & + \varepsilon^2 A_2 \cos\left(2((k^{(0)} + \varepsilon k^{(1)} + \varepsilon^2 k^{(2)}) - \omega t)\right) \\ & + \varepsilon^3 C_1 \left(9\cos(3((k^{(0)} + \varepsilon k^{(1)} + \varepsilon^2 k^{(2)})x - \omega t)) \right. \\ & \left. + \tfrac{1}{9}\cos((k^{(0)} + \varepsilon k^{(1)} + \varepsilon^2 k^{(2)})x - \omega t)\right) \end{aligned} \tag{26}$$

Approximate solutions (26) with their each derivative of the variables x and t are resubstituted to equations (10), (13), and (17) so that we get these following results.

- First-order, $O(\varepsilon)$:
  The result in first order is a new shape of dispersion relation, namely

$$\frac{\omega^2}{(k^{(0)})^2} = gH\left(1 + \frac{(k^{(0)}H)^2}{3}\left(1 + 2k^{(0)}\right)\right) \tag{27}$$

- Second-order, $O(\varepsilon^2)$:
  In this order, it was obtained $k^{(1)} = 0$, so that the value of coefficient A_2 in the equation (14) can be rewritten as below

$$A_2 = \frac{k^{(2)}}{2}\left(\frac{HB_1^2 k^{(0)} + \omega A_1 B_1}{gH(k^{(0)})^2 - \frac{4}{3}gH^3(k^{(0)})^4 - \omega^2}\right) \tag{28}$$

- Third-order, $O(\varepsilon^3)$:
  The value of $k^{(1)} = 0$ obtained in the second-order causes some terms of the third-order, in which each order is still dependent on the value $k^{(1)}$, can be eliminated. Therefore, the new expansion form of k can be rewritten as follows.

$$k = k^{(0)} + \varepsilon^2 k^{(2)} \tag{29}$$

where

$$k^{(2)} = -\frac{C_1}{9}\left(\frac{(k^{(0)})^2 + \omega^2 + \frac{1}{3}gH^3(k^{(0)})^4)}{A_1 k^{(0)}\left(gH(k^{(0)})^2\right) - 2)}\right) \tag{30}$$

Based on all the above results, now a solution of shallow water equations solved using the asymptotic expansion method can be described as the new following form

$$\begin{aligned}\eta(x,t) = \quad & \varepsilon A_1 \cos\Big((k^{(0)} + \varepsilon^2 k^{(2)})x - \omega t\Big) \\ & + \varepsilon^2 A_2 \cos\Big(2((k^{(0)} + \varepsilon^2 k^{(2)}) - \omega t)\Big) \\ & + \varepsilon^3 C_1 \Big(9 \cos(3((k^{(0)} + \varepsilon^2 k^{(2)})x - \omega t)) \\ & + \tfrac{1}{9} \cos((k^{(0)} + \varepsilon^2 k^{(2)})x - \omega t)\Big) \end{aligned} \tag{31}$$

The coefficient of $k^{(2)}$ has an important role in the expansion of the wave number k. It also can be interpreted as a consequence of the use of the asymptotic expansion method to obtain the approximate solution of shallow water equations. In other words, the convergence of the solution will be achieved when the wave number k is corrected by $k^{(2)}$.

5. Summary

Based on the modified result, asymptotic expansion method is very appropriate in approaching Boussinesq analytical solution of high order equation. However, the solution that has been obtained can not be said as a valid solution. This is because of the resonance part, which is the shape that looks like the form of the first-order solution. Therefore, in order for the solution to be valid, further correction is required on other parameters in the Boussinesq wave equation.

Meanwhile, the result of modifying the Boussinesq equation into a form that can be solved by asymptotic expansion method shows that many new coefficients and variables appear. Of course, the emergence of these new coefficients and variables enlarge the calculation time and also increase the amount of computer memory used for computer simulation.

References

[1] Ma G, Kirby J T and dan Shi F 2013 *Ocean Modelling* **69** 146–165
[2] Guo A, Xiao S and Li H 2015 *Pure and Applied Geophysics* **172** 569–587
[3] Groesen E and Klopman 2005 Dispersive effects in tsunami generation *Proceedings of The Indonesian Ocean Forum 2005*
[4] Adytia D 2012 *Coastal zone simulations with variations Boussinesq modelling* Ph.D. thesis eemcs-eprint-22024
[5] Ramli M, Munzir S, Khairuman T and Halfiani V 2014 *Far East Journal of Mathematical Sciences* **90** 97–117
[6] Donahue A S, Zhang Y, Kennedy A B, Westerink J J, Panda N and Dawson C 2015 *Ocean Modelling* **86** 36–57
[7] Walkley M A 1999 A numerical method for extended boussinesq shallow water wave equation *Dissertation* (University of Leeds)
[8] Lannes D and Bonneton P 2009 *Physics of Fluids* **21** 1–9
[9] Appuhamy J M R S 2007 Numerical simulation of tsunami in indian ocean *Master Thesis*
[10] Liu P L F, Woo S B and Cho Y S 1998 Computer programs for tsunami propagation and inundation *Technical Report* (Cornell University)
[11] Bagheri J and Das S K 2013 *Journal of Applied & Computational Mathematics* **2** 1–9

IOP Conf. Series: Materials Science and Engineering **300** (2018) 012050 doi:10.1088/1757-899X/300/1/012050

Mitigating Local Natural Disaster through Social Aware Preparedness Using Complexity Approach

Irwan Supadli[1], Andini Saputri[2] , Herman Mawengkang[3]
[1]Mangement of Natural Resources and The Environmental Departement, Postgraduate University of Sumatera Utara, Medan, Indonesia
[2]Biology Department, Faculty of Engineering, Samudra University, Aceh, Indonesia
[3]Department of Mathematics, University of Sumatera Utara, Medan, Indonesia

[1]supadli.irwan@yahoo.com,[2]puputandini87@gmail.com,[3]hmawengkang@yahoo.com

Abstract. During and after natural disaster, such as, eruption of vulcano, many people have to abandon their living place to a temporary shelter. Usually, there could be several time for the occurrence of the eruption. This situation, for example, happened at Sinabung vulcano, located in Karo district of North Sumatera Province, Indonesia. The people in the disaster area have become indifferent. In terms of the society, the local natural disaster problem belong to a complex societal problem. This research is to find a way what should be done to these society to raise their social awareness that they had experienced serious natural disaster and they will be able to live normally and sustainable as before. Societal complexity approach is used to solve the problems. Social studies referred to in this activity are to analyze the social impacts arising from the implementation of the relocation itself. Scope of social impact assessments include are The social impact of the development program of relocation, including the impact of construction activities and long-term impact of construction activity, particularly related to the source and use of clean water, sewerage system, drainage and waste management (solid waste), Social impacts arising associated with occupant relocation sites and the availability of infrastructure (public facilities, include: worship facilities, health and education) in the local environment (pre-existing). Social analysis carried out on the findings of the field, the study related documents and observations of the condition of the existing social environment Siosar settlements.

Keywords: Disaster, Mitigating, Preparedness, Societal complexity

1. Introduction

Construction of residential relocation area is a planned effort to transform an area that is occupied by a community with a variety of ecological, social, economic and physical limitations to become a region forward with the community quality of life equal or not far behind compared to other Indonesian society. This condition should be the focus of attention of various stakeholders towards refugees Sinabung, which will be placed in the new residential area in the village relocation program Siosar at Merek Sub Region of Karo Regency.

In the construction of the relocation area, the logging is done in the area of pine forest into something that is not inevitable. This leads to the occurrence of landslides if not controlled, so that the necessary effort to keep the preservation of pine forests in addition there are many important fauna endangered birds such as hornbills and large mammals. Pine forest conservation rescue efforts not only to prevent the occurrence of landslides, global warming or changes in land use, but also as an effort to protect Germ Plasma of extinction, especially species that are threatened with extinction.

Relocation after the eruption of Mount Sinabung community requires careful handling effort because it involves many aspects, among others: social, cultural, economic and environment, both natural environment as well as the built environment. It is hoped this study is able to ensure that the management of various aspects of the above to be better than the original, so as to prevent the threat of landslides, flooding and the extinction of some species are considered rare.

2. Scope Of Social Impact Assessment

Social studies referred to in this activity are to analyze the social impacts arising from the implementation of the relocation itself. Scope of social impact assessments include:

- The social impact of the development program of relocation, including the impact of construction activities and long-term impact of construction activity, particularly related to the source and use of clean water, sewerage system, drainage and waste management (solid waste).
- Social impacts arising associated with occupant relocation sites and the availability of infrastructure (public facilities, include: worship facilities, health and education) in the local environment (pre-existing).

3. Research Result

3.1. Integer Settlemnet

Preparation of the master plan in development of Siosar area aimed at creating an integrated development of the residential area with the construction of public facilities and synergy with aspects of the locality in the settlements. In the implementation of the settlement, the processes that are local and which are widely should be accommodated existence and relevance to the function in the process of settlement construction as a whole. Both of these processes are accommodated into the system implementation settlement construction. Unifying settlement objectives locally or widely of and integrate every element or aspect of development will be an integral part of the overall development objectives settlements.

Integrated residential area is an area of physical elements forming the structure of a regular space and blends. However, the issue of settlements that arise can not be separated from the dynamics that occur in the life of society and government policy in managing the settlement area. The importance policy and strategy on the relocation of the Siosar village settlements is an inseparable unity. Policies and strategies should promote settlement of spatial arrangement of space that can harmonize the natural environment and the built environment. The purpose of spatial planning is to achieve integration in the use of natural resources and man-made and can provide protection against preventive function in negative environmental impacts due to the arrangement of space that is not appropriate.

Therefore, in addition to settlements is one of the basic human needs also has a strategic function in the role of family education center, nursery culture, and improving the quality of future generations, as well as a manifestation of identity. The realization of people's welfare can be characterized by increasing the quality of life of a decent and dignified, among others through the fulfillment of housing and settlement. Thus efforts to manage the housing and settlement as one of the priority sector in the development of human beings.

[1] points out that the settlements were created and developed can be a means for a life full of piety and faith, creates a feeling of security and comfort, ensure the physical and spiritual health, increase intimacy and create social relationships and interaction quality, have the basic infrastructure planned or unplanned and have facilities that support life and livelihood.

In accordance with master plan of Siosar settlements, integrated settlements models have become priorities in the future development of the area, because in addition to homes, public facilities will be built several other villages such as; health centers, schools, community centers and so forth.

Masterplan prepared a design that refers to the existing condition of existing ones, as well as data from relevant ministries and institutions. However, during the ongoing assessment of integrated models of settlements are still inadequate. Public facilities are not a single village was built. Only about 103 houses are already built with the condition of the house ready for habitation 53 and 50 home again still in the final stages of completion. Houses are built very simple home category (RSS) by type of building 36 meters. The vast size of the building amounted to 5x8 meters. The house has a rest area on the front and back of the house which is 7 meters and on the right side and left the house amounted to 2 meters.

In 10 relocation homes contained 1 septic tank that serves as the sanitary disposal of solid and liquid waste from the rest of human waste. Building an existing septic tank has met the eligibility standard to build reservoirs of fiber that is not easy to sink into the ground, so that the source of drinking water to be protected from bacterial contamination ecoly. Availability sanitation should be a concern by local governments and refugees who will live here to keep the septic tank's usefulness if the reservoir has a full bath is necessary to normalize the siphon or septic tank emptying back settlements conducted regularly. Local governments prepare the means of transport, such as vacuum truck through the Department of Hygiene and reaching out to the Siosar village. These activities are carried out periodically to perform routine monitoring of the condition of the community septic tank. Until now ongoing assessment is already built 10 septic tanks were ready.

Housing relocation conditions of Sinabung volcano eruption has adequate ethical standards with 1 toilet facilities measuring 2 x 2 m, and equipped with one toilet and tap water and do not have a window. Conditions of cement-floored interior of the house still, and do not have a kitchen. Therefore, it is necessary refugee initiatives that will determine each kitchen space. Air circulation system is like in general where each house has a glass window on the front and a door on the front and one on the rear door.

3.2. Accessibility And Availability

A. Road Condition And Drainage Systems

Concepts design contained in the road network in the master plan, based on existing regional area and topography of the site, then the concept of the road network is made in the form of 'loop' is closed, thus simplifying the circulation and overall affordability of the region. The main road network is set to follow the existing topographic characteristics in order to minimize the 'cut and fill' at roadworks, so friendly to the existing contours. Secondary road network designed an extension of the main network, and aims to connect every function and villages in this region with the main road network.

The road network is able to accommodate the needs of residential infrastructure for refugees from 7 villages and 1 village, which consists of 2,053 plots of houses for disaster victims, as well as public facilities and social aims to provide a livelihood that not only match the life before the disaster, but it gives chance of a better life for the people who relocated to this area. Up to this time, the relocation of the Siosar village settlements was concentrated to prepare a settlement phase 2 and phase 3 will be built, so that the problem of accessibility to the Siosar still requires serious treatment.

[2] says that the drainage water is a series of buildings that serve to reduce or remove excess water from an area or land, so the land can function optimally. Drainage also can control the quality of groundwater in relation to salinity were able to drain, drain, discarding, or diverting unwanted excess water in an area, as well as ways alleviation consequences caused by excess water.

Structuring the drainage network system Siosar village settlement area should have a structured drainage network, comprising: a network of primary, secondary, and tertiary. This network will be able to perform normalization and rehabilitation of the channel in order to create a safe environment and good to inundation, overflowing rivers, flood, or local rain. Primary network serves as a channel that used the river and creeks as the water flow. Secondary network serves as a channel that connects the tertiary channels with the primary channel (built with concrete/stucco). While the tertiary network serves as a conduit for the flow of household waste into secondary channels, such as plastering, plumbing and soil, thus flooding and overflow of water flowing into the house due to high rainfall is not flooded yard.

3.3. Infrastructure and Public Facilities

Infrastructure and public facilities serve as the facilities required community that provision is done simultaneously or bulk. The fulfillment level of the facility is the size of community welfare. The provision of infrastructure and public facilities include roads, electricity, water, and drainage. Facilities of the road connecting the village to another village or the surrounding area and the road network that connects the village part, plays a very important for the activity of the population and the development of the village itself and at the same time as the basic framework that forms the structure of the village [3].

Network utility as a major part of the infrastructure and facilities for basic daily life such as electricity, water and drainage constructed above and below ground. The networks usually follow or ride on the road network form. Once the utility built its presence will be long and will attract people to occupy land and build access the utility.

Electricity network has been built in all the way to the Siosar village. Electricity network built on land along the road to the Siosar village. Electricity network is intended as a source of illumination and energy to people who will live in the Siosar village. From field observations indicate the installation of electricity networks have been built in almost all residential areas with adequate conditions. In addition to the power grid, residential areas of Siosar village have built a network of water flowing into people's homes adequately. The water comes from mountain springs within approximately 2.5 miles from the relocation site. The method used to drain the mountains towards the houses is by gravity methods.

This method utilizes the properties of water that flows into the lower place. Installing water pipes to drain the water from the fountain to get into homes by utilizing the natural properties of water that always flows from a higher to a lower place, so that the availability of clean water in the Siosar village very adequate. The water can be used for 24 hours for a variety of domestic activities. If seen by naked eyes clear water conditions, cold and odorless. This course will benefit those who live later.

The process of water distribution has been prepared properly. While in the Siosar village, we find water pipes with water meter for every home that will meter can be used to control the level of spending citizens of water. With clean water distribution system like this, will not be difficult for residents in getting clean water. This of course will vary with the condition of people in the previous settlement that had to queue to get a supply of clean water.

Clean water is a vital necessity of every human being, so that the availability of water determines the health and welfare of the community. The need for clean water in residential areas have to be addressed on a massive scale in the form of provision of facilities for drinking water pipelines are maintained in quality and quantity. The facility managers are generally in the form of a local company called PDAM (Regional Water Company). The local government is expected to bridge the management of clean water to the taps Siosar settlements in a sustainable manner. However, the selves-help could also be an alternative model in water management in the future.

Pro-actively displaced communities are given instruction in the maintenance and management of communal water supply, by forming groups of administrators managing clean water, so that by itself would be awakened obligation to maintain continuity of supply of clean water for them. Surely displaced water is obliged to pay contributions in accordance with the volume of water consumption every month. The money collected can be used to finance operational in the maintenance and management of clean water to the citizens of the future. It takes a commitment of local governments through the taps to provide technical assistance to the refugee community in the form of training capacity in the control treatment water pipes and water meters in order to keep it functioning properly. Besides, given the understanding on all citizens to always pro-actively maintain the condition of the pipe is always functioning properly.

3.4. Domestic Waste Management System

Until the assessment takes place, domestic waste management for the Siosar village still not well planned. There is no shelter construction (tub) bins for each home. Likewise, the waste temporary location (TPS) is not planned well. Where trash will be disposed of and how waste management, then it is a problem in the future to achieve a clean Siosar village of waste generation.

Karo Government should prepare a variety of policies for domestic waste issue. Starting from lug and garbage collection to prevent scattering and simplify the process of collecting up to transporting waste from waste sources to landfills while (TPSS). Lug is done by providing a barrel (bath) bins for each home, providing workers in charge citing waste and prepare the vehicle (truck) garbage which operates regularly to transport garbage Siosar village every day. Karo Government's commitment to allocate a special budget is needed to support this policy, because the distance is quite far from the city Kaban Jahe will have an impact on the rising operational costs in the handling of waste.

In addition to the government's commitment, it takes creative efforts of communities to manage and utilize domestic waste into economically valuable products. The three products that can be produced from domestic waste management derived from residual organic materials, namely: charcoal, compost and Takakura composting of duck excrement. Third products are the potential of processed products made by refugees, considering the manufacture of these products can be obtained easily and not costly. Processing methods can be learned easily, without having to follow special education. Provide intensive training and periodically by bringing the facilitators who have skills in making briquettes and compost.

Figure 1. Management of manure compost ducks

Briquettes and compost product can be utilized solely by refugees for agricultural purposes as well as for fuel in cooking food that still depend on gas and kerosene. If the processing of organic waste can be developed into a superior product, then this sector could be one of creative economy and be a source of rural income. Community empowerment can be done communally by forming groups of micro-businesses that are able to create a populist economic power competitive. Local governments and other stakeholders are expected to provide massive support to the sustainability of the future development of this business.

For the processing of inorganic waste from domestic waste, can be done by reused economically valuable products, such as: making crafts (souvenirs) derived from processed, plastic, rubber, steorofoam, bottles of mineral water and others. Refugee communities provided with training to recover these wastes into economically valuable souvenir products. This product will be marketed to Berastagi or other tourist destinations that can accommodate and market derived from refugee communities Siosar craft.

4. Conclusions And Recommendations

4.1. Conclusions

1. An integrated model planning of settlement is still not performing well. Only about 103 houses are already built with the condition of the house ready for habitation 53 and 50 home again still in the final stages of completion. No one woke public facilities.
2. The availability of infrastructure is not yet fully adequate. Only the power grid and water network that has been built in almost all residential areas with fairly good condition.
3. The management of domestic waste of the Siosar village is still not well planned. Not having found the construction of shelters (tub) bins for each home. Likewise, the temporary of waste location (TPS) is undeveloped.

4.2. Recommendations

1. Policies and strategies of spatial Siosar settlement should put forward a model of integrated settlements can harmonize the natural environment and the built environment. The goal is to achieve integration in the use of natural resources and man-made and can provide protection against the negative impact of prevention function on the environment and serves as a family education center, nursery culture, improving the welfare and quality of future generations, as well as a manifestation of identity.
2. The local government is expected to bridge the management of clean water to the Siosar settlements with PDAM in a sustainable manner. However, efforts by themselves could also be an alternative model in water management in the future.
3. Create a policy in the provision of infrastructure to cope with the problem of waste. For example: the provision of places (tub) storage bins for each home, providing workers in charge citing waste and prepare the vehicle (truck) garbage which operates regularly to transport garbage of Siosar village every day.
4. Creating the best practices of communities to manage and utilize organic waste and anorganic domestic become creative economic sectors, namely: charcoal briquettes from domestic waste, composting of residual domestic waste (takakura) and compost from dirt ducks, handicrafts from husk and cob corn and crafts of plastic household waste (souvenirs, bags, plastic flowers, flower pots)

5. References

[1] Pramumijoyo, S., & Karnawati, D., 2006. *Pemantauan dan Mitigasi Bencana Alam Longsor. Makalah dalam Se-minar "Pemantauan dan Mitigasi Bencana Alam Banjir, Tanah Long-sor dan Kekeringan.* Balai Peneli-tian dan Pengembangan Teknologi Pengelolaan DAS Wilayah Indone-sia Bagian Barat. Surakarta, 29 Agustus 2006.
[2] Pawitan H., 2009. *Perubahan Penggunaan Lahan dan Pengaruhnya Terhadap Hidrologi Daerah Aliran Sungai (Land Use Changes and Their Impacts on Watershed Hydrology).* Laboratorium Hidrometerologi FMIPA IPB, Bogor.
[3] Bintarto.1983. *Interaksi Desa Kota dan Permasalahannya.*Jakarta : Ghalia Indonesia.

IOP Conf. Series: Materials Science and Engineering **300** (2018) 012051 doi:10.1088/1757-899X/300/1/012051

Sedimentation Optimization on River Dam Flow by Using COMSOL Multiphysics

Tulus[1], **Suriati**[2], **T. J. Marpaung**[1]

[1] Department of Mathematics, Universitas Sumatera Utara, Padang Bulan 20155 USU Medan Indonesia

[2] Department of Informatics, Sekolah Tinggi Teknik Harapan, Medan 20218 Indonesia

E-mail: tulus@usu.ac.id

Abstract. This paper is a study of the application of material sedimentation on a river water dam that does not need to be streamed again. The problem will be shown in three forms of occurrence that is, while the water is flat, curved and steep down. The finite element method is used to solve the problem in the form of Darcy equation in two dimensional case. The solution is performed by using the COMSOL Multiphysics 5.2 for the flow phenomenon. The results obtained will be more of the material sedimented in flat water conditions, but for the water conditions curved and steep to the bottom water that has not yet experienced more sedimentation is discarded.

1. Introduction

Sedimentation sourced from human activities along the river provides many positive effects such as agricultural land clearing, agricultural irrigation, industrial waste, vegetation clearance on the banks of the river and so on. While its negative impact can be seen along the river body and river estuary. These impacts are as follows: death of marine organisms, decreased bideversity, shipping line resistance due to silting, disturbance or loss of habitat, decline in natural seafood stocks, sediment size change change, turbidity increase and depth change. Pattern of sediment deployment along rivers and River mouths can be readily known quantitatively over time through sediment model simulations.

The results of the sediment distribution survey can be analyzed by suspicion of potential sediment sources with modeling using initial conditions scenario. The physical, chemical and biological processes of sedimentation effects in rivers and estuaries can be identified using certain scenarios. Some simulations for sedimentation phenomena have been studied by some authors for some case of considerations, i.e mass fraction in the basin [1, 2], flows in the water treatment [3].

Fluids that move through pore spaces in an aquifer or reservoir can shield the porous medium from stress because they bear part of the load from, for instance, overlying rocks, sediments, fluids, and buildings. Withdrawing fluids from the pore space increases the stress the solids bear, sometimes to the degree that the reservoir measurably compacts. The reduction in the pore space loops back and alters the fluid pressures. The feedback brings about more fluid movement, and the cycle continues. Terzaghi Compaction describes a conventional flow model and uses the results in postprocessing to calculate vertical compaction following Terzaghi theory.

Based on the research by Sucipto [9], erosion and sedimentation is the process of removing the grains of land from its mother in a place and the transport of the material by the movement of water or wind and then followed by the deposition of material found elsewhere [8]. The occurrence of erosion and sedimentation according to [8] depends on several factors such as rainfall characteristics, slope inclination, cover plant and soil ability to absorb and release water into shallow soil layers, the impact of soil erosion can cause sedimentation in the river so as to reduce the capacity river. A number of erosion materials that can be fully exposed from the source until they reach the control point are called sediment yields. The sediment yield is expressed in units of weight (ton) or unit volume (m^3) and is also a function of the area of the drainage region. It can also be said that the sediment yield is the amount of sediment derived from the erosion occurring in the catchment area measured over a certain time period and place [4].

This paper will discuss how the sedimentation process in the channel in two dimentional case. The problem is solved by using of the finite element method. The input data come from the physical properties of river water that contain some sediment. What will be done in this simulation is to see how sedimentation occurs in river water purification. And it will be simulated in the three drawings which are in the first drawing in the absence of the lifting of the river flow cover, the second time when the river flow cover opens, and the third when the cover is fully opened.

2. Finite Element Method with COMSOL 5.2

The basic idea in the finite element method is to find the solution of a complicated problem by replacing it by a simpler one. Since the actual problem is replaced by a simpler one in finding the solution, we will be able to find only an approximate solution rather than the exact solution. The existing mathematical tools will not be sufficient to find the exact solution (and sometimes, even an approximate solution) of most of the practical problems. Thus, in the absence of any other convenient method to find even the approximate solution of a given problem, we have to prefer the finite element method. Moreover, in the finite element method, it will often be possible to improve or refine the approximate solution by spending more computational effort.

In 1943 Courant considerably increased possibilities of the Ritz method by introduction of the special linear functions defined over triangular regions and applied the method for the solution of torsion problems [5]. As unknowns, the values of functions in the node points of triangular regions were chosen. Thus, the main restriction of the Ritz functions a satisfaction to the boundary conditions was eliminated. Some problems of chemical mixture can also be computed by using Comsol [6]. The Ritz method together with the Courant modification is similar with FEM proposed independently by Clough many years later introducing for the first time in 1960 the term finite element in the paper of finite element method in plane stress analysis [7]. The main reason of wide spreading of FEM in 1960 is the possibility to use computers for the big volume of computations required by FEM. However, Courant did not have such possibility in 1943. Although the method has been extensively used previously in the field of structural mechanics, it has been successfully applied now for the solution of several other types of engineering problems like heat conduction, fluid dynamics, electric and magnetic fields, and others.

COMSOL Multiphysics software is a powerful finite element (FEM), partial differential equation (PDE) solution engine. The basic COMSOL Multiphysics software has eight add-on modules that expand the capabilities of the basic software into the following application areas: AC/DC, Acoustics, Chemical Engineering, Earth Science, Heat Transfer, MEMS, RF, and Structural Mechanics. The COMSOL Multiphysics software also has other supporting software, such as the CAD Import Module and the Material Library. Therefore for this simulation we use COMSOL program can simulate water flow and its sedimentation using Finite Element Method.

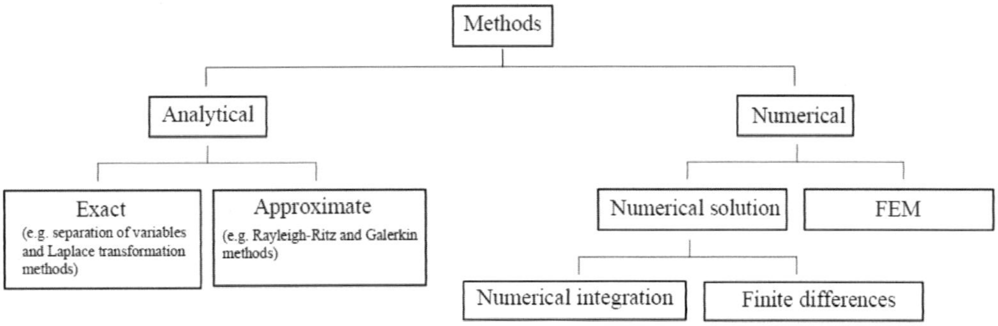

Figure 1. Clasification of common Methods

3. Simulation in two dimension

The flow field is fully described using the Darcy velocity in an equation of continuity

$$S_h \frac{\partial H}{\partial t} + \nabla.(-K\nabla H) = 0$$

where S_h is the storage coefficient (m^{-1}), K equals hydraulic conductivity (m/s), and H represents hydraulic head (m). In most conventional flow models, S_h represents small changes in fluid volume and pore space in that it combines terms describing the fluids compressibility, the solids compressibility, and the reservoirs porosity.

Instead of solving Darcy's law in the hydraulic head formulation, we solve in the pressure formulation.

$$\rho S \frac{\partial p}{\partial t} + \nabla.\rho\left[-\frac{\kappa}{\mu}(\nabla p + \rho g \nabla D)\right] = 0 \tag{1}$$

Here, the storage coefficient $S(1/Pa)$ is related to the fluid density, acceleration of gravity and the storage coefficient given in Darcy velocity equation by the relation $S = \frac{S_h}{\rho g}$. Also, the hydraulic head is related to the fluid pressure and elevation $H = \frac{p}{\rho g} + D$ and the hydraulic conductivity is related to the permeability and dynamic viscosity of the fluid $K = \frac{\kappa \rho g}{\mu}$. In this simulation obtained three images for element volume and Darcy velocity in sedimentation.

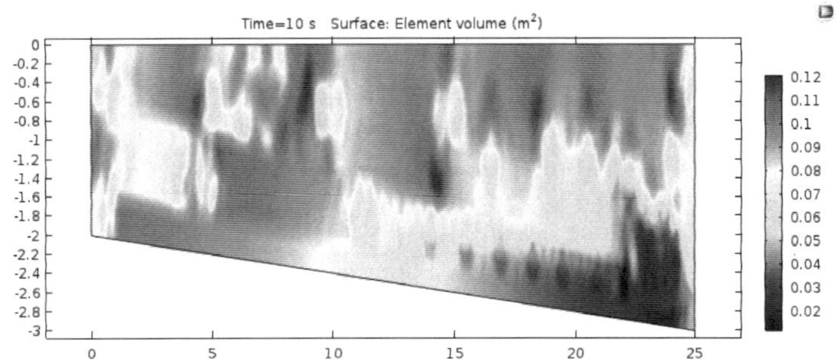

Figure 2. Simulation 1. Element Volume when Closed River Cover

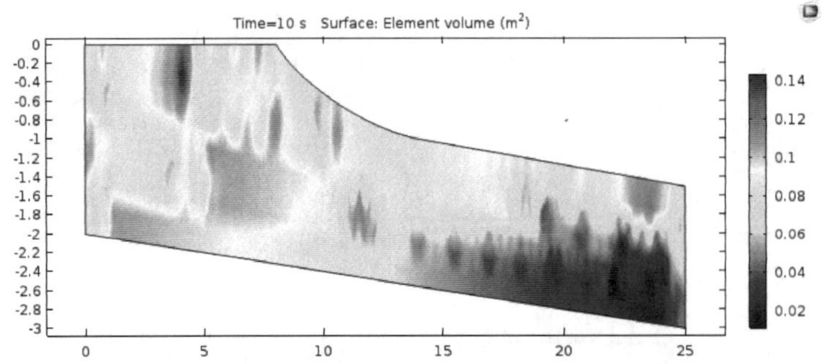

Figure 3. Simulation 2. . Element Volume at Half Open River Flow Covers

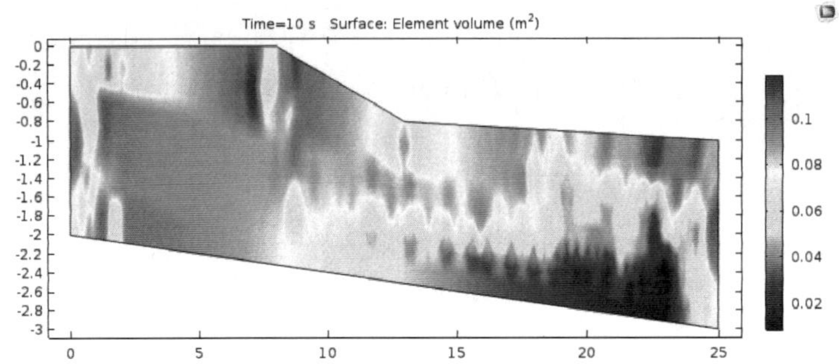

Figure 4. Simulation 3. . Element Volume on Open River Flow

From the simulation we can see how the elemental volume changes (at this time the element in question is water) in the first and second simulations, where the volume of the element decreases shown the less red color and the blue color in the image due to the change in flow shape where the dam cover In half open. While the time is fully open, the volumes of the element tend to be stable and the sedimentation formed (blue color) tends to be less, so for this condition, we can see to optimize the sedimentation process, it is necessary to open the lid of the river dam.

In figure 2, the volume on the sedimentation is stable and looks a bit mixing between water and other solid materials. While the cover will be opened, Figure 3, there is a sudden increase in the volume of solid material in the middle of the section. And when the cover is all open, the solid material is there, almost missing everything, figure 4. It shows the sedimentation process on the water purification where through the three processes we can say, the first process for the collection of solid materials with laminar currents, the second process occurs turbulence So that the solids are unified, and the third process of disposal of solid materials present in the first process

In the three simulations for Darcy Velocity, we can see how the velocity in the second simulation changed drastically because of turbulence when the cover of the dam was opened in half, but the switch was normal after it was fully opened.

4. Conclusion

A 2D numerical computation based on darcy's equation is presented to search the sedimentation phenomenon in the channel of irrigation. Three types of consideration are computed and it is concluded that there are differences between the position for the velocities.

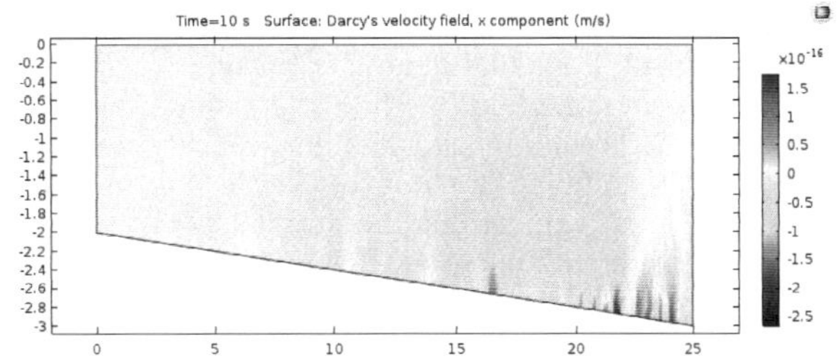

Figure 5. Simulation 1. Darcy Velocity field x when Closed River Cover

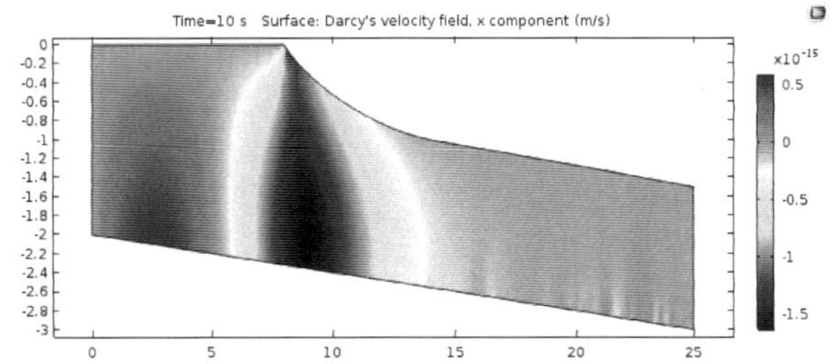

Figure 6. Simulation 2. Darcy Velocity field x at Half Open River Flow Covers

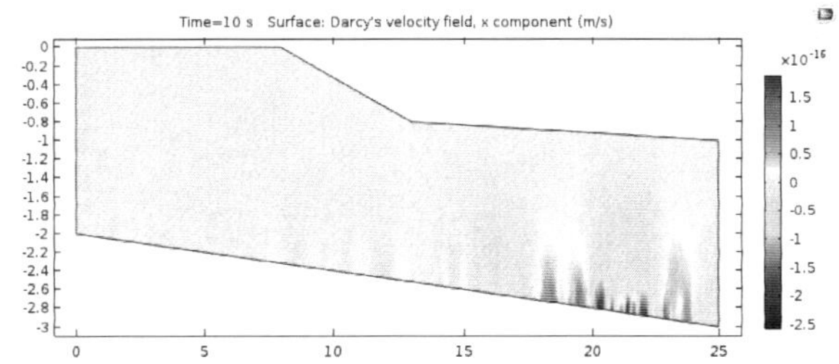

Figure 7. Simulation 3. Darcy Velocity field x on Open River Flow

Acknowledgments
The authors gratefully acknowledge that the present research is supported by Ministry of Research and Technology and Higher Education Republic of Indonesia. The support is under the research grant TALENTA USU of Year 2017.

References
[1] Tulus, and M. Situmorang 2017 Modeling of Sedimentation Process in the Irrigation Channel IOP Conference Series: Materials Science and Engineering **180(1)** 012021
[2] Tulus, Suriati, and M. Situmorang 2017 Computational Analysis of Suspended Particles in the Irrigation Channel *Journal of Physics: Conference Series* **801(1)** 012094
[3] Tulus, Suriati, M. Situmorang, and D. M. Zain 2017 Computational Analysis of Sedimentation Process in the Water Treatment Plant *Journal of Physics Conference Series* **890(1)** 012077.
[4] C. Asdak 2007 Hidrologi dan Pengendalian Daerah Aliran Sungai, Gadjah Mada University Press, Yogyakarta.
[5] R. Courant 1943 Variational methods for the solution of problems of equilibrium and vibrations *Bulletin of the American Mathematical Society* **49** 1-23.

IOP Conf. Series: Materials Science and Engineering **300** (2018) 012051 doi:10.1088/1757-899X/300/1/012051

[6] Tulus, O. S. Sitompul, Sawaluddin, and Mardiningsih 2015 Mathematical Modeling and Simulation of Fluid Dynamic in The Continuous Stirred Tank *Bulletin of Mathematics* **7**(2) 1-12

[7] R. W. Clough 1960 The finite element method in plane stress analysis. *Proc. American Society of Civil Engineers (2nd Conference on Electronic Computation, Pitsburg, Pennsylvania* **23** 345-378.

[8] Suripin 2002 *Pelestarian Sumber Daya Tanah dan Air* Yogyakarta: Penerbit ANDI.

[9] Sucipto 2008 Kajian Sedimentasi di sungai Kaligarang dalam Upaya Pengelolaan Daerah Aliran Sungai Kaligarang-Semarang *Thesis* Universitas Diponegoro Semarang.

IOP Conf. Series: Materials Science and Engineering **300** (2018) 012052 doi:10.1088/1757-899X/300/1/012052

Conceptual Model for Mitigating Human – Wildlife Conflict based on System Thinking

Pindi Patana[1], Herman Mawengkang[2], Maya Silvi Lydia[3]
[1]Graduate School of Environment Management/Faculty of Forestry, University of Sumatera Utara, Medan, Indonesia
[2]Department of Mathematics, University of Sumatera Utara, Medan, Indonesia
[3]Faculty of Computer Science and Information Technology, University of Sumatera Utara, Medan, Indonesia
[1]pindipatana@gmail.com

Email: hmawengkang@yahoo.com

Abstract. In conservation process it is unavoidably that conflict incidents may occur among the people and wild-life in the surrounding of the conservation area. Mitigating conflict between wildlife and people is considered a top conservation priority, particularly in landscapes where high densities of people and wildlife co-occur. This conflict is also happened in Leuser conservation area located in the border of North Sumatra and Aceh province, Indonesia. Easing the conflict problem is very difficult. This paper proposes a conceptual model based on system thinking to explore factors that may have great influence on the conflict and to figure out mitigating the conflict. We show how this conceptual framework can be utilized to analyze the conflict occur and further how it could used to develop a multi-criteria decision model.
Keywords: Conflict, conservation, conceptual model, system thinking

1. Introduction

Whenever a conservation takes place, particularly for a forest, a conflict between human and wildlife (HWC) could happen unavoidably. Analog with the meaning of a conflict, HWC can be defined as interaction between humans and wildlife that results in negative impacts on human social, economic or cultural life, on the conservation of wildlife populations, or on the environment. This issue has negative meaning due to the fact that it could destruct the effective conservation and also to prevent economic development and resource sustainability ([1]; [2]).

A set of global trends relating to human populations, habitat evolution and animal distribution and behaviour has contributed to the escalation of human-wildlife conflict worldwide. While humans and wildlife have co-existed for millennia, the frequency of conflicts involving problem animals has grown in recent decades, mainly because of the exponential increase in human populations and consequential expansion of human activities ([3]; [1]), partition of wildlife distributions ([4]; [5]; [6]), and natural of landscape characteristics ([7]).

Responses to tackle these conflicts are varied. However, when conflicts are handled improperly, they can create continuing public frustration, further reducing the credibility of the board which administers the program and detracting from long-term objectives ([8]; [9]). Some authors are finding that conflict management approaches can be used effectively to manage the HWC ([8]; [9]; [10]).

[11] propose a decision modeling based on multicriteria analysis to resolve human-wildlife conflicts, which requires the participation of local communities and other stakeholder groups. [2] suggests that social factors should be considered in resolving effectively HWC. He points out that

direct wildlife damage is the main driver of conflict. Technical approaches which are usually used to limit the damage ([4]; [12]) cannot be expected to lessen the conflict. However, we argue for his approach could handle HWC effectively, as there are other factors beside social should be involved. HWC can be regarded as a complex system. In order to mitigate the conflict we need to include the whole elements in the system.

[13] use mental models to analyze HWC from the perspective of a social-ecological system in Namibia. They explore the process of mind mapping to get insight the understanding of HWC in a social-ecological system. They show that the model could be used to indicate significant variables in easing the conflict.

[14] also consider the HWC in terms of social dimensions. Therefore they use multi criteria decision analysis (MCDA) as a decision support tool to evaluate the management of how to reduce HWC. [7] point out that although landscape characteristics is crucial in affecting human-wildlife interactions, it is necessary to have a better insight the mechanisms that drive those interactions. As there are many factors involved, they use a conceptual model to integrate those factors in such a way to provide a decision making model at multiple objectives.

Indicators derived from social science also focus on wildlife population size as an independent variable. Metrics such as the sociological carrying capacity [15] have been used to understand the diversity of stakeholder viewpoints related to wildlife management. Regardless of metrics, ecological and social frameworks consistently describe wildlife population size as a primary driver of human-wildlife interactions leading to conflict. Fewer conceptual frameworks have integrated ecological and social factors that affect human-wildlife interactions.

This paper is concerned with HWC occurred in the Leuser Landscape. We address a conceptual framework to mitigate the conflict. We use system thinking to describe inter-relationship among the component of the system human-wildlife as a whole.

2. Leuser Landscape

The Leuser Landscape is an area of forest located in the provinces of Aceh and North Sumatra on the island of Sumatra in Indonesia. Covering more than 2.6 million hectares it is one of the richest expanses of tropical rain forest in Southeast Asia and is the last place on earth where sumatra elephant, sumatran rhinoceros, sumatran tiger and sumatran orangutan are found within one area. It has one of the world's richest yet least-known forest systems, and its vegetation is an important source of Earth's oxygen.It is among the most biodiverse and ancient ecosystems ever documented by science

The ecosystem stretches from the coast of the Indian Ocean to the Malacca Straits. It encompasses two vast mountain ranges including Mount Leuser that reaches 3455m, two major volcanoes, three lakes and more than nine major river systems. As well as providing habitats for a number of endangered wildlife species, the ecosystem acts as a life support for approximately four million people who live around it by providing a steady supply of water, soil fertility, flood control, climate regulation and pest mitigation.

The Leuser Ecosystem comprises one of the remaining examples of Indo-Malayan (Malesian) vegetation communities with an estimated 45% of the approximately 10 000 recorded plant species. In general the ecosystem can be characterised as a montane rainforest community. However, the typical vegetation type up to an altitude of 600 metres is moist tropical lowland forest characterised by multi-layered stories with emergent trees reaching between 45 and 60 metres in height and high densities of fruit tree species. The large variety of tree species found in Leuser represent virtually all life-strategies of trees, from root flowering and trunk flowering to common twig flowering types. Among the most important and impressive trees are the several species of strangling fig. The largest flower on earth; the parasitic Rafflesia is a relatively common in the ecosystem.

3. System Thinking

The approach of systems thinking is to consider of any given issue, such as HWC, as a whole, emphasizing interrelationships between components rather than the components themselves. It does not try to break systems down into parts in order to understand them; instead, it focuses attention on how the parts act together in networks of interactions. Systems thinking is not a discipline, but rather an interdisciplinary conceptual framework used in a wide range of areas.

Despite the absence of a commonly accepted definition for systems thinking, the diverse available definitions clearly yield two main complementary meanings for systems thinking: rising above the separate components to see the whole system, and thinking about each separate component as a part of the whole system. Systems thinking is considered an effective means of facing real-world situations.

Systems thinking is not a new concept, however, it is increasingly being regarded as a new way to understand and manage complex problems at both local or global levels [16] .

[17] use the analogy of an iceberg to illustrate the conceptual model known as the Four Levels of Thinking as a framework for systemic interventions.

In this model, events or symptoms (those issues that are easily identifiable) represent only the visible part of the iceberg above the waterline. Most decisions and interventions currently take place at this level, because 'quick fixes' (treating the symptoms) appear to be the easiest way out, although they do not provide long lasting solutions. However, at the deeper (fourth) level of thinking that hardly ever comes to the surface are the 'mental models of individuals and organisations that influence why things work the way they do. Mental models reflect the beliefs, values and assumptions that a person hold ([17]).

Moving to the third level of thinking is a critical step towards understanding how these mental models can be integrated in a systems structure that reveals how the different components are interconnected and affect one another. Thus, systemic structures unravel the intricate lace of relationships in complex systems.

The second level of thinking is to explore and identify the patterns that become apparent when a larger set of events (or data points) become linked to create a 'history' of past behaviours or outcomes and to quantify or qualify the relationships between the components of the system as a whole.

The systems thinking paradigm and methodology embrace these four levels of thinking by moving decision-makers and stakeholders from the event level to deeper levels of thinking and providing a systemic framework to deal with complex problems ([17]), such as HWC problem.

4. Conceptual Framework

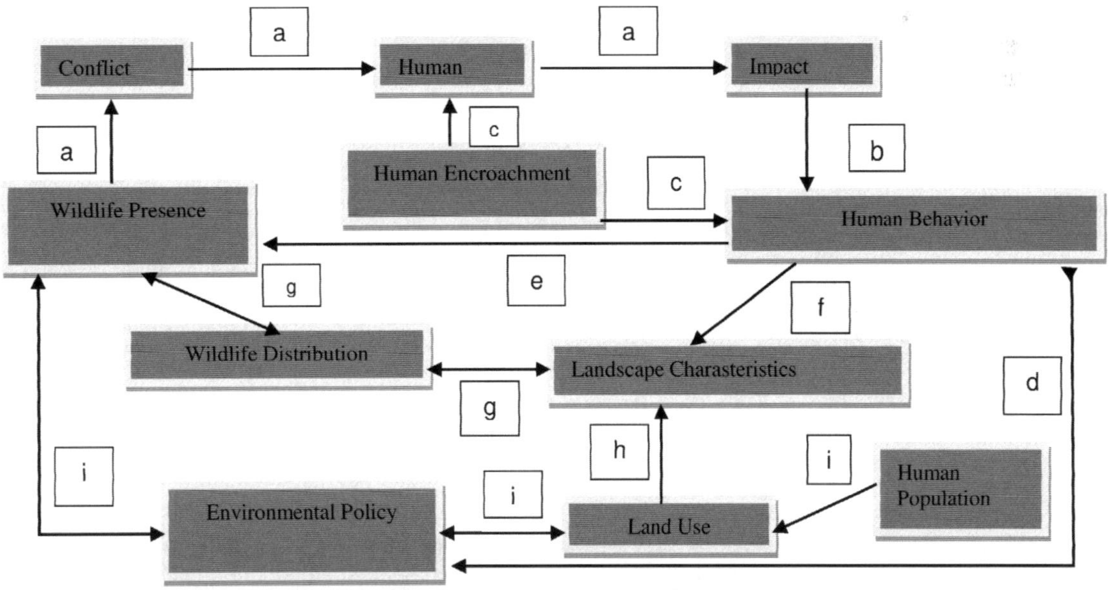

Figure 1. Conceptual framework based on system thinking

Figure 1 shows the conceptual framework model for HWC based on system thinking, which involve interactions and feedbacks within human and natural system. We use the notion of set to describe the model, such as, {conflict, human}would have an impact to {human behaviour}. Set of {human characteristic} could give influence to the set of {human reaction, human behaviour}.

Factors that influence human wildlife conflict can be described as follows.

i. Landscape characteristics
ii. Human encroachment
iii. Human population
iv. Wildlife distribution
v. Leuser board strategies
vi. Government policies

In diagram can be shown as in Figure 2.

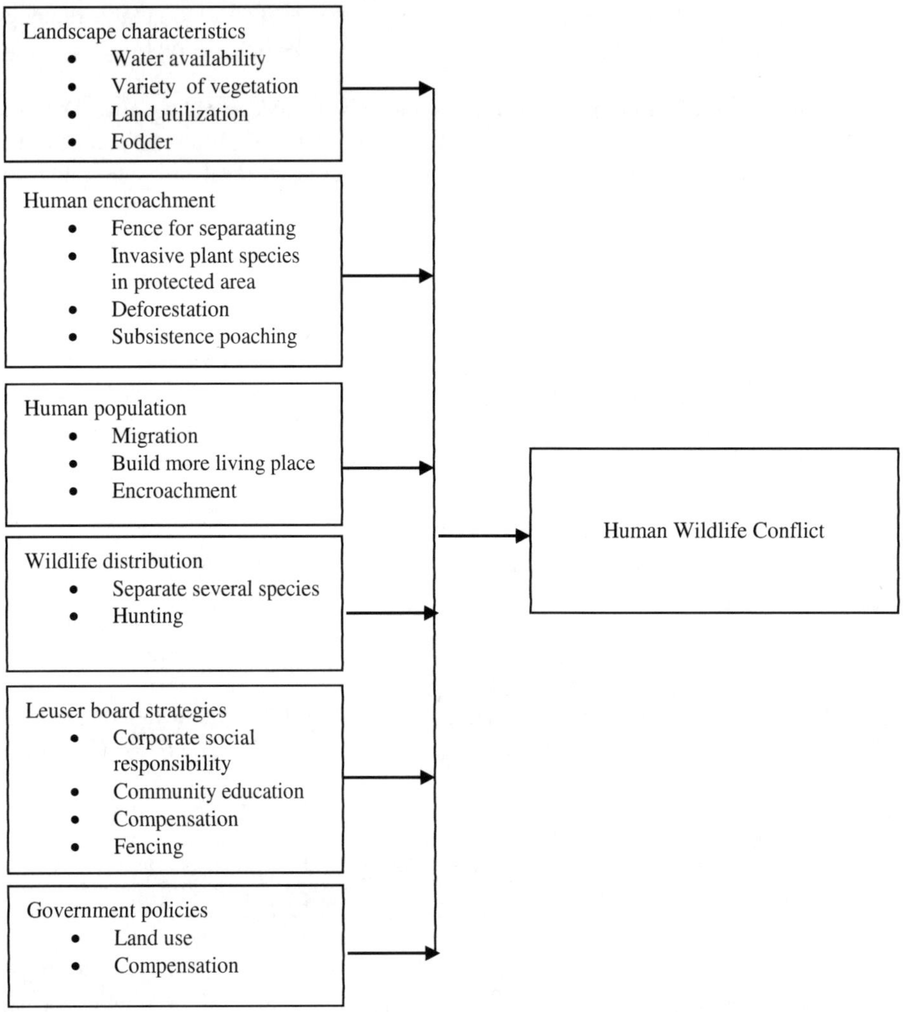

Figure 2. Conceptual framework of factors for HWC

5. Multi-criteria Decision Making

Nowadays it is quite often in the realm of decision making one encounters with problems which contain more than one objectives. These objectives in particular are independent and may have conflicts to one and another. In terms of mathematical programming, this type of problem belongs to

an approach which is called multi-criteria decision making (MCDM). Practically, goal programming (GP) is the most popular among all MCDM techniques ([18]). An interesting characteristic of GP is that a decision maker is allowed to consider the environmental, organizational, and managerial situation into the model via goal levels and priorities.
Mathematically, the lexicographic GP are expressed as follows.

The objective is to minimize deviation from the desired goals which is consistent with the AHP ranking of the alternatives.

$$Minimize \quad z = \sum_{i=1}^{m} \mathrm{P}_i (d_i^+ + d_i^-) \tag{1}$$

Subject to

$$\text{Goal constraints:} \quad \sum_{j=1}^{n} a_{ij} A_j - d_i^+ + d_i^- = b_i, \quad for\ i = 1,...,m \tag{2}$$

$$\text{System constraints:} \quad \sum_{j=1}^{n} a_{ij} A_j = b_i, \quad for\ i = m+1,...,m+k \tag{3}$$

$$\text{with } d_i^+, d_i^-, A_j \geq 0, \quad for\ i = 1,...,m;\ j = 1,...n \tag{4}$$

It can be seen that there are m goals, k system constraints and n decision variables. Where d_i^+ is a deviational variable of overachievement of goal *i*, and d_i^- is a deviational variable of underachievement of goal *i*.

6. Conclusion

We have focused on conflicts between human and wildlife due to conservation concern, particularly happened at Leuser Landscape. The conflict which threatened human living and crops is necessarily to be managed properly. We consider the HWC as a system, and then we put into the concept of system thinking. The linkages between the whole factors involved are described in the conceptual model framework. In order to mitigate the conflict optimally we will use multi-criteria decision making.

References

[1] Woodroffe, R., Thirgood, S., & Rabinowitz, A. (Eds.). 2005. *People and wildlife, conflict or coexistence?* Cambridge: Cambridge University Press.

[2] Dickman, A. J., 2010. *Complexities of conflict: the importance of considering social factors for effectively for resolving human-wildlife conflict.* Anim. Conserv. 13, pp. 458-466.

[3] Gilleland, A. H., 2010. *Human-wildlife conflict across urbanization gradients: Spatial, social, and ecological factors. Graduate Theses and Dissertation.* University of South Florida.

[4] Breitenmoser, U., Angst, C., Landry, J-M., Breitenmoser-Wursten, C., Linnell, J.D.C. & Weber, J-M., 2005. *Nonlethal techniques for reducing depredation. In People and wildlife: conflict or coexistence?:* pp. 49–61. Woodroffe, R., Thirgood, S. & Rabinowitz, A. (Eds). Cambridge: Cambridge University Press.

[5] Muruthi, P., 2005. *Human wildlife conflicts: Lessons learned from AWD's African heartland.* AWF working Papers, July 2015, 12p

[6] White, R. M., Fischer, A., Marshall, K., Travis, J. M. J., Webb, T. J., di Falco, S., Redpath, S. M., van der Wal, R., 2009. *Developing and integrated conceptual framework to understand biodiversity conflicts.* Land Use Policy, 26, pp. 242-253.

[7] Morzillo, A. T., de Beurs, K. M., Martin-Mikle, C. J., 2014. *A conceptual framework to evaluate human-wildlife interactions within coupled human and natural systems.* Ecology and Society, 19(3), pp. 44.

[8] Messmer, T., 2000. *The emergence of human wildlife conflict management: Turning challenges into opportunities.* International Bio-deterioration & Biodegredation, 45 (2000) 971102.

[9] Redpath, Young, J., Evely, A., Adams, W. M., Sutherland, W. J., Whitehouse, A., Amar, A., Lambert, R. A., Linnell, J. D. C., Watt, A., & Gutierrez, R. J., 2013. *Understanding and managing conservation conflicts.* Trends in Ecology and Evolution, 28(2).
[10] Shakeel, T, 2016. *Approaches to human-wildlife conflict management in and around Ayubia National Park.* Khyber Pakhthunkhwa, Pakistan. Discovery, 52(250), pp. 2059-2067.
[11] Redpath, S. M., Arroyo, B. E., Leckie, F. M., Bacon, P., Bayfield, N., Gutierrez, R. J., & Thirgood, S. J., 2004. *Using decision modeling with stakeholders to reduce human-wildlife conflict: A raptor-grouse case study.* Conservation Biology, 18(2), pp. 1-10.
[12] Thirgood, S., Redpath, S., 2005. *Hen herriers and red grouse: the ecology of a conflict.* Conserv. Biol., Ser., 9, pp. 192-208.
[13] Mosimane, A. W., McCool, S., Brown, P., & ingrebretson, J., 2013. *Using mental model in the analysis of human-wildlife conflict from the perspective of a social-ecological system in Namibia.* Fauna & Flora International Oryx, pp. 1-7.
[14] Davies, A. L., Bryce, R., & Redpath, S. M., 2013. *Use of multi-criteria decision analysis to address conservation conflicts.* Conservation Biology, Vol 00(00), pp. 1-9.
[15] Decker D. J., & Purdy, K. G., 1988. *Toward a concept of wildlife acceptance capacity in wildlife management.* Wildlife Society Bulletin, 16, pp. 53–57.
[16] Monat, J. P., & Gannon, T. F., 2015. *What is system thinking? A review of selected literature plus recommendations.* American Journal of Systems Science, 4(1), pp. 11-26.
[17] Maani, K. E., & Cavana, R. Y., 2007. *Systems Thinking, System Dynamic: Managing Change and Complexity (2nd Ed.).* Pearson: Prentice Hall.
[18] Lee, S. M., 1972. *Goal programming for decision analysis.* Aurbach , Philadelphia.

Semantic interpretation of search engine resultant

M K M Nasution

Teknologi Informasi, Fasilkom-TI, Universitas Sumatera Utara, Padang Bulan 20155 USU Medan Indonesia

E-mail: mahyuddin@usu.ac.id

Abstract. In semantic, logical language can be interpreted in various forms, but the certainty of meaning is included in the uncertainty, which directly always influences the role of technology. One results of this uncertainty applies to search engines as user interfaces with information spaces such as the Web. Therefore, the behaviour of search engine results should be interpreted with certainty through semantic formulation as interpretation. Behaviour formulation shows there are various interpretations that can be done semantically either temporary, inclusion, or repeat.

1. Introduction

In the logic of language there are issues that present the motivation for fomulating the phenomena of relationships between entities [1, 2]. The phenomena of this relationship is recognized as semantics such as meronymy [3], holonomy [8], hyponymy [5], synonymy [6], and polysemy [7]. Although the characteristics of this different relationships initially relate only to natural language, but because the language logically runs the technology, the consequences of any technology are directly influenced by language behavior [8, 9, 10]. For example, the disambiguation of entity name not only becomes the social nature of the language [11], but also the special nature of search engine for the social media, like in Web [12].

In general, researchers involve the natural language processing (NLP) and the search engine (SE) to reveal the semantic behavior of language logic [13]. This is done by affirming language needs such as keywords, alias search, grammar, and data structure [14]. But not a lot of efforts to explore the potential of semantic formulation based on the results were returned by a search engine, the resultant. Each search engine gives characteristics to the information it produces [15]. Therefore, search engine resultants become important to interpret through implementation of set theory of the words. This paper will describe some semantic behavior formulations generated by search engines from the information space it accesses.

2. Related Works and Motivation

Search engine resultant is the result of return of each search engine based on query content [16, 17]. Query content is a search term, and the concept of return results is generally based on the similarity between search terms and the content of the information space with which the search engine becomes its interface [18]. If $t = (w_1, w_2, \ldots, w_n)$ is a search term with w_i, $i-1, \ldots, I$ are the words. Suppose Ω is information space with members are in $\{\omega_1, \omega_2, \ldots, \omega_J\}$ with which every ω_j consists of the words w_i, $k = 1, \ldots, I$, and size of t are smaller or equal to

Table 1. Similarity of query and contents of pages in information space

t_i	$\|t\|$	$\|\omega\|$	kecocokan dengan ω	$\|t \in \omega\|$	sim
t_1	5	5	5 same words	5	1.00
t_2	5	5	4 same words	4	0.64
t_3	5	5	3 same words	3	0.36
t_4	5	5	2 same words	2	0.16
t_5	5	5	1 same words	1	0.04
t_6	5	5	0 same words	0	0.00

the size of ω or $|t| \leq |\omega|$ [19, 20]. Suppose that the search term size n with the member of ω gives the largest pseudo size is n, thus based on the measurement of similarity [21]

$$sim = \frac{2|xy|}{|x|^2 + |y|^2} \tag{1}$$

we get Table 1 for $|t| = 5$.

Suppose to apply the similarity, the search engine is modeled as follows [22, 23, 24].

Definition 1. Let Ω contains the ordered pair of the terms t_i and the pages ω_j, or (t_i, ω_j), $i = 1, \ldots, I$, $j = 1, \ldots, J$. (t_i, ω_j) as a relation table of t_i and ω_j such that $\Omega = \{(t, \omega)_{ij}\}$ as presentation of search engine, and $|\Omega|$ is a cardinality of Ω.

Definition 2. Let t_x is a search term and $t_x \in \mathcal{S}$ is a set of singleton search term of search engine. A vector space $\Omega_x \subseteq \Omega$ is a singleton search engine event (or singleton space of event) of pages that contain an occurrence of $t_x \in \omega_x$. The cardinality of Ω_x is denoted by $|\Omega_x|$, $|\Omega_x| \leq |\Omega|$.

In occurrence, the semantic interpretation of Table 1 describes that for a query $q = t_x$, the search engine returns results sequentially based on the correspondence between t_x and contents of information space. In other words, first the page in information space is relevant to a query in logical implication if it implies the query or if $\omega \Rightarrow q$ is true: $\omega \Rightarrow t$ is true for all $\omega \in \Omega$, simply $\omega \Rightarrow t = 1$ [25]. In other hand, for $|t_x| = k$ vocabularies we have $2^k - 2$ anothers of $\{\{t_k^{2^k-2}\} \subset \{w_1, w_2, \ldots, w_k\} = t_k\}$. Thus, $\omega \Rightarrow q$ is true based on Eq. (1) or simply

$$\omega \overset{sim}{\Longrightarrow} q = \omega \overset{sim}{\Longrightarrow} t_k \tag{2}$$

whereby if each the matched content of pages we denoted as search terms also: $t_x, t_y, \ldots, t_z$ then $t_z \subset \ldots \subset t_y \subset t_x$. Therefore, each search engine has a meaning semantics as follows [22].

Theorem 1. Let $t_x, t_y \in \mathcal{S}$ are two different search terms. If Ω_x and Ω_y are the singleton search engine events of t_x and t_y, respectively, then

$$|\Omega_x| = |\Omega_x| + |\Omega_y| \tag{3}$$

Definition 3. Let $t_x, t_y \in \mathcal{S}$, $t_x \neq t_y$. A doubleton search term is $\mathcal{D} = \{\{t_x, t_y\} : t_x, t_y \in \mathcal{S}\}$ and its vector space denoted by $\Omega_x \cap \Omega_y$ is a double search engine event (or doubleton space of eent) of pages that contain a co-occurrence of t_x and t_y such that $t_x, t_y \in \omega_x$ and $t_x, t_y \in \omega_y$ where $\Omega_x, \Omega_y, \Omega_x \cap \Omega_y \subseteq \Omega$.

Similar to occurrence, in co-occurrence each search engine based on doubleton has a meaning semantics as follows [23].

Theorem 2. Let $t_x, t_y \in \mathcal{S}$, $t_x \neq t_y$. If $t_x, t_y \in \Omega_x \cap \Omega_y$, then

$$|\Omega_x \cap \Omega_y| = |\Omega_x \cap \Omega_y| + |\Omega_x \cap \Omega_x| + |\Omega_y \cap \Omega_y| \tag{4}$$

3. An Approach

As adaptive approach we propose some scenarios to phenomena of implementation about set of words. Based on assumption that $t_x, t_y \in \mathcal{S}$ we get three concepts [22]

(i) $t_x \neq t_y$ and $q_1 = t_x$ and $q_2 = t_y$, or two search terms in different queries.
(ii) $t_x \neq t_y$, $t_x \cap t_y = \emptyset$, and $\omega_x \cap \omega_y \neq \emptyset$.
(iii) $t_x \neq t_y$, $t_x \cap t_y \neq \emptyset$ and $|t_y| < |t_x|$.

Lemma 1. If $q_x = t_x$ and $q_y = t_y$, then

$$|\Omega_x \cap \Omega_y| = 0 \tag{5}$$

Proof. For queries q_x and q_y we have $\Omega_x = \{(t_x, \omega_x)_{ij}\}$ and $\Omega_y = \{(t_y, \omega_y)_{ij}\}$, respectively. However, $\Omega_x \cap \Omega_y = \{(t_x, \omega_x)_{ij}\} \cap \{(t_y, \omega_y)_{ij}\} = \{(t_x \cap t_y, \omega_x \cap \omega_y)\}$ or based on Eq. (2) $\Omega_x \cap \Omega_y = \{(q_x \cap q_y, \omega_x \cap \omega_y)\} = \{\omega_x \cap \omega_y \Rightarrow q_x \cap q_y)\} = \{(\omega_x \Rightarrow q_x) \wedge \omega_y \Rightarrow q_y)\} = \emptyset$. In other words, although possible that $t_x \cap t_y \neq \emptyset$ and then not $\Omega_x \cap \Omega_y \neq \emptyset$ because resultants of queries q_1 and q_2 are different, clearly that $|\Omega_x \cap \Omega_y| = 0$.

Lemma 2. Let $t_x, t_y \in \mathcal{S}$ are two terms. If $t_x \neq t_y$, $t_x \cap t_y = \emptyset$, and $\omega_x \cap \omega_y \neq \emptyset$, then $\Omega_x = \Omega_y$, $\Omega_x, \Omega_y \subseteq \Omega$.
Proof. Based on assumption, by Definition 1 and Definition 2 we have $\forall w_x \in t_x$, $w_x \notin t_y$, and $\forall w_y \in t_y$, $w_y \notin t_x$, then $t_x \cap t_y = \emptyset$ or $t_x \cup t_y = t_y \cup t_x$, but $\forall w_x \in \omega_x$, $w_x \in \omega_y$ and $\forall w_y \in \omega_y$, $w_y \in \omega_x$, then $\omega_x \cap \omega_y = \omega_x = \omega_y$, $\omega_x \cup \omega_z = \omega_z \cup \omega_x = \omega_x = \omega_z$. For $\Omega_x = \{(t_x, \omega_x)\} = \{(t_x, \omega_x \cap \omega_y)\} = \{(t_x, \omega_x)\} \cap \{(t_x, \omega_y)\} = \{(t_x, \omega_x)\} \cap \{(t_y, \omega_y)\} = \Omega_x \cap \Omega_y$ we get $\Omega_x = \Omega_x \cap \Omega_y$. Similarly, we get $\Omega_y = \Omega_y \cap \Omega_x$. Thus, $\Omega_x = \Omega_y$ is proved.

Proposition 1. Let $t_x, t_y \in \mathcal{S}$ are two terms. If $t_x \neq t_y$ and $t_x \cap t_y \neq \emptyset$, then $|\Omega_x \cup \Omega_y| = |\Omega_x| + |\Omega_y|$.
Proof. Based on assumption, by Definition 1 and Definition 2 we have $\forall w_x \in t_x, \exists w_x \in t_y$ and $\forall w_y \in t_y$, $\exists w_y \in t_x$, then $\forall w_x \in \omega_x, \exists w_x \in \omega_y$ and $\forall w_y \in \omega_y$, $\exists w_y \in \omega_x$ such that $\{(t_x, \omega_x)\} \cap \{(t_y, \omega_y)\} = \Omega_x \cap \Omega_y \subset \Omega_x = \{(t_x, \omega_x)\}$ or $\Omega_x \cap \Omega_y \subset \Omega_y$ and $\{(t_x, \omega_x)\} \cup \{(t_y, \omega_y)\} = \Omega_x \cup \Omega_y \supset \Omega_x$ or $\Omega_x \cup \Omega_y \supset \Omega_y$. Thus, $|\Omega_x \cup \Omega_y| = |\Omega_x| + |\Omega_y| - |\Omega_x \cap \Omega_y|$, and based on Lemma 1 and Lemma 2, we have $|\Omega_x \cup \Omega_y| = |\Omega_x| + |\Omega_y|$.

Lemma 3. Let $t_x, t_y \in \mathcal{S}$ are terms. If $t_x \neq t_y$, $t_x \cap t_y \neq \emptyset$ and $|t_y| < |t_x|$, then

$$\Omega_x = \Omega_x \cup \Omega_y \tag{6}$$

Proof. Based on assumption, by Definition 1 and Definition 2, we have $\forall w_y \in t_y, w_y \in t_x$, $\exists w_x \in t_x, w_x \notin t_y$ and then $\forall w_y \in \omega_y$, $w_y \in \omega_x$, $\exists w_x \in \omega_x$, $w_x \notin \omega_y$ such that $t_x \cap t_y = t_y$ and $t_x \cup t_x = t_x$ or $\omega_x \cap \omega_y = \omega_y$ and $\omega_x \cup \omega_y = \omega_x$. Let $\Omega_x = \{(t_x, \omega_x)\}$, we have $\Omega_x = \{(t_x, \omega_x)\} = \{(t_x \cup t_y, \omega_x \cup \omega_y)\} = \{(t_x, \omega_x) \cup (t_y, \omega_y)\} = \{(t_x, \omega_x)\} \cup \{(t_y, \omega_y)\} = \Omega_x \cup \Omega_y$. It is proved.

IOP Conf. Series: Materials Science and Engineering **300** (2018) 012053 doi:10.1088/1757-899X/300/1/012053

The next step we use the concept of doubleton [23] as follows.

Proposition 2. Let $t_x, t_y \in \mathcal{S}$ are search terms. If $t_x \neq t_y$, but $\{(t_x, \omega_x)\} \cap \{(t_y, \omega_y)\} \neq \emptyset$, then for a doubleton search engine event of t_x and t_y is the $\Omega_x \cap \Omega_y$, and $\Omega_x, \Omega_y \subseteq \Omega$,

$$|\Omega_x \cap \Omega_y| \leq |\Omega_x| \leq |\Omega| \tag{7}$$

and

$$|\Omega_x \cap \Omega_y| \leq |\Omega_y| \leq |\Omega|. \tag{8}$$

Proof. By using assumption of Lemma 3, $\Omega_x \cap \Omega_y = \{(t_x, \omega_x)\} \cap \{(t_y, \omega_y)\} = \{(t_x \cap t_y, \omega_x \cap \omega_y)\} = \{(t_y, \omega_y)\} = \Omega_y$, or simply

$$\Omega_x \cap \Omega_y = \Omega_y, \tag{9}$$

or for $\Omega_y \subset \Omega_x$, we obtain $\Omega_x \cap \Omega_y \subset \Omega_x$. While based on assumption of Proposition 1, $\Omega_x \cap \Omega_y = \{(t_x, \omega_x)\} \cap \{(t_y, \omega_y)\} = \{(t_x \cap t_y, \omega_x \cap \omega_y)\} = \{(\emptyset, \emptyset)\} = \emptyset$. It means that

$$\Omega_x \cap \Omega_y \subset \Omega_x \tag{10}$$

and

$$\Omega_x \cap \Omega_y \subset \Omega_y \tag{11}$$

By using assumption of Lemma 2, $\Omega_x \cap \Omega_y = \{(t_x, \omega_x)\} \cap \{(t_y, \omega_y)\} = \{(t_x \cap t_y, \omega_x \cap \omega_y)\} = \{(t_x, \omega_x)\} = \Omega_x$, or simply

$$\Omega_x \cap \Omega_y = \Omega_x \tag{12}$$

Thus, Eqs. (9), (10), (11), and (12) give $|\Omega_x \cap \Omega_y| \leq |\Omega_x| \leq |\Omega|$ or $|\Omega_x \cap \Omega_y| \leq |\Omega_y| \leq |\Omega|$.

4. Proof of Theorem

Let Ω_x and Ω_y are two singletons, by using Eq. (5) in Lemma 1 [22], Eq. (6) of Lemma 3 be

$$\begin{aligned} |\Omega_x| &= |\Omega_x \cup \Omega_y| \\ &= |\Omega_x| + |\Omega_y| - |\Omega_x \cap \Omega_y| \\ &= |\Omega_x| + |\Omega_y| \end{aligned} \tag{13}$$

and Theorem 1 is proved.

From Eq. (12) [23] we get

$$\begin{aligned} |\Omega_x \cap \Omega_y| &= |\Omega_x| \\ &= |\Omega_x| + |\Omega_y && \text{Lemma 1} \\ &= |\Omega_x| + |\Omega_x \cap \Omega_y| && \text{Eq. (9)} \\ &= |\Omega_x| + |\Omega_y| + |\Omega_x \cap \Omega_y| && \text{Lemma 1} \\ &= |\Omega_x \cap \Omega_y| + |\Omega_x \cap \Omega_x| + |\Omega_y \cap \Omega_y| \end{aligned} \tag{14}$$

and Theorem 2 is proved.

Proposition 3. Let $t_z, \ldots, t_y, t_x \in \mathcal{S}$ are terms where $t_z \neq \ldots \neq t_y \neq t_x$ and $|t_z| < \ldots |t_y| < |t_x|$, then $\Omega_x = \Omega_x \cup \Omega_y$ holds recursively.

Proof. By the Lemma 3 and an assumption that $|t_x| < \ldots |t_y| < |t_x|$, then for $|t_z| < |t_{z_1}|$ we have $\Omega_{z_1} = \Omega_{z_1} \cup \Omega_z$, for $|t_y| < |t_x|$ we have $\Omega_x = \Omega_x \cup \Omega_y$. Thus, we obtain $\Omega_x = \Omega_x \cup \Omega_y \cup \ldots \cup \Omega_z$, or $\Omega_x = \Omega_x \cup \Omega_y$ be recursive where $\Omega_y \cup \ldots \cup \Omega_z$ is a part of Ω_x.

Corollary 1. Let $t_z, \ldots, t_y, t_x \in \mathcal{S}$. If $t_z \neq \ldots \neq t_y \neq t_x$ and $|t_z| < \ldots < |t_y| < |t_x|$, then

$$|\Omega_x| = |\Omega_x| + |\Omega_y| + \ldots + |\Omega_z| \tag{15}$$

Proof. Based on Theorem 1 and Proposition 2. we have $|\Omega_x| = |\Omega_x| + |\Omega_y| = |\Omega_x| + |\Omega_y \cup \ldots| = |\Omega_x| + |\Omega_y| + \ldots = |\Omega_x| + |\Omega_y| + |\ldots \cup \Omega_z| = |\Omega_x| + |\Omega_y| + \ldots + |\Omega_z|$.

Corollary 2. Let $t_x, t_y, t_a, t_b \in \mathcal{S}$, $t_x \cap t_y = \emptyset$ and $t_a \cap t_b \neq \emptyset$. If $|\Omega_x| = |\Omega_x| + |\Omega_a|$ and $|\Omega_y| = |\Omega_y| + |\Omega_b|$, then $|\Omega_x \cap \Omega_y| \geq 0$.
Proof. This is a direct consequence of Proposition 3 and Lemma 3.

Corollary 3. Let $t_z, \ldots, t_y, t_x \in \mathcal{S}$. If $t_z \neq \ldots \neq t_y \neq t_x$ and $|t_z| < \ldots < |t_y| < |t_x|$, then Eq. (14) holds recursively.
Proof. This is a direct consequence of Corollary 1, i.e

$$|\Omega_x \cap \Omega_y| = |\Omega_x \cap \Omega_y| + |\Omega_x \cap \Omega_x| + |\Omega_y \cap \Omega_y| + |\Omega_y \cap \ldots| + \ldots + |\ldots \cap \Omega_z| + |\Omega_z \cap \Omega_z| \tag{16}$$

5. Conclusion

Interpreting the resultant of search engines is semantically into some formulations is to provide certainty of the relationship between several entities, and it produces a temporary behavior of certainty for the events with changes of information dynamically, the inclusion of other information within an available information, and the repeat information recursively as much as iterations possible. By doing so, it is systematically possible to screen uncertain traits for trusty information relate on other and future works.

References

[1] A A Ferreira, M A Goncalves, and A H F Laender 2012 A brief survey of automatic methods for author name disambiguation *ACM SIGMOD Record* **41(2)**.
[2] M K M Nasution, and O S Sitompul 2017 Enhancing extraction method for aggregating strength relation between social actors *Advances in Intelligent Systems and Computing* **573**.
[3] H T Nguyen and T H Cao 2008 Named entity disambiguation on an ontology enriched by Wikipedia *IEEE*.
[4] M K M Nasution 2015 Extracting keyword for disambiguating name based on the overlap principle *International Conference on Information Technology and Engineering Application* (4-th ICIBA), Book 1. (arXiv: 1602.00104v1 [cs:IR] 30 Jan 2016).
[5] H T Nguyen and T H Cao 2008 Named entity disambiguation: A hybrid statistical and rule-based incremental approach *ASWC*, LNCS **5367**.
[6] L Lloyd, V Bhagwan, D Gruhl, and A Tomskins 2005 Disambiguation of references to individuals *RJI0364 Computer Science*.
[7] Y Song, J Huang, I G Councill, J Li, and C L Giles 2007 Generative models for name disambiguation (poster paper) *WWW 2007*.
[8] M K M Nasution, M Elveny, R Syah, and S A Noah 2015 Behaviour of the resoures in the growth of social network *Proceedings - 5th International Conference on Electrical Engineering and Informatics: Bridging the Knowledge between Academic, Industry, and Community, ICEEI 2015 7352551*.
[9] M K M Nasution, M Hardi, R Sitepu 2016 Using social networks to assess forensic of negative issues *Proceedings of 2016 4th International Conference on Cyber and IT Service Management, CITSM 2016 7577513*.
[10] M K M Nasution, M Hardi, R Sitepu, and E Sinulingga 2017 A method to extract the forensic about negative issues from Web *IOP Conference Series: Materials Science and Engineering* **180(1)**, 012241.
[11] H Schutze 1998 Automatic word sense discrimination *Computational Linguistics* **24(1)**.
[12] R Bekkerman and A McCallum 2005 Disambiguating Web appearances of people in a social network *The Fourteenth International World Wide Web Conference or Proc. of WWW*.
[13] M K M Nasution, R Syah and M Elfida 2018 Information retrieval based on the extracted social network *Applied Computational Intelligence and Mathematical Methods*, Advances in Intelligent Systems and Computing **662**.

[14] M K M Nasution 2014 New method for extracting keyword for the social actor *Lecture Notes in Computer Science (including subseries Lecture Notes in Artificial Intelligence and Lecture Notes in Bioinformatics)* **8397** LNAI (PART 1).
[15] M K M Nasution 2017 Modelling and simulation of search engine *Journal of Physics: Conference Series* **801(1)**, 012078.
[16] M K M Nasution 2013 Superficial method for extracting academic sosial network from the Web *Ph.D Dissertation, Universiti Kebangsaan Malaysia.*
[17] Aaron, O S Sitompul, and R F Rahmat 2015 Distributed autonomous Neuro-Gen Learning Engine for content-based document file type identification, *IEEE International Conference on Cyber and IT Service Management.*
[18] M K M Nasution, 2016, Social network mining: A definition of relation between the resources and SNA, *International Journal on Advanced Science, Engineering and Information Technology* **6(6)**.
[19] M K M Nasution, and S A Noah 2011 Extraction of academic social network from online database *2011 International Conference on Semantic Technology and Information Retrieval* (STAIR), 5995766.
[20] M K M Nasution, S A M Noah, and S Saad 2011 Social network extraction: Superficial method and information retrieval *Proceeding of International Conference on Informatics for Development* (ICID'11), (arXiv:1601.02904v1 [cs.IR] 12 Jan 2016).
[21] K M N Mahyuddin, O S Sitompul, S Nasution, and H Ambarita 2017 New similarity *IOP Conference Series: Materials Science and Engineering* **180(1)**, 012297.
[22] M K M Nasution 2012 Simple search engine model: Adaptive properties *Cornell University Library* (arXiv:1212.3906v1 [cs.IR] 17 Dec 2012).
[23] M K M Nasution 2012 Simple search engine model: Adaptive properties for doubleton *Cornell University Library* (arXiv:1212.4702v1 [cs.IR] 19 Dec 2012).
[24] M K M Nasution 2013 Simple search engine model: Selective properties *Cornell University Library* (arXiv:1303.3964v1 [cs.IR] 16 Mar 2013).
[25] M K M Nasution and S A Noah 2012 Information retrieval model: A social network extraction perspective *Proceedings - 2012 International Conference on Information Retrieval and Knowledge Management,* (CAMP'12), 6204999.

Transmission Dinamics Model Of Dengue Fever

Debora[1], Rendy[2] and Rahmi[3]
Department of Mathematics, Universitas Sumatera Utara, Medan, Indonesia

E-mail: [1]dexasi84@gmail.com , [2]rendyjs@ymail.com, [3]rahmisiregar51@yahoo.com

Abstract. Dengue fever is an endemic disease that is transmitted through the Aedes aegypti mosquito vector. The disease is present in more than 100 countries in America, Africa, and Asia, especially tropical countries. Differential equations can be used to represent the spread of dengue virus occurring in time intervals and model in the form of mathematical models. The mathematical model in this study tries to represent the spread of dengue fever based on the data obtained and the assumptions used. The mathematical model used is a mathematical model consisting of Susceptible (S), Infected (I), Viruses (V) subpopulations. The SIV mathematical model is then analyzed to see the solution behaviour of the system.

Keywords: dengue fever, endemic, SIV model, model simulation.

1. Introduction

Dengue Haemorrhagic Fever (DHF) or dengue haemorrhagic fever is a viral disease transmitted by mosquitoes that is currently the main concern of the international community. DHF is found in parts of the tropical and sub-tropical earth, mostly in urban and semi-urban areas. DHF was first noticed in Southeast Asia in the 1950s but starting in 1975 is now the leading cause of death in children in Asian countries. Even since 1997 DBD has been declared the most important viral disease of dangerous origin and fatal to humans. DHF is transmitted to humans through the bite of Aedes female mosquitoes infected with dengue virus.

According to WHO DHF problems have grown dramatically in the last decade, about 40% of the world's people are at risk of dengue. Based on information from the Health Center Protection website in Indonesia, the disease was first reported in 1968, in the cities of Jakarta and Surabaya. The epidemic of DHF outbreaks outside Java was first reported in West Sumatra and Lampung in 1972. Since then, the disease has become widespread in many parts of Indonesia. DHF transmission can only occur through mosquito bites in the body containing Dengue virus. Until now there has not been found a special drug that can kill dengue virus, therefore the main preventive effort is to avoid mosquito bites (Zeth, 2012).

2. Literature Review

2.1 The Ordinary Differential Equations

A common differential equation is an equation that expresses the relationship between a function with a single independent variable and the total derivative of this function against that independent variable. The dependent variable *(y)* depends on the physical problem being model. The independent variable is usually one of the variables of time *(t)* or space *(x)*. The Order of GDP is the highest-order derivative in a differential equation.

The general form of first order GDP is: $\frac{dy}{dt} = f(t, y)$

where *f (t, y)* is called the derivative function. For simplification of notation, the derivative is usually expressed by a single quotation mark: $y' = \frac{dy}{dt}$, So we get : $y' = f(t, y)$
GDP has a common form:

$$a_n y^{(n)} + a_{n-1} y^{(n-1)} + \dots + a_2 y'' + a_1 y' + a_0 y = F(t)$$

Where superscript (n), (n-1), etc., denotes the order derivatives to n, n-1, and so on.

2.2 Ordinary Differential Equations Linear and Nonlinear

Linear GDP is the GDP of which all derivatives appear in linear form and no coefficients depend on the dependent variable. Coefficients can be a function of independent variables, in which GDP is called linear GDP with coefficients changed.

$y' + \alpha y = F(t)$ (Linear, constant coefficient, one-order GDP)

$y' + \alpha t y = F(t)$ (Linear, changeable coefficient, one-order GDP)

$y' = f(t, y)$ (Common form of linear GDP)

If the coefficient depends on the dependent variable, or derivatives appear in the form of *nonlinear,* its GDP is *nonlinear.*

Example:

$$y\,y' + \alpha y = 0$$
$$(y')^2 + \alpha y = 0$$

$y'' + P(x, y)y' + Q(x, y)y = F(x)$ (GDP common form of *non-linear)*

2.3 Model Mathematics for Transmission Virus Dengue

Model that will Discussed on chapter this is model SIV *(Susceptible, Infectious, viruses)* on transmission virus dengue in body man, with Pay attention Facts And assumptions.

Table 1. List of Variables

Variable	Information	Terms
S (t)	amount cell that susceptible Infected virus on time *T*	$S(t) \geq 0$

$I(t)$	amount cell that Infected virus on time *T*	$I(t) \geq 0$
$V(t)$	Number of Dengue Virus in in body on time T	$V(t) \geq 0$

Table 2. List Parameters

Parameter	Information	Terms
α	Pure birth rate of susceptible cells	$\alpha > 0$
β	The proportion of dengue viruses that move	$\beta > 0$
δ	Pure cell death rate is vulnerable	$\delta > 0$
σ	The pure mortality rate of infected cells	$o > 0$
n	Number of new dengue virus duplications	$n > 0$
μ	The proportion of the number of infected cells that produce the virus	$\mu > 0$
$\gamma 1$	The pure death rate of dengue virus	$\gamma 1 > 0$
$\gamma 2$	Dengue virus death rate caused by T cells	$\gamma 2 > 0$
$\gamma 3$	The rate of death of dengue virus caused by herbal medicine	$\gamma 3 > 0$

Diagram in under this Illustrates process transmission virus dengue in in the body human.

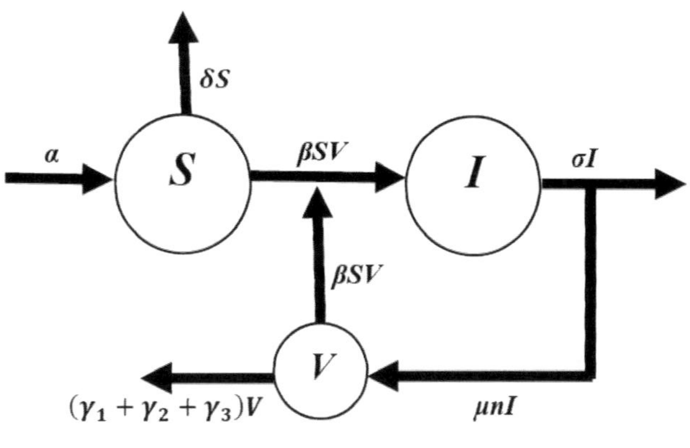

Figure 1. Diagram Process Transmission Virus dengue in Human body

Based on diagram above Obtained Differential equations of dynamic transmission virus dengue that presented in the model Mathematics, namely:

IOP Conf. Series: Materials Science and Engineering **300** (2018) 012054 doi:10.1088/1757-899X/300/1/012054

(1) $\frac{dS}{dt} = \alpha - \beta SV - \delta S$

(2) $\frac{dI}{dt} = \beta SV - \sigma I - \mu nI$

(3) $\frac{dV}{dt} = \mu nI - \gamma_1 V - \gamma_2 V - \gamma_3 V - \beta SV$

3. Conclusion

Based on the graph, the number of cells that are vulnerable to the relative (S) a lot with the passage of time will decrease and toward the value of zero. As for the number of cells that are important (I) at a time will reach the highest point, which indicates the maximum number of infected cells and gradually decreases with the passage of time. Because since S and I are getting less infected, the amount of virus in the body as well (V) decreases. This can be seen based on graphs for S, I, and V.

References:

[1]. Aminah, Ekawati. 2005. Stability Model SEIR, University of Borneo Tarakan.

[2]. Cesar, Liliam, 2011. *Modeling the Dynamic Transmission of Dengue Fever* Vol.05 Issue 1 (2011).

[3]. Hadinegoro, Sri.2004. Dengue Hemorrhagic Fever. Faculty of Medicine, University of Indonesia, Jakarta.

[4]. Leleury, Zeth Arthur. 2012. Optimal control On Dynamic Model Epidemic Dengue Using MISER3. Journal of Barekeng Vol. 6 No. 2nd Hal. 17 -21 (2012)

[5]. Purwanto, Henri ,. Evi Noviani, Muhlasah Novitasari. 2014. Analysis And Simulation of Mathematical Models of Dengue Fever With Disease One Serotif Dengue Virus.Buletin Scientific Mathematics Statistics And Applied (Bimaster) Vol. 03, No. 3 p. 153 - 162 (2014)

[6]. P, Maiti. 2004. Two Stage Non-Parametric Approach For Small Area Estimation. Proceeding of ASA Section on Survey Methods: 4058- 40665.

[7]. Leleury, Zeth Arthur. 2012. Optimal control On Dynamic Model Epidemic Dengue Using MISER3. Journal of Barekeng Vol. 6 No. 2nd Hal. 17 -21 (2012)

IOP Conf. Series: Materials Science and Engineering **300** (2018) 012055 doi:10.1088/1757-899X/300/1/012055

Information Retrieval on social network: An Adaptive Proof

M Elveny, R Syah, M Elfida, M K M Nasution*

Technical Information, Fasilkom-TI, Universitas Sumatera Utara, Padang Bulan 20155 USU Medan Indonesia

E-mail: mahyuddin@usu.ac.id

Abstract. Information Retrieval has become one of the areas for studying to get the trusty information, with which the recall and precision become the measurement form that represents it. Nevertheless, development in certain scientific fields make it possible to improve the performance of the Information Retrieval. In this case, through social networks whereby the role of social actor degrees plays a role. This is an implication of the query in which co-occurrence becomes an indication of social networks. An adaptive approach we use by involving this query in sequence to a stand-alone query, it has proven the relationship among them.

1. Introduction

Information Retrieval (IR) is a part of computer science that systematically examines the relevance of information required with information sources, which mathematically derives and proves the measurement formulation of the relevance of information resources and information required. It is from models to methods [1]. In information era with such a large and change information source dynamically [2], also continue to grow, the role of IR is crucial for generating trusty information [3]. This is based on the work of the search engine, which logically means that the document ω is relevant to the query q if it means query or $\omega \Rightarrow q$, $\omega \in \Omega$, where Ω is a space information [4].

One way of obtaining information is through the extraction of information or data mining, such as the extraction of social networks from the Web, or the mining of social structure from information sources by involving social network analysis (SNA) [5]. On the other hand, social networks obtained either by extracted (semi)-automatically or manually [6], or social networks either by document or in real terms are present in everyday life [7]. It technologically presents IR methods based on social networks [8]. Because after all the source of information such as the Web has always been a shadow of the actual state of the events [20]. However, little interest has been made to prove the existence of social network links with IR [10]. Therefore, this paper aims to reveal an IR formula adaptively based on social networks.

2. Basic Concept and Problem Definition

To be a conceptual bridge toward the problem definition, we disclose the basic concepts and related works of various literature as follows [11, 12].

Definition 1. A document d consists of the words w_k, $k = 1, \ldots, K$ or contains some vocabularies (sometimes it expressed as tokens) i.e. w_l, $l = 1, \ldots, L$ where $L \leq K$ if every word has a weight $|w_l| = \sum_{i=1}^{k_j} p(w_k) = k_i/L$ where $|w_k| = p(w_k) = 1/K$ and k_i the number of the same words for the word w_k or k_i is the word frequency of w_k.

Definition 2. A set of documents D is a collection of documents $d_i, i = 1, \ldots, I$ arranged in such a way that each document has a weight $|d|$ and every vocabulary or word has a weight $|w|$.

2.1. Information Retrieval
To model IR approach required the standard data as the comparison for results returned by the access method toward information source. Therefore, based on Definition 2, a document-set D_r serves as the comparative standard data for document generated by an access tool (search engine) whereby it depends on the query q (we call it as the evaluation document or D_e). The document comparison result will be stated in the mutual document $D_r \cap D_e$ [4, 13].

Definition 3. Document set D_r is set of documents $D = \{d_i | i = 1, \ldots, I\}$ such that each document has a unique identity $id_i = f(d_i)$ where f is a mapping that collects different addresses of the same documents, or $D_r = \langle id_i, d_i \rangle$.

Definition 4. Evaluation document D_e is a collection of documents that is accessed whereby each document has id uniquely based on the information source.

In general, to estimate the trusty information we use the measurement referred to as recall and precision as follows [14].

Definition 5. Recall (rec) is a measurement to the relevant documents is retrieved by the tool, i.e.

$$rec = \frac{|D_r \cap D_e|}{|D_r|} \tag{1}$$

where D_r is a set of relevant documents and D_e is a set of the retrieved documents, and $|D_r \cap D_e|$ is the size of $D_r \cap D_e$ and $|D_r|$ is the size of D_r.

Definition 6. Precision ($prec$) is a measurement to the retrieved documents that is relevant to real documents based on tool, i.e.

$$prec = \frac{|D_r \cap D_e|}{|D_e|} \tag{2}$$

where D_r is a set of relevant documents and D_e is a set of the retrieved documents, and $|D_r \cap D_e|$ is the size of $D_r \cap D_e$ and $|D_e|$ is the size of D_e.

2.2. Social Network
To develop the social network, we have a set of social actors $A = \{a_i | i = 1, \ldots, I\}$ and we determine the relationship between social actors in pairs based on a set of relation clues $C = \{c_j | j = 1, \ldots, J\}$. Therefore, social network can be defined as follows [15, 16].

Definition 7. A social network (SN) is a graph $G(V, E)$, $V = \{v_i | i = 1, \ldots, I\}$ as a set of vertices in G and $E = \{e_k | k = 1, \ldots, K\}$ as a set of edges in G for representing the social relationships between social actors such that

(i) $\gamma_1 : A \stackrel{1:1}{\rightarrow} V$

(ii) $\gamma_2 : R \rightarrow E$

where R is a set of relation between social actors, i.e. $r_j = c(a_k, a_l)$, $r_j \in R$, $j = 1, \ldots, J$, $a_k, a_l \in A$. We notify a social network as $\langle < V, E, A, R, C, \gamma_1, \gamma_2 \rangle$.

Definition 8. A graph $G(V, E)$ is a star graph if one of vertex $v_c \in V$ has degree $d > 1$ is more than another vertices in V and other vertices have degree $d(v_i) = 1$, $v_i \neq v_c$, and v_c as center of star graph.

Lemma 1. If degree of social actors is $d(a) > 1$, then a is a center of star graph.
Proof. For some of social actors we have degrees $d(a_1) \geq d(a_2) \geq \ldots \geq d(a_m) > 1$. Based on it, there are m candidate vertices as center, there are $m - 1$ candidate vertices as leafs, and then we build a star graph by a way we eliminate degrees until $d() = 1$ of all vertices that are not center candidates. This method is done so that all $d(a_i) > 1$ alternately will be the center of the star graph.

Theorem 1. The recall and precision as a presentation of IR can be enhanced on the basis of social network if and only if the social network is optimally shaped star graph.

3. An Approach

To be get information we use cognitive structures as an approach in some of implications as follows [4, 1].

Lemma 2. If D_r and D_e each contains uniquely document id that may be the same between two sets, then the same two id are based on the iteration of $id_i \in D_r$ againts $id_j \in D_e$.
Proof. Suppose $id_i \in D_r$ and $id_j \in D_e$. $id_j \in D_e$ is generated based on the $\omega \Rightarrow q$ implication which is true value if the content q is in the Ω, in other case it is false. While D_r contains a set of documents with specified id, and $\omega \Rightarrow id$ is true if $q \Rightarrow id$. Because D_r contains a set of id_j, so $q \Rightarrow id_j$ has to round every $id_i \in D_e$, and looping $id \in D_r$ done to $id \in D_e$.

Lemma 3. If D_r and D_e each contains a document id that might be the same so as to form $D_r \cap D_e$, looping $id_i \in D_r$ against $id_j \in D_e$ generates a sequance number of D_e.
Proof. Based on the assumptions and consequences of the Lemma 1 and Lemma 2. Suppose $id_j \in D_e$ with $id_1, id_2, \ldots, id_m$, $j = 1, \ldots, m$, as a result sequence of $\omega \Rightarrow q$. Therefore, by doing iteration $id_i \in D_r$ against one by one against from $id_j \in D_e$ from the sequence 1 to m.

Proposition 1. If $D_r \cap D_e$ contains a sequence of id, then the value of $D_r \cap D_e$ is j related to id_j.
Proof. Based on the results of the Lemma 3, the size of $|D_r \cap D_e|$ is smaller than or equal to m, i.e. $1 \leq |D_r \cap D_e| \leq m$.

Proposition 2. If $D_r \cap D_e$ contains a sequence of id, then the value of D_e is the last number of iteration towards $id_j \in D_e$.
Proof. Based on the results of the Lemma 3, the size of $|D_e|$ is smaller than or equal to m, i.e. $1 \leq |D_e| \leq m$, $|D_e|$ is the last number of iteration towards D_e.

By involving the above systematic: Lemma 1, Lemma 2, Lemma 3, Proposition 1 and Proposition 2, we disclose the following ordinances:

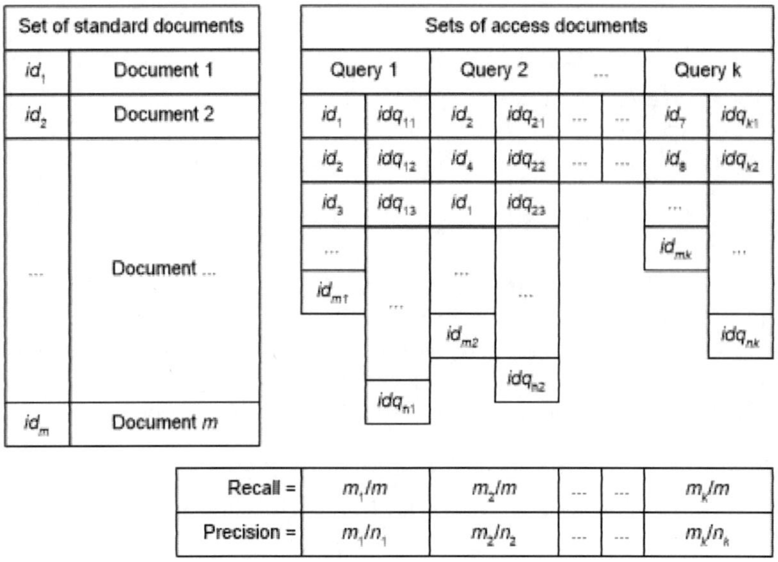

Figure 1. The recall and precision with the standard and the access documents.

Algorithm 1:
INPUT : A set of id_i from D_r
OUTPUT : $|D_r|$, $|D_e|$, $|D_r \cap D_e|$
STEPS :
(1) $id_j \leftarrow (\omega \Rightarrow q)$
(2) Set $j = 1$ to J:
  Set $i = 1$ to I:
    if $id_j = id_i$ then:
      (a) $n = j$
      (b) Collect id_j into $D_r \cap D_e$.
(3) $|D_r| \leftarrow J$, $|D_e| \leftarrow I$, $|D_r \cap D_e| \leftarrow n$.

4. Adaptive Proof

Later in this paper, we reveal the interpretive outlines involving the above approach to prove adaptively Theorem 1.

Assuming that each query stands alone, based on Algorithm 1 the recall and precision calculations can be expressed as Fig. 1, although the query contains the co-occurrence of two names of the social actor [17]. However, since any query involving co-occurrence becomes a clue of the relationship between two social actors, so it is possible that one of the social actor names is the same actor on each query [18]. Thus, the queries produce social networks that are generally the form of star graph with one social actor as the center [19].

Suppose that there is a query sequence $q_1, \ldots, q_k$ whereby each query contains $q_1 \leftarrow a_1, a_2$; $q_2 \leftarrow a_1, a_3$; $\ldots$; $q_k \leftarrow a_1, a_k$. So we get a list of $D_{e_1}, D_{e_2}, \ldots, D_{e_k}$, or a sequence of $D_r \cap D_{e_1}, D_r \cap D_{e_2}, \ldots, D_r \cap D_{e_k}$, and a set of documents is a collection of evaluation documents as follows

$$D_{es} = \cup_{l=1}^{k} D_{e_l}, \tag{3}$$

whereby $|D_{es}| \leq |D_{e_1}| + |E_{e_2}| + \ldots + |D_{e_k}|$. Whereas, a collection of documents comes from the comparison between id of 2 sets of documents is $D_r \cap D_{es} = \cup_{l=1}^{k} D_r \cap D_{e_l}$, and based on Eq.

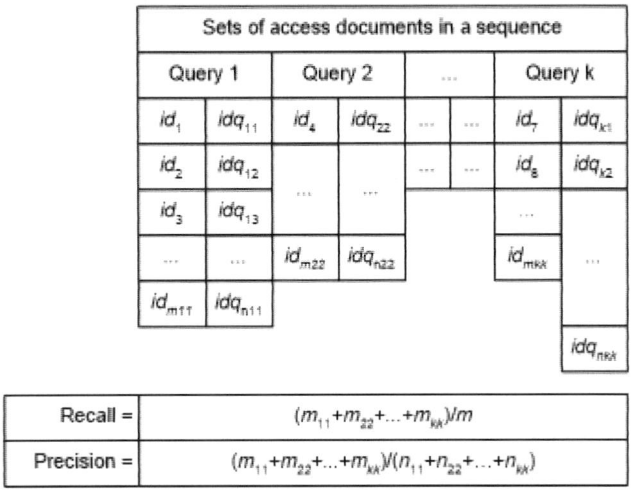

Figure 2. Adaptive approach to the recall and precision.

(3) it be

$$D_r \cap D_{es} = D_r \cap \cup_{l=1}^{k} D_{e_l} \tag{4}$$

whereby $|D_r \cap D_{es}| \leq |D_r \cap D_{e_1}| + |D_r \cap D_{e_2}| + \cdots + |D_r \cap D_{e_k}|$. In general, in Eqs. (1) and (2), the value of $|D_{e_1}| + |D_{e_2}| + \cdots + |D_{e_k}|$ and the value of $|D_r \cap D_{e_1}| + |D_r \cap D_{e_2}| + \cdots + |D_r \cap D_{e_k}|$ each has been reduced to $|\cup_{l=1}^{k} D_{e_l}|$ and $|D_r \cup_{l=1}^{k} D_{e_l}|$. This reduction as a result of merging the set of documents where the same documents to be listed once so that the value of $|D_{es}|$ close to the value of $|D_r|$ or $|D_r \cap D_{es}| \leq |D_r|$, but the number of documents in D_{es} has potential to exceed the number of documents in D_r. This causes a low precision value even if recall value is high. Taking into account that the keyword can reduce unsuitable documents, each query with a form of co-occurrence (one of the social actor names being the keyword for the other) lifts the appropriate document up to the surface [20]. Randomly assigned queries such that every D_{e_l}, $l \neq k$, $|D_{e_l}|$ has the highest value in accordance with $|D_r \cap D_{e_l}|$ in sequence, where in the next sequence the value of $|D_{e_l}|$ does not come from the same document in the previous query, while in the last query or for $|D_{e_k}|$ involves all possible documents, see Fig. 2. Thus by involving the same query as Fig. 1, taking into account the precise measurements consecutively. The query results, except the last query, are considered only to the extent that the last document is appropriate. In other words, if Eqs. (1) and (2) are restated based on Eqs. (3) and (4) as follows [1]

$$rec = \frac{|D_r \cap \cup_{l=1}^{k} D_{e_l}|}{|D_r|} \tag{5}$$

and

$$prec = \frac{|D_r \cap \cup_{l=1}^{k} D_{e_l}|}{|\cup_{l=1}^{k} D_{e_l}|} = \frac{|D_r \cap \cup_{l=1}^{k} D_{e_l}|}{|D_r \cap \cup_{l=1}^{k-1} D_{e_l}| + |D_{e_k}|} \tag{6}$$

As the implementation of the Eqs. (5) and (6) can be seen in Fig. 2. Theorem is proven.

5. Conclusion

The involvement of social actors can be used to improve the performance of recall and precision through effective approaches. An effective approach is made to the use of queries in sequence via a stand-alone query. However, implementation needs to be done by involving data and search engines, in addition to providing more definitive proof of the relation existence between the extracted social networks and IR.

References

[1] M K M Nasution, R Syah and M Elfida 2018 Information retrieval based on the extracted social network *Applied Computational Intelligence and Mathematical Methods*, Advances in Intelligent Systems and Computing **662**.

[2] M K M Nasution, M Elveny, R Syah, and S A Noah 2015 Behaviour of the resoures in the growth of social network *Proceedings - 5th International Conference on Electrical Engineering and Informatics: Bridging the Knowledge between Academic, Industry, and Community, ICEEI 2015, 7352551.*

[3] L Kirchhoff, K Stanoevska-Slabeva, T Nicolai, and M Fleck 2008 Using social network analysis to enhance information retrieval systems. *Social Networks Applicatons Conference.*

[4] M K M Nasution and S A Noah 2012 Information retrieval model: A social network extraction perspective *Proceedings - 2012 International Conference on Information Retrieval and Knowledge Management*, (CAMP'12), 6204999.

[5] M K M Nasution 2016 Social network mining: A definition of relation between the resources and SNA, *International Journal on Advanced Science, Engineering and Information Technology* **6(6)**.

[6] M Hamasaki, Y Matsuo, K Ishida, T Hope, T Nishimura, and H Takeda 2006 An integrated method for social network extraction *Proceeding of the 15th International Conference on World Wide Web* (WWW 2006).

[7] F Benhawi, N M Ali and H M Judi 2012 User engagement attributes and levels in facebook *Journal of Theoretical and Applied Information Technology* **41(1)**.

[8] P Mika 2007 *Social Networks and the Semantic Web* Springer-Verlag: Berlin.

[9] M K M Nasution 2014 New method for extracting keyword for the social actor *Lecture Notes in Computer Science (including subseries Lecture Notes in Artificial Intelligence and Lecture Notes in Bioinformatics)* **8397** LNAI (PART 1).

[10] M K M Nasution, and O S Sitompul 2017 Enhancing extraction method for aggregating strength relation between social actors *Advances in Intelligent Systems and Computing* **573**.

[11] M K M Nasution, and S A Noah 2011 Extraction of academic social network from online database *2011 International Conference on Semantic Technology and Information Retrieval* (STAIR), 5995766.

[12] M K M Nasution, S A M Noah, and S Saad 2011 Social network extraction: Superficial method and information retrieval *Proceeding of International Conference on Informatics for Development* (ICID'11), (arXiv:1601.02904v1 [cs.IR] 12 Jan 2016).

[13] M K M Nasution 2015 Extracting keyword for disambiguating name based on the overlap principle *International Conference on Information Technology and Engineering Application* (4-th ICIBA), Book 1. (arXiv: 1602.00104v1 [cs:IR] 30 Jan 2016).

[14] W B Croft, D Metzler, T Strohman 2010 *Search Engines Information Retrieval in Practice* (New York: Addison Wesley).

[15] M K M Nasution 2016 Social network mining (SNM): A definition of relation between the resources and SNA *International Journal on Advanced Science Engineering Information Technology* **6(6)**.

[16] M K M Nasution, M Hardi, R Syah 2017 Mining of the social network extraction *Journal of Physics: Conference Series* **801 (1)**.

[17] M K M Nasution 2012 Simple search engine model: Adaptive properties for doubleton *Cornell University Library* (arXiv:1212.4702v1 [cs.IR] 19 Dec 2012).

[18] M K M Nasution 2013 Simple search engine model: Selective properties *Cornell University Library* (arXiv:1303.3964v1 [cs.IR] 16 Mar 2013).

[19] M K M Nasution, O S Sitompul, E P Sinulingga, and S A Noah 2016 An extracted social network mining *Proceedings of 2016 SAI Computing Conference* (SAI).

[20] M K M Nasution 2014 New method for extracting keyword for the social actor *Source of the Document Lecture Notes in Computer Science (including subseries Lecture Notes in Artificial Intelligence and Lecture Notes in Bioinformatics)* **8397** LNAI (PART 1).

IOP Conf. Series: Materials Science and Engineering **300** (2018) 012056 doi:10.1088/1757-899X/300/1/012056

Social network extraction based on Web: 1. Related superficial methods

Mahyuddin Khairuddin Matyuso Nasution*

Teknologi Informasi,Fasilkom-TI, Universitas Sumatera Utara, Padang Bulan 20155 USU Medan Indonesia

E-mail: mahyuddin@usu.ac.id

Abstract. Often the nature of something affects methods to resolve the related issues about it. Likewise, methods to extract social networks from the Web, but involve the structured data types differently. This paper reveals several methods of social network extraction from the same sources that is Web: the basic superficial method, the underlying superficial method, the description superficial method, and the related superficial methods. In complexity we derive the inequalities between methods and so are their computations. In this case, we find that different results from the same tools make the difference from the more complex to the simpler: Extraction of social network by involving co-occurrence is more complex than using occurrences.

1. Introduction

The social network extraction is a model for constructing the data structure of the social community from the unstructured data, i.e. $\gamma_1 : A \rightarrow V$ and $\gamma_2 : A \times A \rightarrow E$, whereby A is a set of social actor names, V and E are the set of vertices and the set of edges, respectively, in $G = (V, E)$ as a graph [1]. So, as information source and social media, Web contain data about both the social actors and their relations [2]. The data in Web, over again, are growing very rapidly and dynamically changeable. In that order, there are some methods that have been developed by the researcher, and each has advantages and disadvantages [3]. In general, the methods for extracting social networks from the Web is by exploring the limited potential of search engine [4].

Extracting the social network is to establish a reliable method for revealing the trusted information about social network from dynamic media and large data like Web. Technology that may be able to optimally disclose the trusted information in certain constraint dimensions. This paper intends to analyze some methods of extracting social network from Web.

2. Related Work

The idea of the social network extraction from Web is based on the strength of relations by involving similarity distance. The strength relation comes from the results of the search of two occurrences of social actor names and co-occurrence of the social actor names. Therefore, if there are n social actor names differently, then we obtained $n(n-1)/2$ iterations computationally or exactly is as set $O(n^2)$ [5].

Some of systems have been implemented to generate social network from information sources are related Web. Therefore, the systems use the interrelated methods: The ReferralWeb [6]

Table 1. Iteration for queries and computation

No	Model	Iteration	Queries	Computation
1	Basic	$A = \{a_i \mid i = 1, \ldots, n\}$	n	
	Expanded	m keywords		$n(m(m-1)/2)$
2	Basic	$R = \{r_j \mid j = 1, \ldots, n(n-1)\}$	$n(n-1)/2$	$n(n-1)/2$
	Expanded	m keywords		$n(m(m-1)/2)$
	Threshold			$n(n-1)/2$

involved co-occurrence to find the experts on a sequence of social actors through the social network extraction. Although not disclosed that this system involves similarity distance, but the similarity by using search engine reveal clearly that results based on the implementation of similarity. On different occasion, one system has been developed (i.e. Flink) by a researcher to represent a social network of social actors through the concept similarities in the Web contents [7]. It powered by metadata such as FOAF. In other system, the researchers call it as POLYPHONET, systematically has been defined that to extract social network from Web needs a way of referring to the template, the theory of graph $G(V, E)$, which involves two steps: First define vertices (as social actors), and further define the relationship between them (the strength relations) [8]. To support the system, several measurements of co-occurrences take the decisive roles. One of them is Jaccard coefficient, i.e.

$$J_c = \frac{|X \cap Y|}{|X \cup Y|} \in [0, 1] \tag{1}$$

$|X \cap Y|$ is a cardinality of intersection of a set X and a set Y, and $|X \cup Y|$ is a cardinality of the joint of a set X and a set Y. The co-occurrence analysis is also used to extract the relations in heterogeneous community such as artists and the relations in complex community such as firms [9]. In this case, the analysis is done by adding keywords as appeals parameters in sentences that reveal the relations between firms [10, 11].

3. Kind of Approaches

Considering that the social network extraction from Web is so important [12]: Web as the information source, then be the knowledge, and based on trusted information the knowledge releases the needed wisdom [3]. Thus, the Web have some potentials that may serve as a basis for generating the social network. An extraction method which involves the search engine as the possible approach for crawling webpages [13, 14]. This approach we call as superficial method [15]. The method is to pick up information via the query q and generate hit count and snippet [16, 17, 18]. This is modeled as follows: For a pair of social actor names $a_i, a_j \in A$, query q assigned a_i or/and a_j, and any search engine will generate hit count $|a_i|$ and $|a_j|$ as occurrences and $|a_i \cap a_j|$ as a co-occurrence. So based on the queries, the search engine dedicate sa_i and sa_j as collections of snippets of a_i and a_j, respectively, and $sa_j \cap sa_j$ is a collection of snippets associated with a pair of actors [19].

Each snippet contains URL addresses, web title, and summary of webpage. The canonical form of URL address is

$$u = http(s) : //d_m \ldots d_2\dot{d}_1/p_1/p_2/\cdots/p_{n-2}/x \tag{2}$$

whereby $\{s, d, p, q\}$ = {scheme,authority,path,query}, $x = p_{n-1}$ or $x = p_{n-1}q$. Each URL address has n layers where each layer is separated by a slash [15]. While web title and summary of webpage contain words, and each snippet contains $1, \ldots, max = \pm 50$ words [20].

Thus, the model expresses some methods for extracting social network from Web as follows.

Actors	Extraction Method	Occurrence		Co-occurrence		Scale of	
		Iteration	Content	Iteration	Content	Queries	Computation
n	BSM	n	a_i	$n(n-1)/2$	a_i, a_j	$(n^2+n)/2$	$n(n-1)/2$
	BSMv		a_i, kw		a_i, a_j, kw		
	PSM		"a_i"		"a_i", "a_j"		
	PSMv		"a_i","kw"		"a_i", "a_j","kw"		
	USM		ua_i	-	-	n	
	JUSM	-	-	$n(n-1)/2$	ua_i, ua_j	$(n^2+n)/2$	
	PUSM	n	$ua_{"i"}$	-	-	n	
	JPUSM	-	-	$n(n-1)/2$	$ua_{"i"}, ua_{"j"}$	$(n^2+n)/2$	
	DSM	n	sa_i	-	-	n	
	JDSM	-	-	$n(n-1)/2$	sa_i, sa_j	$(n^2+n)/2$	
	DPSM	n	$sa_{"i"}$	-	-	n	
	JDPSM	-	-	$n(n-1)/2$	$sa_{"i"}, sa_{"j"}$	$(n^2+n)/2$	

Figure 1. Type of methods for extracting social network from Web.

3.1. Basic superficial method

The basic superficial method (BSM) involves 3 (three) times to submit queries to the search engine for a pair of social actors and the computation by involving Eq. (1) [6, 9, 21], $j_c = |a_i \cap a_j|/(|a_i| + |a_j| - |a_i \cap a_j|)$, is done with BSM procedure: $\texttt{BSM}(a_i, a_j)$, (a) $|a_i| \leftarrow q = a_i$, (b) $|a_j| \leftarrow q = a_j$, (c) $|a_i \cap a_j| \leftarrow q = a_i, a_j$, and $\texttt{return}\ j_c$. Therefore, like Table 1, for n social actors, number of queries is $n + n(n-1) = n^2$ or if in symmetric condition the number of queries is $n^2/2 + n/2$, while the computation scale is $n(n-1)/2$ times, where $|a_i \cap a_j| \leq |a_i|$, $|a_i \cap a_j| \leq |a_j|$, $|a_i| + |a_j| - |a_i \cap a_j| \neq 0$.

The BSM produces a strength relation between two social actors $a_i, a_j \in A$. This relation is influenced by social behavior [22], that is if the name of social actor is more than one word, then the other social actor name similar to one and more of the words (phrase) will be counted in, otherwise a name involves the matching pattern will generate exactly corresponding information [23, 24]. In last case, the name of the chosen social actor is a well-define name [25]. For example, the phrase Mahyuddin K. M. Nasution (without quotes) is a social name or in general a_i, "Mahyuddin K. M. Nasution" (in quotes) is a phrase in pattern form or "a_i". Generally there are not many of the same names in the form of patterns, whereas in the form of social names allows the similar names of social actors from anywhere to participate [26]. Therefore, to correctly identify the information source about a social actor, it requires the identity of each social actor, i.e. keywords kw [27]. Like Table 1, it is possible to have m keywords for each social actor, but the keyword serves to lift the related webpages from below of stack and reduces the ambiguity. Thus, BSM procedure can be expanded by adding keyword to each query: $|a_i| \leftarrow q = a_i, kw$; $|a_j| \leftarrow q = a_j, kw$; and $|a_i \cap a_j| \leftarrow q = a_i, a_j, kw$. Or we call it as the expanded basic superficial method or the basic superficial method with keyword (BSMv), see Fig. 1.

In co-occurrence concept, a name of social actor becomes a keyword (keyphrase) for the other name [28]. Therefore, if $|a_i \cap a_j| = 0$ or one actor has nothing to do with the other actor, then Eq. 1 be

$$j_m = \frac{|a_i \cap a_j|}{|a_i| + |a_j|} \in [0, 1] \tag{3}$$

where $j_m \leq j_c$. Moreover, computationally any strength relation generated from social names require a threshold [29]. It is to ensure the trusty of information about social network. In this case, j_m is the approximate value of threshold which is generally the lowest value of j_c. Thus, j_m is a threshold whereby $j_m \neq 0$ is the smallest for $n(n-1)/2$ potential strength relations. The addition of threshold computation on BSM will enhance BSM and is called as the enhanced basic superficial method (EnBSM) [30].

In case of using the name in pattern form and each query contains the couple among co-occurrence with keyword, then social network based on keyword be a community layer. In other words, there are m concentration of social actors based on m keywords, see Fig. 1.

3.2. Pattern superficial method

This method is as development of basic superficial method whereby each query contains entries in quotes. Therefore, complexity of the pattern superficial method (PSM) equal to BSM for n social actors. Similar to BSM, this method have the generated methods such as the pattern underlying superficial method (PUSM), the joint pattern underlying superficial method (JPUSM), the description pattern superficial method (DPSM), and the joint description pattern superficial method (JDPSM), see Fig. 1.

3.3. Underlying superficial method

The underlying superficial method (USM) involves twice submitting queries to the search engine for a pair of social actors [15]. Based on Eq. (1), each query q produces a list of snippets and then a list of URL addresses, or g_a URL addresses for a social actor, i.e. $L(u_a)$. Each URL has c duplication, m is a number of URL parts, thus u to be cm/n_i, n_i is layer of URL. Therefore, we get $|a_i| = \sum_{j=1}^{g_a}(c_j m_j/n_i)^2$. While each same URL of queries q_a and q_b produces the vector $qq_j = (c_{j_a} m_{j_a}/n_a)(c_{j_b} m_{j_b}/n_b)$, thus we get

$$|ab| = \sum_{j=1}^{g_{ab}} (c_{j_a} m_{j_a})(c_{j_b} m_{j_b})/(n_{i_a} n_{i_b}). \tag{4}$$

and then similarity based on USM as follows [31]

$$sim_{usm} = \frac{\sum_{j=1}^{g_{ab}} (c_{j_a} m_{j_a})(c_{j_b} m_{j_b})/(n_{i_a} n_{i_b})}{\sum_{j=1}^{g_a}(c_{j_a} m_{j_a}/n_{i_a})^2 + \sum_{j=1}^{g_b}(c_{j_b} m_{j_b}/n_{i_b})^2} \tag{5}$$

and the computation by involving Eq. (5) is done with USM procedure: $\texttt{USM}(a_i, a_j)$, (a) $L(u_a) \leftarrow q = a$, (b) $L(u_b) \leftarrow q = b$, and **return** sim_{usm}.

The USM can also be expanded and/or enhanced by involving keywords and threshold usage. Keyword inclusion will reduce the number of URL addresses being processed while the involvement of threshold point measurements will reduce the similarity between the lower layers. In general, for n social actors it takes n times the submitting to a search engine, but computationally it takes $n(n-1)/2$ iterations potentially and is charged $g_{ab} = g_a g_b$ iterations for comparison between URL addresses based on a pair of social actors. In another case, co-occurrence also generates a set of URLs for two different social actors, and this will cause in the number of queries being $n^2/2 + n/2$ times submit to the search engine, even though the computation of the Eq. (4) will be simplified into $|ab_i| = \sum_{j=1}^{g_{ab}}(c_j m_j/n_i)^2$. See Fig. 1, for the generated and related methods.

3.4. Description superficial method

Similar to USM, the description superficial method (DSM) also involves 2 (two) queries. Each query produces a list of snippets, or it produces a group of the words sets. Thus, each set of words has probability of sets and probability of word, with which each word has a weight. In other word, based on list of snippets, we have a set of vocabularies with their weights, we call it as the weighted words. The words in snippets are the description of a social actor. By using a similarity measurement towards two lists of snippets, we get the description relation between two social actors [14]. Let W_a and W_b are the sets of weighted vocabularies for social actor $a, b \in A$, respectively. A set of weighted vocabularies for both social actors a and b in A is W_{ab}, then we have a relation based on Eq (1), i.e. $sim_w = W_{ab}/(W_a + W_b - W_{ab})$. To get a weight from two weights of same words of the different social actors we propose using average of two weights, i.e. $w_{ab} = (w_a + w_b)/2$, $w_{ab} \in W_{ab}$, $w_a \in W_a$ and $w_b \in W_b$ [32]. See Fig. 1, for the generated and related methods.

The computational implementation procedures for this method involve the same concepts like USM either involving just occurrence or by involving co-occurrence as the treatment, and likewise about behaviors of complexity of this method is dependent on the use of same treatment.

3.5. Seed based superficial method

This method is developed with the concept of association rule. By the intent that search engine and query are used to obtain a URL address corresponding to the pattern of the social actor name and a name of the referenced online database. The database contains a list of scientific papers: author names, title, year, and events. Social network of authors are formed following a record of online database, and the descriptions of social network refer to titles of scientific paper. Therefore, through the titles that consist of words, we can use the concept of DSM for generating the clue of concentration of the scientists community [25]. We call this method as the seed based superficial method (SSM).

For n social actors this method employs n queries. Each social actor will generate $0, \ldots, m$ others social actors, and SSM will produce $0, \ldots, m$ relations with $0, \ldots, p$ edges, $p \leq m$, because each edge may be consists of more than one relations between two social actors. So, SSM first produces a social network based on the social actors or a social network with one social actor as central, and for n social actors this method generate the social networks. Therefore, the complexity computationally of SSM is $n(m_i)$, $i = 1, \ldots, n$, m_i is number of record rows in online database.

4. Conclusion

Using the same source, Web, there are several methods that can be employed to extract social networks. Each method has a different complexity than others, and they computationally involve multiple iterations. In general, methods involving sources related to snippets will have lighter complexity than others, as well as iterations used. Based on their complexity, the integration of methods is likely to be mutually supportive for generating credible social networks, but this needs the comparison study.

References

[1] P Mika (2007) *Social Networks and the Semantic Web.* Springer-Verlag: Berlin.

[2] M K M Nasution and S A Noah 2012 Information retrieval model: A social network extraction perspective *Proceedings of 2012 International Conference on Information Retrieval and Knowledge Management* (CAMP'12).

[3] M K M Nasution 2016 Social network mining (SNM): A definition of relation between the resources and SNA *International Journal on Advanced Science, Engineering and Information Technology* **6(6)**.

[4] Y Matsuo, J Mori, M Hamasaki, T Nishimura, T Takeda, K Hasida, and M Ishizuka 2007 POLYPHONET: An advanced social networks extraction system from the Web *Journal of Web Semantics: Science, Services and Agents on the World Wide Web* **5**.

[5] M K M Nasution, Marischa Elveny, Rahmad Syah, and S A Noah 2015 Behavior of the resources in the growth of social network *5th International Conference on Electrical Engineering and Informatics: Bridging the Knowledge between Academic, Industry, and Community* (ICEEI).

[6] Henry Kautz, Bart Selman, and Mehul Shah 1997 ReferralWeb: Combining social networks and collaborative filtering *Communication of the ACM* **40(3)**.

[7] Peter Mika 2005 Flink: Semantic Web technology for the extraction and analysis of social networks *Web Semantics: Science, Services and Agent on the World Wide Web* **3 (2-3)**.

[8] Yutaka Matsuo, Junichiro Mori, Masahiro Hamasaki, Takuichi Nishimura, Hideaki Takeda, Koiti Hasida, and Mitsuru Ishizuka 2006 POLYPHONET: An advanced social network extraction system *Proceedings of the 15th International Conference on World Wide Web 2006* (WWW 2006).

[9] YingZi Jin, Y Matsuo, and M Ishizuka 2006 Extracting a social network among entities by web mining *Workshop on Web Content Mining with Human Language* (ISWC 2006).
[10] YingZi Jin, Y Matsuo, and M Ishizuka 2007 Extracting social networks among various entities on the Web *ESWC 2007* LNCS **4519**.
[11] F Benhawi, N Mohamad Ali, and H M Judi 2012 User engagement attributes and levels in facebook *Journal of Theoretical and Applied Information Technology* **41(1)**.
[12] Jason Potts, Stuart Cunningha, John Hartley, and Paul Ormerod 2008 Social network markets: A new definition of the creative industries *Journal of Cultural Economics* **32(3)**.
[13] M K M Nasution 2017 Modelling and Simulation of Search Engine *Journal of Physics: Conference Series* **801(1)**.
[14] M K M Nasution, M Hardi, and R Syah 2017 Mining of the social network extraction *Journal of Physics: Conference Series* **801(1)**.
[15] M K M Nasution and S A Noah 2010 Superficial method for extracting social network for academics using web snipppets *Lecture Notes in Computer Science* **LNAI 6401**.
[16] M K M Nasution 2012 Simple search engine model: Adaptive properties *Cornell University Library,* arXiv:1212.3906v1.
[17] M K M Nasution 2012 Simple search engine model: Selective properties *Cornell University Library,* arXiv:13033964v1.
[18] M K M Nasution 2017 Semantic interpretation of search engine resultant *InteriOR.*
[19] M K M Nasution 2012 Simple search engine model: Adaptive properties for doubleton *Cornell University Library,* arXiv:1212.4702v1.
[20] M K M Nasution 2014 New method for extracting keyword for the social actor *Lecture Notes in Computer Science* **LNAI 8397 (PART 1)**.
[21] M K M Nasution, S A M Noah, and S Saad 2011 Social network extraction: Superficial method and information retrieval *Proceedings of International Conference on Informatics for Development* (ICID11).
[22] M K M Nasution, R Syah, and M Elveny 2017 Studies on behaviour of information to extract the meaning behind the behaviour *Journal of Physics: Conference Series* **801**.
[23] YingZi Jin, Y Matsuo, and M Ishizuka 2007 Extracting inter-Firm networks from World Wide Web *The 9th IEEE International Conference on E-Commerce Technology and the 4th International Conference on Enterprise Computing E-Commerce and E-Services* (CEC-EEE 2007).
[24] YingZi Jin, Y Matsuo, and M Ishizuka 2008 Extracting inter-Firm networks from the World Wide Web using a general-puspose search engine *Online Information Review* **32(2)**.
[25] M K M Nasution and S A Noah 2011 Extraction of academic social network from online database *2011 International Conference on Semantic Technology and Information Retrieval*(STAIR 2011).
[26] Y Matsuo, H Tomobe, and T Nishimura 2007 Robust estimation of Google counts for social networks extraction *AAAI'07 Proceedings of the 22nd National Conference on Artificial Intelligence* **2**.
[27] D Bollegala, Y Matsuo, and M Ishizuka 2007 Measuring semantic similarity between words using web search engine *Proceedings of the 16th International Conference on World Wide Web* (**WWW'07**).
[28] R M Alguliev, R M Aliguliyev, and F S Ganjeliyev 2011 Extracting a heterogeneous social network of academic researchers on the Web based on Information Retrieved from multiple sources *American Journal of Operations Research* **1**.
[29] M K M Nasution and S A Noah 2012 A methodology to extract social network form the Web Snippet *Cornell University Library,* arXiv:1211.5877.
[30] M K M Nasution, R Syah, and M Elfida 2018 Information retrieval based on the extracted social network *Applied Computational Intelligence and Mathematical Method, Advances in Intelligent Systems and Computing* **662**.
[31] M K M Nasution, O S Sitompul, Sawaluddin Nasution, and H Ambarita 2017 New Similarity *IOP Conference Series: Materials Science and Engineering* **180(1)**
[32] M K M Nasution and O S Sitompul 2017 Enhancing extraction method for aggregating strength relation between social actors." *Advances in Intelligent Systems and Computing* (AISC) **573**.

IOP Conf. Series: Materials Science and Engineering **300** (2018) 012057 doi:10.1088/1757-899X/300/1/012057

Enhancing to method for extracting Social network by the relation existence

Maria Elfida, M K Matyuso Nasution*, O S Sitompul

Technical Information, Fasilkom-TI, Universitas Sumatera Utara, Padang Bulan 20155 USU Medan Indonesia

E-mail: mahyuddin@usu.ac.id

Abstract. To get the trusty information about the social network extracted from the Web requires a reliable method, but for optimal resultant required the method that can overcome the complexity of information resources. This paper intends to reveal ways to overcome the constraints of social network extraction leading to high complexity by identifying relationships among social actors. By changing the treatment of the procedure used, we obtain the complexity is smaller than the previous procedure. This has also been demonstrated in an experiment by using the denial sample.

1. Introduction

Future of social network is not depend only on its roles in decision making [1], but information source as a resource for generating social network, whereby the resources of social network today not only consist of: vertices, edges, social actors, traditional relations like family, office, society, nation, and cultural, but also included occurrence, co-occurrence, personal names, text, graphic, image, semantic, corpus, media, and Web [2]. Therefore, the social networks extracted from the Web to be closely related to the treatment of the last mentioned collection of resources for generating trusty information [3]. In this case, the trusty information not just related to extracting social network completely from Web, but identifying correctly the components of a social network builder like the clues of relations, and the labels of vertex and edges [4, 5].

Extracting the social network is a relative approach which is formed through modal relations that depends heavily on the co-occurrence for representing relations between individuals, groups or organization [6]. Some of methods, named with superficial methods, have led to many advances in other fields and natural language processing (NLP) fields involving artificial intelligence (AI) [7]. However, the development of information extraction methods, especially with regard to social networks, is not comparable to the growth of webpages and the dynamics of their change in content. While the method of extraction is always collided with limited access to the overall information available in the source information [8]. Therefore, this paper will reveal a way of reducing the complexity of a method by identifying information related to the relationship between social actors, in addition to overcoming the limitations of access in social network extraction from the Web.

2. Related Works and Motivation

In general, to extract social network from Web formally is as follows [9, 11, 12, 10]

IOP Conf. Series: Materials Science and Engineering **300** (2018) 012057 doi:10.1088/1757-899X/300/1/012057

A1 $\gamma_1 : A \overset{1:1}{\rightarrow} V$, whereby $A = \{a_i|i = 1, \ldots, I\}$ is a set of social actors, and $V = \{v_i|i = 1, \ldots, I\}$ is a set of vertices in graph G.

A2 $\gamma_2 : A \times A \rightarrow E$, $E = \{e_j|j = 1, \ldots, J\}$ is a set of edges in graph G.

A1 means that, there is information about social actors coming from the Web. This is interpreted as submitting a query q to search engine whereby the query contains the name of a social actor, or $q_i \leftarrow a_i$ (the occurrences) [13].

A2 indirectly explains that one of clues about relation between social actors is derived from the application of co-occurrence, i.e. submitting a query q to search engine whereby the query contains a pair of social actor names, or $q_j \leftarrow a_i, a_k$ [14]. However, Cartesian product on A2 states with certainty that for n social actors, in the procedure of social network extraction there is the potential of $n(n-1)$ iterations computationally. In other words, there n times to submit query (occurrence) to search engines and $n(n-1)$ times to submit query (co-occurrence) to search engines, or $n + n(n-1) = n^2$ computationally. This is symmetrically expressed as $n + n(n-1)/2 = n/2 + n^2/2$ times for submitting query to search engine [15].

On the one hand, each search engine has the limitations of receiving query submissions, such as 1000 submissions/day/computer [10]. Meanwhile, as social media the Web constantly changing and so dynamic [16]. Thus, it may be that the old social actors still exist, but the new social actors must have arisen from everyday social activities in society [2]. On the other hand, information about social actors and their relationships must be trusty [12]. For that purpose, in step A1 the method involve keywords to uncover ambiguity or use well defined name in the form of patterns to reduce the bias, nevertheless this is not to reduce the complexity of procedures and the information space like Web [8].

In principle, the method of social network extraction from the Web depend on strength relations sr through the concept of similarity. Generally, the similarity uses three vectors based on two social actors [18], i.e.

$$sr = sim(\Omega_{a_i}, \Omega_{a_k}, \Omega_{a_i} \cap \Omega_{a_k}) \tag{1}$$

whereby $\Omega_{a_i} = q_i \leftarrow a_i$, $\Omega_{a_k} = q_k \leftarrow a_k$, $\Omega_{a_i} \cap \Omega_{a_k} = q_{i,k} \leftarrow a_i, a_k$.
Meanwhile, the last query is a form that involves symmetry of mutual keywords between a social actor and another social actor, and when a name pattern is applied, it will simultaneously be done the disambiguation and the bias reduction [14, 16, 17].

3. An Approach

In many ways, the similarity span plays a role, especially when it involves measurements related to co-occurrence [16, 18] Whether directly or indirectly, the extraction of social networks actually rests on the role of co-occurrence, which generally involves the value of $\Omega_{a_i} \cap \Omega_{a_k}$, i.e. the hit count $|\Omega_{a_i} \cap \Omega_{a_k}|$, which meets the following criteria

$$|\Omega_{a_i} \cap \Omega_{a_k}| \leq |\Omega_{a_i}| \tag{2}$$

and

$$|\Omega_{a_i} \cap \Omega_{a_k}| \leq |\Omega_{a_k}| \tag{3}$$

whereby $|\Omega_{a_i}|$ and $|\Omega_{a_k}|$ are the hit counts based on occurrences for $a_i, a_k \in A$. It means that the existence of the relationship between two actors depends on the value of $|\Omega_{a_i} \cap \Omega_{a_k}|$ for all a_i and a_k in A. Therefore, based Eq. 1 we obtain a rule as follows.

Rule 1. $\forall v_i, v_k \in V, \exists e_j \in E$ if and only if $sr > 0$ between a_i and a_k, $a_i, a_k \in A$.

Table 1. Scale of Iterations

Modification	oc	cs	sr
Before	n	$n(n-1)/2$	$n(n-1)/2$
After	$\leq n$	$n(n-1)/2$	$\leq n(n-1)/2$

Computationally, for example sr depends on formulation of hit counts as follows [18].

$$sr = \frac{2|\Omega_{a_i} \cap \Omega_{a_k}|}{|\Omega_{a_i}| + |\Omega_{a_k}|}. \tag{4}$$

So, let $\mathcal{A}$ as a space for all occurrence for A we have a procedure as follows.

```
Social_Network(A)
  for i ← 1 to n
      𝒜 = [q_i ← a_i]
  for i ← 1 to n
     for k ← i + 1 to n − 1
          function.sr(a_i, a_k, |Ω_{a_i}|, |Ω_{a_k}|)

function.sr(a_i, a_k, |Ω_{a_i}|, |Ω_{a_k}|)
  |Ω_{a_i} ∩ Ω_{a_k}| ≈ q_{i,k} ← a_i, a_k
  if |Ω_{a_i} ∩ Ω_{a_k}| ≤ |Ω_{a_i}| AND |Ω_{a_i} ∩ Ω_{a_k}| ≤ |Ω_{a_k}|:
     return sr(|Ω_{a_i}|, |Ω_{a_k}|, |Ω_{a_i} ∩ Ω_{a_k}|)
  else:
     return 0.
```

Or, briefly the procedure contain steps as follows

(i) Get $|\Omega_{a_i}|$, $i = 1, \ldots, I$.
(ii) Get $|\Omega_{a_i} \cap \Omega_{a_k}|$ in pairs for all social actors.
(iii) If $|\Omega_{a_i} \cap \Omega_{a_k}| \leq |\Omega_{a_i}|$ AND $|\Omega_{a_i} \cap \Omega_{a_k}| \leq |\Omega_{a_k}|$, then compute sr.

Thus, we get a rule for identifying the relationship between two social actors.
Rule 2. $sr > 0$ if and only if $|\Omega_{a_i} \cap \Omega_{a_k}| > 0$.

Under Rule 2, it can be stated that if one social actor has not the hit count about co-occurrence > 0 with all the other social actors, then occurrence should not be disclosed.

Therefore, two first steps in procedure of extracting social network from Web by using search engine change be as follows:

(i) Get $|\Omega_{a_i} \cap \Omega_{a_k}|$ in pairs for all social actors.
(ii) If $|\Omega_{a_i} \cap \Omega_{a_k}| > 0$, then get $|\Omega_{a_i}|$ and $|\Omega_{a_k}|$.
(iii) If $|\Omega_{a_i} \cap \Omega_{a_k}| \leq |\Omega_{a_i}|$ AND $|\Omega_{a_i} \cap \Omega_{a_k}| \leq |\Omega_{a_k}|$, then campute sr.

For which the number of submitting query to search engine as the occurrence may be less than or equal to n social actors. Thus, the complexity of social network extraction $< n/2 + n^2/2$. Computationally, the number of iterations for sr less than or equal to $n(n-1)/2$ with considering that $|\Omega_{a_i} \cap \Omega_{a_k}| = 0$ is not count in computation.

Let us redefine the number of queries for occurrence for n social actors as the occurrence scale (os) and the number of queries for co-occurrence as co-occurrence scale (cs), we obtain the comparison like Table 1 (Scale of Iterations).

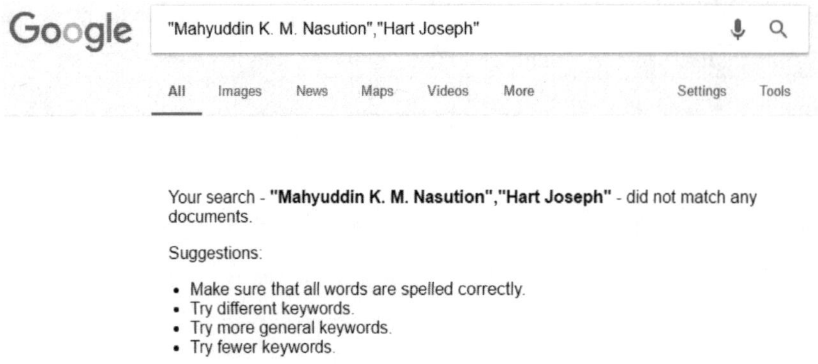

Figure 1. One of 7 names in denial sample: "Hart Joseph"

4. Experiment

Let us implement the last procedure as follows

Social_Network(A)
  identify(A)
  if $\mathcal{I} > 0$:
    for $i \leftarrow 1$ to n
      $\mathcal{A} = [q_i \leftarrow a_i]$
  for $i \leftarrow 1$ to $n_{\mathcal{I}}$
    for $k \leftarrow i + 1$ to $n_{\mathcal{I}} - 1$
      function.sr($\mathcal{A}, \mathcal{A}_{cs}$)

identify(A)
  $\mathcal{I} = [0]$
  for $i \leftarrow 1$ to n
    for $k \leftarrow i + 1$ to $n - 1$
      $|\Omega_{a_i} \cap \Omega_{a_k}| \approx q_{i,k} \leftarrow a_i, a_k$
      if $|\Omega_{a_i} \cap \Omega_{a_k}| > 0$:
        $\mathcal{I} = [1]$
        $\mathcal{A}_{cs} = [|\Omega_{a_i} \cap \Omega_{a_k}|]$

function.sr($\mathcal{A}, \mathcal{A}_{cs}$)
  if $|\Omega_{a_i} \cap \Omega_{a_k}| \leq |\Omega_{a_i}|$ AND $|\Omega_{a_i} \cap \Omega_{a_k}| \leq |\Omega_{a_k}|$ AND
    $|\Omega_{a_i} \cap \Omega_{a_k}| \leq |\Omega_{a_i}| > 0$:
    return $sr(|\Omega_{a_i}|, |\Omega_{a_k}|, |\Omega_{a_i} \cap \Omega_{a_k}|)$

where $\mathcal{I}$ contains binary number $\{0, 1\}$, and the relations between two social actors we call as the binary relationship.

In an experiment involving 469 social actors, there were 7 (seven) names of social actors who at the time of the experiment were not on the Web. 7 names are used as a denial and endorsement for the identification part of procedure, to identify the relationship between social actors. For example, by using Google search engine for query q ="Mahyuddin K. M. Nasution","Hart Joseph", we have

$$|\Omega_{\text{"Mahyuddin K. M. Nasution"}} \cap \Omega_{\text{"Hard Joseph"}}| = 0,$$

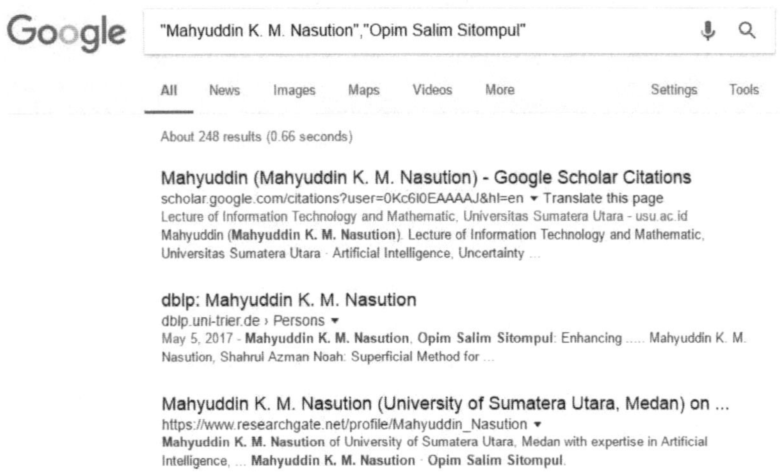

Figure 2. Relation between "Mahyuddin K. M. Nasution" and "Opim Salim Sitompul"

see Fig. 1, while for query $q =$"Mahyuddin K. M. Nasution","Opim Salim Sitompul", we have

$$|\Omega_{\text{"Mahyuddin K. M. Nasution"}} \cap \Omega_{\text{"Opim Salim Sitompul"}}| = 248,$$

see Fig. 2. Tautologically, the method of identification by involving a denial sample behaves correctly if and only the method behaves in a general manner in the population. In other words, if the existence of 7 social actors is not on the Web, then any relationship between one of 7 social actors and one of 462 other social actors does not exist. Thus, there are as many as 3,276 binary relationships cannot be expressed through the information source. Meanwhile, involving 462 names of social actors in the pattern form we obtained 22,836 binary relationships or 21.44% of 106,491 potential relationships. Therefore, the complexity computationally requires 22,836 iterations and 23,988 times generating information for producing the strength relations.

5. Conclusion

In this study have disclosed the key that determines the complexity of social network extraction methods. By changing the treatment of the procedure based on identifying the existence of the relationship between the social actors, the number of queries submissions into the search engine can be reduced, as well as computationally, the iterations number is less than $n/2 + n^2/2$.

References

[1] M K M Nasution 2016 Social network mining (SNM): A definition of relation between the resources and SNA *International Journal on Advanced Science, Engineering and Information Technology* **6(6)**.
[2] M K M Nasution, M Elveny, R Syah, and S A M Noah 2015 Behavior of the resources in the growth of social network *Proceedings of the 5th International Conference on Electrical Engineering and Informatics* (ICEEI).
[3] M K M Nasution, M Hardi, and R Syah 2017 Mining of the social network extraction *Journal of Physics: Conference Series* **801(1)**
[4] M K M Nasution and Shaurul Azman Noah 2011 Extraction of academic social network from online database *Proceeding of 2011 International Conference on Semantic Technology and Information Retrieval* (STAIRS11)).
[5] M K M Nasution, R Syah, and M Elveny 2017 Studies on behaviour of information to extract the meaning behind the behaviour *Journal of Physics: Conference Series* **801(1)**.
[6] M K M Nasution and S A M Noah 2010 Superficial method for extracting social network for academic using web snippets *Rough Set and Knowledge Technology*, **LNAI 6401**.

[7] M K M Nasution, S A M Noah, and S Saad 2011 Social network extraction: Superficial method and information retrieval *Proceeding of International Conference on Informatics for Development* (ICID'11).
[8] M K M Nasution 2017 Modelling and Simulation of Search Engine *Journal of Physics: Conference Series* **801(1)**.
[9] M K M Nasution and Shahrul Azman Noah 2012 Information retrieval model: A social network extraction perspective *International Conference on Information Retrieval & Knowledge Management* (CAMP12).
[10] Y Matsuo, J Mori, M Hamasaki, T Nishimura, H Takeda, K Hasida, M Ishizuka 2007 POLYPHONET: An advanced social network extraction system from the Web *Journal of Web Semantics* **5**.
[11] P Mika 2005 Flink: Semantic web technology for the extraction and analysis of social networks *Journal of Web Semantics* **3**.
[12] P Mika 2007 *Social Networks and the Semantic Web* Heidelberg, Berlin: Springer.
[13] M K M Nasution 2012 Simple Search Engine Model: Adaptive Properties *Cornell University Library* arXiv:1212.3906 [cs.IR].
[14] M K M Nasution 2012 Simple Search Engine Model: Adaptive Properties for Doubleton *Cornell University Library* arXiv:1212.4702 [cs.IR].
[15] M K M Nasution 2013 Superficial method for extracting academic social network from the Web *Ph.D. Dissertation* Universiti Kebangsaan Malaysia.
[16] M K M Nasution 2014 New method for extracting keyword for the social actor *Lecture Notes in Computer Science (including subseries Lecture Notes in Artificial Intelligence and Lecture Notes in Bioinformatics)* **LNAI 8397 (PART 1)**.
[17] M K M Nasution 2015 Extracting keyword for disambiguating name based on the overlap principle *Proceeding of International Conference on Information Technology and Engineering Application* (4-th ICIBA) **Book 1**.
[18] K M N Mahyuddin, O S Sitompul, Sawaluddin Nasution, H Ambarita 2017 New similarity *IOP Conference Series: Materials Science and Engineering* **180(1)**.

CFD Analysis of Evaporation-Condensation Phenomenon In an Evaporation Chamber of Natural Vacuum Solar Desalination

H Ambarita[1,2*], A D Ronowikarto[1,2*], R E T Siregar[1,2] and E Y Setyawan[2,3]

[1]Mechanical Engineering, Faculty of Engineering, University of Sumatera Utara Medan, Indonesia
[2]Sustainable Energy Research Centre, Faculty of Engineering, University of Sumatera Utara Medan, Indonesia
[3]Mechanical Engineering, National Institute of Technology, Malang, Indonesia

Corresponding author: himsar@usu.ac.id

Abstract. Desalination technologies is one of solutions for water scarcity. With using renewable energy, like solar energy, wind energy, and geothermal energy, expected will reduce the energy demand. This required study on the modeling and transport parameters determination of natural vacuum solar desalination by using computational fluid dynamics (CFD) method to simulate the model. A three-dimensional case, two-phase model was developed for evaporation-condensation phenomenon in natural vacuum solar desalination. The CFD simulation results were compared with the avalaible experimental data. The simulation results shows inthat there is a phenomenon of evaporation-condensation in an evaporation chamber. From the simulation, the fresh water productivity is 2.21 litre, and from the experimental is 2.1 litre. This study shows there's an error of magnitude 0.4%. The CFD results also show that, vacuum pressure will degrade the saturation temperature of sea water.

1. Introduction

Water scarcity already affects every continent. Around 1.2 billion people, or almost one-fifth of the world's population, live in areas of physical scarcity, and 500 million people are approaching this situation [1].Water scarcity is among the main problems to be faced by many societies and the World in the XXIst century [1]. Water use has been growing at more than twice the rate of population increase in the last century, and, although there is no global water scarcity as such, an increasing number of regions are chronically short of water [1]. Sharon and Reddy [2] investigated, some regions of the globe are under severe stress due to water scarcity and pollution.The fresh water needs of man kind can be only satisfied if saline water which is available in plenty is converted to portable water by desalination. Desalination tehnologies is one of solution for this problem. Desalination is widely adopted in Middle east, Arab countries, North America, Asia, Europe, Africa, Central America, South America and Australia to meet their fresh water and process water demands [2]. Ambarita [3] investigated, in Indonesia, some areas also use desalination to produce fresh water for their living. There are many methods can be been used in producing fresh water from sea water. The used of renewable energy such as solar, wind, and geothermal energy expected will reduce the energy demand. Eltawil et al. [4]

investigated, 57% of desalination technologies is using solar energy. Ambarita [5] investigated, the higher thermal efficiency is the better performance. Working temperature of solar collector is very important. These facts reveal that the maximum temperature in the evaporator plays an imporant role of the system performance [5].

Thu et al. [6] and KC Ng et al. [7] had researched on renewable adsorption desalination technology and concluded that the introduction of solar significantly reduce energy consumption in water production.The principle of desalination techonolgy in general is actually very simple. Sea water in an evaporation chamber was heated until it evaporates, then the vapor will condensed and streamt down towater storage. The result of vapor condensed is fresh water. In this natural vacuum solar desalination system, is using fluid which heated in copper coil by flat plate type solar collector as a heater. The length of copper coil s 5.5 m. An evaporator as an evaporation chamber is an important component. A solar heating system, an evaporator, a condenser at a height of about 10.34 above ground level, connected via pipes to a sea water and fresh water storage. Ladislav et al. [8] investigated, condensation is associated with the heat transfer from the containment atmosphere to the containment walls and it is one of the factors that determine pressure and steam concentration in the containment. Condensation on the walls can promote containment atmospheremixing and it is a very effective passive cooling mechanism. The vacuum pressure will degrade the saturation temperature of sea water. There is no need electricity to operate this natural vacuum solar desalination. Literature review shows that there has been less attention to evaporation-condensation phenomenon by using CFD. In this study, a two-phase model with 3-dimensional geometry was developed for evaporator as an evaporation chamber of natural vacuum solar desalination. Liquid volume fraction and mass flow rate was determined. Simulation results were compared with experimental data.

2. Method

The research has been conducted at Sustainable Energy Research Centre, Faculty of Engineering, University of Sumatera Utara, Medan with geographic location are $3° 30' - 3° 43'$ North Latitude and $98° 35' - 98° 44'$ East Longitude. Typical of the evaporation chamber and natural vacuum desalination are shown in Figure 1.

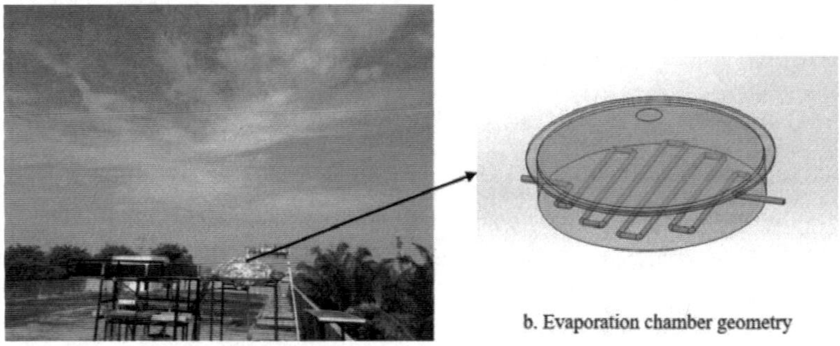

a. Natural Vacuum Solar Desalination

b. Evaporation chamber geometry

Figure 1 Natural vacuum solar desalination and evaporation chamber

2.1 Numerical Method

A two phase with 3-dimensional geometry was developed in Mixture framework for evaporation-condensation phenomenon in a evaporation chamber of natural vacuum solar desalination. Turbulence flow with constant fluid properties and the gravity acts in vertical direction were used.

1. Continutity equations are given by Setoodeh et al [9] :

- Gas phase

$$\nabla \cdot (r_G \rho_G V_G) + S_{LG} = 0 \quad (1)$$

- Liquid phase

$$\nabla \cdot (r_L \rho_L V_L) + S_{LG} = 0 \quad (2)$$

Where S_{LG} is the rate of mass transfer from liquid phase to gas phase and vice versa [9]. Setoodeh et al [9] investigated mass transfer between phases satisfy the local balance condition, so $S_{LG} = -S_{GL}$

2. Pressure constraint and mass transfer equations given by Setoodeh et al [9] :

- Pressure constraint

$$P_G = P_L = P \quad (3)$$

- Gas phase mass transfer

$$\nabla \cdot \left[r_G\left(\rho_G V_G Y_A - \rho_G D_{AG}(\nabla Y_A)\right)\right] - S_{LG} = 0 \quad (4)$$

- Liquid phase mass transfer

$$\nabla \cdot \left[r_L\left(\rho_L V_L Y_A - \rho_L D_{AL}(\nabla X_A)\right)\right] + S_{LG} = 0 \quad (5)$$

3. The total number of evaporation calculated by:

$$m = \dot{m} \times t \quad (6)$$

$$V = \frac{m}{\rho} \quad (7)$$

Where $\dot{m}$ is the mass flow rate and t is the time of evaporation.

2.2 Experimental aspects
Heating fluid by solar collector in a copper coil with a length 5.5 m and diameter of 5/6 inch was used. It transferred to evaporation chamber by a DC pump which generated by photovoltaic. The operation pressure was a vacuum pressure. The vacuum pressure gauge an the thermocouples were used in the experiment to measure an operation pressure and temperature. The measurement tools is shown in Figure 2. In this study, 70°C of the copper coil temperature and 0.46 atm of vacuum pressure were examined. Figure 3. shows the result of measurement.

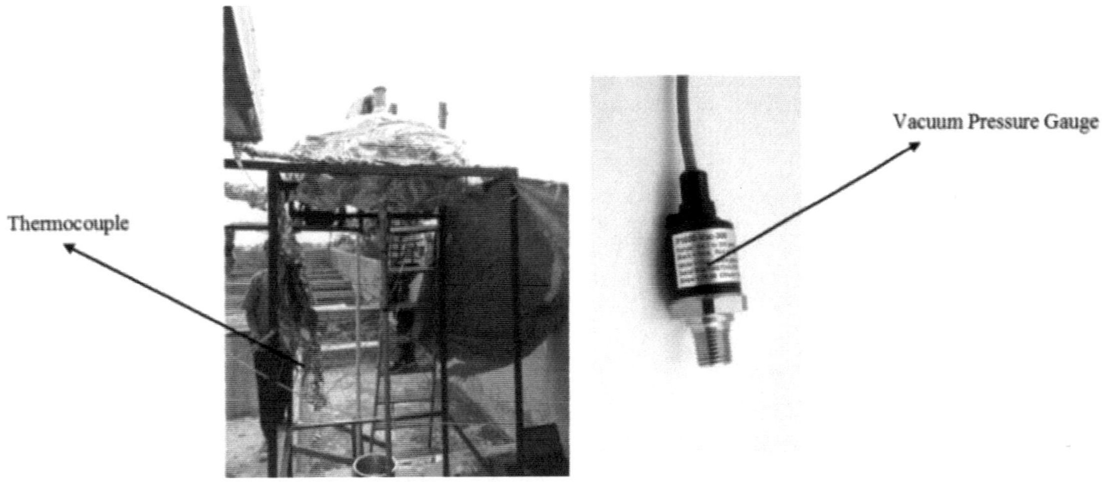

Figure 2 The measurement tools

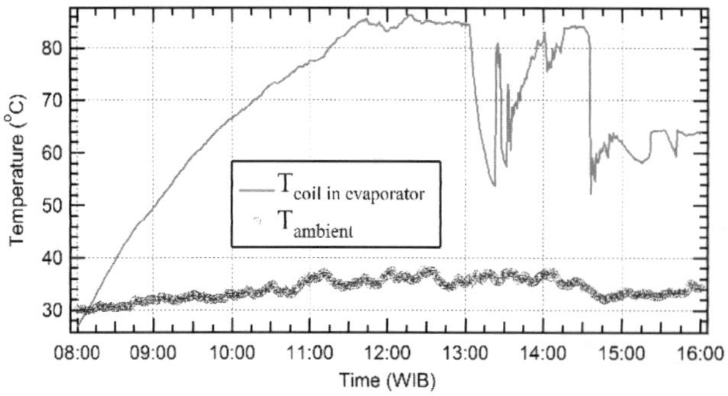

Figure 3 The result of measurement

3. Results and discussions

3.1 Initial and boundary conditions

The recorded experimental data are for 6 hours, but it's impossible to run CFD simulation for 6 hours. In this study, it was assumed in a time period of 1 hour. Boundary conditions for evaporation chamber was divided into two boundary, a heater was solid boundary and an evaporator was fluid boundary. Pressure outlet boundary condition was specified for the vapor at the exit. The heat flux on evaporator was 0. According to the experiments, the value of copper coil temperature 70°C was applied as a fixed values of copper coil in evaporation chamber. The constant operation pressure of 0.46 atm based on the experimental data and gravity acceleration of 9,8 m/s^2 were used.

3.2 Simulation results

The CFD analysis was carried out using the commercial package ANSYS Fluent 16.1 on eight 2,6 GHz CPU processors was used for solving the equations. Building model geometry was done using SolidWorks 2016. Its meshing done using workbench. This study focus on liquid volume fraction and mass flow rate in evaporator. Sea water in the system was evaporated by fluid inside copper coil which heated by flat plate type solar collector. Mass flow rate from simulation results was compared with experimental data. Fig shows the results of simulation in 1 h time period. Fresh water production rate predicted by CFD simulation was compared agains experimental data. The phenomenon of evaporation-condensation in an evaporation chamber has investigated as shown in Figure 4**.** The orange area is shows the liquid phase and the blue area is shows the gas phase.The mass flow rate also investigated by CFD simulation is7.69x10^{-5} kg/s.By using equation (6) and (7) fresh water production has been calculated. The results from the CFD analysis and the experiment are presented in Table 1.The error of magnitude is 0.4%.

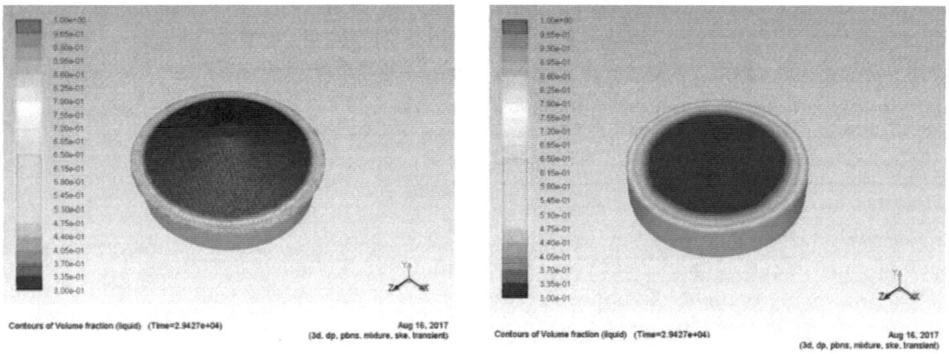

(a) Liquid volume fraction (b) Liquid volume fraction (opened)

Figure 4. The phenomenon of evaporation-condensation

Table 1 Results Comparison

	Mass Flow Rate	Fresh Water Production
Experimental	*unknown*	2.1 litre
CFD Analysis Result	7,69x10^{-5} kg/s	2.21 litre

4. Conclusion

The phenomenon of evaporation-condensation in an evaporation chamber of natural vacuum solar desalination has been studied using CFD simulation. The CFD simulation of evaporation-condensation was carried out on the gas mixture only (as shown in Figure 4.). Mixture framework with turbulence flow were used to the simulations. The results clearly show that water was condensed. The wall thickness and heat flux were neglected. The calculation of mass flow rate was in good agreement with experimental data. An error of magnitude is 0.4%.

References

[1] http://www.un.org/waterforlifedecade/scarcity.shtml (Accesed September 24, 2017)
[2] Sharon H and Reddy K S 2015 *Renewable and Sustainable Energy Reviews* **41**, 1080-1118
[3] Ambarita, H 2016 *Case Studies in Thermal Engineering* **8** 346-358
[4] Eltawil A M, Zhengming Z, and Liqing Y 2008 *Twelfth International Water Technology Conference* 22-23
[5] Ambarita, H 2017 *IOP Conf. Series: Materials Science and Engineering* **180** 012024
[6] Thu K., Yanagi H., Saha B.B, Ng. Kim 2013 *International Journal Heat Mass Transfer* **65** 662e9
[7] Narayan G P, Sharqawy M H, Summers E K, Lienhard J H, Zubair S M, Antar M A 2010 *Renewable Sustain Energy Rev* **14** 1187–201
[8] Ladislav V, Josef S, Jiri M 2014 *Nuclear Engineering and Design* **279** 147-157
[9] Setoodeh N, Rahimi R, and Ameri A 2011 *Journal of Desalination* **268** 103-110
[10] Jun-De Li 2013 *International Journal of Heat and Mass Transfer* **57** 708-721

IOP Conf. Series: Materials Science and Engineering **300** (2018) 012059 doi:10.1088/1757-899X/300/1/012059

Numerical Study on Radiation Effects to Evaporator in Natural Vacuum Solar Desalination System

R E T. Siregar[1,2], A D. Ronowikarto[1,2], E Y Setyawan[3] and H Ambarita[1,2*]

[1]Mechanical Engineering, Faculty of Engineering, Universitas Sumatera Utara, Medan, Indonesia 20155
[2]Sustainable Energy Research Center, Mechanical Engineering, Universitas Sumatera Utara, Medan, Indonesia
[3]Mechanical Engineering, National Institute of Technology Malang

Corresponding author: himsar@usu.ac.id

Abstract. The need for clean water is increasing day by day due to the increasing factor of living standard of mankind, hence designed natural vacuum solar desalination. The natural vacuum Solar desalination is studied experimentally. A small-scale natural vacuum desalination study consists of evaporator and condenser as the main components designed and manufactured. To transfer heat from the solar collector into the evaporator, the fluid transfer system uses a pump powered by a solar cell. Thus, solar collectors are called hybrid solar collectors. The main purpose of this exposure is to know the characteristics of the radiation effects on incoming energy on the evaporator during the process. This system is tested by exposing the unit to the solar radiation in the 4th floor building in Medan. The experiment was conducted from 8.00 to 16.00 local time. The results show that natural vacuum solar desalination with hybrid solar collectors can be operated perfectly. If the received radiation is high, then the incoming energy received by the evaporator will also be high. From measurements with HOBO microstation, obtained the highest radiation 695.6 W/m^2, and the calculation result of incoming energy received evaporator obtained highest result 1807.293 W.

1. Introduction

The need for clean water is increasing day by day due to the factors of industrialization, motorization, and improvement of living standards of mankind. Research shows that the water supply will not be able to meet the needs of the use due to lack of clean water availability. It has been estimated by the United Nations Organization that by 2025, nearly 1800 million people in the world will experience a scarcity of clean water. [1] This condition can be prevented if humans can find other ways to produce clean water. Fortunately, desalination technology has been developed since long resembling a natural hydrological cycle to prevent this problem, but this technology certainly requires a lot of energy and has a negative impact on the environment.

Renewable energy used by the desalination process is generally in the form of solar, wind, and geothermal energy. According to Eltawil et al [2] Among the three, 57% of the desalination system is supplied with solar power as renewable energy. Even countries rich in fossil fuels such as the Middle East and Arab Nations have also turned their attention to solar energy in order to provide clean water without polluting the environment. The conventional desalination systems in services mainly use fossil fuel. According to a study by Kalogirou [3], about 10,000 t of fossil fuel was burn to power desalination systems in the world.

Desalination in principle is a way to get clean water through the process of distillation of dirty water. In general there are various ways that are often used to get clean water that is: boiling, filtering, desalination and others. Boiling is done only to kill germs and harmful bacteria, but the dirt in the form of small solids can not be separated with water. Screening is used only to filter out the dirt in the form of small solids, but the harmful germs and bacteria can not be separated from the water. How desalination is an effective way to produce clean water free of germs, bacteria, and impurities in the form of small solids.

Li et al. [4] proposed a system that combines multi effect distillation desalination with a supercritical organic Rankine Cycle and an ejector. Their system works like a combined heat, power and condensation. The thermal performance of the system is analyzed theoretically. The results show overall exergy efficiency close to 40% for salt concentration of 35 g/kg using a low temperature heat source at 150 °C. Araghi et al. [5] reported an analytical study on the performance of a new combined vacuum desalination and power system as a heat recovery. The system is claimed can run on waste heat and an organic working fluid. The results show that the overall performance of the introduced system is comparable with the discharge thermal energy combined desalination system utilizing ammonia mixture. However in term of desalination, the proposed system produces more fresh water.

One of the promising solar desalination is natural vacuum desalination system. In the system seawater is placed in a container which is vacuumed naturally by using the gravity of the water. The advantage of using the vacuum is that a low grade heat source such as solar energy can be used efficiently. Several researches of natural vacuum desalination have been reported in literature. Gude et al [6] uses a solar collector with 18 m^2 collector area and the system is combined with thermal energy storage. The reported system can produce 100 L/day of fresh water. Gude and Nirmalakandan [7] combined naturally vacuum and combined with solar-powered air cooling systems. This system can produce fresh water at 4.5 kh/hour and 3.25 kW cooling capacity. Maroo and Goswani [8] reported their study of a natural vacuum flash desalination system operated in one stage and two stage modes. This system is supported with 1 m^2 solar collector. This system can produce 5.54 kg and 8.66 kg of fresh water when operated with single stage and stage mode. The performance ratios are 0.748 and 1.35, respectively. Ayhan and Al-Madani [9] reported a study of natural vacuum solar desalination consisting of evaporator columns exposed to solar radiation and shaded condenser columns. To provide better water supply, blowers are used to transfer steam from the evaporator to the condenser. More recently, Ambarita [10] reported numerical and experimental research on the laboratory scale of a natural vacuum desalination unit. For solar energy simulation electric heater is used in evaporator. Ambarita [11] investigated, the higher thermal efficiency is the better performance. Working temperature of solar collector is very important. These facts reveal that the maximum temperature in the evaporator plays an important role of the system performance.

The work described above shows that the study of natural vacuum solar desalination has been in the spotlight. Several modifications are proposed to improve its performance. In this study, a preliminary experimental study on vacuum desalination was done naturally. To improve the thermal performance of the solar collector, a hybrid system is proposed. The goal is to explore the natural vacuum desalination characteristics when exposed to direct solar irradiation. The characteristics and performance of this system is study experimentally and numerically to get better characteristics and performance which has been done before.

The principle of desalination work in general is actually very simple. The sea water is heated until it evaporates, and then the resulting vapor is condensed back and accommodated in a container tank. Water condensate is clean water. While the sea water that does not boil during heating is a salt concentrate. The desalination process that the authors will discuss in this study is the desalination of the vacuum system with the modification of the heat supply sourced from the hot water produced by the heat-catching collector. The concept of this system is to utilize a naturally formed vacuum chamber to be able to evaporate a certain amount of seawater at low pressure so that it can evaporate with less heat energy supply than conventional techniques. A small supply of thermal energy can be extracted from flat plate solar and / or heat collectors. In this study, however, a heating element of a heat exchanger is supplied through a heat-catching collector to keep the heat supply in the evaporator constant. The uniqueness of this system is the way gravity and atmospheric pressure are used in the formation of vacuum conditions. The formation of a vacuum system aims to lower the pressure of the evaporator chamber so that heating can take place with a low heat supply. The atmospheric pressure will be equal to the hydrostatic pressure formed by a water pipe that is about 10.34 meters high. Thus, if the height of the pipe is more than 10.34 meters and is covered from the top with water, and the water is allowed to fall down due to gravity, the water will fall to a height of about 10.34 meters, and form a vacuum chamber above it. The occurrence of this desalination depends not only on the heating system received by the evaporator from the solar collector, but more dependent on the evaporative cooling process.

2. Method

To conduct this research, the experimental system have been designed and manufactured. This system consists of solar collector, evaporator, annulus heat exchanger, condenser and pipe circuit. Evaporator made of stainless steel 304. This is a cylindrical container with the top of the cone. The diameter, height and thickness of each evaporator are 80 cm, 15 cm, and 5 mm. Heat is injected into the evaporator. It provides energy for evaporation. The heat from the solar collector is transferred to the evaporator by using water as a transfer fluid. An electric powered pump is produced by a 100 Wp photovoltaic. The system schematic diagram is shown in Figure 1.

In the first phase of the experiment, the system was first sucked up with seawater on the evaporator side and heat exchangers and fresh water on the condenser side. There is a valve between the evaporator and the condenser to separate fresh water and seawater in the system at first. Once filled with sea water and fresh water, the valve is then opened. On the evaporator side, seawater will drop by gravity and provide a vacuum naturally in the evaporator. The same way is used to provide vacuum to the condenser. The valve between the evaporator and the condenser is open and the system is now ready to be tested.

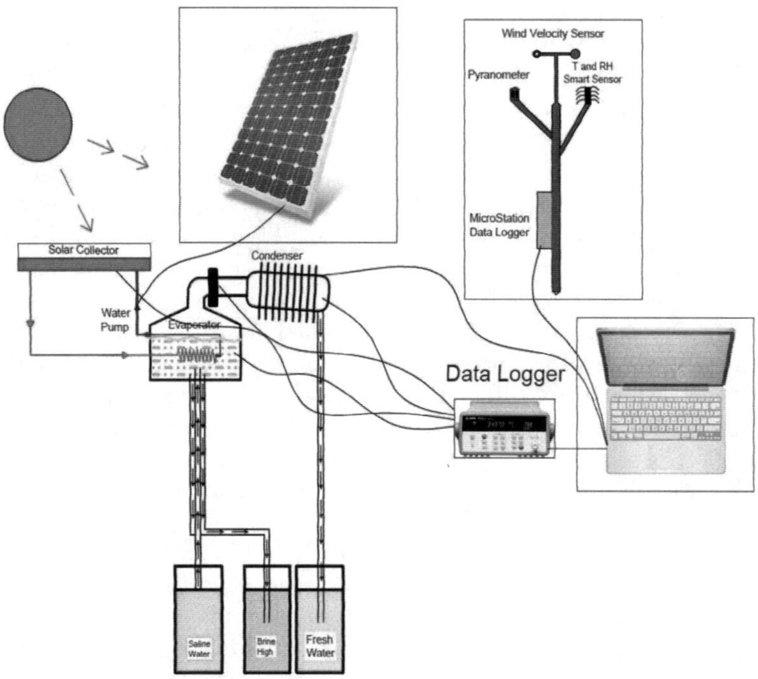

Figure 1. Schematic diagram of natural vacuum solar desalination and data acquisition system

In order to perform experiment, a data acquisition system is used. The system consists of HOBO micro station datalogger and temperature measurement system. There are 20 thermocouples are used to measure temperatures. The sensors are made of *J*-type thermocouples with uncertainly equal to 0.1°C. A datalogger (Agilent 34972) consists of 20 thermocouple channels are used to record temperatures with interval of 1 minute. The solar radiation is measured using HOBO pyranometer smart sensor. The ambient temperature and relative humidity (RH) is measured using HOBO temperature RH smart sensor with an accuracy of 0.2°C and ± 2.5% RH, respectively. The wind speed around the experimental apparatus is measured with HOBO wind speed smart sensor with accuracy ± 1.1 m/s.

The sea water in the evaporator is heated by a heating coil heated with a liquid transfer from the solar collector. Because the pressure in the evaporator is below the atmosphere and approaches the absolute stretch, the water will evaporate. The resulting vapor will flow into a condenser that has a low pressure saturation. In the condenser, the steam will condense. The condenser is cooled with natural convection of ambient air. To avoid the heat from the evaporator transferred to the condenser, the flange made is used. The space between the flanges is filled with rubber sheets that have low conductivity. This system makes the evaporator at a higher temperature while it maintains the condenser at ambient temperature.

In order to provide better heat transfer rates in the condenser, a series of circular fins are mounted on the outer surface. The condensed steam will be collected as a product of the system and fed into freshwater containers. The saline concentration in seawater will increase. The salt water will flow to the salt container through an annulus type heat exchanger. Since the salt water from the evaporator still has a high temperature, the heat will be exchanged for the salt water that enters the evaporator. This is the preheating stage for seawater before entering the evaporator. The resulting freshwater flow in the condenser water and the brine from the evaporator will make the evaporator pressure lower. Low pressure on the evaporator will make the flow of sea water into the evaporator through the sea water pipe bait. This fact shows that all currents in the system are driven naturally without any extra strength. The main component of this system is the evaporator. System performance is mainly influenced by evaporator evaporation rate. In this section an analysis in the evaporator will be provided. Mass conservation in evaporator causes [3]:

$$\frac{d}{dt}(\rho V)_s = \rho_i \dot{V}_i - \rho_w \dot{V}_w - \rho_e \dot{V}_e \tag{1}$$

Where V [m^3] is the volume of sea water in the evaporator, and $\dot{V}$ [m^3/s] the volume flow rate at each inlet side and the evaporator exit side. As a result of evaporation, the salt concentration in the evaporator will increase. If the concentration is expressed by C [%], then the change of salt concentration in the evaporator tube can be expressed by [3]:

$$\frac{d}{dt}(\rho CV)_s = (\rho C)_i \dot{V}_i - (\rho C)_s \dot{V}_w \tag{2}$$

Where "s" is the sea water in the evaporator. The law of conservation of energy in the evaporator can be defined as the amount of incoming heat minus the exit heat will be used to increase the temperature of the fluid in the evaporator. In the form of the following equation [3]:

$$\frac{d}{dt}(\rho C_p VT)_s = Q_{in} + (\rho C_p T)_i \dot{V}_i - (\rho C_p T)_s \dot{V}_w - Q_e - Q_{loss} \tag{3}$$

In this equation C_p [J/kg.K] is a heat type. The parameter Q_{in} is the rate of heat transfer that enters the evaporator from the heating surface. In this study the heater is assumed to be cylindrical and has a constant temperature. The heat transfer is assumed by natural convection. For horizontal pipe, the equation used [3]:

$$\mathrm{Nu} = \left\{ 0.6 + \frac{0.387 Ra_D^{1/6}}{\left[1 + (0.559/\mathrm{Pr})^{9/16}\right]^{8/27}} \right\}^2 \tag{4}$$

By using this equation then the rate of heat transfer going into the evaporator is [3]:

$$\dot{Q}_{in} = \frac{\mathrm{Nu} \times k}{D} A_s (T_{sil} - T_s) \tag{5}$$

Where $A_s = \pi DL$ is the surface area of the heating coil.

3. Results and Discussions

Experiments are tested from 8.00 to 16.00 local time. Solar radiation was measured by HOBO MicroStation and environmental temperature measurements around the study. The solar radiation and the energy received by the evaporator during the experiment are shown in Figure 2. In the picture, theoretically if the solar radiation increases then the energy entering the evaporator will also increase.

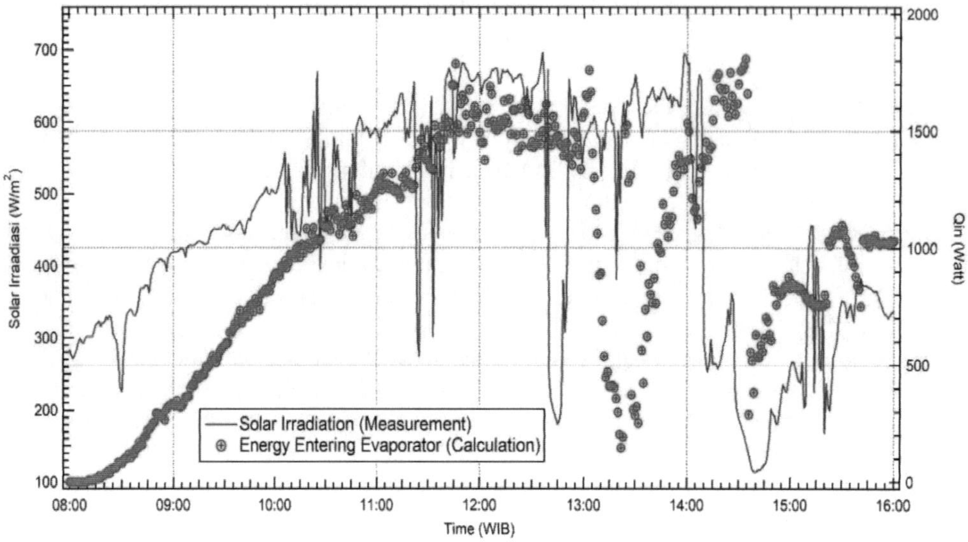

Figure 2. Solar irradiation and energy in evaporator

From the Figure above the results of solar radiation measurements with HOBO microstation measuring the highest obtained 695.6 W / m2 at 12.32 local time, while the received energy received the highest evaporator received 1807.293 W at 14.34 local time. Figure 3 shows the same results in Figure 2, when the ambient temperature increases will also affect the increase in incoming energy received by the evaporator.

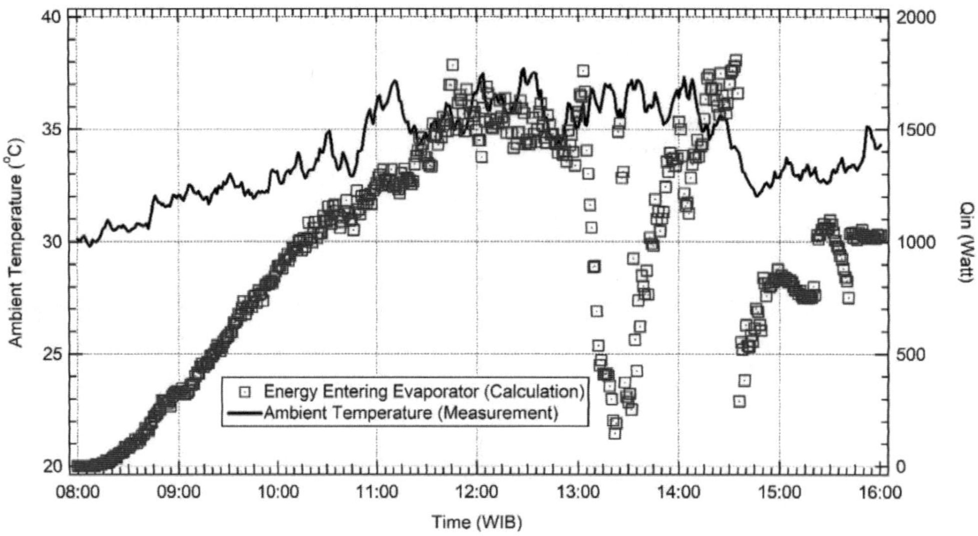

Figure 3. Ambient temperature and energy in evaporator

IOP Conf. Series: Materials Science and Engineering **300** (2018) 012059 doi:10.1088/1757-899X/300/1/012059

4. Conclusion

In this study, small-scale natural vacuum solar desalination using hybrid solar collectors has been designed and manufactured. The area of the solar collector is 3 m^2 as the heat supply to the evaporator and the pump is used to transfer the transfer fluid to the evaporator. This unit has been tested early by exposing the effects of solar radiation in the building of Mechanical Engineering University of North Sumatra city of Medan. The experiment is tested from 8.00 to 16.00. The results show an improvement in previous studies. The conclusions that can be drawn here are as follows. Natural vacuum solar desalination with hybrid solar collectors can be operated perfectly. the result of measurement of solar radiation with HOBO microstation got highest measurement 695.6 W/m^2 at 12.32 local time, while incoming energy received evaporator got highest 1807.293 W at 14.34 local time. In this project same as theory if solar radiation and environmental temperature increases then the incoming energy received evaporator will also increase.

References

[1] http://www.un.org/waterforlifedecade/scarcity.shtml (Accessed September 24, 2017)
[2] Eltawil A. M, Zhengming Z, and Liqing Y 2008 *Twelfth International Water Technology Conference,* Alexandria, Egypt, 22-23.
[3] S. Kalogirou 2005 *Prog. Energy Combust.* Sci. **31** 242–281.
[4] C. Li, D.Y. Goswami, A. Shapiro, E.K. Stefano, G. Demirkaya 2012 *Energy* **46** 582– 595.
[5] A.H. Araghi, M. Khiadani, K. Hooman 2016 *Energy* **98** 215–224.
[6] V.G. Gude 2015 *Appl. Energy* **137** 877–898.
[7] V.G. Gude, N. Nirmalakhadan, S. Deng, A. Maganti 2012 *Energy Convers. Manag.* **56** 192–198.
[8] S.C. Maroo, D.Y. Goswami 2009 Desalination **249** 635–646.
[9] T. Ayhan, H. Al-Madani 2010 Renew. Energy **35** 506–514.
[10] H. Ambarita 2016 Case Studies In Thermal Enginering **8** 346 -358.
[11] Ambarita, H 2017 IOP Conf. Series: *Materials Science and Engineering* **180** 012024 Annual Applied Science and Engineering Co.

Iterative methods for tomography problems: implementation to a cross-well tomography problem

M F Karadeniz[1], G W Weber[1]

[1] Institute of Applied Mathematics, Middle East Technical University, Ankara, Turkey

E-mail: fatih.karadeniz@metu.edu.tr

Abstract. The velocity distribution between two boreholes is reconstructed by cross-well tomography, which is commonly used in geology. In this paper, iterative methods, Kaczmarz's algorithm, algebraic reconstruction technique (ART), and simultaneous iterative reconstruction technique (SIRT), are implemented to a specific cross-well tomography problem. Convergence to the solution of these methods and their CPU time for the cross-well tomography problem are compared. Furthermore, these three methods for this problem are compared for different tolerance values.

1. Introduction

Cross-well seismic tomography, which is used often in geology, mainly deals with reconstructing the velocity structure between two boreholes by measuring travel time for ray paths between them [1]. A cross-well tomography problem can be defined as an inverse problem because we have a set of measurements (travel time), then we try to reconstruct the model (velocity structure) from them [4] in the cross-well tomography problem. Enhanced oil recovery, imaging of geological structures in sediments and crystalline rock are some vital examples in which cross-well tomography is used [1].

Hydrologists should have information regarding the locations of hydraulically conductive fractures if they want to clean up contaminants which are in fractured bedrock as stated in [3]. Herewith, cross-well seismic tomography can be used to obtain the information. There are some methods [3] such as single-borehole hydraulic tests, cross-borehole hydraulic tests to get information about locations of these fractures. However, they have a weakness concerning the hydraulically conductive fractures which are far away from boreholes. So, cross-well seismic tomography is used to overcome this problem. As described in [3], P-waves are used to investigate the bedrock which is located between two wells. After P-waves are used in the bedrock, obtained data are processed by using tomographic methods. The result of this process is known as tomogram [3]; that is, the structure of P-wave velocity between two wells. Then, since P-wave velocity is reduced by fractures, the locations of fractures can be obtained from the low velocity anomalies in the tomogram. Thus, these fractures can be hydraulically conductive ones [3].

In this paper, we apply iterative methods, which are Kaczmarz's method, algebraic reconstruction technique (ART), and simultaneous iterative reconstruction technique (SIRT), to a specific cross-well tomography problem, which was taken from [2].

The problem [2] is explained as follows: there are two vertical wells and the distance between them is 1600 m. A seismic source is placed in one of wells at depths of 50, 150, . . . , 1550 m.

A set of receivers is placed in the other well at same depths. Then, a travel time with standard deviation of 0.5 ms is recorded for every source-receiver pair. Also, ray paths are assumed to be straight. Hence, this paper explains how we obtained the velocity distribution in the two-dimensional plane between the two wells by using three iterative methods for this problem.

This paper compares CPU time and convergence to the solution of Kaczmarz's method, ART, and SIRT for different tolerance values on the cross-well tomography problem.

2. Methods

In cross-well seismic tomography, the travel time for seismic energy [2], which travels through ray path ℓ, can be defined by

$$t = \int_{\ell} s(\mathbf{x}) d\ell, \tag{2.1}$$

if slowness at a point $\mathbf{x}$ and ℓ are known. Slowness is known as inverse of the velocity [9]. Since the model (slowness) and data (travel time) are continuous functions' variables, and we want to determine the model from the data in our cross-well tomography problem, this problem is defined as a continuous inverse problem. We can discretize Equation (2.1) and obtain a discrete inverse problem by using midpoint rule in the cross-well tomography problem. In addition, the reader can analyze [7, 8] to obtain detailed information regarding seismic tomography and slowness.

When we discretize Equation (2.1) into a 16×16 grid of 100 meter by 100 meter uniform blocks in the cross-well tomography problem, 256 model parameters are obtained. Therefore, approximated form of Equation (2.1) [2] can be written as

$$t = \int_{\ell} s(\mathbf{x}) d\ell \approx \sum_{\text{blocks}} s_{\text{block}} \cdot \Delta l_{\text{block}}, \tag{2.2}$$

where Δl_{block} represents lengths of ray paths within corresponding blocks.

Let $\mathbf{m}$ be a model vector with n components which are slowness, $\mathbf{d}$ be a data with m components which are measured travel time and $\mathbf{G}$ be a matrix whose components are lengths of ray paths within corresponding block in our cross-well tomography problem. Then, we can write Equation (2.2) as a linear system of equations [2] by

$$\mathbf{Gm} = \mathbf{d}. \tag{2.3}$$

As stated in [2], if we have a problem which has a large $\mathbf{G}$ matrix, methods such as singular value decomposition (SVD), Tikhonov regularization are not useful for this problem. The reason of this impracticability is that computers need too much memory to store such kind of $\mathbf{G}$ matrix. However, if $\mathbf{G}$ is a sparse matrix; that is, many of its components are zero, the memory problem can be solved by getting rid of zero components of the matrix $\mathbf{G}$. In this way, the nonzero components of the matrix and their locations can be kept [2]. Unlike factorization methods, iterative methods consist of products of vectors with $\mathbf{G}$ and $\mathbf{G}^T$ [5]. In the cross-well tomography problem, which we deal with, the data $\mathbf{d}$ and $\mathbf{G}$ matrix are given.

In this paper, three iterative methods, Kaczmarz, ART, and SIRT, are examined and they are applied to the cross-well tomography problem.

2.1. Kaczmarz's method

Kaczmarz's algorithm is one of the most commonly used methods for solving Equation (2.3) [2]. An essential property of this method is that its implementation is easy. Kaczmarz's method is based on "method of projections" [6]. Tomographic problems are frequently discretized as uniform blocks as we discretized in the cross-well tomography problem. Before explaining algorithm of Kaczmarz, let us write expanded form of Equation (2.3) [6]:

$$\begin{aligned} G_{11}m_1 + G_{12}m_2 + G_{13}m_3 + \ldots + G_{1n}m_n &= d_1 \\ G_{21}m_1 + G_{22}m_2 + G_{23}m_3 + \ldots + G_{2n}m_n &= d_1 \\ &\vdots \\ G_{m1}m_1 + G_{m2}m_2 + G_{m3}m_3 + \ldots + G_{mn}m_n &= d_m. \end{aligned} \tag{2.4}$$

According to [6], a model (or image), which has n degrees of freedom, can be obtained by a grid with n cells. Then, the model, $(m_1, m_2, \ldots, m_n)$ can be defined as a single point in n-dimensional space. So, each of the equations in Equation (2.4) is called a hyperplane in this n-dimensional space. Unique solution of Equation (2.4), if it exists, can be obtained by a single point which is the intersection of all hyperplanes [6]. In Kaczmarz's algorithm, each of m equations $\mathbf{G}_i \cdot \mathbf{m} = d_i$, as we described before, represents an n-dimensional hyperplane in $\mathbb{R}^m$ [2]. Algorithm of Kaczmarz for a system of equations $\mathbf{Gm} = \mathbf{d}$ for a tomography problem [2] is described in Algorithm 1.

Algorithm 1 Kaczmarz's Algorithm

1: $\mathbf{m}^{(0)} \leftarrow \mathbf{0}$
2: **for** $i \leftarrow 0, m-1$ **do**

$$\mathbf{m}^{(i+1)} \leftarrow \mathbf{m}^{(i)} - \frac{\mathbf{G}_{i+1,\cdot}\mathbf{m}^{(i)} - d_{i+1}}{\mathbf{G}_{i+1,\cdot}\mathbf{G}^T_{i+1,\cdot}}$$

3: **end for**
4: If there is no convergence to the solution, go back to the second step.

The algorithm starts with an initial guess $\mathbf{m}^{(0)}$, then, $\mathbf{m}^{(1)}$ is obtained by projecting $\mathbf{m}^{(0)}$ onto the first equation of Equation (2.4), which is determined by the first row in $\mathbf{G}$ [2]. Similar operations are applied for the next solutions until the solution is projected onto all m hyperplanes, which are defined in Equation (2.4) [2].

2.2. Algebraic reconstruction technique

Algebraic reconstruction technique, also known as ART, is an updated form of Kaczmarz's method [2]. In ART, components in row $i+1$ of $\mathbf{G}$, which are not zero, are replaced with ones [2]. The purpose of this replacement is that we can easily determine the total number of cells traversed by ray path. The process of finding cell in the ray path is simplified by this replacement. Approximated travel time [2] through the ray path $i+1$ is defined as

$$q_{i+1} = \sum_{\text{cell } j \text{ in ray path } i+1} \mathbf{m}_j l, \tag{2.5}$$

where l is called cell dimension. Then, $q_{i+1} - d_{i+1}$ is the residual of the estimated travel time of ray $i+1$ [2]. Algorithm of ART [2] for a system of equations $\mathbf{Gm} = \mathbf{d}$ for a tomography problem is explained in Algorithm 2.

Algorithm 2 ART Algorithm

1: $\mathbf{m}^{(0)} \leftarrow \mathbf{0}$
2: **for** $i \leftarrow 0, m$ **do**
  $N_i :=$ # of cells traversed by ray path i,
  $L_i :=$ the length of ray path i.
3: **end for**
4: **for** $i \leftarrow 0, m-1, \quad j \leftarrow 1, n$ **do**

$$m_j^{(i+1)} \leftarrow \begin{cases} m_j^{(i)} + \dfrac{d_{i+1}}{L_{i+1}} - \dfrac{q_{i+1}}{lN_{i+1}}, & \text{if cell } j \text{ in ray path } i+1, \\ m_j^{(i)}, & \text{otherwise.} \end{cases}$$

5: **end for**
6: If convergence of the solution is not obtained yet, let $\mathbf{m}^{(0)} \leftarrow \mathbf{m}^{(m)}$ and turn back to fourth step. Otherwise, the solution $\mathbf{m} \leftarrow \mathbf{m}^{(m)}$ is returned.

2.3. Simultaneous iterative reconstruction technique

Simultaneous iterative reconstruction technique, also known as SIRT, is a variation of ART. SIRT gives better tomographic images than ART, but its convergence is slower than ART [6]. Algorithm of SIRT [2] for a system of equations $\mathbf{Gm} = \mathbf{d}$ is given in Algorithm 3.

Algorithm 3 SIRT Algorithm

1: $\mathbf{m}^{(0)} \leftarrow \mathbf{0}$
2: **for** $j \leftarrow 0, n$ **do**
  $K_j :=$ # of paths which pass through cell j.
3: **end for**
4: **for** $i \leftarrow 0, m$ **do**
  $N_i :=$ # of cells traversed by ray path i,
  $L_i :=$ the length of ray path i.
5: **end for**
6: $\Delta\mathbf{m}^{(i+1)} \leftarrow \mathbf{0}$
7: **for** $i \leftarrow 0, m-1, \quad j \leftarrow 1, n$ **do**

$$\Delta m_j^{(i+1)} \leftarrow \begin{cases} \Delta m_j^{(i+1)} + \dfrac{d_{i+1}}{L_{i+1}} - \dfrac{q_{i+1}}{lN_{i+1}}, & \text{if cell } j \text{ in ray path } i+1, \\ 0, & \text{otherwise.} \end{cases}$$

8: **end for**
9: **for** $j \leftarrow 1, n$ **do**

$$m_j^{(i+1)} \leftarrow m_j^{(i+1)} + \frac{\Delta m_j^{(i+1)}}{K_j}$$

10: **end for**
11: If there is no convergence of the solution, go back to line 6.
  Otherwise, the current solution is returned.

In simultaneous iterative reconstruction technique, firstly, models are calculated if cell j is in ray path $i+1$ [2]. Then, average of these models are computed before the model is updated [10].

3. Results and discussion

This paper focuses on the cross-well tomography problem, which is described in Section 1. In this problem, components of $\mathbf{G}$ are lengths of ray paths within corresponding blocks. In addition, noisy data, **d**, represents measured travel time, and the model, **m**, represents slowness in grid in which row-by-row indexing is used. Thus, the velocity structure can be obtained from this model. Our aim is to find velocity distribution in 2-D plane between two wells by using Kaczmarz's method, ART, and SIRT in this paper. We used MATLAB functions for ART and SIRT, which were produced by [2], by modifying and arranging them for the implementation to the problem.

Our problem is called ill-conditioned and rank deficient since rank($\mathbf{G}$)= 243 is smaller than $n = 256$ (number of model parameters) and cond($\mathbf{G}$)= 1.87^{18} where rank($\mathbf{G}$) is rank of $\mathbf{G}$ matrix and cond($\mathbf{G}$) is condition number of $\mathbf{G}$ matrix.

When we implemented Kaczmarz's algorithm to the cross-well tomography problem, we obtained a model vector

$$\mathbf{m} = [0.0003430, 0.0003424, 0.003419, \ldots, 0.0003404, 0.0003406, 0.0003409]^T. \tag{3.1}$$

In Equation (3.1), there are 256 components which denote slowness in the grid of 100 meter by 100 meter block. For example, first component of **m** in Equation (3.1) is the block which is at first row and first column. Similarly, second component of the same model is the block which is at first row and second column and so on. Therefore, first 16 components (first row) are grids at depth of 50 m, second 16 components (second row) are grids at depth of 150m and so forth. As a result, we obtained the figure of slowness (s/m) vs. depth (m), as shown in Figure 1, for the cross-well tomography problem by using Kaczmarz's algorithm.

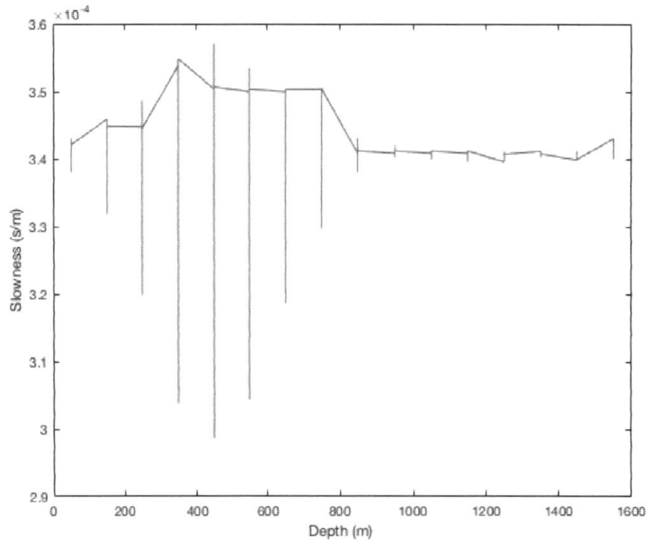

Figure 1: Kaczmarz's method solution for the cross-well tomography problem.

When we applied ART to the cross-well tomography problem, we obtained a slowness model vector

$$\mathbf{m} = [0.0003713, 0.0003519, 0.003444, \ldots, 0.0003148, 0.0003119, 0.0003075]^T, \tag{3.2}$$

which has 256 components as in Equation (3.1). Figure 2, which we obtained, represents slowness (s/m) vs. depth (m) for the cross-well tomography problem by using ART.

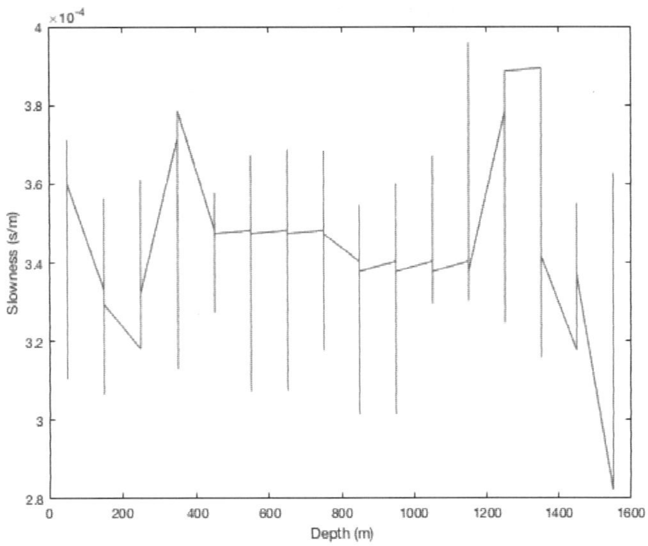

Figure 2: ART solution for the cross-well tomography problem.

Lastly, when we implemented SIRT to the cross-well tomography problem, we obtained a model vector

$$\mathbf{m} = [0.0003747, 0.0003562, 0.003464, \ldots, 0.0003177, 0.0003152, 0.0003097]^T, \qquad (3.3)$$

whose 256 components are slowness at corresponding depths. Then, Figure 3, which we produced, indicates slowness (s/m) vs. depth (m) for the cross-well tomography problem by using SIRT.

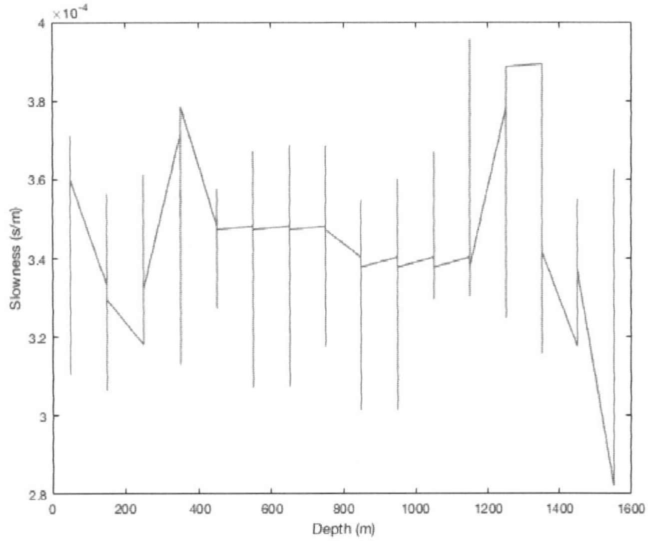

Figure 3: SIRT solution for the cross-well tomography problem.

When we compared the results of the implementation of these three methods to the cross-well problem, we observed that Kaczmarz's algorithm has the fastest convergence to the solution among the methods. When we chose the value of tolerance as 10^{-9}, the solution was converged after 7964, 21597, 61034 iterations by using Kaczmarz's method, ART, and SIRT respectively.

IOP Conf. Series: Materials Science and Engineering **300** (2018) 012060 doi:10.1088/1757-899X/300/1/012060

Moreover, CPU time of the implementation of Kaczmarz's algorithm, ART, and SIRT for the cross-well tomography problem was measured as 14.17 seconds, 32.65 seconds, 96.98 seconds respectively.

We also found that although ART is slower than SIRT until a specific tolerance value (5×10^{-9}), for values greater than that specific value, SIRT is slower than ART for this problem. When we chose tolerance value as 5×10^{-9}, the solution was converged after 10271 iterations with 16.83 seconds and 9432 iterations with 15.38 seconds by using SIRT and ART respectively. When we chose tolerance value as 10^{-8}, which is greater than 5×10^{-9}, the solution was converged after 3563 iterations with 5.94 seconds and 5180 iterations with 8.49 seconds by using SIRT and ART respectively.

Table 1: Convergence to the solution of the iterative methods, when the value of tolerance is 10^{-9}.

	Kaczmarz's algorithm	ART	SIRT
Number of Iterations	7964	21597	61034
CPU Time (seconds)	14.17	32.65	96.98

Table 2: Convergence to the solution of the iterative methods, when the value of tolerance is 5×10^{-9}.

	Kaczmarz's algorithm	ART	SIRT
Number of Iterations	2655	9432	10271
CPU Time (seconds)	4.92	15.38	16.83

Table 3: Convergence to the solution of the iterative methods, when the value of tolerance is 10^{-8}.

	Kaczmarz's algorithm	ART	SIRT
Number of Iterations	1715	5180	3563
CPU Time (seconds)	3.21	8.49	5.94

According to [2], Kaczmarz's algorithm converges quickly, if the hyperplanes are almost orthogonal. If more than two hyperplanes are nearly parallel to each other, then, convergence of the solution could have been immensely slow. Consequently, we can say that hyperplanes in our cross-well tomography problem are almost orthogonal. Moreover, we observed that the solutions of ART and SIRT are noisier than Kaczmarz's method as we expected.

4. Conclusion and outlook

Our major purpose, in the present paper, is to implement Kaczmarz's algorithm, ART, and SIRT to the cross-well tomography problem and compare their convergence to the solution, and

CPU time. As stated in [2], there are some methods, which are based on matrix factorizations, to solve linear system of equations. QR factorization and singular value decomposition (SVD) can be given as examples for these methods. However, they are not enough to solve problems which have sparse matrices [2]. Matrices which are created by using these factorizations are denser than the **G** matrix, which is factorized [2]. Herewith, usage of iterative methods such as Kaczmarz's method, ART, and SIRT for large problems are more appropriate than usage of factorization methods.

The cross-well tomography problem, which was analyzed in this paper, was discretized into a 16×16 grid of 100 meter by 100 meter uniform blocks. This discretization provided 256 model parameters, which are slowness. Then, the velocity distribution in two-dimensional plane between two wells was obtained as the model. Figures of Slowness vs. Depth, which we produced, represent slowness distribution according to the corresponding depths in the cross-well tomography problem. The results we had show that Kaczmarz's method is the method which converges fastest to the solution for the cross-well tomography problem when it is compared to ART and SIRT. Moreover, we found that ART has slower convergence than SIRT when we chose the value of tolerance greater than 5×10^{-9} in the problem.

In the present paper, ray paths were assumed to be straight in the problem, which we examined. Then, we used iterative methods for the linear inverse problem to get velocity structure in the two-dimensional plane between the two wells. If there is ray-path bending because of changing seismic velocity in a cross-well tomography problem, the problem becomes a nonlinear inverse problem [2]. Hence, the implementation of methods for nonlinear inverse problems such as Tikhonov regularization and Occams inversion [2] to the cross-well tomography problem will be future work. Furthermore, in synthetic aperture radar (SAR) imaging [11], a plane or satellite with an antenna moves along a flight track, then, electromagnetic radiation waves are emitted by the antenna. These electromagnetic waves scatter off the land, and the antenna, which emitted the waves, identifies the scattered waves [11]. Herewith, the terrain's image is generated by the received signals. Therefore, SAR imaging can be used in cross-well tomography problems, and this radar imaging application will be our future work.

References

[1] Harris J M 1993 *Crosswell seismic tomography: Newkirk Borehole Test Facility case study* (Stanford, CA: Seismic Tomography Project, Department of Geophysics)
[2] Aster R C, Borchers B and Thurber C H 2013 *Parameter Estimation and Inverse Problems* (Academic Pr)
[3] Ellefsen K J, Hsieh P A and Shapiro A M 2002 Crosswell seismic investigation of hydraulically conductive, fractured bedrock near Mirror Lake, New Hampshire *Journal of Applied Geophysics* **50** 299-317
[4] Snieder R and Trampert J 1999 Inverse Problems in Geophysics *CISM International Centre for Mechanical Sciences Wavefield Inversion* **398** 119-190
[5] Hansen P C 2010 *Discrete inverse problems: insight and algorithms* (Philadelphia, PA: SIAM)
[6] Kak A C and Slaney M 2001 *Principles of computerized tomographic imaging* (New York: SIAM)
[7] Anderson D L and Dziewonski A M 1984 Seismic tomography *Scientific American* **251** 60-68
[8] Nolet G 2008 *A breviary of seismic tomography: imaging the interior of the earth and sun* (Cambridge: Cambridge University Press)
[9] Stein S and Wysession M 2009 *An introduction to seismology, earthquakes, and earth structure* (Wiley)
[10] Sharma P V 1997 *Environmental and engineering geophysics* (Cambridge: Cambridge Univ. Press)
[11] Cheney M 2001 A Mathematical Tutorial on Synthetic Aperture Radar *SIAM Review* **43** 301-312

IOP Conf. Series: Materials Science and Engineering **300** (2018) 012061 doi:10.1088/1757-899X/300/1/012061

Enviromental Health Risks on Community in Coastal Area As a Results The Presence of Pb in Sea Water and Drinking Water.

Sri Malem Indirawati[1], Setiaty Pandia[2],Herman Mawengkang[3],Wirsal Hasan[4]
[1]Graduate School of Environment Mangement, Universitas Sumatera Utara
[2]Department of Chemistry Engineering, Universitas Sumatera Utara
[3]Department of Mathematics, Universitas Sumatera Utara
[4] Department of Public Health, Universitas Sumatera Utara

Email: [1]srimalem_indirawati@yahoo.co.id,[2]setiaty_pandia@yahoo.com, [2]hmawengkang@yahoo.com,[4]wirsal_hasan@yahoo.com

Abstract. The burden of pollution due to industrial waste, ports, community activities and marine intrusion further exacerbate environmental quality. This pollution causes drinking water sources polluted. This study aims to analyze Pb contamination in marine, and drinking water from wellbores and measure the magnitude of health risks This is cross sectional study and quantitative research that analyzes Pb concentrations in marine and drinking water. The sample are 250 people who live in coastal area and drink water from wellbores. Water samples were examined in certified laboratories by using Atomic Absorbstion Spectrophotometer method, health risk was analyzed by the environmental health risk (EHRA) method. Pb concentrations average in marine is 52 μgl-1 . Pb concentration from 92 samples of drinking water average is 4.5 μgl-1 and range 5.4 – 26.2 μgl-1. The amount of health risk RQ <1, which means that it has not shown risk yet. Pb exceeded the environmental quality standard in marine, There are 14.7% of people consuming Pb contaminated drinking water. Community complaints found at the study sites were diarrhea 22.8% and dizziness 17.2% and skin disease 17.2% , upper respiratory tract infection, , rheumatism and hypertension

Keywords: Pb, sea water, drinking water, health risks

.

1. Introduction

Major cities in Indonesia are generally located in coastal areas, the characteristics of coastal areas with characteristics towards the land affected by the physical properties of the ocean such as tidal, wind, wave and intrusion and seaward influenced by natural processes on land such as sedimentation , the flow of fresh water from the stream watersheds, domestic sewage, industrial and agricultural. Environmental pollution in coastal areas make the waters along the shoreline and the beach itself becomes unfit for use by the general public and has become a global health problem[1].

Health quality of the coastal environment is very critical due to sea water pollution by industrial waste, ports, marine transportation, domestic waste and community activities. Polluted sea will have an impact on the environment, especially the quality of community water sources. Belawan is a coastal area in Medan city, heavy metal pollution has taken place in the region's waters of Belawan, as in previous studies which found marine life that has been contaminated with heavy metals Pb, Cd and Cr [2].Heavy metals are persistent elements and can accumulate through the food chain (bioaccumulation), The accumulation of heavy metals (Cd, Pb, Hg and Cr) was measured by atomic absorption spectrophotometry in water, sediment, plankton and fish samples collected from Lake Beyşehir. It was Pb > Cd > Cr > Hg (pv < 0.05), in addition to this, accumulation orders of heavy metals in the food web was also found to be water > plankton > sediment > fish tissues, except for Cr [3].

Another phenomenon the location of this study has been found Pb of freshwater resources of the communities living in coastal areas. Seawater intrusion is suspected as the primary cause in which there has been a process of sea level rise that sea water into the wells of the population. Seawater intrusion assumed that it could cause heavy metal pollution of freshwater resources population that is shallow wells and deep wells. Belawan coastal Presence of heavy metals Pb, As and Mn in drinking water cause effects on the health effects, especially in infants of pregnant women who consume water that has been contaminated with heavy metals [4]. Research on the risk of consuming contaminated drinking water in Kohistan, North Pakistan in measuring the risk of contamination of Cu, Co, Cr, Mn, Ni, Pb, Zn and Cd in drinking water of the population. The pollution comes from activities geogenik the process of weathering, erosion and anthropogenic activities are mining, industrial waste, domestic and agriculture [5] communities still exist that utilize the artesian well water as a source of drinking water [6].

Pollutants such as heavy metals pose a threat to the health of coastal communities, especially Belawan, seawater intrusion phenomenon allows community water sources polluted by heavy metals. People who use the water as drinking water in particular will be exposed to heavy metals contained in their drinking water. Further studies can be done is to analyze the risk in people who consume drinking water sourced from clean water from wells in coastal areas.

Studies of environmental health risk analysis is one method of study environmental effects on health. This method is very suitable for the study of environmental impact on public health in the Environmental Impact Assessment is being able to predict the risk of exposure by the projection period ahead. Enviromental health Risks Assesment (EHRA) consists of four stages of assessment, namely hazard identification, dose-response analysis, exposure analysis and risk characterization. These measures should not be done sequentially, unless the risk characterization as the last stage The study method is the result of research EHRA exposure to arsenic in drinking water in Bangladesh society cause of death based on the cohort study [7].

The phenomenon of environmental quality in coastal Belawan pollution source of some concern to human activities cause pollution both at sea and on community water sources. This condition if it continues to be an impact on public health. Based on these studies will be needed to analyze the presence of heavy metals in the coastal environment and quantify health risks in people who are exposed to heavy metals.

2. Materials and Methods

This study is a quantitative study is to analyze the levels of heavy metals in the coastal environment is by measuring the levels of Pb in the waters, Pb in drinking water sources and the population large measure on the public health risks. The sample was seawater, drinking water that comes from wellbore and the people who live in coastal consume drinking water from the source of the wellbore. The data analysis research variables are as follows:

a. Seawater samples taken 10 points in the territorial waters of the subdistrict of Medan Belawan (see fig 1), Parameters measured were physical quality (odor, taste, clarity), chemical (temperature, DO and PH, Pb), then the results were compared with Water Quality Criteria US EPA [8].

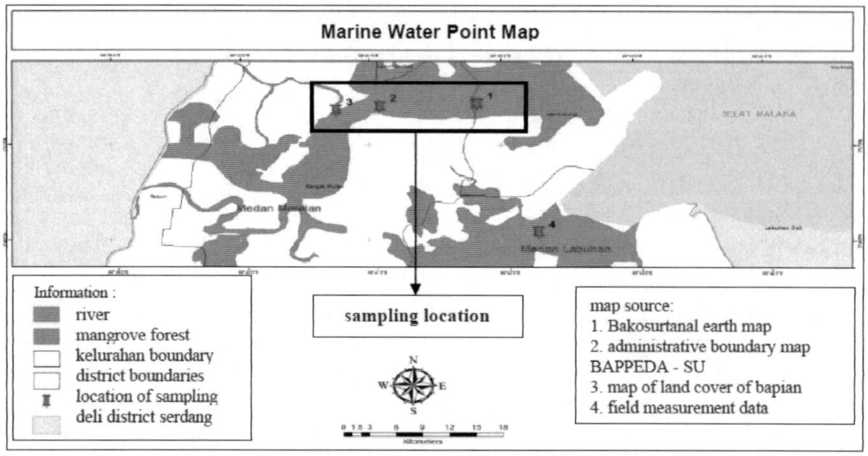

Figure 1. Marine Water Point Map
Sources : Data obtained from the measurement results in the field

b. Samples of drinking water taken from wellbores in the study site there are 92 samples of the wellbore, the sampling technique uses purposive sampling. parameters measured is the concentration of Pb in drinking water The results are then compared with WHO guidelines for drinking water for Pb ; 5 µg/l [9]. Fig. 2 below is the map of wellbore sample point content of Plumbum

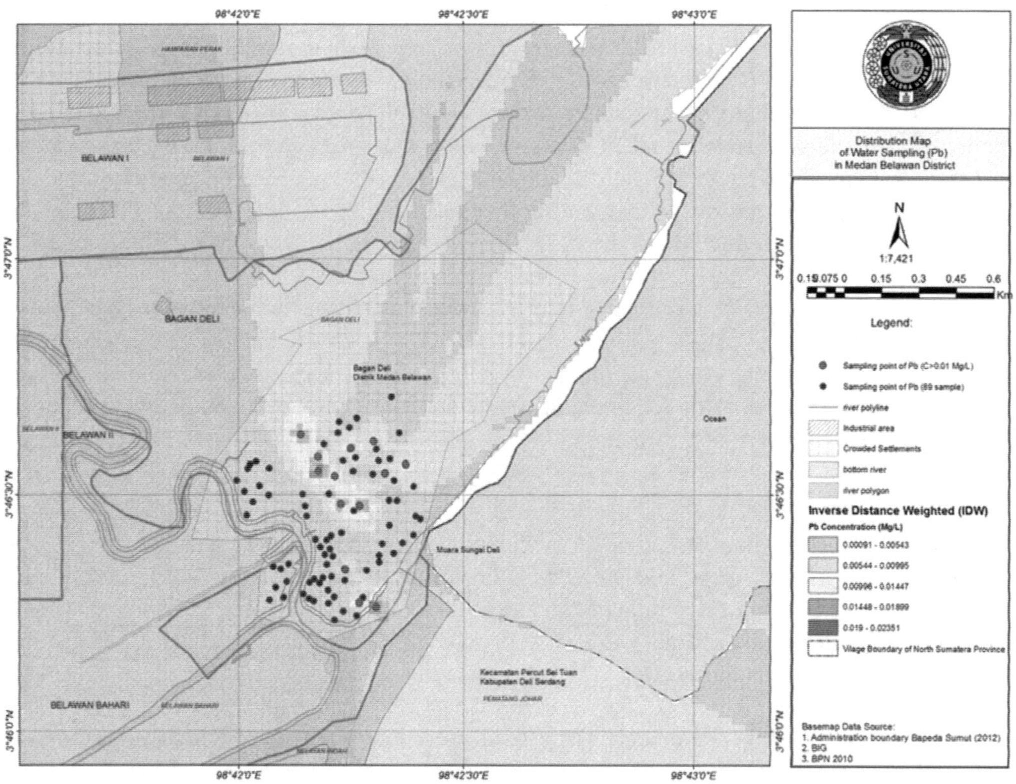

Figure 2. Map of the Wellbore Sample Point
Sources : Data obtained from the measurement results in the field by using GPS

IOP Conf. Series: Materials Science and Engineering **300** (2018) 012061 doi:10.1088/1757-899X/300/1/012061

c. Analysis of environmental health risks conducted the analysis stage of exposure, dose response analysis, risk characterization and risk management. EHRA method by measuring the variables of the public to sample analysis stage exposure is measured variable; concentration of drinking water consumed (mg /l), the rate of intake (l / day), weight (kg), the duration of exposure (years), the time period average (Dt x 365 days) all these variables to measure Intake (intake)[7].

Data collection using questionnaire. The Variabels is described below

I	:	Intake mg / kg / day
C	:	Risk gent concentration, mg / L for drinking water
R	:	Intake or consumption rate, L / day for drinking water
f_E	:	Frequency of exposure, day / year (residential 350 days)
D_t	:	Duration of exposure, (the real time or projection, 30 years for residential default value)
W_b	:	Weight, kg
T_{avg}	:	the average of period time (D_t × 365 days / year for non-carcinogens,70 years × 365 days / year for carcinogens)

Reference Doses of Pb (*RfD* Pb) = 3.5×10^{-4} [7]

The next step is to calculate the risk characterization of numerical risk estimates, by estimating risk by calculating the ratio of intake with reference dose (RfD) with the formula below (1)

$$\text{Risk Quatient (RQ)} = \frac{\text{Intake (mg / kg day)}}{\text{RfD}} \quad (1)$$

3. Results And Discussion

A. Pb and Cd Concentration in The Seawater

Based on laboratory test results to the results obtained tidal conditions that is Pb concentration is above the quality standard of value, this means has been heavily polluted coastal waters. Suspected pollution sources come from domestic sewage, industrial and harbor. This location has 8 major industries, 15 industries and 32 small home industries.

Table 1. Pb Concentration in the Sea Water in Belawan

Sample Location		Pb (μgl^{-1})	pH	DO
MEDAN BELAWAN				
1.N 03.79010⁰ E 098.69701⁰	Tidal	64	7.6	5.0
2. N 03.78963⁰ E 098.68475		32	6.8	5.03
3. N 03.78898⁰ E 098.67907⁰		60	6.6	4.77
Mean:		**52**	**7.0**	**4.93**
1.N 03.79010⁰ E 098.69701⁰	Low Tide	56	8.5	8.0

2. N 03.78963^0 E 098.68475^0	63	8.2	9.1
3. N 03.78898^0 E 098.67907^0	53	8.1	7.9
Mean:	**57**	**8.26**	**8.3**

Data Source : Test Result on Laboratory

Tebel 1. describes the results of laboratory tests of seawater quality that include levels of Pb. The sample is taken from 3 point location with the condition of tide and low tide condition to see difference of metal concentration of Pb. The concentrations of Pb exceeded the environmental quality standard, the threshold value based on the US EPA standard for marine water[8]. The mean of Pb in seawater also exceeds the environmental quality standard in tidal conditions (52 μgl^{-1}.) and lows (57 μgl^{-1}.) the standard is 10 μgl^{-1}. Specifically for PH is still in standard condition pH 6.6 – 8.5 whereas DO level for tidal condition exceedeed the standard of environmental quality standard (NAB> 5). This condition describes the pollution has occurred in marine of Medan Belawan .

B. Metal Pollution in Wellbores

The measurement results Sample some 92 wellbore used by the people as the main source of drinking water in the obtained results mean Pb concentration is 4.5 μgl^{-1}. with a value of Standard Deviation (SD) = 0.057 and a maximum value is 26 μgl^{-1}. There are 13 (14.6%) wellbore have concentrations of Pb exceeds the threshold limit value for Pb based on drinking water quality from WHO is 5 μgl^{-1}. [9]. Pollutant sources are assumed waters, seawater intrusion factor as originator of the contamination of the wellbore which is used by the people as a source of drinking water. Sea water intrusion in the city of Medan had reached 13 Km [10] and the results of the health center reports there are some 48.7% reported unhealthy home, wastewater disposal systems directly into the sea.

C. Enviromental Health Risks Assesment

A total of 34 respondents consume drinking water that has been contaminated with Pb, but poses no risk of this can be explained from the calculation analysis of health risks from exposure to lead in drinking water (see table 3). Results The mean RQ = 0:04, RQ = 0.020 minimum value, maximum value 0.325 with SD = 0.52. The magnitude of health risks (RQ) is still under one (RQ <1) This means that at the time of the study there is no health risk, meaning Pb in drinking water that are resident at the time of measurement do not pose a risk despite being polluted. Pb heavy metal is said to be very dangerous to human health that cannot be destroyed by micro organisms and accumulates in the environment, especially the water-component parts to form a complex compound together with the organic and inorganic compounds by adsorption and combined [4].

The results of the samples of underground water in Bagjata India with resources obtained mining waste contamination Pb levels in drinking water sources 28 μgl^{-1} with RQ) <1[11]. Although the RQ <1 but the exposure will take place throughout the period, the nature of Pb in the body accumulate and accumulate currently not at risk, but a few years later is likely to cause health effects for society. Symptoms of lead poisoning in the community such as abdominal pain, nausea, diarrhea, neuropaty peripheral nerves, muscle weakness, hands and feet, headaches, depression and anemia. Pb is neurotoxic which can accumulate in the body, Pb high in the body causing disorders of hemoglobin synthesis of blood, neurological disorders (nervous), disorders of the kidneys, the reproductive system, acute or chronic nervous system, and impaired function of the lungs. In addition, it can lower the IQ of young children if there are 10-20 μgr / dl in the blood, every increase of 10 μgr / dl in the blood would reduce IQ by 4 points [12][13] .

Table 2. Result Of Risk Assesment Analysis And Risk Quation Of Pb (RQ Pb)

Variable	Mean	Median	Min	Max	SD
Concentration (C) mg/L	0.0045	0.0025	0.0054	0.026	0.056
Consumption Rate (R), L/day	1.8795	2.0	1	5	0.69
Frequency of exposure (f_E) day / year	323	350	48	350	59.56
Duration of exposure (Dt) in year	23.2	20.0	1	70	15.96
Weight (W_b),Kg	56.38	59.00	20	94	16.06
T_{avg} the average of period time (D_t × 365 days / year for non-carcinogens)	8470	7300	365	25550	5826
Intake (I)	8×10^{-5}	5×10^{-5}	5×10^{-6}	5×10^{-4}	8.7×10^{-5}
RQ Pb	0.0419	0.0241	0.02	0.325	0.525

Data Source : Results Field interviews and risk analysis By using formula (I)

Refers to Table 4, the maximum value for Cd 0.78 RQ (RQ <1), this means that the content of cadmium in drinking water do not pose a risk community at the time of measurement lasts. Similarly, the risk Pb at the time of measurement, but still there is no guard against for the foreseeable future.

4. Conclusion

Pb and Cd exceeded the environmental quality standard in marine, There are 14.7% of people consuming Pb contaminated in drinking water but the amount of health risks RQ <1.

Acknowledgements

I would like to extend my sincere gratitude to the chairman of STIKES Sumatera Utara, respondents and my family who in their own way have given various help for my research

References

[1] April Clark, 2003. *Health hazards due to pollution of waters along the coast of Visakhapatnam, east coast of India.* Ecotoxicology and Environmental Safety Volume 56, Issue 3: Pages 390-397.

[2] Siagian, L 2008. Pengaruh Pencemaran Logam Berat Pb, Cd, Cr Terhadap Biota Laut Dan Konsumennya di Kelurahan Bagan Deli Belawan .http://www.usu.Library:Perpustakaan Universitas Sumatera Utara.ac.id, diakses tanggal 4 Agustus 2016

[3] Ahmet A, Sibel 2005. Y. Assessment of heavy metal concentrations in the food web of lake Beyşehir, Turkey. ChemosphereVolume 60, Issue 4, July , Pages 552–556

[4] Jerrold B., Leikin,M . Frank P. Palouce. P, 2008. Poisoning and Toxicology Handbook". Fourth Edition ed.New Yok: Informa Healthcare USA,Inc

[5] IPCS. 2004. Environmental Health Criteria XXX: *Principles for modelling dose-response for the risk assessment of chemicals* (Draft). Geneva: World Health Organization and International Programme on Chemical Safety.

[6] Argos M, Karla T, Rathouz PJ. 2010. *Arsenic exposure from drinking water, and all cause and chronic-disease mortalities in Bangladesh* (HEALS): a prospective cohort study". Lancet pages; 376: 252–5.

[7] US EPA. 2006. *Integreted risk information system (IRIS).*US EPA,office of Research and Development, National Center for Enviromental Assessment.

[8] US EPA. 2016. *Office of Water Policy And Technical Guidelines On Interprestation And Impelementation Of Aquatic Life Metals Area* . Washinton DC

[9] WHO. 2011. Drinking Water Quality Guidelines, WHO Press.

[10] Zaldy, S. 2009. Analisis intrusi air laut dan zona klorida pada sumur bor dalam dan dangkal di kawasan kota Medan dan sekitarnya. Tesis USU, USU Press.

[11] Giri S, Mahato M K, Singh G, Jha V N. 2012. "Risk Assesment due to intake of heavy metals through the ingestion of groundwater around two prpoposes uranium mining areas in Jharkand, India." Enviromen Monit Assess No. 184 March p.1351-1358

[12] Rodigues. 2016. "Neurodevelopmental outcomes among 2- to 3-year-old children in Bangladesh with elevated blood lead and exposure to arsenic and manganese in drinking water". Enviromental Health Journal 1:2-9.

[13] Duruibe, J. O., Ogwuegbu, M. O. C. and Egwurugwu, J. N.2007. *Heavy metal pollution and human biotoxic effects.*International Journal of Physical Sciences Vol. 2 (5), pp. 112-118, May, Available online at http://www.academicjournals.org/IJPS ISSN 1992 - 1950 © 2007 Academic Journals

IOP Conf. Series: Materials Science and Engineering **300** (2018) 012062 doi:10.1088/1757-899X/300/1/012062

Assessment of Seismic Damage on The Exist Buildings Using Fuzzy Logic

Pınar USTA[1], Nihat MOROVA [1], Ahmet EVCİ [2], Serap ERGÜN[3]

[1]Suleyman Demirel University, Faculty of Technology, Civil Engineering Department, 32260 Isparta/Turkey

[2]Afyon Kocatepe University, Faculty of Engineering, Civil Engineering Department, 03200 Afyon/Turkey

[3]Suleyman Demirel University, Faculty of Technology, Software Engineering Department, 32260 Isparta/Turkey

Email : pinarusta@sdu.edu.tr

Abstract. Earthquake as a natural disaster could damage the lives of many people and buildings all over the world. These is micvulnerability of the buildings needs to be evaluated. Accurate evaluation of damage sustained by buildings during natural disaster events is critical to determine the buildings safety and their suitability for future occupancy. The earthquake is one of the disasters that structures face the most. There fore, there is a need to evaluate seismic damage and vulnerability of the buildings to protect them. These days fuzzy systems have been widely used in different fields of science because of its simpli city and efficiency. Fuzzy logic provides a suitable framework for reasoning, deduction, and decision making in fuzzy conditions. In this paper, studies on earthquake hazard evaluation of buildings by fuzzy logic modeling concepts in the literature have been investigated and evaluated, as a whole.

Keywords: Seismic Damage, Earthquake, Historical Buildings, Fuzzy Logic

1. Introduction

Earthquake as a natural disaster could harm the lives of numerous people all over the world. It can impose a hugeeconomical burden on societies, as well. even if earthquake may be unpredictable, proper planning could significantly reduce the damages [1]. Natural disasters such as earthquakes and high winds pose a serious threat to all structures. Recent earthquakes, such as the 1994 Northridge, USA the 1995 Kobe, Japan and the 1999 Kocaeli, Turkey earthquakes caused extensive destructive damage to structures [2]. These earthquake damages in Turkey and all the world take attention for the preserve of the buildings against to earthquake hazard. The earthquake performance evaluation and upgrading of non-standard design existing buildings are among the most significant issues [3]. Especially, Historical buildings preservation, with global respect for cultural heritage, has become a booming trend world-wide due to the emphasis on its benefits, concerning architectural, economic, social, political and spiritual values [4].Exist buildings are vulnerable to earthquakes cause major financial losses and serious disruption on utilities and services for whole community [3-5].Nearly 50 percent of large cities around the world are placed nearby seismic belts.

In developed countries, sensitive technologies used in infrastructural equipment tend to be damaged during earthquakes. However, cities in developing countries are less protected against earthquakes [6]. Global seismic hazard map, All quakes over magnitude 5 since 1973 and Worldwide Earthquakes: 2000 – 2009 are shown Figure 1,2 and 3, respectively.

IOP Conf. Series: Materials Science and Engineering **300** (2018) 012062 doi:10.1088/1757-899X/300/1/012062

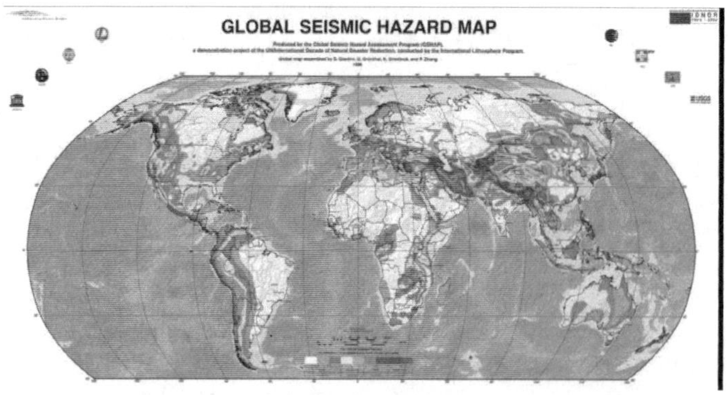

Figure 1. Global seismic hazard map [7]

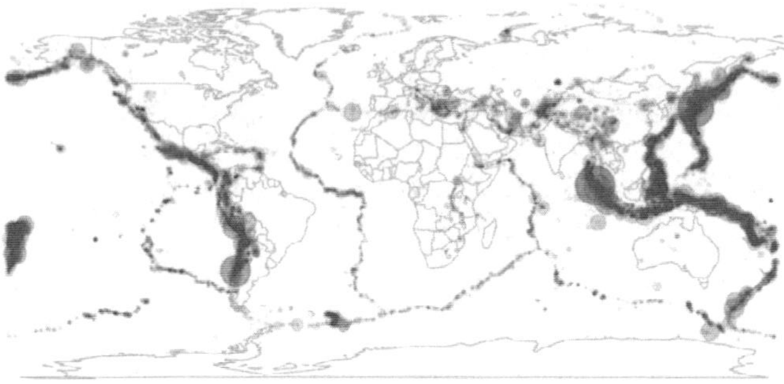

Figure 2. All quakes over magnitude 5 since 1973 [8]

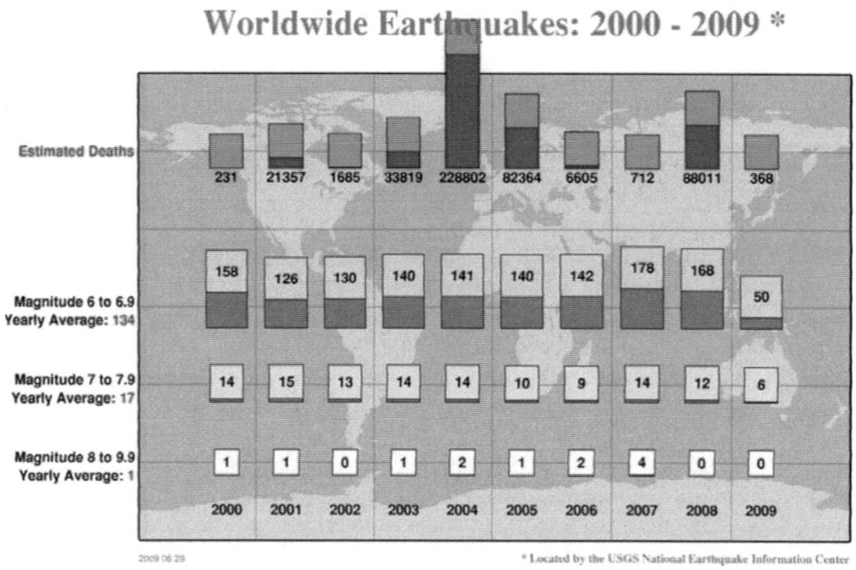

Figure 3. Worldwide Earthquakes: 2000 – 2009 [9]

Approximately 95 percent of all victims of natural disasters around the world are estimated to be from developing countries, and such individuals are twenty times greater in number than the victims of similar catastrophes in the developed World. Figure 4 shows People killed per disaster[6].

Magnitude	2006	2007	2008	2009	2010	2011	2012	2013	2014	2015	*2016*
8.0–9.9	2	4	0	1	1	1	2	2	1	1	0
7.0–7.9	9	14	12	16	21	19	15	17	11	18	17
6.0–6.9	142	178	168	144	151	204	129	125	140	124	126
5.0–5.9	1,712	2,074	1,768	1,896	1,963	2,271	1,412	1,402	1,475	1,413	1,507
4.0–4.9	12,838	12,080	12,292	6,805	10,164	13,303	10,990	9,795	13,494	13,239	12,771
Total	14,703	14,350	14,240	8,862	12,300	15,798	12,548	11,341	15,121	14,795	14,421

Figure 4. Number of earthquakes worldwide for 2006-2016 [10]

Accurate evaluation of damage sustained by buildings during natural disaster events is critical to determine the buildings' safety and their suitability for future occupancy. Time is of the essence in conducting the evaluations as the damaged buildings cannot resume serving their regular purpose until they are deemed safe. The speed with which evaluations are conducted determines the duration for which the potentially damaged buildings remain unusable. The elapsed time directly translates into significant economic losses and to circumstances in which humans are exposed to precarious working and living conditions [11].

Therefore, it is really important to evaluate buildings damage as soon as possible. Nowadays it is possible to be evaluated damage on buildings with computer technologies by using different methods such as fuzzy logic methods.

2. Fuzzy Logic

The fuzzy set was introduced by LotfiZadeh in 1965 since that time Fuzzy logic has come a long way. after first presented to technical society the subject has been the focus of much independent research. Fuzzy set deals with uncertain phenomena often presented in the real-life application. thefuzzy set theory allows the object to have any degrees of membership between zero and 1 but the classical mathematics requires objects to have zero or 1 [12-13-14].

There is our limited understanding of the behavior of the vulnerable elements of the exist buildings under earthquake loads. Due to over the past decade a lot of effort has been devoted to the problem of how to evaluate earthquake damage on buildings by different techniques given the large uncertainties in the pattern of earthquake occurrence. Because of diverse uncertainties and randomness involved both in capacity and seismic demand, evaluation of potential earthquake damage should be carried out based on statistical and probabilistic techniques [15]. In a seismic risk assessment, the subjective judgment of the assessor is always influencing the evaluation and decision making process. For example, the assessor could have totally different opinions about the potential seismic vulnerability of a building if compared to opinions of another assessor, due to the differences in their knowledge and experience. the knowledge given by the assessor is in linguistic terms which aresuch as very high or very low irregularity. Therefore, the available information for this qualitative reasoning, the complexity and the lack of evaluation itself, create ambiguity in the assessment process that can be best handled with fuzzy logic. Fuzzy logic provides a mathematical way to represent fuzziness and imprecision and provides translated for linguistic and quantitative reasoning qualitative features [16].

Fuzzy theory can be applied to establish indicators for earthquake danger or hazard levels. Fuzzy theory is extremely important in building structural analysis. It can indicate that damage conditions or overlaps of different destruction conditions ranging from light structural damage to total collapse. Corresponding earthquake fragility curves can be used to assess the structural risk and damage scale for buildings resulting from an earthquake. Combining fuzzy theory with dynamic risk safety analysis can facilitate safety planning for reconstruction areas following earthquakes [17].

3. Assesment of Seismic Damage and Fuzzy Logic

For the background presented in this section,assessment of seismic damage problems related to buildings type can be addressed through a Fuzzy logic analysis.

Table 1 summarizes the main scientific works available in the literature considering the application of Fuzzy Logic in theassessment of seismic damage problems, putting in evidence the field of application, the objective of the evaluation, the Fuzzy logic technique used and the scientific journal in which the work was published.

Table 1. Some Main works available in the literature concerning Fuzzy Logic applications in assessment of seismic damage on buildings

Authors	Ye ar	Fields of application	Objectives of the evaluation	Journals
Sen and Ekinci [18]	2016	Assessment of Earthquake vulnerability	To evaluate the Earthquake Vulnerability Analysis in Structure Scale with Fuzzy Logic Method in GIS	International Journal of Environment and Geoinformatics
Sen [3]	2010	Earthquake hazard evaluation	To make rapid visual earthquake hazard evaluation of existing buildings by using fuzzy logic	Expert Systems with Applications
Fischer et al. [19]	2002	Structural damage estimation	To advised the use of fuzzy logic in structural damage estimation	Engineering Structures
Demartinosand Dritsos [20]	2006	Pre-earthquake assessment	To discussed the performance of a fuzzy logic based rapid visual screening procedure	Earthquake Spectra
He et al. [21]	2013	Earthquake Damage Assessment	To Assessment Earthquake Damage on RC Structures by Fuzzy logic	Mathematical Problems in Engineering
Alesheikh [1]			To Evaluate earthquake vulnerability with fuzzy logic and GIS	
Elenas et al. [22]	2013	Earthquake Damage Assessment	To classification of seismic damages in buildings	Computational Methods in Structural Dynamics and Earthquake Engineering
Komsari [16]	2011	Seismic risk evaluation	To evaluate seismic risk of wood frame construction with fuzzy based techniques	Phd Thesis
Pakdamarand Guler [23]	2008	Performance evaluation	To evaluate performance of There in forced concrete structures with fuzzy logic approach	The 14th world conference on Earthquake Engineering
Shao et al [24]	2014	Earthquake Damage Assessment	To estimate earthquake damage on buildings using with fuzzy logic application	Journal of Central South University

4. Conclusions

The assessment of seismic damage identification of buildings in pre-earthquake and post- earthquake period presents a very significant task, which must be accomplished in rapid, simple, economic and efficient manner.This paper presents an assessment of seismic damage on the buildings based on Fuzzy techniques.

The search results show that fuzzy logic theory can be use to predict and assessment building damage rates and that estimated results are parallel to real disaster figures. Prediction of disaster damage using building damage rates can provide a reference for immediate disaster response during earthquakes and for regular disaster prevention and rescue planning [24].

As shown in a case study in the United States, the uncertainty of earthquake danger or hazard analysis can be handled with a decay curve obtained using fuzzy theory and regression analysis. the results of this research show that using an appropriate membership function can decrease uncertainty regarding building damage triggered by earthquakes and can be applied to various natural disasters [17].

In any engineering analysis, due to uncertainties which is in structural components and materials, approximation cannot be avoided. Consequently, the fuzzy logic model gives simplicity, speed and flexibility at the deciding buildings' damage assessment.

There is a need to more analyze on exist buildings by conducting experiments and observations.Future work in the area could focus on multiple research directions. The study will be enhanced with real models and fuzzy logic theory will be applied on models.

References

[1] Alesheikh, A. A. Evaluating earthquake vulnerability with the help of fuzzy logic and GIS (Case study: Tehran, District 16).

[2] Guclu, R., & Yazici, H. (2007). Fuzzy logic control of a non-linear structural system against earthquake-induced vibration. Journal of Vibration and Control, 13(11), 1535-1551.

[3] Şen, Z. (2010). Rapid visual earthquake hazard evaluation of existing buildings by fuzzy logic modeling. Expert Systems with Applications, 37(8), 5653-5660.

[4] Perng, Y. H., Juan, Y. K., & Hsu, H. S. (2007). Genetic algorithm-based decision support for the restoration budget allocation of historical buildings. Building and environment, 42(2), 770778.

[5] Tromans, I. J. (2004). Behaviour of buried water supply pipelines in earthquake zones (Doctoral dissertation, Imperial College London).

[6] Meshkini, A., Habibi, K., & Alizadeh, H. (2013). Using fuzzy logic and GIS tools for seismic vulnerability of old fabric in Iranian cities (Case study: Zanjan city). Journal of Intelligent & Fuzzy Systems, 25(4), 965-975.)

[7] http://www.thegeographeronline.net/uploads/2/6/6/2/26629356/5465728_orig.jpg Access date: 3.3.2017.

[8] http://www.thegeographeronline.net/uploads/2/6/6/2/26629356/5079399_orig.png Access date: 3.3.2017.

[9] http://www.thegeographeronline.net/hazards-and-disasters---risk-assessment-and-response.html Access date: 3.3.2017.

[10] https://en.wikipedia.org/wiki/List_of_earthquakes_in_2016 Access date: 3.3.2017.

[11] Kamat, V. R., & El-Tawil, S. (2007). Evaluation of augmented reality for rapid assessment of earthquake-induced building damage. Journal of computing in civil engineering, 21(5), 303310).

[12] M. Ali Lotfollahi Yaghin , M. Reza Bagerzadeh Karimi , B. Bagheri , V. Sadeghi Balkanlou, Vibration control of Multi Degree of Freedom structure under earthquake excitation with TMD control and active control force using fuzzy logic method at the highest and the lowest story of the building, IOSR Journal of Mechanical and Civil Engineering (IOSR-JMCE) e-ISSN: 2278-1684,p-ISSN: 2320-334X, Volume 8, Issue 5 (Sep. - Oct. 2013), PP 07-12);

[13] Sanchez-Silva, M., & Garcia, L. (2001). Earthquake damage assessment based on fuzzy logic and neural networks. Earthquake Spectra, 17(1), 89-112.)

[14] Guclu, R., & Yazici, H. (2007). Fuzzy logic control of a non-linear structural system against earthquake-induced vibration. Journal of Vibration and Control, 13(11), 1535-1551.)

[15] Yücemen, M. S., Özcebe, G., & Pay, A. C. (2004). Prediction of potential damage due to severe earthquakes. Structural Safety, 26(3), 349-366)

[16] Ghorbani Komsari, S. (2014). Seismic Risk Assessment of Wood Frame Construction Using Fuzzy Based Techniques (Doctoral dissertation, Université d'Ottawa/University of Ottawa).

[17] Shao, Y. W., Wu, Y. S., Kao, S. F., Huang, C. J., & Chang, K. Y. (2014). Application of fuzzy theory on earthquake damage rate estimation of buildings. Journal of Central South University, 21, 2454-2459.)

[18] Şen, A., & Ekinci, V. Realization of Earthquake Vulnerability Analysis in Structure Scale with Fuzzy Logic Method in GIS: Kadıköy, Maltepe and Prince Islands Sample.Fischer, T., Alvarez, M., De la Llera, J. C., & Riddell, R. (2002). An integrated model for earthquake risk assessment of buildings. Engineering Structures, 24(7), 979-998.

[19] Demartinos, K., & Dritsos, S. (2006). First-level pre-earthquake assessment of buildings using fuzzy logic. Earthquake spectra, 22(4), 865-885.

[20] He, H., Cong, M., & Lv, Y. (2013). Earthquake damage assessment for RC structures based on fuzzy sets. Mathematical Problems in Engineering, 2013.

[21] Elenas, A., Vrochidou, E., Alvanitopoulos, P., & Andreadis, I. (2013). Classification of seismic damages in buildings using fuzzy logic procedures. In Computational Methods in Stochastic Dynamics (pp. 335-344). Springer Netherlands.

[22] Pakdamar, F., & Güler, K. (2008, October). Fuzzy logic approach in the performance evaluation of reinforced concrete structures (flexible performance). In Proceedings of the 14th world conference on earthquake engineering, Beijing, China (pp. 12-17).

[23] Shao, Y. W., Wu, Y. S., Kao, S. F., Huang, C. J., & Chang, K. Y. (2014). Application of fuzzy theory on earthquake damage rate estimation of buildings. Journal of Central South University, 21, 2454-2459.

SEM Model Medical Solid Waste Hospital Management In Medan City

Verawaty Simarmata[1], Setiaty Pandia[2], Herman Mawengkang[3]
[1]Graduate School, Universitas Sumatera Utara
[2]Department of Natural Resources Management and Environment, Universitas Sumatera Utara
[3]Department of Mathematics, Universitas Sumatera Utara

Email: verawatysimarmata@yahoo.co.id

Abstract. In daily activities, hospitals, as one of the important health care unit, generate both medical solid waste and non-medical solid waste. The occurrence of medical solid waste could be from the results of treatment activities, such as, in the treatment room for a hospital inpatient, general clinic, a dental clinic, a mother and child clinic, laboratories and pharmacies. Most of the medical solid waste contains infectious and hazardous materials. Therefore it should be managed properly, otherwise it could be a source of new infectious for the community around the hospital as well as for health workers themselves. Efforts surveillance of various environmental factors need to be applied in accordance with the principles of sanitation focuses on environmental cleanliness. One of the efforts that need to be done in improving the quality of the environment is to undertake waste management activities, because with proper waste management is the most important in order to achieve an optimal degree of human health. Health development in Indonesian aims to achieve a future in which the Indonesian people live in a healthy environment, its people behave clean and healthy, able to reach quality health services, fair and equitable, so as to have optimal health status, health development paradigm anchored to the healthy. The healthy condition of the individual and society can be influenced by the environment. Poor environmental quality is a cause of various health problems. Efforts surveillance of various environmental factors need to be applied in accordance with the principles of sanitation focuses on environmental cleanliness. This paper proposes a model for managing the medical solid waste in hospitals in Medan city, in order to create healthy environment around hospitals.

Keywords: waste, medical solid waste, non-medical solid waste, infectious and health paradigm.

1. Introduction

Hospital activities will generate a number of by-products in the form of waste, both solids, liquids and gases that contain pathogens, chemicals and medical equipment which are generally dangerous and toxic. In order to improve the quality of services, hospitals should be able to cope with such waste in ([6], [7], [13]). As for the waste treatment facilities in the hospital one of them is to use incinerators in ([4], [8]). Characteristics of medical solid waste produced can be distinguished into two parts, namely, domestic waste and waste hazard and toxicity in ([1], [18]). The latter one are infectious in [9]. An incinerator unit is expected to reduce the volume of waste prior to disposal and also can eliminate the hazard and toxicity [5]. Objective the Purpose of the Research is as Follows :

- To determine the performance of machines and tools in the handling of medical solid waste of the hospital.
- To find out how much the rate of availability of medical solid waste in the activities of medical activity.
- Knowing the extent of the amount of the availability of human resources in the environment that the ability of medical solid waste treatment hospital.

- Generating actions and efforts of the hospital's medical waste management by means of overcoming the impact of medical solid waste of the hospital according to the rules in the Decree of the Minister of Health of the Republic of Indonesia Number 12 Year 2004 Solid Waste Management's Medical Hospital, can be used as input to the Government in enforcing supervision and establish how the actions and efforts for the management and control of solid waste pollution hospital medical until declared safe for the environment and living beings.

The problem of waste is a serious concern of the people and government of Indonesia, especially since the last decade, mainly due to the business development activities of the hospital which is the backbone of economic development for Indonesia ([10], [12]). Waste management is a necessity for the preservation of human health and the environment in general. However, the procurement and operation of medical solid waste processing facility hospital was still considered burdensome for the hospital in ([17], [22], [23]).

2. Problem Formulation

Hazardous and toxic waste resulting from activities at most hospitals contain various elements of medical solid waste ranging from heavy metals that have the nature of accumulative toxic which are harmful to human health ([14], [21], [24]).

Medical waste in the form of gas, liquid or solid generally include the category or the nature of the hazard and toxicity waste ([11], [15]). The operational in activities of the hospital aims to improve the health and well-being, but it also produces waste as an environmental pollutant waters, soil and gas.

From the description of the background of the above in mind that the management of solid waste pollution control medical hospital, is one of the fundamental problems in the city of Medan and application supervision is the main business licensing and control alternatives in accordance with applicable regulations ([2], [20], [21]). However, to determine the extent to which technology and applied that can control solid waste pollution medical hospital in Medan, it is necessary to study the issues as follows:

- Anything that affects the performance of machines and tools in the handling of medical solid waste of the hospital?
- What is the pattern to determine how much the rate of availability of medical solid waste in the activities of medical activity?
- It has been the extent of the amount of the availability of human environment play a role in handling the hospital's medical solid waste according to the rules in the Decree of the Minister of Health of the Republic of Indonesia Number 12 of 2004?
- How do the actions and efforts of the hospital's medical waste management by means of overcoming the impact of hospital medical solid waste right? So there is input to the Government's policy in enforcing supervision and sets out how the actions and efforts for the management and control of solid waste pollution hospital medical until declared safe for the environment and living beings.

For the mathematical description of the model, the following notations are defined.

2.1 Sets and indices

Data from this study were analyzed using statistical techniques that use methods of regression and simple correlation, [25]. Test significance and the truth of the correlation between variables purpose is to determine how big the correlation coefficient between these variables by using correlation and regression formulas [3]. Currently, the research will be obtained comparative data with more than two samples of the variables are interconnected, then to test the statistical parametric (n1) and non parametric (n2) can be used one-way ANOVA or two way ANOVA combined using test method variance or the t test [19].

IOP Conf. Series: Materials Science and Engineering **300** (2018) 012063 doi:10.1088/1757-899X/300/1/012063

2.2 Variables

- Variable performance of machines and tools (incinerator) in the treatment of hospital medical solid waste (X1).
- Variable availability of medical solid waste from the activities of hospital activities (X2).
- Variable of environment variable availability of human resources according to the rules in the Decree of the Minister of Health of the Republic of Indonesia Number 12 Year 2004 (X3).
- Variable overcome the impact of the hospital's medical solid waste (X4).

2.3 Parameters

- X1.1 = Source hospital medical solid waste.
- X1.2 = Type hospital medical solid waste.
- X1.3 = Number of hospital medical solid waste.
- X1.4 = Production of hospital medical solid waste.
- X2.1 = Number of rooms and patient.
- X2.2 = Facilities or Facility Storage While medical solid waste.
- X2.3 = Physical condition of availability of raw materials activities of hospital activities.
- X2.4 = Medical solid waste collection process hospital.
- X3.1 = Condition Availability HR environment.
- X3.2 = Graduates HR environment.
- X3.3 = Ability or expertise HR environment.
- X3.4 = Training officers in the handling of hospital medical solid waste.
- X4.1= Condition of society in the hospital environment.
- X4.2 = Attitude of people in a hospital environment.
- X4.3= The level of awareness of the hospital against medical solid waste treatment.
- X4.4= Level of medical solid waste pollution hospital.

3. Mathematical Formulation

$$\mathbf{AVE} = \frac{\sum_{i=1}^{n} L_i^2}{n} \tag{1}$$

Where : Li = standardized loading estimate
n=value indicator.
While reliability test can be done by calculating construct reliability (CR) with the following formula:

$$\mathbf{CR} = \frac{\left(\sum_{i=1}^{n} L_i\right)^2}{\left(\sum_{i=1}^{n} L_i\right)^2 + \left(\sum_{i=1}^{n} e_i\right)} \tag{2}$$

Dimana: CR = construct reliability
Li = loading factor or standardized loading estimate

IOP Conf. Series: Materials Science and Engineering **300** (2018) 012063 doi:10.1088/1757-899X/300/1/012063

Table 1. Indeks Equilibrium Statistic Equation Models after Modification Model

Criteria	**Value Cut-off for n = 89 ; df = 168; alpha = 0,05**	**Value AMOS**	**Description**
Freedon Degree (db)	> 0	168	Qualify
Chi – Square	< 155.405	53,058	Qualify
P-value	P value ≤ 0.05	0,001	Qualify
CMIN/DF	≤ 2,00	1,196	Qualify
Root Mean Square Error of Approximation (RMSEA)	< 0.08	0,022	Qualify
Goodness of Fit Index (GFI)	≥ 0.90	0,964	Qualify
Adjusted Goodness of Fit Index (AGFI)	≥ 0.09	0,941	Qualify
Tucker Lewis Index (TLI)	≥ 0.95	0,989	Qualify
Comparative Fit Index (CFI)	≥ 0.94	0,993	Qualify

When : n = value sampel; df = freedom degree.

Table 2. Model Direct between Variabel Unobserved Variable

Direct between Variabel Unobserved Variable	**Direct Impact**	**CR**	**Rob (p)**	**Description**
X1 ←Model_Management	1,025	1,02	0,15	Influence
X2 ←Model_ Management	0,927	0,93	0,86	Influence
X3 ←Model_ Management	0,939	0,94	0,88	Influence
X4 ←Model_ Management	0.586	0,59	0,34	Influence

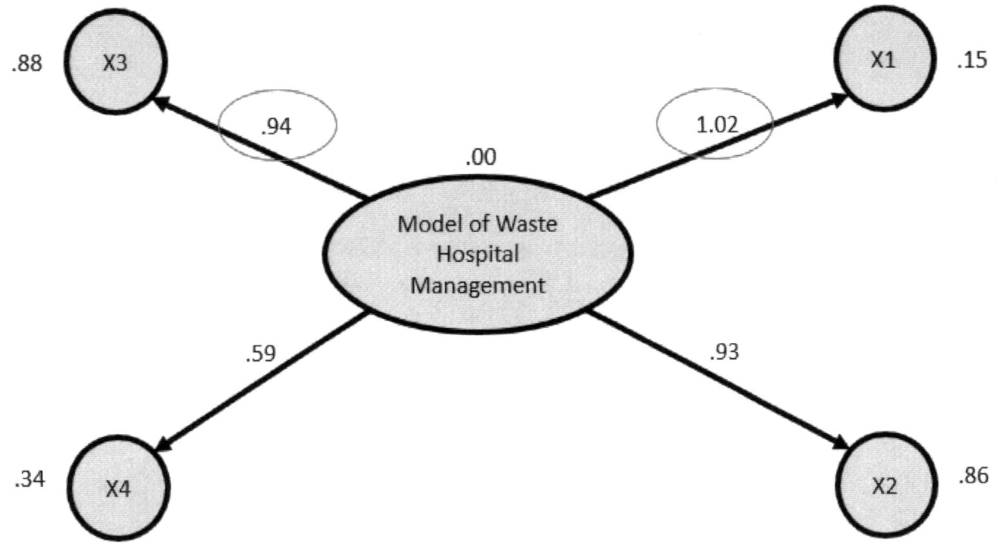

Figure 1. Final Sequence Equation Models (Source: Data Primer 2016)

4. Conclusions

The conclusions of the model SEM medical solid waste hospital management in Medan can be stated as follows.

- Place or health services or medical care and treatment.
- Results of research for procedures special handling solid waste medical hospital, in this case the standard operating procedures or SOPs (Standard Operating Services) acquired point X1 for 1,025 concluded that the procedures for special handling solid waste medical hospital of highly influenced from the source of solid waste medical hospitals, type of hospital medical solid waste, the number of hospital medical solid waste and solid waste collection processes internally medical hospital. In this highly desirable mutilation waste statement whenever there is addition or medical solid waste storage in the storage room. Recording daily or daily logbook which will be reevaluated every 3 (three) months if there is submission to the collector, transportation, destruction and medical solid waste utilizers of the hospital.
- Training and the provision of non-formal education that support the effectiveness of solid waste management medical hospital room starts from action to action to the activities of a doctor or medical personnel. On the basis of this is necessary to add the medical knowledge regarding the handling of solid waste such as hospital medical known to influence. Data obtained with the results of this SEM model test to information how big the accuracy of the analysis of massive data acquisition to rate the availability of medical solid waste in the activities of medical activities with X2 = 0.927. It is concluded that whenever any activity or activities of the hospital medical solid waste will provide a contribution of X2 per activity.
- The results of multiple linear regression analysis provide important information the extent of the amount of the availability of human surroundings ability in handling the hospital's medical solid waste according to the rules in the Decree of the Minister of Health of the Republic of Indonesia Number 12 Year 2004 with X3 figure of 0,939 per activity for medical solid waste management activities. This applies equally to every hospital in the city of Medan because almost all hospitals that have no HR environment capable of handling serious medical solid waste from hospitals.

- Model suitable measures and efforts to become a model Solid Waste Management Medical Hospital in Medan is Y = 1.025 X1 + 0.927 X2 + 0.939 X3 + 0.586 X4.

References

[1] Abah, Obekpa., Stephen and Ige Ohimain, Elijah., 2011, *Healthcare Waste Management in Nigeria: A Case Study*, Journal of Public Health and Epidemiology, Available online at http://www.academicjournals.org/jphe, ISSN:2141-2316 ©2011 Academic Journals, 3(3), 99110.

[2] Al-Shayea, Qeethara., and El-Refea, Ghaleb, 2013, *Predicting the Effects of Medical Waste in the Environment Using Artificial Neural Networks: A Case Study*, IJCSI International Journal of Computer Science Issues, ISSN (Print): 1694-0784, 10(3), ISSN (Online): 16940814, 10(1), 258-261.

[3] Andrés L., Medaglia, Juan, Villegas, G., M., Diana, Rodríguez and Coca, 2009, *Hybrid Biobjective Evolutionary Algoritms for the Design of a Hospital Waste Management*, *J Heuristics*, 15, 153-176.

[4] Brindha. G., and Kishore, Kanchan., 2013, *Bio Medical Waste Management*, International Journal of Health Sciences and Research, ISSN: 2249-9571, 3(8), 69-76.

[5] Chafid Fandeli, 2004, *Analisis Mengenai Dampak Lingkungan Prinsip Dasar Dalam Pembangunan*, Penerbit Liberty, Yogyakarta.

[6] Chowdhary, Asif., and Slathia, Deepika., 2014, *Biomedical Waste Management: A Case Study of Gandhinagar Hospital, Jammu*, International Journal of Environmental Research and Development, ISSN:2249-3131, 4(4), 287-290.

[7] Chudasama, Rajesh K., Rangoonwala, Matib., Sheth, Ankit., Misra, SKC., Kadri and V Patel, Umed., 2013, *Biomedical Waste Management: A Study of Knowledge, Attitude and Practice Among Health Care Personnel At Tertiary Care Hospital in Rajkot*, Journal of Research in Medical and Dental Science, 1(1), 17-22.

[8] Conrardy, Julie., Hillanbrand, Mary., Sandra, Myers., and Nussbaum, George F., 2010, *Reducing Medical Waste,* AORN Journal, 10.1016/j.aorn.2009.12.029, 91(6), 711-721.

[9] Dalal, Parag., 2013, *Management of Infectious Bio-medical Waste of Ujjain City*, International Journal of Advanced Research, ISSN:2320-5407, 1(2), 52-58.

[10] David L. Streiner, PhD, 2006, *Building a Better Model: An Introduction to Structural Equation Modelling*, Research Methods in Psychiatry, Vol. 51, No. 5, pp: 317-324.

[11] Deepak Arora., Mahajan MK., Shekhar Dwive D and Parul GARG, 2015, *Impact Of Training Session On Biomedical Waste Management - An Analysis,* International Journal of Recent Scientific Research, ISSN: 0976-3031, Vol. 6, Issue, 9, pp.6479-6481.

[12] Desi Rahmatina, 2014, *Pemodelan Structural Equation Modelling pada Data Ordinal dengan Menggunakan Metode Weighted Least Square (WLS)*, Prosiding SNM 2014, hal. 1-10.

[13] Dhruv P., Hirani, Krish R., Villaitramani and Snehjit J., Kumbhar, 2014, *Biomedical Waste: An Introduction to its Management*, International Journal of Innovative Research in Advanced Engineering (IJIRAE), ISSN: 2349-2163, 1 (8), 82-87.

[14] Dohare S., V. K., Garg, and B. K., Sarkar, 2013, *A Study Of Hospital Waste Management Status in Health Facilities Of An Urban Area*, International Journal of Pharma and Bio Sciences, ISSN 0975-6299, 4(1): (B) 1107 - 1112.

[15] Jagadeesh, U., Boss, Chandira., Poyyamoli, G., and Roy, Gautam., (2014). *Biomedical Waste Management Practices In Puducherry Region India: A Case Study Of Three Selected Health Care Facilities*, International Journal of Advanced Research, ISSN:2320-5407, 2(8), 780 -799.

[16] Joseph Jr, Hair F., 2001,. *Multivariate Data Analysis*, Seven Edition, New York, PrenticeHall International, Inc.

[17] Kumar Goddu, Vijaya., Duvvuri, Kavita., and Kaumudini Bakki., Vidya, 2007, *A Critical Analysis of Healthcare Waste Management in Developed and Developing Countries: Case Studies from India and England*, Proceedings of the International Conference on Sustainable Solid Waste Management, Chennai, India, 134-141.

[18] Lakbala, P., and Mahesh, T. M., 2011, *Bio Medical Waste Managementin Shiraz City of Iran,* India Journal 8(1), 56-64.

[19] Larwin, Karen., and Harvey., Milton, 2012, *A Demonstration of a Systematic ItemReduction Approach Using Structural Equation Modeling*, Practical Assessment, Research & Evaluation, Item Reduction using SEM, ISSN:1531-7714, 17(8), 1-19.

[20] Prüss, A., Giroult, E., and Rushbrook, P., 2005, *Pengelolaan Aman Limbah Layanan Kesehatan* (Ahli Bahasa: Munaya Fauziah) Safe Management of Wastes From Health Care Activities, World Health Organization. Jakarta. EGC, 3(15), 21-28, 171-175.

[21] Radha, R., 2012, *Assessment of Existing Knowledge, Attitude, and Practices Regarding Biomedical Waste Management among the Health Care Workers in a Tertiary Care Rural Hospital,* International Journal of Health Sciences and Research, ISSN: 2249-9571, 2(7), 1419.

[22] Rajendra, Abhilash and Ramu, 2014, *Bio Medical Waste Management in The Local Planning Area Of Mysore City*, International Journal of Advanced Research, ISSN 2320-5407, 2 (7), 951-955.

[23] Sharma, Shalini., 2010, *Awareness about Bio-Medical Waste Management among Health Care Personnel of Some Important Medical Centers in Agra*, International Journal of Environmental Science and Development, ISSN: 2010-0264, 1(3), 251-255.

[24] Singh, Anshita., Kumari, Reema., Wakhlu, Anupam., Srivastava, Kirti., Wakhlu, Ashish., and Kumar, Sandeep., 2014, *Assessment of Bio-Medical Waste Management in a Government Healthcare Setting of North India*, International Journal of Health Sciences and Research, ISSN:2249-9571, 4(11), 203-208.

[25] Stephan, Levine Krehbiel Berenson., 2001, *Statistics for Managers Using Mocrosoft Excel*, New York, Prentice-Hall International, Inc.

IOP Conf. Series: Materials Science and Engineering **300** (2018) 012064 doi:10.1088/1757-899X/300/1/012064

Developing a feasible neighbourhood search for solving hub location problem in a communication network

Fibri Rakhmawati[1], Herman Mawengkang[2], F. Buulolo[2], Mardiningsih[2]
[1]Graduate School of Mathematics, University of Sumatera Utara
[2] Department of Mathematics, University of Sumatera Utara
[2]hmawengkang@yahoo.com

Abstract. The hub location with single assignment is the problem of locating hubs and assigning the terminal nodes to hubs in order to minimize the cost of hub installation and the cost of routing the traffic in the network. There may also be capacity restrictions on the amount of traffic that can transit by hubs. This paper discusses how to model the polyhedral properties of the problems and develop a feasible neighbourhood search method to solve the model.

Keywords: Communication network, hub location, MILP, neighborhood search.

1. Introduction

Hub facilities can be found in most of the transportation and communication networks. The main role of hubs for this sort of networks usually act as a central point for distribution of communication services. The hub facility will distribute most types of communications services.

Hub location problems arise in all network design, particularly when it is necessary to distribute services. Due to the nature of a location problem, the aim is to find the location of hub nodes (location point) and the allocation of service nodes to these located hub nodes. In standard location problems when the locations of the facilities are determined then logically, each customer node receives service from its nearest facility. However, for hub location problems, it is not necessary to assign each demand node to its nearest hub will give optimal solutions. Therefore the optimal allocations of demand centers to the located hubs must be determined. In term of applications rather than in telecommunication, the hub location problem can also be found in public transportation [1]. parcel delivery [2], [3], and logistic systems [4]. Interesting review for the application of hub location problem can be found in [5] and [6].

This paper considers hab location in telecommunication network. Communication networks are built and designed in order to distribute resources. With the advent of the information age there has been increased interest in the efficient design of communication networks. Network design problems occur in many field of communication, ranging from the old style analogue telephone networks to more modern applications such as wide - area and local – area computer networks, ISDN networks for multi – media transmission, and digital cellular networks for mobile phone communications. They all have a similar intent: how to carry the expected traffic flow from origin to destination at minimum cost. The most important part of network design is to find the best way to layout the components (nodes and arcs) to minimize cost while meeting performance criterion such as transmission delay, throughput or reliability.

If one imagines the communication requirement as a set of nodes representing origin and destinations, the primary problem is to join the nodes together in the most efficient manner. This is not an easy problem; for event a small number of nodes there are many thousands of ways of joining them together. Therefore this type of problem is termed NP – *complete* [7].

Given that we cannot investigate all possible ways of linking the nodes, the challenge is to devise a method for investigating a restricted number that nevertheless obtains a good solution. Further more, communication network design problems are not time critical. Meaning that, most approaches which have been designed to solve these problems based on heuristics, such as, simulated annealing, taboo search, evolutionary computing [8], [9]. [10] proposed a hybrid heuristic based on simulated annealing, tabu list, and improvement procedures for the uncapacitated hub location problem with single allocation. [11] developed a tabu search meta-heuristic for the multiple allocation p-hub median problem. [12] presented a hybrid genetic algorithm embedded with a diagonalization method for the intermodal hub-and-spoke network design problem with multiple stakeholders and multi-type containers. [13] developed two different meta-heuristic procedures that consist of two phases: a solution construction phase and a solution improvement phase based on local search for the intermodal terminal location problem. [14] provided a review for the most recent advances of solution methodologies for the hub location problem

Exact methods are guaranteed to achieve proven optimal solutions under sufficient computation time. [15] proposed a branch-and-cut algorithm for the hub location problem with single assignment. [16] developed a Benders decomposition algorithm for the uncapacitated hub location problem with multiple assignments in telecommunication and transportation systems. [17] proposed a branch-and-price algorithm for the capacitated hub location problem with single assignment, in which Lagrangean relaxation is used to obtain tight lower bounds of the restricted master problem. [18] presented an exact algorithm capable of solving large-scale instances of the uncapacitated hub location problem with multiple assignments. The algorithm applies Benders decomposition to a strong path-based formulation of the problem. [19] proposed a Benders decomposition method for the tree of hub location problems. The exact methods are efficient at solving relatively small problems. However, they are limited on thoroughly searching all possible branches and finding an optimal solution under a reasonable amount of computational time when the problem size is particularly large.

In many applications however, it is necessary to add redundancy into the network to ensure reliability; if one link fails it would still be possible to connect from origin to destination. We have used a combination of linear Programming (LP) and heuristic based on neighbournood search approach to tackle this problem; and these approaches are discussed in Sections 4 and 5.

2. The Network Design Problem

The design problem is to satisfy all traffic requirements at minimum cost. For a tree network synthesis problem involving n nodes, there are n^{n-2} possible tree structures, e.g. one hundred million possibilities for a network as small as 10 nodes. The information required to formulate the problem is the traffic demand between each origin and destination (O – D) pair, and the linier cost function for carrying traffic on each (possible) link (i,j) between nodes i and j.

In order to model the problem, we define first the following notation.

Sets and Indices

N Set of nodes,
T Set of time periods.
M_k Set of modules with different types available for a hub located at node $k \in N$

Traffic communication flow

f_{ij}^{t} Flow from node $i \in N$ that will go to node $j \in N$ at time period $t \in T$

TFE_{i}^{t} Total flow emerged from node $i \in N$ at time period $t \in T$

TFD_{j}^{t} Total flow for node $j \in N$ at time period $t \in T$

Capacities

C_{mk} Capacity of module type $m \in M_k$ available for node $k \in N$

Costs

α_{k}^{t} Setup cost to locate a hub at node $k \in N$ in period $t \in T$

β_{kj}^{t} Costs to operate a hub li9nk between hubs $k \in N$ and $j \in N$ in period $t \in T$

δ_{mk}^{t} Cost for installing a module of type $m \in M_k$ at hub $k \in N$ in period $t \in T$

γ_{k}^{t} Cost to operate per unit of flow for hub $k \in N$ in period $t \in T$

σ_{ij}^{t} Cost for sending flow from node $i \in N$ to node $j \in N$ in period $t \in T$

Decision variables

x_{ij}^{t} Binary variable indicating whether there is a link between node $i \in N$ to hub $j \in N$ in period $t \in T$

z_{kj}^{t} Binary variable indicating whether a hub link between hubs $k \in N$ and $j \in N$ in period $t \in T$

v_{km}^{t} Binary variable indicating whether a module of type $m \in M_k$ is installed at hub $k \in N$ in period $t \in T$

y_{ijk}^{t} Amount of communication flow from node $i \in N$ routed from hub $k \in N$ to hub $j \in N$ in period $t \in T$.

3. The Model

The objective of the communication network problem such that to find the hub location is to minimize the overall costs. From reality it can be understood that the costs which could be involved are as follows: cost to link a node to a hub, cost to install a module, cost to locate a hub and cost relating to the amount of communication flow.. Mathematically the objective can be expressed as follows.

$$\begin{aligned} Minimize & \sum_{t\in T}\sum_{i\in N}\alpha_{t}^{k}(x_{ii}^{t}-x_{ii}^{t-1})+\sum_{t\in T}\sum_{k\in N}\sum_{j\in N,j>k}\beta_{kj}^{t}z_{kj}^{t}+\sum_{t\in T}\sum_{k\in N}\sum_{m\in M_k}\delta_{mk}^{t}v_{mk}^{t} \\ & +\sum_{t\in T}\sum_{k\in N}\gamma_{k}^{t}(\sum_{i\in N}TFE_{i}^{t}x_{ik}^{t}+\sum_{i\in N}\sum_{j\in N}y_{ijk}^{t})+\sum_{t\in T}\sum_{i\in N}\sum_{j\in N}\sigma_{ij}^{t}x_{ij}^{t} \\ & +\sum_{t\in T}\sum_{i\in N}\sum_{j\in N}\sum_{k\in N}\sigma_{ij}^{t}y_{ijk}^{t}+\sum_{t\in T}\sum_{i\in N}\sum_{j\in N}\sigma_{ij}^{t}TFD_{j}^{t}x_{ij}^{t} \end{aligned} \quad (1)$$

There are constraints imposed for the decision variables should be considered. The restrictions are formulated in the following equations.

$$\sum_{j\in N} x_{ij}^t = 1, \quad \forall i \in N, \forall t \in T \tag{2}$$

$$x_{ij}^t \leq x_{jj}^t, \quad \forall i \in N, \forall t \in T \tag{3}$$

$$x_{jj}^t \leq x_{jj}^{t-1}, \quad \forall j \in N, \forall t \in T \tag{4}$$

$$\sum_{m\in M_k} v_{km}^t \leq x_{kk}^t, \ \forall k \in N, \forall t \in T \tag{5}$$

$$\sum_{i\in N} TFE_i^t\, x_{ik}^t + \sum_{j\in N}\sum_{i\in N} y_{ijk}^t \leq \sum_{m\in M_k} C_{km}\, v_{km}^t, \ \forall k \in N, \forall t \in T \tag{6}$$

$$\sum_{j\in N, j\neq k} y_{kji}^t - \sum_{j\in N, j\neq k} y_{jki}^t = TFD_i^t x_{ik}^t - \sum_{j\in N,} f_{ij}^t\, x_{jk}^t, \ \forall i, k \in N, \forall t \in T \tag{7}$$

$$z_{kj}^t \leq x_{kk}^t, \ \forall k, j \in N, k < j, \forall t \in T \tag{8}$$

$$z_{kj}^t \leq x_{jj}^t, \ \forall k, j \in N, k < j, \forall t \in T \tag{9}$$

$$y_{ijk}^t + y_{ikj}^t \leq TFE_i^t z_{kj}^t, \ \forall i, j, k \in N, k < j, \forall t \in T \tag{10}$$

$$x_{ij}^t, z_{kj}^t, v_{km}^t \in \{0, 1\}, \ \forall i, j, k \in N, \forall m \in M_k, \forall t \in T \tag{11}$$

$$x_{ii}^0 = 0, \forall i \in N \tag{12}$$

$$y_{ijk}^t \geq 0, \forall i, j, k \in N, \forall t \in T \tag{13}$$

It can be observed that the model is in the form of large scale mixed integer linear programms.

4. A Suboptimal Search

In integer programming the reduced gradient vector, which is normally used to detect an optimality condition, is not available, even though the problems are convex. Therefore it is necessary to impose a certain condition for the local testing search procedure in order to assure that we have obtained the "best" suboptimal integer feasible solution.

A quantity test to replace the pricing test for optimality in the integer programming problem was proposed by Scarf [18]. The test is conducted by a search through the neighbours of a proposed feasible point to see whether a nearby point is also feasible and yields an improvement to the objective function. However, the result obtained is just a feasible suboptimal solution.

Let $[x_k]$ be the integer feasible point which satisfies the neighborhood conditions $[x_k] + 1 \in N([x_k])$. We could then say if $[x_k] + 1 \in N([x_k])$ implies that the point $[x_k] + 1$ is either infeasible or yields an inferior value to the objective function obtained with respect to $[x_k]$. In this case $[x_k]$ is said to be an "sub-optimal" integer feasible solution to the integer programming problem.

5. The Algorithm

After solving the relaxed problem, the procedure for searching a suboptimal but integer-feasible solution from an optimal continuous solution can be described as follows.

Let

$$x = [x] + f, \quad 0 \le f \le 1$$

be the (continuous) solution of the relaxed problem, $[x]$ is the integer component of non-integer variable x and f is the fractional component.

Stage 1.

Step 1. Obtain $\delta_{i^*} = \min\{f_i, 1 - f_i\}$

Step 2. Find $v_{i^*}^T = e_{i^*}^T B^{-1}$

Step 3. Calculate the maximum movement of nonbasic variable $\sigma_{ij} = v_{i^*}^T a_j$

Otherwise go to next non-integer nonbasic or superbasic j (if available). Eventually the column j^* is to be increased form LB or decreased from UB.

If none go to next i^*.

Step 4.

Solve $B\alpha_{j^*} = \alpha_{j^*}$ for α_{j^*}

Step 5. Do ratio test for the basic variables in order to stay feasible due to the releasing of nonbasic j^* from its bounds.

Step 6. Exchange basis

Step 7. If row $i^* = \{\emptyset\}$ go to Stage 2, otherwise

Repeat from step 1.

Stage 2. Do integer lines search to improve the integer feasible solution

6. Conclusions

The approach described in this paper has been demonstrated to be efficient on a limited range of moderately sized problem. However, the success of these investigations lends hope that it will prove to provide a valuable tool for communication network design. An advantage of the approach described here over other heuristic approaches is the ability to terminate the procedure early and still be assured of a reasonably good solution which still satisfies feasibility requirement.

7. References

[1] Nickel, S., Schöbel, A., & Sonneborn, T., 2001. *Mathematics methods and optimization in transportation systems*. Berlin: Springer. ch. Hub Location Problems in Urban Traffic Networks.

[2] Ernst, A. T., & Krishnamoorthy, M., 1998. *Exact and heuristic algorithms for the uncapacitated multiple allocation p-hub median problem.* European Journal of Operational Research, **104,** pp. 100–112.

[3] Ernst, A., & Krishnamoorthy, M., 1999. *Solution algorithms for the capacitated single allocation hub location problem.* Annals of Operations Research, **86,** pp. 141–159.

[4] Tan, P. Z., & Kara, B. Y., 2007. *A hub covering model for cargo delivery systems.* Networks, **49,** pp. 28–39.

[5] Alumur, S., & Kara, B., 2008. *Network hub location problems: The state of the art.* European Journal of Operational Research, **190,** pp. 1–21.

[6] Campbell, J. F., & O'Kelly, M. E., 2012. *Twenty-five years of hub location research.* Transportation Science, **46,** pp. 153–169.

[7] Alhazov, A., Mart´ın-Vide, C., & Rogozhin, Y., 2006. *On the number of nodes in universal networks of evolutionary processors,* Acta Inf. **43**, pp. 331–339.

[8] Pioro, M., & Medhi, D., 2004. *Routing, Flow and Capacity Design in Communication and Computer Networks*. San Francisco, CA: Morgan Kaufmann.

[9] Resende, M. G. C., & Pardalos, P. M., 2006. *Handbook of Optimization in Telecommunications.* Springer.

[10] Chen, J., 2007. *A hybrid heuristic for the uncapacitated single allocation hub location problem.* Omega 2007, **35(2),** pp. 211–220.

[11] Ishfaq, R., & Sox, C.R., 2011. *Hub location allocation in intermodal logistic networks.* European Journal of Operational Research, **210(2),** pp. 213–230.

[12] Meng, Q., & Wang, X., 2011. *Intermodal hub-and-spoke network design: Incorporating multiple stakeholders and multi-type containers.* Transportation Research Part B: Methodological, **45(4),** pp. 724–742.

[13] Sorensen, K., Vanovermeire, C., & Busschaert, S., 2012. *Efficient metaheuristics to solve the intermodal terminal location problem.* Computers & Operations Research2012; **39(9),** pp. 2079–2090.

[14] Farahani, R.Z., Hekmatfar, M., Arabani, A.B., & Nikbakhsh, E., 2013. *Hub location problems: A review of models, classification, solution techniques, and applications.* Computers & Industrial Engineering; **64(4)**, pp. 1096–1109.

[15] Labbe, M., Yaman, H., & Gourdin, E., 2005. *A branch and cut algorithm for hub location problems with single assignment.* Mathematical Programming, **102(2)**, pp. 371–405.

[16] Camargo, R., Miranda, G., Lokketangen, A., 2013. *A new formulation and an exact approach for the many-to-many hub location routing problem.* Applied Mathematical Modelling, 3**7(12-13)**, pp. 7465–7480.

[17] Contreras, I., Fernández, E., & Marín, A., 2010. *The tree of hubs location problem.* European Journal of Operational Research, **202,** pp. 390–400.

[18] Contreras, I., Cordeau, J.-F., & Laporte, G., 2011. *The dynamic uncapacitated hub location problem.* Transportation Science, **45,** pp. 18–32.

[19] Sa, E.M., Camargo, R., Miranda, G., 2013. *An improved benders decomposition algorithm for the tree of hubs location problem.* European Journal of Operational Research, **226(2)**, pp. 185–202.

Seismic Evaluation of A Historical Structure In Kastamonu - Turkey

Pınar USTA[1], Asuman Işıl ÇARHOĞLU[2], Ahmet EVCİ [3]
[1]Suleyman Demirel University, Faculty of Technology, Civil Engineering Department, 32260 Isparta/Turkey
[2]Suleyman Demirel University, Faculty of Engineering, Civil Engineering Department, 32260 Isparta/Turkey
[3]Afyon Kocatepe University, Faculty of Engineering, Civil Engineering Department, 03200 Afyon/Turkey

Email : pinarusta@sdu.edu.tr, asumancarh@gmail.com, evciahmet@gmail.com

Abstract. The Kastomonu province is a seismically active zone. the city has many historical buildings made of stone-masonry. In case of any probable future earthquakes, existing buildings may suffer substantial or heavy damages. In the present study, one of the historical traditional house located in Kastamonu were structurally investigated through probabilistic seismic risk assessment methodology. In the study, the building was modeled by using the Finite Element Modeling (FEM) software, SAP2000. Time history analyses were carried out using 10 different ground motion data on the FEM models. Displacements were interpreted, and the results were displayed graphically and discussed.

Keywords: Time history analysis, Finite element methods, cultural heritage

1. Introduction

Historical buildings are carrying the thousands of years' cultural wealth and heritage from past to present. These buildings must be preserved for the next generation. They give us historical conception about sociological, economical and political perspective from the past. Historical structures were built based on the masters` knowledge and experience without design standards or scientific studies (Ilerisoy and Soyluk, 2012). Structural performances of historical buildings are in variety with their geometry and defined material properties. Therefore, when modeling historical buildings, defining the material properties and the accurate geometrical shape of the structure are the critical steps in the procedure (Yardim and Mustafaraj, 2016). Historical structures have low ductility because of their brittle structural components. They may have severe damage during the severe earthquakes. Many historical buildings are located in high seismic zone in Turkey (Erdil ve Okuyucu, 2011).

A masonry historical building protection, repair and strengthening is really a difficult and challenging task. cultural value and the desire to preserve historical buildings for future generations, requires a high level of protection against any possible destruction under future actions, like earthquake loads. At the same time, should be avoided to damage on historical building which has survived until these days. To accomplish this duty, a precise understanding of the problems to be faced by the structure, the reasons for them, as well as a sound knowledge of the effect any intervention might have on the structure are needed, so that intervention does not become the cause of future damage to the building under consideration (Asteris et al., 2005).

The many structural systems of historical buildings are masonry elements which are composed of stone, bricks and mortar. For each of the old historic structures (including the monuments) built in high seismic seismic zones, the earthquake has always been the number one enemy, as the earthquake reacted very badly (Asteris, 2008).

The task of protecting historic structures is generally the responsibility of the engineer. For the successful intervention on a monument requires a good understanding of structural behavior under with a static and dynamic (earthquake) loads. For an engineer, taking part to the restoration process of a historical structure, through the analysis of its structural system, means mainly to face the demanding task of equipping the historical structure with the capability to withstand future actions with the minimum possible amount of damage, while bearing in mind the characteristics and values which make this structure unique and worthy of special attention (Asteris, 2008).

2. The Seismicity and Earthquake Hazard of Kastamonu Province

Kastamonu is located on the North Anatolian Fault Zone and total population is 363 700. Kastamonu city center falls into the first-degree earthquake zone in the latest map of 1996 which is still in force. According to Seismic Zoning Map of Turkey, 46 percent of the surface area of Kastamonu province is in the first-degree hazard zone, 22 percent in the second-degree hazard zone, 24 percent in the third degree hazard zone and 8 percent in the fourth degree hazard zone. The city center is the first-degree danger zone but nevertheless, there are very few studies on seismic and seismic hazard for this region and region's buildings (Ozmen, 2001 and Ozmen, 2011). Kastamonu earthquake map, Distribution of earthquakes with a magnitude of $M \geq 4$ and Number of occurrences of earthquakes with $M \geq 4$, which occurred in the instrumental period are given in figure 1, figure 2 and figure 3, respectively.

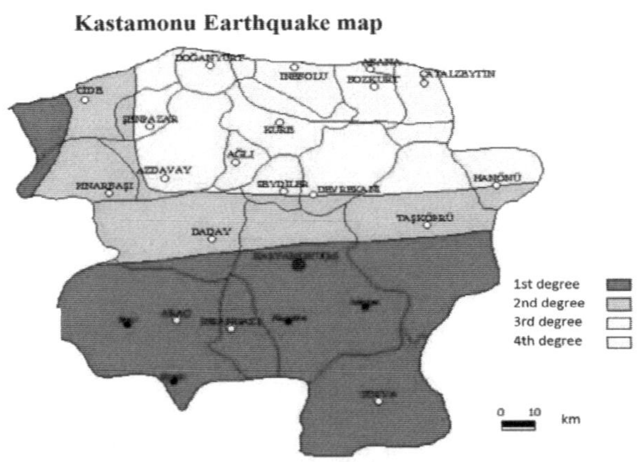

Region	Area (km²)	%	Population (1997)	%
1st Grade	6.007	46	156.391	43
2nd Grade	2.919	22	87.288	24
3rd Grade	3.130	24	72.740	20
4th Grade	1.080	8	47.281	13
Total	13136		363.700	

Figure 1. Earthquake zoning map of Kastamonu province

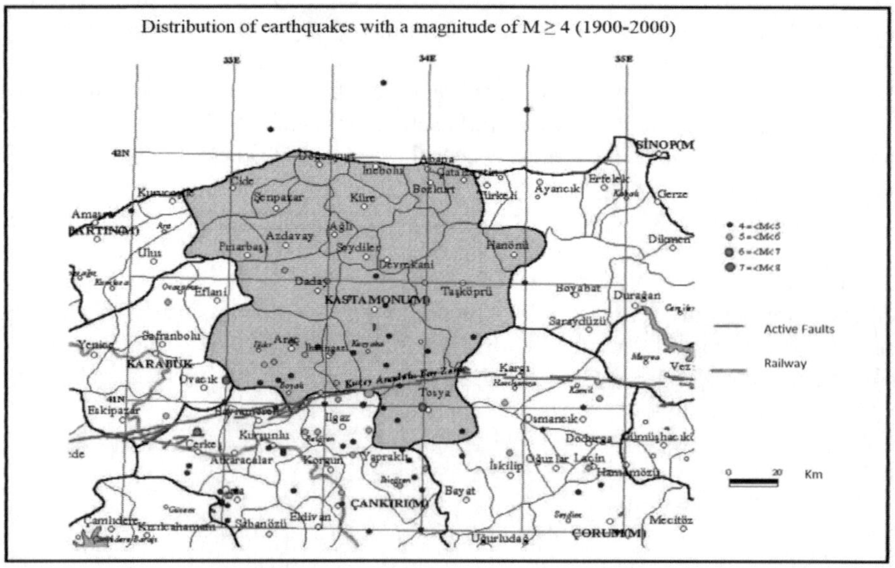

Figure 2. Distribution of earthquakes with a magnitude of M ≥ 4

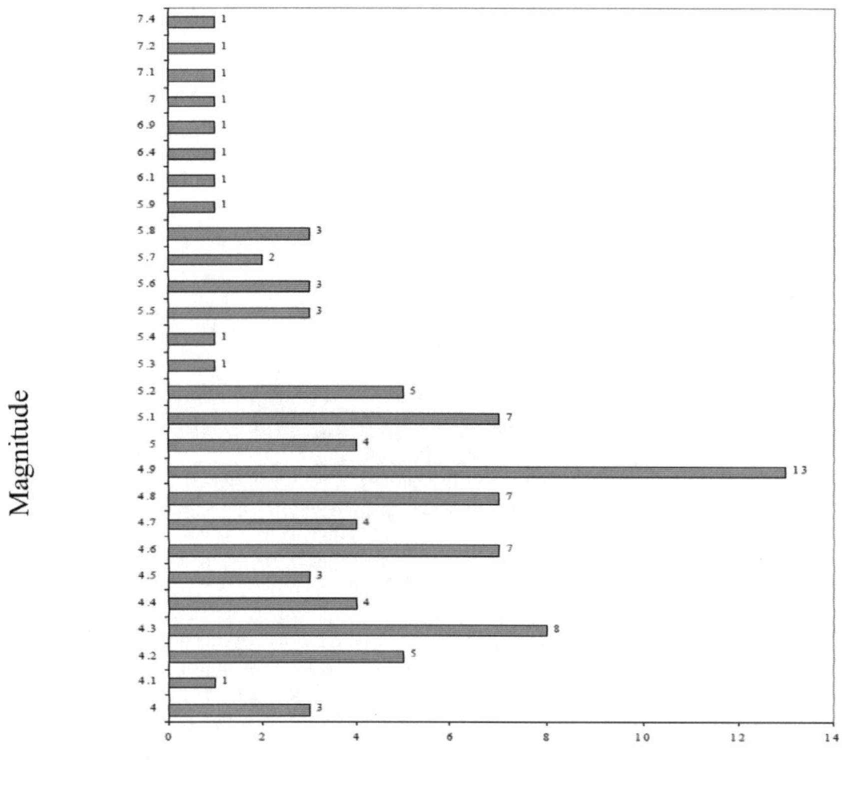

Figure 3. Number of occurrences of earthquakes with M ≥ 4, which occurred in the instrumental period

IOP Conf. Series: Materials Science and Engineering **300** (2018) 012065 doi:10.1088/1757-899X/300/1/012065

Earthquakes which have occurred in historical period (B.C.2100 – 1900) in Kastamonu province is given in Table 1.

Table 1. Earthquakes which have occurred in historical period (B.C.2100 – 1900).

History	Latitude	Longitude	Intensity	Explanation
03.09.968	41.15	34.75	IX	The earthquake that took place at 2 o'clock in Kastamonu, Çorum and Amasya it was caused by damage.
1075	40.60	34.95	VIII	The earthquake that took place in the morning hours was around Çorum caused damage.
1509	40.55	35.00	VIII	It caused damage in Çorum.
10.07.1668	41.30	33.80	VII	It caused damage in Kastamonu and Bolu
18.08.1668	41.20	33.80	VII	
12.05.1844	40.98	34.80	VIII	It caused damage in Osmancık, Çorum and Ankara. 200 people have lost their lives Due to the earthquake.
1845	40.60	33.60	V	It has been effective around Çankırı.
28.03.1880	42.00	35.20	VIII	
28.09.1881	40.60	33.60	VIII	It has been effective around Çankırı. It was stated that the magnitude of the earthquake was 6.1 and 12 people lost their lives.
1882	41.00	34.00	VI	It caused damage in Tosya, Kastamonu, İskilip and Çankırı
1883	41.00	33.70	VI	It caused damage in Kastamonu and Çankırı.
1885	41.30	34.30	VI	It caused damage in Sinop, İnebolu, Taşköprü, Tosya, Çankırı and İskilip.
1890	41.30	33.80	VI	It caused damage in Kastamonu and around.

3. Analizing and Modeling of Structure

After determining of measured drawing of structure and creation of axes, structure has beenmodeled with method of finite element. The three-dimensional analysis model was developed by using SAP2000 structural analysis software (Wilson and Habibullah 1992).

By using Sap2000 finite element program. Shell and solid elements have been used for modeling of structure.8500elements had been used in model such as; 6054-unit solid element, 2446 unit covering element. Numbers of Total crucial point is 14853. All elements have been improved and defined with prismatic elements accordance with characteristic of equipment and behave. Finite element modeling of historical masonry building of Kastamonu is given in Figure 4.

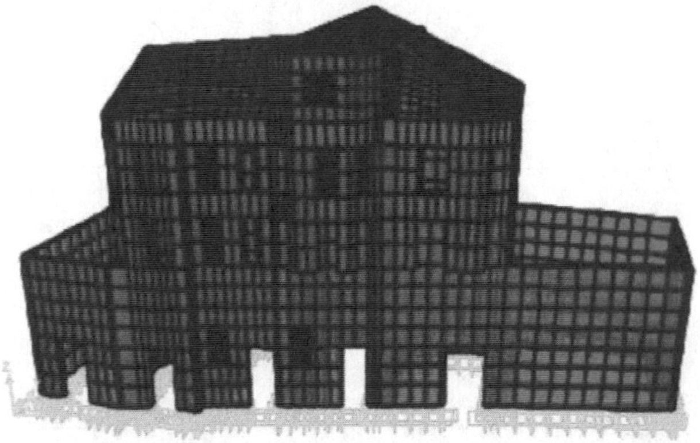

Figure 4. Finite element modeling of historical masonry building of Kastamonu

To determine the structural behavior of masonry building, time history analyses were carried out by using 10 different ground motion records on the masonry building's Finite Element models. Time history analyses are most reliable analysis type in the structural assessment approaches. For time history analyses, ground motion data were taken from Pacific Earthquake Engineering Research Center (PEER) (Yon et all., 2015). Selected ground motion records from Soils used for the dynamic analysis of the historical buildings are given in Tables 2 and 3, respectively.

The building was assumed to be made of a stone material and coverage material. Material values, which are used in model, are given in table 2. These values were determined by literature survey.

Table 2. Characteristics of equipments that used in model

Material Properties **(ton - m)**					
Equipment	Type	Mass	Weight	E	Poisson
Falez stone	Isotrope	0.25	2.50	204000.00	0.20
Brick	Isotrope	0.1835	1.8355	40788.65	0.2

Table 3. Ground motion records of A Soil Class

No	Earthquake	Date	Moment Magnitude (Mw)	Record	Ground Velocity (cm/s)	Ground Acceleration (g)	Epicenter distance (km)	Type
1	Anza (Horse Cany)	25/02/1980	4.9	AZF315	2.6	0.066	12.1	lateral slip
2	Morgan Hill	24/04/1984	6.2	G01320	2.9	0.098	16.2	lateral slip
3	Coyote Lake	06/08/1979	5.7	G01320	8.3	0.132	9.3	lateral slip
4	Landers	28/06/1992	7.3	GRN180	14.1	0.041	141.6	lateral slip
5	Landers	28/06/1992	7.3	ABY090	20	0.146	69.2	lateral slip
6	Landers	28/06/1992	7.3	SIL000	3.8	0.05	51.7	lateral slip
7	Landers	28/06/1992	7.3	29P000	3.7	0.08	42.2	lateral slip
9	Loma Prieta	18/10/1989	6.9	SGI360	8.4	0.06	30.6	Oblique

10	Loma Prieta	18/10/1989	6.9	MCH000	3.5	0.073	44.8	Oblique
11	Loma Prieta	18/10/1989	6.9	PTB297	12.9	0.072	78.3	Oblique

4. Structural Analyses

SAP 2000 software was used for time history analyses to determine the structural behavior of the historical building. 12 mode values were considered in the dynamic time history analyses. Displacements values were interpreted, and the results were displayed graphically and discussed. The values of the displacement obtained from historical masonry building is given in Figures 5, and 6 respectively. In Figures displacements values obtained for earthquakes are seen in the X and Y directions.

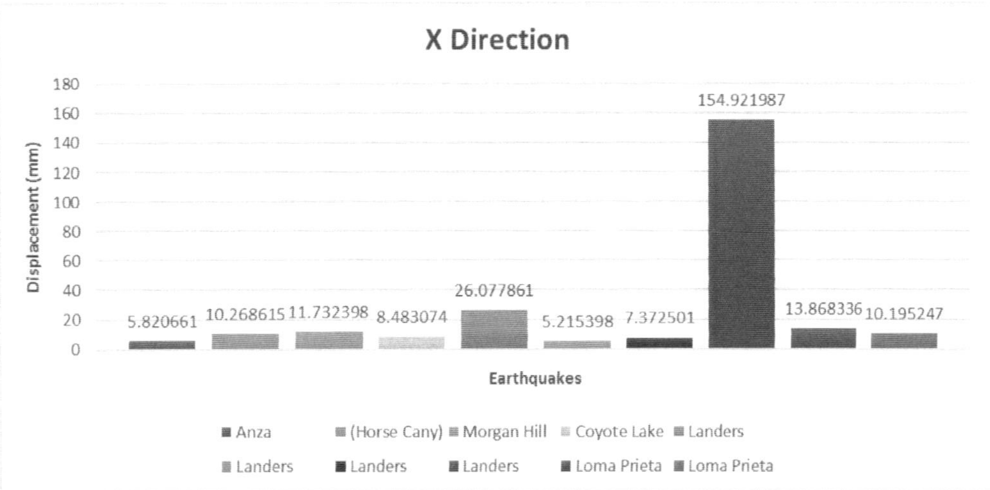

Figure 5. X-direction displacement of masonry building

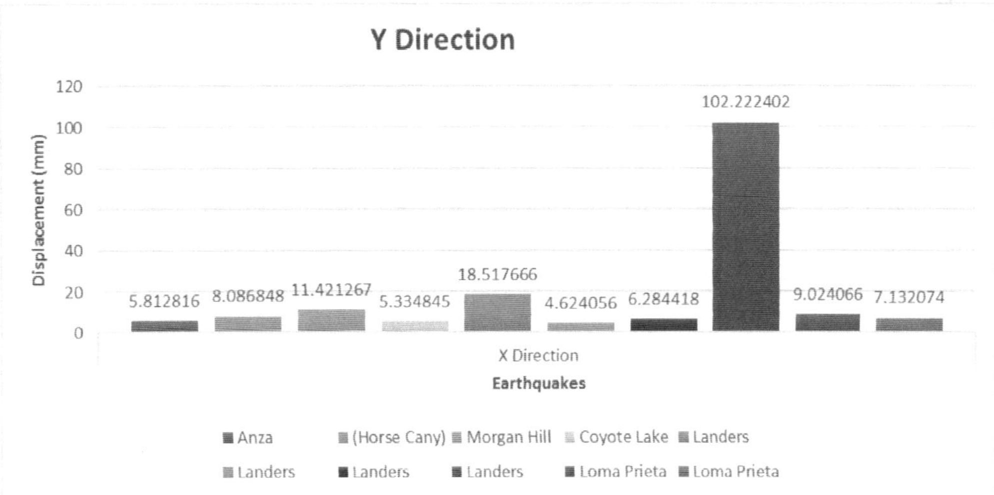

Figure 6. Y-direction displacement of masonry building

5. Result

In this study, structural assessment of masonry building in Kastamonu was discussed with detailed structural assessment approach. The historical building was modelled and analyzed through time history analyses under 10 different ground motion data using SAP2000 Finite Element Analysis software. The dynamic analyses were performed with - 10 ground motion data with soil classes A to the historical building.

When comparing the results of the soil A according to X and Y direction, the highest displacement values for X directionoccurred at Landers (eighth Earthquake in earthquake table)Earthquake. The lowest displacement values for X direction were occurred at Landers (Sixth Earthquake in earthquake table) Earthquake. The highest displacement values for Y direction occurred at Landers (eighth Earthquake in earthquake table) Earthquake. The lowest displacement values for Y direction were occurred at Landers (sixth Earthquake in earthquake table) Earthquake.

References

[1] Asteris, P. G. (2008). On the structural analysis and seismic protection of historical masonry structures. Open construction and building technology journal, 2, 124-133.
[2] Asteris, P. G., Tzamtzis, A. D., Vouthouni, P. P., & Sophianopoulos, D. S. (2005). Earthquake resistant design and rehabilitation of masonry historical structures. Practice Periodical on Structural Design and Construction, 10(1), 49-55.
[3] Erdil B. and Okuyucu D., 2011, Seismic performance evaluation of Adilcevaz Pasha (Tugrul bey) mosque, WCCE-ECCE-TCCE Joint Conference 2, Seismic protection of cultural heritage, October 31- November 1, Antalya, Turkey.
[4] Ilerisoy Z. Y., Soyluk A., 2012, impact of shallow earthquakes on the Sehzade Mehmet Mosque, Gradevinar 9, 735-740.
[5] Ozmen B., 2011, Earthquake Probability Forecasts for Kastamonu and Near Surroundings, Geological Bulletin of Turkey, Volume 54, Number 3, December 2011.
[6] Özmen, B., 2001, The Seismicity and Earthquake Hazard of Kastamonu Province, 54. Türkiye Jeoloji Kurultayı May 7-10, TMMOB Jeoloji Mühendisleri Odası, Ankara.
[7] SAP 2000, Computer Program, Computers and Structures, Berkeley, USA, 2000.
[8] Wilson, E.I., ve Habibullah, A., 1992. Structural Analysis Programs, Computers and Structures Inc., Berkeley, CA, A.B.D.
[9] Yardim Y. and Mustafaraj E., 2016, Selected assessment and retrofitting application techniques for historical unreinforced masonry buildings, Access date:4.23.2016.
[10] Yon B., Oncu M. E. and Calayir Y., 2015, Effects of seismic zones and site conditions on response of RC buildings, Građevinar 67 (2015) 6, 585-596.

IOP Conf. Series: Materials Science and Engineering **300** (2018) 012066 doi:10.1088/1757-899X/300/1/012066

Utilization of Cocoa Peel as Biosorbent for Oil and Color Removal in Palm Oil Mill Effluent (POME)

S Pandia[1], A T Hutagalung[1*], A D Siahaan[1]

[1]Department of Chemical Engineering, University of Sumatera Utara, Medan, Indonesia

*Email: anitatiurmaida94@gmail.com

Abstract. The aim of this study is to discover the effectiveness of cocoa peel as biosorbents for oil and color removal in POME. This study used biosorbent from cocoa peel with variation of particle size which passed through 70 mesh, 100 mesh, and 120 mesh and was activated with the ratio of biosorbent and 0,6 M HNO_3 of 1:2, 1:4, and 1:6 (m/v). The adsorption process was carried out using biosorbent with the highest iodine number in varying biosorbent mass and contact time. The highest iodine number was 596,684 mg/g and obtained at particle size 120 mesh and the ratio of biosorbent : HNO_3 as 1 : 4. The resulting biosorbents were analyzed for their characterization, such as vapor content, ash content, and volatile matter, including FT-IR and SEM-EDX. The POME were analyzed for their oil and color content, using gravimetric method and UV-Vis spectrophotometry. The best removal is were 80,88% for oil and 83.45% for color at 1.5 g biosorbent mass and 3 h contact time. The resullt for oil removal was close to the standard of Indonesian Environment Minister for oil in effluent. Also the adsorption of oil and color behaves as a pseudo-second-order kinetic models.
Keyword: Adsorption, cocoa peel, color, oil, POME (Palm Oil Mill Effluent)

1. Introduction

The needs for palm oil have increased, so it will be one of vegetable oils that has fastest growing every year. The capacity of the palm oil industry will increase too, including the waste, especially the Palm Oil Mill Effluent (POME). Generally, POME is not toxic, but it is a liquid waste which has high volume and an unpleasant odor. The organic content of POME will inhibit the transmission of light into the body of water, so that will be slowing down the process of photosynthesis and probably disrupt the food chain in the ecosystem. The oil content in POME can reach until 6000 mg/L [1]. The color of the POME is due to the presence of plant constituents, such as lignin and phenolic compounds and repolymerization of the color compound after anaerobic treatment [2]. Several ways have been done in POME treatment, such as aerated and facultative anaerobic biological processes, anaerobic biological processes and land applications system, biological processes of anaerobic tank-aeration advanced. However, this liquid waste treatment process takes about 120-140 days, require for large treatment area, high maintenance costs, and the generation of methane gas emissions into the air [3,4].

An alternative method that is quite effective and cheap has been much proposed by the use of biosorbents from agricultural waste and plantations that have high cellulose content. In Indonesia, cocoa fruit is one of the plantation commodities whose role is quite important. The only utilized parts are the seeds, but the skin was limited only for animal feed and raw materials fertilizer. Cocoa peel has a coarse texture, thick and hard. Cocoa peel weighs up to 75% of the total fruit weight. The content of cellulose in cocoa peel is about 23-54%, thus making it potentially biosorbent. Furthermore, the cocoa peel also contains organic compounds such as protein 5.69-9.69%, fat 0.02-0.15%, glucose 1.16-3.92%, sucrose 0.02-0.18%, pectin 5.30-7.08%, crude fiber 33.19-39.45%. These compounds are polymers of carbon elements so that cocoa peel can be made as biosorbents that have a high capasity [5]. Thus, the utilization of cocoa peel as a biosorbent can increase its economic value. Based on that description, the use of cocoa peel as a biosorbent is to be processed as an alternative treatment process.

2. Material and Method

2.1. Material
The main *materials* were the cocoa peel that were colected from the plantations in Medan and Palm Oil Mill Effluent (POME) from Palm Oil Mill in North Sumatera.

2.2. Preparation of the Biosorbent from cocoa peel
Cocoa peels were cutted in small pieces. They were washed with water and dried in oven at 80°C for 24 h. They were crushed to powder form and passed through 70 mesh. Biosorbent was acivated with a solution of 0,6 M HNO_3 at variation of ratio of biosorbent : HNO_3 of 1:2. It was followed by washing several times with distilled water to remove the activating solution. The procedure was repeate for the particle size which passed through 100 mesh and 120 mesh and the ratio of bio sorbent:nitric acid of 1:4 and 1:6 (m/v)

2.3 *Batch Adsorption Process*
Biosorbent (1 g, 1,5 g, 2 g) were poured into 50 mL of POME in a beaker glass and stirred at 200 rpm and pH 2 for 1, 2, 3, 4, and 5 h. Afterwards, POME was filtered with Whatman No. 1 filter paper.

2.4 *Analysis of Oil and Color Residual*
The oil residual was analyzed with gravimetric method. The color residual was analyzed with UV-Vis Spectrophotometer.

3. Results and Discussion

3.1 *Effect of Partcle Size and Ratio of Bio Sorbent : Nitric Acid to Iodine Number*
Before and after activation, an iodine number of the bio sorbent was analyzed which were given in Figure 1 below.

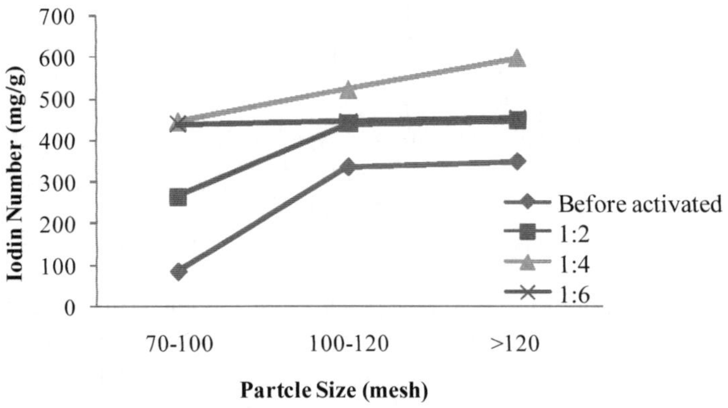

Figure 1. Effect of Particle Size on Iodine Number for Certain Ratio of Biosorbent : Nitric Acid (b/v)

From the Figure 1. it can be seen that iodine number tend to increase by decreasing of the particle size for the same ratio of biosorbent : nitric acid (b/v) and by increasing the ratio of biosorbent : nitric acid (b/v) for the same particle size. However, the iodine number was decrease at the ratio 1:6 of biosorbent : nitric acid (b/v). The surface of biosorbent increase by decreasing of the particle size, hence more percentage removal in the adsorption process. The active site formation depends by the concentration of the activator used. At the higher activator concentration, the larger amount of porous structure was produced by the stronger dehydration or condensation. But perhaps too much activator had blocked a parts of macroporous, so the adsorption capacity decreased [6]. A very high volume of activator may eliminate active biosorbent properties due to the damage caused by dissolution and breakup of pores on the biosorbent [7]. The increasing of the iodine number until the ratio of biosorbent : HNO_3 of 1:4, showed that a very low volume of the activator may cause the active sites formation was incomplete. However, the iodine number deceased at the ratio of biosorbent : HNO_3 of 1:6. These can be happened because the volume of the activator was very high. It can be concluded that the particle size 120 mesh and the ratio of biosorbent : nitric acid (b/v) of 1:4 were the the most favorable conditions to produce the biosorbent from cocoa peel with the highest iodine number.

According to SNI-06-3730-1995, the resulting bio sorbent is close to qualified as activated carbon. It is seen from iodine number 596,684 mg/g, 9,89% water content, 0,52% ash content, and 84% volatile matter content, Where the requirements given by SNI in the table below:

Table 1. Standard of activated carbon from SNI No. 06-3730-1995

Requirements	Content
Iodine Number	>750 mg/g
Vapor Content	< 25%
Ash Content	<10%
Water Content	<15%

3.2 Fourier Transform Infrared Spectroscopy (FT-IR)

Bio sorbent characterization using FTIR Spectrophotometry was conducted fix bio sorbent before and after the activation, and after used in the adsorption process. Functional groups on bio sorbent can be inferred by comparing result on the graph with literature or IR correlation [8]. The graphs are in figure 2 below:

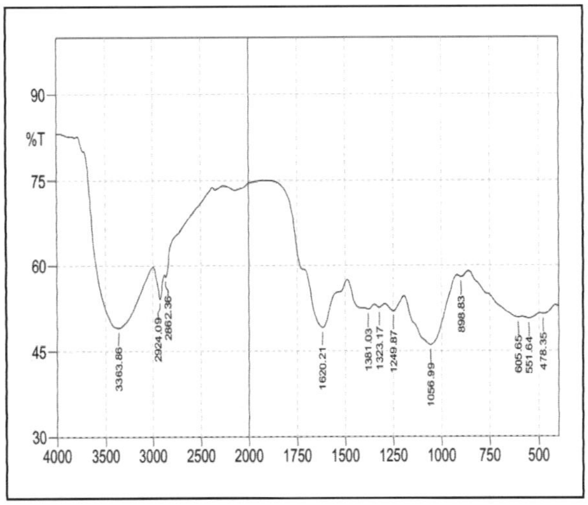

Figure 2(a) FTIR Spectrophotometry Result for Biosorbent Before Activation

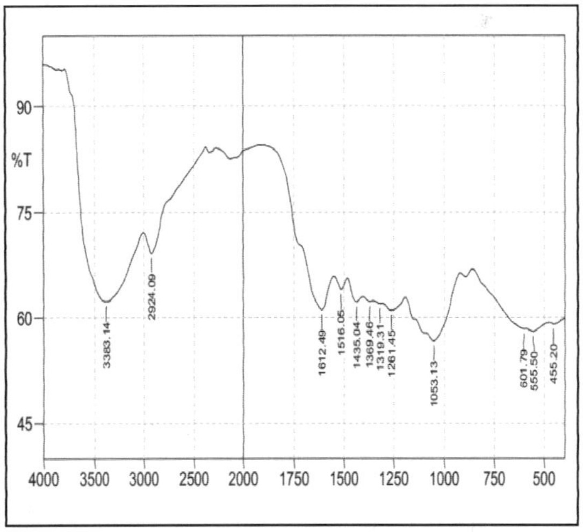

Figure 2(b) FTIR Spectrophotometry Result for Activated Biosorbent

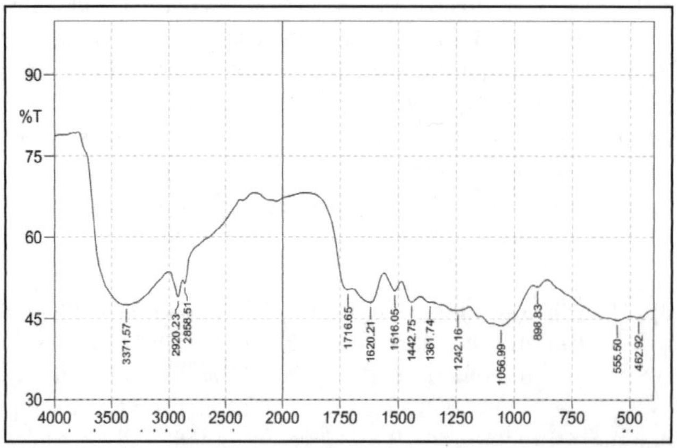

Figure 2(c) FTIR Spectrophotometry Result for Used Biosorbent

Based on Fig 2., the biosorbent before activation generated several peaks of wave numbers in each range of the region, such as -OH group (alcohol), C-H group (alkane), C=C group (alkenes), C-N group (amine), and C-O group (ester). However, after activation there is some wavelength and intensity of the functional groups changes. It may indicate a HNO_3 solution used as an activator that has been decomposed of the mineral salts present in the sample. Thus, the functional groups increasingly formed on the biosorbent, so that can increase the biosorbent capacity in adsorption process by forming the complexs compound [9]. In Figure 2c. showed the biosorbent that has been contacted with POME there are several functional groups attached by the biosorbent after the adsorption. The presence of C=O group in the absorption of 1716.65 cm-1 is the cause of the cloudy appearance [7], and indicates the amount of oil which is the triglyceride. By comparing the result from the HNO_3 activated biosorbent and the used biosorbent, it shows the interaction between the biosorbent functional groups and the molecules present in POME. From the results of IR spectrophotometry, it can be concluded that the functional groups that play a role in oil and color removal in POME are hydroxyl, carbonyl, amine, nitro and alkoxy groups.

3.3 Results of Analysis Morphological Characteristics Biosorbent

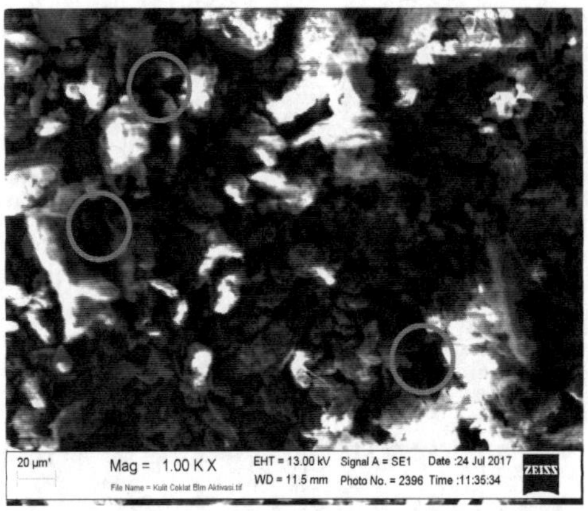

Figure 3(a) Morphology Analysis Report for Biosorbent Before Activation (Magnifation 1000 times)

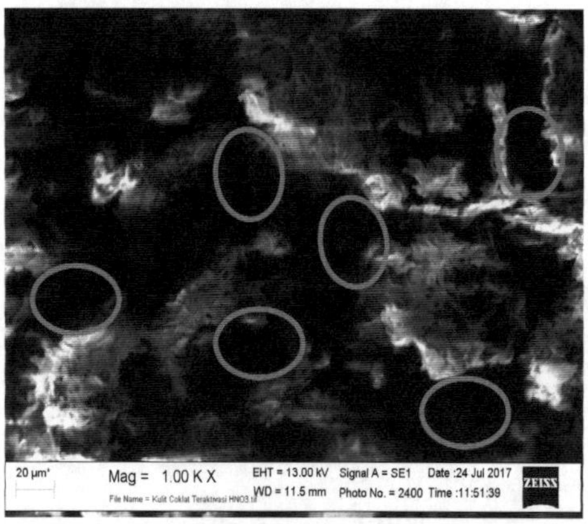

Figure 3(b) Morphology Analysis Report for Activated Biosorbent (Magnifation 1000 times)

Figure 3(c) Morphology Analysis Report for Used Biosorbent (Magnifation 1000 times)

The figure 3., showed that the micropores of activated biosobent are wider and clearer. The figure 3a and 3b showed that the activation process give a significant change of the surface morphology of the biosorbent.This is correspond with the purpose of activation that the ac tivator can eliminated the impurity metals. Thus, the activated cocoa peel biosorbent with HNO_3 has higher capacity adsorption than the biosorbent without activation. Based on the result of SEM analysis on cocoa peel biosorbent before activation there are metal and mineral elements. The presence of these metals and minerals indicates that ash content is usually present in the biosorbent [10].The SEM micrograph in fig7.shows that the micropore structures of the biosorbent after being contacted with POME has been covered with muddy-line substances. These muddy-line substancesperhaps are the components of adsorbate which is indicated by the decreasing color and oil content of POME. The presence of muddy-line substances proves that the components present in POME were adsorbed by the active group on the biosorbent.

3.4 *Adsorption of Oil and Color in POME*

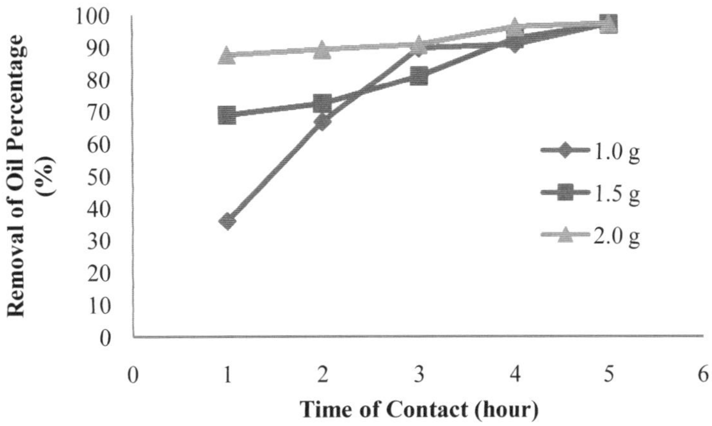

Figure 4. Effect of Biosorbent mass on Removal Oil Percentage for Certain Contact Time

Fig 4. shows the removal oil percentage in POME of intial concentration (2411 mg/L) after adsorption onto biosorbent at different biosorbent mass (1.0 g, 1,5 g, 2.0 g) and at different contact time (1, 2, 3, 4, and 5 h). The results shows that the removal oil percentage increases by increasing the contact time at the same mass of biosorbents. Likewise, the percentage removal of oil tend toincreaseby increasing the biosorbent mass at the same contact time.This is correspond with study wihich has been done by Ahmad, et al. [11], the adsorption of oil will increase since the biosorbent mass increases. However, the percentage removal of oil fluctuates at the contact time of 3 h and 5 h. The adsorption of oil on the condition of strong acid or low pH, can disrupt the oil layer to become unstable. Biosorbents may provoke the physicochemical properties of the oil layer, so that demulsification may occur, which will result in increasing small oil sizes which will increase the adsorption of the oil in solution [12]. At longer mixing times, the demulsification process is increasing. Damage to the oil layer (demulsification) will cause the contact surface area between the biosorbent and the oil to increase, thus increasing the adsorption of residual oil [11]. The discrepancy may be due to the uncontrollable oil content of POME at any initial adsorption condition.

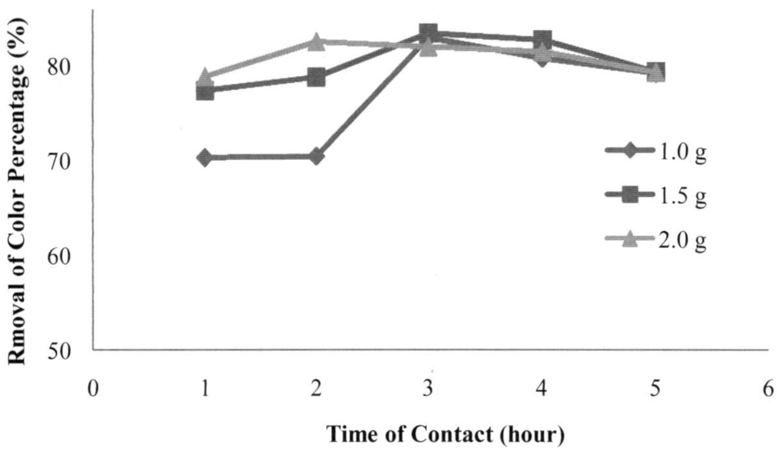

Figure 5. Effect of Biosorbent mass on Removal Color Percentage for Certain Contact Time

Fig 5. shows the removal color percentage in POME of intial concentration (7340 mg/L) after adsorption onto biosorbent at different biosorbent mass (1.0 g, 1,5 g, 2.0 g) and at different contact time (1, 2, 3, 4, and 5 h). From Fig 4, it can be seen that there is an optimum time of each biosorbent mass used. It can prove that the active site contained in the biosorbent takes time to absorb adsorbate. According to Faisol, et al. [13], that the contact time can be as a test of the performance and endurance of the biosorbent before it reaches a condition where the biosorbent can not absorb adsorbate or called saturation point. At contact time of less than 3 hours the adsorption process is not optimum. This can be due to the active site contained in the biosorbent has not reached its saturation point, so the adsorption process can continue to increase. However, at contact time of 4 hours to 5 hours the adsorption power of biosorbent continues to decrease, means that the biosorbent has reached saturation point. According to Hening Purnamawati and Budi Utami [14], this is because the biosorbent has desorbed which is release of ions or molecules that have binded to the active group on the biosorbent, so that a number of dye molecules are released back into the solution. The highest removal percentage for oil and color was obtained at the biosorbent mass of 1.5 g with 3 hours contact time, that is 80, 88 % and 83.45%.

IOP Conf. Series: Materials Science and Engineering **300** (2018) 012066 doi:10.1088/1757-899X/300/1/012066

3.5 Adsorption Kinetic Studies

The adsorption kinetics shows the adsorption rate closely related to the absorbency between the active site and the adsorbate molecule. The adsorption kinetics can be obtained empirically by using the pseudo model of first order and second order [15] as in eq. (1) and (2).

$$\frac{1}{q_t} = \frac{k_1}{(q_e)^t} + \frac{1}{q_e} \quad (1)$$

$$\frac{t}{q_t} = \frac{1}{q_e}t + \frac{1}{(k_2 q_e^2)} \quad (2)$$

Where q_t and q_e are the amount of adsorbate at equilibrium and at any time (t), k_1 and k_2 are the adsorption rate constant at first order and second order. The linear plot as shown in Figure 6.

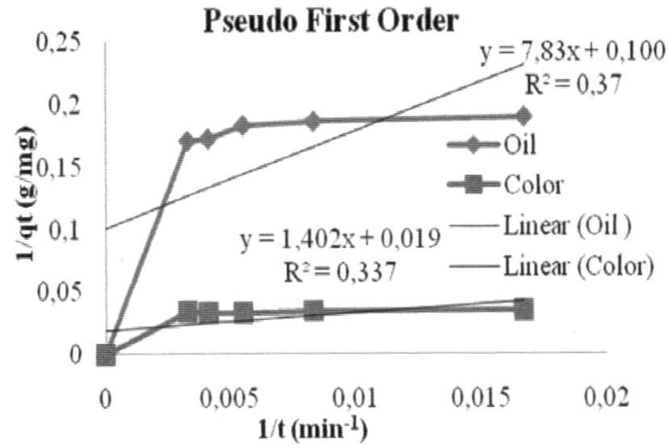

Figure 6 (a). Kinetic model for Pseudo fisrt order

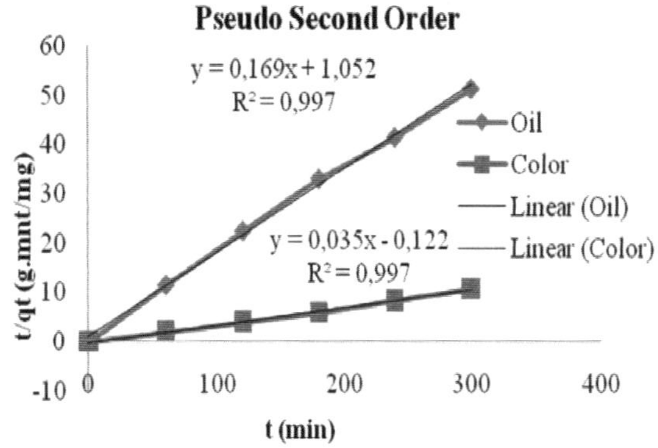

Figure 6 (b). Kinetic model for Pseudo Second order

The graphs obtained in Fig 6a. show that the kinetic model for pseudo first order has a poor correlation coefficient. The graphs obtained in Fig 6b. show that the kinetic model for pseudo second order has a very high correlation coefficient (> 0.99). Second-order equation on oil adsorption obtained value $r^2 = 0.997$ and the first-order equation on the color adsorption obtained value $r^2 = 0.997$. Thus, it can be concluded that the adsorption of oil and color follow the pseudo kinetics mechanism of second order. According to Ho and McKay [16], the pseudo second order kinetic allows chemical reactions to the adsorption process involving the power of valence through division or electron exchange between adsorbent and adsorbate.

3. Conclusion

The highest iodin number of biosorbent from cocoa peel was 596,684 mg/g at particle size which passed through 120 mesh and ratio biosorbent : HNO_3 (b/v) is 1:4. Characteristic of biosorbent of cocoa peel is 9,89% moisture content, 0.52% ash content, and 84% volatile content. The HNO_3 solution used as an activator that has been decomposed of the mineral salts present in the sample which is proved through the FTIR Spectrophotometry and SEM results. The highest removal percentage for oil and color in POME was obtained at the biosorbent mass of 1.5 gram with 3 hours contact time, that is 80,88% and 83.45%. Adsorption kinetic of oil and color in POME follows the pseudo-second order kinetic model. Therefore the cocoa peel could be useful as a biosorbent for removing oil and color.

References

[1] Ahmad A L, Chan C Y, Shukor S R Abd and Mashitah M D 2009 Separation of Oil and Carotenes from Palm Oil Mill Effluent by Adsorption Chromatography with Silica Based Biosorbent *Asia-Pacific Journal of Chemical Engineering* **4** 717–722

[2] Mohammed R R and Chong M F 2014 Treatment and Decolorization of Biologically treated Palm Oil Mill Effluent (POME) Using Banana Peel as Novel Biosorbent *Journal of Environmental Management* **132** 237-249

[3] Said M, Mohammad A W, Tusirin M, Nor M, Abdullah S R, and Hasan H A 2015 Investigation of Three Pre-treatment Methods Prior to Nanofiltration Membrane for Palm Oil Mill Effluent Treatment *Sains Malaysiana* **44** (3) 421–427

[4] Ratna D, Syafruddin, Yunus M, and Suryani 2013 Kombinasi Reaktor Elektrokoagulasi dan Bioadsorber untuk Penyisihan Kontaminan dalam Limbah Cair PKS *Prosiding SNTK*, ISSN. 1907-0500

[5] Masitoh Y F and Sianita M M 2013 Pemanfaatan Arang Aktif Kulit Buah Coklat (*Theobroma cocoa L.*) sebagai Adsorben Logam Berat Cd (II) dalam Pelarut Air *UNESA Journal of Chemistry* **2** (2) 23-28

[6] Li S, Zhang L, Pan B, Xiong G and Cang D 2010 Study on Preparation and Charaterization of The Biosorbents with Sewage Sludge *Srcosmos*,1 – 3

[7] A. S. Pardede, A. M. Siburian, S. Pandia. 2014. Utilization of Biosorbent from Tamarind Seeds for Removing Red and Yellow Color in CPO (Crude Palm Oil). *International Journal of Research in Advent Technology* **2** 10:2321-9637

[8] Stuart B, 2004, *Infrared Spectroscopy: Fundamentals and Aplications*, John Wiley & Sons, Sydney

[9] Safrianti L, Wahyuni N, Zaharah T A 2012 Adsorpsi Timbal (II) oleh Selulosa Limbah Jerami Padi Teraktivasi Asam Nitrat: Pengaruh pH dan Waktu Kontak. JKK **1** (1)

[10] Haura U, Fachrul R and Meilina H 2017 Karakterisasi Adsorben dari Kulit Manggis dan Kinerjanya pada Adsorpsi Logam Pb(II) dan Cr(VI) *BIOPROPAL INDUSTRI* **8** (1)

[11] Ahmad A L, Bhatia S, Ibrahim N, Sumathi S 2005 Adsorption of Residual Oil from Palm Oil Mill Effluent Using Rubber Powder. *Brazilian Journal of Chemical Engineering* **22** (3), pp. 371 - 379

[12] Sokker H H, El-Sawy N M, Hassan M A and El-Anadouli B E 2011 Adsorption of Crude Oil from Aqueous Solution by Hidrogel of Chitosan Based Polycrylamide Prepared by Radiation Induced Graft Polymerization *Jurnal of Hazardous Materials* **190**

[13] Arsip F, Mardhiah R, and Husna 2008 Uji Efektifitas Cangkang Telur dalam Mengadsorpsi Ion Fe dengan Batch *Jurnal Teknik Kimia* **15** (2)

[14] Purnamawati H and Utami B 2014 Pemanfaatan Limbah Kulit Buah Kakao (*Theobroma cacao L.*) sebagai Adsorben Zat Warna *Rhodamin B*. *Prosiding Seminar Nasional Fisika dan Pendidikan Fisika* (SNFPF) Ke-5 **5** (1): 12-18

[15] Liu H, Wang C, Liu J, Wang B and Sun H 2013 Competitive Adsorption of Cd(II), Zn(II), dan Ni(II) from Their Binary and Ternary Acidic Systems Using Tourmaline *Jurnal of Environmental Management* p.128

[16] Ho Y S and McKay G 1999 Pseudo-Second Order Model for Sorption Processes *Process Biochem* **34** 451-465

A Novelty Design Of Minimization Of Electrical Losses In A Vector Controlled Induction Machine Drive

Solly Aryza[1*], M Irwanto[2], Zulkarnain Lubis[3], Andysah Putera Utama Siahaan[4*], Robbi Rahim[5*], Mhd.Furqan[6]

[1]Faculty of Engineering, University Pembangunan Pancabudi, Gatot Subroto street 4,5 Medan North Sumatera, Indonesia.

[2]Faculty Engineering, University Malaysia Perlis, Pauh street Perlis, Malaysia

[2,3]Faculty Engineering, Institute Technology Medan, Gedung Arca street No.52, Medan Kota, North Sumatera 20217.

[4]Faculty Computer Science, University Pembangunan Panca Budi, Gatot Subroto Street 4,5 MedanNorth Sumatera, Indonesia.

[5]Department of Health Information, Akademi Perekam Medik dan Infokes Imelda, Bilal Street Medan 20116, Indonesia

[1,4,5]Postgraduate Student University Malaysia Perlis, Pauh Street Perlis, Malaysia.

[6]Departement of Computer Science, UIN Sumatera Utara, North Sumatera Indonesia.

Abstract. The induction motor has in the industry . More attention has been a focus to develop and design of induction motor drive. With the method of vector control novelty prove the efficiency of induction motor over their entire speed range. In this paper desirable to design a loss minimization controller which can improve the efficiency. Also, this research described Modeling of an induction motor with core loss included. Realization of methods vector control for an induction motor drive with loss element included. The case of the loss minimization condition. The procedure was successful to calculate the gains of a PI controller. Though the problem of obtaining a robust and sensorless induction motor drive is by no means completely solved, the results obtained as part of this work point in a promising direction.

1. Introduction.

IM are critical components in industrial processes. A motor failure may yield an unexpected interruption at the industrial plant, with consequences in costs, product quality, commonly used in adjustable speed drive systems. Induction motors have been widely employed in various industries as actuators or drivers to produce mechanical motions and forces. Since it is estimated that more than 50% of the world electric energy is generated and consumed by electric machines, to improve an efficiency of electric drives are important [1.2]. Induction motors require both full operating range of speed and fast torque response in operational conditions, regardless of load variations. Namely, induction motors have a high efficiency at rated speed and torque.

Its efficient control requires a suitable model with accurate parameters, the minimization of the objective function is carried out using the Particle Swarm Optimization. Particle Swarm Optimization (PSO) is an evolutionary algorithm inspired by social interaction. PSO is an evolutionary once technique (a search method based on a natural system) developed by Kennedy and Eberhart. The basic concept of the PSO technique 'lies in accelerating each particle towards its p best and g best locations, with a random weighted acceleration at each time step. PSO has many parameters, and these are described as follow: V max is the maximum allowable velocity of the particles (i.e. in the case where the velocity of the particle exceeds Vmax, then it is limited to Vmax). Particle swarm optimization (PSO) is one of the modern heuristic algorithms [16], [17]. PSO has attracted great attention due to its features of easy implementation, robustness to control parameters and computation efficiency compared with other existing heuristic algorithms, and has been successful. Particle swarm optimization (PSO) is an evolutionary computation technique the system initially has a population of random solutions. Each potential solution called an electromagnetic torque is given by particle. In this research paper, a new minimum-time minimum-loss control algorithm for induction motors using system particle swarm optimization is suggested to obtain high performance, as well as high efficiency, under practical constraints on voltage and current. The validity of the proposed scheme, which carries out minimum-time speed control in the transient state and minimum-loss supervision in the steady state, will be revealed via simulation, including an induction motor model.

2. Induction Motor Loss Model

This paper described an equivalent circuit which points out the rotor magnetic current is used. An iron loss resistance Rf was added in parallel with magnetic inductance in the rotor flux reference frame which is shown in figure.1 [2,5,7]. Information related to this machine is presented in Appendix

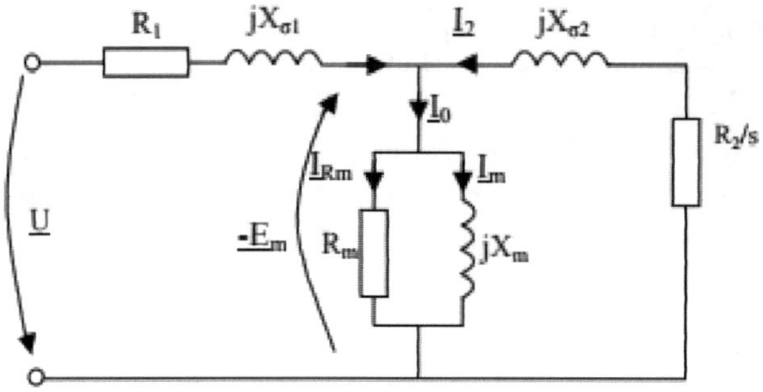

Figure 1. Equivalent circuit of induction motor containing iron loss resistance.

In permanent state, there is no leakage inductance on themotor and the equivalent circuit will be the same as fig.2.

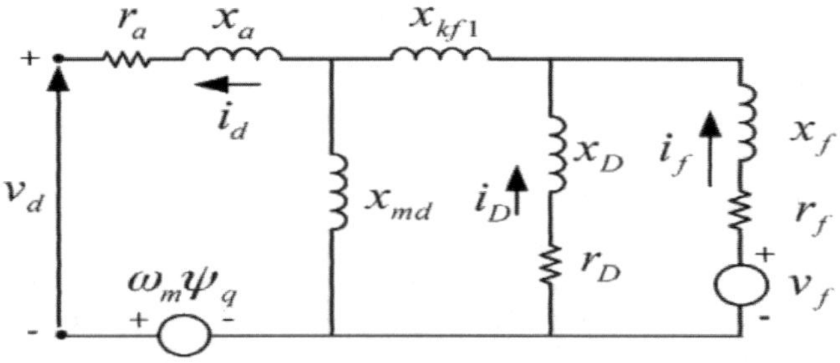

a) Equivalent axis of axis d

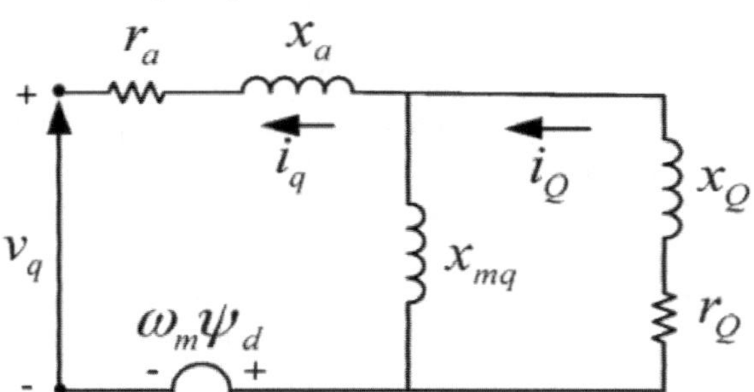

b) Equivalent axis of axis b

Figure 2. Motor equivalent circuit in the permanent state.

To develop the loss model, a typical simple approach has been discussed in the previous kinds of literature[1,2,5].Copper loss of stator, copper loss of rotor and iron loss which are stated as equation 1:

$$\frac{2\pi N_S T_R}{60} - \frac{2\pi N T_R}{60} = \frac{2\pi T_R}{60}(N_s - N) \quad (1)$$

3. Parameter Pso In Induction Motor

We think that a magnetic flux axis d is set up on the magnetic flux electric current vector as shown in figure 1. In the induction motor vector control, the voltage and ampere equations on the d-q axes are following

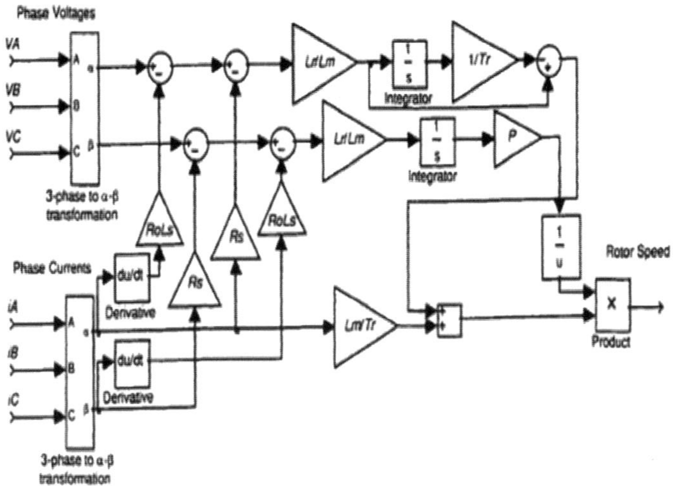

Figure3. Electric Current Vector

Voltage equation:

$$V_{\text{dip}}(r,\theta) = \frac{\hat{\mathbf{r}}\cdot\mathbf{p}}{4\pi\epsilon_0 r^2} = \frac{p\cos\theta}{4\pi\epsilon_0 r^2}. \quad (2)$$

Ampere equation:

$$I_R = \frac{E_R}{(R_R + jX_R)} = \frac{sE_{R0}}{(R_R + jsX_{R0})} \quad (3)$$

In this section, the procedure of PSO in online system parameter identification is described. Here, each particle represents all parameters of estimated model. The proposed algorithm sequentially gives data set by sampling periodically. While starting, in the first period, the best system parameter is found by minimizing the SSE introduced. Here, the simulation for next period does not begin until the fitness of global best becomes lower than a predefined threshold. After that, the estimated parameters will not be updated unless a change in the system parameters are detected. to detect any change. In system parameters, the global optimum in the later period is noticed as a sentry particle. In the beginning of each of the next periods, the sentry reevaluates its fitness and if the fitness changes significantly or it becomes bigger than a predefined threshold, the changes in parameters are confirmed. If no changes are detected, the algorithm leaves this period without changing the positions of particles. In contrast, when any change in parameters occurs, the sentry alerts the swarm to reset their best location memories and then the algorithm runs further to find the new optimum values. For this purpose, the fitness of global optimum particle and personal bests of all particles are evaporated at the rate of a big evaporation constant. As a result, other particles have a chance to find better solutions than those stored on their pervious global and personal memories. Moreover, the velocities of particles are increased to search in a bigger solution space for new best solution

The diagram block of the proposed optimized approach is shown in fig.4. Motor control method was based on rotor flux vector control method (FOC). In this structure, speed and electromagnetic torque of the motor are measured to obtain magnetic current, An FLC is the conversion of Linguistic expressions based on expert knowledge into the control strategy[9]. First , the speed error and its variation derivative are input as FLC variables. Then, FLC output variable is introduced as the reference value. then these numerical variables are converted into the Linguistic variables. Five fuzzy sets are seen in table.1 as NL(Negative large), NS(Negative small), Z(Zero), PS (Positive small) and PL(Positive large)[11-15]. Fuzzy control structure is defined as follow:

1. Five fuzzy sets for each input and output variable
2. Fuzzification by using continuous universe of discourse
3. Using mamdani performance (min)
4. Defuzzification by using centroid.

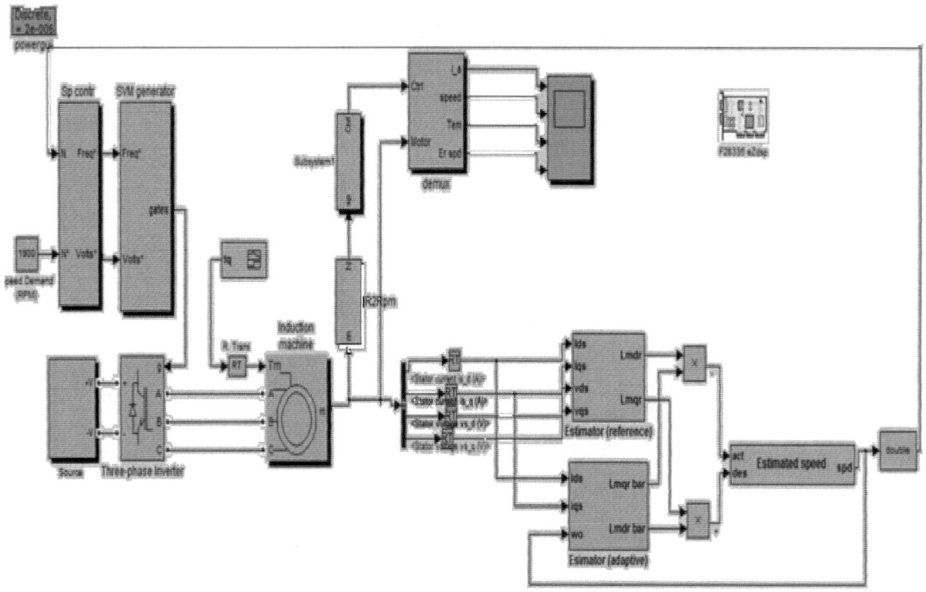

Figure 4. The Proposed approach

Figure 5. The physics Proposed Controlled

IOP Conf. Series: Materials Science and Engineering **300** (2018) 012067 doi:10.1088/1757-899X/300/1/012067

4. Result And Discussion

To certify both, steady state and transient behavior of the proposed algorithm some simulation has been carried out. The three phase induction motor has the following parameters: To know everything, I use a range Rr = 0.0025 ohms, Rs = 0.0015 ohms, f = 60Hz, P = 2, V= 120 V, for my induction motor.

A. Steady State

The steady state of stator flux and Electromagnetic

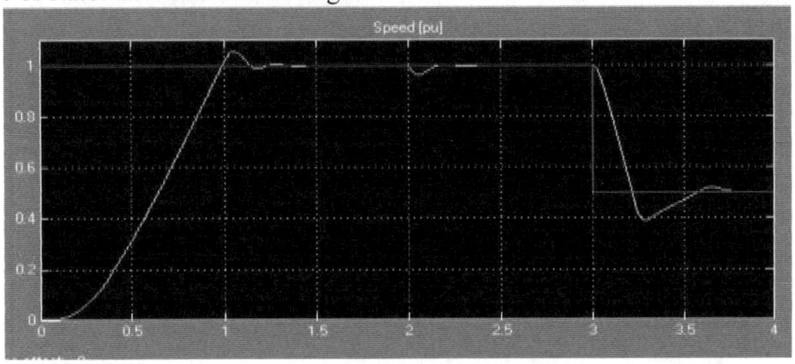

Figure 6. Motor speed and reference speed during using HPSO

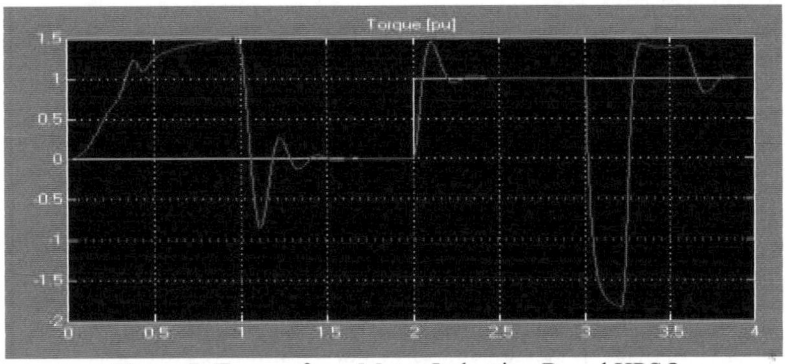

Figure7. Torque from Motor Induction Based HPSO

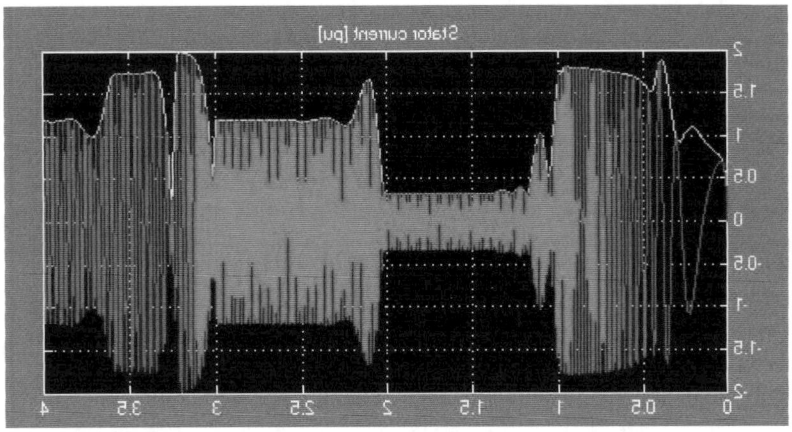

Figure 8. Stator Current From HPSO Based Induction Motor.

In the Figure6, 7, and 8, showed the speed, torque and stator currents of induction motors using system, where everything is quite stable when compared to using other systems.

5. Conclusion

In this research, the system is presented to improve the Efficiency and decrease the losses. In order to increase the motor drive stability, during variations of speed and load, a fuzzy controller was used, which had higher results than a PI controller, the selected voltage vector in the new PSO schemes which produce the lowest start-up stator that may be current and torque ripple is reduced thereby increasing the performance of induction motor. Therefore, the tuning approach simultaneously evaluated in PI controller tuning speed and current PI controllers for vector control of induction motors in the simulation and experiment. Therefore, the following functions speed torque, speed, and efficiencies that can be jointly satisfied. HPSO induction motor model allows estimation of relevant parameters in motor speed, stator flux, rotor flux and torque without using sensors. This means that with a lower input power, the torque of input reference load has been supplied. When the load torque varies between 1 and 20 Nm, the input power using FG algorithm was reduced up to approximately 30%. On the other hand, the amount of loss with FG method had a significant advantage over the other two different torques.

References

[1] Waheeda beevi,Sukesh Kumar," loss minimization of vector controlledinduction motor drive using genetic algorithm",IEEE,2012.

[2] M.Nasir Uddin,Sang Woo Nam," Adaptive backstepping online loss minimization control of an IM drive",IEEE,2007.

[3] Yong Tai, Zhaomian Liu," Efficiency optimization of induction Motor using genetic algorithm and hybrid genetic algorithm",2013,IEEE.

[4] Z.Rouabah,F.Zidani," efficiency optimization of induction motor drive using genetic algorithms", pp. 204 - 208 ,IEEE,April 2008.

[5] M. Nasir Uddin, Sang Woo Nam," New Online Loss-MinimizationBased Control of an Induction Motor Drive", IEEE TRANSACTIONS ON POWER ELECTRONICS, VOL. 23, NO. 2, MARCH 2008.

[6] A.H. Haddoun, J. Ghouili, R. Abdessemed," Efficiency - Minimization Loss Based Fuzzy Logic Control Of Induction Machine",PP. 774 – 777,2005 IEEE.

[7] T.R. Chelliah, J.G. Yadav," Optimal Energy Control of Induction Motor by Hybridization of Loss Model Controller Based on Particle Swarm Optimization and Search Controller ", World Congress on Nature &Biologically Inspired Computing, Pp. 1178 – 1183, 2009.

[8] Turki Y. Abdalla, Haroution Antranik Hairik," Direct Torque Control System for a Three Phase Induction Motor With Fuzzy Logic Based Speed Controller", 1st international confrence on energy power and control ,college of engineering,transection IEEE , 2010, pp.131–138.

[9] Ali M. Eltamaly, A. I. Alolah, and Basem M. Badr," Fuzzy Controller for Three Phases Induction Motor Drives", pp. 1 – 6, Autonomous and Intelligent Systems (AIS), 2010 International Conference .

[10] Bahram-Kimiaghalam, Meisam-Rahmani and Hassan-Halleh," Speed &Torque Vector Control of Induction Motors with Fuzzy Logic Controller", International Conference on Control, Automation and Systems 2008 Oct. 14-17, 2008 in COEX, Seoul, Korea.

[11] Tejavathu Ramesh, A. K. Panda," High Performance Direct Torque and Flux Control of Induction Motor Drive using Fuzzy Logic Based Speed Controller", 2013 International Conference on Circuits, Power and Computing Technologies,pp. 213 – 218.

[12] Tejavathu Ramesh, Anup Kumar Pandl, Y Suresh ," Direct Flux and Torque Control of Induction Motor Drive for Speed Regulator using PI and Fuzzy Logic Controllers", International Conference OnAdvances In Engineering, Science And Management (ICAESM),pp. 288 – 295, 2012.

[13] L. Ramesh, S. P.Chowdhury," Efficiency Optimization of Induction Motor Using a Fuzzy Logic Based Optimum Flux Search Controller", International Conference on Power Electronics, Drives and Energy Systems,pp. 1 – 6, 2006.

[14] Iulian BIROU, Virgil MAIER, Sorin PAVEL," Indirect Vector Control of an Induction Motor with Fuzzy-Logic based Speed Controller ", 3 rd International Symposium on Electrical Engineering and Energy Converters,2009.

[15] T. S. Radwan," Perfect Speed Tracking of Direct Torque ControlledInduction Motor Drive Using Fuzzy Logic",pp.38-43,IEEE 2005.

IOP Conf. Series: Materials Science and Engineering **300** (2018) 012068 doi:10.1088/1757-899X/300/1/012068

LISREL Model Medical Solid Infectious Waste Hazardous Hospital Management In Medan City

Verawaty Simarmata[1], Ungkap Siahaan[2],Setiaty Pandia[3], Herman Mawengkang[4]

[1] Doctor Program Graduate of the University of North Sumatra University of Sumatera Utara.

[2] Master Program Graduate of the University of North Sumatra University of Sumatera Utara.

[3] Lecturers of Doctor Program Natural Resources Management and Environment Departement, University of Sumatera Utara

[4] Lecturers at Mathematics Department and Doctor Program Natural Resources Management and Environment Departement, University of Sumatera Utara

[1] **verawatysimarmata@yahoo.co.id**

[2] **hinalang_siahaan@yahoo.co.id**

[3] **setiatypandia@yahoo.com**

[4] **hmawengkang@yahoo.com**

Abstract. Hazardous and toxic waste resulting from activities at most hospitals contain various elements of medical solid waste ranging from heavy metals that have the nature of accumulative toxic which are harmful to human health.

Medical waste in the form of gas, liquid or solid generally include the category or the nature of the hazard and toxicity waste. The operational in activities of the hospital aims to improve the health and well-being, but it also produces waste as an environmental pollutant waters, soil and gas.

From the description of the background of the above in mind that the management of solid waste pollution control medical hospital, is one of the fundamental problems in the city of Medan and application supervision is the main business licensing and control alternatives in accordance with applicable regulations.

Keywords: toxic waste, medical solid waste, non-medical solid waste, infectious and human health.

1. Introduction

Hospital activities will generate a number of by-products in the form of waste, both solids, liquids and gases that contain pathogens, chemicals and medical equipment which are generally dangerous and toxic ([12], [17]). In order to improve the quality of services, hospitals should be able to cope with such waste ([16], [19]). As for the waste treatment facilities in the hospital one of them is to use incinerators [4].

The problem of waste is a serious concern of the people and government of Indonesia, especially since the last decade, mainly due to the business development activities of the hospital which is the backbone of economic development for Indonesia [2]. Waste management is a necessity for the preservation of human health and the environment in general [1, 7]. However, the procurement and operation of medical solid waste processing facility hospital was still considered burdensome for the hospital [6, 7, 8].

2. Problem Formulation

If the amount of waste collected quite a lot, it is necessary to increase the number of containers. Containers are usually are made from metal or plastic [3]. Medical waste category of medical waste can be classified based on the potential dangers of hanging in it, as well as the volume and persistence properties which cause problems [2] for example:

1. Waste sharp objects such as needles, intravenous equipment, pasteur pipette, broken glass, and others;
2. Infectious waste, has a sense as waste associated with patients who require isolation of infectious diseases (intensive care) and laboratory waste [14];
3. Waste pathology (body tissue) is removed from the body tissue surgery or autopsy process;
4. Waste cytotoxic is material contaminated or possibly contaminated with bath cytotoxic during compounding, transport or treatment measures cytotoxic;
5. Waste pharmaceuticals derived from expired drugs that are not needed;
6. Chemical waste resulting from the use of chemicals in action, medical, veterinary, laboratory, sterilization processes and research;
7. Radioactive waste is material contaminated with radio-isotopes originating from medical or research used to radionucleous [20].

Until now, there are various laws and regulations governing the management of hazardous and toxic chemicals, but not sufficient, especially to prevent pollution or environmental damage [10]. For example, the Government Regulation on the Management of of hazardous and toxic chemicals, the issue focused on the management of hazardous and toxic materials for solid waste hospital medical and other third parties who act as producers, users, transporters, storage, user and collection especially for solid waste home medical in accordance with certain diseased appendix in the regulation [13].

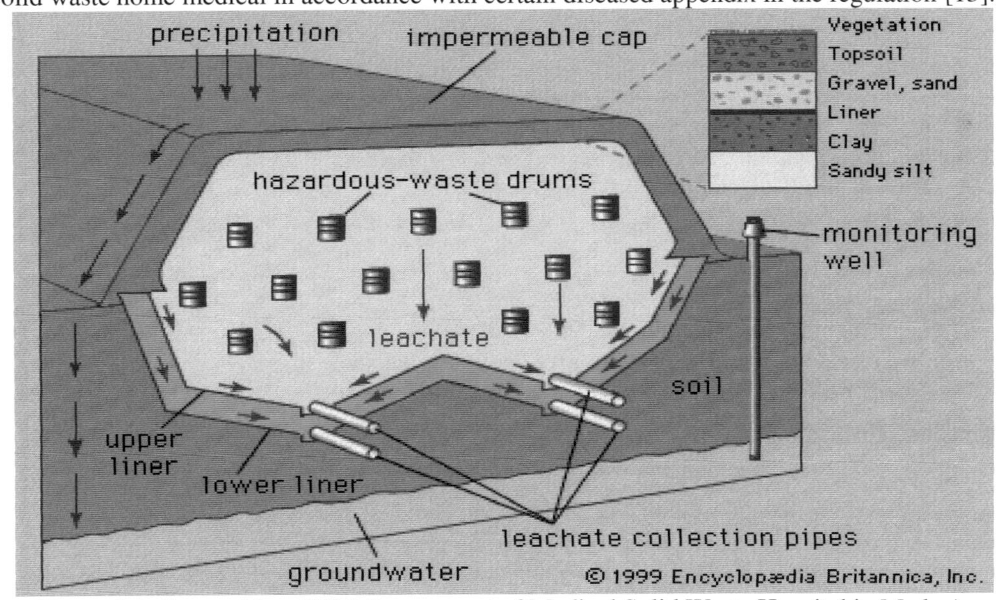

Figure 2.1. *Secured Landfill (*Management of Medical Solid Waste Hospital in Medan).

Sets and indices

One quality that is fairly common in the LISREL model is that the models disregard the means and regard all variables to be centered about their group means. This, in turn, results in

IOP Conf. Series: Materials Science and Engineering **300** (2018) 012068 doi:10.1088/1757-899X/300/1/012068

having the models with zero means. This is done in order to reduce the complexity associated in the analysis [11, 15].

If a multi-group model is being worked on with the help of LISREL, then it will give the same output of that process as is obtained by running a regression with dummy variables in SPSS [18].

LISREL helps the researcher in providing a fairly influential and flexible means for the examination of various group differences. It provides indicative information called modification indices which help the researcher in identifying the equality constraints [5].

LISREL can help the user to identify the interaction effects that need to be included in the model and the ones that do not need to be included in the model. The indicative information can be used in diagnosing the model specification for medical solid infectious waste hospital management in Medan city [9].

Variables

- Variable performance of machines and tools (incinerator) in the treatment of hospital medical solid waste (X1).
- Variable availability of medical solid waste from the activities of hospital activities (X2).
- Variable of environment variable availability of human resources according to the rules in the Decree of the Minister of Health of the Republic of Indonesia Number 12 Year 2004 (X3).
- Variable overcome the impact of the hospital's medical solid waste (X4).

Parameters

- X1.1 = Source hospital medical solid waste.
- X1.2 = Type hospital medical solid waste.
- X1.3 = Number of hospital medical solid waste.
- X1.4 = Production of hospital medical solid waste.
- X2.1 = Number of rooms and patient.
- X2.2 = Facilities or Facility Storage While medical solid waste.
- X2.3 = Physical condition of availability of raw materials activities of hospital activities.
- X2.4 = Medical solid waste collection process hospital.
- X3.1 = Condition Availability HR environment.
- X3.2 = Graduates HR environment.
- X3.3 = Ability or expertise HR environment.
- X3.4 = Training officers in the handling of hospital medical solid waste.
- X4.1 = Condition of society in the hospital environment.
- X4.2 = Attitude of people in a hospital environment.
- X4.3 = The level of awareness of the hospital against medical solid waste treatment.
- X4.4 = Level of medical solid waste pollution hospital.

3. Mathematical Formulation

Scheme dependent and independent variables influence in the determination of the conceptual study, are: $R = Y$, and

$$Y = a_1X_1 \{ (a_1X_{1.1}) + (a_1X_{1.2}) + (a_1X_{1.3}) + (a_1X_{1.4}) \} + a_2X_2 \{ (a_2X_{2.1}) + (a_2X_{2.2}) + (a_2X_{2.3}) + (a_2X_{2.4}) \} + a_3X_3 \{ (a_3X_{3.1}) + (a_3X_{3.2}) + (a_3X_{3.3}) + (a_3X_{3.4}) \} + a_4X_2 \{ (a_4X_{4.1}) + (a_4X_{4.2}) + (a_4X_{4.3}) + (a_4X_{4.4}) \}$$

IOP Conf. Series: Materials Science and Engineering **300** (2018) 012068 doi:10.1088/1757-899X/300/1/012068

Model LISREL medical solid infectious waste hospital management in Medan city for Living protected Environmental Law paying. Applicable Legislation (Y). Whatever *t* formula = statistic test if will n_1 = the total number of samples and n_2 = number of sample groups.

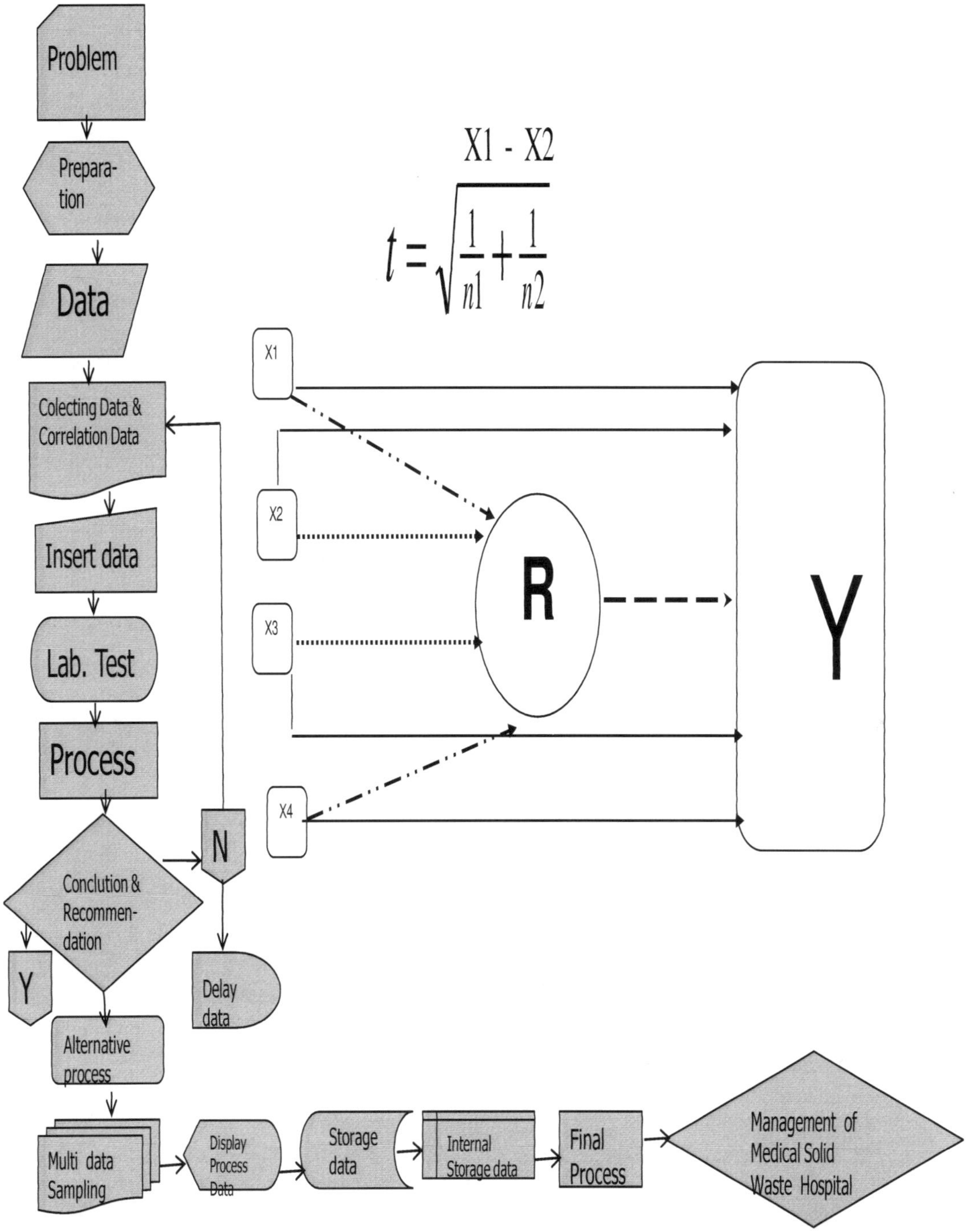

Figure 3.1. Conceptual Framework Scheme.

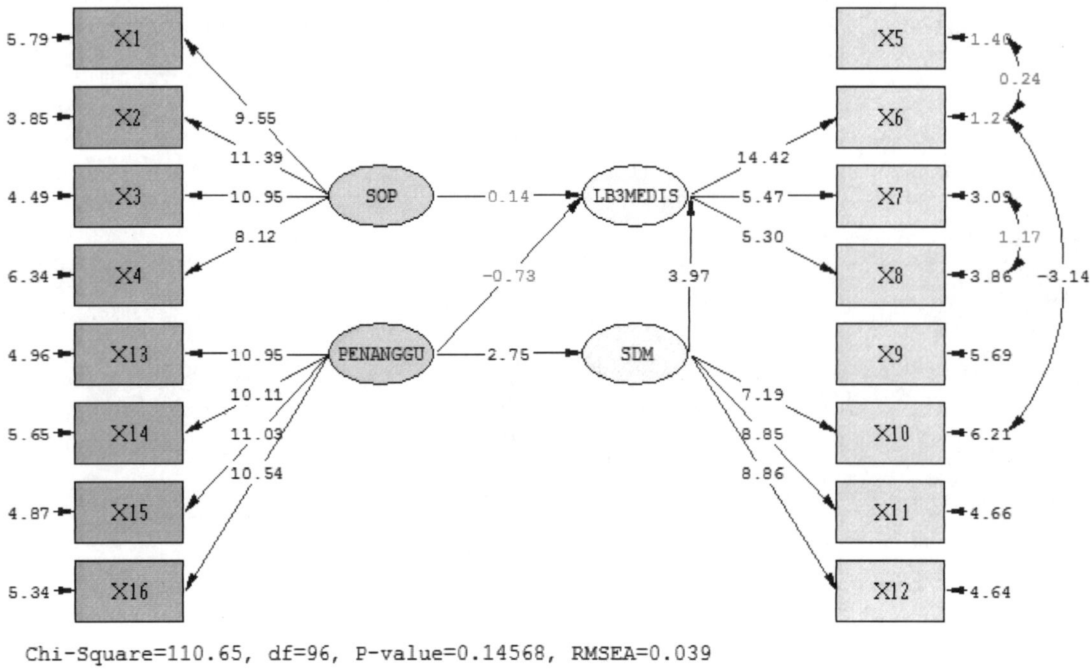

Figure 3.2. Statistical T Test Table LISREL Models

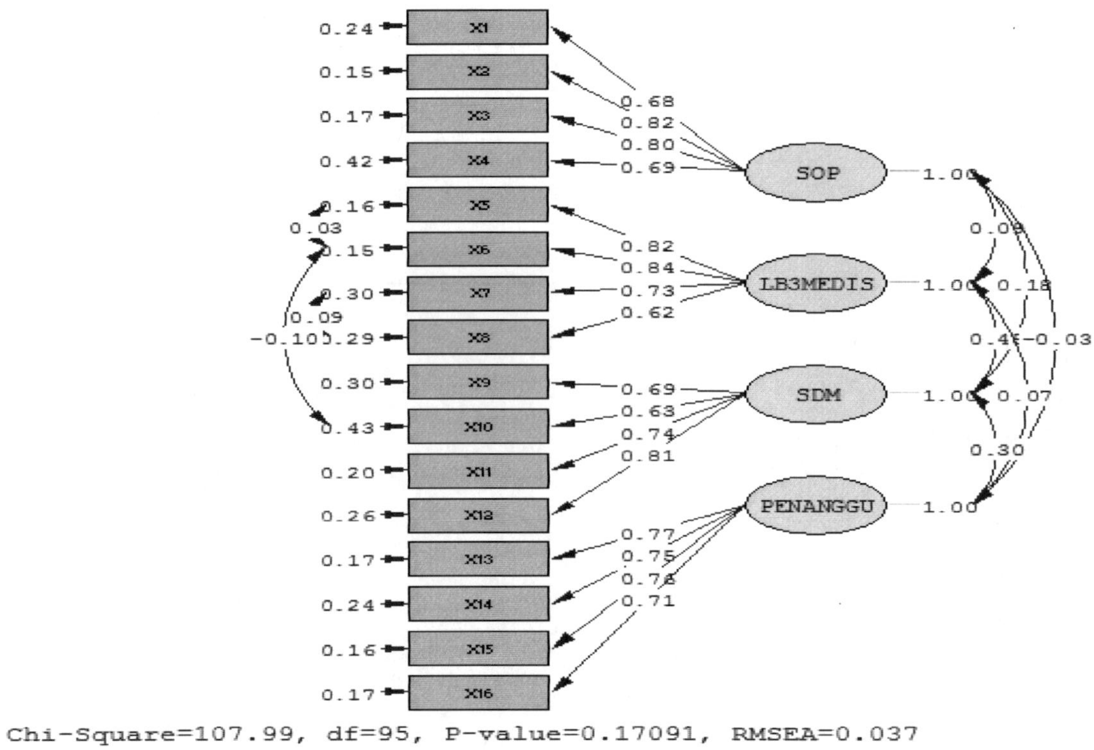

Figure 3.3. LISREL Models

4. Conclusions

The conclusions of the model LISREL medical solid infectious waste hospital management in Medan City, can be stated as follows.

a. Internal hospital solid waste collection process internally. (X1= 0.768).
b. Sources of hospital medical solid waste. (X2 = 0.743).
c. Number of hospital medical solids produced (X3 = 0.738).
d. Type of hospital medical solid waste (X4 = 0.665).
e. Model suitable measures and efforts to become a model LISREL medical solid infectious waste hospital management in Medan City, is Y = 0,768 X1 + 0.743 X2 + 0.738 X3 + 0.665X4.

References

[1] Andrés L., Medaglia, Juan, Villegas, G., M., Diana, Rodríguez and Coca, 2009, *Hybrid Biobjective Evolutionary Algoritms for the Design of a Hospital Waste Management*, *J Heuristics*, 15, 153-176.

[2] Anurag V., Tiwari, and Prashant A., Kadu, 2014, *Assessment of Biomedical waste management in Amravati*, International Journal of Engineering Sciences & Research Technology, ISSN:2277-9655, 3(5), 399-402.

[3] Chudasama, Rajesh K., Rangoonwala, Matib., Sheth, Ankit., Misra, SKC., Kadri and V Patel, Umed., 2013, *Biomedical Waste Management: A Study of Knowledge, Attitude and Practice Among Health Care Personnel At Tertiary Care Hospital in Rajkot*, Journal of Research in Medical and Dental Science, 1(1), 17-22.

[4] Conrardy, Julie., Hillanbrand, Mary., Sandra, Myers., and Nussbaum, George F., 2010, *Reducing Medical Waste,* AORN Journal, 10.1016/j.aorn.2009.12.029, 91(6), 711-721.

[5] David L. Streiner, PhD, 2006, *Building a Better Model: An Introduction to Structural Equation Modelling*, Research Methods in Psychiatry, Vol. 51, No. 5, pp: 317-324.

[6] David N. Ogbonna, Alex Chindah and Ndubuisi Ubani, 2012, *Waste Management Options for Health Care Wastes in Nigeria: A Case Study of Port Harcourt Hospitals*, Journal of Public Health and Epidemiology, ISSN 2141-2316, Academic Journals, Vol. 4(6), pp.156-169.

[7] Dhruv P., Hirani, Krish R., Villaitramani and Snehjit J., Kumbhar, 2014, *Biomedical Waste: An Introduction to its Management*, International Journal of Innovative Research in Advanced Engineering (IJIRAE), ISSN: 2349-2163, 1 (8), 82-87.

[8] Diaz, L.F., Savage, G.M. and Eggerth, L.L., 2005, *Alternatives for the Treatment and Disposal of Healthcare Wastes in Developing Countries*, Waste Management, 25: 626-37.

[9] Joseph Jr, Hair F., 2001, *Multivariate Data Analysis*, Seven Edition, New York, Prentice-Hall International, Inc.

[10] Kumar Goddu, Vijaya., Duvvuri, Kavita., and Kaumudini Bakki., Vidya, 2007, *A Critical Analysis of Healthcare Waste Management in Developed and Developing Countries: Case Studies from India and England*, Proceedings of the International Conference on Sustainable Solid Waste Management, Chennai, India, 134-141.

[11] Larwin, Karen., and Harvey., Milton, 2012, *A Demonstration of a Systematic Item-Reduction Approach Using Structural Equation Modeling*, Practical Assessment, Research & Evaluation, Item Reduction using SEM, ISSN:1531-7714, 17(8), 1-19.

[12] M. Shareefdeen, Zarook., 2012, *Medical Waste Management and Control*, Scientifict Research, Journal of Environmental Protection, 3, 1625-1628.

[13] Muluken, Azage, Haimanot, Gebrehiwot., and Mesafint, Molla., 2013, *Healthcare Waste Management Practices Among Healthcare Workers In Healthcare Facilities of Gondar Town, Northwest Ethiopia*, Health Science Journal, E-ISSN:1791-809, 7(3), 315-326.

[14] OB, Asante., E, Yanful., and EB, Yaokumah., 2014, *Healthcare Waste Management; Its Impact : A Case Study Of The Greater Accra Region, Ghana*, International Journal Of Scientific & Technology Research, www.ijstr.org., ISSN:2277-8616, 3, 106-102.
[15] Prüss, A., Giroult, E., and Rushbrook, P., 2005, *Pengelolaan Aman Limbah Layanan Kesehatan* (Ahli Bahasa: Munaya Fauziah) Safe Management of Wastes From Health Care Activities, World Health Organization. Jakarta. EGC, 3(15), 21-28, 171-175.
[16] Silva, C.E., Hoppe, A.E., Ravanello, M.M. and Mello, N., 2005, Medical Waste Management in the South of Brazil, *Waste Management*, 25, 600-605.
[17] Singh, Anshita., Kumari, Reema., Wakhlu, Anupam., Srivastava, Kirti., Wakhlu, Ashish., and Kumar, Sandeep., 2014, *Assessment of Bio-Medical Waste Management in a Government Healthcare Setting of North India*, International Journal of Health Sciences and Research, ISSN:2249-9571, 4(11), 203-208.
[18] Stephan, Levine Krehbiel Berenson., 2001, *Statistics for Managers Using Mocrosoft Excel*, New York, Prentice-Hall International, Inc.
[19] Tabasi, Rahele., and Marthandan, Govindan., 2013, *Clinical Waste Management: A Review on Important Factors in Clinical Waste Generation Rate*, International Journal of Science and Technology, ISSN:2224-3577, 3(3), 194-200.
[20] Thirumala, S., 2013, *Study Of Bio-Medical Waste Generation and Management in Various Hospitals in Davangere City of Karnataka India*, Nitte University Journal of Health Science, ISSN:2249-7110, 3(3), 22-24.

Perceptron Genetic to Recognize Openning Strategy Ruy Lopez

Zulfian Azmi[1], Herman Mawengkang[2]

[1]Graduate School, Universitas Sumatera Utara, Indonesia
[2]Department of Mathematics, Universitas Sumatera Utara, Indonesia

Email: zulfian.azmi@gmail.com

Abstract. The application of Perceptron method is not effective for coding on hardware based systems because it is not real time learning. With Genetic algorithm approach in calculating and searching the best weight (fitness value) system will do learning only one iteration. And the results of this analysis were tested in the case of the introduction of the opening pattern of chess Ruy Lopez. The Analysis with Perceptron Model with Algorithm Approach Genetics from group Artificial Neural Network for open Ruy Lopez. The data is processed with base open chess, with step eight a position white Pion from end open chess. Using perceptron method have many input and one output process many weight and refraction until output equal goal. Data trained and test with software Matlab and system can recognize the chess opening Ruy Lopez or Not open Ruy Lopez with Real time.
Keywords: Perceptron, Genetic Algorithm, Ruy Lopez

1. Introduction

1.1. Problem Background

The use of perceptron methods on this system is not effective, the learning process is longer and not real time into a smart system. One with the Neurall Network approach and the Genetic Algorithm approach can to prediction a case and more quickly learning or more real time. And in the training, genetic algorithm (GA) can be used to determine the initial weight of Multi Layer Perceptron. ([1] .Ming Zhu and Lipo Wang.2010) Genetic algorithms can be used to optimize the weight, topology or neural network parameters ([4] Stephen Ashmore and Michael Gashler.2015). The proposed exponential multilayer perceptron is proportional to a linear network of similar size. ([2] Exponential-Weight Multilayer Perceptron.2017).Especially for predicting and oxygen needs in ponds as material testing.The approach taken to make the waterwheel by the method of Neural Network.(Lie Jasa.2016 Method).

And the results of this analysis were tested in the case of the introduction of the opening pattern of chess Ruy Lopez. Because the knowledge and practice should be supported chessplayer base which is a collection of chess programs. Based on this, it is necessary to research a system that can recognize patterns based on the positions of chess opening end of a pawn using Perceptron with algorithm genetic. The algorithm use iterate over and over again to find out the possibility of the existing pattern so that the system can recognize patterns chess opening characterized by equal all network output with the desired output targets. Which is expected to improve the performance of existing chess programs and increase knowledge of the chess players preparing for it in every game. And so writer is interested to lift want to examine with the title "Analysis with Perceptron Model with Algorithm Approach Genetic for Open Ruy Lopez.".

2. Information Technology

2.1. Technology Computer

The method discovered by Rosenblatt (1962 and Minsky-Papert (1969) is part of an Artificial Neural Network (ANN), which has characteristics similar to the neural network biology. This network as a generalization of mathematical models of neural network biology, assuming that, information occurs in neurons and transmitted between neurons. Neuron has a weight that will amplify signal. and to determine the input, each neuron is using activation functions to imposed on the amount of input received. the magnitude of this output is then compared with a boundary threshold. And this perceptron algorithm in the method are:

- Initialize all weights and bias.
- Set Learning rate.
- Set threshold value.
- Set activation input.
- Compute response to the output unit.
- Put it in the activation function.
- Compare the value of the network output Y with goal. If not the same then changed weight and bias. Do iteration changes constantly and if you already have an output equal to the goal and iteration is stopped. (T.Sutojo, S.Si., M.Kom.2011).

And the term Genetic Algorithm was first proposed by John Holland in 1975 in his book Adaptation in Natural and Artificial Systems. One application is the path planning on mobile robot so that move efficiently to the destination. Therefore, evolution programming techniques, based on genetic algorithm, apply to many optimization problems, such as functional optics with linear and non-linear constraints, traveling salesman problems and scheduling, partitioning and control issues.

Genetic Algorithm is an adaptive method, starting with a set (represented by chromosomes) called population. Solutions from one population are taken and used to form a new population. This is driven by expectations, that the new population will be better than the old. The solution chosen to form a new solution (heredity) is in accordance with fitness. If it fits, then their chance to reproduce will be more. In theory, Gnome has several genes, and a collection of more than 1 gname becomes a population. Genetic Algorithm is a heuristic method developed based on genetic principles and the process of scientific selection of the theory of Darwinian Evolution Genetic Algorithm is a search method based on the process of natural evolution (Wolfgang Ertel.2011), namely the formation of a random population consisting of individuals with properties that depend on the genes -gen in the chromosome. The computation process that occurs in this algorithm is analogous to a process that may be known as a population. The initial population in the Genetic Algorithm was randomly generated by Genetic Algorithm operators for generations. In each generation, the chromosomes will undergo an evolutionary process using a measuring instrument that uses the so-called fitness function. The chromosomes formed from previous generations of chromosomes are called offspring. As well as parents. Crossovers. In GA, known operators can alter genes in chromosomes, this operator is called a mutation operator (mutation). The Basic Structure of the Genetic Algorithm consists of several steps: a) Initialization of population

- Population evaluation
- Population selection to be known genetic operator
- The process of crossing certain chromososm pairs.
- The process of mutation of a particular chromosome.

- Evaluation of the new population.
- Repeat from step 3 if the stop condition has not been met.

Coding is useful to decode gen so the value does not exceed for predetermined range and also becomes the value of the variable to be searched for as a solution to the problem. If the value of the encoded x variable is changed into rb, ra, ierb = lower boundary, ra = upper boundary, then the way to override the values of the above variables until they are in the new range [rb, ra] is called Decoding [Encoding].
Real Number Encoding: x=rb+(ra-rb)g.

Discrete Encoding Decimal:
$x=rb+(ra-rb)(gx10 +10^{-2}+\ldots+10^{-N}$

Binary Encoding:
$x=rb+(ra-rb)(gx2^{-1}+ gx2^{-2}\ldots+gx2^{-N}$

And the opening, an early stage of the game of chess. In the opening round, chess players must try as soon as possible to master the central plot and develop chess pieces so ready to attack, set up defenses and has a good vision for the game. Type of opening so are manifold, among which:

- The Game of open, such as Spanish Defence, Defence 4 Horse Defense of Russia, Games Philidor. This game is very sharp, and players must be vigilant step by step to make no mistakes.
- Games of Closed such as Opening Gambit Minister, Slavic Defense, Tarrasch Defense. Games that happens very quiet, setting the attack and pertahaan gradually.
- Games of Half-Open as Defense caro Khan, Alekhin, defense Sicilia, playing was calm, but can turn out to be sharp and quick.
- The game Half Closed namely Nimzo Indian Defence, the Indian Gaming king, Benoni Defense, Defense of the Netherlands, Ruy Lopez.
- The game Wings as the Indian Game turning upside down, Retri system, opening Inggris..Dan opening can be expected normally in the step-8 s / d 15, and after it was entered in the middle innings. (BeyMagethi, 2007: 7-9).

By understanding the basic theories and perceptron with algorithm genetic methods then chess opening as a base to approach in this study. Furthermore, the theory will be applied using a variety of referral sources that have been modified to be studied. So identify the problem to determined the pattern based of Ruy Lopez opening on the position of white pawn at the end of the opening (step 15) by the method of Perceptron.

2.2. Formulation of Problem

- How is Perceptron method with Genetic Algorithm approach to find the best fitness value at best weight and bias fast compared to usual Perceptron method?
- The results of this analysis were tested in the case of the introduction of the opening pattern of chess Ruy Lopez with Perceptron Model with Algorithm Approach Genetics ?

2.3. Purpose Theoretical

- As an essential and expanding contribution to the study of computer science in Perceptron methods that can be used as a reference for development different instances in the future.
- Provide an important contribution and expand the study of computer science that involves the chessing.
- Adding a new concept that can be used as reference material for further research for the development of computer science.

2.4. Practical Uses

- The results can be used as their contributions to the development of computer science to enhance software related to chess opening.
- The results of this study can be used as a benchmark the performance of the computer to determine the pattern of chess opening by the end position of the chess pieces.
- The results of the research can promote the development of chess, especially in Indonesia and improve the quality of chess players.

3. Results And Discussion

The variables in this study consisted of independent variables and one dependent variable. The independent variables are the input pattern (x1, x2) and the dependent variable is the output (Y).

3.1. The research design

The research design is a design experiment with using the input vector 8 * 8 = 63 components, according to the regional chessboard 8x8, 8 columns and 8 rows. The figures fill the pattern, namely: the number 1 marked, white pawn chess pawn position and the number 0 in the pattern that marked positions other than the pawns / white pawns or empty areas.

3.2. Research Model

The model used in this study are the Perceptron with algorithm genetic. This method has a network architecture that consists of several units and a bias input that has an output unit. And the activation function is a binary function or bipolar, with a possible value of -1.0 or 1. For a price determined thresold θ: ff (net) = 1 if net> θ, ff (net) = 0 if -θ <net <- θ, ff (net) = 1 if net <- θ.

4. Material And Methods

4.1. Data Analysis Technique

The algorithm make perceptron training for introduction of several patterns at once is as follows;

- Indicate each input pattern as a vector whose elements are 1, -1 and 0
- Give the target value tj = 1 if the pattern resembles the desired opening. Otherwise, provide a target value, tj = -1 (j = 1,2,2, ..., m),
- Give initiation weights, bias, rate of comprehension and trheshold.
- Make perceptron training process for each unit of output, namely: to calculate the output response of the j-th; netj = € xi.wi + bj, f (net) = 1 if net> θ, ff (net) = 0 if -θ <net <- θ, and f (net) = -1 if net <- θ. And fix weighting pattern contains an error output that is not the same as the (target), the equation; new Wji = wji (old) + αtj.xi, Bj = bj (old) + αtj and Perform steps continuously until tj = yj; j = 1, .. m.(T.Sutojo,S.Si.,M.Kom (2011)).

4.2. Chessing

Conditions of chessing

The gamewas held on a board consisting of 8 rows and 8 rows box or plot black and white with alternately. The game starts with 16 pieces on each side, which are arranged specifically lined up on each side of the chess board face to face. One piece can occupy only one plot. At the forefront of each row - there are 8 pawns, followed behind the two forts, two horses, two elephants, one minister and one king.

Before the match, player chose chess pieces that will used. Holders white berries begin the first step, which is followed by black holder interchangeably. The game objective is to achieve a checkmate position. This may happen when the king is threatened and could not escape to another plot. Game not always ended in defeat, because it could happen anyway or mussels series of events in which both sides are no longer able to continue the match because it cannot achieve checkmate. In the chess game winning party usually get a value of 1, the loser 0, while the draw is 0.5.

4.3. Mechanical Chess Games

To be able to master how to play chess well, full of variety and the dynamics there are steps that need to be considered and lived by chess players, Chess players both novice level, as well as an advanced degree or an expert. Technique playing, is divided into three acts namely: the initial round (opening), round them idle, final round.

- Preliminary Round (Preamble).
- Early rounds is the initial stage of the game of chess.
- Round Mid
- It is estimated that after the chess players stepped between 8 to 15 steps, it can be said that the game went in the middle rounds.
- End Game The final

Game can be started at step 20, could step into the 50th but also less than 10. In this round required precision and accuracy in determining step, so as not to make a blunder to be fatal. (Bey Mageth, How Understanding Chess Games Games Opening Ceremony: 1:11).

4.4. Computer chess opening with Fritz 6

Fritz is a computer chess program that always uses the theory of the opening. Hence, if you play without opening theory, then the next step will experience a lot of emphasis. Fritz 6 been aware opening move as much as 10 billion. Therefore, if making strides opening new ones, then Fritz 6 will say: "I have ten million moves and this guy comes up with something new!" (I already know 10 billion steps and this guy comes up with something the new ") .It shows us, bring Fritz 6 already know all the steps of opening and determine if we get out of theory.

4.5. Definition of Neural Network

Before discussing the method Perceptron Artificial, it helps us to study on Artificial Neural Networks is part of the artificial intelligence.Neural Network is an information processing system that has characteristics similar to biological neural networks. Artificial Neural Networks as a generalization of mathematical models of biological neural networks with assuming that, process happened many neurons. Signal information transmitted between neurons Between neurons have weights that will strengthen signal. And to specify the input, each neuron using activation functions to imposed on the amount of input received. The magnitude of this output is then compared with a threshold. (Pitowarno. 2004).

4.6. Perceptron

Perceptron network model was found by frank Ronbalatt late 1950s . He was a psychologist .This technique is a simple modeling of the retina of the human eye.Perceptron single layer are can be regarded as one of the artificial neural network techniques are simple . This technique only have an input layer and an output unit . Training algorithm is used to train the network with a case or patterns until the network managed to recognize the pattern. Each output generated by the network was not consistent with the targets then the weight is changed It is continuously performed until no more changed to weight.The last weight acquired during the training artificial neural network that will be used (Pitowarno . 2004).

4.7. Algorithm Genetic

Coding is useful to decode gen so the value does not exceed for predetermined range and also becomes the value of the variable to be searched for as a solution to the problem. If the value of the encoded x variable is changed into rb, ra, ierb = lower boundary, ra = upper boundary, then the way to override the values of the above variables until they are in the new range [rb, ra] is called Decoding [Encoding]. Real Number Encoding:

x=rb+(ra-rb)g.

Discrete Encoding Decimal: $x=rb+(ra-rb)(gx10 +10^{-2}+\ldots+10^{-N}$

Binary Encoding:

$x=rb+(ra-rb)(gx2^{-1}+ gx2^{-2}\ldots+gx2^{-N}$

The application of Perceptron method is not effective for coding on hardware based systems because it is not realtime learning. With Genetic algorithm approach in calculating and searching the best weight (fitness value) system will do learning only 1 iteration only. And it is realtime.

4.8. Results And Discussion

0	0	0	0	0	0	0	0
-1	-1	-1	0	0	-1	-1	-1
-1	0	0	0	0	0	0	0
0	0	0	0	-1	0	0	0
0	0	0	0	1	0	0	0
0	0	0	0	0	0	0	0
1	1	1	1	0	1	1	1
0	0	0	0	0	0	0	0

Figure 1. The Open Ruy Lopez (Morphy Defence)

0	0	0	0	0	0	0	0
-1	-1	-1	0	0	-1	-1	-1
-1	0	0	0	0	0	0	0
0	0	0	0	-1	0	0	0
0	0	0	0	1	0	0	0
0	0	0	0	0	0	0	0
1	1	1	1	0	1	1	1
0	0	0	0	0	0	0	0

Figure 2. The Open Ruy Lopez (Berlin Defence)

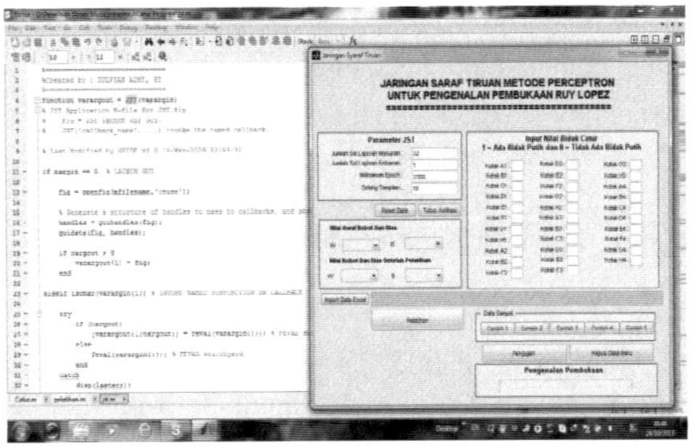

Figure 3. Input matriks 8x8 The Open Ruy Lopez(Berlin Defence)

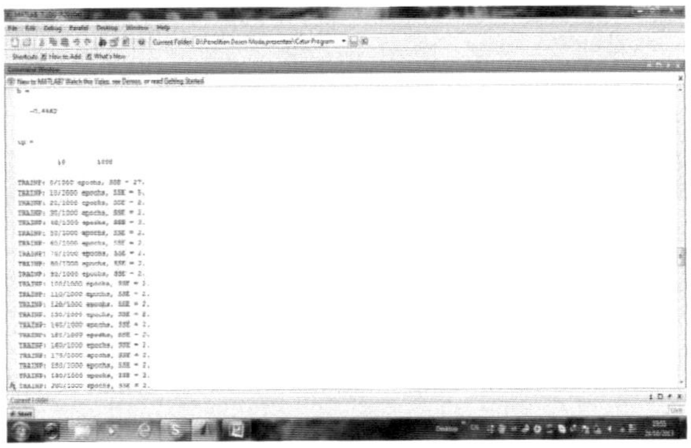

Figure 4. User interface

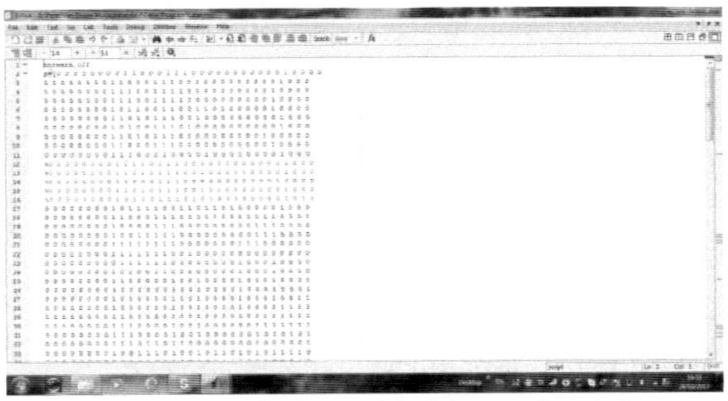

Figure 5. Trainning with bias and weight

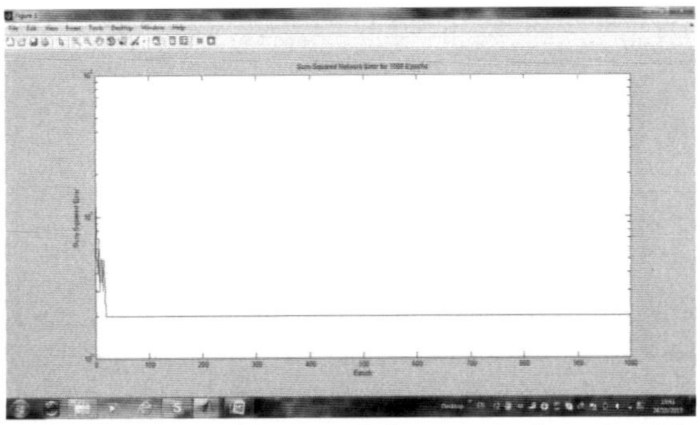

Figure 6. Graphics Software training results by using Matlab.

5. Conclusion

- The system was created by Method Perceptron with Genetic algorithm approach to recognize pattern RuyLopez openning and not the Ruy Lopez. And so it is real time.
- Data trained and test with software Mathlaband system can recognize the chess openingRuy Lopez or Not open Ruy Lopez with Real time.

References

[1] (Strukov Farnood Merrikh Bayat, Xinjie Guo dan Dmitri Perceptron Multilayer Exponential,(ECE Department.Universitas California Santa Barbara CA 93106-9560, AS E-mail: farnoodmb@ece.ucsb.edu).

[2] Andrii Shalaginov.*Evolutionary Optimization of On-line Multilayer Perceptron for SimilarityBased Access Control.*Norwegian Information Security Laboratory Center for Cyber- and Information Security Norwegian University of Science and Technology andrii.shalaginov@ccis.no).

[3] Stephen Ashmore dan Michael Gashler.2015. *Perceptron Multi-Layer dengan Alignment Bipartite Terusan.Arkansas*: Departemen Ilmu Komputer dan Teknik Komputer.Universitas Arkansas.Fayetteville, Arkansas

[4] Mehran Amiri a, Abdollah Ardeshir b,c, Mohammad Hossein Fazel Zarandi (2016).Fuzzy probabilistic expert system for occupational hazard assessmenti construction.

[5] Iran:Journalhomepage:www.elsevier.com/locate/ssci. (Svetlana Filicheva, Alexey Zaikin, Oleg Kanakov.2016.*Dynamical Decision making in a Genetic Perceptron.*Physica D: Nonlinear Phenomena).

[6] Yi-chung Hu.2008. *Nonadditive grey single-layer perceptron with Choquet integral for pattern classification problems usingGenetic algorithms.Neurocomputing,* Volume 72, Issues 1– 3, December 2008, Pages 331-340).

[7] J.J.DGroote,D.M.L.Barbato.2008. Generalization of finite size Boolean perceptrons with genetic algorithms.Neurocomputing, Volume 71, Issues 16– 18, October 2008, Pages 3650-3655.

[8] T.Sutojo,E.Mulyato,V.Suhartono.2010.*Kecerdasan Buatan*.Yogyakarta: Andi.

[9] Rajasekaran, Sdan G.A. Vijayalakshmi Pai. 2003, *Neural Network, Fuzzy*.

[10] T.Sutojo,S.Si.,M.Kom., (2011), *Kecerdasan Buata*n.Yogyakarta: Andi Publishing.

[11] Magethi,Bey.(2007). *Bagaimana Memahami Catur Permainan Pembukaan. Bandung:* Pionir Jaya Publishing.

[12] Didik Wahyudi/Imam Sucahyo.1996.*Pertahanan Spanyol Variant Marshal dan Hindia Menteri*. Jakarta: Titik Terang. Publishing.

[13] Mico Pardosi.2000.Teori PembukaanCatur. Surabaya: Indah Publishing.

[14] Gunaidi Abdia Away.2010.*The Shortcut of Matlab Programming. Bandung*; Informatika Bandung Publishing.

IOP Conf. Series: Materials Science and Engineering **300** (2018) 012070 doi:10.1088/1757-899X/300/1/012070

Triangle Number of Laurent polynomial for the closed braids σ^{2j}

M K M Nasution[1]*, A S Harahap[2], S Nasution[2], E Herawati[2]

[1]Teknologi Informasi, Fasilkom-TI, Universitas Sumatera Utara, Padang Bulan 20155 USU Medan Indonesia

[2]Matematika, FMIPA, Universitas Sumatera Utara, Padang Bulan 20155 USU Medan Indonesia

E-mail: mahyuddin@usu.ac.id

Abstract. In various computations, the triangle numbers help inductively to override a pattern from different structures increasingly. However, different structures of a particular entity effect the structure of presentation, including the order of numbers that characterize it. The structure of uniforms knots based on a particular woven pattern is followed by the behavior of the structure which in this case is represented by the presence of triangle numbers. This paper describes the existence of the triangle number based on the closed braids σ^{2j}.

1. Introduction

Like Pascal's triangle number, there is an arrangement of the triangle number that serves to solve an equation inductively [1]. The arrangement of this triangle number is based on the arrangement of certain object patterns that are mathematically modelled so that they can explicitly explain the pattern arrangement [2]. In everyday life the structure of the object can be considered to reinforce the arrangement of the object pattern, for example the fabric of the rope in addition to beautify also strengthens the relationship between the rope composers such as the hair braid to strengthen the thin hair [3]. Although at a glance, such arrangement appear similar but structurally between one pattern arrangement with another are mutually different or equal, and this difference can be explained mathematically [4].

Mathematically, a braid is the basic pattern of the links of strings or the hair braid. Each braid may turn into a knot in mathematical, i.e. a closed braid, and a braid has a specific identifier formula [5]. When pattern of braid regularly increase, in general the knots will follow increase of the pattern. If the pattern of braid regularly can be formulated based on the number pattern, the arrangement of the knots forms an array of numbers. This paper describe the composition of number pattern of the related closed braids.

2. Related Works and Motivation

A braid has been modelled mathematically [3]: A system called a *woven pattern* is two parallel lines in $\mathcal{R}^3$ in the same direction, each is said to be a top frame and the bottom frame, there are n different points p_i and q_j, $i, j = 1, \ldots, n$, they located on lines L_1 and $L2$, respectively. The adjacent points in a line have the same distance, and then one point of p_i is connected by the curve c_i just to a point q_j, where i and j should not be different [6, 7].

A braid is a weave together with a deformation operation (let's call it intuitively as a sweep), resulting in no weave of the same high such that (i) L_1 and L_2 stays parallel during the sweep, (ii) there are no slices between the curves during the sweep, and (iii) the curve remains normal during sweeping, i.e. the projected curve c_i is moving from p_i to q_j and take the distance from L_1 always increase. Next, intuitively it is stated that σ_i as notation for a woven geometric shape in which the curve c_i via c_{i+1} exactly once on it, whereas c_j for $j \neq i, i+1$ directly connects the points p_j to q_j. Instead, σ_i^{-1} is a webbing where the curve c_i via c_{i-1} exactly once on it, whereas c_j, $j \neq i-1, i$ are direct linking points p_j with q_j [8]. Therefore, for n-order braid, n is positive integer, we have generators $\sigma_1, \sigma_2, \ldots, \sigma_{n-1}$ of braid, and for a collection of n-order braids, we obtain an algebraic definition of the braid group B_n whereby the definition of a group presentation by generators and relations as follows.

Definition 1. The Artin braid group B_n is the group generated by $n-1$ generators σ_i, $i = 1, \ldots, n-1$ and the braid relations $\sigma_i\sigma_j = \sigma_j\sigma_i$ for all $i, j = 1, 2, \ldots, n-1$ with $|i-j| \geq 2$, and $\sigma_i\sigma_{i+1}\sigma_i = \sigma_{i+1}\sigma_i\sigma_{i+1}$ for $i = 1, 2, \ldots, n-2$.

Suppose B_n contains σ_i^{2j}, $j = 1, 2, \ldots, k$. or a group B_2 is generated by a single generator σ_1 only without a relation such that the group B_2 contains σ_1^{2j}. Thus in any braid possible contains the cyclic i.e. $\sigma^2, \sigma^4, \ldots$, or there is an infinite cyclic group of braid.

Let σ^{2j} be basic pattern of the knots, or the closed braids based on σ^{2j}, so there are a normal knots, that is unprime knot, whereby a prime knot is a knot with one string. Thus, σ^{2j} generate the knots possible consist of two or more string. Based on the shape of the cross in braids and trivial knot (the circle) we get

$$\ell P(L_+) + \ell^{-1}P(L_-) + mP(L_0) = 0 \tag{1}$$

where $P(L)$ is a polynomial of oriented link, it has ℓ and m are two variables of integer coeficient, and L_+, L_-, and L_0 are three oriented links corresponding with overlink, underlink, and unlink [9]. For example, for knot s_2 based on pattern σ^2 Eq. (1) be

$$P(s_2) = \ell m^{-1} - \ell m + \ell^3 m^{-1} \tag{2}$$

3. An Approach

The set of numbers for triangular shapes has a triangular frame composition based on entries [1] like Fig. 1. This entry is based on a basic braid arrangement with a pattern $\sigma^{2j} \in B_2$, $j = 1, 2, \ldots, n$, i.e. a finite collection of disjoint curves in $\mathcal{R}^3$. Each closed curve is a component of link, and a link of one component of knot [5]. It is expressed as follows [9].

Definition 2. Let $P(L)$ as a unique way to determine a knot. P is a function as follows

$$P : \cup_n \mathcal{L}_n \to B[\ell^{\pm}, m^{\pm}]$$

where ℓ and m as $u = -(\ell + \ell^{-1})m^{-1}$, $\ell \in \cup_n \mathcal{L}_n$ such that $P(\alpha L) - u^{cL-1}$, L has component cL and B is braid.

Theorem 1. Laurent polynomial $P(L)$ determine a unique way so that equivalent oriented links have same polynomial if and only if $P(unknot) = 1$, and if L_+, L_-, L_0 are three oriented

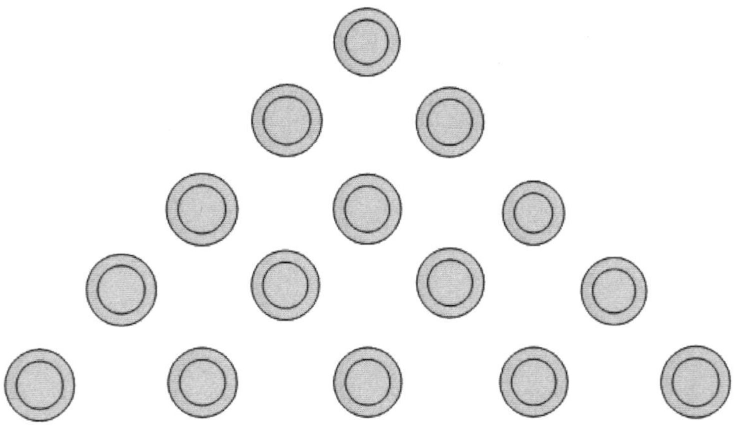

Figure 1. Frame of triangle number entries

link then Eq. (1).

Intuitively, a triple links of L_+, L_-, and L_0 are as two links and one unlink: First two links are the picture of unknot that have positive and negative crosses, respectively, but the last link as based on Theorem 1 is $\ell + \ell^{-1} + mP(L_0) = 0$. Thus, the last of triple links has

$$P(L_0) = -(\ell + \ell^{-1})m^{-1} \tag{3}$$

Thus, to derive the polynomial of a knot is based on Eq. (1), and we have a sequence of polynomial based on braid patterns σ^{2j}, and this inductively will build a general form of polynomial [10].

4. New Triangle Number

Based on Eq. (1) for each knot has triple link of basic patterns in braid σ^{2j}, $j = 1, \ldots, n$. Thus for σ^2 there are knot s_2 with triple links in braids as follows

$$s_2 \to (\sigma^0 = 1), (\sigma^2), (\sigma^1), \tag{4}$$

the knot s_4 has triple links as follows

$$s_4 \to (\sigma^2), (\sigma^4), (\sigma^3), \tag{5}$$

the triple links of knot s_6 be

$$s_6 \to (\sigma^4), (\sigma^6), (\sigma^5), \tag{6}$$

and so on, so for knot s_{2h} we have

$$s_{2h} \to (\sigma^{2(h-1)}), (\sigma^{2h}), (\sigma^{2h-1}). \tag{7}$$

see Table 1.

By using Theorem 1, we have $P(s_0) = P(unknot)$ like Eq. (3) or $P(s_0) = -\ell m^{-1} + \ell^{-1}m^{-1}$, while for $P(s_2)$ is Eq. (2),

$$P(s_4) = -\ell^3 m^{-1} + 3\ell^3 m - \ell^3 m^3 - \ell^5 m^{-1} + \ell^5 m \tag{8}$$

Table 1. Knot with two strings

Knot	$\rightarrow$	$+2$	$+1$	$+$
s_2	σ^0	σ^2	σ^1	2
s_4	σ^2	σ^4	σ^3	2
s_6	σ^4	σ^6	σ^5	2
...	...	...	...	$2 \cdot \ldots$
s_{2h}	$\sigma^{2(h-1)}$	σ^{2h}	σ^{2h-1}	2

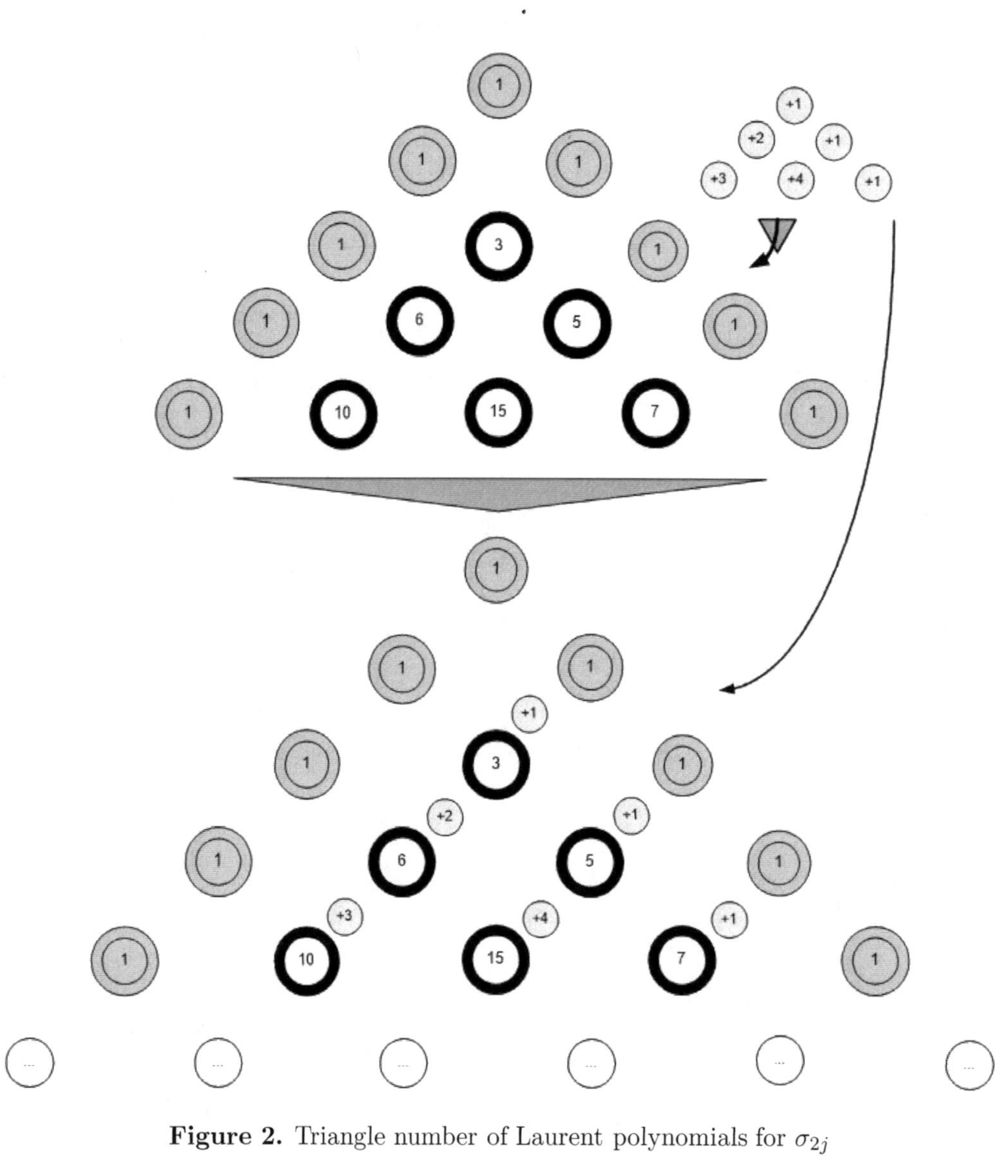

Figure 2. Triangle number of Laurent polynomials for σ_{2j}

and

$$P(s_6) = \ell^5 m^{-1} - 6\ell^5 m + 5\ell^5 m^3 - \ell^5 m^5 + \ell^7 m^{-1} - 3\ell^7 m + \ell^7 m^7. \quad (9)$$

Therefore, based on Eqs. (3), (2), (8), and (9) we have two composition of numbers, i.e. [−1], [−1]-[−1], [−1]-[3]-[−1], [−1]-[−6]-[5]-[−1] and [1], [1], [−1]-[1], [1]-[−3]-[1]. By ignoring the negative sign, the second composition follows the first composition pattern. Thus obtained the initial form of the triangle number as follows

$$\begin{array}{ccccccc} & & & 1 & & & \\ & & 1 & & 1 & & \\ & 1 & & 3 & & 1 & \\ 1 & & 6 & & 5 & & 1 \end{array}$$

The outer numbers of the triangle numbers are 1, while the inner number varies: 3 come from $1+1$ and plus 1, 5 come from $1+3$ and plus 2, while 5 come from $1+3$ plus 1. Then if forwarded to the next line, $1+6$ along with 3 to 10, $6+5$ along with 4 to 15, and $5+1$ along with 1 to 7, and the next line item arrangement is $1, 10, 15, 7$ and 1. So there is a companion triangle number as follows

$$\begin{array}{ccccc} & & +1 & & \\ & +2 & & +1 & \\ +3 & & +4 & & +1 \end{array}$$

with the numbers on the next line are +4, +7, +5, and +1.Thus, generally the number triangle for knot s_{2h} as follows

$$P(s_{2h}) = \sum_{i=1}^{k} (-1)^{|i-k-1|} B_{hi} \ell^{2h-1} m^{2i-3} + \sum_{i=1}^{k-1} (-1)^{|i-(k-1)|} B_{(h-1)i} \ell^{2h+1} m^{2i-3} \quad (10)$$

see Fig. 2. Diagonal from up-left to down-right on right side has an addition value +1, diagonal from up-right to down-left on left side has an addition value +2 increasingly from +1, second diagonal from up-left to down-right on second right has addition value +2 increasingly from +2, while second diagonal from up-right to down-left on second left has addition value +3 increasingly from 2.

5. Conclusion

The triangle numbers of Laurent polynomials that typically illustrates the cyclic knot follows the generators pattern σ^{2j}. The triangle numbers forms certain structures based on the braid structure that regularly their weave increases. Besides these triangle numbers are different from other triangle numbers both from the composition and the structure, which is formed from two arrangement of triangle numbers that support each other's existence. The next study deals with the possibility of developing computationally to determine the number of more strings.

References

[1] S Falcón, and Á Plaza 2007 The k-Fibonacci sequence and the Pascal 2-triangle *Chaos, Solitons and Fractals* **33**.
[2] J B Shearer 1983 A note on the independence number of triangle-free graphs *Discrete Mathematics* **46(1)**.
[3] M K M Nasution 2000 The number triangle of Laurent polynomial on knots *Jurnal Matematika dan Terapannya* **1(2)**.
[4] K M N Mahyuddin, O S Sitompul, S Nasution, and H Ambarita 2017 New similarity *IOP Conference Series: Materials Science and Engineering* **180(1)**, 012297.
[5] M K M Nasution 1992 Studi pembuktian eksistensi simpul normal dari polinomial simpul *Jurusan Matematika, FMIPA USU.*
[6] E Artin 1925 Theorie der Zöpfe *Hamb. Abs.* **4**.

[7] E Artin 1947 Theory of braids *Annals of Mathematics* **48**.
[8] E Artin 1950 The theory of braids *American Scientist* **38**.
[9] K Murasugi 1987 Jones Polynomials and Classical Conjectures in Knot Theory *Topology* **26(2)**.
[10] P Freyd, D Yetter, J Hoste, W B R Lickorish, M Millett, and A Ocneanu 1985 A new polynomial invariant of knots and links *Bulletin American Mathematical Society*, **12(2)**.

IOP Conf. Series: Materials Science and Engineering **300** (2018) 012071 doi:10.1088/1757-899X/300/1/012071

Some structural properties of vector valued φ-function sequence space

[1]S N R Gultom and [2]E Herawati

[1,2]Department of Mathematics, University of Sumatera Utara, Indonesia

E-mail: [1]septiannandarivald@gmail.com, [2]herawaty.elv@gmail.com

Abstract. The sequence space $W(M)$, where M is an *Orlicz function* was introduced by Parashar and Choudhary [1] and Maddox [2]. Let f be *φ-function* and X be a Banach space. In this work, we introduce vector valued sequence space defined by f, denoted by $W(X, f)$. We study some topological properties and inclusion relations of this space.

1. Introduction and Preliminaries

An *Orlicz function* is a continuous, convex, non-decreasing function defined from $[0, \infty)$ to itself such that $M(0) = 0$, $M(x) > 0$ for $x > 0$, and $M(x) \to \infty$ as $x \to \infty$. Lindenstrauss and Tzafriri [3] introduced the sequence space $\ell^{\exists}(M)$ using *Orlicz function* M as follows

$$\ell^{\exists}(M) = \left\{ x = (x_k) : x_k \in \mathbb{R}\ \forall \text{k} \in \mathbb{N} \text{ and } \exists \rho > 0 \text{ such that } \sum_{k=1}^{\infty} M\left(\frac{|x_k|}{\rho}\right) < \infty \right\}$$

The space $\ell^{\exists}(M)$ equipped with the *Luxemburg norm*

$$\|x\| = \inf \left\{ \rho > 0 : \sum_{k=1}^{\infty} M\left(\frac{|x_k|}{\rho}\right) \leq 1 \right\}$$

becomes a Banach space which is called an *Orlicz sequence space*. The space $\ell^{\exists}(M)$ is closely related to the space ℓ_p with $1 \leq p < \infty$,

$$\ell_p = \left\{ x = (x_k) : x_k \in \mathbb{R}\ \forall \text{k} \in \mathbb{N} \ \text{ and } \sum_{k=1}^{\infty} |x_k|^p < \infty \right\}$$

which is an *Orlicz sequence space* with $M(x) = x^p$. In the mathematical literature there exists various modifications of these definitions, where ℓ is replaced by another solid sequence space (see [4–6]). A sequence space X is said to be solid (or normal) if $(\lambda_k x_k) \in X$, whenever $(x_k) \in X$ and for all sequences (λ_k) of scalars with $|\lambda_k| \leq 1$ for all $k \in \mathbb{N}$.

A norm $\|\cdot\|$ on a normal sequence space X is said to be *absolutely monotone norm* if $x = (x_k), y = (y_k) \in X$ and $|x_k| \leq |y_k|$ for all $k \in \mathbb{N}$ implies $\|x_k\| \leq \|y_k\|$. The norm

$$\|x\|_\infty = \sup |x_k|$$

over the classical sequence space $\ell\infty$, c, c_0 and the norm

$$\|x\| = \left(\sum_{k=1}^{\infty} |x_k|^p\right)^{\frac{1}{p}}$$

over ℓ_∞ for $p \geq 1$ are absolutely monotone.

A completed normed space X is said to be a *BK-space* if the function $p_k : X \to \mathbb{R}$ where $p_k(x) = x_k$ is continuous in X for every $x = (x_k) \in X$ and every $k \in \mathbb{N}$. An *AK-space* X with the norm $\|\cdot\|$ is a *BK-space* and $\|x - x^{[n]}\| \to 0$ as $n \to \infty$ for every $x \in X$, where $x^{[n]}$ denotes the $n - th$ section of x.

Let X be a vector space. The collection of all vector valued sequences denoted by $\Omega(X)$. Any vector subspace of $\Omega(X)$ is called vector valued sequence space. The studies on vector valued sequence spaces are done by Rath and Srivastava [7], Das and Choudhary [8], Leonard [9], Srivastava and Srivastava [10] and many others.

A function $f : \mathbb{R} \to [0, \infty)$ which is continuous, vanishing at zero, non-decreasing on $[0, \infty)$ and even is called $\varphi - function$. A $\varphi - function$ f is said to satisfy $\Delta_2 - condition$ (written as $f \in \triangle_2$ for shortly), if there exists $K > 0$ such that $f(2x) \leq Kf(x)$ for every $x \geq 0$.

A functional $\rho : X \to [0, \infty)$ is called a convex modular if $\rho(x) = 0 \Leftrightarrow x = 0$, even, $\rho(\alpha x + \beta y) \leq \alpha\rho(x) + \beta\rho(y)$ if $\alpha + \beta = 1$ and $\alpha, \beta \geq 0$. In this case, we say that X is a convex modular space.

Let $X = (X, \|\cdot\|_X)$ be a *Banach space* with an absolutely monotone norm $\|\cdot\|_X$ and f is a $\varphi - function$. Using convex $\varphi - function$ f, we introduce the following set, denoted by $W(X, f)$.

$$W(X, f) = \left\{x = (x(i)) \in \Omega(X) : \rho_f\left(\frac{x(i) - \ell}{\alpha}\right) \to 0 \text{ as } i \ \to \infty, \text{ for some } \alpha > 0 \text{ and } \ell \in X\right\}$$

where

$$\rho_f(x(i)) = \frac{1}{m}\sum_{i=1}^{m} f(\|x(i)\|_X), \text{ for every } x(i) \in X$$

is a *convex modular.*

A function $g : X \to \mathbb{R}$ is said to be *paranorm* if $g(\theta) = 0$, $g(x) \geq 0$, $g(x+y) \leq g(x)+g(y)$, even and every scalar sequence (λ_n) with $|\lambda_n - \lambda| \to 0$ and every sequence (x_n) with $g(x_n - x) \to 0$ implies $g(\lambda_n x_n - \lambda x) \to 0$ for all $\lambda \in \mathbb{R}$ and $x \in X$, where θ is the zero in the linear space X. The notion of *paranormed sequence space* was introduced by Nakano [11] and Simons [12]. Later on it was further investigated by Rath and Tripathy [13], Tripathy and Sen [14].

In this work, we investigate some of topological properties of the set $W(X, f)$ equipped with a paranorm that we will define and study some inclusion relations of this set.

2. Main Results

In this section we examine some topological properties and inclusion relations of the set $W(X, f)$.
Lemma 1. *If $x, y \in X$ such that $0 \leq x \leq y$, then $\rho_f(x) \leq \rho_f(y)$.*

Theorem 2.1. *If $\varphi - function$ f satisfies the $\Delta_2 - condition$ and convex, then $W(X, f)$ is a linear space.*

Proof. Let $x = (x(i))$, $y = (y(i)) \in W(X, f)$, then there exist $\alpha_1, \alpha_2 > 0$ and $\ell_1, \ell_2 \in X$ such that

$$\rho_f\left(\frac{x(i) - \ell_1}{\alpha_1}\right) \to 0 \text{ and } \rho_f\left(\frac{y(i) - \ell_2}{\alpha_2}\right) \to 0 \text{ as } i \to \infty$$

Take $\alpha = \max\{\alpha_1, \alpha_2\}$ and $\ell = \ell_1 + \ell_2$. Considering Lemma 1 and the property of $\varphi - function$, we have

$$\rho_f\left(\frac{(x(i)+y(i))-\ell}{\alpha}\right) \leq \rho_f\left(\left|\frac{x(i)-\ell_1}{\alpha_1}\right| + \left|\frac{y(i)-\ell_2}{\alpha_2}\right|\right)$$

Since ρ_f is convex and $f \in \Delta_2$, there exists $\frac{K_1}{2}, \frac{K_2}{2} > 0$ such that

$$\rho_f\left(\frac{(x(i)+y(i))-\ell}{\alpha}\right) \leq \frac{K_1}{2}\rho_f\left(\frac{x(i)-\ell_1}{\alpha_1}\right) + \frac{K_2}{2}\rho_f\left(\frac{y(i)-\ell_2}{\alpha_2}\right)$$

Consequently, $\rho_f\left(\frac{(x(i)+y(i))-\ell}{\alpha}\right) \to 0$ as $i \to \infty$. Hence, $x + y \in W(X, f)$. Let $\beta \in \mathbb{R}$ and $x = (x(i)) \in W(X, f)$, then there exists $\alpha > 0$ and $\ell \in X$ such that

$$\rho_f\left(\frac{x(i)-\ell}{\alpha}\right) \to 0 \text{ as } i \to \infty$$

Choose $p = \beta\ell$. For $\beta = 0$, is clear that $\rho_f\left(\frac{\beta x(i)-p}{\alpha}\right) \to 0$ as $i \to \infty$. Now, assume that $\beta \neq 0$. Since $f \in \Delta_2$, then by using *Archimedian* there exists $n_0 \in \mathbb{N}$ and $K > 0$ such that

$$\rho_f\left(\frac{\beta x(i)-p}{\alpha}\right) \leq K^{n_0}\rho_f\left(\frac{x(i)-\ell}{\alpha}\right) \to 0 \text{ as i} \to \infty$$

Therefore, $\beta x \in W(X, f)$ and the proof is complete. □

Theorem 2.2. *A function* $g : W(X, f) \to \mathbb{R}$ *with*

$$g(x) = \inf\left\{\alpha > 0 : \rho_f\left(\frac{x(i)}{\alpha}\right) \leq 1\right\}$$

is a paranorm.

Proof. It is easy to show that $g(\theta) = 0$, $g(x) \geq 0$ and $g(-x) = g(x)$, for every $x \in W(X, f)$, where θ is the zero in the linear space $W(X, f)$. We shall now show the subaddivity of g. Let $x = (x(i))$, $y = (y(i)) \in W(X, f)$, then there exist $\alpha_1, \alpha_2 > 0$ such that

$$\rho_f\left(\frac{x(i)}{\alpha_1}\right) \leq 1 \text{ and } \rho_f\left(\frac{y(i)}{\alpha_2}\right) \leq 1$$

Take $\alpha = \max\{2\alpha_1, 2\alpha_2\}$. Considering Lemma 1 and using the convexity of ρ_f, we have

$$\begin{aligned}\rho_f\left(\frac{x(i)+y(i)}{\alpha}\right) &\leq \frac{1}{2}\rho_f\left(\frac{x(i)}{\alpha_1}\right) + \frac{1}{2}\rho_f\left(\frac{y(i)}{\alpha_2}\right)\\ &\leq \rho_f\left(\frac{x(i)}{\alpha_1}\right) + \rho_f\left(\frac{y(i)}{\alpha_2}\right)\end{aligned}$$

Therefore, $g(x + y) \leq g(x) + g(y)$ for every $x, y \in W(X, f)$. Finally, we show that scalar multiplication is continuous. Let (λ_n) be any scalar sequence and $(x_n(i)) \subset W(X, f)$, with $|\lambda_n - \lambda| \to 0$ and $g(x_n(i) - x(i)) \to 0$ as n $\to \infty$. Considering Lemma 1 and using the the convexity of ρ_f, we have

$$\begin{aligned}\rho_f\left(\frac{\lambda_n x_n(i) - \lambda x(i)}{\alpha}\right) &\leq \rho_f\left(\left|\frac{(\lambda_n-\lambda)x_n(i)}{\alpha}\right| + \left|\frac{\lambda(x_n(i)-x(i))}{\alpha}\right|\right)\\ &\leq \frac{1}{2}\rho_f\left(2|\lambda_n-\lambda|\left|\frac{x_n(i)}{\alpha}\right|\right) + \frac{1}{2}\rho_f\left(2|\lambda|\left|\frac{x_n(i)-x(i)}{\alpha}\right|\right)\\ &\leq \rho_f\left(2|\lambda_n-\lambda|\left|\frac{x_n(i)}{\alpha}\right|\right) + \rho_f\left(2|\lambda|\left|\frac{x_n(i)-x(i)}{\alpha}\right|\right)\end{aligned}$$

Therefore,

$$\begin{aligned}
g(\lambda_n x_n(i) - \lambda x(i)) &= \inf\left\{\alpha > 0 : \rho_f\left(\frac{\lambda_n x_n(i) - \lambda x(i)}{\alpha}\right) \le 1\right\} \\
&\le 2|\lambda_n - \lambda| \inf\left\{\alpha^* = \left(\frac{\alpha}{2|\lambda_n - \lambda|}\right) > 0 : \rho_f\left(\frac{x_n(i)}{\alpha^*}\right) \le 1\right\} \\
&\quad + 2|\lambda| \inf\left\{\alpha^{**} = \left(\frac{\alpha}{2|\lambda|}\right) > 0 : \rho_f\left(\frac{x_n(i) - x(i)}{\alpha^{**}}\right) \le 1\right\} \\
&= 2|\lambda_n - \lambda|\, g(x_n(i)) + 2|\lambda|\, g(x_n(i) - x(i)) \to 0
\end{aligned}$$

Hence, $g(\lambda_n x_n(i) - \lambda x(i)) \to 0$. This completes the proof of the theorem. □

Theorem 2.3. *The linear space $W(X, f)$ is a complete paranormed sequence space.*

Proof. Let (x_n) be any *Cauchy* sequence in $W(X, f)$ where $(x_n) = (x_n(i)) = (x_n(1), x_n(2), ...)$. This implies for any $\epsilon > 0$, there exists $n_0 \in \mathbb{N}$ such that for every $m \ge n \ge n_0$, we get $g(x_m - x_n) < \epsilon$. Consequently, $\rho_f\left(\frac{x_m(i) - x_n(i)}{\epsilon}\right) \le 1$. Since ρ_f is convex, we have $\rho_f(x_m(i) - x_n(i)) \le \epsilon$.

Using the continuity of f, it follows that $\|x_m(i) - x_n(i)\|_X < \epsilon$ for every $\epsilon > 0$. Hence, for every fixed i, the sequence $(x_n(i))$ is a *Cauchy* sequence in X. It converges since X is complete. Say, $x_n(i) \to x(i)$ as $n \to \infty$. Using these limits, we define $x = (x(i))$ and show that $x \in W(X, f)$ and $g(x_n - x) \to 0$. Since $X = (X, \|\cdot\|_X)$ is a Banach space, we get

$$\|x_m(i) - x(i)\|_X = \|x_m(i) - \lim_{n\to\infty} x_n(i)\|_X = \lim_{n\to\infty} \|x_m(i) - x_n(i)\|_X < \epsilon^2$$

Since $(x_n(i)) \in W(X, f)$, there exists $\alpha > 0$ and $\ell \in X$ such that

$$\rho_f\left(\frac{x_n(i) - \ell}{\alpha}\right) \to 0 \text{ as } i \to \infty$$

Using the continuity of f, we obtain

$$\rho_f\left(\frac{x(i) - \ell}{\alpha}\right) = \rho_f\left(\frac{\lim_{n\to\infty} x_n(i) - \ell}{\alpha}\right) = \lim_{n\to\infty} \rho_f\left(\frac{x_n(i) - \ell}{\alpha}\right) \to 0 \text{ as } i \to \infty$$

It follows that $x \in W(X, f)$. We will show that $g(x_n - x) \to 0$. Since f is continuous, then

$$\rho_f\left(\frac{x_n(i) - x(i)}{\alpha}\right) = \rho_f\left(\frac{x_n(i) - \lim_{m\to\infty} x_m(i)}{\alpha}\right) \le 1$$

Therefore, $g(x_n - x) = \inf\left\{\alpha > 0 : \rho_f\left(\frac{x_n(i) - x(i)}{\alpha}\right) \le 1\right\}$. Hence, there exists sequence $\left(\frac{c}{2^n}\right), n \ge 1$, for a real number c with $g(x_n - x) < \frac{c}{2^n}$, for every $n \ge 1$. Therefore, we get $g(x_n - x) \to 0$. We can conclude that $W(X, f)$ is a complete paranormed space. □

Theorem 2.4. *The linear space $W(X, f)$ is an AK space.*

Proof. Let $x = (x(i)) \in W(X, f)$, then there exists $\alpha > 0$ and $\ell \in X$ such that

$$\frac{1}{m}\sum_{i=1}^{m} f\left(\left\|\frac{x(i) - \ell}{\alpha}\right\|_X\right) \to 0 \text{ as i} \to \infty$$

It follows that for every $i = 1, ..., m$, we have $\|x(i) - \ell\|_X \to 0$, as $i \to \infty$. Consequently,

$$\frac{1}{m}\sum_{i=n}^{m} f\left(\left\|\frac{x(i+1)-\ell}{\alpha}\right\|_X\right) \to 0 \text{ as i} \to \infty$$

Hence, $\rho_f\left(\frac{x-x^{[n]}}{\alpha}\right) \to 0$ as $n \to \infty$, where $x^{[n]}$ denotes the $n-th$ section of x. Therefore, for $\epsilon = 1$, there exists $n_0 \in \mathbb{N}$ such that for every $n \geq n_0$, we obtain $\rho_f\left(\frac{x-x^{[n]}}{\alpha}\right) \leq 1$. It follows that $g(x - x^{[n]}) \to 0$ as $n \to \infty$. This completes the proof. □

Theorem 2.5. *Let f and g be two $\varphi-functions$, then*
(i) $W(X, f) \subseteq W(X, g \circ f)$
(ii) $W(X, f) \cap W(X, g) \subseteq W(X, f + g)$

Proof. (i) Let $x = (x(i)) \in W(X, f)$, then there exists $\alpha > 0$ and $\ell \in X$ such that

$$\frac{1}{m}\sum_{i=1}^{m} f\left(\left\|\frac{x(i)-\ell}{\alpha}\right\|_X\right) \to 0 \text{ as i} \to \infty$$

Hence, for every $\epsilon > 0$, there exists $i_0 \in \mathbb{N}$ such that for every $i \geq i_0$, we have

$$\frac{1}{m}\sum_{i=1}^{m} f\left(\left\|\frac{x(i)-\ell}{\alpha}\right\|_X\right) < \epsilon$$

It follows that for every $i = 1, ..., m$, we have $f\left(\left\|\frac{x(i)-\ell}{\alpha}\right\|_X\right) \to 0$ as i $\to \infty$. Since g is a $\varphi-function$, we have $g\left(f\left(\left\|\frac{x(i)-\ell}{\alpha}\right\|_X\right)\right) \to 0$ as i $\to \infty$. Hence, $\rho_{g\circ f}\left(\frac{x(i)-\ell}{\alpha}\right) \to 0$ as i $\to \infty$. This implies $x \in W(X, g \circ f)$. This concludes the proof.
(ii) The result of this point is obvious. □

References

[1] Parashar S and Choudhary B 1994 *Indian J. Pure and App. Math.* **25** 419-28.
[2] Maddox I J 1988 *Elements of Functional Analysis* (Cambridge: Cambridge University Press).
[3] Lindenstrauss J and Tzafriri L 1971 *Israel J. Math.* **10** 379-90.
[4] Ghosh D and Srivastava P D 1999 *Glasnik Matematicki* **34** 253-261.
[5] Malkowsky E and Savas E 2000 *Arch. Math.* **36** 219-228.
[6] Kolk E 1994 *Tartu UI Toimetised* **970** 65-72.
[7] Rath A and Srivastava P D 1996 *Ganita* **47** 1-12.
[8] Das N R and Choudhary 1992 *Bul. Calcutta Math. Soc.* **84** 47-54.
[9] Leonard I E 1976 *J. Math. Anal. App.* **54** 245-65.
[10] Srivastava J K and Srivastava B K 1996 *India J. Pure App. Math.* **27** 73-84.
[11] Nakano 1951 *Proc. Japan Acad.* vol 27 p 508-12.
[12] Simons S 1965 *Proc. London Math. Soc.* p 422-436.
[13] Rath D and Tripathy B C 1994 *Indian J. Pure App. Math.* **25** 381-86.
[14] Tripathy B C and Sen M 2001 *Indian J. Pure App. Math.* **32** 1689-94.
[15] Mahmut Isik 2011 *Tamsui Oxford J. of Information and Mathematical Sciences* **27** 137-48.
[16] Özdemir M K and Solak I 2006 *Thai J. of Math.* **27** 93-105.

IOP Conf. Series: Materials Science and Engineering **300** (2018) 012072 doi:10.1088/1757-899X/300/1/012072

Notes on matrix of Fibonacci numbers

Rendy Jehanshah*[1], Yan Putra[2], Debora E Sirait[3], Mardiningsih[4] and Zuhri[5]

[1,2,3,4]Department of Mathematics, University of Sumatera Utara, Medan 20155, Indonesia
[5] STIM SUKMA Medan, KOPERTIS AREA 1, Medan-Indonesia

E-mail: *rendyjs@ymail.com

Abstract. For integer $m \geq 3$, we discuss m by m matrices of Fibonacci numbers that resemble the simplicity of Binet formula and maintain the calculation using only integer computation. We show that the m by m matrices of Fibonacci numbers will eventually store $m+1$ distinct consecutive Fibonacci numbers.

1. Introduction

One of the most well known sequence is the Fibonacci sequence (Fibonacci numbers). The Fibonacci sequence $\{f_n\}_{n\geq 0}$ is the sequence of numbers of the form

$$\begin{array}{ccccccccc} f_0 & f_1 & f_2 & f_3 & f_4 & f_5 & f_6 & f_7 & \cdots \\ 0 & 1 & 2 & 3 & 5 & 8 & 13 & 21 & \cdots \end{array}$$

Interesting problem regarding Fibonacci numbers is to find the nth term of the sequences. Around 1875 Binet proposed a formula that can calculate the nth term Fibonacci numbers known as Binet formula [1, 4]. For $n \geq 2$, Binet formula can be formulated as follows

$$f_n = \frac{1}{\sqrt{5}}(\lambda_1^n - \lambda_2^n) = \frac{1}{\sqrt{5}}\left[\left(\frac{1+\sqrt{5}}{2}\right)^n - \left(\frac{1-\sqrt{5}}{2}\right)^n\right] \tag{1}$$

with $\lambda_1 = \frac{1+\sqrt{5}}{2}$, $\lambda_2 = \frac{1-\sqrt{5}}{2}$. One difficulty in determining the nth term f_n of Fibonacci numbers is the fact that the formula (1) dealing with calculation of irrational number.

One way to find the Fibonacci numbers, without dealing with irrational number, is using a recurrence relation that based on the property that for $n \geq 2$ each term the Fibonacci numbers is the sum of the previous two terms [4]. So the Fibonacci numbers can be found using recurrence relation

$$f_n = f_{n-1} + f_{n-2}, \quad \text{for} \quad n \geq 2. \tag{2}$$

We note that the recurrence relation (2) will store three consecutive terms of the Fibonacci numbers.

Based on the recurrence relation (2) an attempt has been made to store three terms of Fibonacci numbers using a 2 by 2 matrix [2–4]. In this paper, for $m \geq 3$ we establish an m by m matrix that will eventually store $m+1$ terms of the Fibonacci numbers. We organize the paper as follows. In Section 2, we review the 2 by 2 matrix for Fibonacci numbers. In Section 3, based on the 2 by 2 matrix of Fibonacci numbers, for $m \geq 3$ we discuss an m by m matrix of Fibonacci numbers.

2. Case of 2 by 2 matrix

We review the 2 by 2 matrix of Fibonacci numbers as in [2–4]. We start with the special 2 by 2 matrix $F_2 = \begin{pmatrix} 1 & 1 \\ 1 & 0 \end{pmatrix}$. Let $U_n = \begin{pmatrix} u_{n,1} \\ u_{n,2} \end{pmatrix}$ and define a recurrence relation

$$U_{n+1} = \begin{pmatrix} u_{n+1,1} \\ u_{n+1,2} \end{pmatrix} := F_2 U_n = \begin{pmatrix} u_{n,1} + u_{n,2} \\ u_{n,1} \end{pmatrix}.$$

Notice that $u_{n+1,1} = u_{n,1} + u_{n,2}$ resembles a Fibonacci like recurrences as in (2). Hence if we choose the vector U_n to be $U_n = \begin{pmatrix} f_n \\ f_{n-1} \end{pmatrix}$ for some two consecutive terms of Fibonacci numbers, then the vector U_{n+1} will store the two consecutive terms f_{n+1} and f_n of the Fibonacci numbers.

Fortunately, the entries of each column of

$$F_2 = \begin{pmatrix} 1 & 1 \\ 1 & 0 \end{pmatrix} = \begin{pmatrix} f_2 & f_1 \\ f_1 & f_0 \end{pmatrix} \tag{3}$$

are two consecutive Fibonacci numbers. This implies the power of F_2 for some positive integer $n \geq 1$ will store three consecutive Fibonacci numbers as stated in the following theorem.

Theorem 1. [4] *Let $\{f_n\}_{n\geq 0}$ be the squence of Fibonacci numbers. If $F_2 = \begin{pmatrix} 1 & 1 \\ 1 & 0 \end{pmatrix}$, then $(F_2)^n = \begin{pmatrix} f_{n+1} & f_n \\ f_n & f_{n-1} \end{pmatrix}$ for $n \geq 1$.*

Proof. We prove by mathematical induction on n. If $n = 1$, we have $(F_2)^1 = \begin{pmatrix} f_{1+1} & f_1 \\ f_1 & f_{1-1} \end{pmatrix} = \begin{pmatrix} 1 & 1 \\ 1 & 0 \end{pmatrix}$ as in (3).

Assume that for some positive integer $k \geq 1$, $(F_2)^k = \begin{pmatrix} f_{k+1} & f_k \\ f_k & f_{k-1} \end{pmatrix}$. This implies

$$\begin{aligned} (F_2)^{k+1} &= (F_2)(F_2)^k = \begin{pmatrix} 1 & 1 \\ 1 & 0 \end{pmatrix} \begin{pmatrix} f_{k+1} & f_k \\ f_k & f_{k-1} \end{pmatrix} \\ &= \begin{pmatrix} f_{k+1} + f_k & f_k + f_{k-1} \\ f_{k+1} & f_k \end{pmatrix} = \begin{pmatrix} f_{k+2} & f_{k+1} \\ f_{k+1} & f_k \end{pmatrix} \\ &= \begin{pmatrix} f_{(k+1)+1} & f_{(k+1)} \\ f_{(k+1)} & f_{(k+1)-1} \end{pmatrix}. \end{aligned}$$

Therefore, we conclude that $(F_2)^n = \begin{pmatrix} f_{n+1} & f_n \\ f_n & f_{n-1} \end{pmatrix}$ for $n \geq 1$. □

We note that the statement of Theorem 1 constitutes the beauty of Binet formula (1) of finding Fibonacci numbers using a simple formula and the beauty of the recurrence relation (2) that finds the Fibonacci numbers using only integer computation.

As a direct consequence of Theorem 1 we have the following result.

Corollary 2. *Let $\{f_n\}_{n\geq 0}$ be the sequence of Fibonacci numbers. If $F = \begin{pmatrix} f_{k+2} & f_{k+1} \\ f_{k+1} & f_k \end{pmatrix}$ for some positive integer k, then $F^n = \begin{pmatrix} f_{(k+1)n+1} & f_{(k+1)n} \\ f_{(k+1)n} & f_{(k+1)n-1} \end{pmatrix}$.*

Proof. Notice that $F = \begin{pmatrix} f_{k+2} & f_{k+1} \\ f_{k+1} & f_k \end{pmatrix} = \begin{pmatrix} 1 & 1 \\ 1 & 0 \end{pmatrix}^{k+1}$. Hence,

$$F^n = \left[\begin{pmatrix} 1 & 1 \\ 1 & 0 \end{pmatrix}^{k+1}\right]^n = \begin{pmatrix} 1 & 1 \\ 1 & 0 \end{pmatrix}^{(k+1)n} = \begin{pmatrix} f_{(k+1)n+1} & f_{(k+1)n} \\ f_{(k+1)n} & f_{(k+1)n-1} \end{pmatrix}.$$ □

3. Our results

We will generalize the result in Theorem 1 for m by m matrix for some $m \geq 3$. We start with $m = 3$. Let $F_3 = \begin{pmatrix} 1 & 1 & 0 \\ 1 & 0 & 0 \\ 0 & 1 & 0 \end{pmatrix} = \begin{pmatrix} f_2 & f_1 & 0 \\ f_1 & f_0 & 0 \\ f_0 & 1 & 0 \end{pmatrix}$. Let U_{n+1} and U_n be 3 by 1 matrices and consider the recurrence

$$\begin{pmatrix} u_{n+1,1} \\ u_{n+1,2} \\ u_{n+1,3} \end{pmatrix} := \begin{pmatrix} 1 & 1 & 0 \\ 1 & 0 & 0 \\ 0 & 1 & 0 \end{pmatrix} \begin{pmatrix} u_{n,1} \\ u_{n,2} \\ u_{n,3} \end{pmatrix}.$$

Then $u_{n+1,1} = u_{n,1} + u_{n,2}$ and $u_{n+1,j} = u_{n,j-1}$ for $j = 2, 3$. As in the case of 2 by 2 matrix, if we replace U_n with vector with entries consist of three consecutive Fibonacci numbers then FU_n will result in three consecutive Fibonacci numbers.

Proposition 3. *Let $\{f_n\}_{n\geq 0}$ be the sequence of Fibonacci numbers. If* $F_3 = \begin{pmatrix} 1 & 1 & 0 \\ 1 & 0 & 0 \\ 0 & 1 & 0 \end{pmatrix}$, *then*

$$(F_3)^n = \begin{pmatrix} f_{n+1} & f_n & 0 \\ f_n & f_{n-1} & 0 \\ f_{n-1} & f_{n-2} & 0 \end{pmatrix} \text{ for } n \geq 2.$$

Proof. We proof by induction on n. If $n = 2$, we have

$$\begin{aligned} (F_3)^2 &= \begin{pmatrix} 1 & 1 & 0 \\ 1 & 0 & 0 \\ 0 & 1 & 0 \end{pmatrix}^2 = \begin{pmatrix} 2 & 1 & 0 \\ 1 & 1 & 0 \\ 1 & 0 & 0 \end{pmatrix} = \begin{pmatrix} f_3 & f_2 & 0 \\ f_2 & f_1 & 0 \\ f_1 & f_0 & 0 \end{pmatrix} \\ &= \begin{pmatrix} f_{2+1} & f_2 & 0 \\ f_2 & f_{2-1} & 0 \\ f_{2-1} & f_{2-2} & 0 \end{pmatrix}. \end{aligned}$$

Assume that $(F_3)^k = \begin{pmatrix} f_{k+1} & f_k & 0 \\ f_k & f_{k-1} & 0 \\ f_{k-1} & f_{k-2} & 0 \end{pmatrix}$ for some $k \geq 2$. Hence we have

$$\begin{aligned} (F_3)^{k+1} &= F_3(F_3)^k = \begin{pmatrix} 1 & 1 & 0 \\ 1 & 0 & 0 \\ 0 & 1 & 0 \end{pmatrix} \begin{pmatrix} f_{k+1} & f_k & 0 \\ f_k & f_{k-1} & 0 \\ f_{k-1} & f_{k-2} & 0 \end{pmatrix} \\ &= \begin{pmatrix} f_{k+1} + f_k & f_k + f_{k-1} & 0 \\ f_{k+1} & f_k & 0 \\ f_k & f_{k-1} & 0 \end{pmatrix} = \begin{pmatrix} f_{k+2} & f_{k+1} & 0 \\ f_{k+1} & f_k & 0 \\ f_k & f_{k-1} & 0 \end{pmatrix} \\ &= \begin{pmatrix} f_{(k+1)+1} & f_{k+1} & 0 \\ f_{k+1} & f_{(k+1)-1} & 0 \\ f_{(k+1)-1} & f_{(k+1)-2} & 0 \end{pmatrix}. \end{aligned}$$

Therefore,

$$(F_3)^n = \begin{pmatrix} f_{n+1} & f_n & 0 \\ f_n & f_{n-1} & 0 \\ f_{n-1} & f_{n-2} & 0 \end{pmatrix}$$

for integer $n \geq 2$. □

Let

$$F_4 = \begin{pmatrix} 1 & 1 & 0 & 0 \\ 1 & 0 & 0 & 0 \\ 0 & 1 & 0 & 0 \\ 0 & 0 & 1 & 0 \end{pmatrix} = \begin{pmatrix} f_2 & f_1 & 0 & 0 \\ f_1 & f_0 & 0 & 0 \\ f_0 & 1 & 0 & 0 \\ 0 & 0 & 1 & 0 \end{pmatrix}. \quad (4)$$

Using the same argument as in the case of 3 by 3 , we have the following result.

Proposition 4. *Let $\{f_n\}_{n\geq 0}$ be the sequence of Fibonacci numbers. If F_4 is defined as in (4), then* $(F_4)^n = \begin{pmatrix} f_{n+1} & f_n & 0 & 0 \\ f_n & f_{n-1} & 0 & 0 \\ f_{n-1} & f_{n-2} & 0 & 0 \\ f_{n-2} & f_{n-3} & 0 & 0 \end{pmatrix}$ *for $n \geq 3$.*

We now consider, for some $m \geq 5$, the m by m matrix

$$F_m = \begin{pmatrix} 1 & 1 & 0 & 0 & \cdots & 0 & 0 \\ 1 & 0 & 0 & 0 & \cdots & 0 & 0 \\ 0 & 1 & 0 & 0 & \cdots & 0 & 0 \\ 0 & 0 & 1 & 0 & \cdots & 0 & 0 \\ 0 & 0 & 0 & 1 & \cdots & 0 & 0 \\ \vdots & \vdots & \vdots & \vdots & \ddots & \vdots & \vdots \\ 0 & 0 & 0 & 0 & \cdots & 1 & 0 \end{pmatrix} = \begin{pmatrix} f_2 & f_1 & 0 & 0 & \cdots & 0 & 0 \\ f_1 & f_0 & 0 & 0 & \cdots & 0 & 0 \\ f_0 & 1 & 0 & 0 & \cdots & 0 & 0 \\ 0 & 0 & 1 & 0 & \cdots & 0 & 0 \\ 0 & 0 & 0 & 1 & \cdots & 0 & 0 \\ \vdots & \vdots & \vdots & \vdots & \ddots & \vdots & \vdots \\ 0 & 0 & 0 & 0 & \cdots & 1 & 0 \end{pmatrix}. \quad (5)$$

Then the recurrence relation $U_{n+1} := F_m U_n$ will result in $u_{n+1,1} = u_{n,1} + u_{n,2}$ and for $j = 2, 3, \ldots, m$ we have $u_{n+1,j} = u_{n,j-1}$. Let $F_m(:, k)$ be the kth column of F_m, then

$$(F_m)^{m-2}F_m(:,1) = \begin{pmatrix} f_m \\ f_{m-1} \\ f_{m_2} \\ \vdots \\ f_1 \end{pmatrix}, (F_m)^{m-2}F_m(:,2) = \begin{pmatrix} f_{m-1} \\ f_{m-2} \\ f_{m_3} \\ \vdots \\ f_0 \end{pmatrix} \text{ and } (F_m)^{m-2}F(:,k) = \begin{pmatrix} 0 \\ 0 \\ 0 \\ \vdots \\ 0 \end{pmatrix}$$

for $k = 3, \ldots, m$. This implies

$$(F_m)^{m-1} = \begin{pmatrix} f_m & f_{m-1} & 0 & 0 & \cdots & 0 & 0 \\ f_{m-1} & f_{m-2} & 0 & 0 & \cdots & 0 & 0 \\ f_{m-2} & f_{m-3} & 0 & 0 & \cdots & 0 & 0 \\ f_{m-3} & f_{m-4} & 0 & 0 & \cdots & 0 & 0 \\ f_{m-4} & f_{m-5} & 0 & 0 & \cdots & 0 & 0 \\ \vdots & \vdots & \vdots & \vdots & \ddots & \vdots & \vdots \\ f_1 & f_0 & 0 & 0 & \cdots & 0 & 0 \end{pmatrix}. \quad (6)$$

We now have the following result.

Theorem 5. *Let $\{f_n\}_{n\geq 0}$ be the sequence of Fibonacci numbers. If F_m, for some $m \geq 3$, is the m by m matrices defined in* (5)*, then*

$$(F_m)^n = \begin{pmatrix} f_{n+1} & f_n & 0 & 0 & \cdots & 0 & 0 \\ f_n & f_{n-1} & 0 & 0 & \cdots & 0 & 0 \\ f_{n-1} & f_{n-2} & 0 & 0 & \cdots & 0 & 0 \\ f_{n-2} & f_{n-3} & 0 & 0 & \cdots & 0 & 0 \\ f_{n-3} & f_{m-4} & 0 & 0 & \cdots & 0 & 0 \\ \vdots & \vdots & \vdots & \vdots & \ddots & \vdots & \vdots \\ f_{n-(m-2)} & f_{n-(m-1)} & 0 & 0 & \cdots & 0 & 0 \end{pmatrix}$$

for $n \geq m-1$.

Proof. We prove by induction on n. Equation (6) guarantees that theorem is true for $n = m-1$.

Assume now that $(F_m)^k = \begin{pmatrix} f_{k+1} & f_k & 0 & 0 & \cdots & 0 & 0 \\ f_k & f_{k-1} & 0 & 0 & \cdots & 0 & 0 \\ f_{k-1} & f_{k-2} & 0 & 0 & \cdots & 0 & 0 \\ f_{k-2} & f_{k-3} & 0 & 0 & \cdots & 0 & 0 \\ f_{k-3} & f_{k-4} & 0 & 0 & \cdots & 0 & 0 \\ \vdots & \vdots & \vdots & \vdots & \ddots & \vdots & \vdots \\ f_{k-(m-2)} & f_{k-(m-1)} & 0 & 0 & \cdots & 0 & 0 \end{pmatrix}$ for $k \geq m-1$. Then

$$(F_m)^{k+1} = F_m(F_m)^k = \begin{pmatrix} f_{k+1}+f_k & f_k+f_{k-1} & 0 & 0 & \cdots & 0 & 0 \\ f_{k+1} & f_k & 0 & 0 & \cdots & 0 & 0 \\ f_k & f_{k-1} & 0 & 0 & \cdots & 0 & 0 \\ f_{k-1} & f_{k-2} & 0 & 0 & \cdots & 0 & 0 \\ f_{k-2} & f_{k-3} & 0 & 0 & \cdots & 0 & 0 \\ \vdots & \vdots & \vdots & \vdots & \ddots & \vdots & \vdots \\ f_{k-(m-3)} & f_{k-(m-2)} & 0 & 0 & \cdots & 0 & 0 \end{pmatrix}$$

and hence

$$(F_m)^{k+1} = \begin{pmatrix} f_{(k+1)+1} & f_{k+1} & 0 & 0 & \cdots & 0 & 0 \\ f_{(k+1)} & f_{(k+1)-1} & 0 & 0 & \cdots & 0 & 0 \\ f_{(k+1)-1} & f_{(k+1)-2} & 0 & 0 & \cdots & 0 & 0 \\ f_{(k+1)-2} & f_{(k+1)-3} & 0 & 0 & \cdots & 0 & 0 \\ f_{(k+1)-3} & f_{(k+1)-4} & 0 & 0 & \cdots & 0 & 0 \\ \vdots & \vdots & \vdots & \vdots & \ddots & \vdots & \vdots \\ f_{(k+1)-(m-2)} & f_{(k+1)-(m-1)} & 0 & 0 & \cdots & 0 & 0 \end{pmatrix}.$$

Therefore,

$$(F_m)^n = \begin{pmatrix} f_{n+1} & f_n & 0 & 0 & \cdots & 0 & 0 \\ f_n & f_{n-1} & 0 & 0 & \cdots & 0 & 0 \\ f_{n-1} & f_{n-2} & 0 & 0 & \cdots & 0 & 0 \\ f_{n-2} & f_{n-3} & 0 & 0 & \cdots & 0 & 0 \\ f_{n-3} & f_{m-4} & 0 & 0 & \cdots & 0 & 0 \\ \vdots & \vdots & \vdots & \vdots & \ddots & \vdots & \vdots \\ f_{n-(m-2)} & f_{n-(m-1)} & 0 & 0 & \cdots & 0 & 0 \end{pmatrix}$$

for $n \geq m-1$. □

We note that first two columns the matrix $(F_m)^n$ for some $n \geq m-1$ store $m+1$ distinct consecutive terms of the Fibonacci numbers.

References

[1] Dunlap R A 1997 *The Golden Ratio and Fibonacci Numbers*(World Scientific Publishing : Singapore).
[2] Hawkins K, Johnson U and Mathes B 2015 *Linear Algebra and its Applications*, **475** 80–89.
[3] Luis R and Oliveira H M 2014 *Applied Mathematics and Computation*, **226**, 101–116.
[4] Meinke A M 2011 *Fibonacci Numbers And Associated Matrices* (Thesis Kent State University).

On some new normed sequence spaces

[1]G Pranajaya and [2]E Herawati

[1,2]Department of Mathematics, University of Sumatera Utara, Indonesia

E-mail: [1]gunturpranajaya13@gmail.com, [2]herawaty.elv@gmail.com

Abstract. The sequence spaces $(c_0)_\Lambda$, c_Λ, and $(\ell_\infty)_\Lambda$ was introduced and studied by Mursaleen and Noman [11]. In the present paper, for M is a generalization of Orlicz function, we extend the spaces Mursaleen and Noman's to $[c_0(M)]_\Lambda$, $[c(M)]_\Lambda$, and $[\ell_\infty(M)]_\Lambda$, respectively, and investigate some topological properties of these spaces. Finally, we determine the necessary and sufficient conditions of an infinite matrix A belonging to classes $\big(c_0(M), c_0(M)\big)$, $\big(c(M), c(M)\big)$, and $\big(\ell_\infty(M), \ell_\infty(M)\big)$.

1. Introduction and Preliminaries

By ω, we denote the space of all sequences of real or complex numbers. Any linear subspace of ω is called a *sequence space*. We shall write c_0, c, and ℓ_∞ for the spaces of all convergent to zero, convergent, and bounded sequences, respectively.

A sequence space X is called a *BK space* provided X is a complete normed space and a function $p_k : X \to \mathbb{R}$ defined by $x \mapsto p_k(x) = x_k$ is continuous for all $k \in \mathbb{N}$ (see [5]).

The sequence spaces c_0, c, and ℓ_∞ are BK spaces equipped with sup-norm $\|\cdot\|_\infty$ given by

$$\|x\| = \sup_{k\in\mathbb{N}} |x_k|.$$

Let $A = (a_{nk})$ be an infinite matrix of real or complex numbers a_{nk}, where $n, k \in \mathbb{N}$, and X, Y be the sequence spaces. The map A from X into Y is said to be *matrix transformation* if $Ax = (A_n(x))$ exists in Y where

$$A_n(x) = \sum_{k=0}^{\infty} a_{nk}x_k \text{ converges for all } n \in \mathbb{N} \text{ and all } x \in X. \tag{1}$$

We denote (X, Y) as the class of all infinite matrices that map X into Y. Thus, $A \in (X, Y)$ if and only if (1) hold, and $Ax \in Y$ for all $x \in X$. The theory of matrix transformation deals with establishing necessary and sufficient conditions on the entries of a matrix to map a sequence space X into a sequence space Y.

For a sequence space X, the *matrix domain* of an infinite matrix A in X is a sequence space defined by

$$X_A = \Big\{x = (x_k) \in \omega : Ax \in X\Big\}.$$

The idea of constructing a new sequence space by means of the matrix domain of a particular limitation method has recently been studied by several authors, e.g., Altay and Başar [1],

Mursaleen et al. [2], Mursaleen and Noman [11, 12], Malkowsky [9], Malkowsky and Savaş [10]. Mursaleen and Noman [11] introduced Λ-matrix and constructed the matrix domains on Λ-matrix in the classical sequence spaces c_0, c, and ℓ_∞. They examined some topological properties, established inclusion relations concerning with those spaces, determined their α-, β-, γ-duals, and characterized some relate matrix classes.

On the other side, Lindenstrauss and Tzafriri [7] introduced the sequence space defined by Orlicz function as follows :

$$\ell_M = \left\{ x = (x_k) \in \omega : (\exists \rho > 0) \sum_{k=1}^{\infty} M\left(\frac{|x_k|}{\rho}\right) < \infty \right\}$$

which is called *Orlicz sequence space.* The space ℓ_M equipped with norm

$$\|x\| = \inf\left\{ \rho > 0 : \sum_{k=1}^{\infty} M\left(\frac{|x_k|}{\rho}\right) \leq 1 \right\}$$

becomes a Banach space.

Using the matrix domain X_Λ defined by Mursaleen and Noman [11], in this work, we introduce Λ-matrix domain for the sequences generated by a generalization of Orlicz function M, denoted by $[X(M)]_\Lambda$ where $X \in \{c_0, c, \ell_\infty\}$. Furthermore, we investigate some topological properties of these spaces over the norm spaces, and give the necessary and sufficient conditions on an infinite matrix A belonging to classes $\big(c_0(M), c_0(M)\big)$, $\big(c(M), c(M)\big)$, and $\big(\ell_\infty(M), \ell_\infty(M)\big)$.

2. Results

2.1. The Sequence Space $[X(M)]_\Lambda$

A function $M : \mathbb{R}^+ \to \mathbb{R}^+$ is called a *generalization of Orlicz function* which is vanishing at zero, non decreasing, and continuous. A generalization of Orlicz function M is said to satisfy Δ_2*-condition* for all values of x if there exists a constant $K > 0$ such that $M(2x) \leq KM(x)$ for $x \geq 0$. Furthermore, in [11], Mursaleen and Noman defined the infinite matrix $\Lambda = (\lambda_{nk})$ by

$$\lambda_{nk} = \begin{cases} \dfrac{\lambda_k - \lambda_{k-1}}{\lambda_n} & ; 0 \leq k \leq n \\ 0 & ; k > n \end{cases} \tag{2}$$

where $\lambda = (\lambda_k)$ be a strictly increasing sequence of positive reals tending to infinity, that is, $0 < \lambda_0 < \lambda_1 < \cdots$ and $\lambda_k \to \infty$ as $k \to \infty$. By using (2), in the present section we define the sequence space $[X(M)]_\Lambda$ where $X \in \{c_0, c, \ell_\infty\}$ and M is a generalization of Orlicz function, and prove that these sequence spaces according to its norm are complete normed spaces. These sequence spaces are as follows:

$$[c_0(M)]_\Lambda = \left\{ x = (x_k) \in \omega : (\exists\ \rho > 0)\ M\left(\frac{|\Lambda_n(x)|}{\rho}\right) \to 0, n \to \infty \right\},$$

$$[c(M)]_\Lambda = \left\{ x = (x_k) \in \omega : (\exists\ \rho > 0,\ l \in \mathbb{R})\ M\left(\frac{|\Lambda_n(x)|}{\rho}\right) \to l, n \to \infty \right\}, \text{ and}$$

$$[\ell_\infty(M)]_\Lambda = \left\{ x = (x_k) \in \omega : (\exists\ \rho > 0)\ \sup_{n\in\mathbb{N}} M\left(\frac{|\Lambda_n(x)|}{\rho}\right) < \infty \right\}.$$

Now, we may begin with the following results which is essential in the text.

2.2. Linear Topological Structure of $[X(M)]_\Lambda$

In this section, we examine some topological properties of the sequence spaces defined above.

Theorem 2.1. *If M is a convex function, then the sequence space $[X(M)]_\Lambda$ for $X \in \{c_0, c, \ell_\infty\}$ is linear space over the set of real numbers $\mathbb{R}$.*

Proof. We prove the theorem for $X = \ell_\infty$. Let $x, y \in [\ell_\infty(M)]_\Lambda$ and $\alpha, \beta \in \mathbb{R}$, then there exist some positive ρ_1 and ρ_2 such that

$$\sup_{n\in\mathbb{N}} M\left(\frac{|\Lambda_n(x)|}{\rho_1}\right) < \infty \text{ and } \sup_{n\in\mathbb{N}} M\left(\frac{|\Lambda_n(y)|}{\rho_2}\right) < \infty.$$

Take $\rho = \max(2|\alpha|\rho_1, 2|\beta|\rho_2)$, then for a convex function M we get

$$\begin{aligned}\sup_{n\in\mathbb{N}} M\left(\frac{|\Lambda_n(\alpha x + \beta y)|}{\rho}\right) &\le \sup_{n\in\mathbb{N}} M\left(\frac{|\alpha\Lambda_n(x)|}{\rho} + \frac{|\beta\Lambda_n(y)|}{\rho}\right)\\ &\le \sup_{n\in\mathbb{N}} M\left(\frac{|\alpha||\Lambda_n(x)|}{2|\alpha|\rho_1} + \frac{|\beta||\Lambda_n(y)|}{2|\beta|\rho_2}\right)\\ &\le \frac{1}{2}\sup_{n\in\mathbb{N}} M\left(\frac{|\Lambda_n(x)|}{\rho_1}\right) + \frac{1}{2}\sup_{n\in\mathbb{N}} M\left(\frac{|\Lambda_n(y)|}{\rho_2}\right) < \infty.\end{aligned}$$

This proves that $[\ell_\infty(M)]_\Lambda$ is linear space. □

It is easy to show that $[c_0(M)]_\Lambda$ and $[c(M)]_\Lambda$ are also linear spaces.

Theorem 2.2. *If M satisfy Δ_2-condition, then the space $[X(M)]_\Lambda$ for $X \in \{c_0, c, \ell_\infty\}$ is complete normed space equipped with the norm defined by*

$$\|x\|_{[X(M)]_\Lambda} = \inf\left\{\rho > 0 : \sup_{n\in\mathbb{N}} M\left(\frac{|\Lambda_n(x)|}{\rho}\right) \le 1\right\} \tag{3}$$

Proof. We prove the theorem for $X = \ell_\infty$ and the other cases will follow similarly. Let $x, y \in [\ell_\infty(M)]_\Lambda$. It is easily seen that $\|x\|_{[\ell_\infty(M)]_\Lambda} \ge 0$. Next, if $x = 0$, then obviously $\|x\|_{[\ell_\infty(M)]_\Lambda} = 0$. Conversely, suppose $\|x\|_{[\ell_\infty(M)]_\Lambda} = 0$, then for every $\epsilon > 0$ we get $\|x\|_{[\ell_\infty(M)]_\Lambda} < \epsilon$. This implies there exists some ρ_0 with $0 < \rho_0 < \epsilon$ such that

$$\sup_{n\in\mathbb{N}} M\left(\frac{|\Lambda_n(x)|}{\epsilon}\right) < \sup_{n\in\mathbb{N}} M\left(\frac{|\Lambda_n(x)|}{\rho_0}\right) \le 1.$$

Since M is a generalization of Orlicz function, it follows that for every $\epsilon > 0$ and for every $n \in \mathbb{N}$,

$$|\Lambda_n(x)| = \left|\frac{1}{\lambda_n}\sum_{k=0}^{n}(\lambda_k - \lambda_{k-1})x_k\right| = 0.$$

Under the assumption that $\lambda = (\lambda_k)$ is a strictly increasing sequence of positive real numbers, it is easy to check by mathematical induction that $x_k = 0$ for every $k \in \mathbb{N}$. Thus, $x = 0$.

Furthermore, let $x \in [\ell_\infty(M)]_\Lambda$ and $\alpha \in \mathbb{R}$. If $\alpha = 0$, it is clear that the homogeneous property of the norm holds. Assume $\alpha \neq 0$, we get

$$\|\alpha x\|_{[\ell_\infty(M)]_\Lambda} = |\alpha| \inf\left\{\frac{\rho}{|\alpha|} > 0 : \sup_{n\in\mathbb{N}} M\left(\frac{|\Lambda_n(x)|}{\frac{\rho}{|\alpha|}}\right) \le 1\right\}.$$

This gives $\|\alpha x\|_{[\ell_\infty(M)]_\Lambda} = |\alpha|\|x\|_{[\ell_\infty(M)]_\Lambda}$.

Now, let $x, y \in [\ell_\infty(M)]_\Lambda$, then there exists some $\rho_1, \rho_2 > 0$ such that

$$\sup_{n\in\mathbb{N}} M\left(\frac{|\Lambda_n(x)|}{\rho_1}\right) \leq 1 \text{ and } \sup_{n\in\mathbb{N}} M\left(\frac{|\Lambda_n(y)|}{\rho_2}\right) \leq 1.$$

Hence, if we choose $\rho = \rho_1 + \rho_2$, then by the properties of M, we have $M\left(\frac{|\Lambda_n(x+y)|}{\rho}\right) \leq M\left(\frac{|\Lambda_n(x)|}{\rho_1}\right) + M\left(\frac{|\Lambda_n(y)|}{\rho_2}\right)$ for all $n \in \mathbb{N}$. Consequently, $\|x+y\|_{[\ell_\infty(M)]_\Lambda} \leq \|x\|_{[\ell_\infty(M)]_\Lambda} + \|y\|_{[\ell_\infty(M)]_\Lambda}$. Hence, $[\ell_\infty(M)]_\Lambda$ is a normed space.

Now, suppose that (x^i) be any Cauchy sequence in $[\ell_\infty(M)]_\Lambda$. Then, for each $\epsilon > 0$ there exists $i_0 \in \mathbb{N}$ and ρ_0 where $0 < \rho_0 < \epsilon$ such that

$$\sup_{n\in\mathbb{N}} M\left(\frac{|\Lambda_n(x^j - x^i)|}{\epsilon}\right) < \sup_{n\in\mathbb{N}} M\left(\frac{|\Lambda_n(x^j - x^i)|}{\rho_0}\right) \leq 1.$$

Hence, for every $\epsilon > 0$, $M\left(|\Lambda_n(x^j - x^i)|\right) < \epsilon$. Consequently, $|x_k^j - x_k^i| < \epsilon$ for each $\epsilon > 0$, every $j \geq i \geq i_0$, and every $k \in \mathbb{N}$, where $\lambda = (\lambda_k)$ is a strictly increasing sequence of positive real numbers. We see that (x_k^j) is Cauchy sequences of real numbers. Since $\mathbb{R}$ is complete, there exists $x_k \in \mathbb{R}$ such that $x_k^j \to x_k$ as $j \to \infty$ for all $k \in \mathbb{N}$. Using these limits, we define $x = (x_k)$ and show that $x \in [\ell_\infty(M)]_\Lambda$ and $x^i \to x$ as $i \to \infty$ in $[\ell_\infty(M)]_\Lambda$. From (3), we have for all $i \geq i_0$

$$\sup_{n\in\mathbb{N}} M\left(\frac{|\Lambda_n(x^i - x)|}{\rho_0}\right) = \lim_{j\to\infty} \sup_{n\in\mathbb{N}} M\left(\frac{|\Lambda_n(x^i - x^j)|}{\rho_0}\right) \leq 1.$$

We obtain $\|x^i - x\| < \epsilon$ for every $i \geq i_0$. This shows that

$$x^i \to x \text{ as } i \to \infty \text{ in } [\ell_\infty(M)]_\Lambda.$$

Since $x^i \in [\ell_\infty(M)]_\Lambda$ and M satisfy Δ_2-condition, we have

$$\sup_{n\in\mathbb{N}} M\left(\frac{|\Lambda_n(x)|}{\rho}\right) \leq \frac{K_1}{2} \sup_{n\in\mathbb{N}} M\left(\frac{|\Lambda_n(x - x^i)|}{\rho}\right) + \frac{K_2}{2} \sup_{n\in\mathbb{N}} M\left(\frac{|\Lambda_n(x^i)|}{\rho}\right) < \infty$$

for some $K_1, K_2 > 0$. It is show that $x \in [\ell_\infty(M)]_\Lambda$.

Since (x^i) was an arbitrary Cauchy sequence in $[\ell_\infty(M)]_\Lambda$, this proves completeness of $[\ell_\infty(M)]_\Lambda$. □

Now, the following result is immediate by Theorem 2.2.

Theorem 2.3. *$[X(M)]_\Lambda$ is a BK space where $X \in \{c_0, c, \ell_\infty\}$ and $[X(M)]_\Lambda$ is a AK space where $X = c_0$.*

2.3. The Certain Classes of Matrix Transformations $\big(X(M), X(M)\big)$

Theorem 2.4. *$A \in \big(c(M), c(M)\big)$ if and only if for $\rho > 0$ the following conditions are held :*

(i) $\sup_{n\in\mathbb{N}} M\left(\frac{\sum_{k=0}^{\infty} |a_{nk}|}{\rho}\right) < \infty$,

(ii) $\lim_{n\to\infty} M\left(\frac{|a_{nk}|}{\rho}\right) = \alpha_k$ *exists for each* $k \in \mathbb{N}$, *and*

(iii) $\lim_{n\to\infty} M\left(\frac{|\sum_{k=0}^{\infty} a_{nk}|}{\rho}\right) = \alpha$ *exist.*

Proof. For proving the necessity, suppose that $A \in \big(c(M), c(M)\big)$. If we choose $x = (x_k)$ by $x_k = sgn(a_{nk})$ for all $k \in \mathbb{N}$ and $n \in \mathbb{N}$, then $\sup_{n\in\mathbb{N}} M\left(\frac{\sum_{k=0}^{\infty}|a_{nk}|}{\rho}\right) = \sup_{n\in\mathbb{N}} M\left(\frac{|\sum_{k=0}^{\infty} a_{nk}x_k|}{\rho}\right) < \infty$. This shows that the condition *(i)* holds. Furthermore, if we take $x = (x_j)$ where for all $j \in \mathbb{N}$ and $k \in \mathbb{N}$,

$$x_j = e_j^{[k]} = \begin{cases} 1 & ; j = k \\ 0 & ; j \neq k \end{cases}$$

Since $Ax \in c(M)$ for every $x \in c(M)$, there exists $\rho > 0$ such that $\lim_{n\to\infty} M\left(\frac{|\sum_{j=0}^{\infty} a_{nj}x_j|}{\rho}\right)$ exist. Hence, for all $k \in \mathbb{N}$, $\lim_{n\to\infty} M\left(\frac{|\sum_{j=0}^{\infty} a_{nj}x_j|}{\rho}\right) = \lim_{n\to\infty} M\left(\frac{|a_{nk}|}{\rho}\right)$. It is shown that $\lim_{n\to\infty} M\left(\frac{|a_{nk}|}{\rho}\right)$ exists for each $k \in \mathbb{N}$, and the condition *(ii)* holds. Next, since $x = (x_k) = (1, 1, 1, \cdots)$ belongs to $c(M)$, the condition *(iii)* holds.

For sufficiency, let $x_k \to r$ as $k \to \infty$ and let the conditions *(i)*, *(ii)*, and *(iii)* hold. We write

$$\sum_{k=0}^{\infty} a_{nk}x_k = \sum_{k=0}^{\infty} a_{nk}(x_k - r) + r\sum_{k=0}^{\infty} a_{nk} \text{ for all } n \in \mathbb{N}.$$

Since M satisfy Δ_2-condition, then for some $\rho > 0$ we get

$$M\left(\frac{|\sum_{k=0}^{\infty} a_{nk}x_k|}{\rho}\right) \leq \frac{K_0}{2} M\left(\frac{|\sum_{k=0}^{\infty} a_{nk}(x_k - r)|}{\rho}\right) + \frac{K_1^{m_0+1}}{2} M\left(\frac{|\sum_{k=0}^{\infty} a_{nk}|}{\rho}\right)$$

for all $n \in \mathbb{N}$ and for some $K_0, K_1 > 0$. By *(iii)*, we get $\lim_{n\to\infty} \frac{K_1^{m_0+1}}{2} M\left(\frac{|\sum_{k=0}^{\infty} a_{nk}|}{\rho}\right) = \frac{\alpha K_1^{m_0+1}}{2}$. Further, since $x_k \to r$ as $k \to \infty$, we get

$$\frac{K_0}{2} M\left(\frac{|\sum_{k=0}^{\infty} a_{nk}(x_k - r)|}{\rho}\right) + \frac{K_1^{m_0+1}}{2} M\left(\frac{|\sum_{k=0}^{\infty} a_{nk}|}{\rho}\right) \to \alpha_1 \text{ as } n \to \infty$$

where $\alpha_1 = \frac{\alpha K_1^{m_0+1}}{2}$. It is shown that $M\left(\frac{|\sum_{k=0}^{\infty} a_{nk}x_k|}{\rho}\right) \to \alpha_1$ as $n \to \infty$. Thus, $Ax \in c(M)$. Since for each $x \in c(M)$ implies $Ax \in c(M)$, we conclude that $A \in \big(c(M), c(M)\big)$, which proves the theorem. □

Theorem 2.5. *$A \in \big(c_0(M), c_0(M)\big)$ if and only if for $\rho > 0$ the following conditions are held :*
(i) $\sup_{n\in\mathbb{N}} M\left(\frac{\sum_{k=0}^{\infty}|a_{nk}|}{\rho}\right) < \infty$, *and*
(ii) $\lim_{n\to\infty} M\left(\frac{|a_{nk}|}{\rho}\right) = 0$ *for each* $k \in \mathbb{N}$.

Proof. We first derive the necessary conditions *(i)* and *(ii)*. Since $x = (x_j) = (e_j^{[k]})$ belongs to $c_0(M)$, then

$$\lim_{n\to\infty} M\left(\frac{|a_{nk}|}{\rho}\right) = \lim_{n\to\infty} M\left(\frac{|\sum_{j=0}^{\infty} a_{nj}x_j|}{\rho}\right) = 0 \text{ for each } k \in \mathbb{N}.$$

It is shown that *(ii)* holds. Further, we define $x = (x_k)$ by $x_k = sgn(a_{nk})$ for all $k \in \mathbb{N}$ and $n \in \mathbb{N}$. Thus, for some $\rho > 0$ we have

$$\sup_{n\in\mathbb{N}} M\left(\frac{\sum_{k=0}^{\infty}|a_{nk}|}{\rho}\right) = \sup_{n\in\mathbb{N}} M\left(\frac{|\sum_{k=0}^{\infty}|a_{nk}||}{\rho}\right) = \sup_{n\in\mathbb{N}} M\left(\frac{|\sum_{k=0}^{\infty} a_{nk}x_k|}{\rho}\right).$$

Since $\lim_{n\to\infty} M\left(\frac{|\sum_{k=0}^{\infty} a_{nk}x_k|}{\rho}\right)$ exists, then sequence $\left(M\left(\frac{|\sum_{k=0}^{\infty} a_{nk}x_k|}{\rho}\right)\right)$ is bounded. Thus, $\sup_{n\in\mathbb{N}} M\left(\frac{\sum_{k=0}^{\infty}|a_{nk}|}{\rho}\right) < \infty$, which yields *(i)* holds.

For proving the sufficiency, let us take any $x \in c_0(M)$. Since M is continuous, then $x_k \to 0$ as $k \to \infty$. Thus, $\lim_{n\to\infty} M\left(\frac{|\sum_{k=0}^{\infty} a_{nk}x_k|}{\rho}\right) = 0$. It is shown that $Ax \in c_0(M)$. Since for each $x \in c(M)$ implies $Ax \in c(M)$, then the infinite matrix A belongs to the class $\big(c_0(M), c_0(M)\big)$, which completes the proof. □

Theorem 2.6. *$A \in \big(\ell_\infty(M), \ell_\infty(M)\big)$ if and only if for $\rho > 0$*

$$\sup_{n\in\mathbb{N}} M\left(\frac{\sum_{k=0}^{\infty}|a_{nk}|}{\rho}\right) < \infty.$$

Proof. For proving the necessity, suppose that $A \in \big(\ell_\infty(M), \ell_\infty(M)\big)$, that is, for each $x \in \ell_\infty(M)$ implies $Ax \in \ell_\infty(M)$. Thus, $\sup_{n\in\mathbb{N}} M\left(\frac{|\sum_{k=0}^{\infty} a_{nk}x_k|}{\rho}\right) < \infty$. Then, define $x = (x_k)$ by $x_k = sgn(a_{nk})$ for all $k \in \mathbb{N}$ and $n \in \mathbb{N}$. Thus, for some $\rho > 0$ and for every $n \in \mathbb{N}$, we have

$$M\left(\frac{\sum_{k=0}^{\infty}|a_{nk}|}{\rho}\right) = M\left(\frac{|\sum_{k=0}^{\infty} a_{nk}x_k|}{\rho}\right).$$

Since $\sup_{n\in\mathbb{N}} M\left(\frac{|\sum_{k=0}^{\infty} a_{nk}x_k|}{\rho}\right) < \infty$, then $\sup_{n\in\mathbb{N}} M\left(\frac{\sum_{k=0}^{\infty}|a_{nk}|}{\rho}\right) < \infty$.

For sufficiency, suppose $\sup_{n\in\mathbb{N}} M\left(\frac{\sum_{k=0}^{\infty}|a_{nk}|}{\rho}\right) < \infty$ and take any $x \in \ell_\infty(M)$. Since M is non decreasing, there exists $N_1 > 0$ such that $|x_k| \le N_1 = \rho N_0$ for all $k \in \mathbb{N}$. Hence, by using Hölder inequality [8], we get

$$\sup_{n\in\mathbb{N}} M\left(\frac{|A_n(x)|}{\rho}\right) \le \sup_{n\in\mathbb{N}} M\left(\frac{\sup_{k\in\mathbb{N}}|x_k|\sum_{k=0}^{\infty}|a_{nk}|}{\rho}\right).$$

Further, by using the Archimedean property, since $N_1 \in \mathbb{R}$, then there exists $m_0 \in \mathbb{N}$ such that $N_1 \le 2^{m_0}$. Since M satisfy Δ_2-condition, we have

$$\sup_{n\in\mathbb{N}} M\left(\frac{|A_n(x)|}{\rho}\right) \le K^{m_0} \sup_{n\in\mathbb{N}} M\left(\frac{\sum_{k=0}^{\infty}|a_{nk}|}{\rho}\right) < \infty$$

for some $K > 0$. It is show that $Ax \in \ell_\infty(M)$. Since for each $x \in \ell_\infty(M)$ implies $Ax \in \ell_\infty(M)$, then the infinite matrix $A \in \big(\ell_\infty(M), \ell_\infty(M)\big)$, and the proof is complete. □

References

[1] Altay B and Basar F 2005 *Ukrainian Math. J.* **57** 1:1-17
[2] Altay B, Başar F, Mursaleen M 2006 *Inform. Sci.* **176** 10:1450-62
[3] Aydin C and Basar F 2004 *Hokkaido Math. J.* **33** 2:383-98
[4] Aydin C and Basar F 2005 *Demonstratio Math.* **38** 3:641-56
[5] Kamthan P K and Gupta M 1981 *Sequence Spaces and Series* (New York: Marcel Dekker Inc)
[6] Kreyszig E 1978 *Introductory Functional Analysis With Applications* (America: John Wiley and Sons)
[7] Lindenstrauss J and Tzafriri L 1971 *Israel J. Math.* **10** 379-90
[8] Maddox I J 1970 *Elements of Functional Analysis* 1st ed (Cambridge: The University Press)
[9] Malkowsky E 1997 *Mat. Vesnik* **49** (3-4):187-96
[10] Malkowsky E and Savas E 2004 *Appl. Math. Comput.* **147** 2:333-45
[11] Mursaleen M and Noman A K 2010 *Thai J. Math.* **8** 2:311-29
[12] Mursaleen M and Noman A K 2011 *Filomat* **25** 2:33-51

Generalized Moore Penrose Inverse of Normal Elements in a Ring with Involution

Titi Udjiani SRRM, Harjito, Suryoto, Nikken Prima P
Department of Mathematics, Faculty of Science and Mathematics, Diponegoro University

Email: udjianititi@yahoo.com

Abstract. Based on the definition of a normal element in a ring with involution, it is found that each normal element is commutatively with the product of itself and the involution of itself. On the other hand, if the element of a ring with involution has generalized Moore Penrose inverse, then the element is also commutative with the product of itself and the involution of itself. In this paper, the phenomenon of the similarity properties from normal elements and generalized Moore Penrose inverse is used to establish the relationship between of them with them. .

1. Introduction

Let R be a ring with identity and $a \in R$. An involution "*" in a ring R is unary operation $a \in R \mapsto a^* \in R$ such that [1, 3]

$$(a^*)^* = a \qquad (a+b)^* = a^* + b^* \qquad (ab)^* = b^* a^*$$

for each elements a, b in a ring R.

In ring R with involution "*", element $a \in R$ is called normal if $aa^* = a^*a$. It is an extension of the normal matrix definition in the set of matrices over real number [4] . The set of normal elements in R is symbolized by R^{nor}.

Cvetkovi´c, et.al. [3] explain that the group inverse of $a \in R$ is an element $a^\# \in R$ such that :

$$aa^\# a = a, \qquad a^\# a a^\# = a^\#, \qquad aa^\# = a^\# a.$$

The set of all group invertible elements of R will be denoted by $R^\#$.

Koliha and Patrıcio [2] say that an element $a \in R$ is *-cancellable if for any $x \in R$ the following hold :

$$a^* ax = 0 \Rightarrow ax = 0 \text{ and } xaa^* = 0 \Rightarrow xa = 0.$$

Suppose R is a ring with involution "*". Element $b \in R$ is generalized Moore Penrose inverse of $a \in R$ if the followings are satisfied [5] :

$$aba = a, \qquad (ab)^* = ab, \qquad (ba)^* = ba. \tag{1}$$

Any b that hold (1) is a generalized Moore–Penrose inverse of a. We use a_g^+ and R_g^+, respectively to denote generalized Moore–Penrose inverse of a and the set of generalized Moore Penrose invertible elements in R.

If $a \in R$ is the generalized Moore Penrose invertible element, then a_g^+ is not usually unique. That are $a_g^+ = a_g^+ aa_g^+$ or $a_g^+ \neq a_g^+ aa_g^+$. The set of generalized Moore Penrose invers of $a \in R$ will be denoted by $(a_g^+)^-$.

The next theorem describes the existence of the generalized Moore Penrose inverse on the ring with involution that has been given in [5, Theorem 2.10].

Theorem 1: *For any $a \in R$ the following equivalent conditions holds :*

(1) $a \in R_g^+$,

(2) $a^* \in R_g^+$,

(3) *a is *-cancellable and* $a^*a \in R_g^+$,

(4) *a is *-cancellable and* $aa^* \in R_g^+$,

(5) *a is *-cancellable and* $a^*a \in R^\#$,

(6) *a is *-cancellable and* $aa^* \in R^\#$,

(7) *a is *-cancellable and both $a*a$ and $aa*$ are regular,*

(8) $a \in aa^*R \cap Ra^*a$,

(9) *a is *-cancellable and* a^*aa^* *regular*

Theorem 2 in the following describes the properties of a_g^+ if $a \in R_g^+$, which has been discussed in [5, Theorem 2.5].

Theorem 2: *If* $a \in R_g^+$, *then*

1. $(a^*)_g^+ a^* (a^*)_g^+ = (a_g^+ aa_g^+)^*$ for each $a_g^+ \in (a_g^+)^-$ and $(a^*)_g^+ \in ((a^*)_g^+)^-$.
2. $(a^*a)_g^+ a^*a(a^*a)_g^+ = a_g^+ (a_g^+ aa_g^+)^*$ and $(aa^*)_g^+ aa^*(aa^*)_g^+ = (a_g^+ aa_g^+)^* a_g^+$ for each $a_g^+ \in (a_g^+)^-$, $(aa^*)_g^+ \in ((aa^*)_g^+)^-$ and $(a^*a)_g^+ \in ((a^*a)_g^+)^-$.
3. $a^* = a^*aa_g^+ = a_g^+ aa^*$ for each $a_g^+ \in (a_g^+)^-$.
4. $a_g^+ aa_g^+ = (a^*a)_g^+ a^*a(a^*a)_g^+ a^* = a^*(aa^*)_g^+ aa^*(aa^*)_g^+$ for each $a_g^+ \in (a_g^+)^-$, $(aa^*)_g^+ \in ((aa^*)_g^+)^-$ and $(a^*a)_g^+ \in ((a^*a)_g^+)^-$.
5. $(a^*)_g^+ a^*(a^*)_g^+ = a(a^*a)_g^+ a^*a(a^*a)_g^+ = (aa^*)_g^+ aa^*(aa^*)_g^+ a$ for each $(a^*)_g^+ \in ((a^*)_g^+)^-$, $(aa^*)_g^+ \in ((aa^*)_g^+)^-$ and $(a^*a)_g^+ \in ((a^*a)_g^+)^-$.
6. $(a_g^+ aa_g^+)_g^+ = a$ for each $a_g^+ \in (a_g^+)^-$.

By definition of normal element, if $a \in R^{nor}$ then

$$aa^*a = a^*aa \tag{2}$$

and

$$aaa^* = aa^*a \tag{3}$$

IOP Conf. Series: Materials Science and Engineering **300** (2018) 012074 doi:10.1088/1757-899X/300/1/012074

Equation (2) and (3) show that a is commutative with a^*a and aa^*. According to Theorem 1, if $a \in R_g^+$ then $aa^*, a^*a \in R^\#$. Udjiani, et.al. [5] have observed that

$$(aa^*)^\# = (a\,a^*)_g^+ a\,a^*(a\,a^*)_g^+ \quad (4)$$

and

$$(a^*a)^\# = (a^*a)_g^+ a^*a(a^*a)_g^+ \quad (5)$$

for each $(aa^*)_g^+ \in \left((aa^*)_g^+\right)^-$ and $(a^*a)_g^+ \in \left((a^*a)_g^+\right)^-$. (Existence of $(a^*a)_g^+$ guaranted by Theorem 1)

In other side, Mosic dan Djordjevic [4] say that if $a \in R$ is group invertible then a is double commutative with $a^\#$. So, if $a \in R_g^+$ then aa^* is double commutative with $(aa^*)^\#$ and a^*a is double commutative with $(a^*a)^\#$. From the explanation we can conclude that there are a relation between normal element and generalized Moore Penrose invertible element. Motivated by Mosic dan Djordjevic [4], this paper discusses the properties of elements in $R_g^+ \cap R^{nor}$.

2. Normal Elements and generalized Moore Penrose invertible element.

The In this section normal elements in rings with involution are characterized by conditions involving their group and Generalized Moore–Penrose inverse. Some of these results are proved for set of matrices over real number with transpose involution.

From Theorem 1, if $a \in R_g^+$ then a^*a, $a\,a^* \in R^\#$. Inspired that if $a \in R$ is group invertible then a is double commutative with $a^\#$, so we get Lemma 3 in the following :

Lemma 3: *If $aa^* \in R^\#$, $x \in R$ and $aa^*x = xaa^*$ then $(aa^*)^\# x = x(aa^*)^\#$. Also if $a^*a \in R^\#$, $x \in R$ and $a^*ax = x\,a^*a$ then $(a^*a)^\# x = x(a^*a)^\#$.*

Proof :

$$\begin{aligned}(a\,a^*)^\# x &= ((a\,a^*)^\#)^2\, a\,a^*x = ((a\,a^*)^\#)^2\, xa\,a^* = ((a\,a^*)^\#)^2\, x\,(a\,a^*)^2(a\,a^*)^\# \\ &= ((a\,a^*)^\#)^2 a\,a^*\, xa\,a^*(a\,a^*)^\# = (a\,a^*)^\#\, xa\,a^*(a\,a^*)^\# \\ &= (a\,a^*)^\#\, a\,a^*x(a\,a^*)^\# = (a\,a^*)^\#\, a\,a^*xa\,a^*((a\,a^*)^\#)^2 \\ &= (a\,a^*)^\#\, a\,a^*a\,a^*x((a\,a^*)^\#)^2 = a\,a^*x((a\,a^*)^\#)^2 = xa\,a^*((a\,a^*)^\#)^2 \\ &= x(a\,a^*)^\#.\end{aligned}$$

By the same way we have $a^*a \in R^\#$, $x \in R$ and $a^*ax = xa^*a$ then $(a^*a)^\# x = x(a^*a)^\#$. ■

The next Theorem 4 is inspired by the work of Lemma 1.1. in [4].

Theorem 4: *Suppose $a \in R_g^+$ and $b \in R$. If $ab = ba$ and $a^*b = ba^*$ then $a_g^+aa_g^+b = b\,a_g^+aa_g^+$ for each $a_g^+ \in \left(a_g^+\right)^-$.*

Proof :

Using Theorem 2, if $a \in R_g^+$ then $a_g^+aa_g^+ = a^*(aa^*)_g^+aa^*(aa^*)_g^+$ for each $a_g^+ \in \left(a_g^+\right)^-$ and $(aa^*)_g^+ \in \left((aa^*)_g^+\right)^-$.

Futhermore, using Equation (4) we have

$$a_g^+aa_g^+ = a^*(a\,a^*)^\# \text{ for each } a_g^+ \in \left(a_g^+\right)^- \quad (6)$$

By Lemma [3], we have $(a\,a^*)^\#$ is double commutative with $a\,a^*$. Since b is commutative with a^*, then $(a\,a^*)^\#$ is commutative with *b*.
By Equation (6), we obtain $b\,a_g^+aa_g^+ = b\,a^*(a\,a^*)^\# = a^*b(a\,a^*)^\# = a^*(a\,a^*)^\#b = a_g^+aa_g^+b$ for each $a_g^+ \in (a_g^+)^-$. ■

Normal and generalized Moore Penrose invertible element have the same properties, that is commutative with *a* and a^*. Motivated by Mosic dan Djordjevic [4, Lemma 1.2.], we can construct the existence of normal element in R_g^+ in Theorem 5 as follows :

Theorem 5: *Let* $a \in R_g^+$. *Element* a *is normal if and only if* $aa_g^+ = a_g^+a$ *and* $a^*a_g^+aa_g^+ = a_g^+aa_g^+a^*$ *for each* $a_g^+ \in (a_g^+)^-$.
Proof :
By properties of normal element, if $a \in R^{nor}$ then *a* commutative with aa^* and a^*a. In other side if $a \in R_g^+$, then by Theorem 1 we have $aa^*,\ a^*a \in R^\#$. So, using Lemma 3, we conclude $aa_g^+(a_g^+aa_g^+)^* = a_g^+(a_g^+aa_g^+)^*a$ for each $a_g^+ \in (a_g^+)^-$.
Futhermore $aa_g^+ = aa_g^+\,aa_g^+ = aa_g^+(aa_g^+)^* = aa_g^+(a_g^+aa_g^+)^*a^* = a_g^+(a_g^+aa_g^+)^*aa^* = a_g^+(a_g^+aa_g^+)^*a^*a = a_g^+a$ for each $a_g^+ \in (a_g^+)^-$. In the same way $a^*a_g^+aa_g^+ = a^*a_g^+(aa_g^+)^* = a^*a_g^+(a_g^+aa_g^+)^*a^* = a_g^+(a_g^+aa_g^+)^*a^*a^* = a_g^+aa_g^+a^*$ for each $a_g^+ \in (a_g^+)^-$. Conversely if $aa_g^+ = a_g^+a$ and $a^*a_g^+aa_g^+ = a_g^+aa_g^+a^*$ for each $a_g^+ \in (a_g^+)^-$, then using Theorem 2 we have $aa^* = aa^*aa_g^+ = aa^*a_g^+a = aa^*a_g^+aa_g^+a = aa_g^+aa_g^+a^*a = a^*a$. ■

Acknowledgments
This article is funded by the DIPA Faculty of Science and Mathematics Diponegoro University with contract number: 1645h / UN7.5.8 / PP / 2017. Dated April 3, 2017.

References
[1] Harte R and Mbekhta M 1992 *Mathematica*, **103(1)** 71.
[2] Koliha J J and Patrıcio P 2002 *J. Australian Math. Soc.* **72** 137.
[3] Cvetkovi´c D Djordjevi´c D S and Koliha JJ 2007 *Linear Algebra Appl.* **426** 371.
[4] Mosic D and Djordjevic D S 2009 *Linear Algebra Appl.* **431** 732.
[5] Udjiani T Surodjo B and Wahyuni S 2014 *Far East Journal of Mathematical Sciences*, **92(1)** 29.

The competition index of two-colored non-hamiltonian digraph with two cycles whose lengths differ by three

Hari Sumardi[1], Saib Suwilo[2] and Elly Rosmaini[3]

Department of Mathematics, University of Sumatera Utara, Medan 20155 Indonesia

E-mail: [2]saib@usu.ac.id

Abstract. A two-colored digraph is a digraph each of whose arcs is colored by red or blue. An (h,g)-walk in a two-colored digraph is a walk consisting of h red arcs and g blue arcs. A strongly connected two-colored digraph is primitive provided there exists a positive integer $h+g$ such that for each ordered pair of vertices x and y there is a (h,g)-walk from x to y. The competition index of a primitive two-colored digraph $D^{(2)}$ is the least positive integer $h+g$ over all pairs of nonnegative integers h and g such that for each pair of vertices x and y in $D^{(2)}$ there is a vertex w such that there are (h,g)-walks from x to w and from y to w. For some positive integer $s \geq 7$ ($s \not\equiv 0 \bmod 3$), we discuss the competition index of a class of strongly connected primitive two-colored digraph $D^{(2)}$ consisting of two cycles of lengths s and $s+3$, respectively. We present the competition index of such primitive two-colored digraph $D^{(2)}$ in terms of s and the lengths of two specific paths in $D^{(2)}$.

1. Introduction

Let D be a digraph with vertex set $V = \{v_1, v_2, \ldots, v_n\}$. We follow the notation and terminologies for digraph in [3]. A $v_i \to v_j$ path is a $v_i \to v_j$ walk with no repeated vertices except possibly $v_i = v_j$. A cycle is closed path.

A two-colored digraph $D^{(2)}$ is a digraph D each of whose arcs is colored by red or blue. A $v_i \xrightarrow{(h,g)} v_j$ walk in $D^{(2)}$ is a walk from v_i to v_j consisting of h red arcs and g blue arcs. For an (h,g)-walk W_{v_i,v_j} in $D^{(2)}$ we denote $r(W_{v_i,v_j}) = h$ and $b(W_{v_i,v_j}) = g$. The vector $\begin{bmatrix} r(W_{v_i,v_j}) \\ b(W_{v_i,v_j}) \end{bmatrix}$ is the composition of the walk W_{v_i,v_j}. A strongly connected two-colored digraph $D^{(2)}$ is primitive provided there is a positive integer $h+g$ over all nonnegative integers h and g such that for each pair of ordered vertices v_i and v_j there is a $v_i \xrightarrow{(h,g)} v_j$ walk. The notion of primitive two-colored digraph is a generalization of the notion of primitive digraph.

We generalize the notion of m-competition index of a primitive digraph in [2, 3] into the m-competition index of a primitive two-colored digraph as follows. For $1 \leq m \leq n$, the m-competition index of a primitive two-colored digraph $D^{(2)}$, denoted by $k_m(D^{(2)})$, is the smallest positive integer $h+g$ over all nonnegative integers h and g such that for each pair of vertices v_i and v_j there exist at least m distinct vertices $u_1, u_2, \ldots, u_m$ with the property that there are $v_i \xrightarrow{(h,g)} u_t$ walk and $v_j \xrightarrow{(h,g)} u_t$ for each $t = 1, 2, \ldots, m$. The competition index of a primitive

IOP Conf. Series: Materials Science and Engineering **300** (2018) 012075 doi:10.1088/1757-899X/300/1/012075

two-colored digraph $D^{(2)}$ is defined to be $k_1(D^{(2)})$. In 2015, Mulyono et al. [5] introduced the notion of the scrambling index of a primitive two-colored digraph which is the smallest positive integer $h+g$ over all nonnegative integers h and g such that for each pair of distinct vertices v_i and v_j there is a vertex v_t with property that there is a $v_i \stackrel{(h,g)}{\longrightarrow} v_t$ walk and a $v_j \stackrel{(h,g)}{\longrightarrow} v_t$ walk. The notion of the competition index of a primitive two-colored digraph is the same as the notion of the scrambling index of a primitive two-colored digraph.

Mulyono et al. [5] have discussed the competition index of primitive two-colored digraph consisting of two cycles whose lengths differ by one. Mardinigsih and Pasaribu [4] discussed the competition index of a class of Hamiltonian digraph on n vertices consisting of two cycles of length n and $(n-1)/2$ respectively. In this paper, for $s \geq 7$, we discuss the competition index of a class of primitive two-colored digraph consisting of two cycles of lengths s and $s+3$, respectively, where $s \not\equiv 0 \bmod 3$.

2. Preliminaries

Let $D^{(2)}$ be a two-colored digraph and let $C = \{C_1, C_2, \ldots, C_q\}$ be the set of all cycles in $D^{(2)}$. A cycle matrix of $D^{(2)}$ is a 2 by q matrix M of the form $M = \begin{bmatrix} r(C_1) & r(C_2) & \cdots & r(C_q) \\ b(C_1) & b(C_2) & \cdots & b(C_q) \end{bmatrix}$. That is M is a matrix such that, for $i = 1, 2, \ldots, q$, the ith column of M is the composition of the ith cycle of $D^{(2)}$. If the rank of M is 1, the content of M is 0, and the content of M is defined to be the greatest common divisor of the determinants of 2 by 2 submatrices of M, otherwise. The following is an algebraic charaterization of a primitive two-colored digraph.

Theorem 1. [1] *Let $D^{(2)}$ be a strongly connected two-colored digraph and let M be a cycle matrix of $D^{(2)}$. The strongly connected two-colored digraph $D^{(2)}$ is primitive if and only if the content of M is* 1.

Let v_i and v_j be two distinct vertices in $D^{(2)}$ and let v_t be a vertex in $D^{(2)}$. The local competition index of v_i and v_j at the vertex v_t, denoted by $k_1(v_t : v_i, v_j)$, is the smallest positive integer $h_t + g_t$ over all nonnegative integers h_t and g_t such that there are $v_i \stackrel{(h_t,g_t)}{\longrightarrow} v_t$ walk and $v_j \stackrel{(h_t,g_t)}{\longrightarrow} v_t$ walk. The local competition index of v_i and v_j, denoted by $k_1(D^{(2)} : v_i, v_j)$, is defined to be

$$k_1(D^{(2)} : v_i, v_j) = \min_{v_t \in V}\{k_1(v_t : v_i, v_j)\}.$$

We note that

$$\max_{v_i, v_j \in V}\{k_1(D^{(2)} : v_i, v_j)\} \leq k_1(D^{(2)}).$$

The following lemma gives a lower bound for local competition index.

Lemma 2. [5] *Let $D^{(2)}$ be a primitive two-colored digraph consisting of two cycles with cycle matrix $M = \begin{bmatrix} r(C_1) & r(C_2) \\ b(C_1) & b(C_2) \end{bmatrix}$ and let $\det(M) = 1$. Let v_i and v_j be two distinct vertices in $D^{(2)}$. If $k_1(v_t : v_i, v_j)$ is obtained by (h_t, g_t)-walk, then $\begin{bmatrix} h_t \\ g_t \end{bmatrix} \geq M \begin{bmatrix} b(C_2)r(P_{v_i,v_t}) - r(C_2)b(P_{v_i,v_t}) \\ r(C_1)b(P_{v_j,v_t}) - b(C_1)r(P_{v_j,v_t}) \end{bmatrix}$ for some paths P_{v_i,v_t} and P_{v_j,v_t}.*

Let v_i and v_j be two distinct vertices in $D^{(2)}$ and let v_t be a vertex in $D^{(2)}$. Define $q_1 = b(C_2)r(P_{v_i,v_t}) - r(C_2)b(P_{v_i,v_t})$ and $q_2 = r(C_1)b(P_{v_j,v_t}) - b(C_1)r(P_{v_j,v_t})$. Then by Lemma 2 we have

$$\begin{bmatrix} h_t \\ g_t \end{bmatrix} \geq M \begin{bmatrix} q_1 \\ q_2 \end{bmatrix} \quad \text{and} \quad k_1(v_t : v_i, v_j) \geq \ell(C_1)q_1 + \ell(C_2)q_2$$

where $\ell(C_1)$ and $\ell(C_2)$ are the lengths of the cycles C_1 and C_2, respectively.

3. The Case $s \equiv 2 \bmod 3$

In this section we discuss the competition index of strongly connected primitive two-colored digraph $D^{(2)}$ on n vertices consisting of the cycle $C_1 : v_1 \to v_2 \to \cdots \to v_c \to v_{s+1} \to v_{s+2} \to \cdots \to v_n \to v_1$ of length $s+3$ and the cycle $C_2 : v_1 \to v_2 \to \cdots \to v_c \to v_{c+1} \to v_{c+2} \to \cdots \to v_s \to v_1$ of length s where $s \geq 8$, $s \equiv 2 \bmod 3$, $n = 2s+3-c$, and $1 \leq c \leq s-1$. We note that C_1 and C_2 have c vertices in common and the cycle matrix of $D^{(2)}$ is $M = \begin{bmatrix} r(C_1) & r(C_2) \\ b(C_1) & b(C_2) \end{bmatrix}$. By Theorem 1 we have $\det(M) = \pm 1$. Without loss of generality we may assume that $\det(M) = 1$. This implies $M = \begin{bmatrix} (2s+5)/3 & (2s-1)/3 \\ (s+4)/3 & (s+1)/3 \end{bmatrix}$. Hence the number of blue arcs of $D^{(2)}$ is at least $(s+4)/3$ and at most $(2s+5)/3$. We discuss only the case where $D^{(2)}$ has exactly $(2s+5)/3$ blue arcs, the cycle C_1 contains a $(0,(s+4)/3)$-path and the cycle C_2 contains a $(0,(s+1)/3)$-path.

Let the $(0,(s+4)/3)$-path on C_1 be the path $P_{v_x,v_{x'}}$ from v_x to $v_{x'}$ and the $(0,(s+1)/3)$-path on C_2 be the path $P_{v_y,v_{y'}}$ from v_y to $v_{y'}$. We define $d_1 = d(v_{x'}, v_1)$ and $d_2 = d(v_{y'}, v_1)$. Notice that $\ell(C_1) = s+3$, $r(C_1) = (2s+5)/3$, $b(C_1) = (s+4)/3$, $\ell(C_2) = s$, $r(C_2) = (2s-1)/3$ and $b(C_2) = (s+1)/3$.

Theorem 3. *Let s be a positive integer such that $s \geq 8$ and $s \equiv 2 \bmod 3$. Let $D^{(2)}$ be a strongly connected primitive two-colored digraph consisting of two cycles C_1 and C_2 of lengths $s+3$ and s, respectively. Suppose the cycle C_1 contains a $(0,(s+4)/3)$-path $P_{v_x,v_{x'}}$ and the cycle C_2 contains a $(0,(s+1)/3)$-path $P_{v_y,v_{y'}}$. If the paths $P_{v_x,v_{x'}}$ and $P_{v_y,v_{y'}}$ have no vertices in common, then*

$$k_1(D^{(2)}) = \begin{cases} \frac{2s^3+13s^2+20s}{9} + d_1 & \text{if } d_1 > d_2 \text{ and } d_1 - d_2 \leq 1, \\ \frac{2s^3+s^2(3d_1-3d_2+7)+s(12d_1-12d_2+5)}{9} + d_1 & \text{if } d_1 > d_2 \text{ and } d_1 - d_2 > 1, \\ \frac{2s^3+s^2(3d_2-3d_1+13)+s(12d_2-12d_1+20)}{9} + d_2 & \text{if } d_1 \leq d_2. \end{cases}$$

Proof. We split the proof into three cases. They are the case where $d_1 > d_2$ and $d_1 - d_2 \leq 1$, the case where $d_1 > d_2$ and $d_1 - d_2 > 1$, and the case where $d_1 \leq d_2$. We present a complete proof of the first case and present a sketch for the rest.

Case 1. $d_1 > d_2$ and $d_1 - d_2 \leq 1$.

We first show that $k_1(D^{(2)}) \geq (2s^3+13s^2+20s)/9 + d_1$. For this purpose, for each $t = 1, 2, \ldots, n$, we show $k_1(v_t : v_{x'}, v_x) \geq (2s^3+13s^2+20s)/9 + d_1 + d(v_1, v_t)$. We assume that the $k_1(v_t : v_{x'}, v_x)$ is obtained by (h_t, g_t)-walks and define $q_1 = b(C_2)r(P_{v_{x'},v_t}) - r(C_2)b(P_{v_{x'},v_t})$ and $q_2 = r(C_1)b(P_{v_x,v_t}) - b(C_1)r(P_{v_x,v_t})$.

The vertex v_t lies on the $v_1 \to v_x$ path or $v_1 \to v_y$ path. Using the $(d_1 + d(v_1,v_t), 0)$-path from $v_{x'}$ to v_t we have $q_1 = (s+1)(d_1 + d(v_1,v_t))/3$ and using the $(d_1 + d(v_1,v_t), (s+4)/3)$-path from v_x to v_t we have $q_2 = (2s+5)(s+4)/9 - (s+4)(d_1 + d(v_1,v_t))/3$. By Lemma 2 we have $k_1(v_t : v_{x'}, v_x) \geq (s+3)q_1 + sq_2 = (2s^3+13s^2+20s)/9 + d_1 + d(v_1,v_t)$.

The vertex v_t lies on the $v_y \to v_{y'}$ path. Using the $(d_1-d_2+(2s-1)/3, d_2+d(v_1,v_t)-(2s-1)/3)$-path from $v_{x'}$ to v_t we have $q_1 = (s+1)(d_1-d_2+(2s-1)/3)/3-(2s-1)(d_2+d(v_1,v_t)-(2s-1)/3)/3$ and using the $(d_1 - d_2 + (2s-1)/3, d_2 + d(v_1,v_t) - (s-5)/3)$-path from v_x to v_t we have $q_2 = (2s+5)(d_2 + d(v_1,v_t) - (s-5)/3)/3 - (s+4)(d_1 - d_2 + (2s-1)/3)/3$. Lemma 2 implies that $k_1(v_t : v_{x'}, v_x) \geq (s+3)q_1 + sq_2 = (2s^3+13s^2+20s)/9 + d_1 + d(v_1,v_t)$.

The vertex v_t lies in the $v_{y'+1} \to v_s$ path. Using the $(d_1 - d(v_t,v_1) + (2s-1)/3, (s+1)/3)$-path from $v_{x'}$ to v_t we have $q_1 = (s+1)(d_1 - d(v_t,v_1))/3$ and using the $(d_1 - d(v_t,v_1) + (2s-1)/3, (2s+5)/3)$-path from v_x to v_t we have $q_2 = (2s+5)^2/9 - (s+4)(d_1 - d(v_t,v_1) + (2s-1)/3)/3$. Lemma 2 implies that $k_1(v_t : v_{x'}, v_x) \geq (s+3)q_1 + sq_2 = (2s^3+13s^2+20s)/9 + d_1 + s - d(v_t,v_1)$. Since $s - d(v_t,v_1) = d(v_1,v_t)$, we have $k_1(v_t : v_{x'}, v_x) \geq (2s^3+13s^2+20s)/9 + d_1 + d(v_1,v_t)$.

The vertex v_t lies on the $v_{x+1} \to v_{x'}$ path. Using the $((2s+5)/3, d_1 + d(v_1,v_t) - (2s+5)/3)$-path from $v_{x'}$ to v_t we have $q_1 = s(2s+5)/3 - (2s-1)(d_1 + d(v_1,v_t))/3$ and using the

$(0, d_1+d(v_1,v_t)-(2s+5)/3)$-path from v_x to v_t we have $q_2 = (2s+5)(d_1+d(v_1,v_t)-(2s+5)/3)/3$. Lemma 2 implies that $k_1(v_t : v_{x'}, v_x) \geq (s+3)q_1 + sq_2 = (2s^3+13s^2+20s)/9 + d_1 + d(v_1, v_t)$.

The vertex v_t lies on the $v_{x'+1} \to v_n$ path. Using the $(d_1 - d(v_t, v_1), 0)$-path from $v_{x'}$ to v_t we have $q_1 = (s+1)(d_1 - d(v_t,v_1))/3$ and using the $(d_1 - d(v_t, v_1), (s+4)/3)$-path from v_x to v_t we have $q_2 = (2s+5)(s+4)/9 - (s+4)(d_1 - d(v_t, v_1))/3$. Lemma 2 implies that $h_t + g_t \geq \ell(C_1)q_1 + \ell(C_2)q_2 = (s+3)q_1 + sq_2 = p_1 + p_2$ where $p_1 = (4s^3+24s^2+27s-20)/27 + d_1 - d(v_t, v_1)$ and $p_2 = (2s^3+15s^2+33s+20)/27$.

We examine the existence of the $v_{x'} \overset{(p_1,p_2)}{\longrightarrow} v_t$ walk. Since the $v_{x'} \to v_t$ path is a $(d_1 - d(v_t, v_1), 0)$-path and the solution to the system $Mz + \begin{bmatrix} d_1 - d(v_t, v_1) \\ 0 \end{bmatrix} = \begin{bmatrix} p_1 \\ p_2 \end{bmatrix}$ is $z_1 = 0$ and $z_2 = (2s+5)(s+4)/9$, there is no $v_{x'} \overset{(p_1,p_2)}{\to} v_t$ walk. Hence $k_1(v_t : v_{x'}, v_x) > p_1 + p_2$. Notice that the shortest $v_{x'} \to v_t$ walk with at least p_1 red arcs and at least p_2 blue arcs is a $(p_1 + (2s+5)/3, p_2 + (s+4)/3)$-walk. This implies $k_1(v_t; v_{x'}, v_x) \geq p_1 + p_2 + s + 3 = (s+3)q_1 + sq_2 + s + 3 = (2s^3+13s^2+20s)/9 + d_1 + s + 3 - d(v_t, v_1)$. Since $s+3-d(v_t,v_1) = d(v_1, v_t)$, we have $k_1(v_t : v_{x'}, v_x) \geq (2s^3+13s^2+20s)/9 + d_1 + d(v_1, v_t)$.

Therefore, for each $t = 1, 2, \ldots, n$, we have $k_1(v_t : v_{x'}, v_x) \geq (2s^3+13s^2+20s)/9 + d_1 + d(v_1, v_t)$. This implies $k_1(D^{(2)} : v_{x'}, v_x) \geq (2s^3+13s^2+20s)/9 + d_1$ and hence $k_1(D^{(2)}) \geq (2s^3+13s^2+20s)/9 + d_1$.

We next show that $k_1(D^{(2)}) \leq (2s^3+13s^2+20s)/9 + d_1$. It suffices to show that for each $j = 1, 2, \ldots, n$, there is a $v_j \overset{(h,g)}{\longrightarrow} v_1$ walk with $h = (4s^3+24s^2+27s-20)/27 + d_1$ and $g = (2s^3+15s^2+33s+20)/27$. We first show that, for each $j = 2, 3, \ldots, n$, the system

$$Mz + \begin{bmatrix} r(P_{v_j,v_1}) \\ b(P_{v_j,v_1}) \end{bmatrix} = \begin{bmatrix} (4s^3+24s^2+27s-20)/27 + d_1 \\ (2s^3+15s^2+33s+20)/27 \end{bmatrix} \tag{1}$$

has a nonnegative integer solution for some paths P_{v_j,v_1} from v_j to v_1. The solution to the system (1) is the integer vector $z = \begin{bmatrix} \frac{s+1}{3}(d_1 - r(P_{v_j,v_1})) + \frac{2s-1}{3}b(P_{v_j,v_1}) \\ \frac{2s+5}{3}\left(\frac{s+4}{3} - b(P_{v_j,v_1})\right) + \frac{s+4}{3}(r(P_{v_j,v_1}) - d_1) \end{bmatrix}$. We show that $z_1 \geq 0$ and $z_2 \geq 0$ for some path P_{v_j,v_1} from v_j to v_1.

If the vertex v_j lies on the $v_2 \to v_x$ path, then there is a $(r(P_{v_j,v_1}), (s+4)/3)$-path from v_j to v_1 with $d_1 \leq r(P_{v_j,v_1}) \leq \frac{2s+2}{3}$. Using this path we find $z_1 = \frac{2s-1}{3}\frac{s+4}{3} + \frac{s+1}{3}(d_1 - r(P_{v_j,v_1}))$ and $z_2 = \frac{s+4}{3}(r(P_{v_j,v_1}) - d_1)$. Since $d_1 \geq 1$ and $r(P_{v_j,v_1}) \leq \frac{2s+2}{3}$, we have $z_1 \geq \frac{2s-1}{3}$. Since $d_1 \leq r(P_{v_j,v_1})$, we have $z_2 \geq 0$.

If the vertex v_j lies on the $v_{x+1} \to v_{x'}$ path, there is a $(d_1, b(P_{v_j,v_1}))$-path from v_j to v_1 with $0 \leq b(P_{v_j,v_1}) \leq \frac{s+4}{3}$. Using this path we have $z_1 = \frac{2s-1}{3}b(P_{v_j,v_1})$ and $z_2 = \frac{2s+5}{3}\left(\frac{s+4}{3} - b(P_{v_j,v_1})\right)$. Since $0 \leq b(P_{v_j,v_1}) \leq \frac{s+4}{3}$, we have $z_1 \geq 0$ and $z_2 \geq 0$.

If the vertex v_j lies on the $v_{x'+1} \to v_n$ path, then there is a $(r(P_{v_j,v_1}), 0)$-path from v_j to v_1 with $1 \leq r(P_{v_j,v_1}) \leq d_1 - 1$. Using this path we have $z_1 = \frac{s+1}{3}(d_1 - r(P_{v_j,v_1}))$ and $z_2 = \frac{2s+5}{3}\frac{s+4}{3} + \frac{s+4}{3}(r(P_{v_j,v_1}) - d_1)$. Since $r(P_{v_j,v_1}) \leq d_1 - 1$, we have $z_1 \geq \frac{s+1}{3}$. Since $r(P_{v_j,v_1}) \geq 1$ and $d_1 \leq \frac{2s+5}{3}$ we have $z_2 \geq \frac{s+4}{3}$.

If the vertex v_j lies on the $v_{c+1} \to v_y$ path, then there is a $(r(P_{v_j,v_1}), \frac{s+1}{3})$-path from v_j to v_1 with $d_2 \leq r(P_{v_j,v_1}) \leq \frac{2s-1}{3}$. Using this path we have $z_1 = \frac{s+1}{3}(d_1 - r(P_{v_j,v_1})) + \frac{2s-1}{3}\frac{s+1}{3}$ and $z_2 = \frac{2s+5}{3} + \frac{s+4}{3}(r(P_{v_j,v_1}) - d_1)$. Since $r(P_{v_j,v_1}) \leq \frac{2s-1}{3}$ and $d_1 \geq 1$, we have $z_1 \geq \frac{s+1}{3}$. Since $r(P_{v_j,v_1}) \geq d_2$ and $d_1 - d_2 \leq 1$, we have $z_2 \geq \frac{s+1}{3}$.

If the vertex v_j lies on the $v_{y+1} \to v_{y'}$ path, then there is a $(d_2, b(P_{v_j,v_1}))$-path from v_j to v_1 with $0 \leq b(P_{v_j,v_1}) \leq \frac{s-2}{3}$. Using this path we have $z_1 = \frac{s+1}{3}(d_1 - d_2) + \frac{2s-1}{3}b(P_{v_j,v_1})$ and $z_2 = \frac{2s+5}{3}\left(\frac{s+4}{3} - b(P_{v_j,v_1})\right) + \frac{s+4}{3}(d_2 - d_1)$. Since $b(P_{v_j,v_1}) \geq 0$ and $0 \leq d_1 - d_2 \leq 1$, we have $z_1 \geq 0$. Since $b(P_{v_j,v_1}) \leq \frac{s-2}{3}$ and $d_2 - d_1 \geq -1$, we have $z_2 \geq s+2$.

If the vertex v_j lies on the $v_{y'+1} \rightarrow v_s$, then there is a $(r(P_{v_j,v_1}),0)$-path from v_j to v_1 with $1 \leq r(P_{v_j,v_1}) \leq d_2 - 1$. Using this path we have $z_1 = \frac{s+1}{3}(d_1 - r(P_{v_j,v_1}))$ and $z_2 = \frac{2s+5}{3}\frac{s+4}{3} + \frac{s+4}{3}(r(P_{v_j,v_1}) - d_1)$. Since $r(P_{v_j,v_1}) \leq d_2 - 1$, we have $z_1 \geq \frac{s+1}{3}$. Since $r(P_{v_j,v_1}) \geq 1$ and $d_1 \leq \frac{2s+5}{3}$, we have $z_2 \geq \frac{s+4}{3}$.

Therefore for each $j = 2, \ldots, n$ the system (1) has a nonnegative integer solution for some path P_{v_j,v_1}. This implies, for each $j = 2, 3, \ldots, n$, the walk that starts at v_j, moves to v_1 along the chosen path P_{v_j,v_1}, then moves z_1 and z_2 times around the cycles C_1 and C_2, respectively, and back at v_1 is a $v_j \xrightarrow{(h,g)} v_1$ walk. We note that the walk that starts at the vertex v_1 moves $\frac{s+1}{3}d_1$ and $\frac{2s+5}{3}\frac{s+4}{3} - \frac{s+4}{3}d_1$ times around the cycles C_1 and C_2, respectively, and back at v_1 is a $v_1 \xrightarrow{(h,g)} v_1$ walk. Hence $k_1(D^{(2)}) \leq h + g = (2s^3 + 13s^2 + 20s)/9 + d_1$.

Case 2. $d_1 > d_2$ **and** $d_1 - d_2 > 1$.
As in the proof of case 1, to prove $k_1(D^{(2)}) \geq (2s^3 + s^2(3d_1 - 3d_2 + 7) + s(12d_1 - 12d_2 + 5))/9 + d_1$ we first show, for each $t = 1, 2, \ldots, n$, that $k_1(v_t : v_{x'}, v_y) \geq (2s^3 + s^2(3d_1 - 3d_2 + 7) + s(12d_1 - 12d_2 + 5))/9 + d_1 + d(v_1, v_t)$. For this purpose we set $q_1 = b(C_2)r(P_{v_{x'},v_t}) - r(C_2)b(P_{v_{x'},v_t})$ and $q_2 = r(C_1)b(P_{v_y,v_t}) - b(C_1)r(P_{v_y,v_t})$ and then we employ Lemma 2.

To prove $k_1(D^{(2)}) \leq (2s^3 + s^2(3d_1 - 3d_2 + 7) + s(12d_1 - 12d_2 + 5))/9 + d_1$ we show that, for each $j = 1, 2, \ldots, n$, there is a $v_j \xrightarrow{(h,g)} v_1$ walk with $h = (4s^3 + s^2(6d_1 - 6d_2 + 12) + s(21d_1 - 21d_2 + 3) + 15d_1 + 12d_2 - 5)/27$ and $g = (2s^3 + s^2(3d_1 - 3d_2 + 9) + s(15d_1 - 15d_2 + 12) + 12d_1 - 12d_2 + 5)/27$.

Case 3. $d_1 \leq d_2$.
To prove $k_1(D^{(2)}) \geq (2s^3 + s^2(3d_2 - 3d_1 + 13) + s(12d_2 - 12d_1 + 20))/9 + d_2$ it suffices to show, for each $t = 1, 2, \ldots, n$, $k_1(v_t : v_{y'}, v_x) \geq (2s^3 + s^2(3d_2 - 3d_1 + 13) + s(12d_2 - 12d_1 + 20))/9 + d_2 + d(v_1, v_t)$. For this purpose one sets $q_1 = b(C_2)r(P_{v_{y'},v_t}) - r(C_2)b(P_{v_{y'},v_t})$ and $q_2 = r(C_1)b(P_{v_x,v_t}) - b(C_1)r(P_{v_x,v_t})$ and then employs Lemma 2.

To prove $k_1(D^{(2)}) \leq (2s^3 + s^2(3d_2 - 3d_1 + 13) + s(12d_2 - 12d_1 + 20))/9 + d_2$ it suffices to show that, for each $j = 1, 2, \ldots, n$, there is a $v_j \xrightarrow{(h,g)} v_1$ walk with $h = (4s^3 + s^2(6d_2 - 6d_1 + 24) + s(21d_2 - 21d_1 + 27) + 15d_2 + 21d_1 - 20)/27$ and $g = (2s^3 + s^2(3d_2 - 3d_1 + 15) + s(15d_2 - 15d_1 + 33) + 12d_2 - 12d_1 + 20)/27$. □

4. The Case $s \equiv 1 \bmod 3$

In this section we discuss the competition index of a strongly connected primitive two-colored digraph on n vertices consisting of the cycle $C_1 : v_1 \rightarrow v_2 \rightarrow \cdots \rightarrow v_c \rightarrow v_{c+1} \rightarrow v_{c+2} \rightarrow \cdots \rightarrow v_s \rightarrow v_1$ of length s and the cycle $C_2 : v_1 \rightarrow v_2 \rightarrow \cdots \rightarrow v_c \rightarrow v_{s+1} \rightarrow v_{s+2} \rightarrow \cdots \rightarrow v_n \rightarrow v_1$ of length $s+3$ where $s \geq 7$, $s \equiv 1 \bmod 3$, $n = 2s + 3 - c$ and $1 \leq c \leq s - 1$. Let the cycle matrix of $D^{(2)}$ be $M = \begin{bmatrix} r(C_1) & r(C_2) \\ b(C_1) & b(C_2) \end{bmatrix}$. Since $D^{(2)}$ is primitive, Theorem 1 implies $\det(M) = \pm 1$. We assume that $\det(M) = 1$ and hence the cycle matrix of $D^{(2)}$ is $M = \begin{bmatrix} (2s+1)/3 & (2s+7)/3 \\ (s-1)/3 & (s+2)/3 \end{bmatrix}$. This implies the number of blue arcs of $D^{(2)}$ is at least $(s+2)/3$ and at most $(2s+1)/3$. We discuss only the case where $D^{(2)}$ has exactly $(2s+1)/3$ blue arcs which consists of a $(0, (s-1)/3)$-path that lies on C_1 and a $(0, (s+2)/3)$-path that lies on C_2.

Let the $(0, (s-1)/3)$-path on C_1 be the path $P_{v_x,v_{x'}}$ from v_x to $v_{x'}$ and the $(0, (s+2)/3)$-path on C_2 be the path $P_{v_y,v_{y'}}$ from v_y to $v_{y'}$. We define $d_1 = d(v_{x'}, v_1)$ and $d_2 = d(v_{y'}, v_1)$.

Theorem 4. *Let s be a positive integer such that $s \geq 7$ and $s \equiv 1 \bmod 3$. Let $D^{(2)}$ be a strongly connected primitive two-colored digraph consisting of two cycles C_1 and C_2 of lengths s and $s+3$, respectively. Suppose the cycle C_1 contains a $(0, (s-1)/3)$-path $P_{v_x,v_{x'}}$ and the cycle C_2 contains* *a* $(0,(s+2)/3)$*-path* $P_{v_y,v_{y'}}$*. If the paths* $P_{v_x,v_{x'}}$ *and* $P_{v_y,v_{y'}}$ *have no vertices in common, then*

$$k_1(D^{(2)}) = \begin{cases} \frac{2s^3+11s^2+17s+6}{9} + d_2 & \textit{if } d_1 < d_2 \textit{ and } d_2 - d_1 < 3, \\ \frac{2s^3+s^2(3d_2-3d_1+5)+s(6d_2-6d_1-4)-3}{9} + d_1 & \textit{if } d_1 < d_2 \textit{ and } d_2 - d_1 \geq 3, \\ \frac{2s^3+s^2(3d_1-3d_2+11)+s(6d_1-6d_2+17)+6}{9} + d_2 & \textit{if } d_1 \geq d_2. \end{cases}$$

Sketch of Proof. Since the proof is similar to the proof of Theorem 3 we only present the sketch of the proof.

Case 1. $d_1 < d_2$ and $d_2 - d_1 < 3$.
To prove $k_1(D^{(2)}) \geq (2s^3 + 11s^2 + 17s + 6)/9 + d_2$ we first show, for each $t = 1, 2, \ldots, n$, that $k_1(v_t : v_{y'}, v_y) \geq (2s^3 + 11s^2 + 17s + 6)/9 + d_2 + d(v_1, v_t)$. For this purpose we set $q_1 = b(C_2)r(P_{v_{y'},v_t}) - r(C_2)b(P_{v_{y'},v_t})$ and $q_2 = r(C_1)b(P_{v_y,v_t}) - b(C_1)r(P_{v_y,v_t})$ and then we employ Lemma 2.

To prove $k_1(D^{(2)}) \leq (2s^3 + 11s^2 + 17s + 6)/9 + d_2$ we show that, for each $j = 1, 2, \ldots, n$, there is a $v_j \xrightarrow{(h,g)} v_1$ walk with $h = (4s^3+24s^2+39s+14)/27+d_2$ and $g = (2s^3+9s^2+12s+4)/27$.

Case 2. $d_1 < d_2$ and $d_2 - d_1 \geq 3$.
To prove $k_1(D^{(2)}) \geq (2s^3 + s^2(3d_2 - 3d_1 + 5) + s(6d_2 - 6d_1 - 4) - 3)/9 + d_1$ it suffices to show, for each $t = 1, 2, \ldots, n$, $k_1(v_t : v_{y'}, v_x) \geq (2s^3 + s^2(3d_2 - 3d_1 + 5) + s(6d_2 - 6d_1 - 4) - 3)/9 + d_1 + d(v_1, v_t)$. For this purpose one sets $q_1 = b(C_2)r(P_{v_{y'},v_t}) - r(C_2)b(P_{v_{y'},v_t})$ and $q_2 = r(C_1)b(P_{v_x,v_t}) - b(C_1)r(P_{v_x,v_t})$ and then employs Lemma 2.

To prove $k_1(D^{(2)}) \leq (2s^3 + s^2(3d_2 - 3d_1 + 5) + s(6d_2 - 6d_1 - 4) - 3)/9 + d_1$ it suffices to show that, for each $j = 1, 2, \ldots, n$, there is a $v_j \xrightarrow{(h,g)} v_1$ walk with $h = (4s^3 + s^2(6d_2 - 6d_1 + 12) + s(15d_2 - 15d_1 - 9) - 7)/27 + 7d_1/9$ and $g = (2s^3 + s^2(3d_2 - 3d_1 + 3) + s(3d_2 - 3d_1 - 3) - 2)/27 + 2d_1/9$.

Case 3. $d_1 \geq d2$.
To prove $k_1(D^{(2)}) \geq (2s^3 + s^2(3d_1 - 3d_2 + 11) + s(6d_1 - 6d_2 + 17) + 6)/9 + d_2$ it suffices to show, for each $t = 1, 2, \ldots, n$, $k_1(v_t : v_{x'}, v_y) \geq (2s^3 + s^2(3d_1 - 3d_2 + 11) + s(6d_1 - 6d_2 + 17) + 6)/9 + d_2 + d(v_1, v_t)$. For this purpose one sets $q_1 = b(C_2)r(P_{v_{x'},v_t}) - r(C_2)b(P_{v_{x'},v_t})$ and $q_2 = r(C_1)b(P_{v_y,v_t}) - b(C_1)r(P_{v_y,v_t})$ and then employs Lemma 2.

To prove $k_1(D^{(2)}) \leq (2s^3 + s^2(3d_1 - 3d_2 + 11) + s(6d_1 - 6d_2 + 17) + 6)/9 + d_2$ it suffices to show that, for each $j = 1, 2, \ldots, n$, there is a $v_j \xrightarrow{(h,g)} v_1$ walk with $h = (4s^3 + s^2(6d_1 - 6d_2 + 24) + s(15d_1 - 15d_2 + 39) + 14)/27 + (7d_2 + 2d_1)/9$ and $g = (2s^3 + s^2(3d_1 - 3d_2 + 9) + s(3d_1 - 3d_2 + 12) + 4)/27 + (2d_2 - 2d_1)/9$. □

References

[1] Fornasini E and Valcher M E 1998 *SIAM J. Matrix Anal. Appl.* **19** 71–88.
[2] Kim H K 2010 *Linear Algebra Appl.* **433** 72–79.
[3] Kim H K and Park S G 2012 *Linear Algebra Appl.* **436** 86–98.
[4] Mardiningsih and Pasaribu M L 2016 *AIP Conf. Proc* **1775** 030056.
[5] Mulyono, Sumardi H and Suwilo S 2015 *Far East J. Math. Sci.* **96 (1)** 113–132.

IOP Conf. Series: Materials Science and Engineering **300** (2018) 012076 doi:10.1088/1757-899X/300/1/012076

On the 5-Local Profiles of Trees

Efron Manik[1,2], Saib Suwilo[1], Tulus[1], Opim Salim Sitompul[3]

[1] Department of Mathematics, University of Sumatera Utara, Medan-Indonesia,
[2] Department of Mathematics Education, Nommensen HKBP University, Medan-Indonesia,
[3] Department of Information Technology, University of Sumatera Utara, Medan-Indonesia

E-mail: efmanik@gmail.com

Abstract. For T and U trees, we denote $c(U, T)$ as the number of copies U in T, in other words: the number of homomorphism injective from U to T. The path, star, and y-shaped tree is a list of all the homomorphisms of the tree by 5-point. Furthermore, the 5-profile of the T_n tree is symbolized by P, S, and Y for the paths, star, and y-shaped trees respectively. So the set limit of 5-profile, $\Delta_T(5)$, is a subset of $\boldsymbol{R}^3$. Notation $(p_1, p_2, p_3) \in \Delta_T(5)$ corresponds to P, S, and Y respectively. $\mathscr{P}(5)$ is a projection of $\Delta_T(5)$ on the first two coordinates. Determining the area boundary of $\mathscr{P}(5)$ is a challenging task. The d-millipede tree produces the points of $\mathscr{P}(5)$ whose sum of p_1 and p_2 are very small at some point. The study of whether the point generated by the d-millipede tree is the lower bound of the set $\mathscr{P}(5)$ is a question that requires investigation. The expanded d-millipede tree is defined as the tree produced by adding sides to the leaves of the d-milipede tree. This paper discusses the coordinates of point $\mathscr{P}(5)$ generated by the expanded d-millipede tree. The d-millipede tree is a tree with the least number of sub-trees from the family of trees with the same number of points and degrees. The d-millipede tree produces the lowest point in region $\mathscr{P}(5)$. The points generated by the optimum tree and points generated by the expanding d-millipede tree are above the curve connecting the points generated by the d-millipede tree. So we estimate that the lower boundary region $\mathscr{P}(5)$ is the curve connecting the points generated by the d-millipede tree.

1. Introduction

Image can be viewed as a graph and image segmentation can be generated using the methods in graph [1, 2]. Peter [3] uses a minimum spanning tree to segment an image named with the MSTSSCIMLRO algorithm. The algorithm is run where the first step is to write the image in graph. Determining the minimum spanning tree of the graph is the second step. The next step is to remove all sides that are larger than the threshold that form the tree into several sub-trees. A sub tree is seen as a segment in the image.

The minimum spanning tree of a graph is not unique. Following the Kruskal's algorithm, we can conclude that the difference occurs due to the different sequence of edges of equal weight in the graph. Although the minimum spanning tree is not unique to an image, but the image segmentation produced by the different minimum spanning tree is the same [4].

Determining the same object in a digital image is a challenging task [5]. If segmentation is obtained by cutting edges of a tree produced by the minimum spanning tree method, then determining the same sub-tree is the same case by searching for the same object or segment in an image.

Let U, T tree, we denote $c(U, T)$ as the number of copies U in T, or in other words: the number of homomorphism is injecting from U to T. Suppose T_1^k, ..., $T_{N_k}^k$ are lists of all the homomorphisms of trees with k-points, where T_1^k is the path and T_2^k is the star. A k-profile of the T tree is the $p^{(k)}(T) \in R^{N_k}$ vector in which the i-th coordinate is:

$$\left(p^{(k)}(T)\right)_i = \frac{c(T_i^k, T)}{Z_k(T)}, \text{ where } Z_k(T) = \sum_{j=1}^{N_k} c(T_j^k, T).$$

We are interested in understanding the set limit of k-profiles, namely:

$$\Delta_T(k) = \left\{p \in R^{N_k} : \exists (T_n), |T_n| \xrightarrow[n\to\infty]{} \infty, \text{ and } p^{(k)}(T_n) \xrightarrow[n\to\infty]{} p\right\},$$

where $|T|$ is the number of points in T.

For the case of $k = 5$, we will obtain $N_k = 3$. The path, star, and y-shaped tree are lists of all homomorphism of the 5-point tree. Furthermore, the 5-profile of the T_n tree is symbolized by P, S, and Y for path, star, and y-shaped trees respectively. Thus the limit of the set of 5-profiles, $\Delta_T(5)$, is the subset of $\boldsymbol{R}^3$. The symbols $(p_1, p_2, p_3) \in \Delta_T(5)$ correspond to P, S, and Y respectively. $\mathcal{P}(5)$ is the projection of $\Delta_T(5)$ to the first two coordinates. Determining the lower bounds of $\mathcal{P}(5)$ is a challenging task. The d-millipede tree produces the points of $\mathcal{P}(5)$ whose sum of p_1 and p_2 from that point is very small at some point. Whether the points generated by the d-millipede tree is the lower bound of $\mathcal{P}(5)$ is a question that needs to be investigated. The expanded d-millipede tree is a d-milipede tree on which leaves are followed by several edges. This paper discusses the coordinates of the $\mathcal{P}(5)$ points generated by the expanded d-millipede tree. The problem to be discussed in this paper is to compare the coordinates of the points generated by the d-millipede tree with the coordinates of the points generated by the expanded d-millipede tree on $\mathcal{P}(5)$.

This paper discusses the same sub tree of a tree. Section 2 discusses the properties in the k-profile tree and the set of limits of the k-profile tree. Specifically for $k = 2$ will be discussed more deeply. Section 3 will define an expanded d-millipede. The point coordinates of the limit set of k-profile trees generated by d-milippede and the expanded d-millipede will be compared. Finally, this paper ends with Section of conclusions and suggestions.

2. The 5-Local Profil Tree

In contrast to the set of k-profile limits of the graph, the set of k-profile limits of the tree, $\Delta_T(k)$, has some interesting properties [6]. If two different points are members of $\Delta_T(k)$ then all the points between those points are also members of $\Delta_T(k)$. This is summarized in the following theorem.

Theorem 1 The set $\Delta_T(k)$ is convex.

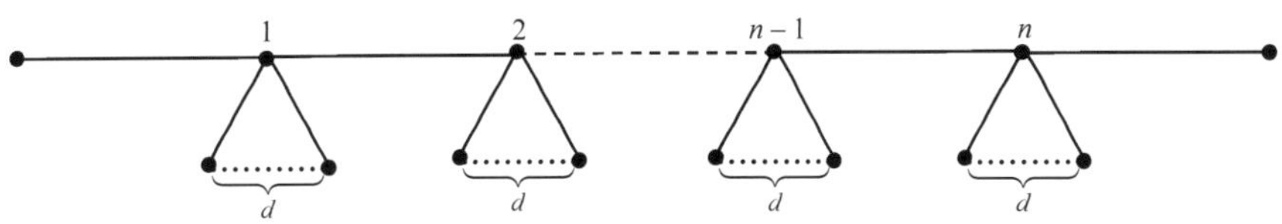

Figure 1. d-millipede

A d-millipede is a tree that all non-leaf points each have d+2 degrees, as shown in Figure 1. The number of non-leaf points, n, is called the length of d-millipede.

Notify $P_k(T) = c(T_1^k, T)$ and $S_k(T) = c(T_2^k, T)$, and T can also not be written for a clear reference. Suppose T_n is (k–4)-millipede with length n, and R_k is $\frac{k-4}{2}$ -millipede with length 2 where k is even. It is easy to see that for $k \geq 6$ apply:

$$z_k(T_n) \geq c(R_k, T_n) \geq 2(n-2)\binom{k-3}{(k-2)/2} \geq (n-2)(3/2)^{k/2},$$

and

$$S_k(T_n) = 0, \quad P_k(T_n) \leq n(k-3)^2.$$

Since $\lim_{k\to\infty} \frac{(k-3)^2}{(3/2)^{k/2}} = 0$ then p1 + p2 can be made close to zero.

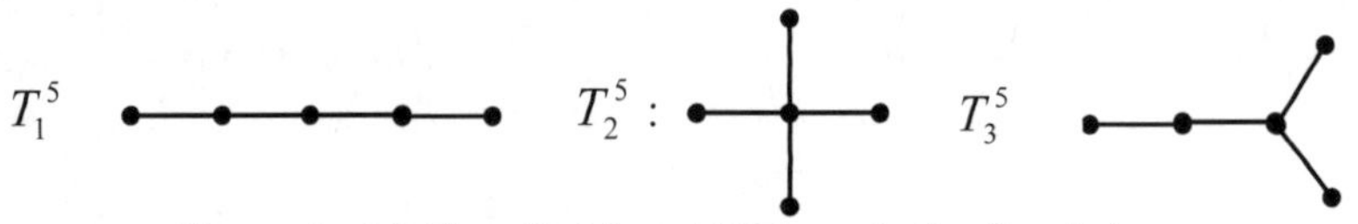

Figure 2. All Tree that is not Homorpic for 5-points

For $k = 5$, we have only three non-homomorphic trees as in Figure 2, so $N_k = 3$. We will denote $P = c(T_1^5, T)$ for path, $S = c(T_2^5, T)$ for star, and $Y = c(T_3^5, T)$ corresponding to its shape. Thus the set of 5-profile limits, $\Delta_T(5)$, is the subset of $\boldsymbol{R}^3$ corresponding to P, S, and Y respectively. Symbols (p_1, p_2, p_3) $\in \Delta_T(5)$ corresponding to P, S, and Y respectively. $\mathscr{P}(5)$ is a projection on the first two coordinates of $\Delta_T(5)$. This set can be written with $\mathscr{P}(5) = \{(p_1, p_2) \in \boldsymbol{R}^2 \mid (p_1, p_2, p_3) \in \Delta_T(5)\}$. By selecting T_n as a path of length n, it is easy to prove that point titik (1, 0) $\in \mathscr{P}(5)$. Likewise, choosing T_n as star n points then it is easy to show that point (0, 1) $\in \mathscr{P}(5)$. Using Theorem 1 it is concluded that the line segment connecting both points (1.0) and (0,1) also lies in $\mathscr{P}(5)$.

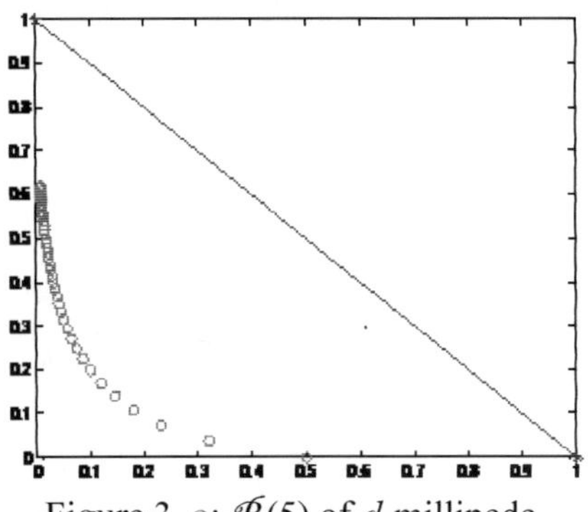

Figure 3. o: $\mathscr{P}(5)$ of d-millipede

Suppose the T_n tree is a d-millipede with length n. It is easy to check that if $d = 1$ then the point (0.5, 0) $\in \mathscr{P}(5)$ which corresponds to T_n. If d = 2, 3, ... then $P = (d+1)^2(n-2)$, $S = nC_4^{d+2}$, and $Y = 2(d+1)(n-2)C_2^{d+1} + 2(d+1)C_2^{d+1}$ .Thus $Z_5(T) = P + S + Y = (d+1)^2(n-2) + nC_4^{d+2} + 2(d+1)(n-2)C_2^{d+1} + 2(d+1)C_2^{d+1}$. By taking the limit obtained:

$$\frac{1}{(d+1)^2 + C_4^{d+2} + 2(d+1)C_2^{d+1}}\left((d+1)^2, C_4^{d+2}, 2(d+1)C_2^{d+1}\right) \in \Delta_T(5)$$

and

$$\frac{1}{(d+1)^2 + C_4^{d+2} + 2(d+1)C_2^{d+1}}\left((d+1)^2, C_4^{d+2}\right) \in \mathscr{P}(5).$$

All points of $\mathscr{P}(5)$ can be seen in Figure 3 which is the same as [6]. Using Theorem 1 it is concluded that all regions between the points corresponding to d-millipede and the line connecting the points (0, 1) and (1, 0) are the points of $\mathscr{P}(5)$.

3. Results and Discussion

The expanded d-millipede tree is a d-millipede tree whose leaves are connected to several edges of the same amount (eg. a number of i) for each leaf, as Figure 4. The length of the expanded d-millipede tree is the same as the length of d-millipede that forms it. The addition of these sides will affect the values of P, S, and Y. Specifically for $i = 1$ or $i = 2$, the value of star S does not change, but the value of P and Y value change.

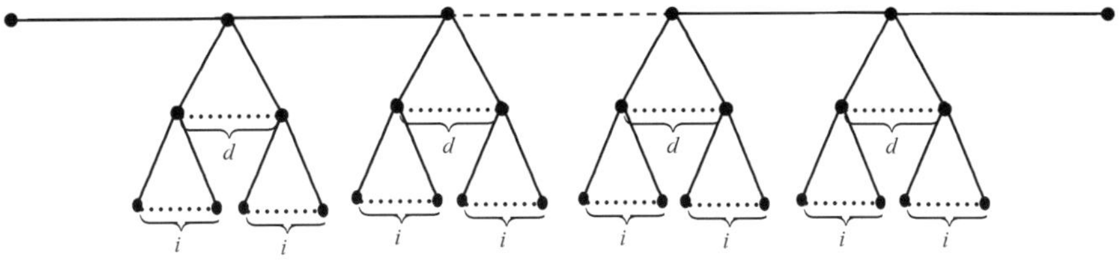

Figure 4. d-millipede with Expansion i

Suppose the T_n tree is a d-millipede with length n with an expansion i. It is easy to check that if i = 1, 2 then $P = (d+1)^2(n-2) + 2id\,(d+1)(n-1) + ni^2 C_2^d$, $S = nC_4^{d+2}$, and $Y = 2(d+1)(n-1)C_2^{d+1} + nd(d+1)\,C_2^i + nid\,C_2^{d+1}$. Thus $Z_5(T) = P + S + Y = (d+1)^2(n-2) + 2id\,(d+1)(n-1) + ni^2 C_2^d + nC_4^{d+2} + 2(d+1)(n-1)C_2^{d+1} + nd(d+1)\,C_2^i + nid\,C_2^{d+1}$. By taking the limit obtained:

$$\frac{1}{Q}\left((d+1)^2 + 2id(d+1) + \tfrac{1}{2}i^2d(d-1), \tfrac{1}{24}(d+2)(d+1)d(d-1), (d+1)^2 d + \tfrac{1}{2}d(d+1)i(i-1) + \tfrac{1}{2}id^2(d+1)\right) \in \Delta_T(5)$$

Where $Q = (d+1)^2 + 2id(d+1) + \tfrac{1}{2}i^2d(d-1) + \tfrac{1}{24}(d+2)(d+1)d(d-1) + (d+1)^2 d + \tfrac{1}{2}d(d+1)i(i-1) + \tfrac{1}{2}id^2(d+1)$, and

$$\frac{1}{Q}\left((d+1)^2 + 2id(d+1) + \tfrac{1}{2}i^2d(d-1), \tfrac{1}{24}(d+2)(d+1)d(d-1)\right) \in \mathscr{P}(5).$$

If $i = 3, 4, \ldots$ then $P = (d+1)^2(n-2) + 2id\,(d+1)(n-1) + ni^2 C_2^d$, $S = nC_4^{d+2} + ndC_4^{i+1}$, and $Y = 2(d+1)(n-1)C_2^{d+1} + nd(d+1)\,C_2^i + nid\,C_2^{d+1}$. Thus $Z_5(T) = P + S + Y = (d+1)^2(n-2) + 2id\,(d+1)(n-1) + ni^2 C_2^d + nC_4^{d+2} + ndC_4^{i+1} + 2(d+1)(n-1)C_2^{d+1} + nd(d+1)\,C_2^i + nid\,C_2^{d+1}$. By taking the limit obtained:

$$\frac{1}{Q}\left((d+1)^2 + 2id(d+1) + \tfrac{1}{2}i^2d(d-1), \tfrac{1}{24}(d+2)(d+1)d(d-1) + \tfrac{1}{24}d(i+2)(i+1)i(i-1), (d+1)^2 d + \tfrac{1}{2}d(d+1)i(i-1) + \tfrac{1}{2}id^2(d+1)\right) \in \Delta_T(5)$$

Where $Q = (d+1)^2 + 2id(d+1) + \tfrac{1}{2}i^2d(d-1) + \tfrac{1}{24}(d+2)(d+1)d(d-1) + \tfrac{1}{24}d(i+2)(i+1)i(i-1) + (d+1)^2 d + \tfrac{1}{2}d(d+1)i(i-1) + \tfrac{1}{2}id^2(d+1)$, and

$$\frac{1}{Q}\left((d+1)^2 + 2id(d+1) + \tfrac{1}{2}i^2d(d-1), \tfrac{1}{24}(d+2)(d+1)d(d-1) + \tfrac{1}{24}d(i+2)(i+1)i(i-1)\right) \in \mathscr{P}(5).$$

For small d and i values, ie. d, i = 1, 2, 3, ..., 30, numerical inference is presented as Figure 5. It was found that the point coordinates by the expanded d-millipede tree lie above the curve connecting the point generated by the d-millipede tree.

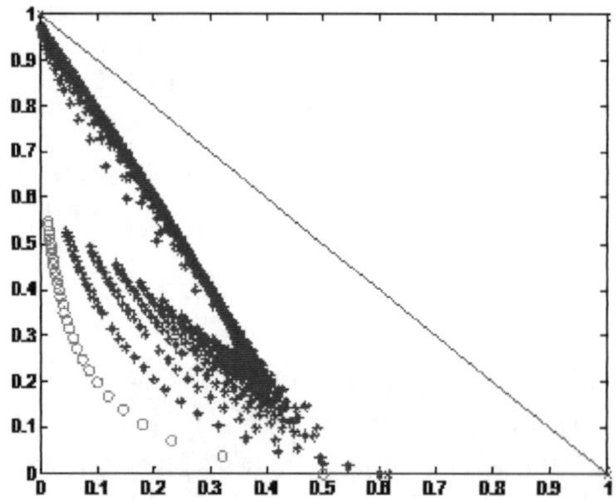

Figure 5. o: $\mathscr{P}(5)$ of *d*-millipede, *: $\mathscr{P}(5)$ of *d*-millipede with extension *i*

For the value of $i > 30$, we will get the value of $S = nC_4^{d+2}$ corresponding to the d-millipede tree much smaller than the value of $S = nC_4^{d+2} + ndC_4^{i+1}$ corresponding to the *d*-millipede tree with the expansion *i*. So it is easy to show that the point coordinates of $\mathscr{P}(5)$ generated by the expanded *d*-millipede tree lie above the curve connecting the point coordinates of $\mathscr{P}(5)$ generated by the *d*-millipede tree.

The optimal tree *T* is the tree that has the most number of subtrees from the family tree where the trees have the number of points and degrees from each point are the same [7]. The formation of such trees can be seen by looking at Figure 6. Note the 2-millipede tree with a length of 20. So this tree has 20 points of degree (2 + 2), and has 20 times 2 and plus 2 (equal to 42) points of one degree. So that the sequence of degrees, ***d***, consists of the first 20 terms of 4 and then the other is 1. The optimum tree can be seen in Figure 6.

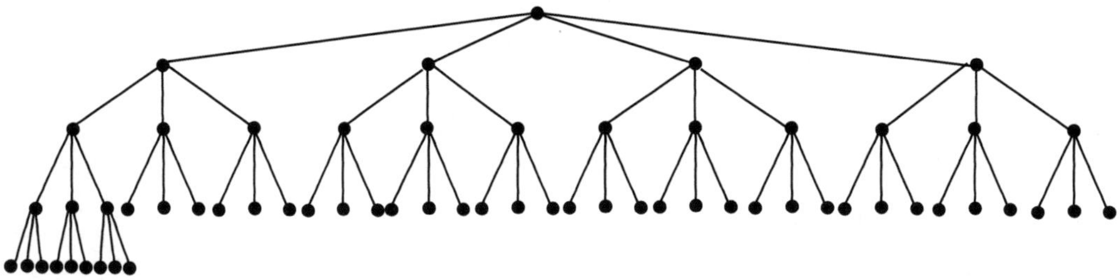

Figure 6. Optimal Tree with 20 Points of Degree 4 and Degree 1 for Other Points

In the optimum tree, each addition of points $(d+1)^2$, at the 5th and subsequent levels, will add the value of *P*, *S*, and *Y* as multiples of its own value. By adding a point of $n(d+1)^2$, the value increases to $P_n = n((d+1)^2 + d(d+1)^2 + (d+1)^2\ C_2^{d+1}) = n((d+1)^2 + d(d+1)^2 + ½(d+1)d(d+1)^2)$. The S_n value becomes $S_n = n(d+1)$, and Y_n becomes $Y_n = n((d+1)\ C_2^{d+1} + dC_2^{d+1}) = ½\, n(d+1)d(2d+1)$. If a sequence is convergent, then the tail of the sequence or sub sequence will converge to the convergence point of the original sequence.

Figure 7 shows that the point generated by the optimum tree for various degrees *d* still remains within the region of $\mathscr{P}(5)$ in Figure 3. The projection area of $\Delta_T(5)$ is always between the segments connecting the points (0, 1) and (1, 0) with a curve connecting the profile points generated by the *d*-millipede tree.

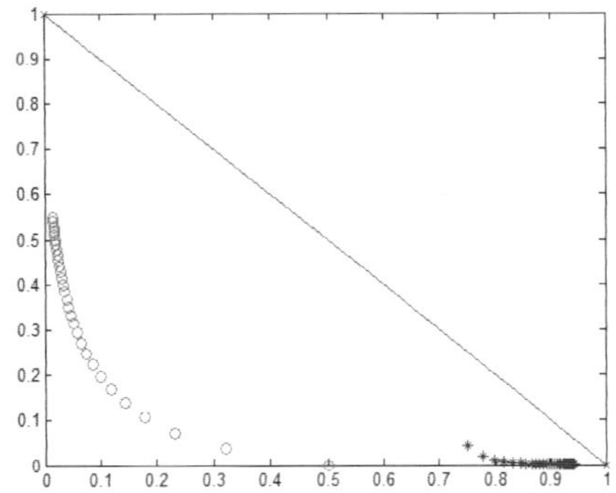

Figure 7. o: $\mathscr{P}(5)$ of *d*-millipede, *: $\mathscr{P}(5)$ of the Optimal Tree

The *d*-millipede tree is a tree with the least number of sub-trees from the family of trees with the same number of points and degrees [8]. The *d*-millipede tree produces the lowest point in region $\mathscr{R}(5)$. The points generated by the optimum tree and points generated by the expanding *d*-millipede tree are above the curve connecting the points generated by the d-millipede tree. So we estimate that the lower boundary region $\mathscr{R}(5)$ is the curve connecting the points generated by the *d*-millipede tree.

4. Conclusions and Suggestions

A *d*-millipede is a tree that all non-leaf points each have *d*+2 degrees. The expanded *d*-millipede tree is a *d*-millipede tree whose leaves are connected to several edges of the same amount (eg. a number of *i*) for each leaf. The addition of these sides will affect the values of *P*, *S*, and *Y*.

The *d*-millipede tree is a tree with the least number of sub-trees from the family of trees with the same number of points and degrees. The *d*-millipede tree produces the lowest point in region $\mathscr{R}(5)$. The points generated by the optimum tree and points generated by the expanding d-millipede tree are above the curve connecting the points generated by the *d*-millipede tree. So we estimate that the lower boundary region $\mathscr{R}(5)$ is the curve connecting the points generated by the *d*-millipede tree.

References

[1] Dessai, R.S.G, S.D.C.S. Araujo, and C. Fernandes. Object Identification using Graph Theory. *International Journal of Engineering Science*, 2017, 10783.

[2] Kim, S., S. Nowozin, P. Kohli, and CD Yoo. Higher-Order Correlation Clustering for Image Segmentation. In: *Advances in neural information processing systems*. 2011. p. 1530-1538.

[3] Peter, S.J., "Minimum Spanning Tree-based Structural Similarity Clustering for Image Mining with Local Region Outliers," *IJCA*, 2010, 8(6), pp. 0975- 8887

[4] Manik, E., S. Suwilo, Tulus, and O.S. Sitompul. The Uniqueness of Image Segmentation Generated by Different Minimum Spanning Tree. *GJPAM,* 2017, 13(7), pp. 2975–2982

[5] Yuan, J. 2009. Image and Video Data Mining. Dissertation

[6] Bubeck, S. and N. Linial. On the local profiles of trees. *Journal of Graph Theory*, 2016, 81.2: 109-119.

[7] Zhang, X. M., Zhang, X. D., Gray, D., & Wang, H. The number of subtrees of trees with given degree sequence. *Journal of Graph Theory*, 2013, 73(3), 280-295.

[8] Sills, A.V. and· H. Wang. The Minimal Number of Subtrees of a Tree. *Graphs and Combinatorics*, 2015, 31:255–264.

Scrambling index of certain primitive graphs consisting of two disjoint odd cycles connected by some paths

Mariana Sitorus, Febryandi Ginting, Putri K Nasution* and Sundari Atikah

Department of Mathematics, Universitas Sumatera Utara, Medan, Indonesia

E-mail: *putrikhairiah09@gmail.com

Abstract. The scrambling index of a primitive graph G is the smallest positive integer k such that for each pair of vertices u and v there is a vertex w such that there exists a uw-walk and a vw-walk of length k. We discuss the scrambling index of primitive graphs G consisting of two disjoint odd cycles each of length s connected by some paths of length ℓ_0. For such primitive graphs G we present formulae for scrambling indices that depend on s and ℓ_0.

1. Introduction

Let $G(V, E)$ be a graph. Let u and v be any vertices in G, a uv-walk W_{uv} of length t connecting u and v is a sequence of vertices $u = u_0, u_1, u_2, \ldots, u_t = v$ and a sequence of edges

$$\{u_0, u_1\}, \{u_1, u_2\}, \ldots \{u_{t-1}, u_t\},$$

where the vertices and the edges are not necessarily distinct. A walk W connecting u and v is denoted by W_{uv}. A walk connecting u and v is closed whenever $u = v$, and is open otherwise. A path P_{uv} connecting u and v is a walk W_{uv} with distinct vertices, except possibly $u = v$. A cycle is a closed path. The length of a walk W_{uv} is the number of edges in W_{uv}, and is denoted by $\ell(W_{uv})$. A walk W_{uv} with sequence of edges $\{u_0, u_1\}, \{u_1, u_2\}, \ldots \{u_{t-1}, u_t\}$ is also denoted by

$$u = u_0 - u_1 - u_2 - \cdots - u_{t-1} - u_t = v.$$

By an odd (even) cycle we mean a cycle of odd (even) length.

A graph G is connected provided that for each pair of vertices u and v in G there is uv-walk in G. The distance between vertices u and v in a connected graph G, denoted $d(u, v)$, is the length of a shortest path connecting u and v. A connected graph G is primitive provided there is a positive integer m such that for each pair of vertices u and v in G, there is a uv-walk of length m. The smallest of such positive integer m is the exponent of G and is denoted by $\exp(G)$. It is well known (see e.g [1]) that a graph G is primitive if only if G has a cycle of odd length.

In 2009, Akelbek and Kirkland [2, 3] introduced the notion of scrambling index of graph. The *scrambling index* of a primitive graph G, denoted by $k(G)$, is the smallest positive integer k such that for any pair of distinct vertices u and v in G there is a vertex $w$such that there is a uw-walk and a vw-walk of length k. Chen and Liu [4] have shown that for a primitive graph G on n vertices with the smallest cycle of length s, then $k(G) \leq n - (s+1)/2$. They also characterize primitive graph that attains the upper bound $n - (s+1)/2$.

This paper discuss the scrambling index of a class of primitive graph consisting of two disjoint odd cycles connected by some paths. For this purpose, we first introduce some preliminaries in setting up a lower and upper bound for scrambling index. Then we discuss the scrambling index of barbell and finally we discuss scrambling index of a prism.

2. Background

For a pair of distinct vertices u and v in G the local scrambling index of u and v is the number

$$k_{u,v}(G) = \min_{w \in V(G)} \{k : \text{there are } uw\text{-walk and } vw\text{-walk of length } k\}.$$

We note that if the local scrambling index of u and v is $k_{u,v}(G)$, then for any positive integer $\ell \geq k_{u,v}(G)$ we can find a vertex w' such that there is a walk uw'-walk and a vw'-walk of length ℓ. This implies

$$k(G) = \max_{u,v \in V(G)} \{k_{u,v}(G)\}.$$

Proposition 1. *Let G be a primitive graph and k' be an even positive integer. If for each pair of vertices u and v in G there is an even walk W_{uv} of length $\ell(W_{uv}) \leq k'$, then $k(G) \leq k'/2$.*

Proof. Let u and v be any two vertices in G and let W_{uv} be an even walk $u = v_0 - v_1 - v_2 - \cdots - v_{2m-1} - v_{2m} = v$ for some positive integer m of length $\ell(W_{uv}) \leq k'$. Let C_2 be the closed walk $v_{2m} - v_{2m-1} - v_{2m}$ of lenght 2. Then the walk W'_{uv} that starts at u, moves to v along the walk W_{uv} and then moves $k' - \ell(W_{uv})$ times around the cycle C_2 is a uv-walk of length k'. Since k' is even, there is a vertex w such that there is a uw-walk of length $k'/2$ and there is a vw-walk of length $k'/2$. Hence $k(G) \leq k'/2$. □

3. Preliminary results

Let C_1 and C_2 be disjoint odd cycles each of length s. Let P be a path of length ℓ_0 with one end vertex on C_1 and the other end on C_2. A connected graph G consisting of two disjoint odd cycles C_1 and C_2 connected by a path P of length ℓ_0 is called an (s, ℓ_0)-barbell. Let x_1 be a vertex in common to C_1 and P and let x_2 be a vertex in common C_2 and P. We note that $d(x_1, x_2) = \ell_0$. For each vertex $u \in C_1$, let P_{ux_1} be the shortest ux_1-path and let P'_{ux_1} be the ux_1-path of length $s - d(u, x_1)$. Similarly, for each vertex $u \in C_2$, let P_{ux_2} be the shortest ux_2-path and let P'_{ux_2} be the ux_2-path of length $s - d(u, x_2)$.

Lemma 2. *Let $s \geq 5$ be an odd integer and G be an (s, ℓ_0)-barbell. If ℓ_0 is odd, then $k(G) = (s + \ell_0)/2$.*

Proof. Let u and v be the two median vertices on P such that $d(u, x_1) = (\ell_0 - 1)/2$ and $d(v, x_2) = (\ell_0 - 1)/2$. Notice that $d(u, v) = 1$. Then the shortest even uv-walk is the walk W_{uv} that starts at u, moves to x_1 along the ux_1-path, then moves one time around the cycle C_1 and back at x_1, finally moves to v along the x_1v-path. We note that $\ell(W_{uv}) = s + \ell_0$. Hence $k(G) \geq (s + \ell_0)/2$

By Proposition 1 it remains to show that for each pair of vertices u and v there is an even walk W_{uv} of length $\ell(W_{uv}) \leq s + \ell_0$. If $d(u, v)$ is even then we are done. So we assume that $d(u, v)$ is odd.

Case 1. Both vertices $u, v \in C_1$ or $u, v \in C_2$. Since u and v both lie on the same cycle, then there is an even uv-path of length $s - d(u, v) < s + \ell_0$.

Case 2. The vertex $u \in C_1$ and $v \in C_2$. We note that $d(u,v) = d(u,x_1) + \ell_0 + d(x_2,v)$. Assume without loss of generality that $d(u,x_1) \leq d(v,x_2)$. Then the uv-walk W_{uv} that starts at u, moves to x_1 along the ux_1-path of length $d(u,x_1)$, then moves to x_2 along the path P, and finally moves to v along the vx_2-path of length $s - d(v,x_2)$ is an even uv-walk of length $s + \ell_0 + d(u,x_1) - d(v,x_2)$. Since $d(u,x_1) \leq d(v,x_2)$, we have $\ell(W_{uv}) \leq s + \ell_0$.

Case 3. The vertex $u \in C_1$ and $v \in P$ or $u \in C_2$ and $v \in P$. We may assume that $u \in C_1$. Notice that $d(u,v) = d(u,x_1) + d(x_1,v)$ is odd. Then the uv-walk W_{uv} that starts at u, moves to x_1 along the ux_1-path of length $s - d(u,x_1)$, and finally moves to v along the x_1v-path is an even uv-walk of length $\ell_0(W_{uv}) = s - d(u,x_1) + d(x_1,v) \leq s + \ell_0$.

Case 4. The vertices $u, v \in P$. We assume that $d(x_1,u) < d(x_1,v)$. If $d(u,x_1) \leq d(v,x_2)$, then $d(u,x_1) \leq (\ell_0 - 1)/2$. The uv-walk W_{uv} that starts at u, moves to x_1 along the ux_1-path of length $d(u,x_1)$, then moves one time around the cycle C_1 and back at x_1, then moves to u along the x_1u-path of length $d(u,x_1)$, and finally moves to v along the uv-path of length $d(u,v)$, is an even uv-walk. Notice that $\ell_0(W_{uv}) = s+2d(u,x_1)+d(u,v)$. Since $\ell_0 = d(x_1,u)+d(u,v)+d(v,x_2)$ and $d(u,x_1) \leq d(v,x_2)$, we have $\ell(W_{uv}) \leq s + \ell_0$.

If $d(u,x_1) > d(v,x_2)$, then $d(v,x_2) < (\ell_0 - 1)/2$. The uv-walk W_{uv} that starts at u, moves to v along the uv-path of length $d(u,v)$, then moves to x_2 along the vx_2-path of length $d(v,x_2)$, then moves one time around the cycle C_2 and back at x_2, and finally moves to v along the x_2v-path of length $d(v,x_2)$, is an even uv-walk. Notice that $\ell(W_{uv}) = d(u,v) + 2d(v,x_2) + s$. Since $\ell_0 = d(x_1,u) + d(u,v) + d(v,x_2)$ and $d(u,x_1) > d(v,x_2)$, we have $\ell(W_{uv}) < s + \ell_0$.

Hence for any pair of distinct vertices u and v there is a uv-walk W_{uv} of length $\ell_0(W_{uv}) \leq s + \ell_0$. Proposition 1 guarantees that $k(G) \leq (s + \ell_0)/2$. □

Lemma 3. *Let $s \geq 5$ be an odd integer and G be an (s,ℓ_0)-barbell. If ℓ_0 is even, then $k(G) = (s + \ell_0 - 1)/2$.*

Proof. Let v be the median vertex on P and let u be the vertex such that $d(u,x_1) = d(v,x_1) - 1$. Then $d(v,x_1) = \ell_0/2$ and $d(u,v) = 1$. Notices that the shortest even uv-walk is the walk that starts at u, moves to x_1 along the path of length $\ell_0/2 - 1$, then moves one time around the cycle C_1 and back at x_1, then moves to u along the path of length $\ell_0/2 - 1$, and finally moves to v along the path of length ℓ_0. Therefore $\ell(W_{uv}) = s + \ell_0 - 1$ and hence $k(G) \geq (s + \ell_0 - 1)/2$.

By Proposition 1 it remains to show that for each pair of vertices u and v there is an even walk W_{uv} of length $\ell(W_{uv}) \leq s + \ell_0 - 1$. If $d(u,v)$ is even then we are done. So we assume that $d(u,v)$ is odd.

Case 1. Both vertices $u, v \in C_1$ or $u, v \in C_2$. As in the proof of Lemma 2 there is a uv-walk of lengths $d(u,v) \leq s + \ell_0 - 1$.

Case 2. The vertex $u \in C_1$ and $v \in C_2$. We note that $d(u,v) = d(u,x_1) + \ell_0 + d(x_2,v)$. Since ℓ_0 is odd, we assume without loss of generality that $d(u,x_1) < d(v,x_2)$. Then the uv-walk W_{uv} that starts at u, moves to x_1 along the ux_1-path of length $d(u,x_1)$, then moves to x_2 along the path P, and finally moves to v along the vx_2-path of length $s - d(v,x_2)$ is an even uv-walk of length $s + \ell_0 + d(u,x_1) - d(v,x_2)$. Since $d(u,x_1) < d(v,x_2)$, we have $\ell(W_{uv}) \leq s + \ell_0 - 1$

Case 3. The vertex $u \in C_1$ and $v \in P$ or $u \in C_2$ and $v \in P$. We may assume that $u \in C_1$. Notice that $d(u,v) = d(u,x_1) + d(x_1,v)$ is odd. Then the uv-walk W_{uv} that starts at u, moves to x_1 along the ux_1-path of length $s - d(u,x_1)$, and finally moves to v along the x_1v-path is an even uv-walk of length $\ell(W_{uv}) = s - d(u,x_1) + d(x_1,v) \leq s + \ell_0$. Since ℓ_0 is even, then $\ell(W_{uv}) \leq s + \ell_0 - 1$.

Case 4. The vertices $u, v \in P$. We assume that $d(x_1,u) < d(x_1,v)$. If $d(u,x_1) \leq d(v,x_2)$, then $d(u,x_1) \leq \ell_0/2 - 1$. The uv-walk W_{uv} that starts at u, moves to x_1 along the ux_1-path of length $d(u,x_1)$, then moves one time around the cycle C_1 and back at x_1, then moves to u along the x_1u-path of length $d(u,x_1)$, and finally moves to v along the uv-path of length $d(u,v)$ is an even uv-walk. Notice that $\ell(W_{uv}) = s+2d(u,x_1)+d(u,v)$. Since $\ell_0 = d(u,x_1)+d(u,v)+d(v,x_2)$, and $d(u,x_1) < d(v,x_2)$, we have $\ell(W_{uv}) \leq s+\ell_0-1$.

If $d(u,x_1) > d(v,x_2)$, then the uv-walk W_{uv} that starts at u, moves to v along the uv-path of length $d(u,v)$, then moves to x_2 along the vx_2-path of length $d(v,x_2)$, then moves one time around the cycle C_2 and back at x_2, and finally moves to v along the x_2v-path of length $d(v,x_2)$, is an even uv-walk. Notice that $\ell(W_{uv}) = d(u,v)+2d(v,x_2)+s$. Since $\ell_0 = d(u,x_1)+d(u,v)+d(v,x_2)$ and $d(u,x_1) > d(v,x_2)$, we have $\ell(W_{uv}) \leq s+\ell_0-1$.

Hence for any pair of distinct vertices u and v there is a uv-walk W_{uv} of length $\ell_0(W_{uv}) \leq s+\ell_0-1$. Proposition 1 guarantees that $k(G) \leq (s+\ell_0-1)/2$. □

Combining Lemma 2 and Lemma 3 we have the following result.

Theorem 4. *Let $s \geq 5$ be an odd integer and G be an (s,ℓ_0)-barbell. Then* $k(G) = \lceil (s+\ell_0-1)/2 \rceil$.

4. Main results

Let C_1 and C_2 be cycles each of odd length s. Let $x_1, y_1 \in C_1$ and $x_2, y_2 \in C_2$ such that $d(x_1,y_1) = d(x_2,y_2)$ and let P_1 be an x_1x_2-path and P_2 be a y_1y_2-path each of length ℓ_0. An (s,ℓ_0)-two bar barbell is a graph consisting of disjoint cycles C_1 and C_2 connected by two paths P_1 and P_2.

Lemma 5. *Let $s \geq 5$ be an odd integer and G be an (s,ℓ_0)-two bar barbell. If ℓ_0 is odd, then* $k(G) = (s+\ell_0)/2$.

Proof. Let u' and v' be the two median vertices on P_1. Then the shortest even $u'v'$-walk is of length $s+\ell_0$. Hence $k(G) \geq (s+\ell_0)/2$. Let u and v be two vertices in G. We may assume that $d(u,v)$ is odd. Considering the proof of Lemma 2 it remains to show that if $u \in P_1$ and $v \in P_2$ then there is an uv-walk of length no more than $s+\ell_0$. Define $\ell_m = \min\{d(x_1,u)+d(y_1,v), d(x_2,u)+d(y_2,v)\}$. Since $d(x_1,u)+d(y_1,v)+d(x_2,u)+d(y_2,v) = 2\ell_0$, then $\ell_m \leq \ell_0$. We may assume that $\ell_m = d(x_1,u)+d(y_1,v)$. Then the uv-walk W_{uv} that starts at u, moves to x_1 along the ux_1-path of length $d(u,x_1)$, then moves to y_1 along the x_1y_1-path of length $s-d(x_1,y_1)$, and finally moves to v along the y_1v-path of length $d(y_1,v)$, is an even uv-walk. Notice that $\ell_0(W_{uv}) = d(u,x_1)+s-d(x_1,y_1)+d(y_1,v)$. Since $d(x_1,u)+d(y_1,v) \leq \ell_0$ and $d(x_1,y_1) \geq 1$, then $\ell(W_{uv}) < s+\ell_0$. □

Lemma 6. *Let $s \geq 5$ be an odd integer and G be an (s,ℓ_0)-two bar barbell. If ℓ_0 is even, then* $k(G) = (s+\ell_0-1)/2$.

Proof. Let v' be the median vertex on P_1 and let u' be the vertex such that $d(u',x_1) = d(v',x_1)-1$. Then $d(v',x_1) = \ell_0/2$ and $d(u',v') = 1$. Notices that the shortest even $u'v'$-walk $W_{u'v'}$ is the walk that starts at u', moves to x_1 along the path of length $\ell_0/2-1$, then moves one time around the cycle C_1 and back at x_1, then moves to u' along the path of length $\ell_0/2-1$, and finally moves to v' along the path of length 1. notice that $\ell_0(W_{u'v'}) = s+\ell_0-1$ and hence $k(G) \geq (s+\ell_0-1)/2$.

Let u and v be two vertices in G and we assume that $d(u,v)$ is odd. Considering the proof of Lemma 3 it remains to show that if $u \in P_1$ and $v \in P_2$, then there is an uv-walk of length no more than $s+\ell_0-1$. Define $\ell_m = \min\{d(x_1,u)+d(y_1,v), d(x_2,u)+d(y_2,v)\}$. Since $d(x_1,u)+d(y_1,v)+d(x_2,u)+d(y_2,v) = 2\ell_0$, then $\ell_m \leq \ell_0$. We may assume that $\ell_m = d(x_1,u)+d(y_1,v)$. Then the uv-walk W_{uv} that starts at u, moves to x_1 along the ux_1-path of length $d(u,x_1)$, then moves to y_1 along the x_1y_1-path of length $s-d(x_1,y_1)$, and finally moves to v along the y_1v-path of length $d(y_1,v)$ is an even uv-walk. Notice that $\ell(W_{uv}) = d(u,x_1)+s-d(x_1,y_1)+d(y_1,v)$. Since $d(x_1,u)+d(y_1,v) \leq \ell_0$ and $d(x_1,y_1) \geq 1$, then $\ell(W_{uv}) \leq s+\ell_0-1$. □

IOP Conf. Series: Materials Science and Engineering **300** (2018) 012077 doi:10.1088/1757-899X/300/1/012077

Combining Lemma 5 and Lemma 6 we have the following result.

Theorem 7. *Let $s \geq 5$ be an odd integer and G be an (s, ℓ_0)-two bar barbell. Then $k(G) = \lceil (s + \ell_0 - 1)/2 \rceil$.*

Let C_1 be the cycle

$$v_1 - v_2 - v_3 - \cdots - v_s - v_1$$

and let C_2 be the cycle

$$v_{s+1} - v_{s+2} - v_{s+3} - \cdots - v_{2s} - v_{s+1}$$

each of length s. For $i = 1, 2, \ldots, s$, we define the path P_i to be a path of length ℓ_0 connecting v_i and v_{s+i}. An (s, ℓ_0)-prism is a graph consisting of disjoint odd cycles C_1 and C_2 each of length s connected by paths P_i, $i = 1, 2, \ldots, s$, each of length ℓ_0. As a consequence and generalization of Theorem 7 we have the following theorem.

Theorem 8. *Let $s \geq 5$ be an odd integer. If G is an (s, ℓ_0)-prism. Then $k(G) = \lceil (s + \ell - 1)/2 \rceil$.*

Proof. The proof is similar to the proof of Theorem 7. □

References

[1] Brualdi R A and Ryser H J 1991 *Combinatorial matrix Theory* (Cambridge: Cambridge University Press)
[2] Akelbek M and Kirkland S 2009 *Linear Algebra Appl.* **430** 1111.
[3] Akelbek M and Kirkland S 2009 *Linear Algebra Appl.* **430** 1099.
[4] Chen S and Liu B 2010 *Linear Algebra Appl.* **433** 1110.

IOP Conf. Series: Materials Science and Engineering **300** (2018) 012078 doi:10.1088/1757-899X/300/1/012078

Scrambling index of two-colored two-cycles whose lengths differ by four

Tantri O D S Putri, Sawaluddin Nasution* and Junior B Pasorong

Departmen of Mathematics, Universitas Sumatera Utara, Medan, Indonesia

E-mail: *`sawal@usu.ac.id`

Abstract. A two-cycles is a digraph consists of two cycles. A two-colored digraph is a digraph whose arcs are colored by red or blue. Scrambling index of a primitive two-colored digraph $D^{(2)}$, denoted by $k(D^{(2)})$, is the smallest positive integer $h+\ell$ over all pairs of nonnegative integers h and ℓ such that for every pair of vertices u and v in $D^{(2)}$ there is a vertex w in $D^{(2)}$ with the property that there exist a walk from u to w and a walk from v to w consisting of h red arcs and ℓ blue arcs. We discuss the scrambling index of a certain class of primitive two-colored two-cycles with cycles of length c and $c+4$, respectively, where $c \equiv 1 (\mathrm{mod}\ 4)$. We present a formula for $k(D^{(2)})$ that depends on c and the distance of two certain vertices to the vertex of indegree 2.

1. Introduction

A two-colored digraph $D^{(2)}$ is a digraph D each of whose arcs is colored by red or blue. For nonnegative integers h and ℓ, an (h,ℓ)-walk in a two-colored digraph is a walk of length $h+\ell$ consisting of h red arcs and ℓ blue arcs. An (h,ℓ)-walk from v_i to v_j denoted by (h,ℓ)-walk W_{v_i,v_j}. A v_iv_j-path P_{v_i,v_j} is a v_iv_j-walk with distinct vertices except possibly $v_i = v_j$. A cycle is a v_iv_j-path with $v_i = v_j$. For a (h,ℓ)-walk W_{v_i,v_j} we denote $r(W_{v_i,v_j}) = h$ and $b(W_{v,v_j}) = \ell$. We note that the length of v_i, v_j-walk W_{v_i,v_j} is $\ell(W_{v_i,v_j}) = r(W_{v_i,v_j}) + b(W_{v_i,v_j})$. The 2 by 1 vector $\begin{bmatrix} r(W_{v_i,v_j}) \\ b(W_{v_i,v_j}) \end{bmatrix}$ is called the composition of the walh W_{v_i,v_j}.

A two-colored digraph $D^{(2)}$ is strongly connected provided for each pair of vertices v_i and v_j in $D^{(2)}$ there is a v_iv_j-walk. By a two-cycles we mean a strongly connected digraph consisting of exactly two cycles. A strongly connected two-colored digraph $D^{(2)}$ is primitive if there is a positive integer $h+\ell$ such that for every ordered pair of vertices v_i and v_j in $D^{(2)}$, there exists an (h,ℓ)-walk W_{v_i,v_j}. The least of such integer $h+\ell$ is called the exponent of $D^{(2)}$ and it is denoted by $\exp(D^{(2)})$. The scrambling index of a primitive two-colored digraph $D^{(2)}$, denoted by $k(D^{(2)})$, is the smallest positive integer $h+\ell$ over all pairs of nonnegative integers h and ℓ such that for every two vertices v_i and v_j in $D^{(2)}$ there is a vertex v_t in $D^{(2)}$ with the property that there exist an (h,ℓ)-walk W_{v_i,v_t} and an (h,ℓ)-walk W_{v_j,v_t}. We note from definitions that for primitive two-colored digraph we have $k(D^{(2)}) \leq \exp(D^{(2)})$. Shader and Suwilo [5] proved that for primitive two-colored digraph $D^{(2)}$ on n vertices $\exp(D^{(2)}) \leq (3n^3 + 2n^2 - 2n)/2$.

Research on scrambling index of two-colored digraphs is initiated by Mulyono et al. [3, 4]. They dicussed the scrambling index of two-colored two-cycles whose lengths differ by one.

Their results show for such primitive two-colored two-cycles $D^{(2)}$ on n vertices the $k(D^{(2)}) \approx n^2$. Mardiningsih and Pasaribu [2] discussed the scrambling index of primitive two-colored hamiltonian two-cycles $D^{(2)}$ on n vertices with cycles of length n and $(n-1)/2$, respectively. Their results also show that $k(D^{(2)}) \approx n^2$. In this paper, we discussed a scrambling index of two-colored two-cycles $D_4^{(2)}$ on n vertices consisting of cycles whose lengths c and $c+4$ respectively, where $c \geq 5$ and $c \equiv 1 (\text{mod } 4)$. We show that the $k(D_4^{(2)}) \approx n^3$.

2. Preliminaries

Let v_i and v_j be two distinct vertices in $D^{(2)}$. The local scrambling index of vertices v_i and v_j at the vertex v_t, denoted by $k_{v_iv_j}(v_t)$, is the smallest positive integer $h+\ell$ over all pairs of nonnegative integers h and ℓ such that there exist an (h,ℓ)-walk $W_{v_iv_t}$ and an (h,ℓ)-walk W_{v_j,v_t}. The local scrambling index of v_i and v_j, denoted by $k_{v_iv_j}(D^{(2)})$, is defined to be $k_{v_iv_j}(D^{(2)}) = \min_{v_t \in V(D^{(2)})}\{k_{v_iv_j}(v_t)\}$. We note that $\max_{v_i,v_i \in V(D^{()})}\{k_{v_i,v_j}(D^{(2)})\} \leq k(D^{(2)})$.

Let $D^{(2)}$ be a two-colored two-cycles with cycles C_1 and C_2. The cycle matrix of $D^{(2)}$ is defined to be the matrix $M = \begin{bmatrix} r(C_1) & r(C_2) \\ b(C_1) & b(C_2) \end{bmatrix}$. If $D^{(2)}$ has at least one arc of each color then it is known that the two-cycle $D^{(2)}$ is primitive if and only if $\det(M) = \pm 1$ [1].

The following lemma is useful for determining lower bound for local scrambling index at a specific vertex.

Lemma 1. [4] *Let $D^{(2)}$ be a primitive two-colored two-cycles with the cycle matriks $M = \begin{bmatrix} r(C_1) & r(C_2) \\ b(C_1) & b(C_2) \end{bmatrix}$ and assume that* $\det(M) = 1$. *Let v_i, v_j and v_t be vertices in $D^{(2)}$. If $k_{v_iv_j}(v_t)$ is obtained from (h_t, ℓ_t)-walk, then $\begin{bmatrix} h_t \\ \ell_t \end{bmatrix} \geq M \begin{bmatrix} b(C_2)r(P_{v_iv_t}) - r(C_2)b(P_{v_iv_t}) \\ r(C_1)b(P_{v_jv_t}) - b(C_1)r(P_{v_jv_t}) \end{bmatrix}$ for some paths P_{v_i,v_t} and P_{v_j,v_t}.*

If we set $f_1 = b(C_2)r(P_{v_iv_t}) - r(C_2)b(P_{v_iv_t})$ and $f_2 = r(C_1)b(P_{v_jv_t}) - b(C_1)r(P_{v_jv_t})$, then Lemma 1 implies that $k_{v_i,v_j}(v_t) \geq \ell(C_1)f_1 + \ell(C_2)f_2$, for some paths $P_{v_iv_t}$ and $P_{v_jv_t}$.

3. Result

Let $D_4^{(2)}$ be a two-colored two-cycles on n vertices consisting of cycles $C_1 : v_1 \rightarrow v_2 \rightarrow \ldots \rightarrow v_s \rightarrow v_{s+1} \rightarrow \ldots v_c \rightarrow v_1$ of length c and $C_2 : v_1 \rightarrow v_2 \rightarrow \ldots \rightarrow v_s \rightarrow v_{c+1} \rightarrow v_{c+2} \rightarrow \ldots \rightarrow v_n \rightarrow v_1$ of length $c+4$. We note that $n = 2c - s + 4$ for some $1 \leq s \leq c$. Let $M = \begin{bmatrix} r(C_1) & r(C_2) \\ b(C_1) & b(C_2) \end{bmatrix}$ be the cycle matrix of $D^{(2)}$. Since $D^{(2)}$ is primitive if and only if $\det(M) = \pm 1$, it is not hard to show the following result.

Corollary 2. *Let $D^{(2)}$ be a two-colored two-cycles with cycles of length c and $c+4$ respectively. Let $c \geq 5$ and $c \equiv 1(\text{mod}4)$. If $D^{(2)}$ is primitive, then cycle matrix of $D^{(2)}$ is*

$$M = \begin{bmatrix} (3c+1)/4 & (3c+13)/4 \\ (c-1)/4 & (c+3)/4 \end{bmatrix} \text{ or } M = \begin{bmatrix} (c-1)/4 & (c+3)/4 \\ (3c+1)/4 & (3c+13)/4 \end{bmatrix}.$$

In this paper, without loss of generality we use the cycle matriks $M = \begin{bmatrix} (3c+1)/4 & (3c+13)/4 \\ (c-1)/4 & (c+3)/4 \end{bmatrix}$. Hence the number of blue arcs of $D^{(2)}$ is at least $(c+3)/4$ and at most $(2c+2)/4$. We discuss only the case where the number of blue arcs is $(c+3)/4$ and $D^{(2)}$ has a blue path of length $(c-1)/4$ that lies on both cycles. Let v_iv_j-path be the $(0, (c-1)/4)$-path lies on the both cycles, where $1 \leq i < j \leq s$ and one blue arc on C_2 is $v_s \rightarrow v_{c+1}$ or $v_k \rightarrow v_{k+1}$ where $c+1 \leq k \leq n = 2c-s+4$. If $v_k = v_n$, then assume that $v_{k+1} = v_1$. Define $d_1 = d(v_{k+1}, v_1)$ and $d_2 = d(v_j, v_1)$ where $d(v_a, v_1)$ means the distance from the vertex v_a to the vertex v_1.

Theorem 3. *Let $D^{(2)}$ is a a two-colored digraph consisting of two cycles with lengths c and $c+4$ respectively. Let $c \geq 5$ and $c \equiv 1(\mathrm{mod}4)$. Suppose $D^{(2)}$ has a $(0,(c-1)/4)$-path that lies on both cycles and one blue arc that lies on C_2 but not on C_1, then*

$$k(D^{(2)}) = \begin{cases} ((d_2-d_1+3)c^2+(3d_2-3d_1+9)c+4d_1+4)/4, & \text{if } d_2-d_1 \geq (3c-11)/4, \\ (3c^3+10c^2-25c-4+16d_2)/16, & \text{if } 1 \leq d_2-d_1 \leq (3c-15)/4, \\ (3c^3+(4d_1-4d_2+10)c^2+(12d_1-12d_2-9)c+16d_2-4)/16, & \text{if } d_1 \geq d_2. \end{cases}$$

Proof. **Case 1.** $d_2 - d_1 \geq (3c-11)/4$.
First, we show that $k(D^{(2)}) \geq ((d_2-d_1+3)c^2+(3d_2-3d_1+9)c+4d_1+4)/4$. That is for every $t = 1,2,\ldots,n$, we show $k_{v_jv_k}(v_t) \geq ((d_2-d_1+3)c^2+(3d_2-3d_1+9)c+4d_1+4)/4+d(v_1,v_t)$. For this purpose we define $f_1 = b(C_2)r(P_{v_jv_t})-r(C_2)b(P_{v_jv_t})$ and $f_2 = r(C_1)b(P_{v_kv_t})-b(C_2)r(P_{v_kv_t})$. If there are two v_jv_t-paths or two v_kv_t-paths, then in determining the lower bound of scrambling index, we take the smallest value of f_1 and f_2. We consider several cases depending on the position of the vertex v_t.

The vertex v_t lies on the v_1v_i-path. There are two v_jv_t-paths, they are a $(d_2+d(v_1,v_t),0)$-path and $(d_2+3+d(v_1,v_t),1)$-path. Using the $(d_2+3+d(v_1,v_t),1)$-path we get $f_1 = ((d_2+d(v_1,v_t))c+3d_2+3d(v_1,v_t)-4)/4$. There is a $(d_1+d(v_1,v_t),1)$-path $P_{v_kv_t}$. Using this path we get $f_2 = ((3-d(v_1,v_t)-d_1)c+d_1+d(v_1,v_t)+1)/4$. From Lemma 1, we get $k_{v_jv_k}(v_t) \geq ((d_2-d_1+3)c^2+(3d_2-3d_1+9)c+4d_1+4)/4+d(v_1,v_t)$.

The vertex v_t lies on the $v_{i+1}v_j$-path. There are two v_jv_t-paths $P_{v_jv_t}$, they are a $(d_2+d(v_1,v_t)-d(v_i,v_t),d(v_i,v_t))$-path and a $(d_2+3+d(v_1,v_t)-d(v_i,v_t),d(v_i,v_t)+1)$-path. Using the $(d_2+3+d(v_1,v_t)-d(v_i,v_t),d(v_i,v_t)+1)$-path we get $f_1 = ((d_2+d(v_1,v_t)-4d(v_i,v_t))c+3d(v_1,v_t)+3d_2-16d(v_i,v_t)-4)/4$. There is a $(d_1+d(v_1,v_t)-d(v_i,v_t),1+d(v_i,v_t))$-path $P_{v_kv_t}$. Using this path we get $f_2 = ((d(v_1,v_t)+4d(v_i,v_t)-3)c+d_1+d(v_1,v_t)+1)/4$. From Lemma 1, we get $k_{v_jv_k}(v_t) \geq ((d_2-d_1+3)c^2+(3d_2-3d_1+9)c+4d_1+4)/4+d(v_1,v_t)$.

The vertex v_t lies on the $v_{j+1}v_s$-path, $v_{s+1}v_c$-path or $v_{c+1}v_k$ path. There is a $(d_2-c+d(v_1,v_t),0)$-path $P_{v_jv_t}$. Using this path we get $f_1 = (-c^2+(d(v_1,v_t)+d_2-3)c+3d(v_1,v_t)+3d_2)/4$. There is a $(d_1+d(v_1,v_t)-(c-1)/4,(c+3)/4)$-path $P_{v_kv_t}$. Using this path we get $f_2 = (c^2+(2-d_1-d(v_1,v_t))c+d_1+d(v_1,v_t)+1)/4$. From Lemma 1, we get $k_{v_jv_k}(v_t) \geq ((d_2-d_1+3)c^2+(3d_2-3d_1+9)c+4d_1+4)/4+d(v_1,v_t)$.

The vertex v_t lies on the $v_{k+1}v_n$ path. There is a $(d_2-c-1+d(v_1,v_t),1)$-path $P_{v_jv_t}$. Using this path we get $f_1 = (-c^2+(d_2+d(v_1,v_t)-7)c+3d_2+3d(v_1,v_t)-16)/4$. There is a $(d_1-c-4+d(v_1,v_t),1)$-path $P_{v_kv_t}$. Using this path we get $f_2 = (c^2+(-d_1-d(v_1,v_t)+6)c+d_1+d(v_1,v_t)-3)/4$. Lemma 1 implies that

$$\begin{bmatrix} h_t \\ \ell_t \end{bmatrix} \geq \begin{bmatrix} ((3d_2-3d_1+9)c^2+(10d_2-10d_1+14)c+3d_2+13d_1-55)/16+d(v_1,v_t) \\ ((d_2-d_1+3)c^2+(2d_2-2d_1+6)c-3d_2+3d_1+7)/16 \end{bmatrix} = \begin{bmatrix} e_1 \\ e_2 \end{bmatrix}.$$

We note that the solution to $M\mathbf{z}+\begin{bmatrix} r(P_{v_kv_t}) \\ b(P_{v_kv_t}) \end{bmatrix} = \begin{bmatrix} e_1 \\ e_2 \end{bmatrix}$ is $z_1 = (3c-3d_1+3d_2-cd_1+cd_2+9)/4$ and $z_2 = 0$. That means there is no (e_1,e_2)-path P_{v_k,v_t}, thus $k_{v_jv_k}(v_t) > e_1+e_2$. Notice that the shortest v_kv_t-walk that contains at least e_1 red arcs and at least e_2 blue arcs is a $(e_1+(3c+13)/4, e_2+(c+3)/4)$-walk. This implies $k_{v_jv_k}(v_t) \geq e_1+e_2+c+3 = ((d_2-d_1+3)c^2+(3d_2-3d_1+9)c+4d_1+4)/4+d(v_1,v_t)$.

Therefore, for each $t = 1,2,\ldots n$, we get $k_{v_jv_k}(v_t) \geq ((d_2-d_1+3)c^2+(3d_2-3d_1+9)c+4d_1+4)/4+d(v_1,v_t)$. This implies that $k_{v_jv_k}(D^{(2)}) = ((d_2-d_1+3)c^2+(3d_2-3d_1+9)c+4d_1+4)/4$, and hence $k(D^{(2)}) \geq ((d_2-d_1+3)c^2+(3d_2-3d_1+9)c+4d_1+4)/4$.

Next we show that $k(D^{(2)}) \leq ((d_2-d_1+3)c^2+(3d_2-3d_1+9)c+4d_1+4)/4$. It is sufficient to show that for each $t=1,2,\dots n$, there are (h,ℓ)-walk W_{v_t,v_1} with $h=((3d_2-3d_1+9)c^2+(10d_2-10d_1+30)c+3d_2+13d_1+9)/16$ and $\ell=((d_2-d_1+3)c^2+(2d_2-2d_1+6)c+3d_2+3d_1+7)/16$. We show that the system

$$M\mathbf{z}+\begin{bmatrix} r(P_{v_tv_1}) \\ b(P_{v_tv_1}) \end{bmatrix} = \begin{bmatrix} h \\ \ell \end{bmatrix} \tag{1}$$

has nonnegative integer solution for some path $P_{v_tv_1}$. The solution of system (1) is $z_1 = ((c+3)(d_2-r(P_{v_tv_1}))+(3c+13)b(P_{v_tv_1})-4)/4$ and $z_2=((c-1)(r(P_{v_tv_1})-d_1)+(-3c-1)b(P_{v_tv_1})+3c+1)/4$.

If the vertex v_t lies on the v_1v_i-path, then there is a $(r(P_{v_tv_1}),(c-1)/4)$-path from v_t to v_1, where $d_2 \leq r(P_{v_tv_1}) \leq (3c+1)/4$. Using this path we get $z_1=((4c+12)d_2-32)/16$. Since $d_2 \geq (3c-11)/4$, we have $z_1 \geq (3c^2-2c-65)/16$. Since $d_2-d_1 \geq (3c-11)/4$, we get $z_2 \geq 1$.

If the vertex v_t lies on the $v_{i+1}v_j$ path, then there is a $(d_2+3, b(P_{v_tv_1}))$-path from v_t to v_1, where $1 \leq b(P_{v_tv_1}) \leq (c-1)/4$. Using this path we get $z_1 \geq 0$. Since $d_2-d_1 \geq (3c-11)/4$, we get $z_2 \geq (3c+1)/4$.

If the vertex v_t lies on the $v_{j+1}v_s$-path or $v_{c+1}v_k$-path, then there is a $(r(P_{v_tv_1}),1)$-path from v_t to v_1, where $d_1 \leq r(P_{v_tv_1}) \leq d_2+2$. Using this path we get $z_1 \geq (c+3)/4$ and $z_2 \geq 0$.

If the vertex v_t lies on the $v_{k+1}v_n$ path, then there is a $(r(P_{v_tv_1}),0)$-path from v_t to v_1, where $1 \leq r(P_{v_tv_1}) \leq d_1$. Using this path and since $d_2-d_1 \geq (3c-11)/4$, we get $z_1=(3c^2-2c-49)/16$. Since $d_2 \leq (3c+1)/4$ and $d_2-d_1 \geq (3c-11)/4$, then $d_1 \leq 3$. We get $z_2 \geq (c+3)/4$.

If the vertex v_t lies on the $v_{s+1}v_c$ path, then there is a $(r(P_{v_tv_1}),0)$-path from v_t to v_1, where $1 \leq r(P_{v_tv_1}) \leq d_2-1$. Using this path we get $z_1=(c-1)/4$. Since $d_2 \leq (3c+1)/4$ and $d_2-d_1 \geq (3c-11)/4$, then $d_1 \leq 3$. We get $z_2 \geq (c+3)/4$.

Therefore, for each $t=1,2,\dots,n$, there exist (h,ℓ)-walk $W_{v_tv_1}$. Thus $k(D^{(2)}) \leq ((d_2-d_1+3)c^2+(3d_2-3d_1+9)c+4d_1+4)/4$.

Case 2: $d_2-d_1 \leq (3c-15)/4$

First, we show that $k(D^{(2)}) \geq (3c^3+10c^2-25c-4+16d_2)/16$. That is for every $t=1,2,\dots,n$, we show $k_{v_jv_i}(v_t) \geq (3c^3+10c^2-25c-4+16d_2)/16+d(v_1,v_t)$. We define $f_1=b(C_2)r(P_{v_jv_t})-r(C_2)b(P_{v_jv_t})$ and $f_2=r(C_1)b(P_{v_iv_t})-b(C_2)r(P_{v_iv_t})$. If there are two v_jv_t-paths or two v_iv_t-paths, then we take the smallest value of f_1 and f_2. We consider several cases.

The vertex v_t lies on the v_1v_i-path. There are two paths $P_{v_jv_t}$, they are a $(d_2+d(v_1,v_t),0)$-path and a $(d_2+3+d(v_1,v_t),1)$-path. Using the $(d_2+3+d(v_1,v_t),1)$-path, we get $f_1=((d_2+d(v_1,v_t))c+3d_2+3d(v_1,v_t)-4)/4$. There are two paths $P_{v_iv_t}$, they are a $(d_2+d(v_1,v_t),(c-1)/4)$-path and a $(d_2+3+d(v_1,v_t),(c+3)/4)$-path. Using the $(d_2+d(v_1,v_t),(c-1)/4)$-path we get $f_2=(3c^2+(-4d_2-4d(v_1,v_t)-2)c+4d_2+4d(v_1,v_t)-1)/16$. From Lemma 1, we get $k_{v_jv_i}(v_t) \geq (3c^3+10c^2-25c-4+16d_2)/16+d(v_1,v_t)$.

The vertex v_t lies on the $v_{i+1}v_j$ path. There are a two paths $P_{v_jv_t}$, they are a $((3c+1)/4, d(v_1,v_t)-(3c+1-4d_2)/4)$-path and a $((3c+13)/4, d(v_1,v_t)-(3c+1-4d_2)/4+1)$-path. Using the $((3c+13)/4, d(v_1,v_t)-(3c+1-4d_2)/4+1)$-path we get $f_1=(12c^2+(-12d_2-12d(v_1,v_t)+52)c-52d_2-52d(v_1,v_t))/16$. There is a $(0,d(v_1,v_t)-(3c+1-4d_2)/4)$-path $P_{v_iv_t}$. Using this path we get $f_2=(-9c^2+(12d_2+12d(v_1,v_t)-6)c+4d_2+4d(v_1,v_t)-1)/16$. From Lemma 1, we get $k_{v_jv_i}(v_t) \geq (3c^3+10c^2-25c-4+16d_2)/16+d(v_1,v_t)$.

The vertex v_t lies on the $v_{j+1}v_s$-path, $v_{s+1}v_c$-path or $v_{c+1}v_k$-path. There is a $(d_2-c+d(v_1,v_t),0)$-path $P_{v_jv_t}$. Using this path we get $f_1=(-c^2+(d(v_1,v_t)+d_2-3)c+3d(v_1,v_t)+3d_2)/4$. There is a $(d_2-c+d(v_1,v_t),(c-1)/4)$-path $P_{v_iv_t}$. Using this path we get $f_2=(7c^2+(-4d_2-4cd(v_1,v_t)-6)c+4d_2+4d(v_1,v_t)-1)/16$. From Lemma 1, we get $k_{v_jv_i}(v_t) \geq (3c^3+10c^2-25c-4+16d_2)/16+d(v_1,v_t)$.

The vertex v_t lies on the $v_{k+1}v_n$ path. There is a $(d_2-c-1+d(v_1,v_t),1)$-path $P_{v_jv_t}$. Using this path we get $f_1=(-c^2+(d_2+d(v_1,v_t)-7)c+3d_2+3d(v_1,v_t)-16)/4$. There is a $(d_2-c-1+d(v_1,v_t),(c+3)/4)$-path of $P_{v_iv_t}$. Using this path we get $f_2=(7c^2(-4d_2-4d(v_1,v_t)+10)c+4d_2+4d(v_1,v_t)-1)/16$. From Lemma 1, we get $k_{v_jv_i}(v_t)\geq(3c^3+10c^2-25c-4+16d_2)/16+d(v_1,v_t)$.

Therefore, for each $t=1,2,\dots n$, we get $k_{v_jv_i}(v_t)\geq(3c^3+10c^2-25c-4+16d_2)/16+d(v_1,v_t)$. This implies that $k_{v_jv_i}(D^{(2)})=(3c^3+10c^2-25c-4+16d_2)/16$, and hence $k(D^{(2)})\geq(3c^3+10c^2-25c-4+16d_2)/16$.

Next we show that $k(D^{(2)})\leq(3c^3+10c^2-25c-4+16d_2)/16$. It is sufficient to show that for each $t=1,2,\dots n$, there are $v_t\xrightarrow{(h,\ell)}v_1$ with

$$M\mathbf{z}+\begin{bmatrix} r(P_{v_tv_1})\\ b(P_{v_tv_1})\end{bmatrix}=\begin{bmatrix} h\\ \ell\end{bmatrix}=\begin{bmatrix}(9c^3+33c^2-77c+64d_2-29)/64\\ (3c^3+7c^2-23c+13)/64\end{bmatrix} \tag{2}$$

has nonnegative integer solution for some path P_{v_t,v_1}. The solution to system (2) is $z_1=((c+3)(d_2-r(P_{v_tv_1}))+(3c+13)b(P_{v_tv_1})-4)/4$ and $z_2=(3c^2-2c+(4c-4)(r(P_{v_tv_1})-d_2)+(-12c-4)b(P_{v_tv_1})-1)/16$.

If the vertex v_t lies on the v_1v_i-path, then there is a $(r(P_{v_tv_1}),(c-1)/4)$-path from v_t to v_1, where $d_2\leq r(P_{v_tv_1})\leq(3c+1)/4$. Using this path we get $z_1=((4c+12)d_2-32)/16$. Since $d_2\geq1$, we have $z_1\geq(4c-20)/16$. Since $d_2\leq r(P_{v_tv_1})\leq(3c+1)/4$, we get $z_2\geq0$.

If the vertex v_t lies on the $v_{i+1}v_j$-path, then there is a $(d_2+3,b(P_{v_tv_1}))$-path from v_t to v_1 where $1\leq b(P_{v_tv_1})\leq(c-1)/4$. Using this path we get $z_1\geq0$ and $z_2\geq(3c-3)/4$

If the vertex v_t lies on the $v_{j+1}v_s$-path or $v_{c+1}v_k$-path, then there is a $(r(P_{v_tv_1}),1)$-path from v_t to v_1 where $d_1\leq r(P_{v_tv_1})\leq d_2+2$. Applying this path we get $z_1\geq(c+3)/4$. Since $d_2-d_1\leq(3c-15)/4$, we get $z_2\geq2c+2$.

If the vertex v_t lies on the $v_{k+1}v_n$-path, then there is a $(r(P_{v_tv_1}),0)$-path from v_t to v_1 where $1\leq r(P_{v_tv_1})\leq d_1$. Since $1\leq d_2-d_1\leq(3c-15)/4$, using this path we get $z_1=(c-1)/4$. Since $d_2\leq(3c+1)/4$, we get $z_2\geq(c-1)/4$.

If the vertex v_t lies on the $v_{s+1}v_c$-path, then there is a $(r(P_{v_tv_1}),0)$-path from v_t to v_1 where $1\leq r(P_{v_tv_1})\leq d_2-1$. Using this path we get $z_1=(c-1)/4$. Since $d_2\leq(3c+1)/4$, we get $z_2\geq(c+3)/4$.

Therefore, for each $t=1,2,\dots,n$ there exist (h,ℓ)-walk $W_{v_tv_1}$. Thus $k(D^{(2)})\leq(3c^3+10c^2-25c-4+16d_2)/16$.

Case 3: $d_1\geq d_2$

First, we show that $k(D^{(2)})\geq(3c^3+(4d_1-4d_2+10)c^2+(12d_1-12d_2-9)c+16d_2-4)/16$. That is for every $t=1,2,\dots,n$, we show $k_{v_{k+1}v_i}(v_t)\geq(3c^3+(4d_1-4d_2+10)c^2+(12d_1-12d_2-9)c+16d_2-4)/16+d(v_1,v_t)$. We define $f_1=b(C_2)r(P_{v_{k+1}v_t})-r(C_2)b(P_{v_{k+1}v_t})$ and $f_2=r(C_1)b(P_{v_iv_t})-b(C_2)r(P_{v_iv_t})$. If there is a two $v_{k+1}v_t$-paths or v_iv_t-paths, then we take the smallest value of f_1 and f_2. We consider several cases.

The vertex v_t lies on the v_1v_i-path. There is a $(d_1+d(v_1,v_t),0)$-path $P_{v_{k+1}v_t}$. Using this path we get $f_1=((d_1+d(v_1,v_t))c+3d_1+3d(v_1,v_t))/4$. There are two paths $P_{v_iv_t}$, they are a $(d_2+d(v_1,v_t),(c-1)/4)$-path and a $(d_2+3+d(v_1,v_t),(c+3)/4)$-path. Using $(d_2+d(v_1,v_t),(c-1)/4)$-path we get $f_2=(3c^2+(-4d_2-4d(v_1,v_t)-2)c+4d_2+4d(v_1,v_t)-1)/16$. From Lemma 1, we get $k_{v_{k+1}v_i}(v_t)\geq(3c^3+(4d_1-4d_2+10)c^2+(12d_1-12d_2-9)c+16d_2-4)/16+d(v_1,v_t)$.

The vertex v_t lies on the $v_{i+1}v_j$-path. There is a $(d_1+(3c+1-4d_2)/4,d(v_1,v_t)-(3c+1-4d_2)/4)$-path $P_{v_{k+1}v_t}$. Using this path we get $f_1=(12c^2+(4d_1-16d_2-12d(v_1,v_t)+52)c+12d_1-64d_2-52d(v_1,v_t)+16)/16$. There is a $(0,d(v_1,v_t)-(3c+1-4d_2)/4)$-path $P_{v_iv_t}$. Using this path we get $f_2=(-9c^2+(12d_2+12d(v_1,v_t)-6)c+4d_2+4d(v_1,v_t)-1)/16$. From Lemma 1, we get $k_{v_{k+1}v_i}(v_t)\geq(3c^3+(4d_1-4d_2+10)c^2+(12d_1-12d_2-9)c+16d_2-4)/16+d(v_1,v_t)$.

The vertex v_t lies on the $v_{j+1}v_s$-path, $v_{s+1}v_c$-path or $v_{c+1}v_k$-path. There is a $(d_1+d(v_1,v_t)-(c-1)/4, (c-1)/4)$-path $P_{v_{k+1}v_t}$. Using this path we get $f_1 = (-4c^2+(4cd_1+4cd(v_1,v_t)-12)c+12d_1+12d(v_1,v_t)+16)/16$. There is a $(d_2-c+d(v_1,v_t),(c-1)/4)$-path $P_{v_iv_t}$. Using this path we get $f_2 = (7c^2+(-4d_2-4cd(v_1,v_t)-6)c+4d_2+4d(v_1,v_t)-1)/16$. From Lemma 1, we get $k_{v_{k+1}v_i}(v_t) \geq (3c^3+(4d_1-4d_2+10)c^2+(12d_1-12d_2-9)c+16d_2-4)/16+d(v_1,v_t)$.

The vertex v_t lies on the $v_{k+1}v_n$-path. There is a $(d_1-c-4+d(v_1,v_t),0)$-path $P_{v_{k+1}v_t}$. Using this path we get $f_1 = (-4c^2+(4d_1+4d(v_1,v_t)-28)c+12d_1+12d(v_1,v_t)-48)/16$. There is a $(d_2-c-1+d(v_1,v_t),(c+3)/4)$-path $P_{v_iv_t}$. Using this path we get $f_2 = (7c^2(-4d_2-4d(v_1,v_t)+10)c+4d_2+4d(v_1,v_t)-1)/16$. From Lemma 1, we get $k_{v_{k+1}v_i}(v_t) \geq (3c^3+(4d_1-4d_2+10)c^2+(12d_1-12d_2-9)c+16d_2-4)/16+d(v_1,v_t)$.

Therefore, for each $t=1,2,\ldots n$, we get $k_{v_{k+1}v_i}(v_t) \geq (3c^3+(4d_1-4d_2+10)c^2+(12d_1-12d_2-9)c+16d_2-4)/16+d(v_1,v_t)$. This implies that $k_{v_{k+1}v_i}(D^{(2)}) = (3c^3+(4d_1-4d_2+10)c^2+(12d_1-12d_2-9)c+16d_2-4)/16$, and hence $k(D^{(2)}) \geq (3c^3+(4d_1-4d_2+10)c^2+(12d_1-12d_2-9)c+16d_2-4)/16$.

Next we show that $k(D^{(2)}) \leq (3c^3+(4d_1-4d_2+10)c^2+(12d_1-12d_2-9)c+16d_2-4)/16$. It is sufficient to show that for each $t=1,2,\ldots n$, the system

$$\begin{aligned} M\mathbf{z} + \begin{bmatrix} r(P_{v_tv_1}) \\ b(P_{v_tv_1}) \end{bmatrix} &= \begin{bmatrix} h \\ \ell \end{bmatrix} \\ &= \begin{bmatrix} (9c^3+(12d_1-12d_2+33)c^2+(40d_1-40d_2-29)c+12d_1+52d_2-13)/64 \\ (3c^3+(4d_2-4d_1+7)c^2+(8d_1-8d_2-7)c+12d_1+12d_2-3)/64 \end{bmatrix} \end{aligned} \qquad (3)$$

has nonnegative integer solution for some path P_{v_t,v_1}. The solution to the system (3) is $z_1 = ((c+3)(d_1-r(P_{v_tv_1}))+(3c+13)b(P_{v_tv_1}))/4$ and $z_2 = (3c^2-2c+(4c-4)(r(P_{v_tv_1})-d_2)+(-12c-4)b(P_{v_tv_1})-1)/16$.

If the vertex v_t lies on the v_1v_i-path, then there is a $(r(P_{v_tv_1}),(c-1)/4)$-path from v_t to v_1, where $d_2 \leq r(P_{v_tv_1}) \leq (3c+1)/4$. Using this path we get $z_1 = ((4c+12)d_1+(4c-4))/16$. Since $d_1 \geq 1$, we have $z_1 \geq (4c-4)/16$. Since $d_2 \leq r(P_{v_tv_1}) \leq (3c+1)/4$, we get $z_2 \geq 0$.

If the vertex v_t lies on the $v_{i+1}v_j$-path, then there is a $(d_2+3, b(P_{v_tv_1}))$-path from v_t to v_1 where $1 \leq b(P_{v_tv_1}) \leq (c-1)/4$. Using this path and since $d_1 \geq d_2$, we get $z_1 \geq (c+7)/4$ and $z_2 \geq (3c-3)/4$

If the vertex v_t lies on the $v_{j+1}v_s$-path or $v_{c+1}v_k$-path, then there is a $(r(P_{v_tv_1}),1)$-path from v_t to v_1 where $d_1 \leq r(P_{v_tv_1}) \leq d_2+2$. Using this path we get $z_1 \geq (c+3)/4$. Since $d_1 \geq d_2$, we get $z_2 \geq (3c^2-14c-5)/16$.

If the vertex v_t lies on the $v_{k+1}v_n$ path, then there is a $(r(P_{v_tv_1}),0)$-path from v_t to v_1 where $1 \leq r(P_{v_tv_1}) \leq d_1$. Using this path we get $z_1 = 0$. Since $d_2 \leq (3c+1)/4$, we get $z_2 \geq (c-1)/4$.

If the vertex v_t lies on the $v_{s+1}v_c$-path, then there is a $(r(P_{v_tv_1}),0)$-path from v_t to v_1 where $1 \leq r(P_{v_tv_1}) \leq d_2-1$. Using this path we get $z_1 = (c+3)/4$. Since $d_2 \leq (3c+1)/4$, we get $z_2 \geq (c+3)/4$.

Therefore, for each $t=1,2,\ldots,n$ there exist (h,ℓ)-walk $W_{v_tv_1}$. Thus $k(D^{(2)}) \leq (3c^3+(4d_1-4d_2+10)c^2+(12d_1-12d_2-9)c+16d_2-4)/16$. □

References

[1] Fornasini E and Valcher E 1998 *SIAM J. Matrix Anal. Appl.* **19** 71-88.
[2] Mardiningsih and Pasaribu M L 2016 *AIP Conf. Proc.* **1775** 030056.
[3] Mulyono and Suwilo S 2014 *Universal Journal of Applied Mathematics* **2** (6) 250-255.
[4] Mulyono, Sumardi H and Suwilo S 2015 *Far East Journal of Mathematical Sciences* **96**(1) 119-132.
[5] Shader B L and Suwilo S 2003 *Linear Algebra App.* **363** 275–293

Exponent and scrambling index of double alternate circular snake graphs

Sri Rahmayanti, Valdo E Pasaribu, Sawaluddin Nasution* and Sishi Liani Salnaz

Department of Mathematics, University of Sumatera Utara, Medan 20155, Indonesia

E-mail: *sawal@usu.ac.id

Abstract. A graph is primitive if it contains a cycle of odd length. The exponent of a primitive graph G, denoted by $\exp(G)$, is the smallest positive integer k such that for each pair of vertices u and v in G there is a uv-walk length k. The scrambling index of a primitive graph G, denoted by $k(G)$, is the smallest positive integer k such that for each pair of vertices u and v in G there is a uv-walk of length $2k$. For an even positive integer n and an odd positive integer r, a (n, r)-double alternate circular snake graph, denoted by $DA(C_{r,n})$, is a graph obtained from a path $u_1u_2 \dots u_n$ by replacing each edge of the form $u_{2i}u_{2i+1}$ by two different r-cycles. We study the exponent and scrambling index of $DA(C_{r,n})$ and show that $\exp(DA(C_{r,n})) = n + r - 4$ and $k(DA(C_{r,n})) = (n + r - 3)/2$.

1. Introduction

Let $G(V, E)$ denote a graph on n vertices. Let u and v be any vertices in G, a *uv-walk* W_{uv} of length t connecting u and v is a sequence of vertices $u = u_0, u_1, u_2, \dots, u_t = v$ and a sequence of edges $\{u_0, u_1\}, \{u_1, u_2\}, \dots, \{u_{t-1}, u_t\}$, where the vertices and the edges are not necessarily distinct. The length of the walk W_{uv} is denoted by $\ell(W_{uv})$. A walk connecting u and v is closed whenever $u = v$, and is open otherwise. A *path* P_{uv} connecting u and v is a walk W_{uv} with distinct vertices, except possibly $u = v$. A *cycle* is a closed path. A walk W_{uv} with sequence of edges $\{u = u_0, u_1\}, \{u_1, u_2\}, \dots, \{u_{t-1}, u_t = v\}$ is also denoted by $W_{uv} : u = u_0 - u_1 - u_2 - \cdots - u_{t-1} - u_t = v$. The distance between vertices u and v, denoted by $d(u, v)$, is the length of a shortest path connecting u and v.

A graph G is connected provided that for each pairs of vertices u and v in G there is a walk connecting u and v. A connected graph G is primitive provided there is a positive integer m such that for each pair of vertices u and v in G, there is a uv-walk of length m. The smallest of such positive integer m is the exponent of G and is denoted by $\exp(G)$. It is well known that a graph G is primitive if and only if G has a cycle of odd length [3].

Alkelbek and Kirkland [1, 2] introduced the notion of scrambling index of primitive graph G for the first time in 2009. The scrambling index of a primitive graph G, denoted by $k(G)$, is the smallest positive integer k such that for any pair of distinct vertices u and v in G there exists a vertex w such that there is a uv-walk of length m and a vw-walk of length m. Chen and Liu [4] discussed the scrambling index of classes of primitive graph with the smallest cycle of length $s \geq 3$ and classes of primitive graphs with loops.

IOP Conf. Series: Materials Science and Engineering **300** (2018) 012079 doi:10.1088/1757-899X/300/1/012079

By an r-double circular DC_r we mean a connected graph consisting of two r-cycles that have one edge in common. The edge in common of the two r-cycles is called the base of the r-double circular or the the base of the r-cycles. Figure 1 presents a DC_5 with base is the edge $\{u_3, u_4\}$.

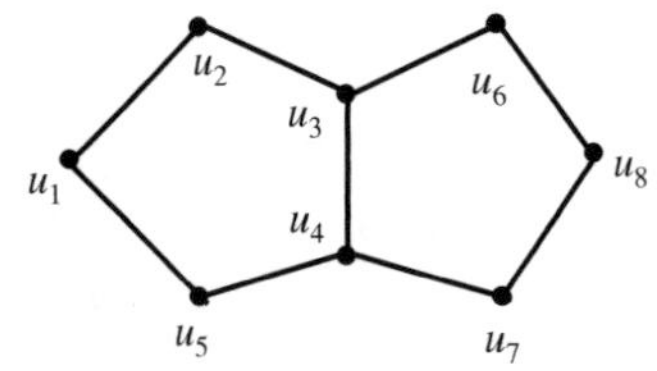

Figure 1. The graph of $DC(5)$

Let n be an even positive integer and r be an odd positive integer. A double alternate circular snake graph ($DA(C_{r,n})$ is a graph obtained from a path $u_1, u_2, \dots, u_n$ by replacing each edges of the form $\{u_{2i}, u_{2i+1}\}$, $1 \leq i \leq \frac{n-2}{2}$ by an r-double circular with base $\{u_{2i}, u_{2i+1}\}$. Figure 2 presents a $DA(C_{5,6})$.

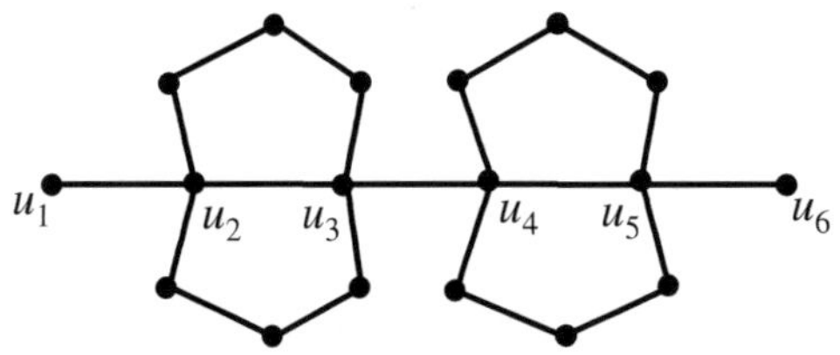

Figure 2. The graph of $DA(C_{5,6})$

Double alternate circular snake graph, especially double alternate triangular snake graph, has attracted many researchers [5, 6]. In this paper, we discussed the exponent and the scrambling index of double alternate circular snake graph, $DA(C_{r,n})$. We present formulae for exponent and scrambling index that depends on n and r.

2. Preliminaries

We discuss some properties of uv-walk in a graph and property of scrambling index.

Proposition 1. *Let G be a graph and let k and ℓ be positive integers such that $\ell \leq k$ and $k \equiv \ell \bmod 2$. If there is a uv-walk of length ℓ in G, then there is a uv-walk of length k in G.*

Proof. Since $k \geq \ell$ and $k \equiv \ell \bmod 2$, there is a nonnegative integer t such that $k = \ell + 2t$. Let W_{uv} be a uv-walk of length ℓ in G and let $\{v, w\}$ be an edge in W_{uv}. Then the uv-walk obtained by moving from u to v along the walk W_{uv} and then moves t times around the closed walk $v - w - v$ is a uv-walk of length k. □

Proposition 2. *Let G be a primitive graph and k' be an even positive integer. If for each pair of vertices u and v in G there is an even uv-walk W_{uv} of length $\ell(W_{uv}) \leq k'$, then $k(G) \leq k'/2$.*

Proof. For each pair of vertices u and v let W_{uv} be an even uv-walk of length $\ell(W_{uv}) \leq k'$. Since $k' \equiv \ell(W_{uv}) \bmod 2$, Proposition 1 implies there is a uv-walk of length k'. Since k' is even, then there exists a vertex w with the property that there is a uw-walk of length $k'/2$ and there is a vw-walk of length $k'/2$. Thus, $k(G) \leq k'/2$. □

For a pair of distinct vertices u and v in G the local scrambling index of u and v is the number

$$k_{u,v}(G) = \min_{w \in V(G)} \{k : \text{there are } uw\text{-walk and } vw\text{-walk of length } k\}.$$

We note that if the local scrambling index of u and v is $k_{u,v}(G)$, then by Proposition 1 for any positive integer $k' \geq k_{u,v}(G)$ we can find a vertex w such that there is a walk $u \overset{k'}{\leftrightarrow} w$ and $v \overset{k'}{\leftrightarrow} w$. This implies

$$k(G) = \max\{k_{u,v}(G) | u, v \in V(G), u \neq v\}.$$

3. Result

We discuss the exponent and scrambling index of double alternate circular snake graph $DA(C_{r,n})$ obtained from a path $u_1, u_2, \ldots u_n$ by replacing each edge of the form $\{u_{2i}, u_{2i+1}\}$, $1 \leq i \leq \frac{n-2}{2}$ by r-double cycle. We note that there are $\frac{n-2}{2}$ double r-cycles. Let P_n be the path $u_1 - u_2 - \cdots - u_n$ of length $n-1$ and let C_r be an r-cycle. For any vertices u_i and u_j in C_r, we define $P_{u_i u_j}$ to be the shortest path connecting u_i and u_j and $P'_{u_i u_j}$ to be the path connecting u_i and u_j of length $r - d(u_i, u_j)$. We note that since r is odd, $\ell(P_{u_i,u_j}) \not\equiv \ell(P'_{u_i,u_j}) \bmod 2$. We first present the result on the exponent of primitive double alternate circular snake head.

Theorem 3. *Let $n \geq 4$ is an even positive integer and $r \geq 3$ is an odd positive integer. Then* $\exp(DA(C_{r,n})) = n + r - 4$.

Proof. We first show that $\exp(DA(C_{r,n})) \geq n+r-4$. We note that for any pair of distinct vertices u and v in $DA(C_{r,n})$ we have $\exp(DA(C_{r,n})) \geq d(u, v)$. Hence $\exp(DA(C_{r,n})) \geq d(u_1, u_n) = n-1$. Notice that the shortest u_1u_{n-1}-walk of length at least $n-1$ is the u_1u_{n-1}-walk of length $n+r-4$. Hence

$$\exp(DA(C_{r,n})) \geq n + r - 4.$$

We now show that $\exp(DA(C_{r,n})) \leq n + r - 4$. For each pair of vertices u_α and u_β it suffices to show that there is a $u_\alpha u_\beta$-walk of length $n + r - 4$. We note if $r \geq 5$, then for any pair of vertices u_α and u_β in $DA(C_{r,n})$ the distance $d(u_\alpha, u_\beta) \leq n + r - 6$. If $d(u_\alpha, u_\beta)$ is odd, then we can extend the path $P_{u_\alpha u_\beta}$ of length $d(u_\alpha, u_\beta)$ into a walk $W_{u_\alpha u_\beta}$ of length exactly $n + r - 4$.

We now assume that $d(u_\alpha, u_\beta)$ is even. Then $d(u_\alpha, u_\beta) < n + r - 4$. We consider four cases depending on the position of the vertices u_α and u_β.

Case 1. *The vertices u_α and u_β lie on the path P_n.* Assume that $\alpha < \beta$. Since $d(u_\alpha, u_\beta)$ is even, we have $d(u_\alpha, u_\beta) \leq n - 2$ and the path P_{u_α,u_β} must contains an edge of the form $\{u_{2i}, u_{2i+1}\}$ for some i. This implies the path that start at u_α moves to u_{2i} along the path of length $d(u_\alpha, u_{2i})$, then moves to v_{2i+1} along the path $P'_{u_{2i},u_{2i+1}}$ of length $r - 1$, then finally moves to u_β along the path of length $d(u_{2i+1}, u_\beta)$ is a $u_\alpha u_\beta$-path of odd length $\ell(P_{u_\alpha,u_\beta}) = d(u_\alpha, u_{2i}) + r - 1 + d(u_{2i+1}, u_\beta) \leq n - 3 + r - 1 = n + r - 4$.

Case 2. *The vertices u_α and u_β lie on a double circular $C_{2,r}$.* If u_α and u_β lie on the same cycle C_r, then there is an odd path P'_{u_α,u_β} of length $r - d(u_\alpha, u_\beta) \leq n + r - 4$. Suppose now u_α and u_β lie on two different cycles C_r. Since $d(u_\alpha, u_\beta)$ is even, we have $d(u_\alpha, u_\beta) \leq r - 1$. We note that the path P_{u_α,u_β} of length $d(u_\alpha, u_\beta)$ can be decomposed into two paths P_{u_α,u_y} and P_{u_y,u_β} for $y = 2i, 2i+1$. Since $d(u_\alpha, u_\beta)$ is even, then $\ell(P_{u_\alpha,u_y}) \equiv \ell(P_{u_y,u_\beta}) \bmod 2$. Assume that $\ell(P_{u_\alpha,u_y}) \leq \ell(P_{u_y,u_\beta})$. Then the walk that starts at u_α moves to u_y along the path P_{u_α,u_y}, then moves to u_β along the path P'_{u_y,u_β} of length $r - \ell(P_{u_\beta,u_y})$ is a $u_\alpha u_\beta$-walk of odd length $r - (\ell(P_{u_\alpha,u_y}) - \ell(P_{u_y,u_\beta})) \leq r < n + r - 4$.

Case 3. *The vertices u_α and u_β lie on different double-cycle.* Suppose u_α lies on the cycle with base $\{u_{2i}, u_{2i+1}\}$ and u_β lies on the cycle with base $\{u_{2j}, u_{2j+1}\}$ with $i < j$. Let P_{u_α,u_β} be the $u_\alpha u_\beta$-path of length $d(u_\alpha, u_\beta)$. We note that the vertices u_{2i+1} and u_{2j} must lie on the path P_{u_α,u_β}.

If $\ell(P_{u_\alpha,u_{2i+1}}) \leq \ell(P_{u_{2j},u_\beta})$, then $\ell(P_{u_\alpha,u_{2i+1}}) + \ell(P'_{u_{2j},u_\beta}) \leq r$. Therefore, the walk W_{u_α,u_β} that consists of the paths $P_{u_\alpha,u_{2i+1}}$, $P_{u_{2i+1},u_{2j}}$, and P'_{u_{2j},u_β} is an odd $u_\alpha u_\beta$-walk of length $\ell(W_{u_\alpha,u_\beta}) \leq n + r - 5$.

If $\ell(P_{u_\alpha,u_{2i+1}}) \geq \ell(P_{u_{2j},u_\beta})$, then $\ell(P'_{u_\alpha,u_{2i+1}}) + \ell(P_{u_{2j},u_\beta}) \leq r$. Therefore, the walk W_{u_α,u_β} that consists of the paths $P'_{u_\alpha,u_{2i+1}}$, $P_{u_{2i+1},u_{2j}}$, and P_{u_{2j},u_β} is an odd $u_\alpha u_\beta$-walk of length $\ell(W_{u_\alpha,u_\beta}) \leq n + r - 5$.

Case 4. *The vertex $u_\alpha = u_1$ and u_β lies on a cycle C_r or the vertex u_α lies on a cycle C_r and $u_\beta = u_n$.* We first consider the case where $u_\alpha = u_1$ and u_β lies on some cycle C_r. Let the base of the cycle C_r be the edge $\{u_{2i}, u_{2i+1}\}$ for some i. Then the walk W_{u_α,u_β} that consist of the paths $P_{u_1,u_{2i}}$ and the path P'_{u_{2i},u_β} is an odd $u_\alpha u_\beta$-walk of length $\ell(W_{u_\alpha,u_\beta}) \leq n + r - 4$. If u_α lies on some cycle C_r with base $\{u_{2j}, u_{2j+1}\}$, the the W_{u_α,u_β} that consists of the paths $P'_{u_\alpha,u_{2j+1}}$ and P_{u_{2j+1},u_β} is an odd $u_\alpha u_\beta$-walk of length $\ell(W_{u_\alpha,u_\beta}) \leq n + r - 4$.

Therefore, from Case 1-4 we conclude that for each pair of vertices u_α and u_β there is an odd $u_\alpha u_\beta$-walk of length $\ell(W_{u_\alpha,u_\beta}) \leq n + r - 4$. Since W_{u_α,u_β} is of odd length, it can be extended to a $u_\alpha u_\beta$-walk of length exactly $n + r - 4$. Therefore, $\exp(DA(C_{r,n}) \leq n + r - 4$. □

We now discuss the scrambling index of the double alternate circular snake graphs.

Theorem 4. *Let $n \geq 4$ is an even positive integer and $r \geq 3$ is an odd positive integer. Then $k(DA(C_{r,n})) = (n + r - 3)/2$.*

Proof. Since $d(u_1, u_n) = n-1$ is odd, the shortest even u_1u_n-walk $W_{u_1u_n}$ is the walk that consists of the paths $P_{v_1,v_{2i}}$, $P'_{u_{2i},u_{2i+1}}$ and P_{u_{2i+1},u_n} of length $\ell(W_{u_1,u_n}) = n - 2 + r - 1 = n + r - 3$. Hence $k_{u_1,u_n}(DA(C_{r,n})) = (n + r - 3)/2$. This implies $k(DA(C_{r,n})) \geq (n + r - 3)/2$.

We now show that $k(DA(C_{r,n})) \leq (n+r-3)/2$. Considering Proposition 2 it suffices to show that for any pair of vertices u_α and u_β in $DA(C_{r,n})$ there is a $u_\alpha u_\beta$-walk of length $n + r - 3$. We have shown in the proof of Theorem 3 that for each pair of vertices u_i and u_j there is a u_iu_j-walk of length exactly $n + r - 4$. Let u_α and u_β be any two vertices in $DA(C_{r,n})$ and let u_γ be a vertex in $DA(C_{r,n})$ such that $\{u_\gamma, u_\beta\}$ is an edge of $DA(C_{r,n})$. By Theorem 3 there is a $u_\alpha u_\gamma$-walk of length $n + r - 4$. Since $\{u_\gamma, u_\beta\}$ is an edge, there is a $u_\alpha u_\beta$-walk of length exactly $n + r - 3$. Thus $k(DA(C_{r,n})) \leq (n + r - 3)/2$. □

References

[1] Alkelbek M and Kirkland S 2009 *Linear Algebra and Appl.* **430** 1111
[2] Alkelbek M and Kirkland S 2009 *Linear Algebra and Appl.* **430** 1099
[3] Brualdi R A and Ryser H J 1991, *Combinatorial matrix Theory* (Cambridge: Cambridge University Press)
[4] Chen S and Liu B 2010 *Linear Algebra and Appl.* **433** 1110.
[5] Ponraj R and Narayanan S S 2015 *Palestine Journal of Mathematics* **4**(2) 439.
[6] Sandhya S S, Somasundaram S and Anusa S 2015 *Journal of Mathematics Research* **7**(1) 72

IOP Conf. Series: Materials Science and Engineering **300** (2018) 012080 doi:10.1088/1757-899X/300/1/012080

Local exponents of a certain class of two-colored Hamiltonian digraphs with two cycles whose lengths n and $n-3$

Mardiningsih[1], Syahriol Sitorus[2], Raidani[3] and Ferry Safriandi[4]

[1,2] Department of Mathematics, Universitas Sumatera Utara, Medan 20155, Indonesia
[3,4] Instansi Kopertis Wilayah I, DPK Universitas Al Azhar Medan, Indonesia

E-mail: mardiningsih.math@gmail.com

Abstract. A two-colored digraph is a digraph for which each of its arc is colored by red or blue. An (h, ℓ)-walk is a walk consisting of h red arcs and ℓ blue arcs. A strongly connected two-colored digraph is primitive provided there exist nonnegative integers h and ℓ such that for each ordered pair of vertices u and v there is a (h, ℓ)-walk from u to v. The local exponent of a vertex v in a primitive two-colored digraph $D^{(2)}$, denoted by $\exp(v, D^{(2)})$ is the least positive integer $h+\ell$ over all nonnegative integers h and ℓ such that for vertex u in $D^{(2)}$ there is an (h, ℓ)-walk from u to v. We discuss local exponent of primitive two-colored Hamiltonian digraph $D_3^{(2)}$ on n vertices consisting of two cycles of lengths n and $n-3$, respectively. For each vertex v in $D_3^{(2)}$, we present an explicit formula for $\exp(v, D_3^{(2)})$.

1. Introduction

Let D be a digraph on n vertices $\{v_1, v_2, \ldots, v_n\}$. We follow digraph terminologies especially the terms walk, path, cycle, strongly connected and distance from Brualdi and Ryser [1]. A two-colored digraph $D^{(2)}$ is a digraph D with the property that each of its arcs is colored by red or blue. An (h, k)-walk in a two-colored digraph $D^{(2)}$ is a walk with exactly h red arcs and k blue arcs. We use the term (h, k)-walk W_{v_i,v_j} to denote an (h, k)-walk from v_i to v_j. For an (h, k)-walk W_{v_i,v_j}, we let $r(W_{v_i,v_j}) = h$ and $b(W_{v_i,v_j}) = k$. The length of W_{v_i,v_j} is $\ell(W_{v_i,v_j}) = r(W_{v_i,v_j}) + b(W_{v_i,v_j})$. The composition of the walk W_{v_i,v_j} is defined to be the vector $(r(W_{v_i,v_j}), b(W_{v_i,v_j}))^T$.

A strongly connected two-colored digraph $D^{(2)}$ is primitive if there exists a positive integer $h+k$ over all nonnegative integers h and k such that for each two vertices v_i and v_j there is a (h, k)-walk W_{v_i,v_j} [2]. Shader and Suwilo [3] introduced the parameter exponent of a primitive two-colored digraph. The *exponent* of a primitive two-colored digraph $D^{(2)}$, denoted by $\exp(D^{(2)})$, is the smallest positive integer $h+k$ over all pairs of nonnegative integer h and k such that for each two vertices v_i and v_j there exists a (h, k)-walk W_{v_i,v_j}. Since then more parameters on primitive two-colored digraph have been introduced. The *local exponent* of a vertex v_t in $D^{(2)}$, denoted by $\exp(v_t, D^{(2)})$, is the smallest positive integer $h_t + k_t$ over all nonnegative integers h_t and k_t such that for each vertex v_i in $D^{(2)}$ there exists an (h_t, k_t)-walk W_{v_i,v_t} [4, 9].

Research on parameters of primitive two-colored digraph focuse on the class of two-colored digraph consisting of two cycles. Results on local exponent can be found on [4–10]. In this paper, we discuss local exponents of a class of primitive two-colored Hamiltonian digraph $D_3^{(2)}$ whose uncolored digraph is the digraph consisting of hamiltonian cycle $v_1 \to v_2 \to \cdots \to v_n \to v_1$ of length n and the cycle $v_1 \to v_2 \to \cdots \to v_{n-3} \to v_1$ of length $n-3$. In Section 2, we discuss the primitivity of $D_3^{(2)}$ and present a way in setting up a lower bound for local exponents of a primitive two-colored digraph consisting of two cycles. Sections 3 and 4 discuss primitive coloring and the local exponents of primitive two-colored digraph $D_3^{(2)}$.

2. Necessary Background

Let $D^{(2)}$ be a strongly connected two-colored digraph consisting of two cycles C_1 and C_2. We define the cycle matrix of $D^{(2)}$ to be the 2 by 2 matrix $M = \begin{bmatrix} r(C_1) & r(C_2) \\ b(C_1) & b(C_2) \end{bmatrix}$.

Theorem 1. [2] *Let $D^{(2)}$ be a strongly connected two-colored digraph consisting of two cycles with at least one arc of each color and let M be a cycle matrix of $D^{(2)}$. The two-colored digraph $D^{(2)}$ is primitive if and only if* $\det(M) = \pm 1$.

The following lemma produces a technique to establish a lower bound for local exponents.

Lemma 2. [9] *Let $D^{(2)}$ be a primitive two-colored digraph consisting of two cycles C_1 and C_2. Suppose the cycle matrix of $D^{(2)}$ is $M = \begin{bmatrix} r(C_1) & r(C_2) \\ b(C_1) & b(C_2) \end{bmatrix}$ and* $\det(M) = 1$. *If* $\exp(v_t : D^{(2)})$ *is achieved by (h_t, k_t)-walks, then $\begin{bmatrix} h_t \\ k_t \end{bmatrix} \geq M \begin{bmatrix} b(C_2)r(P_{v_i,v_t}) - r(C_2)b(P_{v_i,v_t}) \\ r(C_1)b(P_{v_j,v_t}) - b(C_1)r(P_{v_j,v_t}) \end{bmatrix}$ for some paths P_{v_i,v_t} and P_{v_j,v_t}.*

If we define $q_1 = b(C_2)r(P_{v_i,v_t}) - r(C_2)b(P_{v_i,v_t})$ and $q_2 = r(C_1)b(P_{v_j,v_t}) - b(C_1)r(P_{v_j,v_t})$, then from Lemma 2 we have

$$\exp(v_t, D^{(2)}) = h_t + k_t \geq \ell(C_1)q_1 + \ell(C_2)q_2 \tag{1}$$

for some paths P_{v_i,v_t} and P_{v_j,v_t}.

3. The Case where $n \equiv 2 \bmod 3$

Let $D_3^{(2)}$ be a two-colored digraph on $n \geq 8$ vertices where $n \equiv 2 \bmod 3$ whose uncolored digraph is the digraph consisting of the cycle $C_1 : v_1 \to v_2 \to \cdots \to v_n \to v_1$ of length $\ell(C_1) = n$, and the cycle $C_2 : v_1 \to v_2 \to \cdots \to v_{n-3} \to v_1$ of length $\ell(C_2) = n-3$. We discuss the local exponent of $D_3^{(2)}$

As a consequence of Theorem 1 we have the following characterization of the cycle matrix of a primitive two-colored digraph $D_3^{(2)}$.

Corollary 3. *Let $n \geq 8$ and let $M = \begin{bmatrix} r(C_1) & r(C_2) \\ b(C_1) & b(C_2) \end{bmatrix}$ be a cycle matrix of $D_3^{(2)}$. If the strongly connected two-colored digraph $D_3^{(2)}$ is primitive and $n \equiv 2 \bmod 3$, then $M = \begin{bmatrix} (2n-1)/3 & (2n-7)/3 \\ (n+1)/3 & (n-2)/3 \end{bmatrix}$ or $M = \begin{bmatrix} (n+1)/3 & (n-2)/3 \\ (2n-1)/3 & (2n-7)/3 \end{bmatrix}$.*

Without loss of generality we may assume that the cycle matrix of $D_3^{(2)}$ is the matrix $M = \begin{bmatrix} (2n-1)/3 & (2n-7)/3 \\ (n+1)/3 & (n-2)/3 \end{bmatrix}$. This implies that if $D_3^{(2)}$ is primitive, then $r(C_1) = (2n-1)/3$, $b(C_1) = (n+1)/3$, $r(C_2) = (2n-7)/3$, and $b(C_2) = (n-2)/3$. In this paper, we consider only the case where the cycle C_1 contains $((2n-1)/3, 0)$-path and the cycle C_2 contains $((2n-7)/3, 0)$-path. There are four types of such two-colored digraph $D_3^{(2)}$.

(i) The digraph $D_3^{(2)}$ is of Type I if the arc $v_{n-3} \to v_1$ is a red arc and it is the first arc of the $((2n-7)/3, 0)$-path that lies on C_2.

(ii) The digraph $D_3^{(2)}$ is of Type II if the arc $v_{n-3} \to v_1$ is a red arc and it is the last arc of the $((2n-7)/3, 0)$-path that lies on C_2.

(iii) The digraph $D_3^{(2)}$ is of Type III if the arc $v_{n-3} \to v_1$ is a blue arc and it is the first arc of the $(0, (n-2)/3)$-path that lies on C_2.

(iv) The digraph $D_3^{(2)}$ is of Type IV if the arc $v_{n-3} \to v_1$ is a blue arc and it is the last arc of the $(0, (n-2)/3)$-path that lies on C_2.

We next discuss the local exponents of $D_3^{(2)}$. We present a complete proof for Type I and present a sketch of proof for the other types.

Theorem 4. *Let $D_3^{(2)}$ be a primitive two-colored digraph of type I on $n \geq 8$ vertices where $n \equiv 2 \bmod 3$. Then $\exp(v_t, D_3^{(2)}) = (2n^3 - 5n^2 - 4n + 30)/9 + d(v_1, v_t)$ for each $t = 1, 2, \ldots, n$.*

Proof. We first show that $\exp(v_t, D_3^{(2)}) \geq (2n^3 - 5n^2 - 4n + 30)/9 + d(v_1, v_t)$ for each $t = 1, 2, \ldots, n$. Suppose $\exp(v_t, D^{(2)})$ is obtained by (h_t, ℓ_t)-walks. We define the quantities $q_1 = b(C_2)r(P_{v_{n-2},v_t}) - r(C_2)b(P_{v_{n-2},v_t})$ and $q_2 = r(C_1)b(P_{v_{(2n-7)/3},v_t}) - b(C_1)r(P_{v_{(2n-7)/3},v_t})$.

The vertex v_t is on the path $P_{v_1 v_{(2n-7)/3}}$. Applying the $(3+d(v_1, v_t), 0)$-path from v_{n-2} to v_t we have $q_1 = (n-2)(3+d(v_1, v_t))/3$. There are two paths $P_{v_{(2n-7)/3},v_t}$. They are a $(1+d(v_1, v_t), (n-2)/3)$-path and a $(3+d(v_1, v_t), (n+1)/3)$-path. Applying the $(1+d(v_1, v_t), (n-2)/3)$-path, we have $q_2 = (2n^2 - 8n - 1)/9 - (n+1)d(v_1, v_t)/3$, and applying the $(3+d(v_1, v_t), (n+1)/3)$-path we have $q_2 = (2n^2 - 8n - 10)/9 - (n+1)d(v_1, v_t)/3$. We conclude that $q_2 = (2n^2 - 8n - 10)/9 - (n+1)d(v_1, v_t)/3$. By Lemma 2 $\begin{bmatrix} h_t \\ k_t \end{bmatrix} \geq M \begin{bmatrix} q_1 \\ q_2 \end{bmatrix} = \begin{bmatrix} (4n^3 - 12n^2 - 9n + 88)/27 + d(v_1, v_t) \\ (2n^3 - 3n^2 - 3n^2 + 2)/27 \end{bmatrix}$.
Hence $\exp(v_t, D_3^{(2)}) \geq (2n^3 - 5n^2 - 4n + 30)/9 + d(v_1, v_t)$ for each vertex v_t that is on the path $P_{v_1, v_{(2n-7)/3}}$.

The vertex v_t is on the path $P_{v_{(2n-4)/3},v_{n-2}}$. Applying the $((2n-1)/3, d(v_1, v_t) - (2n-10)/3)$-path from v_{n-2} to v_t we have $q_1 = (2n^2 - 13n + 8)/3 - (2n-7)d(v_1, v_t)/3$. Applying the $(0, d(v_1, v_t) - (2n-10)/3)$-path from $v_{(2n-7)/3}$ to v_t we have $q_2 = (2n-1)d(v_1, v_t)/3 - (4n^2 - 22n + 10)/9$. By (1), $\exp(v_t, D_3^{(2)}) \geq \ell(C_1)q_1 + \ell(C_2)q_2 = nq_1 + (n-3)q_2 = (2n^3 - 5n^2 - 4n + 30)/9 + d(v_1, v_t)$ for each vertex v_t that is on the path $P_{v_{(2n-4)/3},v_{n-2}}$.

The vertex v_t is on the path P_{v_{n-2},v_n} path. Applying the $(3-d(v_t, v_1), 0)$-path from v_{n-2} to v_t we have $q_1 = (n-2)(3-d(v_t, v_1))/3$. Applying the $(3-d(v_t, v_1), (n+1)/3)$-path from $v_{(2n-7)/3}$ to v_t we have $q_2 = (2n^2 + n - 1)/9 - (n+1)(3 - d(v_t, v_1))/3$. By Lemma 2 we have $\begin{bmatrix} h_t \\ k_t \end{bmatrix} \geq \begin{bmatrix} (4n^3 - 12n^2 - 9n + 88)/27 - d(v_t, v_1) \\ (2n^2 - 3n^2 - 3n + 2)/27 \end{bmatrix} = \begin{bmatrix} p_1 \\ p_2 \end{bmatrix}$. We next examine the existence of (p_1, p_2)-walk W_{v_{n-2},v_t} in $D_3^{(2)}$. Since the $v_{n-2} \to v_t$ path is a $(3-d(v_t, v_1), 0)$-path and the solution to the system $Mz + \begin{bmatrix} 3 - d(v_t, v_1) \\ 0 \end{bmatrix} = \begin{bmatrix} p_1 \\ p_2 \end{bmatrix}$ is $z_1 = 0$ and $z_2 = (2n^2 + n - 1)/9$, there is no (p_1, p_2)-walk W_{v_{n-2},v_t} in $D_3^{(2)}$. Hence $\exp(v_t, D_3^{(2)}) > p_1 + p_2$. Notice that the shortest $v_{n-2} \to v_t$ walk that contains at least p_1 red arcs and p_2 blue arcs is a $(p_1 + (2n-1)/3, p_2 + (n+1)/3)$-walk. Since $n - d(v_t, v_1) = d(v_1, v_t)$, we now have $\exp(v_t, D_3^{(2)}) \geq p_1 + p_2 + n = (2n^3 - 5n^2 - 4n + 30)/9 + d(v_1, v_t)$ for each vertex v_t that is on the path P_{v_{n-2},v_n}.

Therefore we now conclude that $\exp(v_t, D_3^{(2)}) \geq (2n^3 - 5n^2 - 4n + 30)/9 + d(v_1, v_t)$ for each $t = 1, 2, \ldots, n$.

We next show that $\exp(v_t, D_3^{(2)}) \leq (2n^3 - 5n^2 - 4n + 30)/9 + d(v_1, v_t)$. We first show that for each vertices $v_j, j = 1, 2, \ldots, n$ there is a $v_j \overset{(h_1,\ell_1)}{\longrightarrow} v_1$ walk with $h_1 = (4n^3 - 12n^2 - 9n + 88)/27$ and $\ell_1 = (2n^3 - 3n^2 - 3n + 2)/27$. We show that the system

$$Mz + \begin{bmatrix} r(P_{v_j,v_1}) \\ b(P_{v_j,v_1}) \end{bmatrix} = \begin{bmatrix} h_t \\ \ell_t \end{bmatrix} \tag{2}$$

has a nonnegative integer solution for some path P_{v_j,v_1} from v_j to the vertex v_1. The solution to the system (2) is $z_1 = (n-2)(3 - r(P_{v_j,v_1}))/3 + (2n-7)b(P_{v_j,v_1})/3$ and $z_2 = (n+1)((2n-10)/3 + r(P_{v_j,v_1}))/3 - (2n-1)b(P_{v_j,v_1})/3$.

If $1 \leq j \leq (2n-7)/3$, then there is a $(r(P_{v_j,v_1}), (n+1)/3)$-path from v_j to v_1 with $3 \leq r(P_{v_j,v_1}) \leq (2n-1)/3$. Applying this path and since $r(P_{v_j,v_1}) \leq (2n-1)/3$ we have $z_1 \geq (n-2)(3 - (2n-1)/3)/3 + (2n-7)(n+1)/3 = n - 3$. Applying this path and since $r(P_{v_j,v_1}) \geq 3$, we have $z_2 \geq (n+1)((2n-10)/3 + 3) - (2n-1)(n+1)/9 = 0$.

If $(2n-4)/3 \leq j \leq n-2$, then there is a $(3, b(P_{v_j,v_1}))$-path from v_j to v_1 with $0 \leq b(P_{v_j,v_1}) \leq (n-2)/3$. Applying this path and since $b(P_{v_j,v_1}) \geq 0$, we have $z_1 = (2n-7)b(P_{v_j,v_1})/3 \geq 0$. Applying this path and since $b(P_{v_j,v_1}) \leq (n-2)/3$, we have $z_2 \geq (2n-1)/3$.

If $n-1 \leq j \leq n$, then there is a $(r(P_{v_j,v_1}), 0)$-path from v_j to v_1 with $1 \leq r(P_{v_j,v_1}) \leq 2$. Applying this path and since $r(P_{v_j,v_1}) \leq 2$, we have $z = (n-3)/2$. Applying this path and since $r(P_{v_j,v_1}) \geq 1$, we have $z_2 = (n+1)(2n-7)/9$.

Therefore, for each vertex $v_j, j = 1, 2, \ldots, n$ there is a path P_{v_j,v_1} such that the system (2) has a nonnegative integer solution. This implies the walk that starts at v_j, moves to v_1 along the chosen path P_{v_j,v_1} and along the way moves z_1 times around the cycle C_1 and moves z_2 times around the cycle C_2 is a $v_j \overset{(h_1,\ell_1)}{\longrightarrow} v_1$ walk in $D_3^{(2)}$. Notice that for each $t = 2, 3, \ldots, n$ there is a unique $v_1 \rightarrow v_t$ path P_{v_1,v_t} of length $d(v_1, v_t) = r(P_{v_1,v_t}) + b(P_{v_1,v_t})$. Therefore for each $v_j = 1, 2, \ldots, n$ there is a $(h_1 + r(P_{v_1,v_t}), \ell_1 + b(P_{v_1,v_t})$-walk from v_j to v_t. Hence $\exp(v_t, D_3^{(2)}) \leq h_1 + \ell_1 + d(v_1, v_t) = (2n^3 - 5n^2 - 4n + 30)/9 + d(v_1, v_t)$. □

Theorem 5. *Let $D_3^{(2)}$ be a primitive two-colored digraph of type II on $n \geq 8$ vertices where $n \equiv 2 \bmod 3$. Then $\exp(v_t, D_3^{(2)}) = (2n^3 - 5n^2 - 7n + 9)/9 + d(v_1, v_t)$ for each $t = 1, 2, \ldots, n$.*

Sketch of Proof. To show $\exp(v_t, D_3^{(2)}) \geq (2n^3 - 5n^2 - 7n + 9)/9 + d(v_1, v_t)$, for each vertex $v_t, t = 1, 2, \ldots, n$, we consider the paths $P_{v_{(n+1)/3},v_t}$ and P_{v_n,v_t}. We define the quantities $q_1 = b(C_2)r(P_{v_{(n+1)/3},v_t}) - r(C_2)r(P_{v_{(n+1)/3},v_t})$ and $q_2 = r(C_1)b(P_{v_n,v_t}) - b(C_1)r(P_{v_n,v_t})$.

The vertex v_t is on the path $P_{v_1,v_{(n+1)/3}}$. Applying the $((2n-7)/3, d(v_1, v_t))$-path from $v_{(n+1)/3}$ to v_t we have $q_1 = (2n^2 - 11n + 14)/9 - (2n-7)d(v_1, v_t)/3$. Applying the $(0, 1 + d(v_1, v_t))$-path from v_n to v_t we have $q_2 = (2n-1)d(v_1, v_t)/3$. By (1) we have $k(v_{(n+1)/3}, v_n : v_t) \geq nq_1 + (n-3)q_2 = (2n^3 - 5n^2 - 7n + 9)/9 + d(v_1, v_t)$ for each vertex v_t that is on the path $P_{v_1,v_{(n+1)/3}}$.

The vertex v_t is on the path $P_{v_{(n+4)/3},v_n}$. Applying the $(d(v_1, v_t) - (n-2)/3, 0)$-path form $v_{(n+1)/3}$ to v_t, we have $q_1 = (n-2)d(v_1, v_t)/3 - (n-2)^2/9$. Applying the $(d(v_1, v_t) - (n - 2)/3, (n+1)/3)$-path from v_n to v_t we have $q_2 = (n^2-1)/3 - (n+1)d(v_1, v_t)/3$. By (1) we have $k(v_{(n+1)/3}, v_n : v_t) \geq nq_1 + (n-3)q_2 = (2n^3 - 5n^2 - 7n + 9)/9 + d(v_1, v_t)$ for each vertex v_t that is on the path $P_{v_{(n+4)/3},v_n}$.

Therefore, we now conclude that $\exp(v_t, D_3^{(2)}) \geq (2n^3 - 5n^2 - 7n + 9)/9 + d(v_1, v_t)$ for each $t = 1, 2, \ldots, n$.

As in the proof Theorem 4, in order to show that $\exp(v_t, D_3^{(2)}) \leq (2n^3 - 5n^2 - 7n + 9)/9 + d(v_1, v_t)$, it suffices show that for each vertices $v_j, j = 1, 2, \ldots, n$ there is a $v_j \overset{(h_1,\ell_1)}{\longrightarrow} v_1$ walk with $h_1 = (4n^3 - 12n^2 - 9n + 7)/27$ and $\ell_1 = (2n^3 - 3n^2 - 12n + 20)/27$. That is the system $Mz + \begin{bmatrix} r(P_{v_j,v_1}) \\ b(P_{v_j,v_1}) \end{bmatrix} = \begin{bmatrix} h_1 \\ \ell_1 \end{bmatrix}$ has a nonnegative integer solution for some path P_{v_j,v_1} from v_j to the vertex v_1. □

Theorem 6. *Let $D_3^{(2)}$ be a primitive two-colored digraph of type III on $n \geq 8$ vertices where $n \equiv 2 \bmod 3$. Then $\exp(v_t, D_3^{(2)}) = (2n^3 - 5n^2 - 7n + 18)/9 + d(v_1, v_t)$ for each $t = 1, 2, \ldots, n$.*

Sketch of Proof. To show $\exp(v_t, D_3^{(2)}) \geq (2n^3 - 5n^2 - 7n + 18)/9$ for each $t - 1, 2, \ldots, n$ we consider the $v_{(n-2)/3} \to v_t$ path and the $v_{n-1} \to v_t$ path and we define the quantities $q_1 = b(C_2)r(P_{v_{(n-2)/3},v_t}) - r(C_2)r(P_{v_{(n-2)/3},v_t})$ and $q_2 = r(C_1)b(P_{v_{n-1},v_t}) - b(C_1)r(P_{v_{n-1},v_t})$.

The vertex v_t is on the path $P_{v_1,(n-2)/3}$. Applying the $((2n-7)/3, 1 + d(v_1, v_t))$-path from $v_{(n-2)/3}$ to v_t we have $q_1 = (2n^2 - 17n + 35)/9 - (2n - 7)d(v_1, v_t)/3$. Applying the $(0, 2 + d(v_1, v_t))$-path from v_{n-1} to v_t we have $q_2 = (2n-1)(2 + d(v_1, v_t))/3$. By (1) we have $\exp(v_t, D_3^{(2)}) \geq nq_1 + (n-3)q_2 = (2n^3 - 5n^2 - 7n + 18)/9 + d(v_1, v_t)$ for each vertex v_t that lies on the path $P_{v_1,v_{(n-2)/3}}$.

The vertex v_t is on the $P_{v_{(n+1)/3},v_{n-1}}$. applying the $(d(v_1, v_t) - (n-5)/3, 0)$-path from $v_{(n-2)/3}$ to v_t we have $q_1 = (n-2)d(v_1, v_t)/3 - (n^2 + 7n - 10)/9$. Applying the $(d(v_1, v_t) - (n-5)/3, (n+1)/3)$-path from v_{n-1} to v_t we have $q_2 = (n^2 - n - 2)/3 - (n+1)d(v_1, v_t)/3$. By (1) we have $\exp(v_t, D_3^{(2)}) \geq nq_1 + (n-3)q_2 = (2n^3 - 5n^2 - 7n + 18)/9 + d(v_1, v_t)$ for each vertex v_t that lies on the path $P_{v_{(n+1)/3},v_{n-1}}$.

The vertex $v_t = v_n$. Applying the $((2n-1)/3, 1)$-path from $v_{(n-2)/3}$ to v_t we have $q_1 = (2n^2 - 11n + 23)/9$. Applying the $(0, 1)$-path from v_{n-1} to v_t we have $q_2 = (2n-1)/3$. By (1) we have $\exp(v_t, D_3^{(2)}) \geq nq_1 + (n-3)q_2 = (2n^3 - 5n^2 + 2n + 9)/9 = (2n^3 - 5n^2 - 7n + 18)/9 + n - 1 = (2n^3 - 5n^2 - 7n + 18)/9 + d(v_1, v_t)$ for vertex $v_t = v_n$.

Therefore, we now conclude that $\exp(v_t, D_3^{(2)}) \geq (2n^3 - 5n^2 - 7n + 18)/9 + d(v_1, v_t)$ for each $t = 1, 2, \ldots, n$

As in the proof Theorem 4, in order to show that $\exp(v_t, D_3^{(2)}) \leq (2n^3 - 5n^2 - 7n + 18)/9 + d(v_1, v_t)$, it suffices to show that for each vertices $v_j, j = 1, 2, \ldots, n$ there is a $v_j \overset{(h_1,\ell_1)}{\longrightarrow} v_1$ walk with $h_1 = (4n^3 - 12n^2 - 9n + 7)/27$ and $\ell_1 = (2n^3 - 3n^2 - 12n + 47)/27$. That is the system $Mz + \begin{bmatrix} r(P_{v_j,v_1}) \\ b(P_{v_j,v_1}) \end{bmatrix} = \begin{bmatrix} h_1 \\ \ell_1 \end{bmatrix}$ has a nonnegative integer solution for some path P_{v_j,v_1} from v_j to the vertex v_1. □

Theorem 7. *Let $D_3^{(2)}$ be a primitive two-colored digraph of type IV on $n \geq 8$ vertices where $n \equiv 2 \bmod 3$. Then $\exp(v_t, D_3^{(2)}) = (2n^3 - 5n^2 - 4n + 21)/9 + d(v_1, v_t)$ for each $t = 1, 2, \ldots, n$.*

Skecth of Proof. To show $\exp(v_t, D_3^{(2)}) \geq (2n^3 - 5n^2 - 4n + 21)/9 + d(v_1, v_t)$, for each $t = 1, 2, \ldots, n$, we consider the $v_{(2n-4)/3} \to v_t$ path and the $v_{n-1} \to v_t$ path and we define the quantities $q_1 = b(C_2)r(P_{v_{(n-1)},v_t}) - r(C_2)r(P_{v_{(n-1)},v_t})$ and $q_2 = r(C_1)b(P_{v_{(2n-4)/3},v_t}) - b(C_1)r(P_{v_{(2n-4)/3},v_t})$.

The vertex v_t is on the path $P_{v_1,v_{(2n-4)/3}}$. Applying the $(2 + d(v_1, v_t), 0)$-path from v_{n-1} to v_t we have $q_1 = (n-2)(2 + d(v_1, v_t))/3$. Applying the $(2 + d(v_1, v_t), (n+1)/3)$-path from $v_{(2n-4)/3}$ to v_t we have $q_2 = (2n^2 - 5n - 7)/9 - (n+1)d(v_1, v_t)/3$. By (1) we have $\exp(v_t, D_3^{(2)}) \geq nq_1 + (n-3)q_2 = (2n^3 - 5n^2 - 4n + 21)/9 + d(v_1, v_t)$ for each vertex v_t that lies on the path $P_{v_1,v_{(2n-4)/3}}$.

The vertex v_t is on the path $P_{v_{(2n-1)/3},v_{n-1}}$. Applying $((2n-1)/3, d(v_1, v_t) - (2n-7)/3)$-path from v_{n-1} to v_t we have $q_1 = (2n^2 - 11n + 17)/3 - (2n-7)d(v_1, v_t)/3$. Applying the $(0, d(v_1, v_t) - (2n-7)/3)$-path from $v_{(2n-4)/3}$ to v_t we have $q_2 = (2n-1)d(v_1, v_t)/3 - (4n^2 - 16n + 7)/9$. By

(1), $\exp(v_t, D_3^{(2)}) \geq nq_1 + (n-3)q_2 = (2n^3 - 5n^2 - 4n + 21)/9 + d(v_1, v_t)$ for each vertex v_t that lies on the path $P_{v_{(2n-1)/3}, v_{n-1}}$.

The vertex $v_t = v_n$. Applying the $(1,0)$-path from v_{n-1} to v_t we have $q_1 = (n-2)/3$. Applying the $(1,(n+1)/3)$-path from $v_{(2n-4)/3}$ to v_t we have $q_2 = (2n^2 - 2n - 4)/9$. By Lemma 2 we have $\begin{bmatrix} h_t \\ k_t \end{bmatrix} \geq \begin{bmatrix} (4n^3 - 12n^2 - 9n + 34)/27 \\ (2n^2 - 3n^2 - 3n + 2)/27 \end{bmatrix} = \begin{bmatrix} p_1 \\ p_2 \end{bmatrix}$. Since there is no (p_1, p_2)-walk W_{v_{n-1}, v_t} in $D_3^{(2)}$, we have $\exp(v_t, D_3^{(2)}) \geq p_1 + p_2 + n = (2n^3 - 5n^2 - 4n + 21)/9 + (n-1) = (2n^3 - 5n^2 - 4n + 21)/9 + d(v_1, v_t)$ for vertex $v_t = v_n$ path.

Therefore, we now conclude that $\exp(v_t, D_3^{(2)}) \geq (2n^3 - 5n^2 - 4n + 21)/9 + d(v_1, v_t)$ for each $t = 1, 2, \ldots, n$.

As in the proof Theorem 4, in order to show that $\exp(v_t, D_3^{(2)}) \leq (2n^3 - 5n^2 - 4n + 21)/9 + d(v_1, v_t)$ for each $t = 1, 2, \ldots, n$, it suffices to show that for each vertices $v_j, j = 1, 2, \ldots, n$ there is a $v_j \overset{(h_1,\ell_1)}{\longrightarrow} v_1$ walk with $h_1 = (4n^3 - 12n^2 - 9n + 61)/27$ and $\ell_1 = (2n^3 - 3n^2 - 3n + 2)/27$. That is the system $Mz + \begin{bmatrix} r(P_{v_j,v_1}) \\ b(P_{v_j,v_1}) \end{bmatrix} = \begin{bmatrix} h_1 \\ \ell_1 \end{bmatrix}$ has a nonnegative integer solution for some path P_{v_j,v_1} from v_j to the vertex v_1. □

4. The Case where $n \equiv 1 \bmod 3$

Let $D_3^{(2)}$ be a two-colored digraph on $n \geq 7$ vertices where $n \equiv 1 \bmod 3$ whose uncolored digraph is the digraph consisting of two cycles $C_1 : v_1 \to v_2 \to \cdots \to v_{n-3} \to v_1$ of length $n-3$ and $C_2 : v_1 \to v_2 \to \cdots \to v_n \to v_1$ of length n.

Corollary 8. *Let* $n \geq 7$ *and let* $M = \begin{bmatrix} r(C_1) & r(C_2) \\ b(C_1) & b(C_2) \end{bmatrix}$ *be a cycle matrix of* $D_3^{(2)}$. *If the strongly connected two-colored digraph* $D_3^{(2)}$ *is primitive and* $n \equiv 1 \bmod 3$, *then* $M = \begin{bmatrix} (2n-5)/3 & (2n+1)/3 \\ (n-4)/3 & (n-1)/3 \end{bmatrix}$ *or* $M = \begin{bmatrix} (n-4)/3 & (n-1)/3 \\ (2n-5)/3 & (2n+1)/3 \end{bmatrix}$.

Without loss of generality the cycle matrix of $D_3^{(2)}$ is the matrix $M = \begin{bmatrix} (2n-5)/3 & (2n+1)/3 \\ (n-4)/3 & (n-1)/3 \end{bmatrix}$. In this paper, we consider only the case where the cycle C_1 contains $((2n-5)/3, 0)$-path and the cycle C_2 contains $((2n+1)/3, 0)$-path. There are four types of such two-colored digraph $D_3^{(2)}$.

(i) The digraph $D_3^{(2)}$ is of Type I if the arc $v_{n-3} \to v_1$ is a red arc and it is the first arc of the $((2n-5)/3, 0)$-path that lies on C_1.

(ii) The digraph $D_3^{(2)}$ is of Type II if the arc $v_{n-3} \to v_1$ is a red arc and it is the last arc of the $((2n-5)/3, 0)$-path that lies on C_1.

(iii) The digraph $D_3^{(2)}$ is of Type III if the arc $v_{n-3} \to v_1$ is a blue arc and it is the first arc of the $(0, (n-4)/3)$-path that lies on C_1.

(iv) The digraph $D_3^{(2)}$ is of Type IV if the arc $v_{n-3} \to v_1$ is a blue arc and it is the last arc of the $(0, (n-4)/3)$-path that lies on C_1.

Theorem 9. *Let* $D_3^{(2)}$ *be a primitive two-colored digraph of type I on* $n \geq 7$ *vertices where* $n \equiv 1 \bmod 3$. *Then for each* $t = 1, 2, \ldots, n$, $\exp(v_t, D^{(2)}) = (2n^3 - 7n^2 - 4n + 27)/9 + d(v_1, v_t)$.

Theorem 10. *Let* $D_3^{(2)}$ *be a primitive two-colored digraph of type II on* $n \geq 7$ *vertices where* $n \equiv 1 \bmod 3$. *Then for each* $t = 1, 2, \ldots, n$, $\exp(v_t, D^{(2)}) = (2n^3 - 7n^2 + 2n + 12)/9 + d(v_1, v_t)$.

Theorem 11. *Let $D_3^{(2)}$ be a primitive two-colored digraph of type III on $n \geq 7$ vertices where $n \equiv 1 \bmod 3$. Then for each $t = 1, 2, \ldots, n$, $\exp(v_t, D^{(2)}) = (2n^3 - 7n^2 + 2n + 21)/9 + d(v_1, v_t)$.*

Theorem 12. *Let $D_3^{(2)}$ be a primitive two-colored digraph of type IV on $n \geq 7$ vertices where $n \equiv 1 \bmod 3$. Then for each $t = 1, 2, \ldots, n$, $\exp(v_t, D^{(2)}) = (2n^3 - 7n^2 - 4n + 18)/9 + d(v_1, v_t)$.*

References

[1] Brualdi R A and Ryser H J 1991 *Combinatorial Matrix Theory* (Cambridge University Press: Cambridge).
[2] Fornasini E and Valcher M E 1998 *SIAM J. Matrix Anal. Appl.* **19** 71
[3] Shader B L and Suwilo S 2003 *Linear Algebra Appl.* **263** 275
[4] Gao Y and Shao Y 2009 *Linear Algebra Appl.* **430** 1550
[5] Suwilo S 2011 *AIP Conference Proceedings* **1337** 235
[6] Syahmarani A and Suwilo S 2012 *J. Indones. Math. Soc.* **18** (1) 1
[7] Hussein I, Prasetyo Y D and Suwilo S 2014 *AIP Conference Proceedings* **1635** 430
[8] Perangin-angin M G, Fathoni M, Syahmarani A and Suwilo S 2014 *AIP Conference Proceedings* **1635** 471
[9] Mardiningsih, Fathoni M and Suwilo S 2016 *Malaysian Journal of Mathematical Sciences* **10**(S) 205
[10] Sumardi H and Suwilo S 2016 *AIP Conference Proceedings* **1775** 030055

IOP Conf. Series: Materials Science and Engineering **300** (2018) 012081 doi:10.1088/1757-899X/300/1/012081

Primitive graphs with small exponent and small scrambling index

Ita Y Surbakti, Saib Suwilo* Ericha A Butar-butar, and Helda Oktaviani

Department of Mathematics, Universitas Sumatera Utara, Medan 20155, Indonesia

E-mail: *saib@usu.ac.id

Abstract. A connected graph G is primitive provide there is a positive integer k such that for each pair of vertices u and v there is a uv-walk of length k. The smallest of such positive integer k is the exponent of G and is denoted by $\exp(G)$. The scrambling index of a primitive graph G, denoted by $k(G)$, is the smallest positive integer k such that for each pair of vertices u and v there is a vertex w such that there is a uw-walk and a vw-walk of length k. By an n-chainring $CR(n)$ we mean a graph obtained from an n-cycle by replacing each edge of the n-cycle by a triangle. By a (q,p)-dory, $D(q,p)$, we mean a graph with vertex set $V(D(q,p)) = V(P_q \times P_p) \cup \{w_1, w_2\}$ and edge set $E(D(q,p)) = E(P_q \times P_p) \cup \{w_1 - (u_i, v_1) : i = 1, 2, \ldots, q\} \cup \{w_2 - (u_i, v_p) : i = 1, 2, \ldots, q\}$, where P_n is a path on n vertices. We discus the exponent and scrambling index of an n-chainring and (q,p)-dory. We present formulae for exponent and scrambling index in terms of their diameter.

1. Introduction

Let G be a simple graph. We follow graph terminologies from [1,2]. Let u and v be two vertices in a graph G. A walk connecting u and v is denoted by W_{uv} or uv-walk. A uv-path is a uv-walk without repeated vertices except possibly $u = v$. A cycle is a closed path. The length of a walk W_{uv} is denoted by $\ell(W_{uv})$. By a triangle we mean a cycle of length three. A walk is even or odd if it is of even or odd length respectively. For a connected graph G the distance $d(u,v)$ of vertices u and v in G is defined to be the length of a shortest uv-path in G. The diameter of a connected graph G, denoted by $\mathrm{diam}(G)$, is defined to be

$$\mathrm{diam}(G) = \max_{u,v}\{d(u,v)\}.$$

A connected graph G is said to be primitive if there is a positive integer k such that for each pair of vertices u and v in G there is a uv-walk of length k. The exponent of a primitive graph G, denoted by $\exp(G)$, is the smallest of such positive integer k. The scrambling index of a primitive graph G is the smallest positive integer k such that for each pair of vertices u and v there is a uv-walk of length $2k$ [3–5]. It is known that a connected graph is primitive if and only if it contains an odd cycle [2]. From definition we have $\exp(G) \geq \mathrm{diam}(G)$. Notice also that if $\mathrm{diam}(G)$ of G is even then by definition $k(G) \geq \mathrm{diam}(G)$. If the $\mathrm{diam}(G)$ is odd, then there is a pair of vertices u and v such that the shortest even uv-walk is of length $\mathrm{diam}(G) + 1$. Hence if $\mathrm{diam}(G)$ is odd, we have $k(G) \geq (\mathrm{diam}(G) + 1)/2$. Therefore, for a primitive graph G we find that $k(G) \geq \left\lceil \frac{\text{diam}(G)}{2} \right\rceil$. Therefore, for a primitive graph G with the smallest cycle of odd length $r \geq 3$

$$\text{diam}(G) \leq \exp(G) \text{ and } \left\lceil \frac{\text{diam}(G)}{2} \right\rceil \leq k(G). \tag{1}$$

In this paper we discuss two-classes of primitive graph with small exponent and small scrambling, that is, primitive graphs G with $\exp(G) = \text{diam}(G)$ and $k(G) = \left\lceil \frac{\text{diam}(G)}{2} \right\rceil$. For positif integer $n \geq 3$, an n-chainring is a graph obtained form an n-cycle by replacing each edge of the n-cycle by a triangle. More precisely, an n-chainring $CR(n)$ is a primitive graph on $2n$ vertices consisting of the the n-cycle $v_2 - v_4 - \cdots - v_{2n-2} - v_{2n} - v_2$ and the $2n$-cycle $v_1 - v_2 - v_3 - v_4 - \cdots - v_{2n-1} - v_{2n} - v_1$. The graph of $CR(8)$ is given in Figure 1.

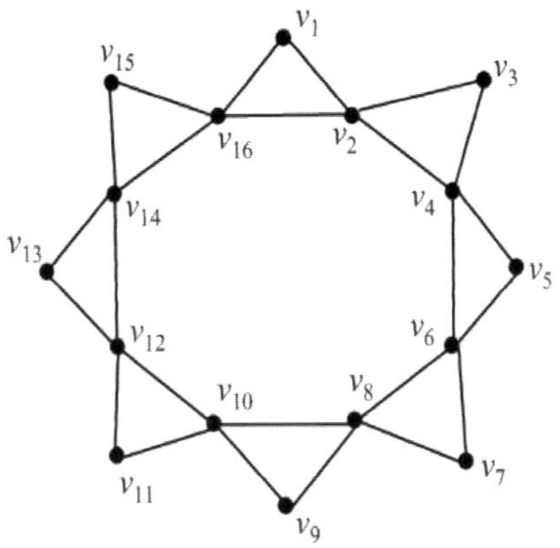

Figure 1. The Graph of $CR(8)$.

For positive integers p and q, let P_q be a path on q vertices $\{u_1, u_2, \ldots, u_q\}$ and P_p be a path on p vertices $\{v_1, v_2, \ldots, v_p\}$. By a (q,p)-dory, $D(q,p)$, we mean a graph with vertex set $V(D(q,p)) = V(P_q \times P_p) \cup \{w_1, w_2\}$ and edge set $E(D(q,p)) = E(P_q \times P_p) \cup \{w_1 - (u_i, v_1) : i = 1, 2, \ldots, q\} \cup \{w_2 - (u_i, v_p) : i = 1, 2, \ldots, q\}$.

In Section 2, we discuss properties of uv-walk especially uv-walk in an n-chainring. In Section 3, we discuss the exponent and scrambling index of n-chainring. In Section 4, we discuss the exponent and scrambling index of (q,p)-dory.

2. Properties of Walks

We discuss some properties of uv-walk necessary for our discussion.

Proposition 1. *Let G be a graph and let W_{uv} be a uv-walk of length $\ell(W_{uv})$ in G. If k is a positive integer such that $k \geq \ell(W_{uv})$ and $k \equiv \ell(W_{uv}) \bmod 2$, then there is a uv-walk of length k in G.*

Proof. Let

$$W_{uv} : u = v_0 - v_1 - v_2 - \cdots - v_i - v_{i+1} - \cdots - v_m = v$$

be a uv-walk of length $\ell(W_{uv})$. Since $k \geq \ell(W_{uv})$ and $k \equiv \ell(W_{uv}) \bmod 2$, there is a nonnegative integer t such that $k - \ell(W_{uv}) = 2t$. Then the walk that starts at $u = v_0$, moves to v_{i+1} along the walk $u = v_0 - v_1 - v_2 - \cdots - v_i - v_{i+1}$ and then moves t times around the closed walk $v_{i+1} - v_i - v_{i+1}$ and finally moves to v along the walk $v_{i+1} - \cdots - v_m = v$ is a uv-walk of length $k = \ell(W_{uv}) + 2t$. □

IOP Conf. Series: Materials Science and Engineering **300** (2018) 012081 doi:10.1088/1757-899X/300/1/012081

Proposition 2. *Let n be integer such that $n \geq 4$. Then for any pair of vertices u and v in $CR(n)$ there is a uv-path of length $d(u,v)+1$.*

Proof. Let u and v be any two vertices in $CR(n)$ and let the path

$$P_{uv} : u = v_0 - v_1 - v_2 - \cdots - v_i - v_{i+1} - \cdots - v_\ell = v$$

be a uv-path of length $d(u,v) = \ell$. Since every edge $v_i - v_{i+1}$ of the path P_{uv} lies on a triangle, then there is a vertex $v_{i'}$ such that the closed path $v_i - v_{i'} - v_{i+1} - v_i$ is a triangle. This implies the uv-path

$$P_{uv} : u = v_0 - v_1 - v_2 - \cdots - v_i - v_{i'} - v_{i+1} - \cdots - v_\ell = v$$

is a uv-path of length $d(u,v)+1$. □

3. Exponent and scrambling index of n-chainring

We discuss the exponent and scrambling index of $CR(n)$. We first present formulae for exponent and scrambling index in term of the diameter of $CR(n)$ and then present formulae that depends on n.

Theorem 3. *Let n be a positive integer with $n \geq 4$. Then* $\exp(CR(n)) = \mathrm{diam}(CR(n))$.

Proof. From (1) we have $\exp(CR(n)) \geq \mathrm{diam}(CR(n))$. It remains to show that $\exp(CR(n)) \leq \mathrm{diam}(CR(n))$. For each pair of vertices u and v we show that there exists a uv-walk of length $\mathrm{diam}(CR(n))$. Notice that for each pair of vertices u and v there is a uv-path P_{uv} of length $d(u,v)$. If $d(u,v) = \mathrm{diam}(CR(n)) \bmod 2$, then by Proposition 1 the path P_{uv} can be extended to a walk W_{uv} of length diam(CR(n)). If $d(u,v) \not\equiv \mathrm{diam}(CR(n)) \bmod 2$, then by Proposition 2 there is a path P'_{uv} of length $\ell(P'_{uv}) = d(u,v)+1$. We now have $\ell(P'_{uv}) \equiv \mathrm{diam}(CR(n))$. Hence Proposition 1 guarantees that the path P'_{uv} can be extended to a walk W_{uv} of length $\mathrm{diam}(CR(n))$. □

Corollary 4. *Let n be a positive integer with $n \geq 4$. Then*

$$\exp(CR(n)) = \begin{cases} (n+1)/2, & \text{if } n \text{ is odd} \\ (n+2)/2, & \text{if } n \text{ is even.} \end{cases}$$

Proof. If n is odd, the $\mathrm{diam}(G) = d(v_1, v_n) = (n+1)/2$. If n is even, the $\mathrm{diam}(G) = d(v_1, v_{n+1}) = (n+2)/2$. □

Theorem 5. *Let n be a positive integer such that $n \geq 3$. Then* $k(CR(n)) = \left\lceil \frac{\mathrm{diam}(CR(n))}{2} \right\rceil$.

Proof. From (1) we have $k(CR(n)) \geq \left\lceil \frac{diam(CR(n))}{2} \right\rceil$. It remains to show that $k(CR(n)) \leq \left\lceil \frac{diam(CR(n))}{2} \right\rceil$.

If the $\mathrm{diam}(CR(n))$ is even, then by Proposition 1 for each pair of vertices u and v there is a uv-walk of length $diam(CR(n))$. Thus we conclude that $k(CR(n)) \leq \mathrm{diam}(CR(n))/2$. If the $\mathrm{diam}(CR(n))$ is odd, then the shortest even walk connecting u_0 and v_0 is of length $\mathrm{diam}(CR(n))+1$. Notice that for every pair of vertices u and v, $d(u,v) \leq \mathrm{diam}(CR(n))$. Proposition 1 and Proposition 2 imply that for each pair of vertices u and v there is an even uv-walk of length $\mathrm{diam}(CR(n))+1$. Hence $k(CR(n)) \leq \frac{\mathrm{diam}(CR(n))+1}{2}$. We now conclude that $k(CR(n)) \leq \left\lceil \frac{\mathrm{diam}(CR(n))}{2} \right\rceil$.

We now conclude that $k(CR(n)) = \left\lceil \frac{diam(CR(n))}{2} \right\rceil$. □

Corollary 6. *For positive integer $n \geq 3$, $k(CR(n)) = \lfloor n/4 \rfloor + 1$.*

Proof. Suppose n is even. Then the $\text{diam}(CR(n)) = (n+2)/2$ and is obtained by the v_1v_{n+1}-path $v_1 - v_2 - v_4 - \cdots - v_n - v_{n+1}$. If $n \equiv 0 \bmod 4$, then $\text{diam}(CR(n)) = (4m+2)/2$ for some positive integer m. By Theorem 5 we have

$$\begin{aligned} k(CR(n)) &= \left\lceil \frac{\text{diam}(CR(n))}{2} \right\rceil = \left\lceil \frac{4m+2}{4} \right\rceil = \lceil m + 2/4 \rceil \\ &= m + 1 = \frac{n}{4} + 1. \end{aligned} \tag{2}$$

If $n \equiv 2 \bmod 4$, then $\text{diam}(CR(n)) = (4m+4)/2$ for some positive integer m. By Theorem 5 we have

$$\begin{aligned} k(CR(n)) &= \left\lceil \frac{\text{diam}(CR(n))}{2} \right\rceil = \left\lceil \frac{4m+4}{4} \right\rceil = \lceil m + 1 \rceil \\ &= m + 1 = \frac{n-2}{4} + 1. \end{aligned} \tag{3}$$

Suppose now that n is odd. Then the $\text{diam}(CR(n)) = (n+1)/2$ and is obtained by the v_1v_n-path $v_1 - v_2 - v_4 - \cdots - v_{n-1} - v_n$. If $n \equiv 1 \bmod 4$, then $\text{diam}(CR(n)) = (4m+2)/2$ for some positive integer m. By Theorem 5 we have

$$\begin{aligned} k(CR(n)) &= \left\lceil \frac{\text{diam}(CR(n))}{2} \right\rceil = \left\lceil \frac{4m+2}{4} \right\rceil = \lceil m + 2/4 \rceil \\ &= m + 1 = \frac{n-1}{4} + 1. \end{aligned} \tag{4}$$

If $n \equiv 3 \bmod 4$, then $\text{diam}(CR(n)) = (4m+4)/2$ for some positive integer m. By Theorem 5 we have

$$\begin{aligned} k(CR(n)) &= \left\lceil \frac{\text{diam}(CR(n))}{2} \right\rceil = \left\lceil \frac{4m+4}{4} \right\rceil = \lceil m + 1 \rceil \\ &= m + 1 = \frac{n-3}{4} + 1. \end{aligned} \tag{5}$$

From (2), (3), (4) and (5) we have $k(CR(n)) = \lfloor n/4 \rfloor + 1$. □

4. Exponent and scrambling index of (q,p)-dory

Let P_q be a path on q vertices $\{u_1, u_2, \ldots, u_q\}$ and P_p be a path on p vertices $\{v_1, v_2, \ldots, v_p\}$. By a (q,p)-dory, $D(q,p)$, we mean a graph with vertex set $V(D(q,p)) = V(P_q \times P_p) \cup \{w_1, w_2\}$ and edge set $E(D(q,p)) = E(P_q \times P_p) \cup \{w_1 - (u_i, v_1) : i = 1, 2, \ldots, q\} \cup \{w_2 - (u_i, v_p) : i = 1, 2, \ldots, q\}$. Since $D(q,p)$ is connected and contains triangles, $D(q,p)$ is primitive. We also note that if $p \geq q$, then $\text{diam}(D(q,p)) = d(w_1, w_2) = p + 1$.

Theorem 7. *Let p and q be positive integers such that $p \geq q$. Then $\exp(D(q,p)) = p + 1$.*

Proof. We note from (1) that $\exp(D(q,p)) \geq \text{diam}(D(q,p)) = p + 1$. It remains to show that $\exp(D(q,p)) \leq p + 1$. We show that for each pair of vertices x_0 and y_0 in $D(q,p))$ there is a walk connecting x_0 and y_0 of length $p + 1$.

If $d(x_0, y_0) \equiv p + 1 \bmod 2$, then by Proposition 1 there is a x_0, y_0-walk of length $p + 1$. It remains to consider the case where $d(x_0, y_0) \not\equiv p + 1 \bmod 2$.

Let P_{x_0,y_0} be the x_0, y_0-path of length $\ell(P_{x_0,y_0}) = d(x_0, y_0)$. If one end vertex of P_{x_0,y_0} is w_1 or w_2, then there is a path P_{x_0,y_0} such that $\ell(P_{x_0,y_0}) = d(x_0, y_0) + 1 \equiv p + 1$. We now assume that the end vertices of P_{x_0,y_0} are not w_1 or w_2. We consider three cases.

Case 1. The vertices $x_0 = (u_i, v_j)$ and $y_0 = (u_i, v_k)$ for some $1 \leq i \leq q$ and $1 \leq j < k \leq p$. We note that $d(x_0, y_0) = k - j \not\equiv p + 1 \bmod 2$. We assume without loss of generality that $p - k \geq j - 1$. The walk that starts at (u_i, v_j) moves to (u_i, v_1) along the path of length $(j-1)$, then moves one times around the triangle $(u_i, v_1) - w_1 - (u_{i-1}, v_1) - (u_i, v_1)$, and finally moves to (u_i, v_k) along the path of length $k-1$ is a x_0, y_0-walk of length $k+j+1$. We note that $k+j+1 = 2(j-1)+k-j+3$. Since $k-j \not\equiv p+1 \bmod 2$, we have $k+j+1 \equiv p+1 \bmod 2$. Moreover since $(j-1) \leq n-k$, we have $k+j+1 \leq p+2$. Since $k+j+1 \equiv p+1 \bmod 2$, then $k+j+1 \leq p+1$.

Case 2. The vertices $x_0 = (u_i, v_k)$ and $y_0 = (u_j, v_k)$ for some $1 \leq k \leq p$ and $1 \leq i < j \leq q$. We note that $d(x_0, y_0) = j - i \not\equiv p+1 \bmod 2$. Assume without loss of generality that $p - k > k - 1$. If $(j - i)$ is even, then the walk that starts at (u_i, v_k) moves to (u_i, v_1) along the path of length $k-1$, then moves to (u_j, v_1) along the path $(u_i, v_1) - (u_{i+1}, v_1) - w_1 - (u_j, v_1)$ of length 3, and finally moves to (u_j, v_k) along the path of length $k-1$, is a x_0, y_0-path of length $2(k-1)+3$. Since $2(k-1)+3 \not\equiv j-i \bmod 2$, we have $2(k-1)+3 \equiv p+1 \bmod 2$. We note that $n-k > k-1$. Therefore $2(k-1)+3 \leq p+2$. Since $2(k-1)+3 \equiv p+1 \bmod 2$, we conclude that $2(k-1)+3 \leq p+1$.

If $(j - i)$ is odd, then the walk that starts at (u_i, v_k) moves to (u_i, v_1) along the path of length $k-1$, then moves to (u_j, v_1) along the path $(u_i, v_1) - w_1 - (u_j, v_1)$ of length 2, and finally moves to (u_j, v_k) along the path of length $k-1$, is a x_0, y_0-path of length $2(k-1)+2$. Since $2(k-1)+2 \not\equiv j - i \bmod 2$, we have $2(k-1)+2 \equiv p+1 \bmod 2$. We note that $n-k > k-1$. Therefore $2(k-1)+2 \leq p+1$.

Case 3. The vertices $x_0 = (u_i, v_j)$ and $y_0 = (u_r, v_s)$ for some $1 \leq j < s \leq p$ and $1 \leq i < r \leq q$. Notice that $d(x_0, y_0) = (r-i) + (s-j) \not\equiv p+1 \bmod 2$. Without loss of generality we assume that $j - 1 \leq p - s$. If $(s-j) \equiv p+1 \bmod 2$, then the walk that starts at (u_i, v_j) moves to (u_i, v_1) along the path of length $(j-1)$, then moves to (u_r, v_1) along the path $(u_i, v_1) - w_1 - (u_r, v_1)$, and finally moves to (u_r, u_s) along the path of length $(j-1)+(s-j)$ is a x_0, y_0-path of length $2(j-1)+2+(s-j) \equiv p+1 \bmod 2$. Since $j-1 \leq p-s$, then $2(j-1)+2+(s-j) \leq p+1$.

If $(s-j) \not\equiv p+1 \bmod 2$, then the walk that starts at (u_i, v_j) moves to (u_i, v_1) along the path of length $(j-1)$, then moves to (u_r, v_1) along the path $(u_i, v_1) - (u_{i+1}, v_1) - w_1 - (u_r, v_1)$, and finally moves to (u_r, u_s) along the path of length $(j-1)+(s-j)$ is a x_0, y_0-path of length $2(j-1)+3+(s-j) \equiv p+1 \bmod 2$. Since $j-1 \leq p-s$, then $2(j-1)+3+(s-j) \leq p+2$. Since $2(j-1)+3+(s-j) \equiv p+1 \bmod 2$, we conclude that $2(j-1)+3+(s-j) \leq p+1$.

Therefore, for each pair of vertices x_0 and y_0 there is a x_0, y_0-walk W_{x_0,y_0} of length $\ell(W_{x_0,y_0}) \leq p+1$. Proposition 1 guarantees that for each for each pair of vertices x_0 and y_0 there is a x_0, y_0-walk W_{x_0,y_0} of length $\ell(W_{x_0,y_0}) = p+1$. Hence $\exp(D(q,p)) \leq p+1$. □

Theorem 8. *Let p and q be positive integers such that $p \geq q$. Then* $k(D(q,p)) = \left\lceil \frac{p+1}{2} \right\rceil$.

Proof. From (1) we have $k(D(q,p)) \geq \left\lceil \frac{diam(D(q,p))}{2} = \frac{p+1}{2} \right\rceil$. It remains to show that $k(D(q,p)) \leq \left\lceil \frac{p+1}{2} \right\rceil$.

If $p+1$ is even, then by Proposition 1 for each pair of vertices x_0 and y_0 there is a x_0, y_0-walk of even length $p+1$. Thus we conclude that $k(D(q,p)) \leq (p+1)/2$.

If the $p+1$ is odd, then the shortest even walk connecting w_1 and w_2 is of length $p+2$. Notice that from Theorem 7 for every pair of vertices x_0 and y_0 there is a x_0y_0-walk of length $p+1$. This implies for each each pair of vertices x_0 and y_0 there is a x_0y_0-walk of even length $p+2$. Hence if $p+1$ is odd, $k(D(q,p)) \leq (p+2)/2$.

Therefore we now conclude that $k(D(q,p)) \leq \left\lceil \frac{p+1}{2} \right\rceil$. Hence we now have $k(D(q,p)) = \left\lceil \frac{p+1}{2} \right\rceil$. □

References

[1] Chartrand G, Lesniak L and Zhang P 2016 *Graphs and Digraphs* (Boca Raton FL: CRC Press)
[2] Brualdi R A and Ryser H J 1991 *Combinatorial matrix Theory* (Cambridge: Cambridge University Press)
[3] Akelbek M and Kirkland S 2009 *Linear Algebra Appl.* **430** 1111
[4] Akelbek M and Kirkland S 2009 *Linear Algebra Appl.* **430** 1099
[5] Chen S and Liu B 2010 *Linear Algebra Appl.* **433** 1110

IOP Conf. Series: Materials Science and Engineering **300** (2018) 012082 doi:10.1088/1757-899X/300/1/012082

The grey Shapley value: an axiomatization

M Ekici[1,*], O Palanci[2] and S Z Alparslan Gök[2]

[1]Usak University, Faculty of education, Department of Mathematics and Science Education, 61 200 Usak, Turkey

[2]Süleyman Demired University, Faculty of Arts and Sciences, Department of Mathematics, 32 260 Isparta, Turkey

E-mail: *mutafa.ekici@usak.edu.tr

Abstract. This study focuses on an interesting class of cooperative games where the coalitional values are interval grey numbers. These cooper- ative games are called cooperative grey games. In this paper, we deal with an axiomatization of the grey Shapley value. We introduce the Banzhaf value and the egalitarian rule by using cooperative grey game theory. Finally, we conclude the paper with a conclusion.

1. Introduction

In this paper, we consider the cooperative games by using grey calculus. The grey systems theory is a new methodology that examines the study of problems involving small samples and poor information. In grey system theory, random variables are regarded as grey numbers, and a stochastic process is referred to as grey process. A grey system is defined as a system containing information presented as grey numbers; and a grey decision is defined as a decision made within a grey system [4,5,13].

This paper focuses on cooperative grey game is where interval grey numbers are taken. This work constructs the framework system of cooperative game theory based on grey data and applies cooperative grey game theory. With the application theory of a grey system, this paper deals with cooperative game issues based on incomplete information and grey uncertainty.

The article is organized as follows. We give basic notions from the theory of grey calculus is Section 2. The grey Shapley value is studied on the class of size monotonic grey games and axiomatic characterization of the grey Shapley value is given in Section 3. In Section 4, we introduce the grey Banzhaf value and the grey egalitarian rule on the special classes of cooperative grey games.

2. Preliminaries

In this section some preliminaries from grey calculus, and grey calculus used for cooperative grey games are given[7,11].

A grey number is such a number whose exact value is known but a range within that the value lies is known. In applications, a grey number in general is a grey or a general set of numbers.

In this paper, we consider the interval grey numbers.

A grey number with both a lower limit $\underline{a}$ and an upper limit $\overline{a}$ is called a grey grey number, denoted as $\otimes \in [\underline{a}, \overline{a}]$.

For example, the weight of a seal is between 30 and 35 kg. A specific person's height is between 1.9 and 2.0 meters. These two grey numbers can be respectively written as

$$\otimes_1 \in [30, 35] \text{ and } \otimes_2 \in [1.9, 2.0].$$

Now, we discuss various operations on grey grey numbers.
Let

$$\otimes_1 \in [a, b], a < b \text{ and } \otimes_2 \in [c, d], c < d.$$

The sum of $\otimes_1$ and $\otimes_2$, written $\otimes_1 + \otimes_2$, is defined as follows:

$$\otimes_1 + \otimes_2 \in [a + c, b + d].$$

For example, let $\otimes_1 \in [2, 7]$ and $\otimes_2 \in [3, 6]$, then,

$$\otimes_1 + \otimes_2 \in [5, 13].$$

Assume that $\otimes \in [a, b], a < b$ and k is a positive real number. The scalar multiplication of k and $\otimes$ is defined as follows:

$$k\otimes \in [ka, kb].$$

We denote by $\mathcal{G}(\mathbb{R})$ the set of grey grey numbers in $\mathbb{R}$. Let $\otimes_1, \otimes_2 \in \mathcal{G}(\mathcal{R})$ with $\otimes_1 \in [a, b], a < b; \otimes_2 \in [c, d], c < d, |\otimes_1| = b - a$ and $\alpha \in \mathbb{R}_+$. Then,

(i) $\otimes_1 + \otimes_2 \in [a + c, b + d]$;
(ii) $\alpha\otimes = [\alpha a, \alpha b]$.

By (i) and (ii) we see that $\mathcal{G}(\mathbb{R})$ has a cone structure.
In general, the difference of $\otimes_1$ and $\otimes_2$ is defined as follows:

$$\otimes_1 \ominus \otimes_2 = \otimes_1 + (-\otimes_2) \in [a - d, b - c],$$

(see[8]).
For example, let $\otimes_1 \in [6, 8]$ and $\otimes_2 \in [2, 5]$, then we have

$$\otimes_1 \ominus \otimes_2 \in [6 - 5, 8 - 2] = [1, 6],$$

$$\otimes_2 \ominus \otimes_1 \in [2 - 8, 5 - 6] = [-6, -1].$$

Different from the above subtraction we use a partial subtraction operator. We define $\otimes_1 - \otimes_2$, only if $|b - a| \geq |d - c|$, by $\otimes_1 - \otimes_2 \in [a - c, b - d]$. Note that $a - c \leq b - d$. We recall that $[a, b]$ is weakly better than $[c, d]$, which denote by $[a, b] \succeq [c, d]$, if and only if $a \geq c$ and $b \geq d$. We also use the reverse notation $[a, b] \preceq [c, d]$ if and only if $a \leq c$ and $b \leq d$ (for details see [1,3]).

Notice that if we make a comparison with the above example, then in our case $[6, 8] - [2, 5]$ is not defined. But, $[2, 5] - [6, 8]$ is defined.

Let $\otimes_1 \in [2, 5]$, and $\otimes_2 \in [6, 8]$, $\otimes_1 - \otimes_2$ is defined since $|5 - 2| \geq |8 - 6|$, but $\otimes_2 - \otimes_1$ is not defined since $|8 - 6| = 2 \ngeq 3 = |5 - 2|$, then we have

$$\otimes_1 - \otimes_2 \in [2 - 6, 5 - 8] = [-4, -3]$$

A cooperative grey game is an ordered pair $< N, w >$ where $N = \{1, \ldots, n\}$ is the set of players, and $w = \otimes : 2^n \to \mathcal{G}(\mathbb{R})$ is characteristic function such that $w(\emptyset) = \otimes_\emptyset \in [0, 0]$, grey payoff function $w(S) = \otimes_S \in [\underline{A_S}, \overline{A_S}]$ refers to the value of the grey expection benefit belonging to coalition $S \in 2^N$, where $\overline{A^S}$ and $\underline{A_S}$ represent the maximum and minimum possible profits of the coalition S. So, a cooperative grey game can be considered as a classical cooperative game with grey profits $\otimes$. Grey solutions are useful to solve reward/cost sharing problems with grey data using cooperative grey game as a tool. Building blocks for grey solutions are grey payoff vectors, i.e., vectors whose components belong to $\mathcal{G}(\mathbb{R})$ we denote by $(\mathcal{G}(\mathbb{R})^{\mathcal{N}}$ the set of all such grey payoff vectors. We denote by $\mathcal{G}G^N$ the family of all cooperative grey games.

We call a game $< N, w >$ *grey size monotonic* if $< N, |w| >$ is monotonic, i.e., $|w|(S) \leq |w|(T)$ for all $S, T \in 2^N$ with $S \subset T$. For further use we denote by $SM\mathcal{G}G^N$ the class of grey size monotonic games with player set N. The grey marginal operators and the grey Shapley value are defined on $SM\mathcal{G}G^N$. Denote by $\Pi(N)$ the set of permutations $\sigma : N \to N$ of N. The *grey marginal operator* $m^\sigma : SM\mathcal{G}G^N \to \mathcal{G}(\mathbb{R})^N$ corresponding to σ, associates with each $w' \in SM\mathcal{G}G^N$ the *grey marginal vector* $m^\sigma(w')$ of w' with respect to σ defined by

$$m_i^\sigma(w') := w'(P^\sigma(i) \cup \{i\}) - w'(P^\sigma(i)) \in [\underline{A_{P^\sigma(i)\cup\{i\}}} - \underline{A_{P^\sigma(i)}}, \overline{A_{P^\sigma(i)\cup\{i\}}} - \overline{A_{P^\sigma(i)}}],$$

for each $i \in N$, where

$$P^\sigma(i) = \{r \in N | \sigma^{-1}(r) < \sigma^{-1}(i)\},$$

and σ^{-1} denotes the entrance number of player i. For grey size monotonic games $< N, w >$, $w(T) - w(S) \in w(T) - w(S)$ is defined for all $S, T \in 2^N$ with $S \subset T$ since $|w(T)| = |w|(T) \geq |w|(S) = |w(S)|$. We notice that for each $w' \in SM\mathcal{G}G^N$ the grey marginal vectors $m^\sigma(w)$ are defined for each $\sigma \in \Pi(N)$, because the monotonicity of $|w|$ implies $\overline{A_{S\cup\{i\}}} - \underline{A_{S\cup\{i\}}} \geq \overline{A_S} - \underline{A_S}$, which can be written as $\overline{A_{S\cup\{i\}}} - \overline{A_S} \geq \underline{A_{S\cup\{i\}}} - \underline{A_S}$. So, $w'(S \cup \{i\}) - w'(S) \in w(S \cup \{i\}) = w(S)$ is defined for each $S \subset N$ and $i \notin S$. Next, we notice that all the grey marginal vectors of a grey size monotonic game are efficient grey payoff vectors.
The *grey Shapley value* $\Phi : SM\mathcal{G}G^N \to \mathcal{G}(\mathbb{R})^N$ is defined by [9]

$$\Phi'(w) := \frac{1}{n!} \sum_{\sigma \in \Pi(N)} m^\sigma(w) \in [\frac{1}{n!} \sum_{\sigma \in \Pi(N)} m^\sigma(\underline{A}), \frac{1}{n!} \sum_{\sigma \in \Pi(N)} m^\sigma(\overline{A})]. \tag{1}$$

Next, we recall that efficiency, null player and additivity axioms for solutions $f : SM\mathcal{G}G^N \to \mathcal{G}(\mathbb{R})^N$.

Efficiency (E): For every $w \in SM\mathcal{G}G^N$ it holds that $\sum_{i \in N} f_i(w) = w(N)$.

Let $w \in SM\mathcal{G}G^N$ and $i \in N$. Then, i is called a *null player* if $w(S \cup \{i\}) = w(S)$, for each $S \in 2^{N \setminus \{i\}}$.

Null Player (NP): If $i \in N$ is a null player in a grey game $w \in SM\mathcal{G}G^N$, then $f_i(w) \in [0, 0]$

Additivity (A): For every $w, v \in SM\mathcal{G}G^N$ it holds that $f(w + v) = f(w) + f(v)$.

Now, we mention the fairness property . Fairness property is introduced in van den Brink (2002).

Let $w \in SM\mathcal{G}G^N$ and $i, j \in N$. Then, i and j are called *symmetric players*, if $w(S \cup \{j\}) - w(S) = w(S \cup \{i\}) - w(S)$, for each S with $i, j \notin S$.

Symmetry (S): If $i, j \in N$ are symmetric players in $w \in SM\mathcal{G}G^N$, then $f_i(w) = f_j(w)$.

Fairness state that if to an grey game $w \in SM\mathcal{G}G^N$ we add an grey game $v \in SM\mathcal{G}G^N$ in which players i and j are symmetric, then the grey payoffs of players i and j change by the same amount.

Fairness (F): If $i, j \in N$ are symmetric players in $w \in SM\mathcal{G}G^N$, then $f_i(w + v) - f_i(w) = f_j(w + v) - f_j(w)$ for all $v \in SM\mathcal{G}G^N$ [12].

3. An axiomatic characterization of grey Shapley value

In this section, we give the characterization of grey Shapley value by using fairness property on the special subclass of cooperative grey games. Firstly, we mention about the relationship between our axioms.

Every grey solution that fulfills symmetry and additivity also satisfies fairness. Further, every grey solution that fulfills the null player property and fairness also satisfies symmetry. These propositions are an extension of the propositions obtained by [12] to cooperative grey games.

Proposition 1. *If $f : SM\mathcal{G}G^N \to \mathcal{G}(\mathbb{R})^N$ fulfills symmetry and additivity, then f also satisfies fairness.*

Proof. We know that $f : SM\mathcal{G}G^N \to \mathcal{G}(\mathbb{R})^N$ fulfill symmetry and additivity. Let $f : SM\mathcal{G}G^N \to \mathcal{G}(\mathbb{R})^N$ satisfy symmetry and additivity. If $i, j \in N$ are symmetric in $w \in SM\mathcal{G}G^N$, then for every $v \in SM\mathcal{G}G^N$ it holds that

$$\begin{aligned} f_i(w+v) - f_i(w) &= f_i(w) + f_i(v) - f_i(w) && \text{(from } \mathbf{A}\text{)} \\ &= f_i(v) && \text{(from } \mathbf{S}\text{)} \\ &= f_j(v) && \\ &= f_j(v) + f_j(w) - f_j(w) && \text{(from } \mathbf{A}\text{)} \\ &= f_j(w+v) - f_j(w). && \end{aligned}$$

Thus fulfills fairness. □

Proposition 2. *If $f : SM\mathcal{G}G^N \to \mathcal{G}(\mathbb{R})^N$ fulfills the null player property and fairness, then f also satisfies symmetry.*

Proof. Let $f : SM\mathcal{G}G^N \to \mathcal{G}(\mathbb{R})^N$ fulfill the null player property and fairness. For the null game $w_0 \in SM\mathcal{G}G^N$ given by $w_0(S) = [0,0]$ for all $S \subset N$, the null player property implies that $f_i(w_0) = [0,0]$ for all $i \in N$. If $i, j \in N$ are symmetric players in $w \in SM\mathcal{G}G^N$, then

$$\begin{aligned} f_i(w) &= f_i(w_0 + w) - f_i(w_0) && \text{(from } \mathbf{F}\text{)} \\ &= f_j(w_0 + w) - f_j(w_0) && \\ &= f_j(w) && \end{aligned}$$

Thus, f fulfills symmetry. □

We know that the grey Shapley value is characterized by efficiency, the null player property, symmetry and additivity. By Proposition 1 it also guarantees fairness. Now, we give the main result of this paper.

Theorem 3. *A grey solution $g : SM\mathcal{G}G^N \to \mathcal{G}(\mathbb{R})^N$ is equal to the grey Shapley value if and only if it fulfills efficiency, the null player property and fairness.*

Proof. The proof is a straight forward generalization from the classical case and can be obtained by following the steps of Theorem 2.5 in [12]). □

Logical independence of the three axioms of Theorem 3 can be illustrated by the following two well-known solutions. Grey Banzhaf value fulfills the null player property and fairness but it does not satisfy efficiency. Grey egalitarian rule fulfills efficiency and fairness but it does not satisfy the null player property.

4. The grey Banzhaf value and the grey egalitarian rule

A cooperative game describes a situation in which a fnite set of n players can generate certain payoffs by cooperation. A one-point solution concept for cooperative games is a function which assigns to every cooperative game a n-dimensional real vector which represents a payoff distribution over the players. The study of solution concepts is central in cooperative game theory. Two well-known solution concepts are the Shapley value as proposed by [10], and the Banzhaf value, initially introduced in the context of voting games by [2]. Now, we introduce the Banzhaf value by using grey uncertainty.

The classical Banzhaf value $\beta : G^N \to \mathbb{R}^N$ given by

$$\beta_i(v) := \frac{1}{2^{|N|-1}} \sum_{i \in S} (v(S) - v(S \backslash \{i\}))$$

for all $i \in N$.

The Banzhaf value considers that every player is equally likely to enter to any coalition whereas the Shapley value assumes that every player is equally likely to join to any coalition of the same size and all coalitions with the same size are equally likely. In addition, the Shapley value is efficient, while the Banzhaf value is not efficient. Thus, the Shapley value distributes the total utility among players while the total amount that players get from Banzhaf's allocation depends on the structure of the TU game.

The grey Banzhaf value is defined for $SM\mathcal{G}G^N$ since the grey marginal operators is defined for $SM\mathcal{G}G^N$.

The grey Banzhaf value $\beta : SM\mathcal{G}G^N \to \mathcal{G}(\mathbb{R})^N$ is defined by

$$\beta'(w) \in \left[\frac{1}{2^{|N|-1}} \sum_{i \in S} (\underline{w(S)} - \underline{w(S \backslash \{i\})}, \frac{1}{2^{|N|-1}} \sum_{i \in S} (\overline{w(S)} - \overline{w(S \backslash \{i\})}) \right],$$

for each $w \in SM\mathcal{G}G^N$.

Example 4. *Let $< N, w >$ be a cooperative grey game with $N = \{1, 2, 3\}$ and $w(1) = w(13) \in [7, 7]$, $w(12) \in [12, 17]$, $w(123) \in [24, 29]$, and $w(S) \in [0, 0]$ otherwise. The grey Shapley value of this game can be calculated as follows: Then the grey marginal vectors are given in the following table, where $\sigma : N \to N$ is identified with $(\sigma(1), \sigma(2), \sigma(3))$. Firstly, for $\sigma_1 = (1, 2, 3)$, we calculate the grey marginal vectors. Then,*

$$\begin{aligned} m_1^{\sigma_1}(w) &= w(1) \in [7, 7], \\ m_2^{\sigma_1}(w) &= w(12) - w(1) \in [5, 10], \\ m_3^{\sigma_1}(w) &= w(123) - w(12) \in [12, 12]. \end{aligned}$$

The others can be calculated similarly, which is shown in Table 1.

Table 1 illustrates the grey marginal vectors of the cooperative grey game in Example 4. The average of the six grey marginal vectors is the grey Shapley value of this game which can be shown as:

$$\Phi'(w) \in ([\frac{27}{2}, 16], [\frac{13}{2}, 9], [4, 4]).$$

The grey Banzhaf value of this game can be calculated as follows: The grey Banzhaf value is defined as

$$\beta'(w) \in \left[\frac{1}{2^{|N|-1}} \sum_{i \in S} (\underline{w(S)} - \underline{w(S \backslash \{i\})}, \frac{1}{2^{|N|-1}} \sum_{i \in S} (\overline{w(S)} - \overline{w(S \backslash \{i\})}) \right].$$

Table 1. Grey marginal vectors

σ	$m_1^\sigma(w')$	$m_2^\sigma(w')$	$m_3^\sigma(w')$
$\sigma_1=(1,2,3)$	$m_1^{\sigma_1}(w')\in[7,7]$	$m_2^{\sigma_1}(w')\in[5,10]$	$m_3^{\sigma_1}(w')\in[12,12]$
$\sigma_2=(1,3,2)$	$m_1^{\sigma_2}(w')\in[7,7]$	$m_2^{\sigma_2}(w')\in[17,22]$	$m_3^{\sigma_2}(w')\in[0,0]$
$\sigma_3=(2,1,3)$	$m_1^{\sigma_3}(w')\in[12,17]$	$m_2^{\sigma_3}(w')\in[0,0]$	$m_3^{\sigma_3}(w')\in[12,12]$
$\sigma_4=(2,3,1)$	$m_1^{\sigma_4}(w')\in[24,29]$	$m_2^{\sigma_4}(w')\in[0,0]$	$m_3^{\sigma_4}(w')\in[0,0]$
$\sigma_5=(3,1,2)$	$m_1^{\sigma_5}(w')\in[7,7]$	$m_2^{\sigma_5}(w')\in[17,22]$	$m_3^{\sigma_5}(w')\in[0,0]$
$\sigma_6=(3,2,1)$	$m_1^{\sigma_6}(w')\in[24,29]$	$m_2^{\sigma_6}(w')\in[0,0]$	$m_3^{\sigma_6}(w')\in[0,0]$

Then,

$$\begin{aligned}\beta_1(w) &= \frac{1}{2^2}\sum_{1\in S}(w(S)-w(S\backslash\{1\}) \\ &\in [12\frac{1}{2},15].\end{aligned}$$

The others can be calculated similarly:

$$\beta_2(w)\in[5\frac{1}{2},8],\quad \beta_3(w)\in[3,3].$$

Then, the grey Banzhaf value can be shown as

$$\beta'(w)\in([12\frac{1}{2},15],[5\frac{1}{2},8],[3,3]).$$

Remark 5. *The grey Banzhaf value satisfies the null player property, and fairness but does not fulfill efficiency.*

In Example 4, the grey Banzhaf value does not satisfy efficiency.

$$\begin{aligned}\sum_{i=1}^{3}\beta_i'(w) &\in \beta_1'(w)+\beta_2'(w)+\beta_3'(w) \\ &= [12\frac{1}{2},15]+[5\frac{1}{2},8]+[3,3] \\ &= [21,26]\neq[24,29]=w(N)\end{aligned}$$

A family of monotonic solutions to general cooperative games (coalitional form games where utility is not assumed to be transferable) is the egalitarian rule. The egalitarian rule is introduced by [6].

The classical egalitarian rule $\gamma: G^N\to\mathbb{R}^N$ is given by

$$\gamma_i(v):=\frac{v(N)}{|N|},$$

for all $i\in N$.

The grey egalitarian rule is defined for $\mathcal{G}G^N$. The grey egalitarian rule $\gamma':\mathcal{G}G^N\to\mathcal{G}(\mathbb{R})^N$ is defined by

$$\gamma'(w)=\frac{w(N)}{|N|},\text{ for each } w\in\mathcal{G}G^N.$$

Example 6. *Consider again the cooperative grey game in Example 1. Then, the grey egalitarian rule of this game is:*

$$\gamma_i'(w) = \frac{w(N)}{|N|} \in \frac{[24, 29]}{3} = [8, 9\frac{2}{3}], \quad i = 1, 2, 3.$$

Remark 7. *The grey egalitarian rule satisfies efficiency and fairness but does not fulfill the null player property.*

Example 8. *Consider the cooperative grey game* $< N, w >$ *with* $N = \{1, 2, 3\}$ *and* $w(2) = w(12) \in [7, 7]$, $w(3) = w(13) \in [12, 17]$, $w(23) = w(N) \in [24, 29]$, *and* $w(S) \in [0, 0]$ *otherwise. The grey egalitarian rule of this game is:*

$$\gamma_i'(w) = \frac{w(N)}{|N|} \in \frac{[24, 29]}{3} = [8, 9\frac{2}{3}], \quad i = 1, 2, 3.$$

Here, the null player is player 1 and $\gamma_1'(w) \in [8, 9\frac{2}{3}]$. *Thus, the grey egalitarian rule does not satisfy the null player property.*

5. Conclusion

The Shapley value and the Banzhaf value are interesting one-point solutions in cooperative game theory. In this context, we give the characterization of the grey Shapley value by using fairness property. In addition the interval Banzhaf value and the egalitarian rule are introduced by using grey systems theory.

The grey Shapley value is axiomatically characterized but the grey Banzhaf value and the grey egalitarian rule are not axiomatically characterized. There exists a gap to be filled by characterizing the grey Banzhaf value and the grey egalitarian rule. It is possible to characterize the grey Banzhaf value and the grey egalitarian rule in the future studies.

References

[1] Alparslan Gök S Z, Branzei O, Branzei R,and Tijs S 2011 Set-valued solution concepts using interval-type payoffs for interval games *Journal of Mathematical Economics***47** 621–626.
[2] Banzhaf J F 1965 Weighted Voting Doesn't Work: A Mathematical Analysis *Rutgers Law Review* **19** 317–343.
[3] Branzei R, Dimitrov D and Tijs S 2003 Shapley-like values for interval bankruptcy games, *Economics Bulletin* **3** 1–8.
[4] Deng J 1982) Control problems of Grey Systems *Systems and Control Letters* **5** 288–294.
[5] Deng J 1985 *Grey System Fundamental Method*(Huazhong University of Science and Technology: Wuhan, China).
[6] Kalai E and Samet D 1985 Monotonic Solutions to General Cooperative Games, *Econometrica* **53** (2) 307–327.
[7] Liu S and Lin Y 2006 *Grey Information: Theory and Practical Applications* (Springer: Germany).
[8] Moore R 1979 *Methods and Applications of Interval Analysis*, (SIAM Studies in Applied Mathematics: Philadelphia).
[9] Palanci O, Alparslan Gök S Z, Ergün S, and Weber G W, 2015 Cooperative grey games and the grey Shapley value, *Optimization: A Journal of Mathematical Programming and Operations Research* **64** (8) 1657–1668.
[10] Shapley L S 1953 A value for n-person games, *Annals of Mathematics Studies* **28** 307–317.
[11] Tijs S 2003 *Introduction to Game Theory* Vol. 23 of Texts and Readings in Mathematics (Hindustan Book Agency: New Delhi, India).
[12] van den Brink R 2002 An axiomatization of the Shapley value using a fairness property, *International Journal of Game Theory* **30** (3) 309–319.
[13] Zhang J J, Wu D S, and Olson D L 2005 The method of grey related analysis to multiple attribute decision making problems with interval numbers, *Mathematical and Computer Modelling* **42** (9-10) 991–998.

Bounds for scrambling index of primitive graphs

Shaflina Izar[1], Mardiningsih[2,*] Zuhri[3], and Sylvi Harleni[4]
[1,2]Department of Mathematics, University of Sumatera Utara, Medan 20155, Indoneisa
[3]STIM SUKMA Medan, KOPERTIS AREA 1, Medan-Indonesia
[4]Fakultas Sains dan Teknologi, Universitas Islam Negeri Sumatera Utara, Medan-Indonesia
E-mail: *mardiningsih@usu.ac.id

Abstract. A connected graph is primitive provided there is a positive integer m such that for each pair of vertices u and v there is a walk of length m connecting u and v. The scrambling index of a primitive graph G is the smallest positive integer k such that for each pair of vertices u and v there is a vertex w such that there exist a walk of length k connecting u and w and a walk of length k connecting v and w. For a primitive graph G with smallest cycle C_s of length s, we present an upper bound on the scrambling index of G that depends on s and the maximum distance between vertices in G and the cycle C_s. We then classify the graphs that satisfy the upper bound.

1. Introduction

Let $G(V, E)$ denote a simple graph on n vertices. We follow [1, 2] for terminologies on graph. A walk W connecting u and v is denoted by W_{uv}. A walk connecting u and v is closed whenever $u = v$, and is open otherwise. A path P_{uv} connecting u and v is a walk W_{uv} with distinct vertices, except possibly $u = v$. A cycle is a closed path. The length of a walk W_{uv} by $\ell(W_{uv})$. A walk W_{uv} is also denoted by $u \overset{W_{uv}}{--} v$. For simplicity a walk of length k connecting u and v is denoted by $u \overset{k}{--} v$ walk. The distance between vertex u and vertex v in a connected graph, denoted $d(u, v)$, is the length of the shortest path connecting u and v. For any set $X \subseteq V(G)$ and a vertex $v \notin X$ the distance between v and X is defined by

$$d(v, X) = \min\{d(v, x) : x \in X\}.$$

If $v \in X$ we define $d(v, X) = 0$.

A connected graph G is primitive if there is a positive integer k such that for each pair of vertices u and v in G, there is a $u \overset{k}{--} v$ walk. The least of such positive integer k is the exponent of G and is denoted by $\exp(G)$. It is well known (see e.g. [1]) that a graph G is primitive if and only if G has a cycle of odd length.

In 2009, Akelbek and Kirkland [3, 4] introduced the notion of scrambling index of graph. The *scrambling index* of a primitive graph G, denoted by $k(G)$, is the least positive integer k such that for any pair of distinct vertices u and v in G there exists a vertex w with the property that there is a $u \overset{k}{--} w$ walk and a $v \overset{k}{--} w$ walk. For a pair of distinct vertices u and v in G the local scrambling index of u and v is the number

$$k_{u,v}(G) = \min_{w \in V(G)} \{k : \text{ there are } u \overset{k}{--} w \text{ walk and } v \overset{k}{--} w \text{ walk}\}.$$

We note that if the local scrambling index of u and v is $k_{u,v}(G)$, then for any positive integer $\ell \geq k_{u,v}(G)$ we can find a vertex w' such that there is a walk $u \overset{\ell}{--} w'$ and $v \overset{\ell}{--} w'$. This implies

$$k(G) = \max_{u,v \in V(G)} \{k_{u,v}(G)\}.$$

Chen and Liu [5] have shown that for a primitive graph G on n vertices with the smallest cycle of length s, $k(G) \leq (s-1)/2 + n - s$. The purpose of this paper is to explore a different upper bound from Chen and Liu's bound, which in most case will be smaller than $(s-1)/2 + n - s$. In Section 2 we discuss some important properties of $u{--}v$ walks. In Section 3 we present an upper bound for the scrambling index of primitive graphs. Finally, in Sections 4 we discuss classes of primitive graphs whose scrambling index achieved the upper bound.

2. Properties of walks

In this section we discuss some properties of $u{--}v$ walk with end vertices u and v in some primitive graph.

Theorem 1. [2] *Let G be a graph. Every $u{--}v$ walk in G contains a $u{--}v$ path.*

Theorem 1 basically saying that we can shorten a $u{--}v$ walk into a shorter $u - -v$ walk. The following result guarantees that we can lengthen a $u{--}v$ walk into a longer $u{--}v$ walk with the same parity.

Proposition 2. *Let G be a graph and u and v be two vertices in G. Every $u \overset{t}{--} v$ walk can be extended to a $u \overset{t+2m}{--} v$ walk for some positive integer m.*

Proof. Let u and v be vertices in G and let

$$W_{uv} : u = v_0 - v_1 - v_2 - \cdots - v_{t-1} - v_t = v$$

be a $u \overset{t}{--} v$ walk in G and let $W_2 : v - v_{t-1} - v$ be a closed walk of length 2. Then the walk

$$W'_{uv} : u \overset{W_{uv}}{--} v \overset{mW_2}{--} v$$

that starts a u, moves to v along the walk W_{uv}, and then moves m times around the closed walk $W_2 : v - v_{t-1} - v$ is a $u \overset{t+2m}{--} v$ walk. □

Proposition 3. *Let G be a graph. Then there is a $u \overset{2t}{--} v$ walk in G if and only if there is vertex w in G such that there are a $u \overset{t}{--} w$ walk and a $v \overset{t}{--} w$ walk in G.*

Proof. Suppose there is a vertex w in G such that there are a $u \overset{t}{--} w$ walk and a $v \overset{t}{--} w$ walk in G. Then there is a $w \overset{t}{--} v$ walk in G. This implies the walk $W_{uv} : u \overset{t}{--} w \overset{t}{--} v$ is a $u \overset{2t}{--} v$ walk.

Assume now that

$$W_{uv} : u = v_0 - v_1 - v_2 - \cdots - v_{2t-1} - v_{2t} = v$$

is a $u \overset{2t}{--} v$ walk in G. If we choose $w = v_t$, then there is a vertex w such that there are $u \overset{t}{--} w$ walk and a $v \overset{t}{--} w$ walk in G. □

3. Bounds for scrambling index

We present a lower bound and an upper for scrambling index of primitive graphs.

Theorem 4. *Let G be a primitive graph and let C_s be a cycle of odd length s. If G does not have odd cyles with length smaller than s, Then*

$$k(G) \leq \frac{s-1}{2} + \max_{v \in V(G)} \{d(v, V(C_s))\}.$$

Proof. For each pair of distinct vertices u and v we show that there is a $u -- v$ walk of length $(s-1) + 2\max_{v\in V(G)}\{d(v, V(C_s))\}$.

First, we claim that for any distinct vertices u and v in G, there is a path P_{uv} such that $\ell(P_{uv}) \leq (s-1)/2 + 2\max_{v\in V(G)}\{d(v, V(C_s))\}$. If both u and v lie on the cycle C_s, then there is a path P_{uv} with $\ell(P_{uv}) \leq (s-1)/2$. If v lies in $V(G) \setminus V(C_s)$ and u lies on C_s, then there is a path P_{uv} with $\ell(P_{uv}) \leq \max_{v\in V(G)}\{d(v, V(C_s))\}$. Suppose u and v lie on $V(G) \setminus v(C_s)$. Assume without loss of generality that $d(u, V(C_s)) \geq d(v, V(C_s))$ and that $d(u, V(C_s))$ is obtained by a path P_{uc} for some vertex c in C_s. If the vertex v lies on the path P_{uc}, then $\ell(P_{uv}) \leq \max_{v\in V(G)}\{d(v, V(C_s))\}$. Otherwise, there is a vertex y in C_s such that $d(v, y) = d(v, V(C_s))$. Since $d(u, c), d(v, y) \leq \max_{v\in v(G)}\{d(v, V(C_s))\}$, then the walk

$$W_{uv} : u \overset{P_{u,c}}{--} c \overset{P_{c,y}}{--} y \overset{P_{y,v}}{--} v$$

is a $u--v$ walk of length $\ell(W_{uv}) \leq (s-1)/2 + 2\max_{v\in V(G)}\{d(v, V(C_s))\}$. By Theorem 1 there is a path P_{uv} of length $\ell(P_{uv}) \leq (s-1)/2 + 2\max_{v\in V(G)}\{d(v, V(C_s))\}$.

We now show that the path P_{uv} can be extended to a $u--v$ walk of length exactly $(s-1) + 2\max_{v\in V(G)}\{d(v, V(C_s))\}$. If there is a $u--v$ path P_{uv} of even length, then Proposition 2 guarantees that we can extend the path P_{uv} to a $u--v$ walk of length exactly $(s-1) + 2\max_{v\in V(G)}\{d(v, V(C_s))\}$. We now assume that all $u--v$ paths are of odd lengths. Notice that u and v cannot be both on C_s. We show that we can extend a $u--v$ path into a $u--v$ walk of length exactly $(s-1) + 2\max_{v\in V(G)}\{d(v, V(C_s))\}$. We consider two cases.

Case 1: *There exists a path P_{uv} that has vertices in common with the cycle C_s*

We claim that the path P_{uv} and the cycle C_s have exactly one vertex in common. Suppose on the contrary that P_{uv} and C_s have more than one vertex in common. Let x_0 and y_0 be two vertices on C_s that lie on the path P_{uv}. We note that there are two paths on C_s say P_{x_0,y_0} and P'_{x_0,y_0} connecting u_0 and y_0. Since C_s is of odd length, $\ell(P_{x_0,y_0}) \not\equiv \ell(P'_{x_0,y_0}) \bmod 2$. This implies either the path $P_{uv} : u--x_0 \overset{P_{x_0,y_0}}{--} y_0--v$ or the path $P_{uv} : u--x_0 \overset{P'_{x_0,y_0}}{--} y_0--v$ is a $u--v$ path of even length. This contradicts the fact that all $u--v$ paths in G are of odd lengths.

We show that there is a $u--v$ walk W_{uv} of even length $\ell(W_{uv}) \leq (s-1) + 2\max_{v\in V(G)}\{d(v, V(C_s))\}$. Suppose the path P_{uv} and the cycle C_s have a vertex in common at v_0. If $v_0 = u$, then the walk $W_{uv} : u = v_0 \overset{C_s}{--} u = v_0 \overset{P_{u,v}}{--} v$ is a $u--v$ walk of even length $\ell(W_{uv}) \leq (s-1) + 2\max_{v\in V(G)}\{d(v, V(C_s))\}$. Similarly, if $v_0 = v$, then the walk $W_{uv} : u \overset{P_{u,v}}{--} v = v_0 \overset{C_s}{--} v = v_0$ is a $u--v$ walk of even length $\ell(W_{uv}) \leq (s-1) + 2\max_{v\in V(G)}\{d(v, V(C_s))\}$.

If $u \neq v_0$ and $v \neq v_0$, then the $u--v$ path P_{uv} can be decomposed into $u--v_0$ path P_{u,v_0} and v_0--v path $P_{v_0,v}$. Since all paths P_{uv} are of odd length, $\ell(P_{u,v_0}) + \ell(P_{v_0,v}) \leq 2\max_{v\in V(G)}\{d(v, V(C_s))\} - 1$. This implies the walk $W_{uv} : u \overset{P_{u,v_0}}{--} v_0 \overset{C_s}{--} v_0 \overset{P_{v_0,v}}{--} v$ is a $u--v$ walk of even length $\ell(W_{uv}) \leq (s-1) + 2\max_{v\in V(G)}\{d(v, V(C_s))\}$. Proposition 2 guarantees that we can extend the $u--v$ walk W_{uv} into a $u--v$ walk with length exactly

$(s-1)+2\max_{v\in V(G)}\{d(v,V(C_s))\}$.

Case 2: *All paths P_{uv} and the cycle C_s have no vertices in common*
There is a path P_{uv} and a vertex x_0 on P_{uv} and a vertex y_0 on C_s such that

$$d(x_0,y_0)=\min\{d(x,y): x \text{ on } P_{uv} \text{ and } y \text{ on } C_s\}.$$

Notice that the path P_{uv} can be decomposed into a path P_{u,x_0} connecting u and x_0, and a path $p_{x_0,v}$ connecting x_0 and v. Since $\ell(P_{uv})$ is odd, the walk

$$u \overset{P_{u,x_0}}{--} x_0 \overset{P_{x_0,y_0}}{--} y_0 \overset{C_s}{--} y_0 \overset{P_{y_0,x_0}}{--} x_0 \overset{P_{x_0,v}}{--} v$$

is a $u--v$ walk of even length. Notice also that $\ell(P_{u,x_0}) \not\equiv \ell(P_{x_0,v}) \bmod 2$. This implies $\ell(P_{u,y_0}) \not\equiv \ell(P_{y_0,v}) \bmod 2$. Since $\ell(P_{u,y_0}),\ell(P_{y_0,v}) \le \max_{v\in V(G)}\{d(v,V(C_s))\}$, we have $\ell(P_{u,y_0})+\ell(P_{y_0,v}) \le 2\max_{v\in V(G)}\{d(v,V(C_s))\}-1$. Thus $\ell(W_{uv}) \le s-1+2\max_{v\in V(G)}\{d(v,V(C_s))\}$. Proposition 2 guarantees that there is a $u--v$ walk W_{uv} with length equals $(s-1)+2\max_{v\in V(G)}\{d(v,V(C_s))\}$

Now Proposition 3 guarantees that for each pair of vertices u and v there is a vertex w such that there is a $u \overset{t}{--} w$ walk and there is a $v \overset{t}{--} w$ walk with $t=(s-1)/2+\max_{v\in V(G)}\{d(v,V(C_s))\}$. Thus the scrambling index $k(G) \le (s-1)/2+\max_{v\in V(G)}\{d(v,V(C_s))\}$. □

We note that for a primitive graph G with the smallest ood cycle of length s, $\max_{v\in V(G)}\{d(v,V(C_s))\} \le n-s$. Hence the bound given in Theorem 4 is smaller than or equal to the bound $(s-1)/2+(n-s)$.

4. Primitive graphs achieving the upper bound

In this section we discuss classes of primitive graphs that satisfy the upper bound given in Theorem 4. We first characterize and then discuss instances of such primitive graphs.

Corollary 5. *Let G be a primitive graph and let C_s be a smallest odd cycle of length s in G. The scrambling index*

$$k(G)=\frac{s-1}{2}+\max_{v\in V(G)}\{d(v,V(C_s))\}$$

if and only if there are vertices u_0 and v_0 such that the shortest even u_0--v_0 walk is a walk of length $(s-1)+2\max_{v\in V(G)}\{d(v,V(C_s))\}$.

Proof. Assume that there are two distinct vertices u_0 and v_0 in G such that the shortest even walk connecting u_0 and v_0 is of length $s-1+2\max_{v\in V(G)}\{d(v,V(C_s))\}$. Then by Proposition 3 there exists a vertex w such that there is a $u_0 \overset{t}{--} w$ walk and a $v_0 \overset{t}{--} w$ walk with $t=(s-1)/2+\max_{v\in V(G)}\{d(v,V(C_s))\}$. We claim that $k_{u_0,v_0}(G)=(s-1)/2+\max_{v\in V(G)}\{d(v,V(C_s))\}$. Suppose on the contrary that $k_{u_0,v_0}(G)=\ell$ for some positive integer $\ell<(s-1)/2+\max_{v\in V(G)}\{d(v,C_s)\}$. Then there exists a vertex w' with the property that there is a $u_0 \overset{\ell}{--} w'$ walk and a $v_0 \overset{\ell}{--} w'$ walk. But this implies the u_0--v_0 walk $W_{u_0,v_0}: u_0 \overset{\ell}{--} w' \overset{\ell}{--} v_0$ is a u_0--v_0 walk of even length $2\ell<(s-1)+2\max_{v\in V(G)}\{d(v,V(C_s))\}$. This contradicts the fact that the shortest even u_0--v_0 walk is of length $(s-1)+2\max_{v\in V(G)}\{d(v,V(C_s))\}$. Therefore,

$$k_{u_0,v_0}(G)=(s-1)/2+\max_{v\in V(G)}\{d(v,V(C_s))\}$$

and hence

$$k(G) \geq k_{u_0,v_0}(G) = (s-1)/2 + \max_{v \in V(G)} \{d(v, V(C_s))\}.$$

By Theorem 4 we conclude that $k(G) = (s-1)/2 + \max_{v\in V(G)}\{d(v, V(C_s))\}$.

We now assume that $k(G) = (s-1)/2+\max_{v\in V(G)}\{d(v, V(C_s))\}$. Then there are two distinct vertices u_0 and v_0 such that

$$k_{u_0,v_0}(G) = \min_{w\in V(G)} \{t : \text{there are } u_0 \overset{t}{--} w \text{ and } v_0 \overset{t}{--} w \text{ walks}\} = k(G).$$

This implies the walk $u_0 \overset{k(G)}{--} w \overset{k(G)}{--} v_0$ is the shortest walk of even length $(s-1) + 2\max_{v\in V(G)}\{d(v, V(C_s))\}$ connecting u_0 and v_0. □

We next discuss the class of primitive graph with smallest cycle of length s that satisfies the bound in Theorem 4. For that purpose we need the following definition. An open path P of length $\ell(P) = \max_{v\in V(G)}\{d(v, V(C_s))\}$ with one end vertex in C_s is called to be *special* if it does not have vertices in common with cycle of odd length other than C_s.

Corollary 6. *Let G be a primitive graph with the shortest odd cycle of length s. If G has a special path, then $k(G) = (s-1)/2 + \max_{v\in V(G)}\{d(v, V(C_s))\}$.*

Proof. Let P_{v_0,u_0} be a special path in G and let $u_0 \in V(C_s)$ and $v_0 \in V(G)$. Let $v_0 - y_0$ be an edge of P_{v_0,u_0} such that $d(v_0, u_0) > d(y_0, u_0)$. Then the $v_0 -- y_0$ walk $W_{v_0,y_0} : v_0 \overset{P_{v_0,u_0}}{--} u_0 \overset{C_s}{--} u_0 \overset{P_{u_0,y_0}}{--} y_0$ is the shortest walk connecting v_0 and y_0 of even length $(s-1) + 2\max_{v\in V(G)}\{d(v, V(C_s))\}$. Corollary 5 implies that $k(G) = (s-1)/2 + \max_{v\in V(G)}\{d(v, V(C_s))\}$. □

Corollary 7. *Let G be a primitive graph containing a unique cycle of odd length. Let C_s be the odd cycle in G say of length s. Then $k(G) = (s-1)/2 + \max_{v\in V(G)}\{d(v, V(C_s))\}$.*

Proof. Since G has only one cycle of odd length, G must contain a special path. The conclusion follows from Corollary 6. □

Let G be a loopless primitive graphs on n vertices with the smallest cycle of length s. Since $s \geq 3$ and $\max_{v\in V(G)}\{d(v, V(C_s))\} \leq n-3$, then by Theorem 4 we have $k(G) \leq n-2$. Let SI_n denote the set of positive integers t for which there exists a loopless primitive graph on n vertices with scrambling index equals t.

Corollary 8. *For any positive integer $n \geq 3$, $SI_n = \{1, 2, \ldots, n-2\}$.*

Proof. For positive integer t, $3 \leq t \leq n-1$, we define a primitive graph G_t on n vertices $\{v_1, v_2, \ldots, v_n\}$ to be the graph with edge set

$$\begin{aligned} E(G_t) &= \{v_1 - v_2 - v_3 - v_1\} \cup \{v_3 - v_4 - \cdots - v_{t-1} - v_t\} \\ &\quad \cup \{v_t - v_{t+i} : i = 1, 2, \ldots, n-t\} \end{aligned}$$

as shown in Figure 1.

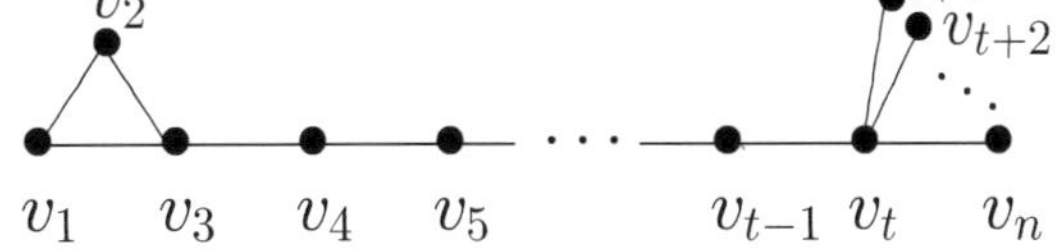

Figure 1. The graph G_t

The graph G_t has a unique odd cycle of length 3. Moreover, G_t has special paths of length $t-2$. Corollary 6 implies that $k(G_t) = t-1$. Since $3 \leq t \leq n-1$ and $k(K_n) = 1$, then for each positive integer $1 \leq p \leq n-2$, there exists primitive graph $k(G) = p$. Hence $SI_n = \{1, 2, \ldots, n-2\}$. □

References

[1] Brualdi R A and Ryser H J 1991 *Combinatorial Matrix Theory* (Cambridge:Cambridge University Press).
[2] Chartrand G. and Lesniak L 1986 *Graphs and Digraphs*(Monterey CA: Wadsworth and Brooks/Cole)
[3] Akelbek M and Kirkland S 2009 *Linear Algebra Appl.*, **430** 1111
[4] Akelbek M and Kirkland S 2009 *Linear Algebra Appl.*, **430** 1099
[5] Chen S and Liu B 2010 *Linear Algebra Appl.*, **433** 1110

An Interactive Strategy for Solving Multi-Criteria Decision Making of Sustainable Land Revitalization Planning Problem

Ruth Mayasari[1], Herman Mawengkang[2], Ronal Gomar Purba[1]
[1]Graduate School of Mathematics, University of Sumatera Utara, Medan, Indonesia
[2]Department of Mathematics, University of Sumatera Utara, Medan, Indonesia

Email: [2]*hmawengkang@yahoo.com*

Abstract. Land revitalization refers to comprehensive renovation of farmland, waterways, roads, forest or villages to improve the quality of plantation, raise the productivity of the plantation area and improve agricultural production conditions and the environment. The objective of sustainable land revitalization planning is to facilitate environmentally, socially, and economically viable land use. Therefore it is reasonable to use participatory approach to fullfil the plan. This paper addresses a multicriteria decision aid to model such planning problem, then we develop an interactive approach for solving the problem.

Keywords: Multi-criteria decision analysis, Modeling, Land revitalization, Interactive approach

1. Introduction

Land is used to meet a multiplicity and variety of human needs and to serve numerous, diverse purposes. Land revitalization is the sustainable redevelopment of abandoned a large piece of land. The main objective of the program is to encourage communities and land owners to reuse and redevelop land that was previously abandoned and turns it into something useful, such as, public parks, restored wetlands, and new businesses. Therefore revitalizing has a rich meaning to a community as to make it safer, greener, and offers more jobs to its residents.

In the revitalization planning program, we examine the effects of uncertainty and irreversibility in valuing and timing conversion and development projects involving land areas or natural resources. This topic has been addressed first by [2, 15], who treat the complete conversion problem as an optimal stopping problem, while the gradual conversion problem was first introduced by [6]. Our analysis is focused on a more general problem that is finding an optimal land/resources portfolio composition through time, in the presence of future market uncertainty. In fact, it is often more realistic to assume that an optimal land management program will involve a gradual sequence of conversion decisions through time, evolving as each land allocation/resource value becomes known more accurately.

In this paper we formulate a scenario-based multi-stage stochastic programming model, which takes into account the uncertainty related to the market value of revenues accruing from the land in different states. As most of the real options literature assumes, whenever the riskiness of a project is diversifiable, as it is in the case of market uncertainty, it is possible to compute the value of the project applying the "risk-neutral" probability distribution and using the risk-free interest rate. In order to take into account the non-constant incremental benefits accruing from different land allocations, we consider piece-wise linear land-use value functions. In fact, in realistic settings, where quantity-dependence is admitted, the incremental value of developed land is contingent on the size of the conversion, see [26].

The second issue investigated is environmental uncertainty. In fact, when we allow for the option of converting an area into a natural park, or, more generally, we include in the model preservation investments, we have to deal, not only with market uncertainty, but also with the more complex and domain specific uncertainty about environmental quality, see [7, 4]. The latter issue is a consequence of the fact that policies are usually concerned with long time horizon decisions, their effects on the environment may be partly unknown and also that environmental processes are stochastic. For example, we typically lack information on the amount of ultimately recoverable resources, characteristics of future technologies and their arrival dates, tastes of people in the future, and so on. Historical data series on fluctuations of natural land value shows discontinuities and frequently a greater volatility when compared to other commercial development opportunities. This often implies that markets are incomplete, thence the riskiness of the environmental investment cannot be hedged by a replicating portfolio and the "risk-neutral" approach cannot be applied.

Nevertheless, option pricing can be profitably integrated to decision analysis methods in order to deal with this domain specific risk. In particular, option pricing techniques can be used to simplify decision analysis when some risks can be hedged by trading and, conversely, decision analysis techniques can be used to extend option pricing techniques to problems with incomplete securities markets, as shown in [25]. Supposing that the decision maker can either choose to sharpen knowledge about the initial value accruing from the preserved land, or take decisions without any further inquiry, we model the gradual conversion problem using a decision tree framework, in [3].

2. Problem formulation

A. Model Notation

This model is developed using a grid of cells which represent portion of land

$j = 1, 2, \dots, n$; where n is the total number of cells (portion of land)in the area

$l, m = 1, 2, \dots, k$ types of land uses

$u =$ undeveloped land use type or status quo (SQ)

$D_l =$ set of cells that already have land use l

$D_M =$ set of cells that already have land use M

$U =$ set of cells of SQ land

$e_j =$ existing land use of cell j

$k_l =$ number of cells that initially have land use $l, l = 1, 2, \dots, k$

$v_l =$ estimated demand for land use l

$$X_{juM}(t) = \begin{cases} 1 \text{ if SQ land at location } j \text{ is converted to } M \text{ at time } t \\ 0 \text{ otherwise} \end{cases}$$

$$X_{jejM}(t) = \begin{cases} 1 \text{ if land use } e_j \text{ at location } j \text{ is converted to } M \text{ at time } t;\ M \neq e_j \\ 0 \text{ otherwise} \end{cases}$$

$$X_{jlM}(t) = \begin{cases} 1 \text{ if land use } l \text{ at location } j \text{ is converted to } M \text{ at time } t;\ M \neq l \\ 0 \text{ otherwise} \end{cases}$$

$W_{juM}(t) \in [0,1]$ portion of SQ land at location j is converted to M at time t

$W_{jejM}(t) \in [0,1]$ portion of land use e_j at location j is converted to M at time t

$W_{jlM}(t) \in [0,1]$ portion of land use l at location j is converted to M at time t

$U(t) \in [0,1]$ the portion of undeveloped land at time t

$D(t) \in [0,1]$ the portion of land use l at time t

$D_M(t) \in [0,1]$ the portion of land use M at time t

We define the benefits and curt accruing from alternative land allocation:

$\pi(t, U(t))$ the discounted values of revenue arising at time t for the portion of undeveloped land

$K(t, D_l(t))$ the discounted values of revenue arising at time t for the portion of land use l

$v(t, D_M(t))$ the discounted values of revenue arising at time t for the portion land use M

$\alpha(t)$ represents the discounted values of variable curt deriving from covension SQ land to M at location j

$\beta(t)$ represents the discounted values of variable curt deriving from convention of land use e_j to M at location j

$\gamma(t)$ represents the discounted values of variable curt deriving from covension of land use l to M at location j

$I_\alpha(t)$ the initial sunk costs necessary to start the convension of the area SQ to M at location j

$I_\beta(t)$ the initial sunk costs necessary to start the convension of the area land use e_j to M at location j

$I_\gamma(t)$ the initial sunk costs necessary to start the convension of the land use l to M at location j

The decision maker objective is to maximize the Net Present Value (NPV) deriving from the managed area, considered both revenues and variable and fixed *conversion* costs:

$$\begin{aligned} \max \ & \sum_{t=1}^{T}[\pi(t,u(t))u(t) + k(t,D_l(t))D_l(t) + v(t,D_M(t))D_M(t) + \\ & -\sum_{t=1}^{T}[\alpha(t)w_{jum}(t) + \beta(t)w_{jejm}(t) + \gamma(t)w_{jlm}(t)] + \\ & -\sum_{t=1}^{T}[I_\alpha(t)x_{jum}(t) + I_\beta(t)x_{jejm}(t) + I_\gamma(t)x_{jlm}(t)] \end{aligned} \tag{1}$$

To include fixed initial investment costs in the model, we have introduced three binary variables each assuming the value one (zero) when the correspondent conversion activity has (has not) been undertaken. $x_{jum}(t) \in \{0, 1\}$, $x_{jum}(t) = 1$ if conversion from SQ into m starts at time t; $x_{jum}(t) = 0$, otherwise; analogously we introduce $x_{jejm}(t) \in \{0, 1\}$, relating to conversion from m into l and $x_{jlm}(t) \in \{0, 1\}$, relating to conversion from SQ into l.

The maximization problem is constrained to investment constraints, formalizing the presence of fixed costs when conversion starts, for $i = \alpha, \beta, \gamma$, with $t \in \{1, \dots, T\}$:

$$\begin{aligned} & w_i(t) \le \sum_{t'=1}^{t} x_i(t') \\ & \sum_{t=1}^{T} x_i(t) \le 1 \end{aligned} \tag{2}$$

3. The Stochastic Programming Model

When uncertainty concerning benefit flows is included in the model, the decision made at each period should take into account all future uncertainties and future decisions. For the sake of simplicity we denote by ξ the random vector defined as $\xi = (\xi_\pi, \xi_v, \xi_\kappa)$.

We can now set out the stochastic linear programming problem as follows:

$$\begin{aligned}\max\ & E_{\xi} \sum_{t=1}^{T}[\pi(t,u(t);\xi)u(t,\xi)+k(t,D_l(t);\xi)D_l(t,\xi)+v(t,z(t);\xi)D_M(t,\xi)+ \\ & -E_{\xi} \sum_{t=1}^{T}[\alpha(t)w_{jum}(t,\xi)+\beta(t)w_{jejm}(t,\xi)+\gamma(t)w_{jlm}(t,\xi)]+ \\ & -E_{\xi} \sum_{t=1}^{T}[I_{\alpha}(t)x_{jum}(,\xi t)+I_{\beta}(t)x_{jljm}(t,\xi)+I_{\gamma}(t)x_{jlm}(t,\xi)]\end{aligned} \quad (3)$$

where E_{ξ} represents the expectation operator relative to the random vector ξ. Again the maximization problem is subject to the following constraints:

1) Initial conditions, for t = 0:

$$u(t,\xi) = 1 \text{ and } D_l(t,\xi) = D_M(t,\xi) = w_{jum}(t,\xi) = w_{jejm}(t,\xi) = w_{jlm}(t,\xi) = 0. \quad (4)$$

2) Investment constraint, for i = α, β, γ , and t ∈ {1, . . . , T },

$$w_i(t,\xi) \le \sum_{t'=1}^{t} x_i(t',\xi),\ a.s.$$
$$\sum_{t=1}^{T} x_i(t,\xi) \le 1 \quad (5)$$

3) Development constraints, for $t \in \{1, \dots, T\}$,

$$u(t,\xi) = u(t-1,\xi) - w_{jum}(t,\xi) - w_{jlm}(t,\xi),$$
$$D_l(t,\xi) = D_l(t-1,\xi) + w_{jum}(t,\xi) - w_{jejm}(t,\xi),$$
$$D_M(t,\xi) = D_M(t-1,\xi) + w_{jejm}(t,\xi) + w_{jlm}(t,\xi),$$
$$u_{\beta}(t,\xi) \le y(t-1,\xi), \text{ a.s.} \quad (6)$$

4) Conservation constraints, for $t \in \{1, \dots, T\}$,

$$u(t,\xi) + D_l(t,\xi) + D_M(t,\xi) = 1\ . \quad (7)$$

5) Information constraints, i.e., for $t \in \{1, \dots, T\}$, ui is the $\Im_t$ -measurable, where $\Im_t$ is the σ-field generated by the observations, i.e.,

$$\Im_t = \sigma\{\xi_{\tau} | \tau \le t\}, \quad (8)$$

where ξ_t , represents the realization of the random vector at time t, for $i = \alpha, \beta, \gamma$.

An important prerequisite, in order to solve the maximization problem, is the discretization of the stochastic processes representing the evolution of the random data, ξ. The aggregation of the discrete processes can be represented by a scenario (event) tree that defines the possible sequences of realizations over the whole planning horizon. Note that nodes in the event tree are associated with decision points while arcs represent realizations of random variables. In particular, the root is associated with the first stage decision variables while leaves are related to all the possible last stage ones. If we denote with N_t the set of nodes at the t-th level, then each node $n \in N_t$ represents a particular realization sequence $\{\xi_{\tau}\}_{\tau}^{t}=1$ of the data process and it can be thought as a particular state of the system at a given time. A probability p_n can be associated with each node n at level t such that $p_n = p\{\xi_t | \xi_{t-1}, \dots, \xi_1\}$.

4. The Deterministic Equivalent Model

Given a set Z of cancelled route, a route $(v_1, v_2, v_3, \dots, v_l)$ is said to avoid E if $(v_i, v_i + 1, \dots, v_j) \notin E$ for all i, j such that $1 \le i < j \le$ l. A route Q from s to t is called a shortest E-avoiding route if the length of Q is the shortest among all E-avoiding route from s to t. We will use the term "exception avoiding" instead of "X-avoiding" when E is equal to Z, the set of all cancelled paths in G.

After the discretization process using a scenario tree approach we can write the deterministic equivalent model as follows.

$$\begin{aligned}
\max \ & \sum_{s\in S} p(s) \sum_{t>0} [\pi(t,s)u(t,s)] + \\
& + \sum_{s\in S} p(s) \sum_{t>0} \sum_{m=1}^{M} [K(m,t,s)dD_l(m,t,s)] + \\
& + \sum_{s\in S} p(s) \sum_{m=1}^{M} m[V(m,t,s)dD_M(m,t,s)] + \\
& - \sum_{s\in S} p(s) \sum_{t>0} [\alpha(t)w_{jum}(t,s) + \beta(t)w_{jejm}(t,s) + \gamma(t)w_{jlm}(t,s)] + \\
& - \sum_{s\in S} p(s) \sum_{t>0} [I_\alpha(t)x_{jum}(t,s) + I_\beta(t)x_{jejm}(t,s) + I_\gamma(t)x_{jlm}(t,s)]
\end{aligned} \tag{9}$$

The objective function is constrained to:

1) Initial conditions, for each $s \in S$ and for $t = 0$,

$$u(t, x) = 1 \text{ and } D_l(t, s) = D_M(t, s) = w_{jum}(t, s) = w_{jejm}(t, s) = w_{jlm}(t, s) = 0. \tag{10}$$

2) Investment constraints, for each $s \in S$, for $i = \alpha, \beta, \gamma$, and $t \in \{1, . . . , T\}$,

$$\begin{aligned}
& w_i(t,s) \le \sum_{t'\le t} x_i(t',s), \\
& \sum_t x_i(t,s) \le 1.
\end{aligned} \tag{11}$$

3) Development constraints, for each $s \in S$ and $t \in \{1, . . . , T\}$,

$$\begin{aligned}
& u(t, s) = u(t-1, s) - w_{jum}(t, s) - w_{jejm}(t, s), \\
& D_l(t, s) = D_l(t-1, s) + w_{jum}(t, s) - w_{jejm}(t, s), \\
& D_l(t, s) = D_M(t-1, s) + w_{jejm}(t, s) - w_{jlm}(t, s), \\
& w_{jejm}(t, s) \le D_l(t-1, s).
\end{aligned} \tag{12}$$

4) Conservation constraint, for any $s \in S$ and for $t \in \{1, . . . , T\}$,

$$D_M(t, s) + D_l(t, s) + D_M(t, s) = 1. \tag{13}$$

5) Non-anticipativity constraints, defined in such a way that the dependencies implied by the scenario tree are satisfied. Defining Bn the bundle of scenarios passing through node n, then

$$\begin{aligned}
& u(t, s_i) = u(t, s_j),\ i \ne j,\ \in s_i,\ s_j \in B_n, \text{ with } n \in N_t,\ t = 1, . . . , T-1, \\
& D_l(t, s_i) = D_l(t, s_j),\ i \ne j,\ \in s_i,\ s_j \in B_n, \text{ with } n \in N_t,\ t = 1, . . . , T-1, \\
& D_M(t, s_i) = D_M(t, s_j),\ i \ne j,\ \in s_i,\ s_j \in B_n, \text{ with } n \in N_t,\ t = 1, . . . , T-1.
\end{aligned} \tag{14}$$

6) Piece-wise linear value of land constraints, defined for each $s \in S$, with $t \in \{0, ..., T\}$.

7) Constraints to ensure that we can allocate maximally one land use to each cell j

$$\begin{aligned}
& \sum_{m\ne e_j} x_{je_jm}(t) \le 1;\ \forall j \in D, t = 1,...,T \\
& \sum_m x_{jum}(t) \le 1;\ \forall j \in U, t = 1,...,T
\end{aligned} \tag{15}$$

8) Constraint that guarantees the demand for land use l is satisfied

$$t_l - \sum_{j\in D_l} \sum_{m\ne l} x_{jlm} + \sum_{j\in(D-D_l)} x_{je_jl} + \sum_{j\in U} x_{jul} \ge v_l;\ \forall l \tag{16}$$

IOP Conf. Series: Materials Science and Engineering **300** (2018) 012084 doi:10.1088/1757-899X/300/1/012084

5. Conclusions

This paper presents a model to plan land revitalization in order to redevelop the abandoned a large piece of land. We formulate a scenario-based multi-stage stochastic programming model, which takes into account the uncertainty related to the market value of revenues accruing from the land in different states.

References

[1] Arrow, K. J., & Fisher, I., 1974. *Environmental Preservation, Uncertainty, and Irreversibility.* Quarterly Journal of Economics 88, pp. 312–319.
[2] Henry, C., 1974. *Investment Decisions Under Uncertainty: The Irreversibility Effect.* American Economic Review 64, pp. 1006–1012.
[3] Clarke, H. R., & Reed, W. J., 1990. *Land Development and Wilderness Conservation Policies Under Uncertainty: A Synthesis.* Natural Resource Modelling, 4, pp. 11–37.
[4] Mulvey, J.M., & Ruszczyński, A., 1995. *A new scenario Decomposition method for large-scale stochastic optimization.* Operations Research, 43(2), pp. 477- 490.
[5] Conrad, J., 2000. *Wilderness: Options to Preserve, Extract, or Develop.* Resources and Energy Economics 22, pp. 205–219.
[6] Bosetti, V., & Messina, V., 2002. *Modelling Flexibility in Land Allocation Problems.* to appear in Annals of Operations Research.
[7] Mawengkang, H., Guno, M. M., Hartama, D., Siregar, A.S., Adam, H. A., & Alfina, O., 2012. *An improved direct search approach for solving mixed-integer nonlinear programming problems.* Accepted to be published in Global Journal of Technology and Optimization.
[8] Bosetti, V., & Messina, V., 2002. *A Contingent Valuation Survey of Development Options for Ginostra and the Resulting Data.* www.life.disco.unimib.it.

Integer Optimization Model for a Logistic System based on Location-Routing Considering Distance and Chosen Route

Joni Mulyasari[1], Herman Mawengkang[2], Syahril Efendi[3]
[1]Graduate School of Mathematics, University of Sumatera Utara, Medan, Indonesia
[2]Department of Mathematics, University of Sumatera Utara, Medan, Indonesia
[3]Faculty of Computer Science and Information Technology, University of Sumatera Utara

Email: [2]*hmawengkang@yahoo.com*

Abstract..In a distribution network it is important to decide the locations of facilities that impacts not only the profitability of an organization but the ability to serve customers.Generally the location-routing problem is to minimize the overall cost by simultaneously selecting a subset of candidate facilities and constructing a set of delivery routes that satisfy some restrictions. In this paper we impose restriction on the route that should be passed for delivery. We use integer programming model to describe the problem. A feasible neighbourhood search is proposed to solve the result model.

Keywords: Logistics, distribution network, location problem, integer programming

1. Introduction

The location-routing problem for a logistic system considered in this paper is to decide where to locate the facilities and how to allocate customers to the selected facilities. The decision can be obtained using a model, which are based on the assumption that customers are served individually on out-and-back routes. However, when customers have demands that are less-than-vehicle capacity and thus can receive service from routes making multiple stops, the assumption of individual routes will not accurately capture the transportation cost. Therefore, it is necessary to integrate location-allocation and routing decisions which could give more accurate and cost-effective solutions.

Location-Routing Problems (LRPs) combine two basic planning tasks in logistics, viz., to decide the locations of facilities and the routing of vehicles to visit customers. It is well-known that making these types of decisions independently of one another may lead to highly suboptimal planning results, even if the location decisions must be made for the long term [1]. Therefore the location-routing problem (LRP) can be defined as a mathematical optimization problem where at least the following two types of decisions must be made interdependently; which facilities should be chosen and which route should be passed by the vehicles ([2]; [3]).

The main objective of the LRP is to minimize the overall operations costs with the following constraints:

(i) Customer demands are satisfied without exceeding vehicle or facility capacities.
(ii) The number of vehicles, the route lengths and the route durations do not exceed the specified limits and
(iii) Each route begins and ends at the same facility.

IOP Conf. Series: Materials Science and Engineering **300** (2018) 012085 doi:10.1088/1757-899X/300/1/012085

In this paper, we consider a LRP with distance and cancelled route constraints.

Cancelled route involving pairs of edges occur frequently in everyday life and can occur dynamically due to rush hour constraints, lane closures, construction, etc. Longer forbidden subpaths are less common, but can arise, for example if heavy traffic makes it impossible to turn left soon after entering a multi-lane roadway from the right. If we are routing a single vehicle it is more natural to find a detour from the point of failure when a forbidden path is discovered.

Location-routing problems are clearly related to both the classical location problem and the vehicle routing problem. In fact, both of the latter problems can be viewed as special cases of the LRP. If we require all customers to be directly linked to a depot, the LRP becomes a standard location problem. If, on the other hand, we fix the depot locations, the LRP reduces to a VRP. From a practical viewpoint, location-routing forms part of *distribution management*, while from a mathematical point of view, it can usually be modeled as a *combinatorial optimization* problem. We note that this is an NP-hard problem, as it encompasses two NP-hard problems (facility location and vehicle routing). In the first instance, the reader is referred to ([4]; [5]) for excellent reviews of various formulations.

Most of the research to date has focused on heuristic methods since LRPs merge two NP-hard problems. The heuristics generally decompose the problem into its three components, facility location, customer allocation to facilities and vehicle routing, and solve a series of well-known problems such as *p*-median, location-allocation and vehicle routing. Exact methods have been developed for a small number of LRP models that are derived from two-index flow formulations for the vehicle routing problem (VRP). Laporte and Nobert [6] solve a single depot model by a constraint relaxation method.

Laporte [4] develops an equivalent model and also extends the model to the case where the number of vehicles used is a variable in the model. Laporte et al. [7] solve a multi-depot capacitated LRP using a constraint relaxation method. In their work, the largest problem solved to optimality has eight candidate facilities and 20 customers. Laporte et al. [7] use a branch and-bound procedure to solve asymmetric LRPs that include as many as three candidate facilities and 80 customers. An adaptive large neighborhood search algorithm was proposed by [8] for two echelon vehicle routing and the LRP. Koc et al. [9] use a heuristic algorithm, called hybrid evolutionary search algorithm, to solve LRP with mix fleet size.

Success in developing exact methods for solving larger instances of LRPs is likely to come from leveraging the advances in exact methods for solving VRPs and other difficult combinatorial optimization problems. Motivated by the success of set partitioning formulations for a variety of transportation problems, such as the VRP with time windows (e.g. [10]), the pickup and delivery problem with time windows (e.g. [11]) and the crew scheduling problem (e.g. [12]), we investigate the effectiveness of set partitioning formulations and branch-and-price algorithms in the context of developing exact algorithms for LRPs. The two main contributions of this paper is to present a new formulation for the LRP with distance constraints and we identify an alternative set of constraints that dramatically improves the linear programming (LP) relaxation bound. Guerra et al. [13] propose a heuristic algorithm for solving LRP in a logistic system. Toyoglu et al. [14] consider the LRP using a combination of facility location and vehicle routing problems. The main objective of their paper is to develop LRP with fewer constraints and variables.

2. Problem formulation

In this section, we present a new set-partitioning-based formulation of the LRP with distance constraints. The objective of the LRP with distance constraints is to select a set of locations and to construct a set of associated delivery routes in such a way as to minimize facility costs plus routing costs. The set of routes must be such that each customer is visited exactly once by one route and that the length of each route does not exceed the maximum distance.

3. Initial Model

Consider a LRP, we define I as the set of customer locations and J as the set of facility locations to be selected. Let graph $G = (V, E)$, where $V = I \cup J$ is the set of nodes and E is the set of arcs, such that $E = VxV$. Define d_{ij} for all $(i, j) \in E$ as the distance between nodes i and j. In this case, the distances satisfy the triangle inequality. We say $d_{ij} = 0, \quad \forall (i, j) \in E$ if $i = j$.

A feasible route k associated with facility j is defined as a simple circuit that begins at facility j, visits one or more customer nodes and returns to facility j and that has a total distance of at most the maximum distance, denoted M. Let R_j be the set of all feasible routes associated with the facility $j, \forall j \in J$. The cost of a route $k \in R_j$ is the sum of the costs of the arcs in the route. The cost within an arc $(i, j) \in E$ is proportional to the distance d_{ij} to reflect distance related operating costs.

We define some notation to be used in the model.

Parameter

$$a_{ijk} = \begin{cases} 1, & \text{if customer i is visited from facility j using route k}, \forall i \in I, \forall j \in J, \forall k \in R_j \\ 0, & \text{otherwise} \end{cases}$$

c_{jk} cost per unit distance of route k associated with facility $j, \forall j \in J, \forall k \in R_j$
f_j fixed cost associated with selecting facility $j, \forall j \in J$
γ object weighted factor

Decision varables

$$X_j = \begin{cases} 1, & \text{if facility } j \text{ is selected}, \forall j \in J \\ 0, & \text{otherwise} \end{cases}$$

$$Y_{jk} = \begin{cases} 1, & \text{if route } k \text{ is used to visit facility } j, \forall j \in J, \forall k \in R_j \\ 0, & \text{otherwise} \end{cases}$$

$$\text{(LRP-DC) Minimize} \quad \gamma . \sum_{j \in J} f_j X_j + \sum_{j \in J} \sum_{k \in R_j} c_{jk} d_{jk} Y_{jk} \qquad (1)$$

s.t.

$$\sum_{j \in J} \sum_{k \in R_j} a_{ijk} Y_{ijk} = 1 \quad \forall i \in I \qquad (2)$$

$$X_j - Y_{jk} \geq 0 \quad \forall j \in J, \forall k \in P_j \qquad (3)$$

$$\sum_{j \in J} \sum_{k \in R_j} d_{ijk} Y_{ijk} \leq M, \quad \forall i \in I \qquad (4)$$

$$X_j \in \{0,1\} \quad \forall j \in J \qquad (5)$$

$$Y_{jk} \in \{0,1\} \quad \forall j \in J, \forall k \in R_j \qquad (6)$$

The objective function (1) is to minimize the weighted sum of the facility costs and the routing costs which depends on distance. Constraints (2) are the set partitioning constraints that require each customer i be served by exactly one of the selected routes. Constraints (3) require that facility j be selected if a route k associated with facility j is selected. Constraints (4) is to ensure that the total distance do not exceed the maximum distance. Constraints (5) and (6) are standard binary restrictions. The LRP with distance constraints is NP-hard. By placing very large costs on the arcs connecting two customer nodes, we obtain a special case of the model in which the selected routes contain exactly one customer.

As presented, the formulation LRP-DC potentially contains an exponential number of variables y_{jk} and an exponential number of constraints (3). Thus, for instances of practical size, enumerating all of the feasible routes and solving the resulting integer program is unlikely to be effective. Instead, we will use feasible neighbourhood search for solving the model.

4. LRP model with selected route

Given a set Z of cancelled route, a route $(v_1, v_2, v_3, \ldots, v_l)$ is said to avoid E if $(v_i, v_i + 1, \ldots, v_j) \notin E$ for all i, j such that $1 \leq i < j \leq l$. A route Q from s to t is called a shortest E-avoiding route if the length of Q is the shortest among all E-avoiding route from s to t. We will use the term "exception avoiding" instead of "X-avoiding" when E is equal to Z, the set of all cancelled paths in G.

It is necessarily to assume that costumers are not in the cancelled route. The model for LRP with selected route can be written as:
(LRP-SR)

$$\text{Minimize} \quad \gamma.\sum_{j\in J} f_j X_j + \sum_{j\in J}\sum_{k\in R_j, k\notin Z}\sum_{i\in I} c_{ijk} d_{ijk} Y_{ijk} \tag{7}$$

$$\text{s.t.} \quad \sum_{j\in J}\sum_{k\in R_j, k\notin Z} a_{ijk} Y_{ijk} = 1, \ \forall i \in I \tag{8}$$

$$X_j - Y_{jk} \geq 0 \quad \forall j \in J, \forall k \in P_j, k \notin Z \tag{9}$$

$$\sum_{j\in J}\sum_{k\in R_j, k\notin Z} d_{ijk} Y_{ijk} \leq M, \quad \forall i \in I \tag{10}$$

$$X_j \in \{0,1\} \quad \forall j \in J \tag{11}$$

$$Y_{jk} \in \{0,1\} \quad \forall j \in J, \forall k \in R_j \tag{12}$$

The objective function (7) is to minimize the weighted sum of the facility costs and the travelling costs without using the cancelled route, which depends on distance Constraints (8) are the set partitioning constraints that require each customer i be served by exactly one of the selected routes. Constraints (9) require that facility j be selected if a route k associated with facility j is selected. Constraints (10) is to ensure that the total distance do not exceed the maximum distance. Constraints (5) and (6) are binary restrictions.

5. Neighbourhood Search

It should be noted that, generally, in integer programming the reduced gradient vector, which is normally used to detect an optimality condition, is not available, even though the problems are convex. Thus we need to impose a certain condition for the local testing search procedure in order to assure that we have obtained the "best" suboptimal integer feasible solution.

Scarf [15] has proposed a quantity test to replace the pricing test for optimality in the integer programming problem. The test is conducted by a search through the neighbours of a proposed feasible point to see whether a nearby point is also feasible and yields an improvement to the objective function.

Let $[\beta]_k$ be an integer point belongs to a finite set of neighbourhood $N([\beta]_{k)}$. We define a neighbourhood system associated with $[\beta]_k$, that is, if such an integer point satisfies the following two requirements

1. If$[\beta]_j \in N([\beta]_k)$ then $[\beta]_k \in [\beta]_j), j \neq k$.
2. $N([\beta]_k) = [\beta]_k + N(0)$

With respect to the neighbourhood system mentioned above, the proposed integerizing strategy can be described as follows.

Given a non-integer component, x_k, of an optimal vector, x_B. The adjacent points of x_k, being considered are $[x_k]$ and $[x_k] + 1$. If one of these points satisfies the constraints and yields a minimum deterioration of the optimal objective value we move to another component, if not we have integer-feasible solution.

Let $[x_k]$ be the integer feasible point which satisfies the above conditions. We could then say if $[x_k] + 1 \in N([x_k])$ implies that the point $[x_k] + 1$ is either infeasible or yields an inferior value to the objective function obtained with respect to $[x_k]$. In this case $[x_k]$ is said to be an "optimal" integer feasible solution to the integer programming problem. Obviously, in our case, a neigbourhood search is conducted through proposed feasible points such that the integer feasible solution would be at the least distance from the optimal continuous solution.

6. The Basic Approach

Before we proceed to the case of MINLP problems, it is worthwhile to discuss the basic strategy of process for linear case, i.e., Mixed Integer Linear Programming (MILP) problems.

Consider a MILP problem with the following form

$$\text{Minimize } P = c^T x \tag{13}$$

$$\text{Subject to } Ax \leq b \tag{14}$$

$$x \geq 0 \tag{15}$$

$$x_j \text{ integer for some } j \in J \tag{16}$$

A component of the optimal basic feasible vector $(x_B)_k$, to MILP solved as continuous can be written as

$$(x_B)_k = \beta_k - \alpha_{k1}(x_N)_1 - \cdots - \alpha_{kj}(x_N)_j - \cdots \alpha_{kn} - m(x_N)_N n \tag{17}$$

Note that, this expression can be found in the final tableau of Simplex procedure. If $(x_B)_k$ is an integer variable and we assume that β_k is not an integer, the partitioning of β_k into the integer and fractional components is that given

$$[\beta_k] + f_k, \quad 0 \leq f_k \leq 1 \tag{18}$$

suppose we wish to increase $(x_B)_k$ to its nearest integer, $([\beta] + 1)$. Based on the idea of suboptimal solutions we may elevate a particular nonbasic variable, say $(x_N)_{j^*}$, above its bound of zero, provided α_{kj^*}, as one of the element of the vector α_{j^*}, is negative. Let Δ_{j^*} be amount of movement of the non variable $(x_N)_{j^*}$, such that the numerical value of scalar $(x_B)_k$ is integer. Referring to Eqn. (17), Δ_{j^*} can then be expressed as

$$\Delta_{f^*} = \frac{1-f_k}{-\alpha_{kj^*}} \tag{19}$$

while the remaining nonbasic stay at zero. It can be seen that after substituting (18) into (19) for $(x_N)_{j^*}$ and taking into account the partitioning of β_k given in (18), we obtain

$$(x_B)_k = [\beta] + 1 \tag{20}$$

Thus, $(x_B)_k$ is now an integer.

It is now clear that a nonbasic variable plays an important role to integerized the corresponding basic variable. Therefore, the following result is necessary in order to confirm that must be a non-integer variable to work with in integerizing process.

Theorem 1. *Suppose the MILP problem* (13)-(16) *has an optimal solution, then some of the nonbasic variables.* $(x_N)_j, j = 1, \dots, n$, *must be non-integer variables.*

Proof:

Solving problem as a continuous of slack variables (which are non-integer, except in the case of equality constraint). If we assume that the vector of basic variables consists of all the slack variables then all integer variables would be in the nonbasic vector x_N and therefore integer valued.

References

[1] Nagy, G., & Salhi, S., 2007. *Location-routing: Issues, models and methods.* European Journal of Operational Research, 177, pp. 649-67.

[2] Drexl, M., & Schneider, M., 2014. *A survey of the standard location-routing proble.* Working paper, Logistic Planning and Information Systems, TU Darmstadt, Germany.

[3] Albareda Sambola, M., 2015. *Location-routing and location-arc-routing.* In Laporte G., Nickel S., Saldanha da Gama F. (Eds.), Location Science, Springer, Berlin-heidelberg.

[4] Laporte, G., 1988. *Location routing problems in golden B,* Assad A (eds) Vehicle Routing: Methods and Studies (North-Holland: Amsterdam) pp. 293–318.

[5] Prodhon, C., & Prins, C., 2014. *A survey of recent research on location-routing problems.* European Journal of Operational Research, 238, pp. 1-17.

[6] Laporte, G., & Nobert, Y., 1981. *An exact algorithm for minimizing routing and operating cost in depot location.* European Journal of Operational Research **6,** pp. 224–226.

[7] Laporte, G., Nobert, Y., & Arpin, D., 1986. *An exact algorithm for solving a capacitated location-routing problem.* Annals of Operations Research **6,** pp. 293–310.

[8] Hemmelmayr, V. C., Cordeua, J. F., & Crainic, T. G., 2012. *An adaptive large neighborhood search heuristic for two echelon vehicle routing problems arising in city logistics.* Computer & Operations Research 39, pp. 3215-3228.

[9] Koc, C., Bektas, T., Jabali, O., & Laporte, G., 2015. *The fleet size and mix location-routing problem with time windows: Formulations and Heuristic algorithm.* CIRRELT-2015-04.

[10] Desrosiers, J., Dumas, Y., Solomon, M. M., & Soumis, F., 1995. *Time Constrained Routing and Scheduling.* M O Ball, T L Magnati, C L Monma, and G L Nemhauser, eds. *Handbooks in Operation Research and Management Science: Network Routing* vol 8 (Nort-Holland: Amsterdam, The Netherlands) pp 35-139.

[11] Sol, M., and Savelsbergh, M. W. P., 1994. *A branch-and-price algorithm for the pickup and delivery problem with time windows.* Memorandum COSOR pp. 94-22 (Department of Mathematics and Computing Science, Eindoven University of Technology: Eindoven, The Netherlands).

[12] Hoffman, K. L., & Padberg, M. W., 1993. *Solving airline crew scheduling problems by branch-and-cut.* Management Sci **39,** pp. 657-683.

[13] Guerra, L., Murino, T., & Romano, E., 2007. *A Heuristic Algorithm for the Constrained Location-Routing Problem* Int. J. of syit. Appl. Eng & Dev. **4** (1), pp. 146-154.

[14] Toyoglu, H., Karasan, O. E., & Kara, B. Y., 2012. *A new formulation approach for location-routing problems.* Netw. Spat. Ecom. DOI 10.1007/511067-011-9170-y. Published on line: 12 Januari 2012.

[15] Scarf, H. E., 1986. *Testing for optimality in the absence of convexity* in Walter P Heller, Ross M Starr and David A Starrett (Eds) (Cambridge University Press) pp. 117-134.

Multi-objective model of waste transportation management for crude palm oil industry

Meslin Silalahi[1], Herman Mawengkang[2], Nenna Irsa Syahputri[3]

[1]University Tapanuli, Siborong-borong, Indonesia
[2] Department of Mathematics, Universitas Sumatera Utara, Indonesia
[3]Teknik Informatika, STT-Harapan Medan, Indonesia

hmawengkang@yahoo.com

Abstract. The crude palm oil industry is an agro-industrial commodity. The global market of this industry has experienced rapid growth in recent years, such that it has a strategic value to be developed for Indonesian economy. Despite these economic benefits there are a number of environmental problems at the factories, such as high water consumption, the generation of a large amount of wastewater with a high organic content, and the generation of a large quantity of solid wastes and air pollution. In terms of waste transportation, we propose a multiobjective programming model for managing business environmental risk in a crude palm oil manufacture which gives the best possible configuration of waste management facilities and allocates wastes to these facilities. Then we develop an interactive approach for tackling logistics and environmental risk production planning problem for the crude palm oil industry.
Keywords: Crude-Palm oil, Environmental Production Planning, Multi-objective programming, Modeling, sustainable optimization.

1. Introduction

Originally, the oil palm was found in the tropical rain forest of West Africa and was used as a source of oil and vitamins. Since then the oil palm tree had spread in many tropical countries in Asia, Africa and Latin America. Oil palm seeds of the Dura variety were introduced to Indonesia and Malaysia in 1848 and 1875, respectively. The oil palm cultivation in South East Asia, such as, in Indonesia, Malaysia and Thailand, produce about 80% of the world's palm oil [1].

Palm oil is used in various food products, such as cooking oil, margarine, frying fats, shortenings, vanaspati, non-dairy creamer, etc. Palm oil is also used in non-food products. It can be substitute products derived from petrochemicals. Due to an increasing environmental awareness, these products have a bright future, particularly, in producing biodiesel.

Due to the increasing market demands and to stay competitive, palm oil companies should have production planning strategies. A plan which can be defined as the acquisition of resources needed and raw materials, as well as planning of the production activities and handling the waste. All of the above, meeting customer demand in the most efficient or economical way possible, i.e. minimizing total costs [2]. Determination of the optimal number of production became one of the key component for a successful production planning [3].

It is no doubt that the crude palm oil (CPO) industry plays an important role for economic development However, it contributes to environmental degradation from both input and output sides of its production activities. From the input side, CPO mill uses much water in production process and needs high energy. The output side , the production process generates large quantity of wastewater (POME), solid waste/ by-product and air pollution.

Due to the involvement of waste in the CPO production system, it is necessary to include waste management in the system planning. How to manage waste is a socially important topic, as it deals with the unavoidable day-to-day needs of people, particularly who live in surrounding industrial area . It possesses a special importance in urbanized areas [4] due to population densities and growing rates of waste generation.

Waste management systems can be divided into several processes, such as waste generation, transportation, treatment, and disposal [4], [5], each of which is associated with both socio-economic and environmental considerations [6].

Waste can be classified by its source, which includes municipal solid waste (MSW), waste waters, specific industrial waste, and medical waste. In addition, it can also be distinguished by the degree of its hazardousness, ranging from organic residues of agricultural industry to highly dangerous nuclear waste.

This paper considers about waste management planning for the transportation and disposal of CPO industrial waste. Industrial waste handling is the final and critical step for industrial pollution control. It is also an important issue to cleaner production and sustainable development. Industrial eco-systems are the environmental friendly systems for industrial waste recycling, resembling the food chains, food webs and the nutrient recycles in natural environment [7]. In scientific terms it can be stated as, what is the trade-offs between environment burden and economic activities

On the normative and quantitative terms, these questions have led to the concept of trade-offs and efficient frontiers for business and the environment [8], [9]. The main idea is to determine the set of solutions in which it is not possible to decrease environmental burden or to increase total environmental quality of each environmental category, unless increasing the costs. From a methodological perspective, however, there is not much developed on determining such a frontier or assessing the trade-offs in sustainable logistics networks, despite the extensive existing literature in the field of multi-objective programming. In this paper we address an approach that is sounded to capitalize the decision maker's most effective cognitive capabilities:visual representation. In order to explore the efficient frontier in feasible time (for the intractability of determining all extreme efficient solutions in a multi-objective linear program, see [10], [11], [12]. Management of waste has become an issue of critical importance during last decades mostly due to the complexity of waste streams and the steadily increasing produced volumes. Most often, the decision-making process towards efficient waste management requires the consideration of a significant number of usually conflicting criteria in order to come up with the optimal solution among alternative scenarios.

Multi-criteria decision-making (MCDM) or multiple-criteria decision analysis is a sub-discipline of operations research that explicitly considers multiple criteria in decision-making environments. MCDM is concerned with theory and methodology that can treat complex problems encountered in business, engineering, and other areas of human activity. A complex problem is characterized by incommensurate and conflicting criteria or objectives such as cost, performance, reliability, safety, productivity and affordability. In the presence of multiple criteria, a unique optimal decision for the problem does not exist but rather many or even infinitely many decisions are suitable [13].

Multi-criteria decision aid (MCDA) is an advanced field of operations research, which provides decision-makers (DMs) and analysts a wide range of methodologies, which are well suited to the complexity of decision problems [14]

Several MCDA methods have been proposed in recent years to help in selecting the best compromise alternatives rather than taking decisions based only on personal thoughts, views or experiences [15], [16]. Moreover, due to a rapid increase in existing technologies towards confronting waste management problems, decision-makers are forced to select among a wide spectrum of available alternatives. The growing environmental consciousness of society has also raised focus from a local to regional or global scale. In addition, the decision-making process towards efficient waste management requires the consideration of different parameters. To that end, MCDA methods have been proposed in recent years as a means for helping decision-makers in selecting the best compromise among alternatives, as well as providing them a powerful tool towards convincing the public over the optimal waste management strategy. Applications of such multi-criteria methods have gained wide acceptance in the last few years over quantitative models, as the former embody many variables, quantitative as well as qualitative in their analysis [17]. MCDA provides a tool towards meeting the decision-makers' needs, as well as gaining the widest possible social acceptance.

Waste management requires not only sound technical assessment of risk but also public participation, consultation and stakeholder dialogue on proposed solutions and associated risks. This is more evident for waste streams that need special treatment and certainly a wide public acceptance and consensus among stakeholders, such as radioactive waste management [18]. The usefulness of using one MCDA method in combination with life cycle analysis (LCA) modelling as a decision aid tool is considered most critical in the selection of an optimal management strategy in waste management problems [19].

MCDA methods present significant merits over other alternative decision-making tools, as well as a number of drawbacks. The pros and cons of the MCDA techniques are analytically addressed in the work of [20]. In brief, the key feature of MCDA techniques is its flexibility on the judgement of the decision-making team, which explores the optimal option by assigning performance scores and weights. In addition, MCDA techniques are able to evaas qualitative) units [21]. Furthermore, MCDA techniques are used to identify either the single most preferred alternative, short-list alternatives for subsequent detailed analysis, rank different alternatives or to distinguish acceptable from unacceptable possibilities. Another important feature of MCDA is the performance matrix, where each row represents an option and each column describes performance of the option against each criterion. The performance matrix can be used when moving towards a final decision by comparing alternatives' performances on a pair-wise basis for all different selected criteria.

On the contrary, there are also some drawbacks of the MCDA methods. One of the major drawbacks is those that are sensitive to uncertainties. For instance, general sources of individual uncertainties could come from data series uncertainties, uncertainty about the future, synergies and idiosyncrasies in the interpretation of ambiguous or incomplete information.

The objective of this paper is to propose a multiobjective programming model for managing business environmental risk in a crude palm oil manufacture which gives the best possible configuration of waste management facilities and allocates wastes to these facilities, to achieve the required objectives (minimization of cost, minimization of perceived risk (PR), minimization of environmental impact (EI) or a compromise between cost, PR and EI in such a way to limit its environmental consequences while increasing profitability.

2. Crude Palm Oil Manaufacturing

2.1. Production Process

The two main products derived from the oil palm fruit are crude palm oil (CPO) and crude palm kernel oil (CPKO). CPO is obtained from the mesocarp (fiber) and CPKO is obtained from the endosperm (kernel). Each oil mill applies a conventional oil milling process, beginning with the steaming of fresh fruit bunches (FFB) under high pressure (sterilization) for a prescribed period of time to condition the fruits. The sterilized bunches are then threshed to separate the fruits from the bunch stalks. The fruits are subsequently pressed to obtain the crude oil. This oil–water mixture undergoes a separation process before the oil is purified and dried prior to storage. The water phase forms the bulk of the raw palm oil mill effluent, which is treated in a waste water treatment plant or a treatment pond.

2.2. Environmental Problems

The entire crude palm oil process does not need any chemicals as a processing aid. However, there are a number of environmental problems at the factories, such as high water consumption, the generation of a large amount of wastewater with a high organic content, and the generation of a large quantity of solid wastes and air pollution. The waste generation (per ton FFB production) from the crude palm oil industry shows that only 22.8% of the raw material input consists of valuable products (CPO and CPKO). Palm oil mills also produce significant quantities of by-products/solid waste, such as empty fruit bunch, fibres, shell, decanter cake and ash from the boiler. Only 23% of raw materials are products, the rests are waste/ by-products. Most of the by-products can be reused in the production process or in other industries. Fibres (14%) are used as fuel in boilers to generate steam and energy, required for the mill operation. Shell (6%) and empty fruit brunch (EFB) (24%) are sold for use in other industries. However, there is a lot of solid waste that has to be treated before disposal. These wastes include 0.03 million ton/ year of decanter sludge and 0.05 million ton/ year of ash. The problems of solid waste management in factories are improper storage and handling of solid waste material and improper land application techniques or practices for solids waste. These wastes consequently cause bad smell and dust that affect the surrounding communities.

2.3. Existing Industrial Ecosystems in the Crude Palm Oil Industry

The crude palm oil industry has developed a number of industrial ecosystem practices for its waste recycling. The nature of these practices can be divided into in-plant industrial ecosystem (clean technology) options and possibilities for external waste exchange, which includes recycling of wastes between the industrial sector and other sectors such as agriculture. There are various technical options for an industrial ecosystem approach.

Environmental impact of palm oil industry can occur due to the following event. In order to produce CPO there are five stages involve: plantation, CPO mills, palm kernel oil mills, refinery factories, and CPO. Get the FFBs from palm oil plantation using transportation to a CPO mill. In here, kernel, CPO and bio-wastes are produced. Afterward kernel and CPO are transported to a place for processing kernel and CPO to get palm oil products. In market people buy this products and use them. Some part of the palm oil products are wasted as municipal waste [22].

3. Model Formulation

We define the notation used as follows.

Ju_{ip}	Amount of waste type i at source p
KLR	Risk factor
$Kupt$	Cost of segregation per unit quantity of waste in time step (t)
Ku_{itp}	Cost of processing unit quantity of waste I at source p in time step (t)
Kru_{it}	Cost recovered from the sale of unit quantity of processable waste type i in time step (t)
Ku_{it}	Cost recovered from the sale of unit quantity of reusable waste type i in time step (t)
$Ksit$	Cost of storage per unit quantity of waste in time step (t)
Bmb	Capital cost for locating disposal facility (b)
$BMfo$	Capital Cost for locating processing facility *fo*
b	Disposal facility
$D_{(p\text{-}fo)}$	Distance between the source node (p) and processing facility (fo)
$D_{(fo\text{-}b)}$	Distance between the processing facility (fo) and disposal facility (b)
$D_{(p\text{-}b)}$	Distance between the source node (p) and disposal facility (b)
$D_{(p\text{-}p^1)}$	Distance between the source node (p) and reuse facility (p^1)
e	Total number of time steps
e^1	Total number of time steps in which primary waste i can arrive back as waste after a cycle of reuse
p	Source node
p^1	Reuse facility
Vc	Importance factor for chemical/component (c)
t	Time step
t^1	Time step in which primary waste (u) going for reuse in time step (t) arrives back as waste on source nodes

$FRdis.b_i$	Risk multiplication factor for disposal of waste type i at disposal facility (*b*)
$FRseg.p_i$	Risk multiplication factor for segregation of waste type i at source node (*p*)
$Fro.fo_i$	Risk Multiplication Factor for processing of waste type i at processing facility (*fo*)
$FRpi.p^1{}_i$	Risk Multiplication Factor for reuse of waste type i at reuse facility (p^1)
$FRsi.p_i$	Risk multiplication factor for storage of waste type i at source node (*p*)
FRa_i	Risk multiplication factor for transportation of waste type i
n	Total number of source nodes
n^1	Total number of reuse facilities
P_{it}	Period in units of time step for which waste type i is stored in time step (*t*)
NG_{itp}	Quantity of new primary waste type i generated in time step (*t*) at source node (*p*)
N_{itp}	Quantity of primary waste type i generated at source node (*p*) in time step (*t*)
N_{it^1p}	Quantity of primary waste type i coming after a cycle of reuse in time step (t^1)
$N_{it(p-b)}$	Quantity of primary waste type i at source node (*p*) in time step (*t*) going directly to disposal site *b*
$N_{it(p-p^1)}$	Quantity of primary waste type i generated at source node (*p*) in time step (*t*) going to reuse facility p^1
$N_{it(p-fo)}$	Quantity of processable waste type i (generated after segregation of primary waste types) at source node '*p*' in time step '*t*' going to processing facility *fo*
$N_{it(p-b)}$	Quantity of processable waste type i (generated after segregation of primary waste types) at source node '*p*' in time step (*t*) going to disposal facility (*b*)
$N_{it(fo-b)}$	Quantity of processable waste type i (generated after segregation of primary waste types) left as residue at processing facility in time step (*t*) going to disposal facility (*b*)
$N_{it(p-b)}$	Quantity of reusable secondary waste type i (generated after segregation of primary waste types) at source node (*p*) in time step (*t*) going to disposal facility (*b*)
$N_{it(p-p^1)}$	Quantity of reusable secondary waste type i (generated after segregation of primary waste types) at source node (*p*) in time step (*t*) going to reuse facility (p^1)
$N_{it(p-b)}$	Quantity of non-reusable, non-processable secondary waste type i (generated after segregation of primary waste types) at source node (*p*) in time step (*t*) going to disposal facility (*b*)
J_{fo}	Total number of processing facilities
$Rsit_{(p)}$	Ratio of stored waste to waste arriving at source node (*p*) in time step (*t*)
R_{it}	Ratio of waste type u^1 which could be processed w.r.t. its total quantity
fo	Processing facility
Nb	Total number of disposal facility
K_{it}	Cost of transportation of waste type i per unit weight per unit distance in time step (*t*)
K_{bt}	Cost of transportation for disposal waste type I per unit weight per unit distance in time step (t)
w_i	Total number of waste type i
Zb	Logical variable associated with disposal facility (*b*)
Zfo	Logical variable associated with processing facility *fo*

The model for the management of logistic CPO with life cycle based approach. There are three objectives to be attained within multi-time step. We define 'time step' as period of time, for which the waste generation and associated costs per unit quantity of weight, for a management of logistic activity (transportation, storage, disposal, etc.) remain constant. The time step could be year, months or any other unit of time for which the waste generation and associated management costs are assumed constant. The three objectives addressed consist of: (i) minimization of total logistic operational cost, (ii) minimization of waste environmental impact (EI), and (iii) minimization of total risk perceived (PR) by the people. Each of these objectives can be minimized individually or a compromise solution using goal programming by assigning different weighting to each objective. The decision maker can assign different weightings to cost, PR and EI, depending on the governing socio-economic scenario. Various scenarios of different weightings to each of the objectives can also be analyzed to arrive at the various tradeoffs between these objectives.

The problem is subjected to the following constraints:

a. Mass balance of wastes at each node
b. Allowable capacities at various facilities.
c. Logical constraints at processing and disposal sites.

4. The Model

The problem is formulated as a multiobjective integer programming model

There are three objectives involve, i.e., Total of logistic operational cost, the cost related to risk, called perceived risk, and environmental impact of waste.

(A) Total of logistic operational cost.

$$TC_1 = \sum_{t=1}^{e} \sum_{p=1}^{n} \sum_{u=1}^{w} \left[\left\{ N_{itp} - \sum_{b=1}^{Nb} N_{it(p-b)} \right\} \times K_{itp} \right] \tag{1}$$

This expression represents the cost of segregation at source node

$$TC_2 = \sum_{t=1}^{e} \sum_{p=1}^{n} \sum_{i=1}^{w} \left[Nut_{(p)} \times Ksit \times Rsit_{(p)} \times R_{it} \right] \tag{2}$$

Cost of storage at source node

$$\begin{aligned} TC_3 = {} & \sum_{t=1}^{e} \sum_{b=1}^{n} \sum_{i=1}^{w} \sum_{fo=1}^{J_{fo}} \left[N_{it(p-fo)} \times K_{it} \times D_{(p-fo)} \right] \\ & + \sum_{t=1}^{e} \sum_{b=1}^{N_b} \sum_{i=1}^{w} \sum_{fo=1}^{J_{fo}} \left[N_{it(fo-b)} \times (1 - R_{it}) \times Ku_{it} \times D_{(fo-b)} \right] \end{aligned} \tag{3}$$

Cost of transportation of waste from source node to processing facilities and residue from processing facilities to disposal facilities

$$TC_4 = \sum_{t=1}^{e} \sum_{p=1}^{n} \sum_{i=1}^{w} \sum_{fo=1}^{Jfo} \left[N_{it(p-fo)} \times K_{it} \right] \tag{4}$$

Representing the cost of processing waste at processing facilities

$$TC_5 = \sum_{fo=1}^{J_{fo}} \left[BMfo \times Zfo \right] \tag{5}$$

This is for Capital cost needed for locating processing facilities

$$TC_6 = \sum_{k=1}^{e} \sum_{p=1}^{n} \sum_{p^1=1}^{n^1} \left[\left\{ \sum_{i=1}^{w} N_{it(p-p^1)} \times Ku_{itp} + \sum_{u^2=1}^{w^2} N_{it(p-p^1)} \times Ku_{it} \right\} \times D_{(p-p^1)} \right] \tag{6}$$

The transportation cost of reusable waste types to reuse facilities

$$TC_7 = \sum_{t=1}^{e} \sum_{i=1}^{w} \sum_{p=1}^{n} \sum_{b=1}^{nb} \left[N_{it(p-b)} \times Ku_{it} \times D_{(p-b)} \right] \tag{7}$$

Another transportation cost for some portion of waste from source nodes to disposal facilities

$$TC_8 = \sum_{b=1}^{Nb} \left[BMb \times Zb \right] \tag{8}$$

Capital cost for locating disposal facilities

$$TC_9 = \sum_{t=1}^{e} \sum_{b=1}^{Nb} \left[\sum_{p=1}^{n} \left\{ \sum_{i=1}^{w} N_{it(p-b)} \right\} \times K_{bt} \right] + \sum_{t=1}^{e} \sum_{b=1}^{Nb} \left[\sum_{i=1}^{w} \left\{ \sum_{p=1}^{n} N_{it(p-b)} + \sum_{fo=1}^{J_{fo}} N_{it(fo-b)} \right\} \times Kbt \right] \tag{9}$$

Representing the cost of waste disposal

$$TC_{10} = -\sum_{t=1}^{e} \sum_{i=1}^{w} \sum_{p=1}^{n} \sum_{fo=1}^{J_{fo}} \left[N_{it(p-fo)} \times Kru_{it} \times R_{it} \right] \tag{10}$$

This is for the cost recovered from the sale of recyclable portion of generated waste, therefore there is a negative sign.

$$TC_{11} = -\sum_{t=1}^{e} \sum_{i=1}^{w} \sum_{p=1}^{n} \sum_{p^1=1}^{n^1} \left[N_{it(p-p^1)} \times Ku_{it} \right] \tag{11}$$

Another cost recovered from the sale of reusable portion of generated waste.

(B) Total perceived risk (PR)

$$PR_1 = \sum_{t=1}^{e} \sum_{i=1}^{w} \sum_{p=1}^{n} \sum_{fo=1}^{J_{fo}} \left[N_{it(p-fo)} \times FRa_i \times KLR \right] \tag{12}$$

This is an expression for risk due to transportation of waste from generation nodes to processing facilities

$$PR_2 = \sum_{t=1}^{e} \sum_{p=1}^{n} \sum_{p^1=1}^{n^1} \left[\left\{ \sum_{i=1}^{w} N_{it(p-p^1)} \times FRa_{(u)} \right\} \times KLR \right] \tag{13}$$

Risk due to transportation of reusable portion of waste to reuse facilities

$$PR_3 = \sum_{t=1}^{e} \sum_{i=1}^{w} \sum_{b=1}^{Nb} \sum_{p=1}^{n} \left[N_{it(p-b)} \times FRa_i \times KLR \right] \tag{14}$$

Risk due to transportation of non-reusable waste from source nodes to disposal

$$PR_4 = \sum_{t=1}^{e} \sum_{i=1}^{w} \sum_{b=1}^{nb} \sum_{p=1}^{n} \left[N_{it(p-b)} \times FRdis.b_i \times KLR \right] \tag{15}$$

PR due to transportation of waste directly going to disposal without segregation

$$PR_5 = \sum_{t=1}^{e} \sum_{p=1}^{n} \sum_{i=1}^{w} \left[\left\{ N_{it(p)} - \sum_{b=1}^{Nb} N_{it(p-b)} \right\} \times FRseq.p_i \times KLR \right] \tag{16}$$

Risk at source nodes due to segregation

$$PR_6 = \sum_{t=1}^{e} \sum_{p=1}^{n} \sum_{i=1}^{w} \left[N_{it(p)} \times Rsit_{(p)} \times FRsi.p_i \times KLR \right] \tag{17}$$

Risk occurring at source nodes due to storage

$$PR_7 = \sum_{t=1}^{e}\sum_{b=1}^{Nb}\left[\sum_{p=1}^{n}\left\{\sum_{i=1}^{w} N_{it(p-b)} \times FRdis.b_i \times KLR\right]\right.$$
$$+\sum_{t=1}^{e}\sum_{b=1}^{Nb}\left[\sum_{i=1}^{w}\left\{\sum_{p=1}^{n} N_{it(p-b)} + \sum_{fo=1}^{J_{fo}} N_{it(fo-b)} \times FRdis.b_i \times KLR\right\}\right] \tag{18}$$

Risk at disposal facilities due to waste

$$PR_8 = \sum_{t=1}^{e}\sum_{p=1}^{n}\sum_{i=1}^{w}\sum_{fo=1}^{J_{fo}}\left[N_{it(p-fo)} \times FRo.fo_i \times KLR\right] \tag{19}$$

Calculating risk at processing facilities

$$PR_9 = \sum_{t=1}^{e}\sum_{p^1=1}^{n^1}\sum_{p=1}^{n}\left[\left\{\sum_{i=1}^{w} N_{it(p-p^1)} \times FRpu.p^1_i\right\} \times KLR\right] \tag{20}$$

Risk at reuse facilities

(C) Total impact or risk of the environment

$$EI = \sum_{t=1}^{e}\sum_{b=1}^{Nb}\left[\sum_{p=1}^{n}\left\{\sum_{i=1}^{w}\left|N_{it(p-b)} \times \sum_{c=1}^{cn} Ju_{ip} \times V_c\right|\right\} \times Pk_{(b)}\right]$$
$$+\sum_{t=1}^{e}\sum_{b=1}^{Nb}\left[\sum_{i=1}^{w}\left\{\sum_{p=1}^{n} N_{it(p-b)} + \sum_{fo=1}^{J_{fo}} N_{it(fo-b)}\right\} \times \sum_{c=1}^{cn}(Ju_{ip} \times V_c) \times Pk_{(b)}\right] \tag{21}$$

Constraints

(a) Mass balance for primary waste type going for reuse in time step t and arriving at source nodes in time step t^1

$$\sum_{p=1}^{n}\sum_{p^1=1}^{n^1} N_{it(p-p^1)} = \sum_{t^1=1}^{e^1}\sum_{p=1}^{n} N_{it^1(p)} \qquad \forall t, \mathrm{i} \tag{22}$$

(b) Mass balance for waste arriving at source nodes

$$NG_{itp} + \sum_{t^1=1}^{e^1} N_{it^1p} = N_{itp} \qquad \forall t^1 = t,\ \forall p, \mathrm{i} \tag{23}$$

(c) Mass balance at source nodes

$$\sum_{t=1}^{e}\sum_{i=1}^{w} N_{itp} = \sum_{t=1}^{e}\left[\sum_{b=1}^{Nb}\left\{\sum_{i=1}^{w} N_{it(p-b)}\right\}\right.$$
$$\left. + \sum_{p^1=1}^{n^1}\sum_{i=1}^{w} N_{it(p-p^1)} + \sum_{fo=1}^{J_{fo}}\sum_{i=1}^{w} Nu^1t_{(p-fo)}\right] \quad \forall p \tag{24}$$

(d) Mass balance at processing facilities

$$\sum_{p=1}^{n}\sum_{i=1}^{w}\left[N_{it(p-fo)} \times (1 - R_{it})\right] = \sum_{i=1}^{w}\sum_{b=1}^{Nb}\left[N_{it(fo-b)}\right] \quad \forall fo,\ t \tag{25}$$

(e) Capacity constraint at processing facilities

$$\sum_{p=1}^{n}\sum_{i=1}^{w}\left[N_{it(p-fo)}\right]\leq Cap.fo.t \qquad \forall fo,\ t \tag{26}$$

(f) Logical constraint at processing facilities

$$\left[\sum_{t=1}^{e}\sum_{p=1}^{n}\sum_{i=1}^{w}\left[N_{it(p-fo)}\right]\Big/\sum_{t=1}^{e}\sum_{i=1}^{w}N_{it(p)}\right]\leq Zfo\leq\sum_{t=1}^{e}\sum_{p=1}^{n}\sum_{i=1}^{w}\left[N_{it(p-fo)}\right] \qquad \forall fo \tag{27}$$

(g) Capacity constraint at disposal facilities

$$\sum_{fo=1}^{J_{fo}}\sum_{i=1}^{w}N_{it(fo-b)}+\sum_{p=1}^{n}\left[\sum_{i=1}^{w}N_{it(p-b)}\right]\times Cap.b.t \qquad \forall b,\ t \tag{28}$$

(h) Logical constraint at disposal facilities

$$\left[\sum_{t=1}^{e}\left[\sum_{fo=1}^{J_{fo}}\sum_{i=1}^{w}N_{it(fo-b)}+\sum_{p=1}^{n}\left\{\sum_{i=1}^{w}N_{it(p-b)}\right\}\right]/\sum_{t=1}^{e}\sum_{i=1}^{w}\left[N_{it(p)}\right]\right]$$
$$\leq Zb\leq\sum_{t=1}^{e}\left[\sum_{fo=1}^{J_{fo}}\sum_{i=1}^{w}N_{it(fo-b)}+\sum_{p=1}^{n}\left\{\sum_{i=1}^{w}N_{it(p-b)}\right\}\right] \qquad \forall b \tag{29}$$

(i) Capacity constraint at reuse facilities

$$\sum_{p=1}^{n}\sum_{i=1}^{w}\left[N_{it(p-p^{1})}\right]\leq cap.p^{1}.t \qquad \forall p^{1},t \tag{30}$$

(j) Logical constraint

$$Zfo=0 \quad or \quad 1$$
$$Zb=0 \quad or \quad 1$$

5. The Algorithm

The proposed interactive algorithm consists of the following three steps.

Steps 1. Determine an initial (weak) efficient solution.

Steps 2. Show the solution to the decision maker (DM). If DM is satisfied with the solution, Stop: otherwise, ask the DM to specify a new reference point $\bar{f}_k$, using Analytic Hierarchy Process (AHP) and go to step 3.

Steps 3. Based on the values of $\bar{f}_k$ and f_k (the last solution), solve

$$\max\ y$$

subject to

$$f_k(x)-(\bar{f}_k-f_k)y\geq f_k,\ k\in H,$$
$$f_k(x)-\bar{f}_k+a(\bar{f}_k-f_k),\ k\in H,$$
$$f_k(x)=\bar{f}_k,\ k\in E$$
$$x\in X$$
$$y\geq 0$$

where y is a scalar.

and find a new intermediate weak efficient (or efficient) solution $f_k(x)$; go to Step 2.

6. Conclusions

Managing business environmental risk in agro-industry consists of making the production process more efficient in such a way as to limit its environmental consequences while increasing profitability.

IOP Conf. Series: Materials Science and Engineering **300** (2018) 012086 doi:10.1088/1757-899X/300/1/012086

In this paper we present a multi-objective integer programming model for managing business environmental risk in a crude palm oil manufacture which consists of making the production process more efficient in such a way to limit the impact of environmental consequences and to meet the investment risk (perceived risk), and then we propose an interactive algorithm for solving the model.

References

[1] Gopal,J., 2001. *The Development of Malaysia's Palm Oil Refining Industry:Obstacles, Policy, and Performance.* Ph.D Thesis of the University of London and Diploma of Imperial College, England.

[2] Wolsey, L. A., & Pochet, Y., 2006.. *Production Planning by Mixed Integer Programming.* (T. V. Mikosch, S. I. Resnick, & S. M. Robinson, Eds.) New York, NY, USA: Springer Science and Business Media, Inc. 2006.

[3] Mathelinea, D., & Mawengkang, H., 2016. *Optimization model for planning and scheduling of perishable fish product supply chain.* International Journal of Trend in Research and Development, **3(2)**, pp. 68-73.

[4] Xu, Y., Huang, G. H., Qin, X. S., Cao, M. F. & Sun, Y., 2010. *An interval-parameter stochastic robust optimization model for supporting municipal solid waste management under uncertainty.* Waste Management, **30(2)**, pp. 316–327.

[5] Wilson, D. C., 1985. *Long-term planning for solid waste management.* Waste Management & Research, **3(3),** pp. 203–216.

[6] Chang, N.-B., & Wang, S., 1997. *A fuzzy goal programming approach for the optimal planning of metropolitan solid waste management systems.* European Journal of Operational Research, **99(2),** pp. 303-321.

[7] Liu, K., & Shyng, Q., 1999, *Eco-system in the steel industry.* Published in Proceeding of International Conference on Cleaner Production and Sustainable Development' 99. December 1999.Taipei,Taiwan.

[8] Huppes, G., & Ishikawa, M., 2005. *A framework for quantified eco-efficiency analysis.* Journal of Industrial Ecology; **9(4),** pp. 25-41.

[9] Bloemhof-Ruwaard, J. M., Krikk, H., & Van Wassenhove, L. N., 2004. *OR models for eco-eco closed-loop supply chain optimization., volume 1 of Reverse Logistics: Quantitative Models for Closed-Loop Supply Chains.* Springer-Verlag, Berlin/Heiderberg, 1 edition.

[10] Steuer,R.E., 1994. *Random problem generation and the computation of efficient extreme points in multiple objective linear programming.* Computational Optimization and Applications, **3**,pp. 333-347.

[11] Steuer, R.E., & Piercy, C. A., 2005. *A regression study of the number of efficient extreme points in multiple objective linear programming.* European Journal of Operational Research, **162(2),** pp. 484-496.

[12] Boyer, O., Tang Sai Hong, Pedram, A., Mohd Yusuff, R., & Zulkifli N., 2013. *A mathematical model for the industrial hazardous waste location-routing problem.* Journal of Applied Mathematics, Article ID 435272.

[13] Achillas, Ch., Moussiopoulos, N., Karagiannidis, A., Banias, G., & Perkoulidis, G., 2013. *The use of multi-criteria decision analysis to tackle waste management problems: a literature review.* Waste Management & Research, **3(12),** pp. 115-129.

[14] Zopounidis, C., & Doumpos, M., 2002. *Multi-criteria decision aid in financial decision making: methodologies and literature review.* Journal of Multi-criteria Decision Analysis **11**, pp. 167–186.

[15] Kami, E., & Werczberger, E., 1995. *The compromise criterion in MCDM: Interpretation and sensitivity to the p parameter.* Environment and Planning B: Planning and Design **22(4)**, pp. 407–418.

[16] Opricovic, S., 2007. *A fuzzy compromise solution for multicriteria problems.* International Journal of Uncertainty, Fuzziness and Knowledge-based Systems **15(3)**, pp. 363–380.

[17] Queiruga, D., Walther, G., & Gonzalez-Benito, J., et al. 2008. *Evaluation of sites for the location of WEEE recycling plants in Spain.* Waste Management **28**, pp. 181–190.

[18] Kemp, R. V., Bennett, D.G., & White, M. J., 2006. *Recent trends and developments in dialogue on radioactive waste management: Experience from the UK.* Environment International **32**, 10212–1032.

[19] Hanandeh, A. E., & El-Zein A., 2010. *The development and application of multi-criteria decision-making tool with consideration of uncertainty: The selection of a management strategy for the bio-degradable fraction in the municipal solid waste*. Bioresource Technology 101: 555–561.

[20] Bhagtani, N., 2008. *A Better Tool for Environmental Decision Making: Comparing MCDA with CBA*. Master Thesis, School of Environmental Sciences, University of East Anglia, UK.

[21] Hajkowicz, S., & Higgins, A., 2006. *A comparison of multiple criteria analysis techniques for water resource management.* European Journal of Operational Research **184(1)**, 255–265.

[22] Hayashi K., 2007. *Environmental impact of palm oil industry in Indonesia*. Proceed. Of Int. Symposium on EcoTopia Science, ISETS07.